physics

for scientists and engineers

Paul A. Tipler

physics
for scientists and engineers

Fourth Edition

Volume 1

Mechanics

Oscillations and Waves

Thermodynamics

advance edition

W.H. FREEMAN AND COMPANY/WORTH PUBLISHERS

Physics for Scientists and Engineers, *Advance Edition*
Fourth Edition, Volume 1
Paul A. Tipler

Copyright © 1999 by W.H. Freeman and Company
Copyright © 1990, 1982, 1976 by Worth Publishers, Inc.
All rights reserved
Manufactured in the United States of America
Library of Congress Catalog Card Number: 98-60168
Volume 1 (Chapters 1–21) ISBN: 1-57259-491-8
Volume 2 (Chapters 22–35) ISBN: 1-57259-492-6
Volume 3 (Chapters 36–41) ISBN: 1-57259-490-X
Volumes 1 and 2, ISBN: 1-57259-614-7
Volumes 1, 2, and 3, ISBN: 1-57259-615-5

Printing: 1 2 3 4 5 02 01 00 99 98

Executive Editor: Anne C. Duffy
Development Editors: Steven Tenney and Morgan Ryan, with Richard Mickey
Marketing Managers: Kimberly Manzi and John Britch
Design: Malcolm Grear Designers
Art Director: George Touloumes
Production Editor: Margaret Comaskey
Production Manager: Patricia Lawson
Layout: Fernando Quiñones and Lee Mahler
Picture Editor: Elyse Rieder
Graphic Arts Manager: Demetrios Zangos
Three-dimensional art by DreamLight Incorporated
Illustrations: DreamLight Incorporated and Mel Erikson Art Services
Composition: Compset, Inc.
Separations: Creative Graphic Services
Printing and Binding: R. R. Donnelley and Sons
Cover Image: Sand atop a vertically driven shaker table spontaneously
forms a roughly sinusoidal outline. Image by Max Aguilera-Hellweg.

Illustration credits begin on page IC-1 and constitute
an extension of the copyright page.

W.H. Freeman and Company
41 Madison Avenue
New York, NY   10010   U.S.A.

For Claudia

preface

In this fourth edition I have worked toward four goals:

1. To help students increase their experience and ability in problem solving
2. To make the reading of the text easier and more fun for students
3. To bring the presentation of physics up to date to reflect the importance of the role of quantum theory
4. To make the text more flexible for the instructor in a wide variety of course formats

Enhanced Problem Solving

To help students learn how to solve problems, the number of worked *Examples* that correspond to intermediate-level problems has been greatly increased. Especially notable is a new two-column side-by-side example format that has been developed to better display the text and equations in worked examples. Care has been taken to show the students a logical method of solving problems. Examples begin with strategies, and often diagrams, in a *Picture the Problem* prologue. When possible, the first step gives an equation relating the quantity asked for to other quantities. This is usually followed by a statement of the general physical principle that applies. For example, this step may be "Apply Newton's second law" or "Use conservation of energy." Examples usually conclude with *Remarks* that discuss the problem and solution, and in many cases there are additional *Check the Result* sections that teach the student how to check the answer, as well as *Exercises* that present additional related problems, which students can solve on their own.

Also new are innovative, interactive types of examples, each labeled *Try it yourself*. In these, students are told in the left column how to proceed with each step of the problem-solving process, but in the right column are given only the answer. Thus, students are guided through the problem, but must independently work through the actual derivations and calculations.

A *Problem-Solving Guide* appears at the end of each chapter in the form of a summary of the worked examples in the chapter. The Problem-Solving Guide is designed to help students recognize types of problems and find the right conceptual strategy for solving them. Here again, general principles such as applying Newton's second law or the conservation of energy are emphasized.

Concluding each chapter is a selection of approximately one hundred *Problems*. The problems are grouped by type, which may or may not coincide with the section titles in the chapter. Each problem is designated easy, intermediate, or challenging. Qualitative questions and problems are integrated with quantitative problems within each group, in the hope that this organization will elevate the stature of qualitative problems in the minds of students (and instructors). At the back of the book, *Answers* are given to the odd-numbered problems. Preceding the answers for each chapter is a *Problem Map* that charts which odd-numbered intermediate-level problems correspond with worked examples in the text. Complete solutions to every other odd-numbered problem, worked out in the two-column example format, are available in the *Solutions Manual for Students.*

I do not believe that students can be given too much help in solving problems. Students learn best when they are successful at the tasks they are given. The hierarchy of worked examples, "Try it yourself" examples, Problem-Solving Guide, and Problem Map gives the student and the instructor maximum flexibility by leading the student through progressive levels of independence. "Try it yourself" problems take students step by step through a problem without doing the math for them. The Problem-Solving Guide gives an overview of the techniques that have been demonstrated in the chapter. The Problem Map shows students who are having difficulty where help may lie in the chapter but gives no other assistance.

Student Interest

Much effort has gone into making the written text more lively and informal. Students build their understanding of physics on the physics they've already learned, each concept serving as a building block that will provide the foundation for further inquiry. Over one hundred enthusiastic student reviews indicate that the changes in the fourth edition will successfully reach the widest range of students and will help them to enjoy learning and doing physics rather than focusing on the difficulty of the subject. To further stimulate the interest of students, supplemental, brief *"Exploring ..."* sections offer essays on various topics of interest to science and engineering undergraduates.

Modern Physics in the Introductory Course

Although quantum theory revolutionized the way we describe the physical world more than 70 years ago, we have been slow to integrate it into our introductory physics courses. To make physics more relevant to today's students, the mass–energy relationship and energy quantization sections are included in the conservation of energy chapter, and the quantization of angular momentum is discussed in the chapter on the conservation of angular momentum. These ideas are then used throughout the text, for example, in Chapter 19 to explain the failure of the equipartition theorem.

In addition, two optional chapters, "Wave–Particle Duality and Quantum Physics" (Chapter 17) and "The Microscopic Theory of Electrical Conduction" (Chapter 27), have been written so that instructors who choose to do so can integrate them into a two-semester course along with the usual topics in classical physics. These chapters offer something completely new—support for professors who choose to introduce quantum physics earlier in the course. Chapter 17 on the wave–particle duality of nature is the concluding chapter in Part II, immediately following the chapter on superposition and standing waves. This chapter introduces the idea of the wave–particle duality of light and matter and uses the frequency quantization of standing waves, just studied in the previous chapter, to introduce energy quantization of confined systems. Many students have heard of quantum theory and are curious about it. Having just studied frequency quantization that arises in standing waves, students can easily grasp energy quantization from standing electron waves, once they have seen from diffraction and interference patterns that electrons have wave properties. Because there is little time to cover even the usual material in the introductory course, some instructors are reluctant to consider adding even one more chapter such as Chapter 17. I would argue that quantum physics is at least as important as many of the other topics we teach.

Chapter 27 on the quantum explanation for electrical conduction is positioned so that it can be covered immediately after the discussion of electric current and dc circuits. The classical model of conduction is developed, concluding with the relation between resistivity and the average speed v_{av} and mean free path λ of electrons. The classical and quantum interpretations of v_{av} and λ are then discussed using the particle-in-a-box problem, discussed in the optional Chapter 17, to introduce the Fermi energy. Simple band theory is discussed to show why materials are conductors, insulators, or semiconductors. My hope in offering these optional chapters is that, given the choice, instructors will take advantage of the means to incorporate simple quantum theory into their elementary physics course.

Flexibility

To accommodate professors in a wide variety of course formats and to respond to the preferences of previous users of this text, there has been some revision in the order of material. With this new edition, instructors can give their students a brief exposure to modern physics integrated with the classical topics, or they can choose to skip the optional chapters on quantum physics entirely, perhaps returning to them in the final part of the course when this material is traditionally taught. To make room for these optional quantum chapters, some traditional material may be deleted from the course. To aid the instructor, material that can be skipped without jeopardizing coverage in other sections has been placed in optional sections. There are also two optional chapters in addition to Chapters 17 and 27. Chapter 12, "Static Equilibrium and Elasticity," and Chapter 21, "Thermal Properties and Processes," gather material that instructors sometimes choose to skip over or offer as added reading. The "optional" labeling of sections and chapters enables the instructor to pick and choose among topics with confidence that no material in nonoptional sections depends on previous coverage of an optional topic. Optional sections and chapters are clearly marked by gray borders down the side of the page. Some optional material, such as numerical methods and the use of complex numbers to solve the driven oscillator equation, is presented in "Exploring ..." essays.

Acknowledgments

Many people have contributed to this edition. I would like to thank everyone who used the earlier editions and offered comments and suggestions.

Gene Mosca, James Garland, Robert Lieberman, and Murray Scureman provided detailed reviews of nearly every chapter. Gene Mosca also wrote the student study guide along with Ron Gatreau. Robert Leiberman and Brooke Pridmore class-tested parts of the book, and assisted in obtaining student reviews and feedback. Howard McAllister was instrumental in the development of a standard approach to problem solving in the examples.

Many new problems were provided by Frank Blatt and Boris Korsunsky. Frank Blatt also provided the end-of-volume answers, wrote the solutions manuals, and offered many helpful suggestions. Jeff Culbert helped to enliven the problem sets with his story problems. Several of the graphs at the ends of the examples were provided by Robert Hollebeek.

I would particularly like to thank the more than one hundred students who read and studied from various chapters and provided detailed and valuable comments. Many instructors have provided extensive and invaluable reviews of one or more chapters. They have all made fundamental contributions to the quality of this revision. I would therefore like to thank:

Michael Arnett, *Iowa State University*

William Bassichis, *Texas A&M*

Joel C. Berlinghieri, *The Citadel*

Frank Blatt, *Retired*

John E. Byrne, *Gonzaga University*

Wayne Carr, *Stevens Institute of Technology*

George Cassidy, *University of Utah*

I. V. Chivets, *Trinity College, University of Dublin*

Harry T. Chu, *University of Akron*

Jeff Culbert, *London, Ontario*

Paul Debevec, *University of Illinois*

Robert W. Detenbeck, *University of Vermont*

Bruce Doak, *Arizona State University*

John Elliott, *University of Manchester, England*

James Garland, *Retired*

Ian Gatland, *Georgia Institute of Technology*

Ron Gautreau, *New Jersey Institude of Technology*

David Gavenda, *University of Texas at Austin*

Newton Greenburg, *SUNY Binghamton*

Huidong Guo, *Columbia University*

Richard Haracz, *Drexel University*

Michael Harris, *University of Washington*

Randy Harris, *University of California at Davis*

Dieter Hartmann, *Clemson University*

Robert Hollebeek, *University of Pennsylvania*

Madya Jalil, *University of Malaya*

Monwhea Jeng, *University of California, Santa Barbara*

Ilon Joseph, *Columbia University*

David Kaplan, *University of California, Santa Barbara*

John Kidder, *Dartmouth College*

Boris Korsunsky, *Northfield Mt. Hermon School*

Andrew Lang (graduate student), *University of Missouri*

David Lange, *University of California, Santa Barbara*

Isaac Leichter, *Jerusalem College of Technology*

William Lichten, *Yale University*

Robert Lieberman, *Cornell University*

Fred Lipschultz, *University of Connecticut*

Graeme Luke, *Columbia University*

Howard McAllister, *University of Hawaii*

M. Howard Miles, *Washington State University*

Matthew Moelter, *University of Puget Sound*

Eugene Mosca, *United States Naval Academy*

Aileen O'Donughue, *St. Lawrence University*

Jack Ord, *University of Waterloo*

Richard Packard, *University of California*

George W. Parker, *North Carolina State University*

Edward Pollack, *University of Connecticut*

John M. Pratte, *Clayton College & State University*

Brooke Pridmore, *Clayton State College*

David Roberts, *Brandeis University*

Lyle D. Roelofs, *Haverford College*

Larry Rowan, *University of North Carolina at Chapel Hill*

Lewis H. Ryder, *University of Kent, Canterbury*

Bernd Schuttler, *University of Georgia*

Cindy Schwarz, *Vassar College*

Murray Scureman, *Amdahl Corporation*

Scott Sinawi, *Columbia University*

Wesley H. Smith, *University of Wisconsin*

Kevork Spartalian, *University of Vermont*

Kaare Stegavik, *University of Trondheim, Norway*

Jay D. Strieb, *Villanova University*
Martin Tiersten, *City College of New York*
Oscar Vilches, *University of Washington*
Fred Watts, *College of Charleston*
John Weinstein, *University of Mississippi*
David Gordon Wilson, *MIT*
David Winter, *Columbia University*
Frank L. H. Wolfe, *University of Rochester*
Roy C. Wood, *New Mexico State University*
Yuriy Zhestkov, *Columbia University*

Focus Group Participants

Cherry Hill, New Jersey, July 15, 1997

John DiNardo, *Drexel University*
Eduardo Flores, *Rowan College*
Jeff Martoff, *Temple University*
Anthony Novaco, *Lafayette College*
Jay Strieb, *Villanova University*
Edward Whittaker, *Stevens Institute of Technology*

Denver, Colorado, August 15, 1997

Edward Adelson, *Ohio State University*
David Bartlett, *University of Colorado at Boulder*
David Elmore, *Purdue University*
Colonel Rolf Enger, *United States Air Force Academy*
Kendal Mallory, *University of Northern Colorado*
Samuel Milazzo, *University of Colorado at Colorado Springs*
Anders Schenstrom, *Milwaukee School of Engineering*
Daniel Schroeder, *Weber State University*
Ashley Schultz, *Fort Lewis College*

Student Reviewers

For this edition we invited the input of student reviewers at all stages of manuscript development. A number of the student reviews were blind submissions. The reviews of the following students were especially helpful:

Jesper Anderson, *Haverford College*
Anthony Bak, *Haverford College*
Luke Benes, *Cornell University*
Deborah Brown, *Northwestern University*
Andrew Burgess, *University of Kent, Canterbury*
Sarah Burnett, *Cornell University*
Sara Ellison, *University of Kent, Canterbury*
Ilana Greenstein, *Haverford College*
Sharon Hovey, *Northwestern University*
Samuel LaRoque, *Cornell University*
Valerie Larson, *Northwestern University*
Jonathan McCoy, *Haverford College*
Aaron Todd, *Cornell University*
Katalin Varju, *University of Kent, Canterbury*
Ryan Walker, *Haverford College*
Matthew Wolpert, *Haverford College*
Julie Zachiariadis, *Haverford College*

I would also like to thank the reviewers of previous editions, whose contributions are part of the foundation of this edition:

Walter Borst, *Texas Technological University*
Edward Brown, *Manhattan College*
James Brown, *The Colorado School of Mines*
Christopher Cameron, *University of Southern Mississippi*
Roger Clapp, *University of South Florida*
Bob Coakley, *University of Southern Maine*
Andrew Coates, *University College, London*
Miles Dresser, *Washington State University*
Manuel Gómez-Rodríguez, *University of Puerto Rice, Río Piedras*
Allin Gould, *John Abbott College C.E.G.E.P., Canada*
Dennis Hall, *University of Rochester*
Grant Hart, *Brigham Young University*
Jerold Izatt, *University of Alabama*
Alvin Jenkins, *North Carolina State University*
Lorella Jones, *University of Illinois, Urbana-Champaign*

Michael Kambour, *Miami-Dade Junior College*

Patrick Kenealy, *California State University at Long Beach*

Doug Kurtze, *Clarkson University*

Lui Lam, *San Jose State University*

Chelcie Liu, *City College of San Francisco*

Robert Luke, *Boise State University*

Stefan Machlup, *Case Western Reserve University*

Eric Matthews, *Wake Forest University*

Konrad Mauersberger, *University of Minnesota, Minneapolis*

Duncan Moore, *University of Rochester*

Elizabeth Nickles, *Albany College of Pharmacy*

Harry Otteson, *Utah State University*

Jack Overley, *University of Oregon*

Larry Panek, *Widener University*

Malcolm Perry, *Cambridge University, United Kingdom*

Arthur Quinton, *University of Massachusetts, Amherst*

John Risley, *North Carolina State University*

Robert Rundel, *Mississippi State University*

John Russell, *Southeastern Massachusetts University*

Michael Simon, *Housatonic Community College*

Jim Smith, *University of Illinois, Urbana-Champaign*

Richard Smith, *Montana State University*

Larry Sorenson, *University of Washington*

Thor Stromberg, *New Mexico State University*

Edward Thomas, *Georgia Institute of Technology*

Colin Thomson, *Queens University, Canada*

Gianfranco Vidali, *Syracuse University*

Brian Watson, *St. Lawrence University*

Robert Weidman, *Michigan Technological University*

Stan Williams, *Iowa State University*

Thad Zaleskiewicz, *University of Pittsburgh, Greensburg*

George Zimmerman, *Boston University*

Finally, I would like to thank everyone at Worth and W. H. Freeman Publishers for their help and encouragement. I was fortunate to work with two talented developmental editors. Steve Tenney worked on the beginning phases of the book and is responsible for many of the innovative ideas, such as the example format, summary format, problem-solving guide, and problem map. Morgan Ryan worked on the final stages, including the entire art program, and made significant improvements in the entire book. I am grateful also for the contributions of Kerry Baruth, Anne Duffy, Margaret Comaskey, Elizabeth Geller, Yuna Lee, Sarah Segal, Patricia Lawson, and George Touloumes.

Berkeley, California
December 1997

Paul Tipler

supplements

For Students

Study Guide

Volume 1 (Chapters 1–21) ISBN: 1-57259-511-6
Volumes 2 and 3 (Chapters 22–41) ISBN: 1-57259-512-4

Each chapter contains a description of key ideas, potential pitfalls, true-false questions that test essential definitions and relations, questions and answers that require qualitative reasoning, and problems and solutions.

Solutions Manual for Students

Volume 1 (Chapters 1–21) ISBN: 1-57259-513-2
Volumes 2 and 3 (Chapters 22–41) ISBN: 1-57259-524-8

The *Solutions Manual for Students* provides answers to every other odd end-of-chapter problem, presented in the same format and with the same level of detail as the *Instructor's Solutions Manual* (see below).

Tipler PLUS⊕ CD-ROM

The *Tipler PLUS⊕* CD-ROM is specifically designed to complement the learning process started in the text. On the CD-ROM, students will find a wealth of features to enhance the learning process. Interactive solution-builder exercises based on additional example problems build problem-solving skills. Video clips of lab demonstrations and applied physics bring main objectives to life. Animated quizzes based on the 3D graphics in the text test concepts from each chapter. And Web links lead the student to the sprawling world of physics on the Web. The student version of *Tipler PLUS⊕*, like the instructor's version, can be updated via the Web.

For Instructors

Instructor's Solutions Manual

Volume 1 (Chapters 1–21) ISBN: 1-57259-514-0
Volumes 2 and 3 (Chapters 22–41) ISBN: 1-57259-515-9

Complete solutions to all problems in the text are worked out in the same two-column format as the examples.

Test Bank

Approximately 3500 multiple-choice questions span all sections of the text. Each question is identified by topic and noted as factual, conceptual, or numerical. ISBN: 1-57259-517-5

Computerized Test-Generation System

A database comprises the questions in the *Test Bank.* Instructors can custom design their tests with the *Computerized Test Bank.* For Windows, ISBN: 1-57259-519-1; for Macintosh, ISBN: 1-57259-520-5

Instructor's Resource Manual

Demonstrations and a film and video cassette guide are included. ISBN: 1-57259-516-7

Transparencies

Approximately 150 full-color acetates of figures and tables from the text are included, with type enlarged for projection. Volume 1, ISBN: 1-57259-521-3; Volumes 2 and 3, ISBN: 1-57259-674-0

Instructor's Tipler PLUS⊕ CD-ROM

The instructor's version of the *Tipler PLUS*⊕ CD-ROM includes everything on the student CD as well as syllabus-making software in one easy-to-navigate environment. Just indicate what part of the book you are teaching and when and *Tipler PLUS*⊕ will link your syllabus to a wealth of CD-ROM and Web content. You can click the update button for new Web links, exercises, and updated content or create your own annotated study and lecture aids. *Tipler PLUS*⊕ can even create an e-mail list of your students and fellow instructors. In addition to the syllabus-maker software and the material on the student CD-ROM, the instructor's CD-ROM also features selected items from the *Study Guide* and the *Instructor's Resource Manual.*

about the author

Paul Tipler was born in the small farming town of Antigo, Wisconsin, in 1933. He graduated from high school in Oshkosh, Wisconsin, where his father was superintendent of the Public Schools. He received his B.S. from Purdue University in 1955 and his Ph.D. at the University of Illinois in 1962, where he studied the structure of nuclei. He taught for one year at Wesleyan University in Connecticut while writing his thesis, then moved to Oakland University in Michigan, where he was one of the original members of the physics department, playing a major role in developing the physics curriculum. During the next 20 years, he taught nearly all the physics courses and wrote the first and second editions of his widely used textbooks *Modern Physics* (1969, 1978) and *Physics* (1976, 1982). In 1982, he moved to Berkeley, California, where he now resides, and where he wrote *College Physics* (1987) and the third edition of *Physics* (1991). In addition to physics, his interests include music, hiking, and camping, and he is an accomplished jazz pianist and poker player.

The author as a student, 1954

Tipler = Quality

Tipler *Physics for Scientists and Engineers*, 4/e continues to be the best resource a student can have for learning physics. Dynamic features like these guide the student to mastery . . .

EXAMPLES

- Text includes a greater number of intermediate-level worked examples.

- Each example has a **"Picture the Problem"** section that teaches students how to solve the problem conceptually before solving it mathematically. By learning how to find and organize the relevant information in a problem, students learn to think like a physicist.

- A major innovation is the potent **two-column side-by-side format** for the solutions to examples. Concepts are explained on the left, and the math is presented on the right. This format allows students to make the connections between the equation and what it means.

- Most examples conclude with a **"Remark"** that supplies additional information, discussion of common errors, and advice on solving problems as a physicist would.

Example 6-8

You ski downhill on waxed skis that are nearly frictionless. (*a*) What work is done on you as you ski a distance *s* down the hill? (*b*) What is your speed on reaching the bottom of the run? Assume the length of the ski run is *s*, its angle of incline is θ, and your mass is *m*. The height of the hill is then $h = s \sin \theta$.

Figure 6-15a   **Figure 6-15b**

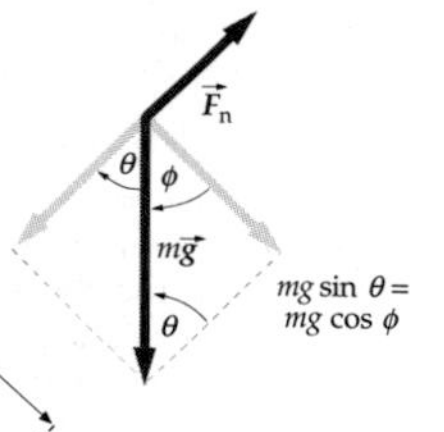

Picture the Problem We assume that you are a particle. Two forces act on you: gravity, $m\vec{g}$, and the normal force exerted by the hill, $\vec{F}_n$ (Figure 6-15*a*). Only gravity does work on you, because the normal force is perpendicular to the hill, and hence has no component in the direction of your motion. The work–kinetic energy theorem with $v_i = 0$ gives the final speed *v*.

Figure 6-15*b* shows a free-body diagram for you on skies. The net force is $mg \sin \theta$, which is the component of the weight in the direction of the displacement Δs.

(*a*)1. The work done by gravity as you traverse the slope is $m\vec{g} \cdot \vec{s}$:

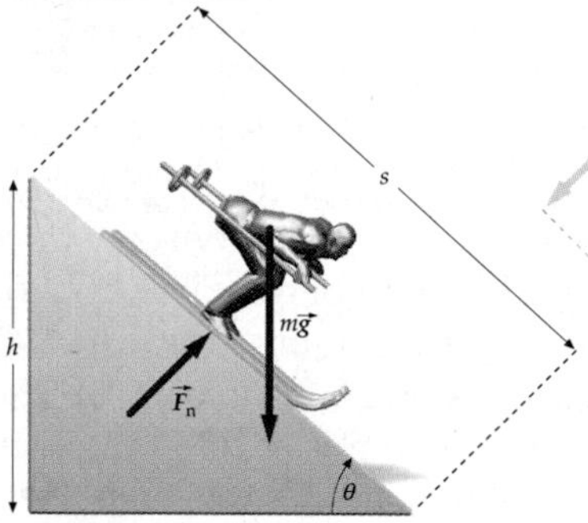

$$W = m\vec{g} \cdot \vec{s} = mgs \cos \phi = mgs \sin \theta$$

2. From Figure 6-15*a*, the angle θ is related to *h* and *s*:

$$\sin \theta = \frac{h}{s}$$

3. Substitute *h* for *s* sin θ:

$$W = mgh$$

(*b*)   Apply the work–kinetic energy theorem to find the final speed *v*:

$$W = mgh = \frac{1}{2}mv^2 - 0 \quad \text{or} \quad v = \sqrt{2gh}$$

Remarks $mg \sin \theta = mg \cos \phi$ is the component of the weight in the direction of the displacement. This is the component that does work on you. The final speed is independent of the angle θ, and the same as if the skier had dropped vertically a height *h* with acceleration *g*. If θ were smaller, the skier would travel a greater distance to drop the same vertical distance *h*, but the component of the force of gravity in the direction of motion would be less. The two effects cancel, and the work done by gravity is *mgh* independent of the angle of the slope. Figure 6-16 shows that for a hill of arbitrary shape, the work done by the earth on the skier is *mgh*.

$\Delta h = \Delta s \cos \phi$

Figure 6-16 Skier skiing down a hill of arbitrary shape. The work done by the earth during a displacement $\Delta \vec{s}$ is $m\vec{g} \cdot \Delta \vec{s} = mg \, \Delta s \cos \phi = mg \, \Delta h$, where Δh is the vertical distance dropped. The total work done by the earth when the skier skis down a vertical distance *h* is $W = \int_0^s m\vec{g} \cdot d\vec{s} = mg \int_0^s \cos \phi \, ds = mg \int_0^h dh = mgh$, independent of the shape of the hill.

- When appropriate, **"Check the Result"** sections teach students how to check their own work.

- Many examples are followed by one or more related **exercises**. Answers are given, but it is up to the student to relate the exercise to the worked-out example.

Check the Result The component of *A* along *B* is *A* cos $\phi = (\sqrt{13}$ m) cos 70.6° = 1.2 m.

Exercise (*a*) Find $\vec{A} \cdot \vec{B}$ for $\vec{A} = (3 \text{ m})\hat{i} + (4 \text{ m})\hat{j}$ and $\vec{B} = (2 \text{ m})\hat{i} + (8 \text{ m})\hat{j}$. (*b*) Find *A*, *B*, and the angle between $\vec{A}$ and $\vec{B}$ for these vectors. (*Answers*   (*a*) 38 m², (*b*) A = 5 m, B = 8.25 m, ϕ = 23°)

Example 2-15

A car is speeding at 25 m/s ($\approx$ 90 km/h $\approx$ 56 mi/h) in a school zone. A police car starts from rest just as the speeder passes and accelerates at a constant rate of 5 m/s^2. (a) When does the police car catch the speeding car? (b) How fast is the police car traveling when it catches up with the speeder?

Picture the Problem To determine when the two cars will be at the same position, we write the positions x_s of the speeder and x_p of the police car as functions of time and solve for the time t when $x_s = x_p$.

(a) 1. Write the position functions for the speeder and the police car:
$$x_s = v_s t \quad \text{and} \quad x_p = \tfrac{1}{2}a_p t^2$$

2. Set $x_s = x_p$ and solve for the time t:
$$v_s t = \tfrac{1}{2}a_p t^2; \quad t = 0 \quad \text{(initial condition)}$$
$$t = \frac{2v_s}{a_p} = \frac{2(25 \text{ m/s})}{5 \text{ m/s}^2} = 10 \text{ s}$$

(b) The velocity of the police car is given by $v = v_0 + at$ with $v_0 = 0$:
$$v_p = a_p t = (5 \text{ m/s}^2)(10 \text{ s}) = 50 \text{ m/s}$$

Remark The final speed of the police car in (b) is exactly twice that of the speeder. Since the two cars covered the same distance in the same time, they must have had the same average speed. The speeder's average speed, of course, is 25 m/s. For the police car to start from rest and have an average speed of 25 m/s, it must reach a final speed of 50 m/s.

Exercise How far have the cars traveled when the police car catches the speeder? (*Answer* 250 m)

Remark In Figure 2-13 the solid lines depict the speeder and the police car in this example. The dashed lines are variations on the example. The smaller acceleration depicted by the lower dashed line means the police car takes longer to reach the speeder. In the higher dashed line, the acceleration is the same as in the example, but the police car does not start accelerating until 4 s after the speeder passes by.

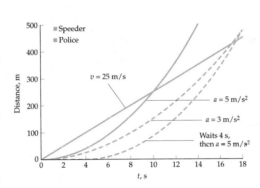

Figure 2-13

Example 6-7 *try it yourself*

A particle is given a displacement $\Delta\vec{s} = 2$ m $\hat{i} - 5$ m $\hat{j}$ along a straight line. During the displacement, a constant force $\vec{F} = 3$ N $\hat{i} + 4$ N $\hat{j}$ acts on the particle (Figure 6-14). Find (a) the work done by the force, and (b) the component of the force in the direction of the displacement.

Picture the Problem The work W is found by computing $W = \vec{F}\cdot\Delta\vec{s} = F_x\Delta x + F_y\Delta y + F_z\Delta z$. Since $\vec{F}\cdot\Delta\vec{s} = F\cos\phi|\Delta\vec{s}|$, we can find the component of $\vec{F}$ in the direction of the displacement from
$$F\cos\phi = \frac{(\vec{F}\cdot\Delta\vec{s})}{|\Delta\vec{s}|} = \frac{W}{|\Delta\vec{s}|}$$

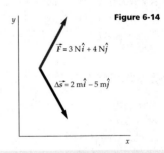

Figure 6-14

Cover the column to the right and try these on your own before looking at the answers.

Steps	Answers
(a) Compute the work done W.	$W = -14$ N·m
(b) 1. Compute $\Delta\vec{s}\cdot\Delta\vec{s}$ and use your result to find the distance $|\Delta\vec{s}|$.	$|\Delta\vec{s}| = \sqrt{29}$ m
2. Compute $F\cos\phi = W/|\Delta\vec{s}|$.	$F\cos\phi = -2.60$ N

Remark The component of the force in the direction of the displacement is negative, so the work done is negative.

Exercise Find the magnitude of $\vec{F}$, and the angle ϕ between $\vec{F}$ and $\Delta\vec{s}$. (*Answer* $F = 5$ N, $\phi = 121°$)

Numerical Methods: Euler's Method

If a particle moves under the influence of a *constant* force, its acceleration is constant and we can find its velocity and position from the constant-acceleration formulas in Chapter 2. But consider a particle moving through space where the force on it, and therefore its acceleration, depends on its position and velocity. The velocity and acceleration of the particle at one instant determine its position and velocity at the next instant, which then determines its acceleration at that instant. The actual position, velocity, and acceleration of an object all change continuously with time. We can approximate this by replacing the continuous time variations with small time steps of duration Δt. The simplest approximation is to assume constant acceleration during each step. This approximation is called **Euler's method**. If the time interval is sufficiently short, the change in acceleration during the interval will be small and can be neglected.

Let x_0, v_0, and a_0 be the known position, veloc-

$$x_2 = x_1 + v_1 \Delta t$$

In general, the connection between the position and velocity at time t_n and time $t_{n+1} = t_n + \Delta t$ is given by

$$v_{n+1} = v_n + a_n \Delta t \qquad 3$$

and

$$x_{n+1} = x_n + v_n \Delta t \qquad 4$$

To find the velocity and position at some time t, we therefore divide the time interval $t - t_0$ into a large number of smaller intervals Δt and apply Equations 3 and 4, beginning at the initial time t_0. This involves a large number of simple, repetitive calculations that are easily done on a computer. The technique of breaking the time interval into small steps and computing the acceleration, velocity, and position at each step using the values from the previous step is called numerical integration.

Drag Forces

To illustrate the use of numerical methods, let us consider a problem in which a sky diver is dropped from rest at some height under the influences of gravity and a drag force that is proportional to the square of the speed. We will find the velocity v and the distance traveled x as functions of time.

The equation describing the motion of an object of mass m dropped from rest is Equation 5-7 with $n = 2$:

$$\sum F_y = mg - bv^n = ma_y$$

Summary

1. Work, kinetic energy, potential energy, and power are important derived dynamic quantities.
2. The work–kinetic energy theorem is an important relation derived from Newton's laws applied to a particle.
3. The dot product of vectors is a mathematical definition that is useful throughout physics.

Topic	Remarks and Relevant Equations
1. Work	
Constant force	The work done by a constant force is the product of the component of the force in the direction of motion and the displacement of the force: $$W = F \cos \theta \, \Delta x = F_x \, \Delta x \qquad 6\text{-}1$$
Variable force	$$W = \int_{x_1}^{x_2} F_x \, dx = \text{area under the } F_x\text{-versus-}x \text{ curve} \qquad 6\text{-}9$$
Force in three dimensions	$$W = \int_1^2 \vec{F} \cdot d\vec{s} \qquad 6\text{-}14$$
Units	The SI unit of work and energy is the joule (J): $$1 \, \text{J} = 1 \, \text{N} \cdot \text{m} \qquad 6\text{-}2$$
2. Kinetic Energy	$$K = \frac{1}{2} mv^2 \qquad 6\text{-}6$$

Problems

In a few problems, you are given more data than you actually need; in a few other problems, you are required to supply data from your general knowledge, outside sources, or informed estimates.

Conceptual Problems

Problems from Optional and Exploring sections

- • Single-concept, single-step, relatively easy
- •• Intermediate-level, may require synthesis of concepts
- ••• Challenging, for advanced students

Conditions for Equilibrium

1 • True or false:
(a) $\Sigma \vec{F} = 0$ is sufficient for static equilibrium to exist.
(b) $\Sigma \vec{F} = 0$ is necessary for static equilibrium to exist.
(c) In static equilibrium, the net torque about any point is zero.
(d) An object is in equilibrium only when there are no forces acting on it.

2 • A seesaw consists of a 4-m board pivoted at the center. A 28-kg child sits on one end of the board. Where should a 40-kg child sit to balance the seesaw?

3 • In Figure 12-23, Misako is about to do a push-up. Her center of gravity lies directly above point P on the floor, which is 0.9 m from her feet and 0.6 m from her hands. If her mass is 54 kg, what is the force exerted by the floor on her hands?

Figure 12-23
Problem 3

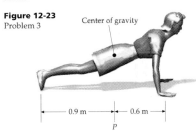
Center of gravity

0.9 m — 0.6 m
P

4 • Juan and Bettina are carrying a 60-kg block on a 4-m board as shown in Figure 12-24. The mass of the board is 10 kg. Since Juan spends most of his time reading cookbooks, whereas Bettina regularly does push-ups, they place the block 2.5 m from Juan and 1.5 m from Bettina. Find the force in newtons exerted by each to carry the block.

Figure 12-24 Problem 4

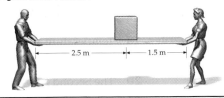
2.5 m — 1.5 m

Figure 12-25 Problem 5

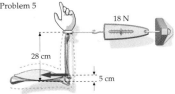
18 N

28 cm

5 cm

the pivot point. If the scale reads 18 N when she exerts her maximum force, what force is exerted by the biceps muscle?

6 • A crutch is pressed against the sidewalk with a force $\vec{F}_c$ along its own direction as in Figure 12-26. This force is balanced by a normal force $\vec{F}_n$ and a frictional force $\vec{f}_s$. (a) Show that when the force of friction is at its maximum value, the coefficient of friction is related to the angle θ by $\mu_s = \tan\theta$. (b) Explain how this result applies to the forces on your foot when you are not using a crutch. (c) Why is it advantageous to take short steps when walking on ice?

Figure 12-26 Problem 6

θ

$\vec{F}_c$

$\vec{f}_s$

$\vec{F}_n$

The Center of Gravity

7 • True or false: The center of gravity is always at the geometric center of a body.

8 • Must there be any material at the center of gravity of an object?

9 • If the acceleration of gravity is not constant over an object, is it the center of mass or the center of gravity that is the pivot point when the object is balanced?

10 • Two spheres of radius R rest on a horizontal table with their centers a distance $4R$ apart. One sphere has twice the weight of the other sphere. Where is the center of gravity of this system?

11 • An automobile has 58% of its weight on the front wheels. The front and back wheels are separated by 2 m.

General Problems

66 • If the net torque about some point is zero, must it be zero about any other point? Explain.

67 • The horizontal bar in Figure 12-52 will remain horizontal if
(a) $L_1 = L_2$ and $R_1 = R_2$.
(b) $L_1 = L_2$ and $M_1 = M_2$.
(c) $R_1 = R_2$ and $M_1 = M_2$.
(d) $L_1 M_1 = L_2 M_2$.
(e) $R_1 L_1 = R_2 L_2$.

R_1 R_2

L_1

L_2

M_1

M_2

68 • Which of the following could not have units of N/m^2?
(a) Young's modulus
(b) Shear modulus
(c) Stress
(d) Strain

69 •• Sit in a chair with your back straight. Now try to stand up without leaning forward. Explain why you cannot do it.

70 • A 90-N board 12 m long rests on two supports, each 1 m from the end of the board. A 360-N block is placed on the board 3 m from one end as shown in Figure 12-53. Find the force exerted by each support on the board.

Figure 12-53
Problem 70

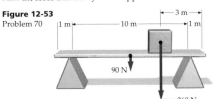
1 m — 10 m — 1 m — 3 m

90 N

$w = 360$ N

CHAPTER 1

Systems of Measurement

Tycho Brahe (1546–1601) with his large brass quadrant for measuring the positions of planets and stars.

We have always been curious about the world around us. Since the beginnings of recorded thought, we have sought ways to impose order on the bewildering diversity of events that we observe. This search for order has taken a variety of forms, and has given birth to religion, art, and science. Although the word "science" has its origins in a Latin verb meaning "to know," science has come to mean not simply knowledge in general, but more specifically, knowledge of the natural world. Most importantly, science is a body of knowledge organized in a specific and rational way.

Today we think of science divided into separate fields, although this division occurred only in the last century or so. The separation of complex systems into smaller categories that can be more easily studied is one of the great successes of science. Biology, for example, is the study of living organisms. Chemistry deals with the interaction of elements and compounds. Geology is the study of the earth. Astronomy is the study of the solar system, the stars and galaxies, and the universe as a whole. Physics is the science of matter and energy, and includes the principles that govern the motion of particles and waves, the interactions of particles, and the properties of molecules, atoms, and atomic nuclei, as well as larger-scale systems such as gases, liquids, and solids. Some consider physics the most fundamental science because its principles supply the foundation of the other scientific fields.

Physics is the science of the exotic and the science of everyday life. At the exotic extreme, black holes boggle the imagination. In everyday life, engineers, musicians, architects, chemists, biologists, doctors, and many others

1

routinely command such subjects as heat transfer, fluid flow, sound waves, radioactivity, and stresses in buildings or bones to perform their daily work. Countless questions about our world can be answered with a basic knowledge of physics. Why must a helicopter have two rotors? Why do astronauts float in space? Why does sound travel around corners while light appears to travel in straight lines? Why does an oboe sound different from a flute? How do CD players work? Why is there no hydrogen in the atmosphere? Why do metal objects feel colder than wood objects at the same temperature? Why is copper an electrical conductor while wood is an insulator? Why is lithium, with its three electrons, extremely reactive, whereas helium, with two electrons, is chemically inert?

Classical and Modern Physics

The earliest recorded efforts to systematically assemble knowledge concerning motion came from ancient Greece. In the system of natural philosophy set forth by Aristotle (384–322 B.C.), explanations of physical phenomena were deduced from assumptions about the world, rather than derived from experimentation. For example, it was a fundamental assumption that every substance had a "natural place" in the universe. Motion was ruled to be the result of a substance trying to reach its natural place. Because of the agreement between the deductions of Aristotelian physics and the motions observed throughout the physical universe, and because there was no tradition of experimentation that could overturn the ancient physics, the Greek view was accepted for nearly 2000 years. It was the Italian scientist Galileo Galilei (1564–1642) whose brilliant experiments on motion established for all time the absolute necessity of experimentation in physics and initiated the disintegration of Aristotelian physics. Within 100 years, Isaac Newton had generalized the results of Galileo's experiments into his three spectacularly successful laws of motion, and the natural philosophy of Aristotle was gone.

Experimentation during the next 200 years brought a flood of discoveries, inspiring the development of physical theories to explain them. By the end of the nineteenth century, Newton's laws for the motions of mechanical systems had been joined by equally impressive laws from Maxwell, Joule, Carnot, and others to describe electromagnetism and thermodynamics. The subjects that occupied physical scientists through the end of the nineteenth century—mechanics, light, heat, sound, electricity, and magnetism—are usually referred to as *classical physics*.

The remarkable success of classical physics led many scientists to believe that the description of the physical universe was complete. However, the discoveries of X rays by Roentgen in 1895 and of nuclear radioactivity by Becquerel in 1896 seemed to be outside the framework of classical physics. The theory of special relativity proposed by Albert Einstein in 1905 contradicted the ideas of space and time of Galileo and Newton. In the same year, Einstein suggested that light energy is quantized; that is, that light comes in discrete packets rather than being wavelike and continuous as had been assumed in classical physics. The generalization of this insight to the quantization of all types of energy is a central idea of quantum mechanics, one that has many amazing and important consequences. The application of special relativity and, particularly, quantum theory to such microscopic systems as atoms, molecules, and nuclei has led to a detailed understanding of solids, liquids, and gases and is often referred to as *modern physics*.

Except for the interiors of atoms and for motions at speeds near the speed of light, classical physics correctly and precisely describes the behavior of the physical world. It is classical physics we must master to understand the macroscopic world we live in, and classical physics is the main subject of this book. Modern physics is itself built on the concepts of classical physics. It is not possible to understand quantum theory without a knowledge of such classical concepts as energy, momentum, angular momentum, wave functions, and standing waves. We thus begin our study of physics with the study of classical topics. However, we will look ahead from time to time and note the relationship between classical and modern physics. When we discuss velocity in Chapter 2, for example, we will take a moment to consider velocities near the speed of light as we cross over to the relativistic universe first imagined by Einstein. After discussing the conservation of energy in Chapter 7, we will discuss the quantization of energy and Einstein's famous relation between mass and energy, $E = mc^2$. After introducing classical waves, we show that light, which is treated as a wave in classical physics, also has particle properties, and electrons, which are particles in the classical view, also have wave properties. Building on the classical concepts of wave functions, interference, diffraction, and standing waves, we show in Chapter 17 that the application of these concepts to electron waves leads to the quantization of energy. With a rudimentary knowledge of quantum theory, we are in position to understand the world around us. We are also prepared to understand when and why classical physics applies in our later studies of physics, and at what times it must be augmented or replaced by quantum physics.

1-1   Units

We all know of things that cannot be measured—the beauty of a flower or of a Bach fugue. As certain as our knowledge of these things may be, we readily admit that this knowledge is not science. The ability not only to define but also to measure is a requisite of science, and in physics, more than in any other field of knowledge, the precise definition of terms and the accurate measurement of quantities have led to great discoveries. We begin our study of physics by establishing a few basic definitions, introducing units, and showing how units are dealt with in equations. The fun comes later.

Measurement of any quantity involves comparison with some precisely defined unit value of the quantity. For example, to measure the distance between two points, we need a standard unit, such as the meter. The statement that a certain distance is 25 meters means that it is 25 times the length of the unit meter. That is, a standard meterstick fits into that distance 25 times. It is important to include the unit, in this case meters, along with the number 25 in expressing this distance, because there are other units of distance such as kilometers or miles in common use. To say that a distance is 25 is meaningless. The magnitude of any physical quantity must include both a number and a unit.

The International System of Units

A small number of fundamental units are sufficient to express all physical quantities. Many of the quantities that we shall be studying, such as velocity, force, momentum, work, energy, and power, can be expressed in terms of three fundamental measures: length, time, and mass. The choice of standard units for these fundamental quantities determines a system of units. The system used universally in the scientific community is called SI (for *Système Internationale*). The standard SI unit for length is the meter, the standard unit of time is the second, and the standard mass is the kilogram. Complete definitions of the SI units are given in Appendix A.

The standard unit of length, the meter (abbreviated m), was originally indicated by two scratches on a bar made of a platinum–iridium alloy kept at the International Bureau of Weights and Measures in Sèvres, France. This length was chosen so that the distance between the equator and the North Pole along the meridian through Paris would be 10 million meters (Figure 1-1). The meter is now defined in terms of the speed of light—the meter is the distance light travels through empty space in 1/299,729,458 second. (This makes the speed of light exactly 299,792,458 m/s.) 3.0×10^8 m/s

The unit of time, the second (s), was originally defined in terms of the rotation of the earth and was equal to $(\frac{1}{60})(\frac{1}{60})(\frac{1}{24})$ of the mean solar day. The second is now defined in terms of a characteristic frequency associated with the cesium atom. All atoms, after absorbing energy, emit light with wavelengths and frequencies characteristic of the particular element. There is a set of wavelengths and frequencies for each element, with a particular frequency and wavelength associated with each energy transition within the atom. As far as we know, these frequencies remain constant. The second is defined so that the frequency of the light from a certain transition in cesium is 9,192,631,770 cycles per second. With these definitions, the fundamental units of length and time are accessible to laboratories throughout the world.

The unit of mass, the kilogram (kg), which equals 1000 grams (g), is defined to be the mass of a standard body, also kept at Sèvres. A duplicate of the standard 1-kg body is kept at the National Institute of Standards and Technology in Gaithersburg, Maryland. We shall discuss the concept of mass in detail in Chapter 4, where we will see that the weight of an object at a given point on earth is proportional to its mass. Thus, masses of ordinary size can be compared by weighing them.

In our study of thermodynamics and electricity, we shall need three more fundamental physical units: one for temperature, the kelvin (K) (formerly the degree Kelvin); one for the amount of a substance, the mole (mol); and one for electrical current, the ampere (A). There is another fundamental unit, the candela (cd) for luminous intensity, which we shall have no occasion to use in this book. These seven fundamental units, the meter (m), second (s), kilogram (kg), kelvin (K), ampere (A), mole (mol), and candela (cd), constitute the international system of units, or SI units.

The unit of every physical quantity can be expressed in terms of the fundamental SI units. Some frequently used combinations are given special names. For example, the SI unit of force, kg·m/s^2, is called a newton (N). Similarly, the SI unit of power, 1 kg·m^2/s^3 = N·m/s, is called a watt (W).

Prefixes for common multiples and submultiples of SI units are listed in Table 1-1. These multiples are all powers of 10. Such a system is called a deci-

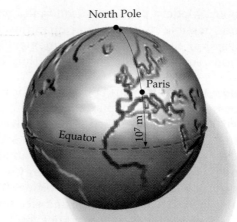

Figure 1-1 The meter was originally chosen so that the distance from the equator to the North Pole along the meridian through Paris would be 10^7 m.

(a)

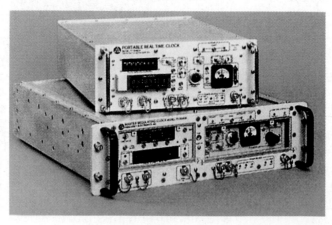

(b)

(a) Water clock used to measure time intervals in the thirteenth century. (b) Modern portable cesium clock.

mal system. The decimal system based on the meter is called the metric system. The prefixes can be applied to any SI unit; for example, 0.001 second is 1 millisecond (ms); 1,000,000 watts is 1 megawatt (MW).

Other Systems of Units

Another decimal system still in use but gradually being replaced by SI units is the cgs system, based on the centimeter, gram, and second. The centimeter is defined as 0.01 m. The gram is now defined as 0.001 kg. Originally the gram was defined as the mass of one cubic centimeter of water. (The kilogram is then the mass of 1000 cubic centimeters or one liter of water.)

In another system of units, the U.S. customary system, a unit of force, the pound, is chosen to be a fundamental unit. In this system, the unit of mass is then defined in terms of the fundamental unit of force. The pound is defined in terms of the gravitational attraction of the earth at a particular place for a standard body. The fundamental unit of length in this system is the foot and the unit of time is the second, which has the same definition as the SI unit. The foot is defined as exactly one-third of a yard, which is now defined in terms of the meter:

$$1 \text{ yd} = 0.9144 \text{ m} \tag{1-1}$$

$$1 \text{ ft} = \tfrac{1}{3} \text{ yd} = 0.3048 \text{ m} \tag{1-2}$$

making the inch exactly 2.54 cm. This scheme is not a decimal system. It is less convenient than the SI or other decimal systems because common multiples of the unit are not powers of 10. For example, 1 yd = 3 ft and 1 ft = 12 in. We will see in Chapter 4 that mass is a better choice for a fundamental unit than force because mass is an intrinsic property of an object independent of its location. Relations between the U.S. customary system and SI units are given in Appendix A.

Table 1-1

Prefixes for Powers of 10

Multiple	Prefix	Abbreviation
10^{18}	exa	E
10^{15}	peta	P
10^{12}	tera	T
10^{9}	giga	G
10^{6}	mega	M
10^{3}	kilo	k
10^{2}	hecto	h
10^{1}	deka	da
10^{-1}	deci	d
10^{-2}	centi	c
10^{-3}	milli	m
10^{-6}	micro	μ
10^{-9}	nano	n
10^{-12}	pico	p
10^{-15}	femto	f
10^{-18}	atto	a

The prefixes hecto (h), deka (da), and deci (d) are not multiples of 10^3 or 10^{-3} and are rarely used. The other prefix that is not a multiple of 10^3 or 10^{-3} is centi (c), now used only with the meter, as in $1 \text{ cm} = 10^{-2} \text{ m}$.

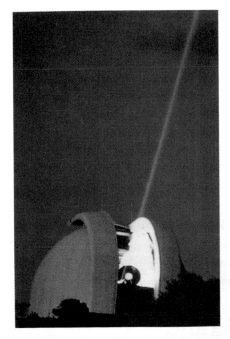

(a)

(b)

(a) Laser beam from the Macdonald Observatory used to measure the distance to the moon. The distance can be measured to within a few centimeters by measuring the time required for the beam to go to the moon and back after reflecting off a mirror (b) placed on the moon by the Apollo 14 astronauts.

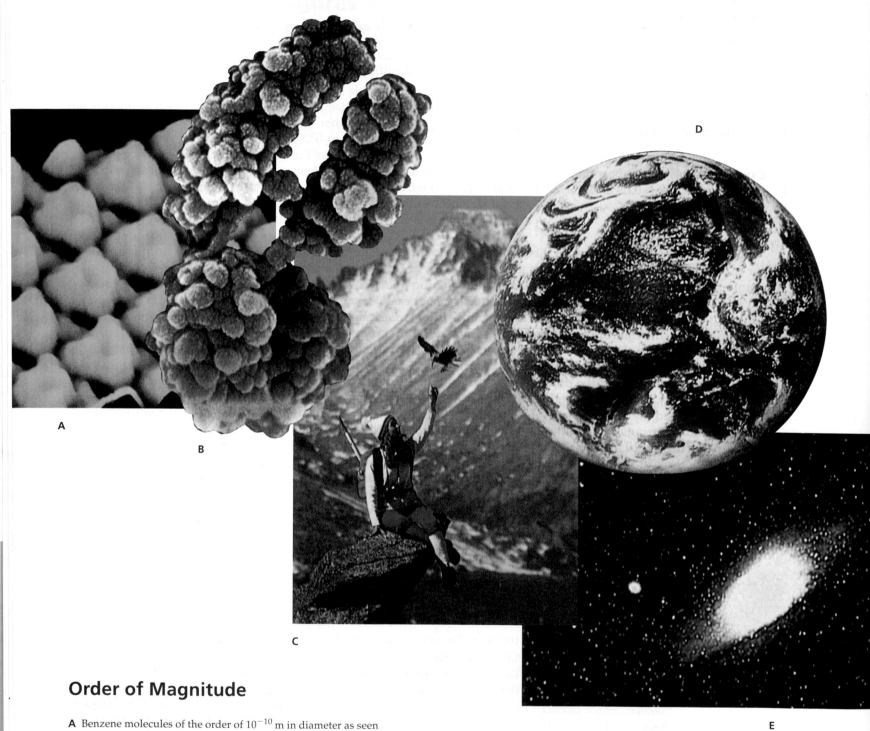

Order of Magnitude

A Benzene molecules of the order of 10^{-10} m in diameter as seen in a scanning electron microscope.

B Chromosomes measuring of the order of 10^{-6} m across as seen in a scanning electron microscope.

C Distances familiar in our everyday world. The height of the girl is of the order of 10^{0} m and that of the mountain is of the order of 10^{4} m.

D Earth with a diameter of the order of 10^{7} m as seen from space.

E The diameter of the Andromeda galaxy is of the order of 10^{21} m.

In many cases the order of magnitude of a quantity can be estimated using reasonable assumptions and simple calculations. The physicist Enrico Fermi was a master at using cunning order-of-magnitude estimations to generate answers for questions that seemed impossible to calculate because of lack of information. Problems like these are often called **Fermi problems**. The following is an example of a Fermi problem.

Example 1-6

What thickness of rubber tread is worn off the tire of an automobile as it travels 1 km (0.6 mi)?

Picture the Problem We assume the tread thickness of a new tire is 1 cm. This may be off by a factor of 2 or so, but 1 mm is certainly too small and 10 cm is too large. Since tires have to be replaced after about 60,000 km (about 37,000 mi), we will assume that the tread is completely worn off after 60,000 km. In other words, the rate of wear is 1 cm of tire per 60,000 km of travel.

Use 1 cm wear per 60,000 km travel to compute the thickness worn after 1 km of travel:

$$\frac{1 \text{ cm wear}}{60,000 \text{ km travel}} = \frac{1.7 \times 10^{-5} \text{ cm wear}}{1 \text{ km travel}}$$

$$\approx 0.2 \ \mu\text{m wear/km of travel}$$

Summary

Topic	Remarks and Relevant Equations
1. Units	The magnitude of a physical quantity (for example, length, time, force, and energy) is expressed as a number times a unit.
Fundamental units	The fundamental SI units (short for *Système Internationale*) are the meter (m), the second (s), the kilogram (kg), the kelvin (K), the ampere (A), the mole (mol), and the candela (cd). The unit of every physical quantity can be expressed in terms of these fundamental units.
Units in equations	Units in equations are treated just like any other algebraic quantity.
Conversion	Conversion factors, which are always equal to 1, provide a convenient method for converting from one kind of unit to another.
2. Dimensions	The two sides of an equation must have the same dimensions.
3. Scientific Notation	For convenience, very small and very large numbers are generally written as a factor times a power of 10.
4. Exponents	
Multiplication	When multiplying two numbers, the exponents are added.
Division	When dividing two numbers, the exponents are subtracted.
Raising to a power	When a number containing an exponent is itself raised to an exponent, the exponents are multiplied.

5. Significant Figures	
Multiplication and division	The number of significant figures in the result of multiplication or division is no greater than the least number of significant figures in any of the numbers.
Addition and subtraction	The result of addition or subtraction of two numbers has no significant figures beyond the last decimal place where both of the original numbers had significant figures.
6. Order of Magnitude	A number rounded to the nearest power of 10 is called an order of magnitude. The order of magnitude of a quantity can often be estimated using reasonable assumptions and simple calculations.

Problem-Solving Guide

Summary of Worked Examples

Type of Calculation	Procedure and Relevant Examples	
1. Units		
Convert a quantity from one set of units to another.	Multiply by a conversion factor that has the value 1.	**Examples 1-1, 1-3**
2. Dimensions		
Analyze the dimensions of quantities to determine how one quantity depends on two or more other quantities.	Inspect the dimensions and try various combinations.	**Example 1-2**
3. Scientific Notation		
Multiply two numbers.	Write the numbers in powers of 10 notation, then add the exponents.	**Example 1-3**
Divide two numbers.	Write the numbers in powers of 10 notation, then subtract the exponents.	**Example 1-3**
Add or subtract two numbers.	Write both numbers so they have the same power of 10.	
4. Significant Figures		
Multiply or divide two numbers.	Round off the answer to the lesser number of significant figures of either factor.	
Add or subtract two numbers.	Round off the answer to the lesser number of significant figures beyond the decimal point of either of the factors.	**Example 1-5**
5. Fermi Problems	Supply needed data by order-of-magnitude estimates.	**Example 1-6**

Problems

In a few problems, you are given more data than you actually need; in a few other problems, you are required to supply data from your general knowledge, outside sources, or informed estimates.

• Single-concept, single-step, relatively easy
•• Intermediate-level, may require synthesis of concepts
••• Challenging, for advanced students

Units

1 • Which of the following is *not* one of the fundamental physical quantities in the SI system?

(*a*) mass (*b*) length (*c*) force (*d*) time
(*e*) All of the above are fundamental physical quantities.

2 • In doing a calculation, you end up with m/s in the numerator and m/s^2 in the denominator. What are your final units?

(*a*) m^2/s^3 (*b*) $1/s$ (*c*) s^3/m^2 (*d*) s (*e*) m/s

3 • Write the following using the prefixes listed in Table 1-1 and the abbreviations listed on the inside cover; for example, 10,000 meters = 10 km. (*a*) 1,000,000 watts, (*b*) 0.002 gram, (*c*) 3×10^{-6} meter, (*d*) 30,000 seconds.

4 • Write each of the following without using prefixes: (*a*) 40 μW, (*b*) 4 ns, (*c*) 3 MW, (*d*) 25 km.

5 • Write out the following (which are not SI units) without using any abbreviations. For example, 10^3 meters = 1 kilometer. (*a*) 10^{-12} boo, (*b*) 10^9 low, (*c*) 10^{-6} phone, (*d*) 10^{-18} boy, (*e*) 10^6 phone, (*f*) 10^{-9} goat, (*g*) 10^{12} bull.

6 •• In the following equations, the distance x is in meters, the time t is in seconds, and the velocity v is in meters per second. What are the SI units of the constants C_1 and C_2?

(*a*) $x = C_1 + C_2 t$ (*b*) $x = \frac{1}{2} C_1 t^2$ (*c*) $v^2 = 2C_1 x$
(*d*) $x = C_1 \cos C_2 t$ (*e*) $v = C_1 e^{-C_2 t}$

(*Hint:* The arguments of trigonometric functions and exponentials must be dimensionless. The "argument" of $\cos \theta$ is θ and that of e^x is x.)

7 •• If x is in feet, t is in seconds, and v is in feet per second, what are the units of the constants C_1 and C_2 in each part of Problem 6?

Conversion of Units

8 • From the original definition of the meter in terms of the distance from the equator to the North Pole, find in meters (*a*) the circumference of the earth, and (*b*) the radius of the earth. (*c*) Convert your answers for (*a*) and (*b*) from meters into miles.

9 • The speed of sound in air is 340 m/s. What is the speed of a supersonic plane that travels at twice the speed of sound? Give your answer in kilometers per hour and miles per hour.

10 • A basketball player is 6 ft $10\frac{1}{2}$ in tall. What is his height in centimeters?

11 • Complete the following:
(*a*) 100 km/h = _____ mi/h. (*b*) 60 cm = _____ in.
(*c*) 100 yd = _____ m.

12 • The main span of the Golden Gate Bridge is 4200 ft. Express this distance in kilometers.

13 • Find the conversion factor to convert from miles per hour into kilometers per hour.

14 • Complete the following:
(*a*) 1.296×10^5 km/h² = _____ km/h·s.
(*b*) 1.296×10^5 km/h² = _____ m/s².
(*c*) 60 mi/h = _____ ft/s. (*d*) 60 mi/h = _____ m/s.

15 • There are 1.057 quarts in a liter and 4 quarts in a gallon. (*a*) How many liters are there in a gallon? (*b*) A barrel equals 42 gallons. How many cubic meters are there in a barrel?

16 • There are 640 acres in a square mile. How many square meters are there in one acre?

17 •• A right circular cylinder has a diameter of 6.8 in and a height of 2 ft. What is the volume of the cylinder in (*a*) cubic feet, (*b*) cubic meters, (*c*) liters?

18 •• In the following, x is in meters, t is in seconds, v is in meters per second, and the acceleration a is in meters per second squared. Find the SI units of each combination:

(*a*) v^2/x (*b*) $\sqrt{x/a}$ (*c*) $\frac{1}{2} at^2$

Dimensions of Physical Quantities

19 • What are the dimensions of the constants in each part of Problem 6?

20 •• The law of radioactive decay is $N(t) = N_0 e^{-\lambda t}$, where N_0 is the number of radioactive nuclei at $t = 0$, $N(t)$ is the number remaining at time t, and λ is a quantity known as the decay constant. What is the dimension of λ?

21 •• The SI unit of force, the kilogram-meter per second squared ($kg \cdot m/s^2$) is called the newton (N). Find the dimensions and the SI units of the constant G in Newton's law of gravitation $F = Gm_1 m_2 / r^2$.

22 •• An object on the end of a string moves in a circle. The force exerted by the string has units of ML/T^2 and depends on the mass of the object, its speed, and the radius of the circle. What combination of these variables gives the correct dimensions?

23 •• Show that the product of mass, acceleration, and speed has the dimension of power.

24 •• The momentum of an object is the product of its velocity and mass. Show that momentum has the dimension of force multiplied by time.

25 •• What combination of force and one other physical quantity has the dimension of power?

26 •• When an object falls through air, there is a drag force that depends on the product of the surface area of the object and the square of its velocity, i.e., $F_{air} = CAv^2$, where C is a constant. Determine the dimension of C.

27 •• Kepler's third law relates the period of a planet to its radius r, the constant G in Newton's law of gravitation ($F = Gm_1m_2/r^2$), and the mass of the sun M_S. What combination of these factors gives the correct dimensions for the period of a planet?

Scientific Notation and Significant Figures

28 • The prefix giga means _____.
(a) 10^3 (b) 10^6 (c) 10^9 (d) 10^{12} (e) 10^{15}

29 • The prefix mega means _____.
(a) 10^{-9} (b) 10^{-6} (c) 10^{-3} (d) 10^6 (e) 10^9

30 • The prefix pico means _____.
(a) 10^{-12} (b) 10^{-6} (c) 10^{-3} (d) 10^6 (e) 10^9

31 • The number 0.0005130 has _____ significant figures.
(a) one   (b) three   (c) four   (d) seven   (e) eight

32 • The number 23.0040 has _____ significant figures.
(a) two   (b) three   (c) four   (d) five   (e) six

33 • Express as a decimal number without using powers of 10 notation:
(a) 3×10^4 (b) 6.2×10^{-3} (c) 4×10^{-6} (d) 2.17×10^5

34 • Write the following in scientific notation.
(a) 3.1 GW = _____ W.
(b) 10 pm = _____ m.
(c) 2.3 fs = _____ s.
(d) 4 μs = _____ s.

35 • Calculate the following, round off to the correct number of significant figures, and express your result in scientific notation.
(a) $(1.14)(9.99 \times 10^4)$
(b) $(2.78 \times 10^{-8}) - (5.31 \times 10^{-9})$
(c) $12\pi/(4.56 \times 10^{-3})$
(d) $27.6 + (5.99 \times 10^2)$

36 • Calculate the following, round off to the correct number of significant figures, and express your result in scientific notation.
(a) (200.9)(569.3)
(b) $(0.000000513)(62.3 \times 10^7)$
(c) $28,401 + (5.78 \times 10^4)$
(d) $63.25/(4.17 \times 10^{-3})$

37 • A cell membrane has a thickness of about 7 nm. How many cell membranes would it take to make a stack 1 in high?

38 • Calculate the following, round off to the correct number of significant figures, and express your result in scientific notation.
(a) $(2.00 \times 10^4)(6.10 \times 10^{-2})$
(b) $(3.141592)(4.00 \times 10^5)$
(c) $(2.32 \times 10^3)/(1.16 \times 10^8)$
(d) $(5.14 \times 10^3) + (2.78 \times 10^2)$
(e) $(1.99 \times 10^2) + (9.99 \times 10^{-5})$

39 • Perform the following calculations and round off the answers to the correct number of significant figures:
(a) $3.141592654 \times (23.2)^2$
(b) $2 \times 3.141592654 \times 0.76$
(c) $4/3\pi \times (1.1)^3$
(d) $(2.0)^5/(3.141592654)$

40 •• The sun has a mass of 1.99×10^{30} kg and is composed mostly of hydrogen, with only a small fraction being heavier elements. The hydrogen atom has a mass of 1.67×10^{-27} kg. Estimate the number of hydrogen atoms in the sun.

General Problems

41 • What are the advantages and disadvantages of using the length of your arm for a standard length?

42 • A certain clock is known to be consistently 10% fast compared with the standard cesium clock. A second clock varies in a random way by 1%. Which clock would make a more useful secondary standard for a laboratory? Why?

43 • True or false:
(a) Two quantities to be added must have the same dimensions.
(b) Two quantities to be multiplied must have the same dimensions.
(c) All conversion factors have the value 1.

44 • On many of the roads in Canada the speed limit is 100 km/h. What is the speed limit in miles per hour?

45 • If one could count $1 per second, how many years would it take to count 1 billion dollars (1 billion = 10^9)?

46 • Sometimes a conversion factor can be derived from the knowledge of a constant in two different systems. (a) The speed of light in vacuum is 186,000 mi/s = 3×10^8 m/s. Use this fact to find the number of kilometers in a mile. (b) The weight of 1 ft^3 of water is 62.4 lb. Use this and the fact that 1 cm^3 of water has a mass of 1 g to find the weight in pounds of a 1-kg mass.

47 •• The mass of one uranium atom is 4.0×10^{-26} kg. How many uranium atoms are there in 8 g of pure uranium?

48 •• During a thunderstorm, a total of 1.4 in of rain falls. How much water falls on one acre of land? (1 acre = 640 mi^2.)

49 •• The angle subtended by the moon's diameter at a point on the earth is about 0.524° (Figure 1-2). Use this and the fact that the moon is about 384 Mm away to find the diameter of the moon. (The angle subtended by the moon θ is approximately D/r_m, where D is the diameter of the moon and r_m is the distance to the moon.)

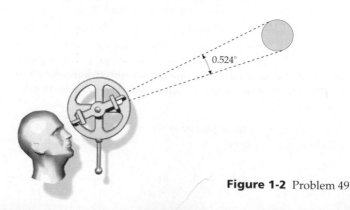

Figure 1-2 Problem 49

50 •• The United States imports 6 million barrels of oil per day. This imported oil provides about one-fourth of our total energy. A barrel fills a drum that stands about 1 m high. (a) If the barrels are laid end to end, what is the length in kilometers of barrels of oil imported each day? (b) The largest tankers hold about a quarter-million barrels. How many tanker loads per year would supply our imported oil? (c) If oil costs $20 a barrel, how much do we spend for oil each year?

51 •• Every year the United States generates 160 million tons of municipal solid waste and a grand total of 10 billion tons of solid waste of all kinds. If one allows one cubic meter of volume per ton, how many square miles of area at an average height of 10 m is needed for landfill each year?

52 •• An iron nucleus has a radius of 5.4×10^{-15} m and a mass of 9.3×10^{-26} kg. (a) What is its mass per unit volume in kg/m^3? (b) If the earth had the same mass per unit volume, what would its radius be? (The mass of the earth is 5.98×10^{24} kg.)

53 •• Evaluate the following expressions.

(a) $(5.6 \times 10^{-5})(0.0000075)/(2.4 \times 10^{-12})$
(b) $(14.2)(6.4 \times 10^{7})(8.2 \times 10^{-9}) - 4.06$
(c) $(6.1 \times 10^{-6})^2(3.6 \times 10^{4})^3/(3.6 \times 10^{-11})^{1/2}$
(d) $(0.000064)^{1/3}/[(12.8 \times 10^{-3})(490 \times 10^{-1})^{1/2}]$

54 •• The astronomical unit is defined in terms of the distance from the earth to the sun, namely 1.496×10^{11} m. The parsec is the radial length that one astronomical unit of arc length subtends at an angle of 1 s. The light-year is the distance that light travels in one year. (a) How many parsecs are there in one astronomical unit? (b) How many meters are in a parsec? (c) How many meters in a light-year? (d) How many astronomical units in a light-year? (e) How many light-years in a parsec?

55 •• If the average density of the universe is at least 6×10^{-27} kg/m^3, then the universe will eventually stop expanding and begin contracting. (a) How many electrons are needed in a cubic meter to produce the critical density? (b) How many protons per cubic meter would produce the critical density? (The mass of an electron and of a proton can be found on the inside cover sheets.)

56 •• Observational estimates of the density of the universe yield an average of about 2×10^{-28} kg/m^3. (a) If a 100-kg football player had his mass uniformly spread out in a sphere to match the estimate for the average mass density of the universe, what would be the radius of the sphere? (b) Compare this radius with the earth–moon distance of 3.82×10^{8} m.

57 •• Beer and soft drinks are sold in aluminum cans. The mass of a typical can is about 0.018 kg. (a) Estimate the number of aluminum cans used in the United States in one year. (b) Estimate the total mass of aluminum in a year's consumption from these cans. (c) If aluminum returns $1/kg at a recycling center, how much is a year's accumulation of aluminum cans worth?

58 •• An aluminum rod is 8.00024 m long at 20.00°C. If the rod's temperature increases, it expands such that it lengthens by 0.0024% per degree temperature rise. Determine the rod's length at 28.00°C and at 31.45°C.

59 ••• The table below gives experimental results for a measurement of the period of motion T of an object of mass m suspended on a spring versus the mass of the object. These data are consistent with a simple equation expressing T as a function of m of the form $T = Cm^n$, where C and n are constants and n is not necessarily an integer. (a) Find n and C. (There are several ways to do this. One is to guess the value of n and check by plotting T versus m^n on graph paper. If your guess is right, the plot will be a straight line. Another is to plot log T versus log m. The slope of the straight line on this plot is n.) (b) Which data points deviate the most from a straight-line plot of T versus m^n?

Mass m, kg	0.10	0.20	0.40	0.50	0.75	1.00	1.50
Period T, s	0.56	0.83	1.05	1.28	1.55	1.75	2.22

60 ••• The table below gives the period T and orbit radius r for the motions of four satellites orbiting a dense, heavy asteroid. (a) These data can be fitted by the formula $T = Cr^n$. Find C and n. (b) A fifth satellite is discovered to have a period of 6.20 y. Find the radius for the orbit of this satellite, which fits the same formula.

Period T, y	0.44	1.61	3.88	7.89
Radius r, Gm	0.088	0.208	0.374	0.600

61 ••• The period T of a simple pendulum depends on the length L of the pendulum and the acceleration of gravity g (dimensions L/T^2). (a) Find a simple combination of L and g which has the dimensions of time. (b) Check the dependence of the period T on the length L by measuring the period (time for a complete swing back and forth) of a pendulum for two different values of L. (c) The correct formula relating T to L and g involves a constant that is a multiple of π, and cannot be obtained by the dimensional analysis of part (a). It can be found by experiment as in (b) if g is known. Using the value $g = 9.81$ m/s^2 and your experimental results from (b), find the formula relating T to L and g.

62 ••• The weight of the earth's atmosphere pushes down on the surface of the earth with a force of 14.7 lb for each square inch of earth's surface. (a) What is the weight in pounds of the earth's atmosphere? (The radius of the earth is about 6370 km.)

63 ••• Each binary digit is termed a bit. A series of bits grouped together is called a word. An eight-bit word is called a byte. Suppose a computer hard disk has a capacity of 2 gigabytes. (a) How many bits can be stored on the disk? (b) Estimate the number of typical books that can be stored on the disk.

mechanics

Modern version of Galileo's legendary experiment in which he dropped a heavy ball and a light ball from the Leaning Tower of Pisa. In this demonstration, a feather and an apple in a large vacuum chamber fall with the same acceleration due to gravity.

CHAPTER 2

Motion in One Dimension

The most coveted records in top fuel dragstrip racing are the elapsed time to complete a quarter mile from a standing start, and the top speed, measured over the last 66 feet of a quarter mile sprint. The record for elapsed time as of early 1998 is 4.564 seconds, and the top speed is 321.78 miles per hour.

We begin our study of the physical universe by examining objects in motion. The study of motion, whose measurement more than 400 years ago gave birth to physics, is called **kinematics.** We start with the simplest case, the motion of a particle along a straight line, like that of a car moving along a flat, straight, narrow road. A particle is an object whose position can be described by a single point. Anything can be considered to be a particle—a molecule, a person, or a galaxy—as long as we can reasonably ignore its internal structure.

2-1 Displacement, Velocity, and Speed

Figure 2-1 shows a car at position x_1 at time t_1 and at position x_2 at a later time t_2. The change in the car's position, called the **displacement,** is given by $x_2 - x_1$. We use the Greek letter Δ (uppercase delta) to indicate the change in a quantity; thus, the change in x is written Δx:

$$\Delta x = x_2 - x_1 \qquad \text{2-1}$$

Definition—Displacement

The notation Δx (read "delta x") stands for a single quantity, the change in x. It is not a product of Δ and x any more than $\cos \theta$ is a product of cos and θ. By convention, the change in a quantity is always its final value minus its initial value.

Velocity is the rate at which the position changes. The **average velocity** of the particle is defined as the ratio of the displacement Δx to the time interval $\Delta t = t_2 - t_1$:

$$v_{av} = \frac{\Delta x}{\Delta t} = \frac{x_2 - x_1}{t_2 - t_1} \qquad 2\text{-}2$$

Definition—Average velocity

Displacement and average velocity may be positive or negative. A positive value indicates motion in the positive x direction. The SI unit of velocity is meters per second (m/s), and the U.S. customary unit is feet per second (ft/s).

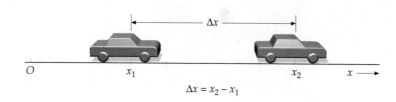

$\Delta x = x_2 - x_1$

Figure 2-1 A car is moving in a coordinate system consisting of a line with a point chosen to be the origin O. Other points on the line are assigned a number x. The value of x depends on its distance from O. Points to the right of O are positive and points to the left are negative. When the car travels from point x_1 to point x_2, its displacement is $\Delta x = x_2 - x_1$.

Example 2-1

An errant space probe is traveling directly toward the sun. At time t_1 it is at $x_1 = 3.0 \times 10^{12}$ m relative to the sun (Figure 2-2). Exactly one year later, it is at $x_2 = 2.1 \times 10^{12}$ m. Find its displacement and average velocity.

Picture the Problem We are given x_1 and x_2. If we choose $t_1 = 0$, then $t_2 = 1$ y $= 3.16 \times 10^7$ s. The average velocity is $\Delta x/\Delta t$.

Figure 2-2

1. The displacement is found from its definition:

$$\Delta x = x_2 - x_1 = 2.1 \times 10^{12}\,\text{m} - 3.0 \times 10^{12}\,\text{m} = -9 \times 10^{11}\,\text{m}$$

2. The average velocity is the displacement divided by the time interval:

$$v_{av} = \frac{\Delta x}{\Delta t} = \frac{-9 \times 10^{11}\,\text{m}}{3.16 \times 10^7\,\text{s}}$$

$$= -2.85 \times 10^4\,\text{m/s} = -28.5\,\text{km/s}$$

Remark Both displacement and average velocity are negative, because the probe moved toward smaller values of x. Note that the units, m for Δx, and m/s or km/s for v_{av}, are essential parts of the answers. It is meaningless to say "the displacement is -9×10^{11}" or "the average velocity of a particle is -28.5."

Exercise A jet plane leaves the gate in Detroit at 2:15 P.M. It arrives at the gate in Chicago, 438 km away, having completed the journey with an average velocity of 500 km/h. What is the arrival time of the flight in Chicago? (*Answer* 3:13 P.M. Detroit time, which is actually 2:13 P.M. Chicago time.)

Example 2-2 *try it yourself*

In a 100-m footrace, you cover the first 50 m with an average velocity of 10 m/s and the second 50 m with an average velocity of 8 m/s. What is your average velocity for the entire 100 m?

Picture the Problem The total displacement is $\Delta x = 100$ m. Find the total time by adding the times for each part of the race. Then compute v_{av} from its definition, $v_{av} =$ (total displacement)/(total time) $= \Delta x / \Delta t$.

Cover the column to the right and try these on your own before looking at the answers.

Steps	Answers
1. Write the average velocity in terms of the total displacement and total time interval.	$v_{av} = \dfrac{\Delta x}{\Delta t}$
2. Compute the time for the first 50 m.	$\Delta t_1 = 5$ s
3. Compute the time for the second 50 m.	$\Delta t_2 = 6.25$ s
4. Calculate the total time.	$\Delta t = \Delta t_1 + \Delta t_2 = 11.25$ s
5. Use your result for the total time to compute the average velocity.	$v_{av} = \dfrac{\Delta x}{\Delta t} = 8.89$ m/s

Check the Result Given the statement of the problem, the answer should be between 8 m/s and 10 m/s, which it is. Since more time is spent running at 8 m/s than at 10 m/s, the average velocity is closer to the lower value.

Exercise A car travels 80 km in a straight line. If the first 40 km is covered with an average velocity of 80 km/h, and the total trip takes 1.2 h, what was the average velocity during the second 40 km? (*Answer*   57.1 km/h)

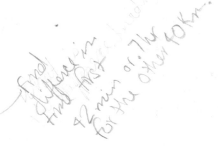

The **average speed** of a particle is the ratio of the total distance traveled to the total time from start to finish:

$$\text{Average speed} = \frac{\text{total distance}}{\text{total time}} = \frac{\Delta s}{\Delta t} \qquad 2\text{-}3$$

Since the total distance and total time are both always positive, the average speed is always positive.

Example 2-3

You run 100 m in 12 s, then turn around and jog 50 m back toward the starting point in 30 s (Figure 2-3). Calculate (*a*) your average speed, and (*b*) your average velocity for the total trip.

Picture the Problem We use the definitions of average speed and average velocity, noting that average *speed* is the total *distance* divided by Δt, whereas average *velocity* is the *net displacement* divided by Δt.

Figure 2-3

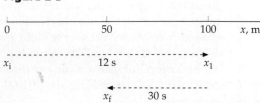

(a) 1. Your average speed equals the total distance divided by the total time:

$$\text{Average speed} = \frac{\Delta s}{\Delta t}$$

2. Calculate the total distance traveled and the total time:

$$\Delta s = 100 \text{ m} + 50 \text{ m} = 150 \text{ m}$$
$$\Delta t = 12 \text{ s} + 30 \text{ s} = 42 \text{ s}$$

3. Use s and t to find your average speed:

$$\text{Average speed} = \frac{150 \text{ m}}{42 \text{ s}} = 3.57 \text{ m/s}$$

(b) 1. Your average velocity is the ratio of the net displacement Δx to the time interval Δt:

$$v_{av} = \frac{\Delta x}{\Delta t}$$

2. Your net displacement is $x_f - x_i$, where $x_i = 0$ is your initial position and $x_f = 50$ m is your final position:

$$\Delta x = x_f - x_i = 50 \text{ m} - 0 = 50 \text{ m}$$

3. Use Δx and Δt to find your average velocity:

$$v_{av} = \frac{\Delta x}{\Delta t} = \frac{50 \text{ m}}{42 \text{ s}} = 1.19 \text{ m/s}$$

Check the Result The world record for a 100-m race is just under 10 s, so 10 m/s is about the maximum speed obtainable. The result of 3.57 m/s for the average speed in part (a) is reasonable, given that you merely jogged for one-third of the distance. If you had obtained 35.7 m/s for the average speed, that would have been a clue that something was wrong with the calculation.

Remark Note that your average speed is greater than your average velocity because the total distance traveled is greater than the total displacement.

Example 2-4

Two trains 75 km apart approach each other on parallel tracks, each moving at 15 km/h. A bird flies back and forth between the trains at 20 km/h until the trains pass each other. How far does the bird fly?

Picture the Problem This problem seems difficult at first, but viewed in the right way it is actually quite simple. We approach it by first writing an equation for the quantity to be found, the total distance Δs flown by the bird.

1. The total distance equals the average speed times the time:

$$\Delta s = \text{average speed} \times \Delta t = (20 \text{ km/h})\Delta t$$

2. The time that the bird is in the air is the time taken for the trains to meet. Since they are moving toward each other at 15 km/h, the distance between them decreases at 30 km/h. Use that to calculate how long it will take for the distance between them to go from 75 km to zero:

$$\Delta t = \frac{75 \text{ km}}{30 \text{ km/h}} = 2.5 \text{ h}$$

3. The total distance traveled by the bird is therefore:

$$\Delta s = (20 \text{ km/h})(2.5 \text{ h}) = 50 \text{ km}$$

Remark Some try to solve this problem by finding and summing the distances flown by the bird each time it moves from one train to the other. This makes a relatively easy problem quite difficult. It is important to develop a thoughtful, systematic approach to solving problems. Begin by writing an equation for the unknown quantity in terms of other quantities. Then proceed by determining the values for each of the other quantities in the equation.

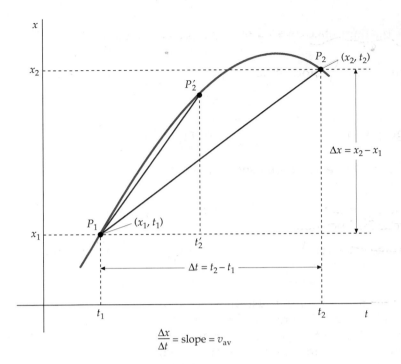

$$\frac{\Delta x}{\Delta t} = \text{slope} = v_{av}$$

Figure 2-4 Graph of x versus t for a particle moving in one dimension. Each point on the curve represents the location x at a particular time t. We have drawn a straight line between positions P_1 and P_2. The displacement $\Delta x = x_2 - x_1$ and the time interval $\Delta t = t_2 - t_1$ between these points are indicated. The straight line between P_1 and P_2 is the hypotenuse of the triangle having sides Δx and Δt, and the ratio $\Delta x / \Delta t$ is its slope. In geometric terms, the slope is a measure of the line's steepness.

Figure 2-4 depicts average velocity graphically. A straight line connects points P_1 and P_2 and forms the hypotenuse of the triangle having sides Δx and Δt. The ratio $\Delta x / \Delta t$ is the line's **slope**, which gives us a geometric interpretation of average velocity:

> The average velocity is the slope of the straight line connecting the points (t_1, x_1) and (t_2, x_2).

Generally, average velocity depends on the time interval on which it is based. In Figure 2-4, for example, the smaller time interval indicated by t_2' and P_2' gives a larger average velocity, as shown by the greater steepness of the line connecting points P_1 and P_2'.

Instantaneous Velocity

On first consideration, defining the velocity of a particle at a single instant seems impossible. At a given instant, a particle is at a single point. If it is at a single point, how can it be moving? If it is not moving, how can it have a velocity? This age-old paradox is resolved when we realize that observing and defining motion requires that we look at the position of the object at more than one time. Consider Figure 2-5. As we consider successively shorter time intervals beginning at t_1, the average velocity for the interval approaches the slope of the tangent at t_1. We define the slope of this tangent as the **instantaneous velocity** at t_1. This tangent is the limit of the ratio $\Delta x / \Delta t$ as Δt, and therefore Δx, approaches zero. So we can say,

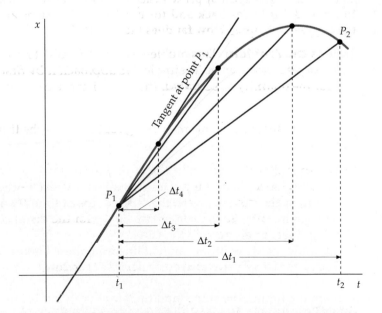

Figure 2-5 Graph of x versus t. Note the sequence of successively smaller time intervals, $\Delta t_1, \Delta t_2, \Delta t_3, \ldots$. The average velocity of each interval is the slope of the straight line for that interval. As the time intervals become smaller, these slopes approach the slope of the tangent to the curve at point t_1. The slope of this line is defined as the instantaneous velocity at time t_1.

The instantaneous velocity is the limit of the ratio $\Delta x/\Delta t$ as Δt approaches zero:

$$v(t) = \lim_{\Delta t \to 0} \frac{\Delta x}{\Delta t}$$

$$= \text{slope of the line tangent to the } x\text{-versus-}t \text{ curve*} \qquad \text{2-4}$$

Definition—Instantaneous velocity

This limit is called the **derivative** of x with respect to t. In the usual calculus notation, the derivative is written dx/dt:

$$v(t) = \lim_{\Delta t \to 0} \frac{\Delta x}{\Delta t} = \frac{dx}{dt} \qquad \text{2-5}$$

A line's slope may be positive, negative, or zero; consequently, instantaneous velocity (in one-dimensional motion) may be positive (x increasing), negative (x decreasing), or zero (no motion). The magnitude of the instantaneous velocity is the **instantaneous speed**.

---

* The slope of the line tangent to a curve is often referred to more simply as the "slope of the curve."

Example 2-5   *try it yourself*

The position of a particle as a function of time is given by the curve shown in Figure 2-6. Find the instantaneous velocity at time $t = 2$ s. When is the velocity greatest? When is it zero? Is it ever negative?

Picture the Problem In the figure, we have sketched the line tangent to the curve at $t = 2$ s. The tangent line's slope is the instantaneous velocity of the particle at the given time. You can use this figure to measure the slope of the tangent line.

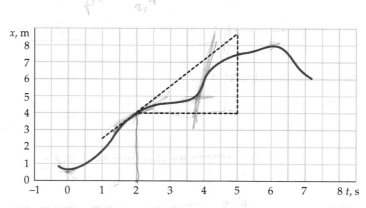

Figure 2-6

Cover the column to the right and try these on your own before looking at the answers.

Steps   **Answers**

1. Find the values x_1 and x_2 on the tangent line at times $t_1 = 2$ s $x_1 \approx 4$ m, $x_2 \approx 8.5$ m
   and $t_2 = 5$ s.

2. Compute the slope of the tangent line from these values. This $v = $ slope $= 1.5$ m/s
   slope equals the instantaneous velocity at $t = 2$ s.

3. From the figure, the slope (and therefore velocity) is greatest
   at about $t = 4$ s. The slope and velocity are zero at $t = 0$ and
   $t = 6$ s and are negative before 0 and after 6 s.

Exercise What is the average velocity of this particle between $t = 2$ s and
$t = 5$ s? (*Answer*   1.17 m/s)

Example 2-6

The position of a stone dropped from a cliff is described approximately by $x = 5t^2$, where x is in meters measured downward from the original position at $t = 0$, and t is in seconds. Find the velocity at any time t. (We omit explicit indication of units to simplify the notation.)

Picture the Problem We can compute the velocity at some time t by computing the derivative dx/dt directly from the definition in Equation 2-4. The corresponding curve giving x versus t is shown in Figure 2-7. Tangent lines are drawn at times t_1, t_2, and t_3. The slopes of these tangent lines increase steadily, indicating that the instantaneous velocity increases steadily with time.

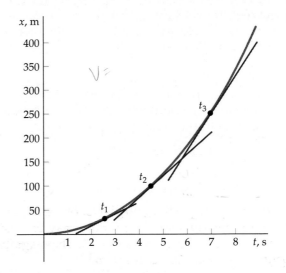

Figure 2-7

1. By definition the instantaneous velocity is:

$$v(t) = \lim_{\Delta t \to 0} \frac{\Delta x}{\Delta t} = \lim_{\Delta t \to 0} \frac{x(t + \Delta t) - x(t)}{\Delta t}$$

2. We compute the displacement Δx from the position function $x(t)$:

$$x(t) = 5t^2$$

3. At a later time $t + \Delta t$ the position is $x(t + \Delta t)$, given by:

$$x(t + \Delta t) = 5(t + \Delta t)^2 = 5[t^2 + 2t\Delta t + (\Delta t)^2]$$
$$= 5t^2 + 10t\Delta t + 5(\Delta t)^2$$

4. The displacement for this time interval is thus:

$$\Delta x = x(t + \Delta t) - x(t)$$
$$= [5t^2 + 10t\Delta t + 5(\Delta t)^2] - 5t^2$$
$$= 10t\Delta t + 5(\Delta t)^2$$

5. Divide Δx by Δt to find the average velocity for this time interval:

$$v_{av} = \frac{\Delta x}{\Delta t} = \frac{10t\,\Delta t + 5(\Delta t)^2}{\Delta t} = 10t + 5\Delta t$$

6. As we consider shorter and shorter time intervals, Δt approaches zero and the second term, $5\,\Delta t$, approaches zero, though the first term, $10t$, remains unchanged. The instantaneous velocity at time t is thus:

$$v(t) = \lim_{\Delta t \to 0} \frac{\Delta x}{\Delta t} = 10t$$

Remark If we had set $\Delta t = 0$ in steps 4 and 5, the displacement would be $\Delta x = 0$, in which case the ratio $\Delta x/\Delta t$ would be undefined. Instead, we leave Δt as a variable until the final step, when the limit $\Delta t \to 0$ is well defined.

To find derivatives quickly, we use rules based on this limiting process (see Appendix Table D-4). A particularly useful rule is

$$\text{If } x = Ct^n, \quad \text{then} \quad \frac{dx}{dt} = Cnt^{n-1} \qquad\qquad 2\text{-}6$$

where C and n are any constants. Using this rule in Example 2-6, we have $x = 5t^2$, and $v = dx/dt = 10t$, in agreement with our previous results.

Relative Velocity

If a particle moves with velocity v_{pA} relative to a coordinate system A, which is in turn moving with velocity v_{AB} relative to another coordinate system B, the velocity of the particle relative to B is

$$v_{pB} = v_{pA} + v_{AB}$$ 2-7a

For example, if you swim in a river parallel to the direction of the flow, your velocity relative to the shore, v_{ys}, equals your velocity relative to the water, v_{yw}, plus the velocity of the water relative to the shore, v_{ws}:

$$v_{ys} = v_{yw} + v_{ws}$$ *approximation*

The velocities add or subtract depending on whether you are swimming with the current or against it. For example, if you are swimming at 2 m/s against the current, and the water speed is 1.2 m/s relative to the shore, then your velocity relative to the shore is $v_{ys} = -2$ m/s $+ 1.2$ m/s $= -0.8$ m/s, where we have chosen the direction of the water flow to be the positive direction. Other common instances when we might want to know the relative velocity are an airplane flying with or against the wind, or a traveler walking on a moving sidewalk at an airport.

 A great surprise of twentieth-century physics was the discovery that Equation 2-7a is only an approximation. A study of the theory of relativity shows that the exact expression for relative velocities is

$$v_{ys} = \frac{v_{yw} + v_{ws}}{1 + v_{yw}v_{ws}/c^2}$$ 2-7b

where $c = 3 \times 10^8$ m/s is the velocity of light in a vacuum. In all everyday cases with macroscopic objects, v_{yw} and v_{ws} are both much smaller than c, so Equations 2-7a and 2-7b are the same, but for very high speeds, such as the speed of an electron or the speed at which distant galaxies are receding from the earth, the difference between these two equations becomes significant. Equation 2-7b has the interesting property that if $v_{yw} = c$, then v_{ys} also equals c, which is a postulate of relativity, namely that the speed of light is the same in all reference frames moving with constant velocity relative to each other (Figure 2-8).

Midair refueling. Each aircraft is nearly at rest relative to the other, though both are moving with very large velocities relative to the earth.

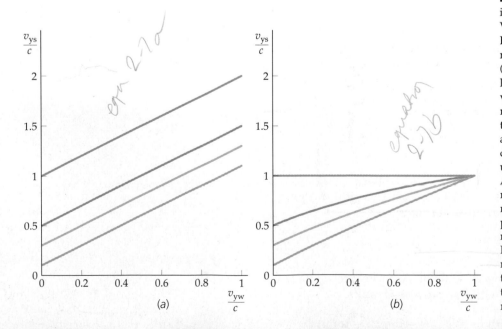

(a)

(b)

Figure 2-8 An boat moving down a river in the everyday and relativistic worlds. (*a*) Velocity addition in our everyday world. Each line shows the boat's velocities relative to shore (*y* coordinate) and water (*x* coordinate). Each of the four different lines corresponds to different relative velocities of water to shore. The uppermost line denotes water flowing relative to the shore at the speed of light, *c*. Note that all velocities are expressed as velocities divided by *c*—that is, as dimensionless units. (*b*) Relativistic velocity addition. Again, each line shows a boat's velocities relative to shore and water for a different relative velocity of water to shore. Unlike part (*a*), the lines are not straight, since the relative velocities combine according to Equation 2-7*b* rather than 2-7*a*. A boat moving at the speed of light *c* relative to the shore also moves at speed *c* relative to the river independent of the velocity of the river relative to the shore.

2-2   Acceleration

Acceleration is the rate of change of the instantaneous velocity. The **average acceleration** for a particular time interval $\Delta t = t_2 - t_1$ is defined as the ratio $\Delta v / \Delta t$, where $\Delta v = v_2 - v_1$:

$$a_{\text{av}} = \frac{\Delta v}{\Delta t}$$

2-8

Definition—Average acceleration

Acceleration has dimensions of length divided by time squared; the SI unit is meters per second squared (m/s^2). For example, if a particle at rest accelerates at 5.1 m/s^2, its velocity after 1 s is 5.1 m/s, its velocity after 2 s is 10.2 m/s, and so on.

   Instantaneous acceleration is the limit of the ratio $\Delta v / \Delta t$ as Δt approaches zero. On a plot of velocity versus time, the instantaneous acceleration at time t is the slope of the line tangent to the curve at that time:

$$a = \lim_{\Delta t \to 0} \frac{\Delta v}{\Delta t}$$

2-9

   = slope of the line tangent to the v versus t curve

Definition—Instantaneous acceleration

Thus, acceleration is the derivative of velocity with respect to time, dv/dt. Since velocity is the derivative of the position x with respect to t, acceleration is the second derivative of x with respect to t, d^2x/dt^2. We can see the reason for this notation when we write the acceleration as dv/dt and replace v with dx/dt:

$$a = \frac{dv}{dt} = \frac{d(dx/dt)}{dt} = \frac{d^2x}{dt^2}$$

2-10

If acceleration is zero, there is no change in velocity over time—velocity is constant. In this case, the curve of x versus t is a straight line. If acceleration is nonzero and constant, as in Example 2-4, then velocity varies linearly with time and the curve of x versus t is quadratic in t.

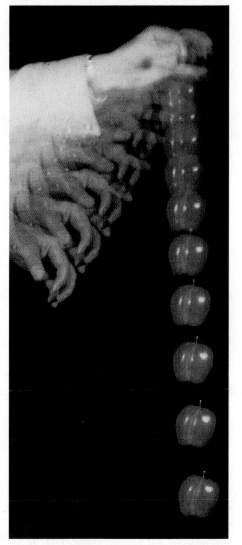

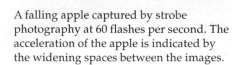

A falling apple captured by strobe photography at 60 flashes per second. The acceleration of the apple is indicated by the widening spaces between the images.

Example 2-7

A cheetah can accelerate from 0 to 96 km/h (60 mi/h) in 2 s, whereas a Corvette requires 4.5 s. Compute the average accelerations for the cheetah and Corvette and compare them with the free-fall acceleration due to gravity, $g = 9.81$ m/s^2.

1. Find the average acceleration from the information given:

Cheetah $a_{av} = \dfrac{\Delta v}{\Delta t} = \dfrac{96 \text{ km/h} - 0}{2 \text{ s}} = 48 \text{ km/h·s}$

Corvette $a_{av} = \dfrac{\Delta v}{\Delta t} = \dfrac{96 \text{ km/h} - 0}{4.5 \text{ s}} = 21.3 \text{ km/h·s}$

2. Convert to m/s^2 using 1 h = 3600 s = 3.6 ks:

Cheetah $\dfrac{48 \text{ km}}{\text{h·s}} \times \dfrac{1 \text{ h}}{3.6 \text{ ks}} = 13.3 \text{ m/s}^2$

Corvette $\dfrac{21.3 \text{ km}}{\text{h·s}} \times \dfrac{1 \text{ h}}{3.6 \text{ ks}} = 5.92 \text{ m/s}^2$

3. To compare the result with the acceleration due to gravity, multiply each by the conversion factor $g/(9.81 \text{ m/s}^2)$:

Cheetah $13.3 \text{ m/s}^2 \times \dfrac{g}{9.81 \text{ m/s}^2} = 1.36g$

Corvette $5.92 \text{ m/s}^2 \times \dfrac{g}{9.81 \text{ m/s}^2} = 0.60g$

Remark Note that by expressing the time in kiloseconds, the k's in km and ks cancel.

Exercise A car is traveling at 45 km/h at time $t = 0$. It accelerates at a constant rate of 10 km/h·s. (a) How fast is it traveling at $t = 2$ s? (b) At what time is the car traveling at 70 km/h? (*Answers*   (a) 65 km/h, (b) 2.5 s)

Exercise in Dimensional Analysis If a car starts from rest at $x = 0$ with constant acceleration a, its velocity v depends on a and the distance traveled x. Which of the following equations has the correct dimensions and therefore could be a possible equation relating x, a, and v?
(a) $v = 2ax$
(b) $v^2 = 2a/x$
(c) $v = 2ax^2$
(d) $v^2 = 2ax$

(*Answer*   Only (d) has the same dimensions on both sides of the equation. Although we cannot obtain the exact equation from dimensional analysis, we can often obtain the functional dependence.)

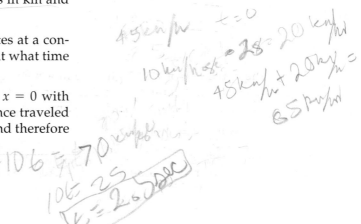

Example 2-8

The position of a particle is given by $x = Ct^3$, where C is a constant having units of m/s^3. Find the velocity and acceleration as functions of time.

1. We find the velocity by applying $x = Ct^3$
$dx/dt = Cnt^{n-1}$ (Equation 2-6):

$$v = \dfrac{dx}{dt} = 3Ct^2$$

2. The time derivative of velocity gives the acceleration:

$$a = \dfrac{dv}{dt} = 6Ct$$

Check the Result We can check the units of our answers. For velocity, $[v] = [C][t^2] = (\text{m/s}^3)(\text{s}^2) = \text{m/s}$. For acceleration, $[a] = [C][t] = (\text{m/s}^3)(\text{s}) = \text{m/s}^2$.

2-3   Motion With Constant Acceleration

The motion of a particle that has constant acceleration is common in nature. For example, near the earth's surface all unsupported objects fall vertically with constant acceleration (provided air resistance is negligible).

If a particle has a constant acceleration a, it follows that the average acceleration for any time interval is also a. Thus,

$$a_{av} = \frac{\Delta v}{\Delta t} = a \quad \text{constant}$$

If the velocity is v_0 at time $t = 0$, and v at some later time t, the corresponding acceleration is

$$a = \frac{\Delta v}{\Delta t} = \frac{v - v_0}{t - 0} = \frac{v - v_0}{t}$$

Rearranging yields v as a function of time.

$$v = v_0 + at \qquad \text{straight line} \tag{2-11}$$

"It goes from zero to 60 in about 3 seconds."
© Sydney Harris

Constant acceleration, v versus t

This is the equation for a straight line in a v-versus-t plot (Figure 2-9). The line's slope is the acceleration a, and its v intercept is the initial velocity v_0.

The displacement $\Delta x = x - x_0$ in the time interval $\Delta t = t - 0$ is

$$\Delta x = v_{av}\Delta t = v_{av}t \tag{2-12a}$$

For constant acceleration, the velocity varies linearly with time, and the average velocity is the mean value of the initial and final velocities. (This relation, which will be derived in Section 2-4, holds only if the acceleration is constant.) If v_0 is the initial velocity and v is the final velocity, the average velocity is

$$v_{av} = \tfrac{1}{2}(v_0 + v) \qquad \frac{v + v}{2} \text{ when } a \text{ is constant} \tag{2-12b}$$

Constant acceleration, v_{av}

The displacement is then

$$\Delta x = x - x_0 = v_{av}t = \tfrac{1}{2}(v_0 + v)t \tag{2-13}$$

We can eliminate v by substituting $v = v_0 + at$ from Equation 2-11:

$$\Delta x = \tfrac{1}{2}(v_0 + v)t = \tfrac{1}{2}(v_0 + v_0 + at)t = v_0t + \tfrac{1}{2}at^2$$

The displacement is thus

$$\Delta x = x - x_0 = \boxed{v_0t + \tfrac{1}{2}at^2} \tag{2-14}$$

Constant acceleration, Δx versus t

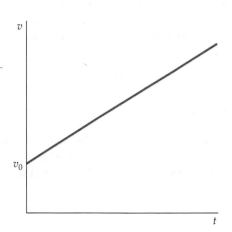

v

v_0

t

Figure 2-9 Graph of velocity versus time for constant acceleration.

The first term on the right, v_0t, is the displacement that would occur if a were zero, and the second term, $\tfrac{1}{2}at^2$, is the additional displacement due to the constant acceleration.

Let's eliminate t from Equations 2-11 and 2-12a and find a relation between Δx, a, v, and v_0. From Equation 2-11, $t = (v - v_0)/a$. Substituting this into Equation 2-12a, we get

$$\Delta x = v_{av}t = \frac{1}{2}(v_0 + v)t = \frac{1}{2}(v_0 + v)\frac{v - v_0}{u} = \frac{v^2 - v_0^2}{2a}$$

or

$$v^2 = v_0^2 + 2a\,\Delta x \qquad\qquad \text{2-15}$$

Constant acceleration, v versus x

Equation 2-15 is useful, for example, if we want to find the final velocity of a ball dropped from rest at some height x and we are not interested in the time the fall takes.

Problems With One Object

Many practical problems deal with objects falling freely due to gravity. The magnitude of the acceleration caused by gravity is designated by g, which has the approximate value

$$g = 9.81 \text{ m/s}^2 = 32.2 \text{ ft/s}^2$$

By convention, g is always positive. If downward is the positive direction, the acceleration due to gravity is $a = g$; if upward is positive, then $a = -g$.

Example 2-9

Upon graduation, a joyful physics student throws his cap upward with an initial speed of 14.7 m/s (Figure 2-10). Given that its acceleration is 9.81 m/s^2 downward (we neglect air resistance), (*a*) how long does it take to reach its highest point? (*b*) What is the distance to the highest point? (*c*) What is the total time the cap is in the air?

Figure 2-10

Picture the Problem When the cap is at its highest point, its instantaneous velocity is zero. Thus we translate the statement "highest point" into the mathematical condition $v = 0$. Similarly, "total time in the air" means the time t following the toss when $x = x_0$. We choose the origin to be at the initial position of the cap, and we designate upward as the positive direction. Then $x_0 = 0$, $v_0 = 14.7$ m/s, and the acceleration, which is downward, is $a = -g = -9.81 \text{ m/s}^2$.

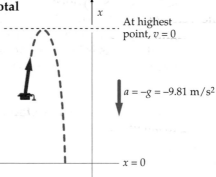

(*a*) 1. The time is related to the velocity and acceleration:

$$v = v_0 + at$$

    2. To find the time at which the cap reaches its greatest height, set $v = 0$, and solve for t:

$$t = \frac{v - v_0}{a} = \frac{0 - 14.7 \text{ m/s}}{-9.81 \text{ m/s}^2} = 1.50 \text{ s}$$

(*b*) We can find the distance traveled from the time t and the average velocity:

$$\Delta x = v_{av}t = \tfrac{1}{2}(v_0 + v)t = \tfrac{1}{2}(14.7 \text{ m/s})(1.50 \text{ s}) = 11.0 \text{ m}$$

(*c*) 1. To find total time, set $x - x_0$ in Equation 2-14 and solve for t:

$$x_0 - x_0 = v_0 t + \tfrac{1}{2}at^2$$

$$0 = t(v_0 + \tfrac{1}{2}at)$$

    2. There are two solutions for t when $x = x_0$. The first corresponds to the time at which the cap is thrown, and the second corresponds to the time at which the cap lands:

$$t = 0 \qquad\qquad\qquad\qquad \text{(first solution)}$$

$$t = -\frac{2v_0}{a} = -\frac{2(14.7 \text{ m/s})}{-9.80 \text{ m/s}^2} = 3 \text{ s} \qquad \text{(second solution)}$$

Remark The $t = 3$ s solution also follows from a symmetry in the system: It takes the same time for the cap to fall from its greatest height as to rise to that height. In reality, however, the cap will not have a constant acceleration because air resistance has a significant effect on a light object like a cap.

Exercise Find Δx using (*a*) Equation 2-13 and (*b*) Equation 2-14. (*c*) Find the velocity of the cap when it returns to its starting point. (*Answers* (*a*) and (*b*) $\Delta x = 11.0$ m, (*c*) -14.7 m/s; notice that the final speed is the same as the initial speed.)

Exercise What is the velocity of the cap (*a*) 0.1 s before it reaches its highest point and (*b*) 0.1 s after it reaches its highest point? (*c*) Compute $\Delta v / \Delta t$ for this 0.2-s time interval. (*Answers* (*a*) 0.981 m/s, (*b*) -0.981 m/s, (*c*) $(-0.981$ m/s $- 0.981$ m/s)/ $(0.2$ s) $= -9.81$ m/s^2)

Exercise A car accelerates from rest at a constant rate of 8 m/s^2. (*a*) How fast is it going after 10 s? (*b*) How far has it gone after 10 s? (*c*) What is its average velocity for the interval $t = 0$ to $t = 10$ s? (*Answers* (*a*) 80 m/s, (*b*) 400 m, (*c*) 40 m/s)

Remark Figure 2-11 shows (*a*) x versus t and (*b*) v versus t for two tosses of a graduate's cap with different initial velocities. Notice that for both tosses the velocity is zero when the cap is at its maximum height. Notice also that the two velocity curves are parallel. This is because the slope of each velocity curve is equal to $-g = -9/81$ m/s^2.

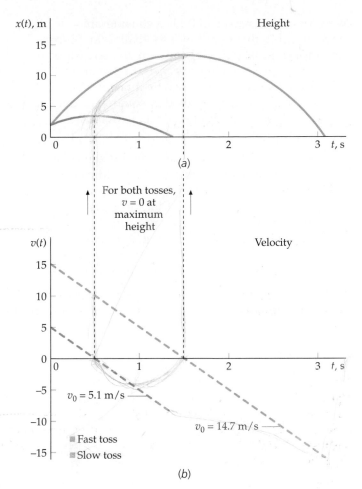

(*a*)

For both tosses, $v = 0$ at maximum height

(*b*)

Figure 2-11

The next example concerns a car's **stopping distance**—how far it travels while coming to a halt.

Example 2-10

On a highway at night you see a stalled vehicle and brake your car to a stop with an acceleration of magnitude 5 m/s^2. (An acceleration that reduces the speed is often called a deceleration.) What is the car's stopping distance if its initial speed is (*a*) 15 m/s (about 34 mi/h) or (*b*) 30 m/s?

Picture the Problem If we choose the direction of motion to be positive, the stopping distance and the initial velocity are positive, but the acceleration is negative. Thus, the initial velocity is $v_0 = 15$ m/s, the final velocity is $v = 0$, and the acceleration is $a = -5$ m/s^2. We seek the distance traveled, Δx. We do not need to know the time it takes for the car to stop, so Equation 2-15 is the most convenient formula to use.

(*a*) 1. Set $v = 0$ in Equation 2-15:

$$v^2 = v_0^2 + 2a\,\Delta x = 0$$

2. Solve for Δx:

$$\Delta x = \frac{v^2 - v_0^2}{2a} = \frac{-v_0^2}{2a} = \frac{-(15 \text{ m/s})^2}{2(-5 \text{ m/s}^2)} = 22.5 \text{ m}$$

(*b*) Use the above result for Δx, but replace v_0 with $v_0' = 2v_0$:

$$\Delta x' = \frac{-(2v_0)^2}{2a} = \frac{-4v_0^2}{2a} = 4\Delta x = 4(22.5 \text{ m}) = 90 \text{ m}$$

for 30 m/sec

Remarks The answer to (b) is a considerable distance, roughly the length of a football field. Since the stopping distance depends on the square of the initial speed, changing v_0 by a factor of 2 changes the stopping distance by a factor $2^2 = 4$. The practical implication of this squared dependence is that even modest increases in speed cause significant increases in stopping distance.

Remark Figure 2-12 shows stopping distance as a function of the initial velocity. The middle curve shows the case where the acceleration is $a = -5.0$ m/s^2; the rightmost and leftmost points on the middle curve are the solutions to parts (a) and (b). Also shown are cases of larger deceleration ($a = -9.81$ m/s^2) with a shorter stopping distance (bottom curve), and of smaller deceleration with a larger stopping distance (top curve).

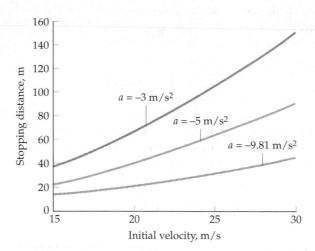

Figure 2-12

Example 2-11   *try it yourself*

In Example 2-10, (a) how long does it take for the car to stop if its initial velocity is 30 m/s? (b) How far does the car travel in the last second?

Picture the Problem (a) You can find the total time t from $v_f = v_0 + at = 0$, with $v_0 = 30$ m/s and $a = -5$ m/s. (b) Since the velocity decreases by 5 m/s each second, the velocity 1 s before the car stops must be 5 m/s. Find the average velocity during the last second and use that to find the distance traveled.

Cover the column to the right and try these on your own before looking at the answers.

Steps

Answers

(a)   Find the total stopping time t.

$t = 6$ s

(b) 1. Find the average velocity during the last second.

$v_{av} = 2.5$ m/s

   2. Compute the distance traveled from $\Delta x = v_{av}\Delta t$.

$\Delta x_1 = v_{av}\Delta t = 2.5$ m

Sometimes valuable insight can be gained about the motion of an object by assuming that the constant-acceleration formulas still apply even when the acceleration is not constant.

Example 2-12

In a crash test, a car traveling 100 km/h (about 62 mi/h) hits an immovable concrete wall. (a) How soon does the car stop? (b) What is its acceleration?

Picture the Problem In this example, it is not accurate to treat the car as a particle because different parts of the vehicle will have different accelerations as the car crumples to a halt. Moreover, the accelerations are not constant. Nevertheless, we can approximate an answer by assuming constant acceleration for a point particle located in the center of the car. We need more information to solve this problem—either the stopping distance or the time to

stop. We can estimate the stopping distance using common sense. Upon impact, the center of the car will certainly move forward less than half the length of the car. We'll choose 0.75 m as a reasonable estimate of the stopping distance.

(a) 1. The time needed for the car to stop is related to the stopping distance and the average velocity:

$$\Delta x = v_{av}\Delta t$$

2. The average velocity is found from the initial and final velocities (since we are only estimating, two significant figures are sufficient):

$$v_{av} = \frac{1}{2}(v_0 + v) = \frac{1}{2}(100 \text{ km/h} + 0) = 50 \text{ km/h}$$
$$= 14 \text{ m/s}$$

3. The time taken to stop the car is thus:

$$\Delta t = \frac{\Delta x}{v_{av}} = \frac{0.75 \text{ m}}{14 \text{ m/s}} = 0.054 \text{ s}$$

(b) 1. The average acceleration equals the ratio of the change in velocity and the time interval:

$$a = \frac{\Delta v}{\Delta t}$$

2. Since the car is brought from $v_0 = 100 \text{ km/h} = 28 \text{ m/s}$ to rest in this time, the average acceleration is:

$$a = \frac{\Delta v}{\Delta t} = \frac{0 - 28 \text{ m/s}}{0.054 \text{ s}} = -520 \text{ m/s}^2$$

Remark The magnitude of this acceleration is greater than $50g$.

Example **2-13**	*try it yourself*

An electron in a cathode-ray tube accelerates from rest with a constant acceleration of 5.33×10^{12} m/s^2 for 0.15 μs (1 μs = 10^{-6} s). The electron then drifts with constant velocity for 0.2 μs. Finally, it comes to rest with an acceleration of -2.67×10^{13} m/s^2. How far does the electron travel?

Picture the Problem The equations for constant acceleration do not apply to this problem directly because the acceleration of the electron varies with time. Divide the electron's motion into three intervals, each with a different constant acceleration, and use the final position and velocity for one interval as the initial conditions for the next interval. Choose the origin to be at the electron's starting position, and the positive direction to be the direction of motion.

Cover the column to the right and try these on your own before looking at the answers.

Steps	**Answers**
1. Find the displacement and final velocity for the first 0.15-μs interval.	$\Delta x_1 = 6.00$ cm; $v_1 = 8.00 \times 10^5$ m/s
2. Use this final velocity as the constant velocity to find the displacement while it drifts at constant velocity.	$\Delta x_2 = 16$ cm
3. Use this same velocity as the initial velocity and Equation 2-15 with $v = 0$ to find the displacement for the third interval in which the electron slows down.	$\Delta x_3 = 1.20$ cm
4. Add the displacements found in steps 1, 2, and 3 to find the total displacement.	$\Delta x = 23.2$ cm

Remark In your television, electrons are accelerated from the cathode to the anode, after which they are focused and passed through deflecting plates. They then drift toward the screen and crash into it, coming abruptly to rest.

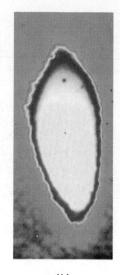

(a) The two-mile-long linear accelerator at Stanford University, used to accelerate electrons and positrons in a straight line to nearly the speed of light. (b) Cross section of the accelerator's electron beam as shown on a video monitor.

(a)

(b)

Example 2-14   *try it yourself*

John climbs a tree to get a better view of the speaker at an outdoor graduation ceremony. Unfortunately, he leaves his binoculars behind. Marsha throws them up to John, but her strength is greater than her accuracy. The binoculars pass John's outstretched hand after 0.69 s and again 1.68 s later. How high is John?

Picture the Problem There are two unknowns in this problem, John's height h and the initial velocity of the binoculars v_0. We know that $x = h$ at $t_1 = 0.69$ s and $x = h$ at $t_2 = 0.69$ s + 1.68 s = 2.37 s. Expressing h as a function of time t gives us two equations from which the two unknowns can be determined.

Cover the column to the right and try these on your own before looking at the answers.

Steps	Answers
1. Write $x(t)$ for times t_1 and t_2, noting that $x_0 = 0$ and $x = h$ in each case.	$h = v_0 t_1 - \frac{1}{2}g t_1^2$ and $h = v_0 t_2 - \frac{1}{2}g t_2^2$
2. Eliminate h from these two equations and solve for v_0 in terms of the times t_1 and t_2.	$v_0 = \dfrac{\frac{1}{2}g(t_1^2 - t_2^2)}{t_1 - t_2} = \frac{1}{2}g(t_1 + t_2)$
3. Substitute your result for v_0 into either of the equations for h.	$h = \frac{1}{2}g t_1 t_2 = 8.02$ m

Remark We have two unknowns, h and v_0, but are given two times, t_1 and t_2, so we can write two equations and solve them for the two unknowns.

Exercise Find the initial velocity of the binoculars and the velocity as they pass John on the way down. (*Answers* $v_0 = 15.0$ m/s and $v_2 = -8.24$ m/s)

Problems With Two Objects

We now give some examples of problems involving two objects moving with constant acceleration.

Example 2-15

A car is speeding at 25 m/s ($\approx$ 90 km/h $\approx$ 56 mi/h) in a school zone. A police car starts from rest just as the speeder passes and accelerates at a constant rate of 5 m/s^2. (*a*) When does the police car catch the speeding car? (*b*) How fast is the police car traveling when it catches up with the speeder?

Picture the Problem To determine when the two cars will be at the same position, we write the positions x_s of the speeder and x_p of the police car as functions of time and solve for the time t when $x_s = x_p$.

(*a*) 1. Write the position functions for the speeder and the police car:

$$x_s = v_s t \quad \text{and} \quad x_p = \tfrac{1}{2}a_p t^2$$

2. Set $x_s = x_p$ and solve for the time t:

$$v_s t = \tfrac{1}{2}a_p t^2; \quad t = 0 \quad \text{(initial condition)}$$

$$t = \frac{2v_s}{a_p} = \frac{2(25 \text{ m/s})}{5 \text{ m/s}^2} = 10 \text{ s}$$

(*b*)   The velocity of the police car is given by $v = v_0 + at$ with $v_0 = 0$:

$$v_p = a_p t = (5 \text{ m/s}^2)(10 \text{ s}) = 50 \text{ m/s}$$

Remark The final speed of the police car in (*b*) is exactly twice that of the speeder. Since the two cars covered the same distance in the same time, they must have had the same average speed. The speeder's average speed, of course, is 25 m/s. For the police car to start from rest and have an average speed of 25 m/s, it must reach a final speed of 50 m/s.

Exercise How far have the cars traveled when the police car catches the speeder? (*Answer*   250 m)

Remark In Figure 2-13 the solid lines depict the speeder and the police car in this example. The dashed lines are variations on the example. The smaller acceleration depicted by the lower dashed line means the police car takes longer to reach the speeder. In the higher dashed line, the acceleration is the same as in the example, but the police car does not start accelerating until 4 s after the speeder passes by.

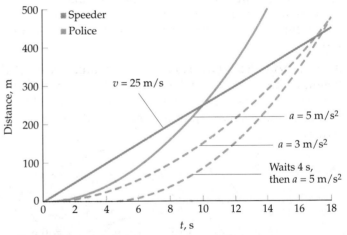

Figure 2-13

Example 2-16   *try it yourself*

How fast is the police car in Example 2-15 traveling when it is 25 m behind the speeding car?

Picture the Problem The speed is given by $v_p = at_1$, where t_1 is the time at which $x_s - x_p = 25$ m.

Cover the column to the right and try these on your own before looking at the answers.

Steps	Answers
1. Using the equations for x_p and x_s from Example 2-15, solve for t_1 when $x_s - x_p = 25$ m.	$t_1 = 5 \text{ s} \pm \sqrt{15} \text{ s} = 1.13 \text{ s}$ or 8.87 s
2. Use $v_p = a_p t$ to compute the speed of the police car at $t = t_1$.	$v = 5.65 \text{ m/s at } 1.13 \text{ s and } 44.4 \text{ m/s at } t = 8.87 \text{ s}$

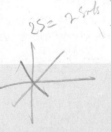

Remark We see from Figure 2-13 that the distance between the cars starts at zero, increases to a maximum value, and then decreases. The separation at any time is $D = x_s - x_p = (25 \text{ m/s})t - \frac{1}{2}(5 \text{ m/s}^2)t^2$. At maximum separation, $dD/dt = 0$, which occurs at $t = 5$ s. At equal intervals before and after $t = 5$ s, the separations are the same.

Example 2-17

Figure 2-14

While standing in an elevator, you see a screw fall from the ceiling. The ceiling is 3 m above the floor (Figure 2-14). (*a*) If the elevator is moving upward with a constant speed of 2.2 m/s, how long does it take for the screw to hit the floor? (*b*) How long is the screw in the air if the elevator starts from rest when the screw falls, and moves upward with a constant acceleration of $a_e = 4.0 \text{ m/s}^2$?

3 m

$v = 2.2$ m/s

Picture the Problem Write the position as a function of time for both the screw, x_s, and the floor, x_f. The screw hits the floor when $x_s = x_f$. Choose the origin to be the initial position of the floor, and designate upward as the positive direction.

(*a*) 1. Write the position functions for the elevator floor and screw:

$$x_f - x_{0f} = v_{0f}t + \frac{1}{2}a_f t^2$$

$$x_s - x_{0s} = v_{0s}t + \frac{1}{2}a_s t^2$$

2. Identify the initial conditions and the accelerations:

$$x_{0f} = 0, \quad v_{0f} = 2.2 \text{ m/s}, \quad a_f = 0$$

$$x_{0s} = h = 3 \text{ m}, \quad v_{0s} = 2.2 \text{ m/s}, \quad a_s = -g$$

3. Substitute these values into the position functions:

$$x_f = (2.2 \text{ m/s})t$$

$$x_s = h + (2.2 \text{ m/s})t - \frac{1}{2}gt^2$$

4. Set $x_s = x_f$ and solve for t:

$$h + (2.2 \text{ m/s})t - \frac{1}{2}gt^2 = (2.2 \text{ m/s})t$$

$$t = \sqrt{\frac{2h}{g}} = \sqrt{\frac{2(3 \text{ m})}{9.81 \text{ m/s}^2}} = 0.78 \text{ s}$$

Note that this result does not depend on the velocity of the elevator.

(*b*) 1. Now the elevator floor moves upward from rest with constant acceleration. The initial conditions are then:

$$x_{0f} = 0, \quad v_{0f} = 0, \quad a_f = 4.0 \text{ m/s}^2$$

$$x_{0s} = 3 \text{ m}, \quad v_{0s} = 0, \quad a_s = -g$$

2. Use the initial conditions to write the position functions for this case:

$$x_f = \frac{1}{2}a_f t^2 = \frac{1}{2}(4.0 \text{ m/s}^2)t^2$$

$$x_s = 3 \text{ m} - \frac{1}{2}gt^2$$

3. Set $x_s = x_w$ and solve for t:

$$\frac{1}{2}a_f t^2 = 3 \text{ m} - \frac{1}{2}gt^2$$

$$t = \sqrt{\frac{2(3 \text{ m})}{g + a_f}} = \sqrt{\frac{2(3 \text{ m})}{(9.81 + 4.0) \text{ m/s}^2}} = 0.66 \text{ s}$$

Remark The time in the air is independent of the speed of the elevator, as long as the elevator does not accelerate. If the elevator has acceleration a_f you and the screw experience an "effective gravity" with acceleration $g' = g + a_f$. For the case in which the elevator accelerates downward with $a_f = -g$, the time of fall becomes infinite and the screw appears weightless.

| Example 2-18 | *try it yourself* |

A probe launches vertically from the surface of Mars and reaches a height of 320 m and a velocity of 80 m/s at time $t = 0$ when its thrusters cut out. It then continues moving upward under the influence of Martian gravity, which is approximately constant and equal to $g_m = 3.72$ m/s^2. At the moment that the probe's thrusters cut out, $t = 0$, you are in a spacecraft 1500 m from the Martian surface, approaching the probe nearly head on, moving downward at 25 m/s and slowing down at a rate of 0.80 m/s^2. (*a*) When do you reach the probe? (*b*) How high above the planet's surface will the first rendezvous occur? (*c*) What is the velocity of each craft when they meet, assuming there are no course adjustments? (*d*) What is the velocity of the probe relative to the ship?

Figure 2-15

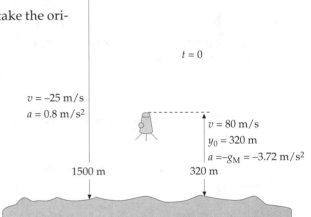

Picture the Problem Let upward be the positive direction, and take the origin to be at the surface of Mars (Figure 2-15). Then for the probe,

$y_0 = 320$ m

$v_0 = 80$ m/s

$a = -3.72$ m/s^2

and for your ship,

$y_0 = 1500$ m

$v_0 = -25$ m/s

$a = +0.8$ m/s^2

The ship reaches the probe when $y_p = y_s$.

Cover the column to the right and try these on your own before looking at the answers.

Steps   **Answers**

(*a*) 1. Write equations for y_p and y_s as functions of time.

$y_p = 320 + 80t - 1.86t^2$, $y_s = 1500 - 25t + 0.4t^2$

2. Set $y_p = y_s$ and solve for the time t. Note that you get two solutions.

$t_1 = 19.0$ s, $t_2 = 27.4$ s

(*b*) Substitute the result for t_1 into either equation in step 1 and solve for y.

$y_p(t_1) = y_s(t_1) = 1.17$ km

(*c*) 1. Write general equations for the velocities and substitute the time(s) found from part (*a*), step 2.

$v_p(t_1) = 9.32$ m/s

$v_s(t_1) = -9.80$ m/s

$v_p(t_2) = -21.9$ m/s

$v_s(t_2) = -3.08$ m/s

(*d*) Find the relative velocity $v_p - v_s$.

$v_r(t_1) = 19.1$ m/s; $v_r(t_2) = -18.8$ m/s

Remark Your ship intercepts the probe twice, once when the probe is moving up and once when it is moving down, as shown in Figure 2-16.

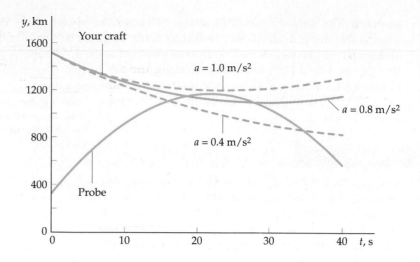

Figure 2-16 The solid curves depict the situation in Example 2-18. Also shown are two other cases: (lower dashed line) A spacecraft with $a = 0.4$ m/s^2 crashes into the probe! (higher dashed line) A spacecraft with $a = 1.0$ m/s^2 narrowly misses the probe. *Suggestion:* Find the conditions for the optimum encounter in which relative velocity is at a minimum.

2-4   Integration

To find the velocity from a given acceleration, we note that the velocity is the function $v(t)$ whose derivative is the acceleration $a(t)$:

$$\frac{dv(t)}{dt} = a(t)$$

If the acceleration is constant, the velocity is that function of time which, when differentiated, equals this constant. One such function is

$$v = at, \quad a = \text{constant}$$

More generally, we can add any constant to at without changing the time derivative. Calling this constant v_0, we have

$$v = at + v_0$$

When $t = 0$, $v = v_0$. Thus, v_0 is the initial velocity.

Similarly, the position function $x(t)$ is that function whose derivative is the velocity:

$$\frac{dx}{dt} = v = v_0 + at$$

We can treat each term separately. The function whose derivative is a constant v_0 is v_0t plus any constant. The function whose derivative is at is $\frac{1}{2}at^2$ plus any constant. Writing x_0 for the combined arbitrary constants, we have

$$x = x_0 + v_0t + \frac{1}{2}at_2$$

When $t = 0$, $x = x_0$. Thus, x_0 is the initial position.

Whenever we find a function from its derivative, we must include an arbitrary constant in the general function. Since we go through the integration process twice to find $x(t)$ from the acceleration, two constants arise. These constants are usually determined from the velocity and position at some given time, which is usually chosen to be $t = 0$. They are therefore called the **initial conditions.** A common problem, called the **initial-value problem,** takes the form "given $a(t)$ and the initial values of x and v, find $x(t)$." This problem is particularly important in physics because the acceleration of a particle is determined by the forces acting on it. Thus, if we know the forces acting on a particle and the position and velocity of the particle at some particular time, we can find its position at all other times.

A function $F(t)$ whose derivative (with respect to t) equals a function $f(t)$ is called the **antiderivative** of $f(t)$. Finding the antiderivative of a function is related to the problem of finding the area under a curve. Consider motion with a constant velocity v_0. The change in position Δx during an interval Δt is

$$\Delta x = v_0 \Delta t$$

This is the area under the v-versus-t curve (Figure 2-17). The geometric interpretation of the displacement as the area under the v-versus-t curve is true not only for constant velocity, but also in general, as illustrated in Figure 2-18. There, the area under the curve is approximated by first dividing the time interval into several smaller intervals, Δt_1, Δt_2, and so on, and drawing a set of rectangles. The area of the rectangle corresponding to the ith time interval Δt_i (shaded in the figure) is $v_i \Delta t_i$, which is approximately equal to the displacement Δx_i during the interval Δt_i. The sum of the rectangular areas is therefore approximately the sum of the displacements during the time intervals and is approximately equal to the total displacement from time t_1 to t_2. Mathematically, we write this as

$$\Delta x \approx \sum_i v_i \Delta t_i$$

where the Greek letter Σ (uppercase sigma) stands for "sum." We can make the approximation as accurate as we wish by choosing enough rectangles under the curve, each having a small value for Δt. For the limit of smaller and smaller time intervals, the resulting sum equals the area under the curve, which equals the displacement. This limit is called the **integral** and is written

$$\Delta x = x(t_2) - x(t_1) = \lim_{\Delta t \to 0} \sum_i v_i \Delta t_i = \int_{t_1}^{t_2} v\, dt \qquad \text{2-16}$$

It is helpful to think of the integral sign $\int$ as an elongated S indicating a sum. The limits t_1 and t_2 indicate the initial and final values of the variable t. The displacement is thus the area under the v-versus-t curve. Figure 2-19 demonstrates that the average velocity has a simple geometric interpretation in terms of the area under a curve.

The process of computing an integral is called **integration**. In Equation 2-16, v is the derivative of x, and x is the antiderivative of v. This is an example of the fundamental theorem of calculus, whose formulation in the seventeenth century greatly accelerated the mathematical development of physics:

$$\text{If } f(t) = \frac{dF(t)}{dt}, \quad \text{then} \quad F(t_2) - F(t_1) = \int_{t_1}^{t_2} f(t)\, dt \qquad \text{2-17}$$

Fundamental theorem of calculus

The antiderivative of a function is also called the indefinite integral of the

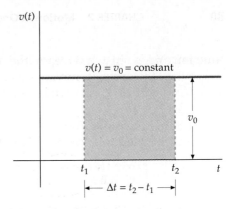

Figure 2-17 The displacement Δx during the time interval $\Delta t = t_2 - t_1$ is equal to the area of the shaded region. By the definition of average velocity, $\Delta x = v_{av} \Delta t$. This is just the area of a rectangle of height v_{av} and width Δt. Thus, the rectangular area $v_{av} \Delta t$ and the area under the v-versus-t curve must be equal.

Shaded area $= v_0 \Delta t = \Delta x$

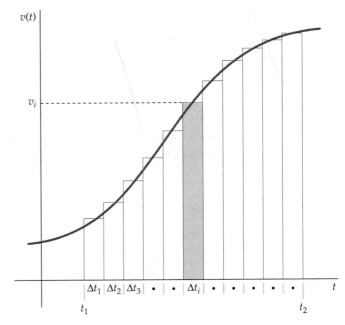

Figure 2-18 Graph of a general $v(t)$-versus-t curve. The total displacement from t_1 to t_2 is the area under the curve for this interval, which can be approximated by summing the areas of the rectangles.

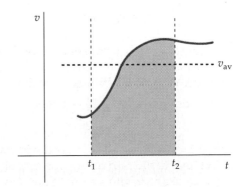

Figure 2-19 The displacement Δx during the time interval $\Delta t = t_2 - t_1$ is equal to the area of the shaded region. We know from the definition of average velocity that $\Delta x = v_{av} \Delta t$. This is just the area of a rectangle of height v_{av} and width Δt. Thus, the rectangular area $v_{av} \Delta t$ and the area under the v-versus-t curve must be equal.

function and is written without limits on the integral sign:

$$x = \int v \, dt$$

Finding the function x from the derivative v (that is, finding the antideriva-tive) is also called integration. For example, if $v = v_0$, a constant, then

$$x = \int v_0 \, dt = v_0 t + x_0$$

where x_0 is the arbitrary constant of integration. We can find a general rule for the integration of a power of t from Equation 2-6, which gives the general rule for the derivative of a power. The result is

$$\int t^n \, dt = \frac{t^{n+1}}{n+1} + C, \qquad n \neq -1 \qquad\qquad \text{2-18}$$

where C is an arbitrary constant. This can easily be checked by differentiating the right side using the rule of Equation 2-6. (For the special case $n = -1$, $\int t^{-1} \, dt = \ln t + C$, where $\ln t$ is the natural logarithm of t.)

The change in velocity for some time interval can similarly be interpreted as the area under the a-versus-t curve for that interval. This is written

$$\Delta v = \lim_{\Delta t \to 0} \sum_i a_i \Delta t_i = \int_{t_1}^{t_2} a \, dt \qquad\qquad \text{2-19}$$

We can now derive the constant-acceleration equations by computing the indefinite integrals of the acceleration and velocity. If a is constant, we have

$$v = \int a \, dt = v_0 + at \qquad\qquad \text{2-20}$$

where we have written the constant of integration, v_0, first. Integrating again and writing x_0 for the constant of integration gives

$$x = \int (v_0 + at) \, dt = x_0 + v_0 t + \frac{1}{2} at^2 \qquad\qquad \text{2-21}$$

Having derived the constant-acceleration equations without any reference to average velocity, we can now show that the average velocity for the special case of constant acceleration is the mean value between the initial and final velocities as given by Equation 2-12. Let v_0 be the initial velocity at $t = 0$, and let v be the final velocity at time t. According to the definition of average ve-locity, the displacement is

$$\Delta x = v_{av} \Delta t = v_{av}(t - 0) = v_{av} t \qquad\qquad \text{2-22}$$

Also, from Equation 2-21, we have

$$\Delta x = v_0 t + \frac{1}{2} at^2$$

We can eliminate the acceleration using $a = (v - v_0)/t$ from Equation 2-20. Then

$$\Delta x = v_0 t + \frac{1}{2}\left(\frac{v - v_0}{t}\right) t^2 = v_0 t + \frac{1}{2} vt - \frac{1}{2} v_0 t = \frac{1}{2}(v + v_0) t \qquad \text{2-23}$$

Comparing this with the definition of average velocity (Equation 2-22), we have

$$v_{av} = \frac{1}{2}\left(v_0 + v_f\right)$$

which is Equation 2-12b.

Example 2-19

A ferry boat moves with constant velocity $v_0 = 8$ m/s for 60 s. It then shuts off its engines and coasts. Its coasting velocity is given by $v = v_0 t_1^2/t^2$, where $t_1 = 60$ s. What is the displacement of the boat from $t = 0$ to $t \to \infty$?

Picture the Problem This velocity function is shown in Figure 2-20. The total displacement is calculated as the sum of the displacement Δx_1 from $t = 0$ to $t = 60$ s and the displacement Δx_2 from $t = 60$ s to $t \to \infty$.

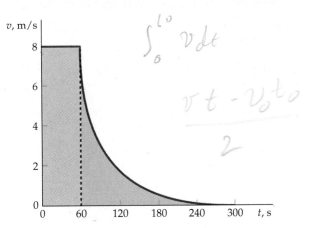

Figure 2-20

1. The velocity of the boat is constant during the first 60 seconds; thus, the displacement is simply the velocity times the elapsed time:

$$\Delta x_1 = v\Delta t = (8 \text{ m/s})(60 \text{ s}) = 480 \text{ m}$$

2. The remaining displacement is given by the integral of the velocity from $t = 60$ s to $t = \infty$. We use Equation 2-18 to calculate the integral:

$$\Delta x_2 = \int_{60 \text{ s}}^{\infty} v \, dt = \int_{60 \text{ s}}^{\infty} \frac{v_0 t_1^2}{t^2} \, dt = v_0 t_1^2 \int_{60 \text{ s}}^{\infty} t^{-2} \, dt = v_0 t_1^2 \left. \frac{t^{-1}}{-1} \right|_{60 \text{ s}}^{\infty}$$

$$= \frac{v_0 t_1^2}{60 \text{ s}} = \frac{(8 \text{ m/s})(60 \text{ s})^2}{60 \text{ s}} = 480 \text{ m}$$

3. The total displacement is the sum of the displacements found above:

$$\Delta x = \Delta x_1 + \Delta x_2 = 480 \text{ m} + 480 \text{ m} = 960 \text{ m}$$

Remark Note that the area under the v-versus-t curve is finite. Thus, even though the boat never stops moving, it travels only a finite distance. A better representation of the velocity of a coasting boat might be an exponentially decreasing function. In that case, the boat would also coast a finite distance in the interval 60 s $\leq t \leq \infty$.

Summary

Displacement, velocity, and acceleration are important *defined* kinematics quantities.

Topic	Remarks and Relevant Equations	
I. General Relations		
1. **Displacement**	$\Delta x = x_2 - x_1$	2-1
Graphical interpretation	Displacement is the area under the v-versus-t curve.	
2. **Velocity**		
Average velocity	$v_{\text{av}} = \dfrac{\Delta x}{\Delta t}$	2-2

Instantaneous velocity	$v(t) = \lim\limits_{\Delta t \to 0} \dfrac{\Delta x}{\Delta t} = \dfrac{dx}{dt}$	2-5
Graphical interpretation	The instantaneous velocity is represented graphically as the slope of the x-versus-t curve.	
Relative velocity	If a particle moves with velocity v_{pA} relative to a coordinate system A, which is in turn moving with velocity v_{AB} relative to another coordinate system B, the velocity of the particle relative to B is $$v_{pB} = v_{pA} + v_{AB}$$	2-7

3. Speed

Average speed	$\text{Average speed} = \dfrac{\text{total distance}}{\text{total time}} = \dfrac{\Delta s}{\Delta t}$	2-3
Instantaneous speed	Instantaneous speed is the magnitude of the instantaneous velocity.	

4. Acceleration

Average acceleration	$a_{av} = \dfrac{\Delta v}{\Delta t}$	2-8
Instantaneous acceleration	$a = \dfrac{dv}{dt} = \dfrac{d^2x}{dt^2}$	2-10
Graphical interpretation	The instantaneous acceleration is represented graphically as the slope of the v-versus-t curve.	
Acceleration due to gravity	The acceleration of an object near the surface of the earth in free fall under the influence of gravity is directed downward and has the magnitude $$g = 9.81 \text{ m/s}^2 = 32.2 \text{ ft/s}^2$$	

5. **Displacement and Velocity as Integrals**	Displacement is represented graphically as the area under the v-versus-t curve. This area is the integral of v over time from some initial time t_1 to some final time t_2 and is written $$\Delta x = \lim\limits_{\Delta t \to 0} \sum_i v_i \Delta t_i = \int_{t_1}^{t_2} v \, dt$$	2-16
	Similarly, the change in velocity for some time is represented graphically as the area under the a-versus-t curve: $$\Delta v = \lim\limits_{\Delta t \to 0} \sum_i a_i \Delta t_i = \int_{t_1}^{t_2} a \, dt$$	2-17

II. Constant-Acceleration Equations

Velocity	$v = v_0 + at$	2-12b
Displacement in terms of v_{av}	$\Delta x = x - x_0 = v_{av} t = \frac{1}{2}(v_0 + v)t$	2-13
Displacement in terms of a	$\Delta x = x - x_0 = v_0 t + \frac{1}{2}at^2$	2-14
v in terms of a and Δx	$v^2 = v_0^2 + 2a\,\Delta x$	2-15

Problem-Solving Guide

The following is applicable to all types of problems:

1. Begin by drawing a neat diagram that includes the important features of the problem.
2. Choose a convenient coordinate system and indicate it on your diagram. Show the origin and positive directions. When possible, choose the origin to be the location of the particle at $t = 0$ so that $x_0 = 0$.
3. Show known quantities on your diagram.
4. When possible, write an equation for the quantity to be found in terms of other quantities that are known or that can be found. Then proceed to find the other quantities in your equation.
5. When possible, solve the problem two different ways to check your solution.
6. Examine your answer to see if it is reasonable.

Summary of Worked Examples

Type of Calculation	Procedure and Relevant Examples
1. Displacement, Velocity, and Acceleration	
Find the average velocity for some time interval.	Find the displacement Δx for that time interval; then $v_{av} = \Delta x / \Delta t$. **Examples 2-1, 2-2, 2-3**
Find the total distance traveled.	Find the total time Δt and the average speed. The total distance is the average speed times the total time. **Example 2-3**
Find the instantaneous velocity from a graph of x versus t.	Draw the tangent line at the point in question. The slope of this line is v. **Example 2-5**
Find the instantaneous velocity and acceleration from a given function $x(t)$.	Compute the derivatives $v = dx/dt$ and $a = dv/dt = d^2x/dt^2$. **Examples 2-6, 2-8**
Find the average acceleration for some time interval.	Find the total change in velocity Δv; then $a_{av} = \Delta v / \Delta t$. **Example 2-7**
2. Constant Acceleration—One Object	
Find the greatest height reached by a thrown object and the time needed to reach it.	Set $v = 0$ in the constant acceleration equation $v^2 = v_0^2 + 2a\Delta x$ and solve for Δx. To find the rise time, set $v = 0$ in $v = v_0 + at$, and solve for t. **Example 2-9**
Find the stopping distance for a braking automobile.	Use $v^2 = v_0^2 + 2a\Delta x$ with $v = 0$. **Example 2-10**
Find the speed after a given time.	Use $v = v_0 + at$. **Example 2-11**
Find the distance traveled during a given time interval.	Find the initial and final time and compute $x_2 - x_1$, or find the average velocity for the interval and compute $\Delta x = v_{av} \Delta t$. **Example 2-11**
Estimate the stopping time and average acceleration for a car crashing into a wall.	Estimate the stopping distance, then use $\Delta t = \Delta x / v_{av}$ and $a = \Delta v / \Delta t$ **Examples 2-10, 2-12**
Find the total distance when the acceleration is different for different intervals.	Find Δx_i for each interval using the final position and velocity for the previous interval as initial conditions for the next interval. **Example 2-13**

3.   Constant Acceleration—Two Objects

Find the time of collision of two moving objects.	Write $x(t)$ for each object, then set $x_1(t) = x_2(t)$ and solve for t. **Examples 2-15, 2-16, 2-17**
Find the speed of one object given the separation of two objects.	Write $x(t)$ for both objects and solve for t for the given separation. Then find v at that time. **Example 2-18**

4.   Integration

Find the displacement of a particle given $v(t)$.	Compute the integral of $v(t)\, dt$. **Example 2-19**

Problems

Conceptual Problems

Problems from Optional and Exploring sections

In a few problems, you are given more data than you actually need; in a few other problems, you are required to supply data from your general knowledge, outside sources, or informed estimates.

- •   Single-concept, single-step, relatively easy
- ••   Intermediate-level, may require synthesis of concepts
- •••   Challenging, for advanced students

For all problems, use g = 9.81 m/s² for the acceleration due to gravity and neglect friction and air resistance unless instructed to do otherwise.

Speed, Displacement, and Velocity

1   • What is the approximate average velocity of the race cars during the Indianapolis 500?

2   • Does the following statement make sense? "The average velocity of the car at 9 A.M. was 60 km/h."

3   • Is it possible for the average velocity of an object to be zero during some interval even though its average velocity for the first half of the interval is not zero? Explain.

4   • The diagram in Figure 2-21 tracks the path of an object moving in a straight line. At which point is the object farthest from its starting point?

(*a*) A
(*b*) B
(*c*) C
(*d*) D
(*e*) E

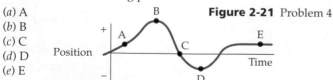

Figure 2-21 Problem 4

5   • (*a*) An electron in a television tube travels the 16-cm distance from the grid to the screen at an average speed of 4×10^7 m/s. How long does the trip take? (*b*) An electron in a current-carrying wire travels at an average speed of 4×10^{-5} m/s. How long does it take to travel 16 cm?

6   • A runner runs 2.5 km in 9 min and then takes 30 min to walk back to the starting point. (*a*) What is the runner's average velocity for the first 9 min? (*b*) What is the average velocity for the time spent walking? (*c*) What is the

average velocity for the whole trip? (*d*) What is the average speed for the whole trip?

7   • A car travels in a straight line with an average velocity of 80 km/h for 2.5 h and then with an average velocity of 40 km/h for 1.5 h. (*a*) What is the total displacement for the 4-h trip? (*b*) What is the average velocity for the total trip?

8   • One busy air route across the Atlantic Ocean is about 5500 km. (*a*) How long does it take for a supersonic jet flying at 2.4 times the speed of sound to make the trip? Use 340 m/s for the speed of sound. (*b*) How long does it take a subsonic jet flying at 0.9 times the speed of sound to make the same trip? (*c*) Allowing 2 h at each end of the trip for ground travel, check-in, and baggage handling, what is your average speed door to door when traveling on the supersonic jet? (*d*) What is your average speed taking the subsonic jet?

9   • As you drive down a desert highway at night, an alien spacecraft passes overhead, causing malfunctions in your speedometer, wristwatch, and short-term memory. When you return to your senses, you can't tell where you are, where you are going, or even how fast you are traveling. The passenger sleeping next to you never woke up during this incident. Although your pulse is racing, hers is a steady 55 beats per minute. (*a*) If she has 45 beats between the mile markers posted along the road, determine your speed. (*b*) If you want to travel at 120 km/h, how many heartbeats should there be between mile markers?

10   • The speed of light, c, is 3×10^8 m/s. (*a*) How long does it take for light to travel from the sun to the earth, a distance of 1.5×10^{11} m? (*b*) How long does it take light to travel from the moon to the earth, a distance of 3.84×10^8 m? (*c*) A light-year is a unit of distance equal to that traveled by light in 1 year. Convert 1 light-year into kilometers and miles.

11 • The nearest star, Proxima Centauri, is 4.1×10^{15} km away. From the vicinity of this star, Gregor places an order at Tony's Pizza in Hoboken, New Jersey, communicating via light signals. Tony's fastest delivery craft travels at $10^{-4}c$ (see Problem 10). (*a*) How long does it take for Gregor's order to reach Tony's pizza? (*b*) How long does Gregor wait between sending the signal and receiving the pizza? If Tony's has a 1000-years-or-it's-yours-free delivery policy, does Gregor have to pay for the pizza?

12 • A car making a 100-km journey travels 40 km/h for the first 50 km. How fast must it go during the second 50 km to average 50 km/h?

13 •• John can run 6.0 m/s. Marcia can run 15% faster than John. (*a*) By what distance does Marcia beat John in a 100-m race? (*b*) By what time does Marcia beat John in a 100-m race?

14 •• Figure 2-22 shows the position of a particle versus time. Find the average velocities for the time intervals *a*, *b*, *c*, and *d* indicated in the figure.

Figure 2-22
Problem 14

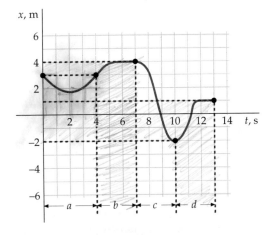

15 •• It has been found that galaxies are moving away from the earth at a speed that is proportional to their distance from the earth. This discovery is known as Hubble's law. The speed of a galaxy at distance r from the earth is given by $v = Hr$, where H is the Hubble constant, equal to 1.58×10^{-18} s^{-1}. What is the speed of a galaxy (*a*) 5×10^{22} m from earth and (*b*) 2×10^{25} m from earth? (*c*) If each of these galaxies has traveled with constant speed, how long ago were they both located at the same place as the earth?

16 •• Cupid fires an arrow that strikes St. Valentine, producing the usual sounds of harp music and bird chirping as Valentine swoons into a fog of love. If Cupid hears these telltale sounds exactly one second after firing the arrow and the average speed of the arrow was 40 m/s, what was the distance separating them? Take 340 m/s for the speed of sound.

Instantaneous Velocity

17 • If the instantaneous velocity does not change, will the average velocities for different intervals differ?

18 • If $v_{av} = 0$ for some time interval Δt, must the instantaneous velocity v be zero at some point in the interval? Support your answer by sketching a possible x-versus-t curve that has $\Delta x = 0$ for some interval Δt.

19 •• An object moves along the x axis as shown in Figure 2-23. At which point or points is the magnitude of its velocity at a minimum?

(*a*) A and E
(*b*) B, D, and E
(*c*) C only
(*d*) E only
(*e*) None of these is correct.

Figure 2-23 Problem 19

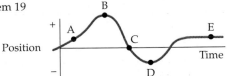

20 •• For each of the four graphs of x versus t in Figure 2-24, answer the following questions. (*a*) Is the velocity at time t_2 greater than, less than, or equal to the velocity at time t_1? (*b*) Is the speed at time t_2 greater than, less than, or equal to the speed at time t_1?

Figure 2-24 Problem 20

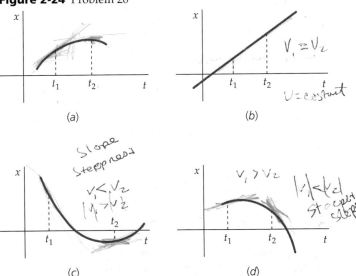

21 • Using the graph of x versus t in Figure 2-25, (*a*) find the average velocity between the times $t = 0$ and $t = 2$ s. (*b*) Find the instantaneous velocity at $t = 2$ s by measuring the slope of the tangent line indicated.

Figure 2-25 Problem 21

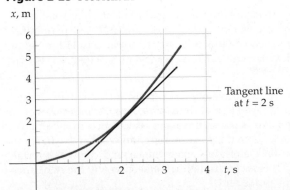

22 • Using the graph of x versus t in Figure 2-26, find (*a*) the average velocity for the time intervals $\Delta t = t_2 - 0.75$ s when t_2 is 1.75, 1.5, 1.25, and 1.0 s; (*b*) the instantaneous ve-

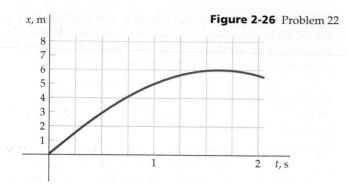

Figure 2-26 Problem 22

locity at $t = 0.75$ s; (c) the approximate time when the instantaneous velocity is zero.

23 •• The position of a certain particle depends on the time according to $x = (1 \text{ m/s}^2)t^2 - (5 \text{ m/s})t + 1$ m. (a) Find the displacement and average velocity for the interval 3 s $\leq t \leq 4$ s. (b) Find a general formula for the displacement for the time interval from t to $t + \Delta t$. (c) Use the limiting process to obtain the instantaneous velocity for any time t.

24 •• The height of a certain projectile is related to time by $y = -5(t - 5)^2 + 125$, where y is in meters and t is in seconds. (a) Sketch y versus t for $0 \leq t \leq 10$ s. (b) Find the average velocity for each of the 1-s time intervals between integral time values from $0 \leq t \leq 10$ s. Sketch v_{av} versus t. (c) Find the instantaneous velocity as a function of time.

25 ••• The position of a body oscillating on a spring is given by $x = A \sin \omega t$, where A and ω are constants with values $A = 5$ cm and $\omega = 0.175$ s^{-1}. (a) Sketch x versus t for $0 \leq t \leq 36$ s. (b) Measure the slope of your graph at $t = 0$ to find the velocity at this time. (c) Calculate the average velocity for a series of intervals beginning at $t = 0$ and ending at $t = 6, 3, 2, 1, 0.5$, and 0.25 s. (d) Compute dx/dt and find the velocity at time $t = 0$.

Relative Velocity

26 • To avoid falling too fast during a landing, an airplane must maintain a minimum airspeed (the speed of the plane relative to the air). However, the slower the ground speed (speed relative to the ground) during a landing, the safer the landing. Is it safer for an airplane to land with the wind or against the wind?

27 •• Two cars are traveling along a straight road. Car A maintains a constant speed of 80 km/h; car B maintains a constant speed of 110 km/h. At $t = 0$, car B is 45 km behind car A. How far will car A travel from $t = 0$ before it is overtaken by car B?

28 •• A car traveling at a constant speed of 20 m/s passes an intersection at time $t = 0$, and 5 s later another car traveling 30 m/s passes the same intersection in the same direction. (a) Sketch the position functions $x_1(t)$ and $x_2(t)$ for the two cars. (b) Determine when the second car will overtake the first. (c) How far from the intersection will the two cars be when they pull even?

29 •• Margaret has just enough gas in her speedboat to get to the marina, an upstream journey that takes 4.0 h. Finding it closed for the season, she spends the next 8.0 h floating back downstream to her shack. The entire trip took 12.0 h;

how long would it have taken if she had bought gas at the marina?

30 •• Joe and Sally tend to argue when they travel. Just as they reached the moving sidewalk at the airport, their struggle for itinerary-making powers peaked. Though they stepped on the moving belt at the same time, Joe chose to stand and ride, while Sally opted to keep walking. Sally reached the end in 1 min, while Joe took 2 min. How long would it have taken Sally if she had walked twice as fast?

Acceleration

31 • Walk across the room in such a way that, after getting started, your velocity is negative, but your acceleration is positive. (a) Describe how you did it. (b) Sketch a graph of v versus t for your motion.

32 • Give an example of a motion for which both the acceleration and the velocity are negative.

33 • Is it possible for a body to have zero velocity and nonzero acceleration?

34 • True or false:

(a) If the acceleration is zero, the body cannot be moving.
(b) If the acceleration is zero, the x-versus-t curve must be a straight line.

35 •• State whether the acceleration is positive, negative, or zero for each of the position functions $x(t)$ in Figure 2-27.

Figure 2-27
Problem 35

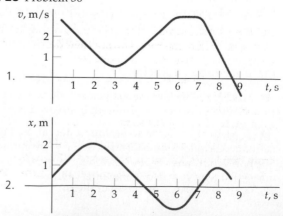

36 •• Answer the following questions for each of the graphs in Figure 2-28: (a) At what times are the accelerations of the objects positive, negative, and zero? (b) At what times are the accelerations constant? (c) At what times are the instantaneous velocities zero?

Figure 2-28 Problem 36

37 • A BMW M3 sports car can accelerate in third gear from 48.3 km/h (30 mi/h) to 80.5 km/h (50 mi/h) in 3.7 s. (*a*) What is the average acceleration of this car in m/s²? (*b*) If the car continued at this acceleration for another second, how fast would it be moving?

38 • At *t* = 5 s, an object at *x* = 3 m is traveling at 5 m/s. At *t* = 8 s, it is at *x* = 9 m and its velocity is −1 m/s. Find the average acceleration for this interval.

39 •• A particle moves with velocity *v* = 8*t* − 7, where *v* is in meters per second and *t* is in seconds. (*a*) Find the average acceleration for the one-second intervals beginning at *t* = 3 s and *t* = 4 s. (*b*) Sketch *v* versus *t*. What is the instantaneous acceleration at any time?

40 •• The position of an object is related to time by *x* = *At*² − *Bt* + *C*, where *A* = 8 m/s², *B* = 6 m/s, and *C* = 4 m. Find the instantaneous velocity and acceleration as functions of time.

Motion With Constant Acceleration

41 • Identical twin brothers standing on a bridge each throw a rock straight down into the water below. They throw the rocks at exactly the same time, but one hits the water before the other. How can this occur if the rocks have the same acceleration?

42 • A ball is thrown straight up. What is the velocity of the ball at the top of its flight? What is its acceleration at that point?

43 • An object thrown straight up falls back to the ground *T* seconds later. Its maximum height is *H* meters. Its average velocity during those *T* seconds is (*a*) *H*/*T*, (*b*) 0, (*c*) *H*/2*T*, (*d*) 2*H*/*T*.

44 • For an object thrown straight up, which of the following is true while it is in the air? (*a*) The acceleration is always opposite to the velocity. (*b*) The acceleration is always directed downward. (*c*) The acceleration is always in the direction of motion. (*d*) The acceleration is zero at the top of the trajectory.

45 • An object projected up with initial velocity *v* attains a height *H*. Another object projected up with initial velocity 2*v* will attain a height of (*a*) 4*H*, (*b*) 3*H*, (*c*) 2*H*, (*d*) *H*.

46 • A ball is thrown upward. While it is in the air, its acceleration is (*a*) decreasing, (*b*) constant, (*c*) zero, (*d*) increasing.

47 • At *t* = 0, object A is dropped from the roof of a building. At the same instant, object B is dropped from a window 10 m below the roof. During their descent to the ground the distance between the two objects

(*a*) is proportional to *t*.
(*b*) is proportional to *t*².
(*c*) decreases.
(*d*) remains 10 m throughout.

48 •• A Porsche accelerates uniformly from 80.5 km/h (50 mi/h) at *t* = 0 to 113 km/h (70 mi/h) at *t* = 9 s. Which graph in Figure 2-29 best describes the motion of the car?

49 •• An object is dropped from rest. If the time during which it falls is doubled, the distance it falls will (*a*) double, (*b*) decrease by one-half, (*c*) increase by a factor of four, (*d*) decrease by a factor of four, (*e*) remain the same.

50 •• A ball is thrown upward with an initial velocity *v*₀. Its velocity halfway to its highest point is (*a*) 0.5*v*₀, (*b*) 0.25*v*₀, (*c*) *v*₀, (*d*) 0.707*v*₀, (*e*) cannot be determined from the information given.

51 • A car starting at *x* = 50 m accelerates from rest at a constant rate of 8 m/s². (*a*) How fast is it going after 10 s? (*b*) How far has it gone after 10 s? (*c*) What is its average velocity for the interval 0 ≤ *t* ≤ 10 s?

52 • An object with an initial velocity of 5 m/s has a constant acceleration of 2 m/s². When its speed is 15 m/s, how far has it traveled?

53 • An object with constant acceleration has velocity *v* = 10 m/s when it is at *x* = 6 m and *v* = 15 m/s when it is at *x* = 10 m. What is its acceleration?

54 • An object has constant acceleration *a* = 4 m/s². At *t* = 0, its velocity is 1 m/s and it is at *x* = 7 m. How fast is it moving when it is at *x* = 8 m? What is *t* at that point?

55 • If a rifle fires a bullet straight up with a muzzle speed of 300 m/s, how high will the bullet rise? (Ignore air resistance.)

56 • A test of the prototype of a new automobile shows that the minimum distance for a controlled stop from 98 km/h to zero is 50 m. Find the acceleration, assuming it to be constant, and express your answer as a fraction of the free-fall acceleration due to gravity. How long does the car take to stop?

57 •• A ball is thrown upward with initial velocity of 20 m/s. (*a*) How long is the ball in the air? (*b*) What is the greatest height reached by the ball? (*c*) When is the ball 15 m above the ground?

58 •• A particle moves with a constant acceleration of 3 m/s². At *t* = 4 s, it is at *x* = 100 m; at *t* = 6 s, it has a velocity *v* = 15 m/s. Find its position at *t* = 6 s.

59 •• A bullet traveling at 350 m/s strikes a telephone pole and penetrates a distance of 12 cm before stopping. (*a*) Estimate the average acceleration by assuming it to be constant. (*b*) How long did it take for the bullet to stop?

60 •• A plane landing on an aircraft carrier has just 70 m to stop. If its initial speed is 60 m/s, (*a*) what is the acceleration of the plane during landing, assuming it to be constant? (*b*) How long does it take for the plane to stop?

61 •• An automobile accelerates from rest at 2 m/s² for 20 s. The speed is then held constant for 20 s, after which there is an acceleration of −3 m/s² until the automobile stops. What is the total distance traveled?

Figure 2-29
Problem 48

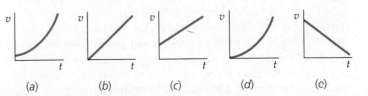

(*a*)   (*b*)   (*c*)   (*d*)   (*e*)

62 •• In the Blackhawk landslide in California, a mass of rock and mud fell 460 m down a mountain and then traveled 8 km across a level plain on a cushion of compressed air. Assume that the mud dropped with the free-fall acceleration due to gravity and then slid horizontally with constant deceleration. (*a*) How long did the mud take to drop the 460 m? (*b*) How fast was it traveling when it reached the bottom? (*c*) How long did the mud take to slide the 8 km horizontally?

63 •• A load of bricks is being lifted by a crane at a steady velocity of 5 m/s when one brick falls off 6 m above the ground. (*a*) Sketch $x(t)$ to show the motion of the free brick. (*b*) What is the greatest height the brick reaches above the ground? (*c*) How long does it take to reach the ground? (*d*) What is its speed just before it hits the ground?

64 •• An egg with a mass of 50 g rolls off a table at a height of 1.2 m and splatters on the floor. Estimate the average acceleration of the egg while it is in contact with the floor.

65 •• To win publicity for her new CD release, Sharika, the punk queen, jumps out of an airplane without a parachute. She expects a stack of loose hay to break her fall. If she reaches a speed of 120 km/h prior to impact, and if a 35g deceleration is the greatest deceleration she can withstand, how high must the stack of hay be for her to survive? Assume uniform acceleration while she is in contact with the hay.

66 •• A bolt comes loose from underneath an elevator that is moving upward at a speed of 6 m/s. The bolt reaches the bottom of the elevator shaft in 3 s. (*a*) How high up was the elevator when the bolt came loose? (*b*) What is the speed of the bolt when it hits the bottom of the shaft?

67 •• An object is dropped from a height of 120 m. Find the distance it falls during its final second in the air.

68 •• An object is dropped from a height H. During the final second of its fall, it traverses a distance of 38 m. What was H?

69 •• A stone is thrown vertically from a cliff 200 m tall. During the last half-second of its flight the stone travels a distance of 45 m. Find the initial velocity of the stone.

70 •• An object in free fall from a height H traverses $0.4H$ during the first second of its descent. Determine the average speed of the object during free fall.

71 •• A bus accelerates at 1.5 m/s² from rest for 12 s. It then travels at constant speed for 25 s, after which it slows to a stop with an acceleration of −1.5 m/s². (*a*) How far did the bus travel? (*b*) What was its average velocity?

72 •• A basketball is dropped from a height of 3 m and rebounds from the floor to a height of 2 m. (*a*) What is the velocity of the ball just as it reaches the floor? (*b*) What is its velocity just as it leaves the floor? (*c*) Estimate the magnitude and direction of its average acceleration during this interval.

73 •• A rocket is fired vertically with an upward acceleration of 20 m/s². After 25 s, the engine shuts off and the rocket continues as a free particle until it reaches the ground. Calculate (*a*) the highest point the rocket reaches, (*b*) the total time the rocket is in the air, (*c*) the speed of the rocket just before it hits the ground.

74 •• A flowerpot falls from the ledge of an apartment building. A person in an apartment below, coincidentally holding a stopwatch, notices that it takes 0.2 s for the pot to fall past his window, which is 4 m high. How far above the top of the window is the ledge from which the pot fell?

75 •• Sharika arrives home late from a gig, only to find herself locked out. Her roommate and bass player Chico is practicing so loudly that he can't hear Sharika's pounding on the door downstairs. One of the band's props is a small trampoline, which Sharika places under Chico's window. She bounces progressively higher trying to get Chico's attention. Propelling herself furiously upward, she miscalculates on the last bounce and flies past the window and out of sight. Chico sees her face for 0.2 s as she moves a distance of 2.4 m from the bottom to the top of the window. (*a*) How long until she reappears? (*b*) What is her greatest height above the top of the window? (Treat Sharika as a point-particle punk.)

76 •• In a classroom demonstration, a glider moves along an inclined air track with constant acceleration a. It is projected from the start of the track ($x = 0$) with an initial velocity v_0. At time $t = 8$ s, it is at $x = 100$ cm and is moving along the track at velocity $v = -15$ cm/s. Find the initial speed v_0 and the acceleration a.

77 •• A rock dropped from a cliff falls one-third of its total distance to the ground in the last second of its fall. How high is the cliff?

78 ••• A typical automobile has a maximum deceleration of about 7 m/s²; the typical reaction time to engage the brakes is 0.50 s. A school board sets the speed limit in a school zone to meet the condition that all cars should be able to stop in a distance of 4 m. (*a*) What maximum speed should be allowed for a typical automobile? (*b*) What fraction of the 4 m is due to the reaction time?

Constant Acceleration With Two Objects

79 •• Two trains face each other on adjacent tracks. They are initially at rest 40 m apart. The train on the left accelerates rightward at 1.4 m/s². The train on the right accelerates leftward at 2.2 m/s². How far does the train on the left travel before the two trains pass?

80 •• Two stones are dropped from the edge of a 60-m cliff, the second stone 1.6 s after the first. How far below the cliff is the second stone when the separation between the two stones is 36 m?

81 •• A motorcycle policeman hidden at an intersection observes a car that ignores a stop sign, crosses the intersection, and continues on at constant speed. The policeman starts off in pursuit 2.0 s after the car has passed the stop sign, accelerates at 6.2 m/s² until his speed is 110 km/h, and then continues at this speed until he catches the car. At that instant, the car is 1.4 km from the intersection. How fast was the car traveling?

82 •• At $t = 0$, a stone is dropped from a cliff above a lake; 1.6 s later another stone is thrown downward from the same point with an initial speed of 32 m/s. Both stones hit the water at the same instant. Find the height of the cliff.

83 •• A passenger train is traveling at 29 m/s when the engineer sees a freight train 360 m ahead traveling on the same track in the same direction. The freight train is moving at a speed of 6 m/s. If the reaction time of the engineer is 0.4 s, what must be the deceleration of the passenger train if a collision is to be avoided? If your answer is the maximum deceleration of the passenger train but the engineer's reaction time is 0.8 s, what is the relative speed of the two trains at the instant of collision and how far will the passenger train have traveled in the time between the sighting of the freight train and the collision?

84 •• After being forced out of farming, Lou has given up on trying to find work locally and is about to "ride the rails" to look for a job. Running at his maximum speed of 8 m/s, he is a distance d from the train when it begins to accelerate from rest at 1.0 m/s². (a) If $d = 30$ m and Lou keeps running, will he be able to jump into the train? (b) Sketch the position function $x(t)$ for the train, with $x = 0$ at $t = 0$. On the same graph, sketch $x(t)$ for various distances d, including $d = 30$ m and the critical separation distance d_c, the distance at which he just catches the train. (c) For the situation $d = d_c$, what is the speed of the train when Lou catches it? What is the train's average speed for the time interval between $t = 0$ and the moment Lou catches the train? What is the exact value of d_c?

85 •• A train pulls away from a station with a constant acceleration of 0.40 m/s². A passenger arrives at the track 6.0 s after the end of the train has passed the very same point. What is the slowest constant speed at which she can run and catch the train? Sketch curves for the motion of the passenger and the train as functions of time.

86 ••• Lou applies for a job as a perfume salesman. He tries to convince the boss to try his daring, aggressive promotional gimmick: dousing perspective customers as they wait at bus stops. A hard ball is to be thrown straight upward with an initial speed of 24 m/s. A thin-skinned ball filled with perfume is then thrown straight upward along the same path with a speed of 14 m/s. The balls are to collide when the perfume ball is at the high point of its trajectory, so that it breaks open and everyone gets a free sample. If $t = 0$ when the first ball is thrown, find the time when the perfume ball should be thrown.

87 ••• Ball A is dropped from the top of a building at the same instant that ball B is thrown vertically upward from the ground. When the balls collide, they are moving in opposite directions, and the speed of A is twice the speed of B. At what fraction of the height of the building does the collision occur?

88 ••• Solve Problem 87 if the collision occurs when the balls are moving in the same direction and the speed of A is 4 times that of B.

89 ••• The Sprint missile, designed to destroy incoming ballistic missiles, can accelerate at $100g$. If an ICBM is detected at an altitude of 100 km moving straight down at a constant speed of 3×10^4 km/h and the Sprint missile is launched to intercept it, at what time and altitude will the interception take place? (*Note:* You can neglect the acceleration due to gravity in this problem. Why?)

90 ••• When a car traveling at speed v_1 rounds a corner, the driver sees another car traveling at a slower speed v_2 a distance d ahead. (a) If the maximum acceleration the driver's brakes can provide is a, show that the distance d must be greater than $(v_1 - v_2)^2/2a$ if a collision is to be avoided. (b) Evaluate this distance for $v_1 = 90$ km/h, $v_2 = 45$ km/h, and $a = 6$ m/s². (c) Estimate or measure your reaction time and calculate the effect it would have on the distance found in part (b).

Integration

91 • The velocity of a particle is given by $v = 6t + 3$, where t is in seconds and v is in meters per second. (a) Sketch $v(t)$ versus t, and find the area under the curve for the interval $t = 0$ to $t = 5$ s. (b) Find the position function $x(t)$. Use it to calculate the displacement during the interval $t = 0$ to $t = 5$ s.

92 • Figure 2-30 shows the velocity of a particle versus time. (a) What is the magnitude in meters of the area of the rectangle indicated? (b) Find the approximate displacement of the particle for the one-second intervals beginning at $t = 1$ s and $t = 2$ s. (c) What is the approximate average velocity for the interval from 1 s $\leq t \leq$ 3 s?

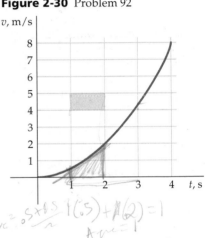
Figure 2-30 Problem 92

93 •• The velocity of a particle is given by $v = 7t^2 - 5$, where t is in seconds and v is in meters per second. Find the general position function $x(t)$.

94 •• The equation of the curve shown in Figure 2-30 is $v = 0.5t^2$ m/s. Find the displacement of the particle for the interval 1 s $\leq t \leq$ 3 s by integration, and compare this answer with your answer for Problem 92. Is the average velocity equal to the mean of the initial and final velocities for this case?

95 •• Figure 2-31 shows the acceleration of a particle versus time. (a) What is the magnitude of the area of the rectangle indicated? (b) The particle starts from rest at $t = 0$. Find the velocity at $t = 1$ s, 2 s, and 3 s by counting the rectangles under the curve. (c) Sketch the curve $v(t)$ versus t from your results for part (b), and estimate how far the particle travels in the interval $t = 0$ to $t = 3$ s.

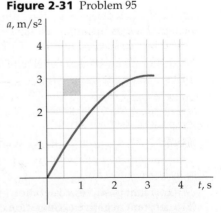
Figure 2-31 Problem 95

96 •• Figure 2-32 is a graph of v versus t for a particle moving along a straight line. The position of the particle at time $t = 0$ is $x_0 = 5$ m. (a) Find x for various times t by counting squares, and sketch x versus t. (b) Sketch the acceleration a versus t.

Figure 2-32 Problem 96

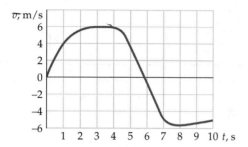

97 ••• Figure 2-33 shows a plot of x versus t for a body moving along a straight line. Sketch rough graphs of v versus t and a versus t for this motion.

Figure 2-33 Problem 97

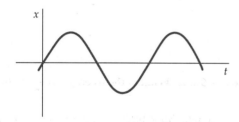

General Problems

98 • True or false:

(a) The equation $\Delta x = v_0 t + \frac{1}{2}at^2$ is valid for all particle motion in one dimension.
(b) If the velocity at a given instant is zero, the acceleration at that instant must also be zero.
(c) The equation $\Delta x = v_{av}\,\Delta t$ holds for all motion in one dimension.

99 • If an object is moving at constant acceleration in a straight line, its instantaneous velocity halfway through any time interval is

(a) greater than its average velocity.
(b) less than its average velocity.
(c) equal to its average velocity.
(d) half of its average velocity.
(e) twice its average velocity.

100 • On a graph showing position on the vertical axis and time on the horizontal axis, a straight line with a negative slope represents

(a) a constant positive acceleration.
(b) a constant negative acceleration.
(c) zero velocity.
(d) a constant positive velocity.
(e) a constant negative velocity.

101 •• On a graph showing position on the vertical axis and time on the horizontal axis, a parabola that opens upward represents

(a) a positive acceleration.
(b) a negative acceleration.
(c) no acceleration.
(d) a positive followed by a negative acceleration.
(e) a negative followed by a positive acceleration.

102 •• On a graph showing velocity on the vertical axis and time on the horizontal axis, zero acceleration is represented by

(a) a straight line with a positive slope.
(b) a straight line with a negative slope.
(c) a straight line with zero slope.
(d) either (a), (b), or (c).
(e) none of the above.

103 •• On a graph showing velocity on the vertical axis and time on the horizontal axis, constant acceleration is represented by

(a) a straight line with a positive slope.
(b) a straight line with a negative slope.
(c) a straight line with zero slope.
(d) either (a), (b), or (c).
(e) none of the above.

104 •• Which graph of v versus t in Figure 2-34 best describes the motion of a particle with positive velocity and negative acceleration?

Figure 2-34 Problems 104 and 105

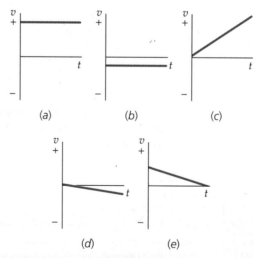

105 •• Which graph of v versus t in Figure 2-34 best describes the motion of a particle with negative velocity and negative acceleration?

106 •• A graph of the motion of an object is plotted with the velocity on the vertical axis and time on the horizontal axis. The graph is a straight line. Which of these quantities *cannot* be determined from this graph?

(a) The displacement from time $t = 0$
(b) The initial velocity at $t = 0$
(c) The acceleration of the object
(d) The average velocity of the object
(e) None of the above

107 •• Figure 2-35 shows the position of a car plotted as a function of time. At which of the times t_0 to t_7 is the velocity (*a*) negative? (*b*) positive? (*c*) zero?

At which times is the acceleration

(*a*) negative?
(*b*) positive?
(*c*) zero?

Figure 2-35 Problem 107

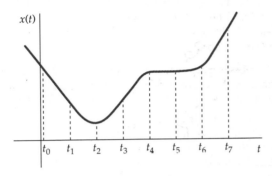

108 •• Sketch *v*-versus-*t* curves for each of the following conditions:

(*a*) Acceleration is zero and constant while velocity is not zero.
(*b*) Acceleration is constant but not zero.
(*c*) Velocity and acceleration are both positive.
(*d*) Velocity and acceleration are both negative.
(*e*) Velocity is positive and acceleration is negative.
(*f*) Velocity is negative and acceleration is positive.
(*g*) Velocity is zero but acceleration is not.

109 •• Figure 2-36 shows nine graphs of position, velocity, and acceleration for objects in linear motion. Indicate the graphs that meet the following conditions:

(*a*) Velocity is constant.
(*b*) Velocity has reversed its direction.
(*c*) Acceleration is constant.
(*d*) Acceleration is not constant.
Which graphs of velocity and acceleration are mutually consistent?

110 • Two cars are being driven at the same speed *v*, one behind the other, with a distance *d* between them. The first driver jams on her brakes and decelerates at a rate $a = 6$ m/s². The second driver sees the brake lights of the first driver and reacts, decelerating at the same rate starting 0.5 s later. (*a*) What is the minimum distance *d* such that the two cars do not collide? (*b*) Express this answer in meters for $v = 100$ km/h (62 mi/h).

111 • The velocity of a particle in meters per second is given by $v = 7 - 4t$, where *t* is in seconds. (*a*) Sketch *v*(*t*) versus *t*, and find the area between the curve and the *t* axis from $t = 2$ s to $t = 6$ s. (*b*) Find the position function *x*(*t*) by integration, and use it to find the displacement during the interval

Figure 2-36 Problem 109

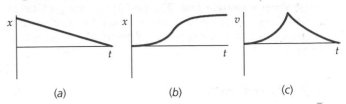

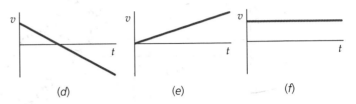

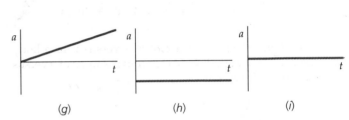

$t = 2$ s to $t = 6$ s. (*c*) What is the average velocity for this interval?

112 •• Estimate how high a ball or small rock can be thrown if it is thrown straight up.

113 •• The cheetah can run as fast as $v_1 = 100$ km/h, the falcon can fly as fast as $v_2 = 250$ km/h, and the sailfish can swim as fast as $v_3 = 120$ km/h. The three of them run a relay with each covering a distance *L* at maximum speed. What is the average speed *v* of this triathlon team?

114 •• In 1997, the men's world record for the 50-m freestyle was held by Tom Jager of the United States, who covered $d = 50$ m in $t = 21.81$ s. Suppose Jager started from rest at a constant acceleration *a*, and reached his maximum speed in 2.00 s, which he then kept constant until the finish line. Find Jager's acceleration *a*.

115 •• The click beetle can project itself vertically with an acceleration of about $a = 400g$ (an order of magnitude more than a human could stand). The beetle jumps by "unfolding" its legs, which are about $d = 0.6$ cm long. How high can the click beetle jump? How long is the beetle in the air? (Assume constant acceleration while in contact with the ground, and neglect air resistance.)

116 •• The one-dimensional motion of a particle is plotted in Figure 2-37. (*a*) What is the acceleration in the intervals AB, BC, and CE? (*b*) How far is the particle from its starting point after 10 s? (*c*) Sketch the displacement of the particle as a function of time; label instants A, B, C, D, and E on your figure. (*d*) At what time is the particle traveling most slowly?

Figure 2-37 Problem 116

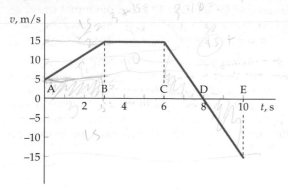

117 •• Consider the velocity graph in Figure 2-38. Assuming $x = 0$ at $t = 0$, write correct algebraic expressions for $x(t)$, $v(t)$, and $a(t)$ with appropriate numerical values inserted for all constants.

Figure 2-38 Problem 117

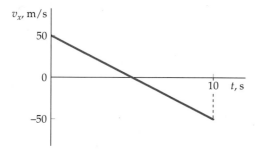

118 •• Starting at one station, a subway train accelerates from rest at a constant rate of 1.0 m/s^2 for half the distance to the next station, then slows down at the same rate for the second half of the journey. The total distance between stations is 900 m. (a) Sketch a graph of the velocity v as a function of time over the full journey. (b) Sketch a graph of the distance covered as a function of time over the full journey. Place appropriate numerical values along both axes.

119 •• The acceleration of a certain rocket is given by $a = Ct$, where C is a constant. (a) Find the general position function $x(t)$. (b) Find the position and velocity at $t = 5$ s if $x = 0$ and $v = 0$ at $t = 0$ and $C = 3 \text{ m/s}^3$.

120 •• A physics professor demonstrates his new "antigravity parachute" by exiting from a helicopter at an altitude of 1500 m with zero initial velocity. For 8 s he falls freely. Then he switches on the "parachute" and falls with a constant upward acceleration of 15 m/s^2 until his downward speed reaches 5 m/s, whereupon he adjusts his controls to maintain that speed until he reaches the ground. (a) On a single graph, sketch his acceleration and velocity as functions of time. (Take upward to be positive.) (b) What is his speed at the end of the first 8 s? (c) For how long does he maintain the constant upward acceleration of 15 m/s^2? (d) How far does he travel during the upward acceleration in part (c)? (e) How many seconds are required for the entire trip from the heli-

copter to the ground? (f) What is his average velocity for the entire trip?

121 •• Without telling Sally, Joe made travel arrangements that include a stopover in Toronto to visit Joe's old buddy. Sally doesn't like Joe's buddy and wants to change their tickets. She hops on a courtesy motor scooter and begins accelerating at 0.9 m/s^2 toward the ticket counter to make the arrangements. As she begins moving, Joe is 40 m behind her, running at a constant speed of 9 m/s. (a) How long does it take for Joe to catch up with her? (b) What is the time interval during which Sally remains ahead of Joe?

122 •• A speeder races past at 125 km/h. A patrol car pursues from rest with a constant acceleration of 8 km/h·s until it reaches its maximum speed of 190 km/h, which it maintains until it catches up with the speeder. (a) How long until the patrol car catches the speeder if it starts moving just as the speeder passes? (b) How far does each car travel? (c) Sketch $x(t)$ for each car.

123 •• When the patrol car in Problem 122 (traveling at 190 km/h) pulls to within 100 m behind the speeder (traveling at 125 km/h), the speeder sees the police car and slams on his brakes, locking the wheels. (a) Assuming that each car can brake at 6 m/s^2 and that the driver of the police car brakes instantly as she sees the brake lights of the speeder (reaction time = 0 s), show that the cars collide. (b) At what time after the speeder applies his brakes do the two cars collide? (c) Discuss how reaction time affects this problem.

124 •• The speed of a good base runner is 9.5 m/s. The distance between bases is 26 m, and the pitcher is about 18.5 m from home plate. If a runner on first base edges 2 m off the base and takes off for second the instant the ball leaves the pitcher's hand, what is the likelihood that the runner will steal second base safely?

125 •• Repeat Problem 124, but with the runner attempting to steal third base, starting from second base with a lead of 3 m.

126 •• Urgently needing the cash prize, Lou enters the Rest-to-Rest auto competition, in which each contestant's car begins and ends at rest, covering a distance L in as short a time as possible. The intention is to demonstrate mechanical and driving skills, and to consume the largest amount of fossil fuels in the shortest time possible. The course is designed so that maximum speeds of the cars are never reached. If Lou's car has a maximum acceleration of a, and a maximum deceleration of $2a$, then at what fraction of L should Lou move his foot from the gas pedal to the brake? What fraction of the time for the trip has elapsed at that point?

127 ••• The acceleration of a badminton birdie falling under the influence of gravity and a resistive force, such as air resistance, is given by $a = dv/dt = g - bv$, where g is the free-fall acceleration due to gravity and b is a constant that depends on the mass and shape of the birdie and on the properties of the medium. Suppose the birdie begins with zero velocity at time $t = 0$. (a) Discuss qualitatively how the speed v varies with time from your knowledge of the rate of change dv/dt given by this equation. What is the velocity when the acceleration is zero? This is called the *terminal velocity*. (b)

Sketch the solution $v(t)$ versus t without solving the equation. This can be done as follows: At $t = 0$, v is zero and the slope is g. Sketch a straight-line segment, neglecting any change in slope for a short time interval. At the end of the interval, the velocity is not zero, so the slope is less than g. Sketch another straight-line segment with a smaller slope. Continue until the slope is zero and the velocity equals the terminal velocity.

128 ••• Suppose acceleration is a function of x, where $a(x) = 2x \, \text{m/s}^2$. (*a*) If the velocity at $x = 1$ m is zero, what is the speed at $x = 3$ m? (*b*) How long does it take to travel from $x = 1$ m to $x = 3$ m?

129 ••• Suppose that a particle moves in a straight line such that, at any time t, its position, velocity, and acceleration all have the same numerical value. Give the position x as a function of time.

130 ••• An object moving in a straight line doubles its velocity each second for the first 10 s. Let the initial speed be 2 m/s. (*a*) Sketch a smooth function $v(t)$ that gives the velocity. (*b*) What is the average velocity over the first 10 s?

131 ••• In a dream, you find that you can run at superhuman speeds, but there is also a resistant force that reduces your speed by one-half for each second that passes. Assume that the laws of physics still hold in your dreamworld, and that your initial speed is 1000 m/s. (*a*) Sketch a smooth function $v(t)$ that gives your velocity. (*b*) What is your average velocity over the first 10 s?

CHAPTER 3

Motion in Two and Three Dimensions

Illuminated fountains, St. Louis, Missouri. The jets follow parabolic paths like those followed by projectiles.

3-1   The Displacement Vector

When motion occurs in two or three dimensions, the displacement of a particle has a direction in space as well as a magnitude. The quantity that gives the direction and the straight-line distance between two points in space is a line segment called the **displacement vector.** It is represented graphically by an arrow whose direction is the same as the direction of the displacement and whose length is proportional to the magnitude of the displacement. We denote vectors by boldface italic letters with an overhead arrow, $\vec{A}$. The magnitude of $\vec{A}$ is written $|\vec{A}|$ or simply A.

Addition of Displacement Vectors

Figure 3-1 shows the path of a particle that moves from point P_1 to a second point P_2 and then to a third point P_3. The displacement from P_1 to P_2 is represented by the vector $\vec{A}$, and the displacement from P_2 to P_3 is represented by $\vec{B}$. Note that the displacement vectors depend only on the endpoints and not on the

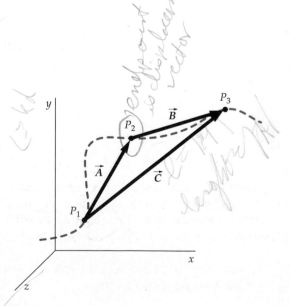

Figure 3-1

actual path of the particle. The *resultant* displacement from P_1 to P_3, labeled $\vec{C}$, is the sum of the two successive displacements $\vec{A}$ and $\vec{B}$:

$$\vec{C} = \vec{A} + \vec{B} \qquad\qquad 3\text{-}1$$

Two displacement vectors are added graphically by placing the tail of one at the head of the other (Figure 3-2). The resultant vector extends from the tail of the first to the head of the second. Note that C does not equal $A + B$ unless $\vec{A}$ and $\vec{B}$ are in the same direction. That is, $\vec{C} = \vec{A} + \vec{B}$ does not imply that $C = A + B$.

An equivalent way of adding vectors, called the **parallelogram method**, is to move $\vec{B}$ so that it is tail to tail with $\vec{A}$. The diagonal of the parallelogram formed by $\vec{A}$ and $\vec{B}$ then equals $\vec{C}$. From Figure 3-3 we can see that it makes no difference in which order we add two vectors; that is, $\vec{A} + \vec{B} = \vec{B} + \vec{A}$.

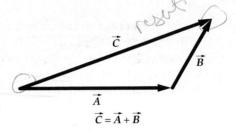

$$\vec{C} = \vec{A} + \vec{B}$$

Figure 3-2 Vector addition.

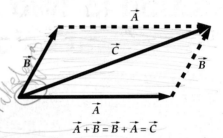

$$\vec{A} + \vec{B} = \vec{B} + \vec{A} = \vec{C}$$

Figure 3-3 Parallelogram method for adding vectors.

Example 3-1

You walk 3 km east and then 4 km north. What is your resultant displacement?

Picture the Problem The two displacements $\vec{A}$ and $\vec{B}$ and the resultant displacement $\vec{C}$ are shown in Figure 3-4. $\vec{A}$ and $\vec{B}$ are at right angles to each other, and $\vec{C} = \vec{A} + \vec{B}$ is the hypotenuse of the corresponding right triangle. The magnitude C can be found from the Pythagorean theorem. The direction of $\vec{C}$ is found using trigonometry.

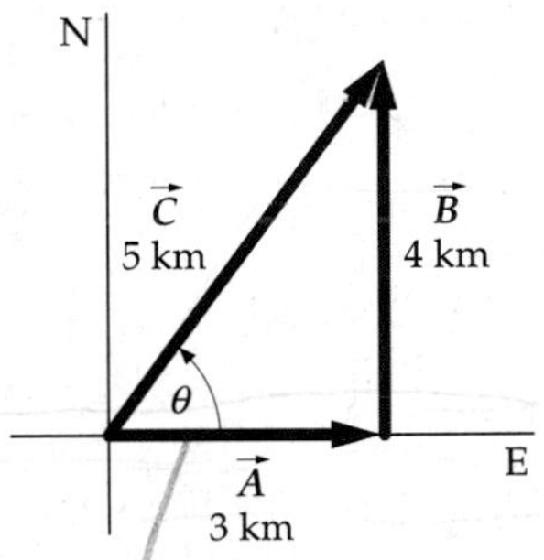

1. The magnitude of the resultant displacement is related to the magnitudes of the two displacements by the Pythagorean theorem:

$$C^2 = A^2 + B^2$$
$$= (3 \text{ km})^2 + (4 \text{ km})^2$$
$$= 25 \text{ km}^2$$

$$C = \sqrt{25 \text{ km}^2} = 5 \text{ km}$$

2. Let θ be the angle from the east axis to the resultant displacement $\vec{C}$. From the figure we find $\tan \theta$; using a calculator with trigonometric functions yields θ:

$$\tan \theta = \frac{4 \text{ km}}{3 \text{ km}} = 1.33$$
$$\theta = \tan^{-1} 1.33 = 53.1°$$

Figure 3-4

Remarks A vector is described by its magnitude and its direction. Your resultant displacement is a vector of length 5 km in a direction 53.1° north of east.

3-2 General Properties of Vectors

Many quantities in physics have magnitude and direction, and add like displacements. Examples include velocity, acceleration, momentum, and force. Such quantities are called **vectors**. Quantities with magnitude but no associated direction—for example, distance and speed—are called **scalars**.

> Vectors are quantities with magnitude and direction that add and subtract like displacements.

Definition—Vectors

A vector is represented graphically by an arrow whose direction is the same as the direction of the vector and whose length is proportional to the magnitude of the vector. When the magnitude of a vector is given, its units must also be given. The magnitude of a velocity vector, for example, requires units such as meters per second. Two vectors are defined to be equal if they have the same magnitude and the same direction. Graphically, this means they have the same length and are parallel to one another. A consequence of this definition is that moving a vector so that it remains parallel to itself does not change it. Thus, all the vectors in Figure 3-5 are equal. If we translate or rotate the coordinate system, all the vectors in Figure 3-5 remain equal. A vector does not depend on the coordinate system used to represent it.

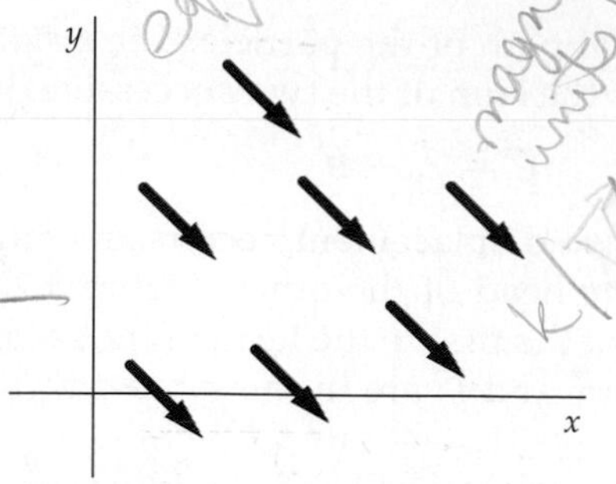

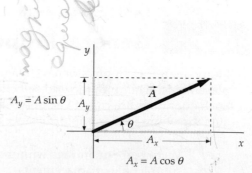

Figure 3-5 Vectors are equal if their magnitudes and directions are the same. All vectors in this figure are equal.

Multiplying a Vector by a Scalar

A vector $\vec{A}$ multiplied by a scalar s is the vector $\vec{B} = s\vec{A}$, which has magnitude $|s|A$ and is parallel to $\vec{A}$ if s is positive and antiparallel if s is negative. Thus the vector $-\vec{A}$ has the same magnitude as $\vec{A}$ but points in the opposite direction so that $\vec{A} + (-\vec{A}) = 0$. The dimensions of $s\vec{A}$ are those of s multiplied by those of A.

Subtracting Vectors

We subtract vector $\vec{B}$ from vector $\vec{A}$ by adding $-\vec{B}$, which has the same magnitude as $\vec{B}$. The result is $\vec{C} = \vec{A} + (-\vec{B}) = \vec{A} - \vec{B}$ (Figure 3-6a). An equivalent way of subtracting $\vec{B}$ from $\vec{A}$ is to draw them tail to tail and then draw a vector $\vec{C}$ from $\vec{B}$ to $\vec{A}$. That is, $\vec{C}$ is the vector that must be added to $\vec{B}$ to obtain the resultant vector $\vec{A}$ (Figure 3-6b). The rules for adding or subtracting any two vectors, such as two velocity vectors or two acceleration vectors, are the same as for adding displacement vectors.

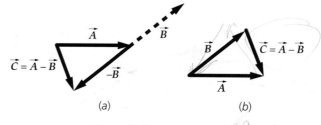

Figure 3-6

Components of Vectors

The component of a vector along a line in space is the length of the projection of the vector on that line. It is found by dropping a perpendicular from the head of the vector to the line, as shown in Figure 3-7. The components of a vector along the x, y, and z directions, illustrated in Figure 3-8 for a vector in the xy plane, are called rectangular components. Note that the components of a vector *do* depend on the coordinate system used to represent the vector, although the vector itself does not.

Rectangular components are useful for the addition or subtraction of vectors. If θ is the angle between $\vec{A}$ and the x axis, then

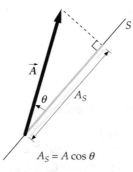

Figure 3-7 Definition of the component of a vector. The component of the vector $\vec{A}$ along the line S is A_S.

$$A_x = A \cos \theta \qquad \text{3-2}$$

The x component of a vector

and

$$A_y = A \sin \theta \qquad \text{3-3}$$

The y component of a vector

where A is the magnitude of $\vec{A}$.

Figure 3-8 The rectangular components of a vector. $A_x = A \cos \theta$, $A_y = A \sin \theta$.

If we know A_x and A_y, we can find the angle θ from

$$\tan\theta = \frac{A_y}{A_x}, \qquad \theta = \tan^{-1}\frac{A_y}{A_x} \qquad\qquad 3\text{-}4$$

and the magnitude A from the Pythagorean theorem:

$$A = \sqrt{A_x^2 + A_y^2} \qquad\qquad 3\text{-}5a$$

In three dimensions,

$$A = \sqrt{A_x^2 + A_y^2 + A_z^2} \qquad\qquad 3\text{-}5b$$

Components can be positive or negative. For example, if $\vec{A}$ points in the negative x direction, A_x is negative. Consider two vectors $\vec{A}$ and $\vec{B}$ that lie in the xy plane. The rectangular components of each vector and those of the sum $\vec{C} = \vec{A} + \vec{B}$ are shown in Figure 3-9. We see that $\vec{C} = \vec{A} + \vec{B}$ is equivalent to both

$$C_x = A_x + B_x \qquad\qquad 3\text{-}6a$$

and

$$C_y = A_y + B_y \qquad\qquad 3\text{-}6b$$

Figure 3-9

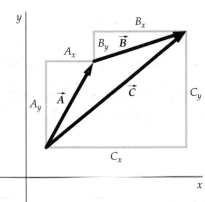

Exercise A car travels 20 km in a direction 30° north of west. Let the x axis point east and the y axis point north as in Figure 3-10. Find the x and y components of the displacement vector of the car. (*Answer* $A_x = -17.3$ km, $A_y = +10$ km)

Figure 3-10

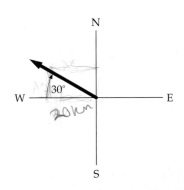

Example 3-2

You walk 3 km west and then 4 km headed 60° north of east (Figure 3-11). Find your resultant displacement (*a*) graphically and (*b*) using vector components.

Picture the Problem The triangle formed by the three vectors is not a right triangle, so the magnitudes of the vectors are not related by the Pythagorean theorem. We find the resultant graphically by drawing each of the displacements to scale and measuring the resultant displacement.

Figure 3-11

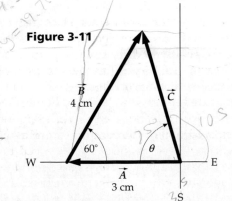

(*a*) If we draw the first displacement vector 3 cm long and the second one 4 cm long, we find the resultant vector to be about 3.5 cm long. Thus, the magnitude of the resultant displacement is 3.5 km. The angle θ made between the resultant displacement and the west direction can then be measured with a protractor. It is about 75°.

(*b*)1. Let $\vec{A}$ be the first displacement and choose the x axis to be in the easterly direction. Compute A_x and A_y from Equations 3-2 and 3-3: $A_x = -3$ km and $A_y = 0$

2. Similarly, compute the components of the second displacement $\vec{B}$:

$$B_x = (4 \text{ km})\cos 60° = 2 \text{ km}$$

$$B_y = (4 \text{ km})\sin 60° = 3.46 \text{ km}$$

3. The components of the resultant displacement $\vec{C} = \vec{A} + \vec{B}$ are found by addition:

$$C_x = A_x + B_x = -3 \text{ km} + 2 \text{ km} = -1 \text{ km}$$

$$C_y = A_y + B_y = 0 + 3.46 \text{ km} = 3.46 \text{ km}$$

4. The Pythagorean theorem gives the magnitude of $\vec{C}$:

$$C^2 = C_x^2 + C_y^2 = (-1 \text{ km})^2 + (3.46 \text{ km})^2 = 13.0 \text{ km}^2$$

$$C = \sqrt{13.0 \text{ km}^2} = 3.60 \text{ km}$$

5. The ratio of C_y to C_x gives the tangent of the angle θ between $\vec{C}$ and the x axis:

$$\tan \theta = \frac{C_y}{C_x} = \frac{3.46 \text{ km}}{-1 \text{ km}} = -3.46$$

$$\theta = \tan^{-1} -3.46 = -74°$$

Remarks Since the displacement (which is a vector) was asked for, the answer must include either the magnitude *and* direction, or both components. In (*b*) we could have stopped at step 3 because the x and y components completely define the displacement vector. We converted to the magnitude and direction to compare with the answer to part (*a*). Note that in step 5 of (*b*), a calculator gives the angle as $-74°$. But the calculator can't distinguish whether the x or y component is negative. We noted on the figure that the resultant displacement makes an angle of about 75° with the negative x axis and an angle of about 105° with the positive x axis. This agrees with the results in (*a*) within the accuracy of our measurement.

Unit Vectors

A **unit vector** is a *dimensionless* vector with unit magnitude. The vector $A^{-1}\vec{A}$ is an example of a unit vector that points in the direction of $\vec{A}$. Unit vectors are often written boldface italic with an overhead caret as in $\hat{A} = A^{-1}\vec{A}$. Unit vectors that point in the x, y, and z directions are convenient for expressing vectors in terms of their rectangular components. They are usually written $\hat{i}$, $\hat{j}$, and $\hat{k}$, respectively. Then the vector $A_x\hat{i}$ has a magnitude A_x and points in the positive x direction (or negative x direction if A_x is negative). A general vector $\vec{A}$ can be written as the sum of three vectors, each of which is parallel to a coordinate axis (Figure 3-12):

$$\vec{A} = A_x\hat{i} + A_y\hat{j} + A_z\hat{k} \qquad 3\text{-}7$$

The addition of two vectors $\vec{A}$ and $\vec{B}$ can be written in terms of unit vectors as

$$\vec{A} + \vec{B} = (A_x\hat{i} + A_y\hat{j} + A_z\hat{k}) + (B_x\hat{i} + B_y\hat{j} + B_z\hat{k})$$

$$= (A_x + B_x)\hat{i} + (A_y + B_y)\hat{j} + (A_z + B_z)\hat{k} \qquad 3\text{-}8$$

The general properties of vectors are summarized in Table 3-1.

Exercise Given two vectors

$$\vec{A} = (4 \text{ m})\hat{i} + (3 \text{ m})\hat{j} \qquad \text{and} \qquad \vec{B} = (2 \text{ m})\hat{i} - (3 \text{ m})\hat{j}$$

find (*a*) A, (*b*) B, (*c*) $\vec{A} + \vec{B}$, and (*d*) $\vec{A} - \vec{B}$. (*Answers* (*a*) $A = 5$ m, (*b*) $B = 3.61$ m, (*c*) $\vec{A} + \vec{B} = (6 \text{ m})\hat{i}$, (*d*) $\vec{A} - \vec{B} = (2 \text{ m})\hat{i} + (6 \text{ m})\hat{j}$)

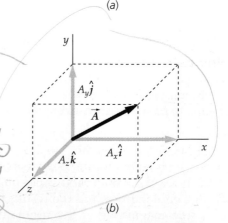

Figure 3-12 (*a*) The unit vectors $\hat{i}$, $\hat{j}$, and $\hat{k}$ in a rectangular coordinate system. (*b*) The vector $\vec{A}$ in terms of unit vectors: $\vec{A} = A_x\hat{i} + A_y\hat{j} + A_z\hat{k}$.

Table 3-1

Properties of Vectors

Property	Explanation	Figure	Component representation
Equality	$\vec{A} = \vec{B}$ if $|\vec{A}| = |\vec{B}|$ and their directions are the same		$A_x = B_x$ $A_y = B_y$ $A_z = B_z$
Addition	$\vec{C} = \vec{A} + \vec{B}$		$C_x = A_x + B_x$ $C_y = A_y + B_y$ $C_z = A_z + B_z$
Negative of a vector	$\vec{A} = -\vec{B}$ if $|\vec{B}| = |\vec{A}|$ and their directions are opposite		$A_x = -B_x$ $A_y = -B_y$ $A_z = -B_z$
Subtraction	$\vec{C} = \vec{A} - \vec{B}$		$C_x = A_x - B_x$ $C_y = A_y - B_y$ $C_z = A_z - B_z$
Multiplication by a scalar	$\vec{B} = s\vec{A}$ has magnitude $|\vec{B}| = s|\vec{A}|$ and has the same direction as $\vec{A}$ if s is positive or $-\vec{A}$ if s is negative		$B_x = sA_x$ $B_y = sA_y$ $B_z = sA_z$

3-3   Position, Velocity, and Acceleration

Position and Velocity Vectors

The **position vector** of a particle is a vector drawn from the origin of a reference frame to the xy position of the particle. For a particle at the point (x, y), its position vector $\vec{r}$ is

$$\vec{r} = x\hat{i} + y\hat{j}$$ 3-9

Definition—Position vector

Figure 3-13 shows the actual path or trajectory of the particle. (Don't confuse the trajectory with the x-versus-t plots of the previous chapter.) At time t_1, the particle is at P_1, with position vector $\vec{r}_1$; by t_2, the particle has moved to P_2, with position vector $\vec{r}_2$. The particle's change in position is the displacement vector $\Delta\vec{r}$:

$$\Delta\vec{r} = \vec{r}_2 - \vec{r}_1$$ 3-10

Definition—Displacement vector

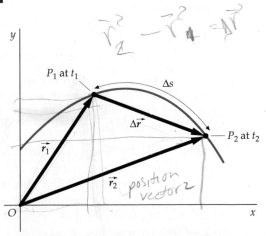

Figure 3-13 The displacement vector $\Delta\vec{r}$ is the difference in the position vectors, $\Delta\vec{r} = \vec{r}_2 - \vec{r}_1$. Equivalently, $\Delta\vec{r}$ is the vector that, when added to $\vec{r}_1$, yields the new position vector $\vec{r}_2$.

The ratio of the displacement vector to the time interval $\Delta t = t_2 - t_1$ is the **average-velocity vector:**

$$\vec{v}_{av} = \frac{\Delta \vec{r}}{\Delta t} \qquad\qquad 3\text{-}11$$

Definition—Average-velocity vector

This vector points in the direction of the displacement.

The magnitude of the displacement vector is less than the distance traveled along the curve unless the particle moves in a straight line. However, if we consider smaller and smaller intervals, the magnitude of the displacement approaches the distance along the curve, and the direction of $\Delta \vec{r}$ approaches the tangent to the curve at the beginning of the interval (Figure 3-14). We define the **instantaneous-velocity vector** as the limit of the average-velocity vector as Δt approaches zero:

$$\vec{v} = \lim_{\Delta t \to 0} \frac{\Delta \vec{r}}{\Delta t} = \frac{d\vec{r}}{dt} \qquad\qquad 3\text{-}12$$

Definition—Instantaneous-velocity vector

The instantaneous-velocity vector is the derivative of the position vector with respect to time. Its magnitude is the speed ds/dt, and its direction is the direction of motion of the particle along the line tangent to the curve.

To calculate the derivative in Equation 3-12, we write the position vector in terms of its components:

$$\Delta \vec{r} = \vec{r}_2 - \vec{r}_1 = (x_2 - x_1)\hat{i} + (y_2 - y_1)\hat{j} = \Delta x \hat{i} + \Delta y \hat{j}$$

Then

$$\vec{v} = \lim_{\Delta t \to 0} \frac{\Delta \vec{r}}{\Delta t} = \lim_{\Delta t \to 0} \frac{\Delta x \hat{i} + \Delta y \hat{j}}{\Delta t} = \lim_{\Delta t \to 0} \frac{\Delta x}{\Delta t} \hat{i} + \lim_{\Delta t \to 0} \frac{\Delta y}{\Delta t} \hat{j}$$

or

$$\vec{v} = \frac{dx}{dt} \hat{i} + \frac{dy}{dt} \hat{j} = v_x \hat{i} + v_y \hat{j} \qquad\qquad 3\text{-}13$$

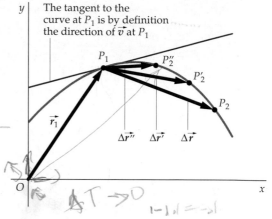

Figure 3-14 As the time interval is made smaller, the direction of the displacement vector approaches the tangent to the curve.

Example 3-3

A sailboat has coordinates $(x_1, y_1) = $ (110 m, 218 m) at $t_1 = $ 60 s. Two minutes later, at time t_2, it has the coordinates $(x_2, y_2) = $ (130 m, 205 m). (*a*) Find the average velocity for this two-minute interval. Express v_{av} in terms of its rectangular components. (*b*) Find the magnitude and direction of this average velocity. (*c*) For $t \geq 20$ s, the position of the sailboat as a function of time is $x(t) = $ 100 m + $(\frac{1}{6}$ m/s$)t$ and $y(t) = $ 200 m + (1080 m·s)t^{-1}. Find the instantaneous velocity at a general time $t \geq 20$ s.

Picture the Problem The initial and final positions of the sailboat are shown in Figure 3-15. (*a*) The average velocity vector points from the initial to the final position. (*b*) The instantaneous velocity components are calculated from Equation 3-13: $v_x = dx/dt$ and $v_y = dy/dt$.

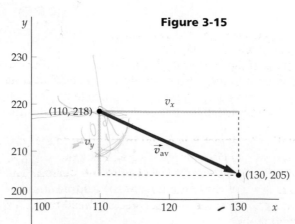

Figure 3-15

(a)   The x and y components of the average velocity $\vec{v}_{av}$ are calculated directly from their definitions:

$$v_{x,av} = \frac{x_2 - x_1}{\Delta t} = \frac{130 \text{ m} - 110 \text{ m}}{120 \text{ s}} = 0.167 \text{ m/s}$$

$$v_{y,av} = \frac{y_2 - y_1}{\Delta t} = \frac{205 \text{ m} - 218 \text{ m}}{120 \text{ s}} = -0.108 \text{ m/s}$$

*(b)*1.   The magnitude of $\vec{v}_{av}$ is found from the Pythagorean theorem:

$$v_{av} = \sqrt{(v_{x,av})^2 + (v_{y,av})^2} = 0.199 \text{ m/s}$$

   2.   The ratio of $v_{y,av}$ to $v_{x,av}$ gives the tangent of the angle θ between $\vec{v}_{av}$ and the x axis:

$$\tan\theta = \frac{v_{y,av}}{v_{x,av}} = \frac{-0.108 \text{ m/s}}{0.167 \text{ m/s}} = -0.65$$

$$\theta = \tan^{-1}(-0.65) = -33.0°$$

(c)   We find the instantaneous velocity $\vec{v}$ by calculating dx/dt and dy/dt:

$$\vec{v} = \frac{dx}{dt}\hat{i} + \frac{dy}{dt}\hat{j} = \left(\frac{1}{6}\text{ m/s}\right)\hat{i} - (1080 \text{ m·s})t^{-2}\hat{j}$$

Remark The magnitude of $\vec{v}$ can be found from $v = \sqrt{v_x^2 + v_y^2}$ and its direction can be found from $\tan\theta = v_y/v_x$.

Exercise Find the x and y components and the magnitude and direction of the instantaneous velocity of the sailboat at time $t_1 = 60$ s. (*Answers* $\vec{v}_1 = (\frac{1}{6}\text{ m/s})\hat{i} - (0.30 \text{ m/s})\hat{j}$, $v_1 = 0.34$ m/s, $\theta_2 = -60.9°$)

Relative Velocity

Relative velocities in two and three dimensions combine just as they do in one dimension, except that the velocity vectors are not necessarily along the same line. If a particle moves with velocity $\vec{v}_{pA}$ relative to a coordinate system A, which is in turn moving with velocity $\vec{v}_{AB}$ relative to another coordinate system B, the velocity of the particle relative to B is

$$\vec{v}_{pB} = \vec{v}_{pA} + \vec{v}_{AB} \qquad\qquad 3\text{-}14$$

Relative velocity

For example, if you are on a railroad car moving with velocity $\vec{v}_{cg}$ relative to the ground (Figure 3-16*a*), and you start walking with a velocity relative to the car of $\vec{v}_{pc}$ (Figure 3-16*b*), then your velocity relative to the ground is the sum of these two velocities: $\vec{v}_{pg} = \vec{v}_{pc} + \vec{v}_{cg}$ (Figure 3-16*c*).

Figure 3-16 Relative velocity in two dimensions.

(a)

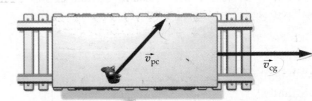

(b)

The velocity of object A relative to object B is equal in magnitude and opposite in direction to the velocity of object B relative to object A. For example, $\vec{v}_{pc} = -\vec{v}_{cp}$, where $\vec{v}_{cp}$ is the velocity of the car relative to the person. The addition of relative velocities is done in the same way as the addition of displacements; either graphically, by placing the velocity vectors head to tail, or analytically, using vector components.

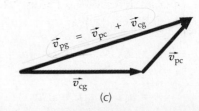

(c)

Example 3-4

A plane is to fly due north. The speed of the plane relative to the air is 200 km/h, and the wind is blowing from west to east at 90 km/h. (*a*) In which direction should the plane head? (*b*) How fast does the plane travel relative to the ground?

Picture the Problem Since the wind is blowing toward the east, the plane must head west of north as shown in Figure 3-17. The velocity of the plane relative to the ground $\vec{v}_{pg}$ will be the sum of the velocity of the plane relative to the air $\vec{v}_{pa}$ and the velocity of the air relative to the ground $\vec{v}_{ag}$.

Figure 3-17

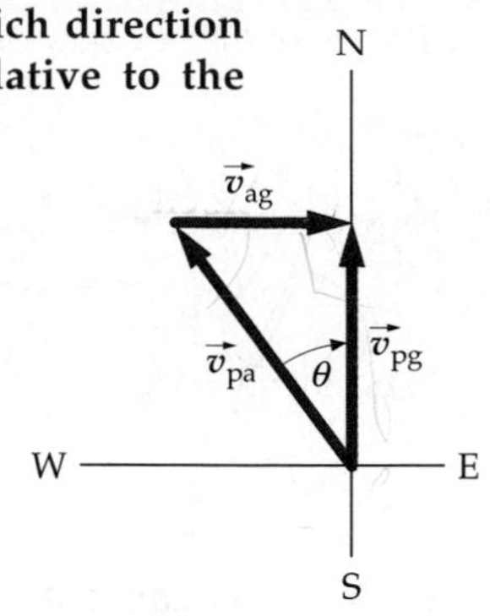

(*a*) 1. The velocity of the plane relative to the ground is given by Equation 3-14:

$$\vec{v}_{pg} = \vec{v}_{pa} + \vec{v}_{ag}$$

2. The sine of the angle θ between the velocity of the plane and north equals the ratio of v_{ag} and v_{pa}:

$$\sin \theta = \frac{v_{ag}}{v_{pa}} = \frac{90 \text{ km/h}}{200 \text{ km/h}} = 0.45$$

$$\theta = 26.7°$$

(*b*) Since v_{ag} and v_{pg} are perpendicular, we can use the Pythagorean theorem to find the magnitude of $\vec{v}_{pg}$:

$$v_{pa}^2 = v_{ag}^2 + v_{pg}^2$$

$$v_{pg} = \sqrt{v_{pa}^2 - v_{ag}^2}$$

$$= \sqrt{(200 \text{ km/h})^2 - (90 \text{ km/h})^2} = 179 \text{ km/h}$$

The Acceleration Vector

The **average-acceleration vector** is the ratio of the change in the instantaneous-velocity vector $\Delta\vec{v}$ to the time interval Δt:

$$\vec{a}_{av} = \frac{\Delta\vec{v}}{\Delta t} \qquad\qquad 3\text{-}15$$

Definition—Average-acceleration vector

The **instantaneous-acceleration vector** is the limit of this ratio as Δt approaches zero; in other words, it is the derivative of the velocity vector with respect to time:

$$\vec{a} = \lim_{\Delta t \to 0} \frac{\Delta\vec{v}}{\Delta t} = \frac{d\vec{v}}{dt} \qquad\qquad 3\text{-}16$$

Definition—Instantaneous-acceleration vector

To calculate the instantaneous acceleration, we express $\vec{v}$ in rectangular coordinates:

$$\vec{v} = v_x\hat{i} + v_y\hat{j} + v_z\hat{k} = \frac{dx}{dt}\hat{i} + \frac{dy}{dt}\hat{j} + \frac{dz}{dt}\hat{k}$$

Then

$$\vec{a} = \frac{dv_x}{dt}\hat{i} + \frac{dv_y}{dt}\hat{j} + \frac{dv_z}{dt}\hat{k} = \frac{d^2x}{dt^2}\hat{i} + \frac{d^2y}{dt^2}\hat{j} + \frac{d^2z}{dt^2}\hat{k}$$

$$= a_x\hat{i} + a_y\hat{j} + a_z\hat{k} \qquad\qquad 3\text{-}17$$

Example 3-5

The position of a thrown baseball is given by $\vec{r} = 1.5\ \text{m}\ \hat{i} + (12\ \text{m/s}\ \hat{i} + 16\ \text{m/s}\ \hat{j})t - 4.9\ \text{m/s}^2\ \hat{j}\ t^2$. **Find its velocity and acceleration.**

1. The x and y components of the velocity are found by differentiating x and y:

$$v_x = \frac{dx}{dt} = \frac{d}{dt}[1.5\ \text{m} + (12\ \text{m/s})t] = 12\ \text{m/s}$$

$$v_y = \frac{dy}{dt} = \frac{d}{dt}[(16\ \text{m/s})t - (4.9\ \text{m/s}^2)t^2]$$

$$= 16\ \text{m/s} - 2(4.9\ \text{m/s}^2)t = 16\ \text{m/s} - (9.8\ \text{m/s}^2)t$$

2. We differentiate again to obtain the components of the acceleration:

$$a_x = \frac{dv_x}{dt} = 0$$

$$a_y = \frac{dv_y}{dt} = -9.8\ \text{m/s}^2$$

3. In vector notation, the velocity and acceleration are:

$$\vec{v} = (12\ \text{m/s})\hat{i} + [16\ \text{m/s} + (9.8\ \text{m/s}^2)t]\hat{j}$$

$$\vec{a} = -9.8\ \text{m/s}^2\ \hat{j}$$

Remark This is an example of projectile motion, a topic we study in the next section.

For a vector to be constant, both its magnitude and direction must remain constant. If either changes, the vector changes. Thus, if a car rounds a curve in the road at constant speed, it is accelerating because the velocity is changing due to the change in direction of the velocity vector.

Example 3-6

A car is traveling east at 60 km/h. It rounds a curve, and 5 s later it is traveling north at 60 km/h. Find the average acceleration of the car.

Picture the Problem The initial and final velocity vectors are shown in Figure 3-18. We choose the unit vector $\hat{i}$ to be east and $\hat{j}$ to be north, and we calculate the average acceleration from its definition, $\vec{a} = \Delta\vec{v}/\Delta t$. Note that $\Delta\vec{v}$ is the vector that, when added to $\vec{v}_i$, results in $\vec{v}_f$.

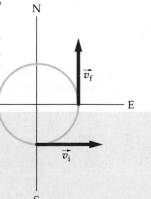

1. The average acceleration is the ratio of the velocity change to the time interval:

$$\vec{a}_{\text{av}} = \frac{\Delta\vec{v}}{\Delta t}$$

2. The change in velocity is related to the initial and final velocities:

$$\Delta\vec{v} = \vec{v}_f - \vec{v}_i$$

3. Express the initial and final velocities as vectors:

$$\vec{v}_i = (60\ \text{km/h})\hat{i}$$

$$\vec{v}_f = (60\ \text{km/h})\hat{j}$$

Figure 3-18

4. Substitute the above results to find the average acceleration:

$$\vec{a}_{\text{av}} = \frac{\vec{v}_f - \vec{v}_i}{\Delta t} = \frac{(60\ \text{km/h})\hat{j} - (60\ \text{km/h})\hat{i}}{5\ \text{s}}$$

$$= -(12\ \text{km/h·s})\hat{i} + (12\ \text{km/h·s})\hat{j}$$

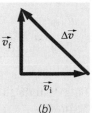

Remark Note that the car accelerates, even though its speed does not change.

Exercise Find the magnitude and direction of the average acceleration vector. (*Answers* $a = 17.0 \text{ km/h·s}$, $\theta = 135°$).

The motion of an object traveling in a circle is a common example of motion in which the velocity of an object changes even as its speed remains constant. We discuss circular motion in Chapter 5.

3-4   Projectile Motion

Figure 3-19 shows a particle launched with initial speed v_0 at angle θ with the horizontal axis. Let the launch point be at (x_0, y_0); y is positive upward, and x is positive to the right. The initial velocity then has components

$$v_{0x} = v_0 \cos \theta_0 \qquad\qquad \text{3-18a}$$

$$v_{0y} = v_0 \sin \theta_0 \qquad\qquad \text{3-18b}$$

In the absence of air resistance, the acceleration is that of gravity, vertically downward:

$$a_x = 0 \qquad\qquad \text{3-19a}$$

and

$$a_y = -g \qquad\qquad \text{3-19b}$$

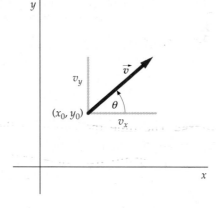

Figure 3-19

Since the acceleration is constant, we can use the kinematics equations discussed in Chapter 2. The x component of the velocity is constant because there is no horizontal acceleration:

$$v_x = v_{0x}$$

The y component varies with time according to Equation 2-11, with $a = -g$:

$$v_y = v_{0y} - gt$$

Notice that v_x does not depend on v_y and vice versa: *The horizontal and vertical components of projectile motion are independent.* This can be demonstrated by dropping a ball from a desktop and projecting a second ball horizontally at the same time. Both balls strike the floor at the same time. The displacements x and y are given by

$$x(t) = x_0 + v_{0x}t \qquad\qquad \text{3-20a}$$

$$y(t) = y_0 + v_{0y}t - \frac{1}{2}gt^2 \qquad\qquad \text{3-20b}$$

Equations of motion for a projectile

(See Equation 2-14.) The notation $x(t)$ and $y(t)$ simply emphasizes that x and y are functions of time. If the y component of the initial velocity is known, the time t for which the particle is at height y can be found from Equation 3-20b. The horizontal position at that time can then be found using Equation 3-20a. The total horizontal distance a projectile travels is called its **range.**

Example 3-7

Another joyful physics student throws his cap into the air with an initial velocity of 24.5 m/s at 36.9° from the horizontal. Find (*a*) the total time the cap is in the air, and (*b*) the total horizontal distance traveled.

Picture the Problem We choose the origin to be the initial position of the cap so that $x_0 = y_0 = 0$. The total time the cap is in the air is found by setting $y = 0$ in Equation 3-20*b*. We can then use this result in Equation 3-20*a* to find the total horizontal distance traveled.

(*a*) 1. Set $y = 0$ in Equation 3-20*b* and solve for *t*:

$$y = v_{0y}t - \tfrac{1}{2}gt^2 = t(v_{0y} - \tfrac{1}{2}gt) = 0$$

2. There are two solutions for *t*:

$$t = 0 \quad \text{(initial conditions)}$$

$$t = \frac{2v_{0y}}{g}$$

3. Compute the vertical component of the initial velocity vector:

$$v_{0y} = (24.5 \text{ m/s})\sin 36.9° = 14.7 \text{ m/s}$$

4. Use this result to find the total time *t*:

$$t = \frac{2v_{0y}}{g} = \frac{2(14.7 \text{ m/s})}{9.81 \text{ m/s}^2} = 3.0 \text{ s}$$

(*b*) Use this value for the time to calculate the total horizontal distance traveled:

$$x = v_{0x}t = (v_0 \cos\theta)t = (24.5 \text{ m/s})\cos 36.9°(3 \text{ s})$$

$$= (19.6 \text{ m/s})(3 \text{ s}) = 58.8 \text{ m}$$

Remarks The time the cap is in the air is the same as in Example 2-7, where the cap was thrown straight up with $v_0 = 14.7$ m/s. Figure 3-20 shows the height *y* versus *t* for the cap. This curve is identical to Figure 2-11 (Example 2-7) because the caps each have the same vertical acceleration and vertical velocity. The figure can be reinterpreted as a graph of *y* versus *x* if its time scale is converted to a distance scale. This can be done by multiplying the time values by 19.6 m/s, because the cap moves 19.6 m/s horizontally. The curve *y* versus *x* is a parabola.

Figure 3-21 shows graphs of the vertical heights versus the horizontal distances for projectiles with an initial speed of 24.5 m/s and several different initial angles. The angles drawn are 45°, which has the maximum range, and pairs of angles of equal amounts above and below 45°. Notice that the paired angles have the same range. The blue curve has an initial angle of 36.9° (0.64 rad), as in this example.

Figure 3-20

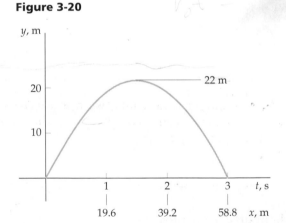

Figure 3-21

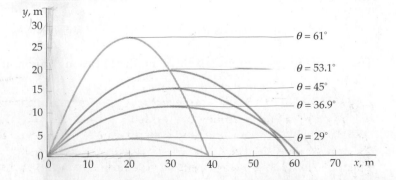

The general equation for the path $y(x)$ can be obtained from Equations 3-20a and 3-20b by eliminating the variable t. Choosing $x_0 = 0$ and $y_0 = 0$, we obtain $t = x/v_{0x}$ from Equation 3-20a. Substituting this into Equation 3-20b gives

$$y(x) = v_{0y}\left(\frac{x}{v_{0x}}\right) - \frac{1}{2}g\left(\frac{x}{v_{0x}}\right)^2 = \left(\frac{v_{0y}}{v_{0x}}\right)x - \left(\frac{g}{2v_{0x}^2}\right)x^2$$

Writing out the velocity components yields

$$y(x) = (\tan \theta_0)x - \left(\frac{g}{2v_0^2 \cos^2\theta_0}\right)x^2$$

3-21
Path of a projectile

for the projectile's path. This is of the form $y = ax + bx^2$, the equation for a parabola passing through the origin. Figure 3-22 shows the path of a projectile with its velocity vector and components at several points.

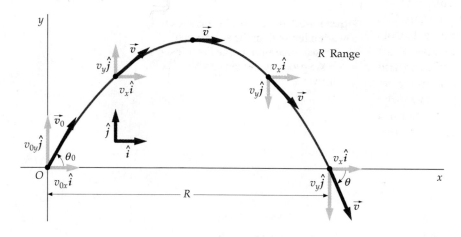

Figure 3-22 Path of a projectile showing velocity vectors.

If the initial and final elevations are equal, the range of a projectile can be written in terms of its initial speed and the angle of projection. As in the above examples, we find the range by multiplying the x component of the velocity by the total time that the projectile is in the air. The total flight time T is obtained by setting $y_0 = 0$ and $y = 0$ in Equation 3-20b:

$$y = v_{0y}t - \frac{1}{2}gt^2 = 0$$

$$t\left(v_{0y} - \frac{1}{2}gt\right) = 0$$

The flight time of the projectile is thus

$$T = \frac{2v_{0y}}{g} = \frac{2v_0}{g}\sin \theta_0$$

and the range is

$$R = v_{0x}T = (v_0 \cos \theta_0)\left(\frac{2v_0}{g}\sin \theta_0\right) = \frac{2v_0^2}{g}\sin \theta_0 \cos \theta_0$$

This can be further simplified by using the following trigonometric identity:

$$\sin 2\theta = 2 \sin \theta \cos \theta$$

Thus,

$$R = \frac{v_0^2}{g}\sin 2\theta_0$$

3-22
Range of a projectile for equal initial and final elevations

Exercise Use Equation 3-21 for the path to derive Equation 3-22. (*Answer* Set $y(x) = 0$ and solve for x.)

Equation 3-22 is useful if you want to find the range for many projectiles with equal initial and final elevations. More importantly, this equation shows how the range depends on θ. Since the maximum value of $\sin 2\theta$ is 1 when $2\theta = 90°$ or $\theta = 45°$, the range is greatest when $\theta = 45°$. In many practical applications, the initial and final elevations may not be equal, and other considerations are important. For example, in the shot put, the ball ends its flight when it hits the ground, but it is projected from an initial height of about 2 m above the ground. This causes the range to be maximum at an angle somewhat lower than 45°, as shown in Figure 3-23. Studies of the best shot-putters show that maximum range occurs at an initial angle of about 42°. When calculating the range of artillery shells, air resistance must be taken into account to predict the range accurately. As expected, air resistance reduces the range for a given angle of projection. It also decreases the optimum angle of projection slightly.

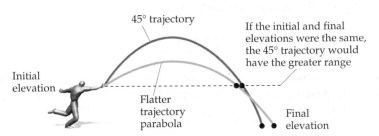

Figure 3-23 If a projectile lands at an elevation lower than the elevation of projection, maximum range is achieved when the projection angle is somewhat lower than 45°.

Example 3-8   *try it yourself*

A policeman chases a master jewel thief across city rooftops. They are both running at 5 m/s when they come to a gap between buildings that is 4 m wide and has a drop of 3 m (Figure 3-24). The thief, having studied a little physics, leaps at 5 m/s and at 45° and clears the gap easily. The policeman did not study physics and thinks he should maximize his horizontal velocity, so he leaps at 5 m/s horizontally. (*a*) Does he clear the gap? (*b*) By how much does the thief clear the gap?

Picture the Problem The time in the air depends only on the vertical motion. Choose the origin at the launch point, with upward positive so that Equations 3-20a and 3-20b apply. Use Equation 3-20b for $y(t)$ and solve for the time when $y = -3$ m. The horizontal distance traveled is the value of x at this time. (*a*) For the policeman, $\theta_0 = 0$, so the equations of motion are $x(t) = v_0 t$ and $y(t) = -\frac{1}{2}gt^2$. (*b*) For the thief, $\theta_0 = 45°$, so $x(t) = v_0 \cos 45° t$ and $y(t) = v_0 \sin 45° t - \frac{1}{2}gt^2$.

Figure 3-24

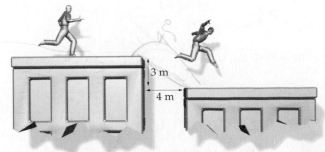

3 m

4 m

Cover the column to the right and try these on your own before looking at the answers.

Steps   **Answers**

(*a*) 1. Write $y(t)$ for the policeman and solve for t when $y = -3$ m.

$$y(t) = -\frac{1}{2}gt^2 = -3 \text{ m}$$

$$t = 0.782 \text{ s}$$

2. Find the horizontal distance traveled during this time.

$$x = v_0 t = 3.91 \text{ m}$$

Since this is less than 4 m, the policeman fails to make it across the gap between buildings.

(*b*) 1. Write $y(t)$ for the thief and set $y = -3$ m.

$$y(t) = v_{0y}t - \frac{1}{2}gt^2 = -3 \text{ m}$$

or

$$\frac{1}{2}gt^2 - v_{0y}t - 3 \text{ m} = 0$$

2. Find the two solutions for t.

$$t = \frac{v_{0y}}{g} + \frac{1}{g}\sqrt{v_{0y}^2 - (-6\text{m})(g)}$$

$$t = -0.5 \text{ s} \quad \text{or} \quad t = 1.22 \text{ s}$$

3. Find the horizontal distance covered for the positive value of t.

$$x = v_{0x}t = 4.31 \text{ m}$$

4. Subtract 4.0 m from this distance.

0.31 m

Remark The thief probably knew that he should jump at slightly less than 45°, but he didn't have time to solve the problem exactly.

Example 3-9

A helicopter drops a supply package to soldiers in a jungle clearing. When the package is dropped, the helicopter is 100 m above the clearing and flying at 25 m/s at an angle $\theta_0 = 36.9°$ above the horizontal (Figure 3-25). (*a*) Where does the package land? (*b*) If the helicopter flies at constant velocity, where is it when the package lands?

Picture the Problem The horizontal distance traveled by the package is given by Equation 3-20*a*, where t is the time the package is in the air. The value of t can be found from Equation 3-20*b*. Choose the origin to be directly below the helicopter when the package is dropped. The initial velocity of the package is the initial velocity of the helicopter.

Figure 3-25

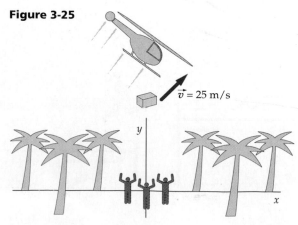

$\vec{v} = 25$ m/s

(a) 1. The point of impact for the package, x, is given by horizontal velocity times the time:

$x = v_{0x}t$

2. Find the horizontal velocity of the package:

$v_{0x} = v_0 \cos \theta = (25 \text{ m/s})\cos 36.9° = 20 \text{ m/s}$

3. Write $y(t)$ and solve for t when $y = 0$:

$y(t) = y_0 + v_{0y}t - \frac{1}{2}gt^2$

$= 100 \text{ m} + (25 \text{ m/s})(\sin 36.9°)t - \frac{1}{2}(9.81 \text{ m/s}^2)t^2$

$= 100 \text{ m} + (15 \text{ m/s})t - 4.9t^2$

$y = 0$ at $t = 6.30$ s and $t = -3.24$ s

4. Use the positive time to find the range x:

$x = v_{0x}t = (20 \text{ m/s})(6.30 \text{ s}) = 126 \text{ m}$

(b) The coordinates of the helicopter at the time of impact are:

$x_h = v_{0x}t = (20 \text{ m/s})(6.30 \text{ s}) = 126 \text{ m}$

$y_h = y_0 + v_{0y}t = 100 \text{ m} + (15 \text{ m/s})(6.30 \text{ s})$

$= 100 \text{ m} + 94.5 \text{ m} = 194.5 \text{ m}$

Remark The positive time is appropriate because it corresponds to a time after the package is dropped (which occurs at $t = 0$). The negative time is when the package would have been at $y = 0$ if its motion had started earlier, as shown in Figure 3-26. Note that the helicopter is directly above the package when the package hits the ground (and at all other times before then).

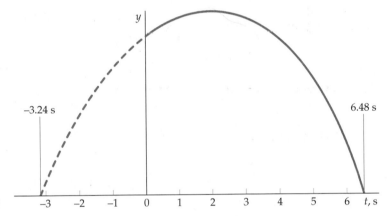

Figure 3-26

Remark Figure 3-27 shows a graph of y versus x for supply packages dropped at various initial angles and with an initial speed of 25 m/s. The green curve is the initial angle of 36.9° given in this example. Note that the maximum range no longer occurs at 45°.

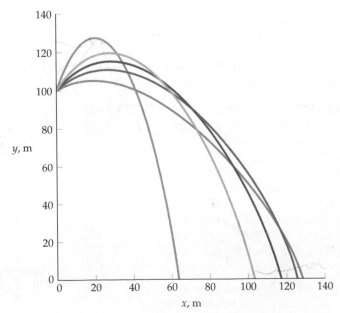

Figure 3-27

Example 3-10   *try it yourself*

In Example 3-9, (*a*) find the time t_1 for the package to reach its greatest height h, (*b*) find its greatest height h, and (*c*) find the time t_2 for the package to fall from its greatest height.

Cover the column to the right and try these on your own before looking at the answers.

Steps	*Answers*
(*a*) 1. Write $v_y(t)$ for the package.	$v_y(t) = v_{y0} - gt = 15 \text{ m/s} - (9.81 \text{ m/s}^2)t$
2. Set $v_y(t_1) = 0$ and solve for t_1.	$t_1 = 1.53$ s
(*b*) 1. Find $v_{y,\text{av}}$ during the time the package is moving up.	$v_{y,\text{av}} = 7.5$ m/s
2. Use $v_{y,\text{av}}$ to find the distance traveled up. Then find h.	$\Delta y = 11.5$ m, $h = 111.5$ m
(*c*) Find the time for the package to fall a distance h.	$t_2 = 4.77$ s

Remark Note that $t_1 + t_2 = 6.3$ s, in agreement with Example 3-9.

Example 3-11

A park ranger with a tranquilizer dart gun intends to shoot a monkey hanging from a branch (Figure 3-28). The ranger aims directly at the monkey, not realizing that the dart will follow a parabolic path that will pass below the present position of the creature. The monkey, seeing the gun discharge, lets go of the branch and drops out of the tree, expecting to avoid the dart. Show that the monkey will be hit regardless of the initial speed of the dart so long as it is great enough for the dart to travel the horizontal distance to the tree before hitting the ground. Assume that the reaction time of the monkey is negligible.

Picture the Problem We choose the origin to be at the muzzle of the gun and let $\vec{r}_0$ be the initial position vector of the monkey. Since the gun is aimed at the original position of the monkey, the initial velocity of the dart $\vec{v}_0$ is parallel to $\vec{r}_0$. We find the position vectors for both the monkey and dart as functions of time, then solve for the time t when they are equal.

Figure 3-28

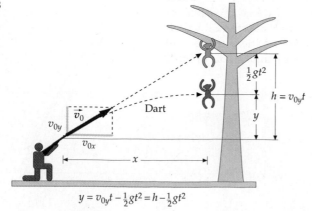

$$y = v_{0y}t - \tfrac{1}{2}gt^2 = h - \tfrac{1}{2}gt^2$$

1. Write the position vector of the monkey at time t:

$$\vec{r}_m = \vec{r}_0 - \frac{1}{2}gt^2\hat{j}$$

2. Write the position vector of the dart at some time t in terms of the initial velocity $\vec{v}_0$:

$$\vec{r}_d = \vec{v}_0 t - \frac{1}{2}gt^2\hat{j}$$

3. When the dart hits the monkey, the position vectors are equal to each other:

$$\vec{r}_m = \vec{r}_d$$

$$\vec{r}_0 - \frac{1}{2}gt^2\hat{j} = \vec{v}_0 t - \frac{1}{2}gt^2\hat{j}$$

$$\vec{r}_0 = \vec{v}_0 t$$

4. We can solve for t in terms of the distance x and initial speed v_0 by taking the x component of the above equation:

$$r_{0x} = x = v_{0x}t$$

$$t = \frac{x}{v_{0x}}$$

Remarks According to these equations, the dart always hits the monkey (Figure 3-29). However, if the dart (and monkey) hits the ground at some time $t < t_1$, the equations for $\vec{r}_m$ and $\vec{r}_d$ are no longer valid. In a familiar lecture demonstration, a target is suspended by an electromagnet. When the dart leaves the gun, the circuit to the magnet is broken and the target falls. The initial velocity of the dart is varied so that for large v_0 the target is hit very near its original height and for small v_0 it is hit just before it reaches the floor.

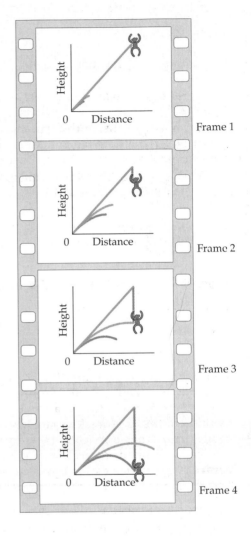

Frame 1

Frame 2

Frame 3

Frame 4

Figure 3-29 In the classic ranger and monkey problem, the dart and the monkey both fall with the same acceleration g. As a result they will meet, independent of the speed of the dart as long as the dart was aimed at the monkey. In *frame 1*, the dart is traveling so fast that it hits the monkey before the monkey has fallen very far. In *frame 2*, the monkey has fallen about a half meter from the tree branch and the dart has fallen about a half meter from a straight line of flight. In *frame 3*, the monkey has fallen about half the distance to the ground before the dart hits. The slower moving dart has also fallen away from a straight line of flight by exactly the same amount. In *frame 4*, the slowest dart and the monkey reach a point just above the ground at the same time.

Example 3-12   *try it yourself*

Your slapshot in hockey is wickedly fast but not very accurate. The puck, struck at ice level, misses the net and just clears the top of the Plexiglas wall of height $h = 2.80$ m. The flight time at the moment the puck clears the wall is $t_1 = 0.650$ s, and the horizontal distance is $x_1 = 12.0$ m. (*a*) Find the initial speed and direction of the puck. (*b*) When does the puck reach its maximum height? (*c*) What is the maximum height of the puck?

Cover the column to the right and try these on your own before looking at the answers.

Steps   **Answers**

(*a*)1. Find the horizontal component of the initial velocity.

$$v_{0x} = \frac{x_1}{t_1} = 18.5 \text{ m/s}$$

2. Write the equation for $y(t)$ and solve for v_{0y} using $y = h$ and $t = t_1$.

$$y = v_{0y}t - \frac{1}{2}gt^2, \qquad v_{0y} = 7.49 \text{ m/s}$$

3. Find v from the components and θ_0 from $\tan \theta_0 = v_{0y}/v_{0x}$.

$$v_0 = \sqrt{v_{0x}^2 + v_{0y}^2} = 20.0 \text{ m/s}, \qquad \theta_0 = 22.0°$$

(b) Write the general equation for $v_y(t)$ and solve for t when $v_y = 0$.

$$v_y = v_{0y} - gt, \qquad t = 0.764 \text{ s}$$

(c) Find the maximum height from $\Delta y = v_{y,av}\, t$.

$$\Delta y = v_{y,av}t = 2.86 \text{ m}$$

Remark In this example, a hockey puck clears a glass wall 2.8 m high and 12 m distant. The puck reaches its maximum height after clearing the wall. Figure 3-30 shows several other cases of initial velocity and angle for which the puck would also just clear the wall.

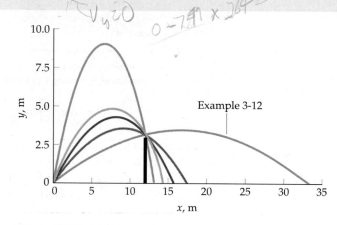

Figure 3-30

Summary

Topic	Remarks and Relevant Equations
1. Vectors	
Definition	Vectors are quantities that have both magnitude and direction. Vectors add like displacements.
Components	The component of a vector along a line in space is its projection on the line. If $\vec{A}$ makes an angle θ with the x axis, its x and y components are

$$A_x = A \cos \theta \qquad \text{3-2}$$

$$A_y = A \sin \theta \qquad \text{3-3}$$

Magnitude	$A = \sqrt{A_x^2 + A_y^2}$ 3-5
Adding vectors graphically	Any two vectors whose magnitudes have the same units may be added graphically by placing the tail of one arrow at the head of the other.

Adding vectors using components	If $\vec{C} = \vec{A} + \vec{B}$, then

$$C_x = A_x + B_x \qquad \text{3-6a}$$

and

$$C_y = A_y + B_y \qquad \text{3-6b}$$

Unit vectors	A vector $\vec{A}$ can be written in terms of unit vectors $\hat{i}$, $\hat{j}$, and $\hat{k}$, which have unit magnitude and lie along the x, y, and z axes, respectively

$$\vec{A} = A_x\hat{i} + A_y\hat{j} + A_z\hat{k} \qquad \text{3-7}$$

Position vector	The position vector $\vec{r}$ points from the origin of the coordinate system to the particle's position.

Instantaneous-velocity vector	The velocity vector $\vec{v}$ is the rate of change of the position vector. Its magnitude is the speed and it points in the direction of motion.

$$\vec{v} = \lim_{\Delta t \to 0} \frac{\Delta \vec{r}}{\Delta t} = \frac{d\vec{r}}{dt} \qquad \text{3-12}$$

Instantaneous-acceleration vector	

$$\vec{a} = \lim_{\Delta t \to 0} \frac{\Delta \vec{v}}{\Delta t} = \frac{d\vec{v}}{dt} \qquad \text{3-16}$$

2. Relative Velocity

If a particle moves with velocity $\vec{v}_{pA}$ relative to a coordinate system A, which is in turn moving with velocity $\vec{v}_{AB}$ relative to another coordinate system B, the velocity of the particle relative to B is

$$\vec{v}_{pB} = \vec{v}_{pA} + \vec{v}_{AB} \qquad \text{3-14}$$

3. Projectile Motion

Independence of motion	In projectile motion, the horizontal and vertical motions are independent. The horizontal motion has constant velocity. The vertical motion is the same as motion in one dimension with constant acceleration due to gravity g downward.

Equations	

$$v_x(t) = v_{0x}$$
$$x(t) = x_0 + v_{0x}t = x_0 + (v_0 \cos \theta)t \qquad \text{3-20a}$$
$$v_y(t) = v_{0y} - gt = v_0 \sin \theta - gt$$

$$y(t) = y_0 + v_{0y}t - \frac{1}{2}gt^2 = y_0 + (v_0 \sin \theta)t - \frac{1}{2}gt^2 \qquad \text{3-20b}$$

Path	

$$y(x) = (\tan \theta_0)x - \left(\frac{g}{2v_0^2 \cos^2 \theta_0}\right)x^2 \qquad \text{3-21}$$

Range	The range is found by multiplying v_x by the total time the projectile is in the air.

Range when initial and final elevations are equal	

$$R = \frac{v_0^2}{g}\sin 2\theta_0 \qquad \text{3-22}$$

Problem-Solving Guide

The following are applicable to all types of problems:

1. Begin by drawing a neat diagram that includes the important features of the problem.
2. Choose a convenient coordinate system and indicate it on your sketch.
3. Indicate the given information on your sketch.

Summary of Worked Examples

Type of Calculation	Procedure and Relevant Examples
1. Vectors	
Add (or subtract) vectors.	Add (or subtract) the components of individual vectors to find the components of the resultant vector. **Example 3-2**
Find the direction of a resultant vector.	The angle made with the positive x direction is found from $\tan \theta = v_y/v_x$. **Examples 3-1, 3-2, 3-3**
Take derivatives of vectors.	Express the vector in component form using unit vectors, and take the derivative of each component separately. **Examples 3-3, 3-5**
2. Relative Velocity	
Express the velocity of a particle relative to a coordinate system that is itself moving relative to another coordinate system.	Use $\vec{v}_{pB} = \vec{v}_{pA} + \vec{v}_{AB}$. **Example 3-4**
3. Projectile Motion	
Find a projectile's time of flight to various positions.	Calculate the time at which a vertical position is reached by using constant acceleration formulas. Calculate the time at which a horizontal position is reached by using constant velocity formulas. **Examples 3-7, 3-8, 3-9, 3-10, 3-11**
Find speeds and angles along a projectile's trajectory	Find the x and y components of the velocity from the constant acceleration formulas. Find the speed from $v = \sqrt{v_x^2 + v_y^2}$. The angle of a projectile's trajectory is the angle of its resultant velocity vector at that moment. **Examples 3-8, 3-12**
Find the position of a projectile	Use $x = v_{0x} t$ and $y = y_0 + v_{0y}t - gt^2$ where t is the time of flight. **Examples 3-7, 3-8, 3-9, 3-12**

Problems

 Conceptual Problems

 Problems from Optional and Exploring sections

In a few problems, you are given more data than you actually need; in a few other problems, you are required to supply data from your general knowledge, outside sources, or informed estimates.

- • Single-concept, single-step, relatively easy
- •• Intermediate-level, may require synthesis of concepts
- ••• Challenging, for advanced students

For all problems, use $g = 9.81$ m/s² for the acceleration due to gravity and neglect friction and air resistance unless instructed to do otherwise.

Vectors and Vector Addition

1 • Can the magnitude of the displacement of a particle be less than the distance traveled by the particle along its path? Can its magnitude be more than the distance traveled? Explain.

2 • Give an example in which the distance traveled is a significant amount yet the corresponding displacement is zero.

3 • The magnitude of the displacement of a particle is _____ the distance the object has traveled.

(a) larger than
(b) smaller than
(c) either larger or smaller than
(d) the same as
(e) smaller than or equal to

4 • A bear walks northeast for 12 m and then east for 12 m. Show each displacement graphically, and find the resultant displacement vector.

5 • (a) A man walks along a circular arc from the position $x = 5$ m, $y = 0$ to a final position $x = 0$, $y = 5$ m. What is his displacement? (b) A second man walks from the same initial position along the x axis to the origin and then along the y axis to $y = 5$ m and $x = 0$. What is his displacement?

6 • A circle of radius 8 m has its center on the y axis at $y = 8$ m. You start at the origin and walk along the circle at a steady speed, returning to the origin exactly 1 min after you started. (a) Find the magnitude and direction of your displacement from the origin 15, 30, 45, and 60 s after you start. (b) Find the magnitude and direction of your displacement for each of the four successive 15-s intervals of your walk. (c) How is your displacement for the first 15 s related to that for the second 15 s? (d) How is your displacement for the second 15-s interval related to that for the last 15-s interval?

7 • For the two vectors $\vec{A}$ and $\vec{B}$ in Figure 3-31, find the following graphically: (a) $\vec{A} + \vec{B}$, (b) $\vec{A} - \vec{B}$, (c) $2\vec{A} + \vec{B}$, (d) $\vec{B} - \vec{A}$, (e) $2\vec{B} - \vec{A}$.

Figure 3-31 Problem 7

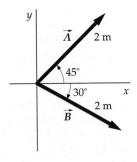

8 • A scout walks 2.4 km due east from camp, then turns left and walks 2.4 km along the arc of a circle centered at the campsite, and finally walks 1.5 km directly toward camp. (a) How far is the scout from camp at the end of his walk? (b) In what direction is the scout's position relative to the campsite? (c) What is the ratio of the final magnitude of the displacement to the total distance walked?

Adding Vectors by Components

9 • Can a component of a vector have a magnitude greater than the magnitude of the vector? Under what circumstances can a component of a vector have a magnitude equal to the magnitude of the vector?

10 • Can a vector be equal to zero and still have one or more components not equal to zero?

11 • Are the components of $\vec{C} = \vec{A} + \vec{B}$ necessarily larger than the corresponding components of either $\vec{A}$ or $\vec{B}$?

12 • The components of a vector are $A_x = -10$ m and $A_y = 6$ m. What angle does this vector make with the positive x axis?

(a) 31°
(b) −31°
(c) 180° − 31°
(d) 180° + 31°
(e) 90° − 31°

13 • A velocity vector has an x component of +5.5 m/s and a y component of −3.5 m/s. Which diagram in Figure 3-32 gives the direction of the vector?

Figure 3-32 Problem 13

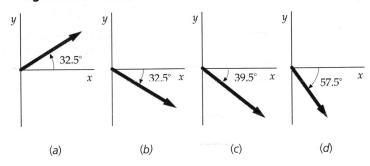

(a)   (b)   (c)   (d)

(e) None of the above.

14 • Three vectors $\vec{A}$, $\vec{B}$, and $\vec{C}$ have the following x and y components:

	$\vec{A}$	$\vec{B}$	$\vec{C}$
x component	+6	−3	+2
y component	−3	+4	+5

The magnitude of $\vec{A} + \vec{B} + \vec{C}$ is _____.

(a) 3.3
(b) 5.0
(c) 11
(d) 7.8
(e) 14

15 • Find the rectangular components of the following vectors $\vec{A}$, which lie in the xy plane, and make an angle θ with the x axis (Figure 3-33) if (a) $A = 10$ m, $\theta = 30°$; (b) $A = 5$ m, $\theta = 45°$; (c) $A = 7$ km, $\theta = 60°$; (d) $A = 5$ km, $\theta = 90°$; (e) $A = 15$ km/s, $\theta = 150°$; (f) $A = 10$ m/s, $\theta = 240°$; and (g) $A = 8$ m/s², $\theta = 270°$.

Figure 3-33 Problem 15

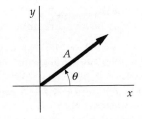

16 • Vector $\vec{A}$ has a magnitude of 8 m at an angle of 37° with the x axis; vector $\vec{B} = 3$ m $\hat{i} - 5$ m $\hat{j}$; vector $\vec{C} = -6$ m $\hat{i} + 3$ m $\hat{j}$. Find the following vectors: (a) $\vec{D} = \vec{A} + \vec{C}$; (b) $\vec{E} = \vec{B} - \vec{A}$; (c) $\vec{F} = \vec{A} - 2\vec{B} + 3\vec{C}$; (d) A vector $\vec{G}$ such that $\vec{G} - \vec{B} = \vec{A} + 2\vec{C} + 3\vec{G}$.

Unit Vectors

17 • Find the magnitude and direction of the following vectors: (a) $\vec{A} = 5\hat{i} + 3\hat{j}$; (b) $\vec{B} = 10\hat{i} - 7\hat{j}$; (c) $\vec{C} = -2\hat{i} - 3\hat{j} + 4\hat{k}$.

18 • Find the magnitude and direction of $\vec{A}$, $\vec{B}$, and $\vec{A} + \vec{B}$ for (a) $\vec{A} = -4\hat{i} - 7\hat{j}, \vec{B} = 3\hat{i} - 2\hat{j}$, and (b) $\vec{A} = 1\hat{i} - 4\hat{j}, \vec{B} = 2\hat{i} + 6\hat{j}$.

19 • Describe the following vectors using the unit vectors $\hat{i}$ and $\hat{j}$: (a) a velocity of 10 m/s at an angle of elevation of 60°; (b) a vector $\vec{A}$ of magnitude $A = 5$ m and $\theta = 225°$; (c) a displacement from the origin to the point $x = 14$ m, $y = -6$ m.

20 • For the vector $\vec{A} = 3\hat{i} + 4\hat{j}$, find any three other vectors $\vec{B}$ that also lie in the xy plane and have the property that $A = B$ but $\vec{A} \neq \vec{B}$. Write these vectors in terms of their components and show them graphically.

21 • If $\vec{A} = 5\hat{i} - 4\hat{j}$ and $\vec{B} = -7.5\hat{i} + 6\hat{j}$, write an equation relating $\vec{A}$ to $\vec{B}$.

22 •• The faces of a cube of side 3 m are parallel to the coordinate planes with one corner at the origin. A fly begins at the origin and walks along three edges until it is at the far corner. Write the displacement vector of the fly using the unit vectors $\hat{i}$, $\hat{j}$, and $\hat{k}$, and find the magnitude of this displacement.

Velocity and Acceleration Vectors

23 • For an arbitrary motion of a given particle, does the direction of the velocity vector have any particular relation to the direction of the position vector?

24 • Give examples in which the directions of the velocity and position vectors are (a) opposite, (b) the same, and (c) mutually perpendicular.

25 • How is it possible for a particle moving at constant speed to be accelerating? Can a particle with constant velocity be accelerating at the same time?

26 • If an object is moving toward the west, in what direction is its acceleration?

(a) North
(b) East
(c) West
(d) South
(e) May be any direction.

27 •• Consider the path of a particle as it moves in space. (a) How is the velocity vector related geometrically to the path of the particle? (b) Sketch a curved path and draw the velocity vector for the particle for several positions along the path.

28 •• A dart is thrown straight up. After it leaves the player's hand, it steadily loses speed as it gains altitude until it lodges in the ceiling of the game room. Draw the dart's velocity vector at times t_1 and t_2, where $\Delta t = t_2 - t_1$ is small. From your drawing find the direction of the change in velocity $\Delta\vec{v} = \vec{v}_2 - \vec{v}_1$, and thus the direction of the acceleration vector.

29 •• As a bungee jumper approaches the lowest point in her drop, she loses speed as she continues to move downward. Draw the velocity vectors of the jumper at times t_1 and t_2, where $\Delta t = t_2 - t_1$ is small. From your drawing find the direction of the change in velocity $\Delta\vec{v} = \vec{v}_2 - \vec{v}_1$, and thus the direction of the acceleration vector.

30 •• After reaching the lowest point in her jump at time t_{low}, the bungee jumper in the previous problem then moves upward, gaining speed for a short time until gravity again dominates her motion. Draw her velocity vectors at times t_1 and t_2, where $\Delta t = t_2 - t_1$ is small and $t_1 < t_{low} < t_2$. From your drawing find the direction of the change in velocity $\Delta\vec{v} = \vec{v}_2 - \vec{v}_1$, and thus the direction of the acceleration vector.

31 • A stationary radar operator determines that a ship is 10 km south of him. An hour later the same ship is 20 km southeast. If the ship moved at constant speed and always in the same direction, what was its velocity during this time?

32 • A particle's position coordinates (x, y) are (2 m, 3 m) at $t = 0$; (6 m, 7 m) at $t = 2$ s; and (13 m, 14 m) at $t = 5$ s. (a) Find v_{av} from $t = 0$ to $t = 2$ s. (b) Find v_{av} from $t = 0$ to $t = 5$ s.

33 • A particle moving at 4.0 m/s in the positive x direction is given an acceleration of 3.0 m/s² in the positive y direction for 2.0 s. The final speed of the particle is _____.

(a) −2.0 m/s
(b) 7.2 m/s
(c) 6.0 m/s
(d) 10 m/s
(e) None of the above

34 • A ball is thrown directly upward. Consider the 2-s time interval $\Delta t = t_2 - t_1$, where t_1 is 1 s before the ball reaches its highest point and t_2 is 1 s after it reaches its highest point. For the time interval Δt, find (a) the change in speed, (b) the change in velocity, and (c) the average acceleration.

35 • Initially, a particle is moving due west with a speed of 40 m/s; 5 s later it is moving north with a speed of 30 m/s. (a) What was the change in the magnitude of the particle's velocity during this time? (b) What was the change in the direction of the velocity? (c) What are the magnitude and direction of $\Delta\vec{v}$ for this interval? (d) What are the magnitude and direction of $\vec{a}_{av}$ for this interval?

36 • At $t = 0$, a particle located at the origin has a velocity of 40 m/s at $\theta = 45°$. At $t = 3$ s, the particle is at $x = 100$ m and $y = 80$ m with a velocity of 30 m/s at $\theta = 50°$. Calculate (a) the average velocity and (b) the average acceleration of the particle during this interval.

37 •• A particle moves in an xy plane with constant acceleration. At time zero, the particle is at $x = 4$ m, $y = 3$ m, and has velocity $\vec{v} = 2$ m/s $\hat{i} - 9$ m/s $\hat{j}$. The acceleration is given by the vector $\vec{a} = 4$ m/s² $\hat{i} + 3$ m/s² $\hat{j}$. (a) Find the velocity vector at $t = 2$ s. (b) Find the position vector at $t = 4$ s. Give the magnitude and direction of the position vector.

38 •• A particle has a position vector given by $\vec{r} = 30t\hat{i} + (40t - 5t^2)\hat{j}$, where r is in meters and t in seconds. Find the instantaneous-velocity and instantaneous-acceleration vectors as functions of time t.

39 •• A particle has a constant acceleration of $\vec{a} = (6\hat{i} + 4\hat{j})\ \text{m/s}^2$. At time $t = 0$, the velocity is zero and the position vector is $\vec{r}_0 - (10\ \text{m})\ \hat{i}$. (a) Find the velocity and position vectors at any time t. (b) Find the equation of the particle's path in the xy plane, and sketch the path.

40 ••• Mary and Robert decide to rendezvous on Lake Michigan. Mary departs in her boat from Petoskey at 9:00 A.M. and travels due north at 8 mi/h. Robert leaves from his home on the shore of Beaver Island, 26 mi 30° west of north of Petoskey, at 10:00 A.M. and travels at a constant speed of 6 mi/h. In what direction should Robert be heading to intercept Mary, and where and when will they meet?

Relative Velocity

41 • A river is 0.76 km wide. The banks are straight and parallel (Figure 3-34). The current is 5.0 km/h and is parallel to the banks. A boat has a maximum speed of 3 km/h in still water. The pilot of the boat wishes to go on a straight line from A to B, where AB is perpendicular to the banks. The pilot should

(a) head directly across the river.
(b) head 68° upstream from the line AB.
(c) head 22° upstream from the line AB.
(d) give up—the trip from A to B is not possible with this boat.
(e) do none of the above.

Figure 3-34 Problem 41

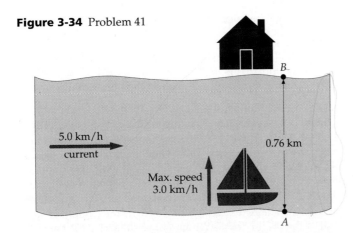

42 •• A plane flies at a speed of 250 km/h relative to still air. There is a wind blowing at 80 km/h in the northeast direction at exactly 45° to the east of north. (a) In what direction should the plane head so as to fly due north? (b) What is the speed of the plane relative to the ground?

43 •• A swimmer heads directly across a river, swimming at 1.6 m/s relative to still water. She arrives at a point 40 m downstream from the point directly across the river, which is 80 m wide. (a) What is the speed of the river current? (b) What is the swimmer's speed relative to the shore? (c) In what direction should the swimmer head so as to arrive at the point directly opposite her starting point?

44 •• A small plane departs from point A heading for an airport at point B 520 km due north. The airspeed of the plane is 240 km/h and there is a steady wind of 50 km/h blowing northwest to southeast. Determine the proper heading for the plane and the time of flight.

45 •• Two boat landings are 2.0 km apart on the same bank of a stream that flows at 1.4 km/h. A motorboat makes the round trip between the two landings in 50 min. What is the speed of the boat relative to the water?

46 •• A model airplane competition has the following rules: Each plane must fly to a point 1 km from the start and then back again. The winner is the plane with the shortest round-trip time. The contestants are free to launch their planes in any direction, so long as the plane travels exactly 1 km out and then returns. On the day of the race, a steady wind blows from the north at 5 m/s. Your plane can maintain an airspeed (speed relative to the air) of 15 m/s, and you know that starting, stopping, and turning times will be negligible. The question: Should you plan to fly into the wind and against the wind on your round-trip, or across the wind flying east and west? Make a reasoned choice by working out the following round-trip times: (1) The plane goes 1 km due north and then back; (2) the plane goes to point 1 km due east of the start, and then back.

47 •• Car A is traveling east at 20 m/s. As car A crosses the intersection shown in Figure 3-35, car B starts from rest 40 m north of the intersection and moves south with a constant acceleration of 2 m/s². (a) What is the position of B relative to A 6 s after A crosses the intersection? (b) What is the velocity of B relative to A for $t = 6$ s? (c) What is the acceleration of B relative to A for $t = 6$ s?

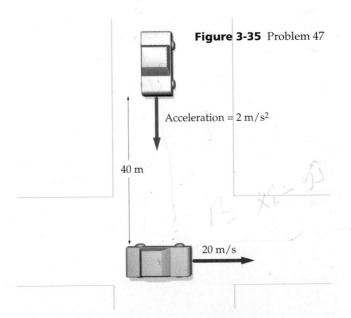

Figure 3-35 Problem 47

Acceleration = 2 m/s²

40 m

20 m/s

48 •• Bernie is showing Margaret his new boat and its autonavigation feature, of which he is particularly proud. "That island is 1 km east and 3 km north of this dock. So I just punch in the numbers like this, and we get ourselves a refreshment and enjoy the scenery." Forty-five minutes later, they find themselves due east of the island. "OK, something went wrong. I'll just reverse the instructions, and we'll go back to the dock and try again." But 45 min later, the boat is 6 km east of their original position at the dock. "Did you allow for the current?" asks Margaret. "For the what?" (a) What is the velocity of the current in the waterway where Bernie and Margaret are boating? (b) What is the velocity of the boat, relative to the water, for the first 45 min? (*c*) What is the velocity of the boat relative to the island for the first 45 min?

49 ••• Airports A and B are on the same meridian, with B 624 km south of A. Plane P departs airport A for B at the same time that an identical plane, Q, departs airport B for A. A steady 60 km/h wind is blowing from the south 30′ east of north. Plane Q arrives at airport A 1 h before plane P arrives at airport B. Determine the airspeeds of the two planes (assuming that they are the same) and the heading of each plane.

Projectiles

50 • What is the acceleration of a projectile at the top of its flight?

51 • True or false: When a bullet is fired horizontally, it takes the same amount of time to reach the ground as a bullet dropped from rest from the same height.

52 • A golfer drives her ball from the tee a distance of 240 yards down the fairway in a high arcing shot. When the ball is at the highest point of its flight,

(*a*) its velocity and acceleration are both zero.
(*b*) its velocity is zero but its acceleration is nonzero.
(*c*) its velocity is nonzero but its acceleration is zero.
(*d*) its velocity and acceleration are both nonzero.
(*e*) insufficient information is given to answer correctly.

53 • A projectile was fired at 35° above the horizontal. At the highest point in its trajectory, its speed was 200 m/s. The initial velocity had a horizontal component of

(*a*) 0.
(*b*) 200 cos(35°) m/s.
(*c*) 200 sin(35°) m/s.
(*d*) (200 m/s)/cos(35°).
(*e*) 200 m/s.

54 • Figure 3-36 represents the parabolic trajectory of a ball going from A to E. What is the direction of the acceleration at point B?

(*a*) Up and to the right
(*b*) Down and to the left
(*c*) Straight up
(*d*) Straight down
(*e*) The acceleration of the ball is zero.

Figure 3-36 Problems 54 and 55

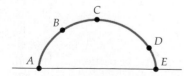

55 • Referring to Figure 3-36, (*a*) at which point(s) is the speed the greatest? (*b*) At which point(s) is the speed the lowest? (*c*) At which two points is the speed the same? Is the velocity the same at those points?

56 • A bullet is fired horizontally with an initial velocity of 245 m/s. The gun is 1.5 m above the ground. How long is the bullet in the air?

57 • A pitcher throws a fastball at 140 km/h toward home plate, which is 18.4 m away. Neglecting air resistance (not a good idea if you are the batter), find how far the ball drops because of gravity by the time it reaches home plate.

58 • A projectile is launched with speed v_0 at an angle of θ_0 with the horizontal. Find an expression for the maximum height it reaches above its starting point in terms of v_0, θ_0, and g.

59 • A projectile is fired with an initial velocity of 30 m/s at 60° above horizontal. At the projectile's highest point, what is its velocity? Its acceleration?

60 •• A projectile is fired with initial speed v at an angle 30° above the horizontal from a height of 40 m above the ground. The projectile strikes the ground with a speed of $1.2v$. Find v.

61 •• If the tree in Example 3-11 is 50 m away and the monkey hangs from a branch 10 m above the muzzle position, what is the minimum initial speed of the dart if it is to hit the monkey before hitting the ground?

62 •• A projectile is fired with an initial speed of 53 m/s. Find the angle of projection such that the maximum height of the projectile is equal to its horizontal range.

63 •• A ball thrown into the air lands 40 m away 2.44 s later. Find the direction and magnitude of the initial velocity.

64 •• Show that if an object is thrown with speed v_0 at an angle θ above the horizontal, its speed at some height h is independent of θ.

65 •• At half its maximum height, the speed of a projectile is three-fourths its initial speed. What is the angle of the initial velocity vector with respect to the horizontal?

66 ••• Wally and Luke advertise their circus act as "The Human Burrs—Trapeze Artists for the New Millennium." Their specialty involves wearing padded Velcro suits that cause them to stick together when they make contact in midair. While working on their act, Wally is shot from a cannon with a speed of 20 m/s at an angle of 30° above the horizontal. At the same moment, Luke drops from a platform having (*x*, *y*) coordinates of (8 m, 16 m), if the cannon is taken to sit at the origin. (*a*) Will they make contact? (*b*) What is the minimum distance separating Wally and Luke during their flight paths? (*c*) At what time does this minimum separation occur? (*d*) Give the coordinates of each daredevil at that time.

Projectile Range

67 • A cargo plane is flying horizontally at an altitude of 12 km with a speed of 900 km/h when a battle tank falls out of the rear loading ramp. (*a*) How long does it take the tank to hit the ground? (*b*) How far horizontally is the tank from where it fell off when it hits the ground? (*c*) How far is the tank from the aircraft when the tank hits the ground, assuming that the plane continues to fly with constant velocity?

68 • A cannon is elevated at an angle of 45°. It fires a ball with a speed of 300 m/s. (*a*) What height does the ball reach? (*b*) How long is the ball in the air? (*c*) What is the horizontal range of the cannon?

69 •• A stone thrown horizontally from the top of a 24-m tower hits the ground at a point 18 m from the base of the tower. (*a*) Find the speed at which the stone was thrown. (*b*) Find the speed of the stone just before it hits the ground.

70 •• A projectile is fired into the air from the top of a 200-m cliff above a valley (Figure 3-37). Its initial velocity is 60 m/s at 60° above the horizontal. Where does the projectile land?

Figure 3-37 Problem 70

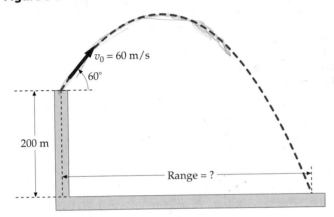

71 •• The range of a projectile fired horizontally from a cliff is equal to the height of the cliff. What is the direction of the velocity vector when the projectile strikes the ground?

72 •• Find the range of the projectile of Problem 60.

73 •• Compute $dR/d\theta$ from Equation 3-22 and show that setting $dR/d\theta = 0$ gives $\theta = 45°$ for the maximum range.

74 •• A rock is thrown from the top of a 20-m building at an angle of 53° above the horizontal. If the horizontal range of the throw is equal to the height of the building, with what speed was the rock thrown? What is the velocity of the rock just before it strikes the ground?

75 •• A stone is thrown horizontally from the top of an incline that makes an angle ϕ with the horizontal. If the stone's initial speed is v, how far down the incline will it land?

76 •• A flock of seagulls has decided to mount an organized response to the human overpopulation of their favorite beach. One tactic popular among the innovative radicals is bombing the sunbathers with clams. A gull dives with a speed of 16 m/s, at an angle of 40° below the horizontal. He releases a projectile when his vertical distance above his target, a sunbather's bronzed tummy, is 8.5 m, and scores a bull's-eye. (*a*) Where is the sunbather in relation to the gull at the instant of release? (*b*) How long is the projectile in the air? (*c*) What is the velocity of the projectile upon impact?

77 ••• A girl throws a ball at a vertical wall 4 m away (Figure 3-38). The ball is 2 m above the ground when it leaves the girl's hand with an initial velocity of $\vec{v}_0 = (10\hat{i} + 10\hat{j})$ m/s. When the ball hits the wall, the horizontal component of its velocity is reversed; the vertical component remains unchanged. Where does the ball hit the ground?

Jumping Gaps; Hitting Targets; Clearing Fences

78 • A boy uses a slingshot to project a pebble at a shoulder-height target 40 m away. He finds that to hit the target he must aim 4.85 m above the target. Determine the velocity of the pebble on leaving the slingshot and the time of flight.

79 •• The distance from the pitcher's mound to home plate is 18.4 m. The mound is 0.2 m above the level of the field. A pitcher throws a fast ball with an initial speed of 37.5 m/s. At the moment the ball leaves the pitcher's hand, it is 2.3 m above the mound. What should the angle between $\vec{v}$ and the horizontal be so that the ball crosses the plate 0.7 m above ground?

80 •• Suppose the puck in Example 3-12 is struck in such a way that it just clears the Plexiglas wall when it is at its highest point. Find v_{0y}, the time t to reach the wall, and v_{0x}, v_0, and θ_0 for this case.

81 •• The coach throws a baseball to a player with an initial speed of 20 m/s at an angle of 45° with the horizontal. At the moment the ball is thrown, the player is 50 m from the coach. At what speed and in what direction must the player run to catch the ball at the same height at which it was released?

82 •• Carlos is on his trail bike, approaching a creek bed that is 7 m wide. A ramp with an incline of 10° has been built for daring people who try to jump the creek. Carlos is traveling at his bike's maximum speed, 40 km/h. (*a*) Should Carlos attempt the jump or emphatically hit the brakes? (*b*) What is the minimum speed a bike must have to make this jump? (Assume equal elevations on either side of the creek.)

83 •• It's the bottom of the ninth with two outs and the winning runs on base. You hit a knee-high fastball that just clears the leaping third baseman's glove. He is standing 28 m from you and his glove reaches to 3.2 m above the ground. The flight time to that point is 0.64 s. Assume that the ball's initial height was 0.6 m. Find (*a*) the initial speed and direction of the ball; (*b*) the time at which the ball reaches its maximum height; (*c*) the maximum height of the ball.

Figure 3-38 Problem 77

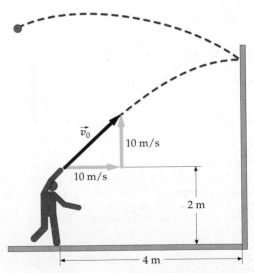

84 •• Noobus is a death-defying squirrel with miraculous jumping abilities. Running to the edge of a flat rooftop, she leaps horizontally with a speed of 6 m/s. If she just clears the 3-m gap between the houses and lands on the neighbor's roof, what is her speed upon landing?

85 ••• If a bullet that leaves the muzzle of a gun at 250 m/s is to hit a target 100 m away at the level of the muzzle, the gun must be aimed at a point above the target. How far above the target is this point?

86 ••• A baseball just clears a 3-m wall that is 120 m from home plate. If the ball leaves the bat at 45° and 1.2 m above the ground, what must its initial speed be?

87 ••• A baseball is struck by a bat, and 3 s later it is caught 30 m away. (a) If the baseball was 1 m above the ground when it was struck and caught, what was the greatest height it reached above the ground? (b) What were the horizontal and vertical components of its velocity when it was struck? (c) What was its speed when it was caught? (d) At what angle with the horizontal did it leave the bat?

88 ••• A baseball player hits a baseball that drops into the stands 22 m above the playing field. The ball lands with a velocity of 50 m/s at an angle of 35° below the horizontal. (a) If the batter contacted the ball 1.2 m above the playing field, what was the velocity of the ball upon leaving the bat? (b) What was the horizontal distance traveled by the ball? (c) How long was the ball in the air?

General Problems

89 • True or false:

(a) The magnitude of the sum of two vectors must be greater than the magnitude of either vector.
(b) If the speed is constant, the acceleration must be zero.
(c) If the acceleration is zero, the speed must be constant.

90 • The initial and final velocities of an object are as shown in Figure 3-39. Indicate the direction of the average acceleration.

Figure 3-39 Problem 90   **Figure 3-40** Problem 91

$\vec{v}_i$
$\vec{v}_f$
$\vec{v}_A$ $\vec{v}_B$

91 • The velocities of objects A and B are shown in Figure 3-40. Draw a vector that represents the velocity of B relative to A.

92 •• A vector $\vec{A}(t)$ has a constant magnitude but is changing direction in a uniform way. Draw the vectors $\vec{A}(t + \Delta t)$ and $A(t)$ for a small time interval Δt, and find the difference $\Delta \vec{A} = \vec{A}(t + \Delta t) - \vec{A}(t)$ graphically. How is the direction of $\Delta \vec{A}$ related to $\vec{A}$ for small time intervals?

93 •• The automobile path shown in Figure 3-41 is made up of straight lines and arcs of circles. The automobile starts from rest at point A. After it reaches point B, it travels at constant speed until it reaches point E. It comes to rest at point F. (a) At the middle of each segment (AB, BC, CD, DE, and EF), what is the direction of the velocity vector? (b) At which of these points does the automobile have an acceleration? In those cases, what is the direction of the acceleration? (c) How do the magnitudes of the acceleration compare for segments BC and DE?

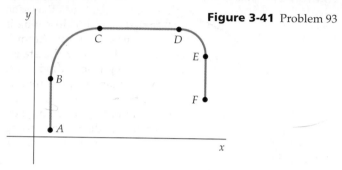

Figure 3-41 Problem 93

94 • The displacement vectors $\vec{A}$ and $\vec{B}$ in Figure 3-42 both have a magnitude of 2 m. (a) Find their x and y components. (b) Find the components, magnitude, and direction of the sum $\vec{A} + \vec{B}$. (c) Find the components, magnitude, and direction of the difference $\vec{A} - \vec{B}$.

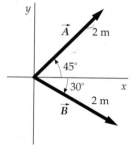

Figure 3-42 Problem 94

95 • A plane is inclined at an angle of 30° from the horizontal. Choose the x axis pointing down the slope of the plane and the y axis perpendicular to the plane. Find the x and y components of the acceleration of gravity, which has the magnitude 9.81 m/s² and points vertically down.

96 • Two vectors $\vec{A}$ and $\vec{B}$ lie in the xy plane. Under what conditions does the ratio A/B equal A_x/B_x?

97 • The position vector of a particle is given by $\vec{r} = 5t\hat{i} + 10t\hat{j}$, where t is in seconds and $\vec{r}$ is in meters. (a) Draw the path of the particle in the xy plane. (b) Find $\vec{v}$ in component form and then find its magnitude.

98 • Off the coast of Chile, a spotter plane sees a school of tuna swimming at a steady 5 km/h northwest (Figure 3-43). The pilot informs a fishing trawler located 100 km due south of the fish. The trawler sails at full steam along the best straight-line course and intercepts the tuna after 4 h. How fast did the trawler move?

Figure 3-43 Problem 98

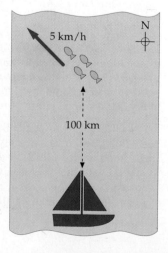

99 •• A worker on the roof of a house drops her hammer, which slides down the roof at a constant speed of 4 m/s. The roof makes an angle of 30° with the horizontal, and its lowest point is 10 m from the ground. What is the horizontal distance traveled by the hammer after it leaves the roof of the house and before it hits the ground?

100 •• A freight train is moving at a constant speed of 10 m/s. A man standing on a flatcar throws a ball into the air and catches it as it falls. Relative to the flatcar, the initial velocity of the ball is 15 m/s straight up. (a) What are the magnitude and direction of the initial velocity of the ball as seen by a second man standing next to the track? (b) How long is the ball in the air according to the man on the train? According to the man on the ground? (c) What horizontal distance has the ball traveled by the time it is caught according to the man on the train? According to the man on the ground? (d) What is the minimum speed of the ball during its flight according to the man on the train? According to the man on the ground? (e) What is the acceleration of the ball according to the man on the train? According to the man on the ground?

101 •• Estimate how far you can throw a ball if you throw it (a) horizontally while standing on level ground; (b) at $\theta = 45°$ while standing on level ground; (c) horizontally from the top of a building 12 m high; (d) at $\theta = 45°$ from the top of a building 12 m high.

102 •• A stunt motorcyclist wants to jump over 10 cars parked side by side below a horizontal launching ramp, as shown in Figure 3-44. With what minimum horizontal speed v_0 must the cyclist leave the ramp in order to clear the top of the last car?

Figure 3-44 Problem 102

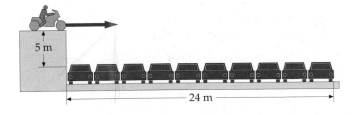

103 •• In 1978, Geoff Capes of Great Britain threw a heavy brick a horizontal distance of 44.5 m. Find the velocity of the brick at the highest point of its flight.

104 •• In 1940, Emanuel Zacchini flew about 53 m as a human cannonball, a record that remains unbroken. His initial velocity was 24.2 m/s at an angle θ. Find θ and the maximum height h Emanuel achieved during the record flight.

105 •• A particle moves in the xy plane with constant acceleration. At $t = 0$ the particle is at $\vec{r}_1 = 4\,m\,\hat{i} + 3\,m\,\hat{j}$, with velocity $\vec{v}_1$. At $t = 2$ s the particle has moved to $\vec{r}_2 = 10\,m\,\hat{i} - 2\,m\,\hat{j}$, and its velocity has changed to $\vec{v}_2 = 5\,m/s\,\hat{i} - 6\,m/s\,\hat{j}$. (a) Find $\vec{v}_1$. (b) What is the acceleration of the particle? (c) What is the velocity of the particle as a function of time? (d) What is the position vector of the particle as a function of time?

106 •• A small steel ball is projected horizontally off the top landing of a long rectangular staircase (Figure 3-45). The initial speed of the ball is 3 m/s. Each step is 0.18 m high and 0.3 m wide. Which step does the ball strike first?

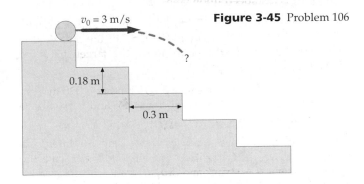

Figure 3-45 Problem 106

107 •• As a car travels down a highway at 25 m/s, a passenger flips out a can at a 45° angle of elevation in a plane perpendicular to the motion of the car. The initial speed of the can relative to the car is 10 m/s. The can is released at a height 1.2 m above the road. (a) Write the initial velocity of the can (relative to the road) in terms of the unit vectors $\hat{i}$, $\hat{j}$, and $\hat{k}$. (b) Where does the can land?

108 •• Suppose you can throw a ball a distance x_0 when standing on level ground. How far can you throw it from a building of height $h = x_0$ if you throw it at (a) 0°? (b) 30°? (c) 45°?

109 •• A baseball hit toward center field will land 72 m away unless it is caught first. At the moment the ball is hit, the center fielder is 98 m away. He uses 0.5 s to judge the flight of the ball, then races to catch it. The ball's speed as it leaves the bat is 35 m/s. Can the center fielder catch the ball before it hits the ground?

110 ••• Darlene is a stunt motorcyclist in a traveling circus. For the climax of her show, she takes off from the ramp at angle θ, clears a fiery ditch of width x, and lands on an elevated platform (height h) on the other side (Figure 3-46). Darlene notices, however, that night after night, the circus owner keeps raising the height of the platform and the flames to make the jump more spectacular. She is beginning to worry about how far this trend can be taken before she becomes a spectacular casualty, so she decides that it is time for some calculations. (a) For a given angle θ and distance x, what is the upper limit h_{max} such that the bike can make the jump? (b) For h less than h_{max}, what is the minimum takeoff speed necessary for a successful jump? (Neglect the size of the bike.)

Figure 3-46 Problem 110

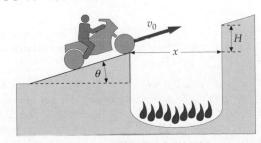

111 ••• A small boat is headed for a harbor 32 km north-west of its current position when it is suddenly engulfed in heavy fog. The captain maintains a compass bearing of northwest and a speed of 10 km/h relative to the water. Three hours later, the fog lifts and the captain notes that he is now exactly 4.0 km south of the harbor. (*a*) What was the average velocity of the current during those three hours? (*b*) In what direction should the boat have been heading to reach its destination along a straight course? (*c*) What would its travel time have been if it had followed a straight course?

112 ••• Galileo showed that, if air resistance is neglected, the ranges for projectiles whose angles of projection exceed or fall short of 45° by the same amount are equal. Prove Galileo's result.

113 ••• Two balls are thrown with equal speeds from the top of a cliff of height H. One ball is thrown upward at an angle α above the horizontal. The other ball is thrown downward at an angle β below the horizontal. Show that each ball strikes the ground with the same speed, and find that speed in terms of H and the initial speed v_0.

CHAPTER 4

Newton's Laws

Isaac Newton

Classical mechanics is a theory of motion based on mass and force. It describes phenomena using Newton's three laws, which relate an object's acceleration to its mass and the forces acting on it. A modern wording of Newton's laws follows.

First law. An object at rest stays at rest unless acted on by an external force. An object in motion continues to travel with constant velocity unless acted on by an external force.

Newton's first law

Second law. The acceleration of an object is in the direction of the net external force acting on it. It is proportional to the net external force, and is inversely proportional to the mass of the object:

$$\vec{a} = \frac{\vec{F}_{net}}{m}$$

or

$$\vec{F}_{net} = m\vec{a}$$

The net force acting on an object, also called the resultant force, is the vector sum of all the forces acting on it: $\vec{F}_{net} = \Sigma\vec{F}$. Thus,

$$\sum \vec{F} = \vec{F}_{net} = m\vec{a} \qquad\qquad \text{4-1}$$

Newton's second law

Third law. Forces always occur in equal and opposite pairs. If object A exerts a force on object B, an equal but opposite force is exerted by object B on object A.

Newton's third law

Friction is greatly reduced by a cushion of air that supports the hovercraft.

4-1 Newton's First Law: The Law of Inertia

Push a piece of ice on a counter top: It slides, then stops. If the counter is wet, the ice will travel farther before stopping. A piece of dry ice riding on a cushion of carbon dioxide slides quite far with little change in velocity. Before Galileo, it was thought that a force, such as a push or pull, was always needed to keep an object moving with constant velocity. Galileo, and later Newton, recognized that the slowing of objects in everyday experience is due to friction. If friction is reduced, the change in velocity is reduced. A water slick or carbon dioxide cushion is especially effective at reducing friction, allowing the object to slide a great distance with little change in velocity. Remove all external forces on an object, Galileo reasoned, and its velocity will never change—a property of matter he described as its **inertia.** This conclusion, restated by Newton as his first law, is also called the **law of inertia.**

Inertial Reference Frames

Newton's first law makes no distinction between an object at rest and an object moving with constant velocity. Whether an object is at rest or is moving with constant velocity depends on the reference frame in which the object is observed. A **reference frame** is a set of coordinate systems at rest relative to each other. Consider a ball sitting in the aisle of an airplane cruising along a horizontal path. In a coordinate system attached to the plane (that is, in the reference frame of the plane) the ball is at rest. It will remain at rest relative to the plane as long as the plane flies with constant velocity. In a coordinate system attached to the earth, the ball is moving with the velocity of the plane. According to Newton's first law, the ball will continue to move with constant velocity in the reference frame of the earth, and will remain at rest in the reference frame of the plane unless it is acted on by a net force.

A reference frame in which the law of inertia holds exactly is called an **inertial reference frame.** Both the cruising plane and the ground are, to a good approximation, inertial reference frames. Any reference frame moving with constant velocity relative to an inertial reference frame is also an inertial reference frame.

Now suppose that the plane accelerates forward relative to the ground. The ball will roll backward, accelerating relative to the plane even though there is no net force acting on it. The ball accelerates *in the plane's frame of ref-*

erence despite there being no net external force acting on it. Also, the back of your seat will exert a horizontal forward force on you, but you do not accelerate *relative to the plane*. The law of inertia does not hold in the reference frame of the accelerating plane. A reference frame accelerating relative to an inertial reference frame is not an inertial reference frame. *Newton's first law thus gives us the criterion for determining if a reference frame is an inertial frame.*

A reference frame attached to the surface of the earth is not quite an inertial reference frame because of the small acceleration of the surface of the earth (relative to the center of the earth) due to the rotation of the earth, and the small acceleration of the earth itself due to its revolution around the sun. However, these accelerations are of the order of 0.01 m/s^2 or less, so to a good approximation, a reference frame attached to the surface of the earth is an inertial reference frame.

4-2   Force, Mass, and Newton's Second Law

Newton's first and second laws allow us to define force. A **force** is an external influence on an object that causes it to accelerate relative to an inertial reference frame. (We assume there are no other forces acting.) The direction of the force is the direction of the acceleration it causes if it is the only force acting on the object. The magnitude of the force is the product of the mass of the object and the magnitude of its acceleration. This definition of force is in accord with our intuitive idea of a force as a push or pull like that exerted by our muscles.

Mass is an intrinsic property of an object that measures its resistance to acceleration. That is, it is a measure of the object's inertia. The ratio of two masses is defined quantitatively by applying the same force to each and comparing their accelerations. If a force F produces acceleration a_1 when applied to an object of mass m_1, and the same force produces acceleration a_2 when applied to an object of mass m_2, then the ratio of the masses is defined by

$$\frac{m_2}{m_1} = \frac{a_1}{a_2}$$

4-2

Definition—Mass

This definition agrees with our intuitive idea of mass. If the same force is applied to two objects, the object with more mass will accelerate less. The ratio a_1/a_2 produced by an identical force acting on two objects is found experimentally to be independent of the magnitude, direction, or type of force used. Mass is an intrinsic property of an object that does not depend on its location—it remains the same whether the object is on the earth, on the moon, or in outer space.

If a direct comparison shows that $m_2/m_1 = 2$ and $m_3/m_1 = 4$, then m_3 will be twice m_2 when the two objects are compared with each other. We can therefore establish a mass scale by choosing a standard object and assigning it a mass of 1 unit. As we noted in Chapter 1, the object chosen as the international standard for mass is a platinum–iridium alloy cylinder carefully preserved at the International Bureau of Weights and Measures at Sèvres, France. The mass of the standard object is 1 **kilogram**, the SI unit of mass.* The force required to produce an acceleration of 1 m/s^2 on the standard object is defined to be 1 **newton** (N). The force that produces an acceleration of 2 m/s^2 on the standard object is 2 N, and so on.

---

* The standard kilogram was originally intended to be equal to the mass of $1000 \text{ cm}^3 = 1$ liter of water.

Example 4-1

A given force produces an acceleration of 5 m/s^2 on the standard object of mass m_1. When the same force is applied to a carton of ice cream of mass m_2, it produces an acceleration of 11 m/s^2. (*a*) What is the mass of the carton of ice cream? (*b*) What is the magnitude of the force?

(*a*)1. The ratio of the masses varies inversely as the ratio of the accelerations under the same applied force:

$$\frac{m_2}{m_1} = \frac{a_1}{a_2} = \frac{5 \text{ m/s}^2}{11 \text{ m/s}^2}$$

2. Solve for m_2 in terms of m_1, which is 1 kg:

$$m_2 = \frac{5}{11} m_1 = \frac{5}{11} (1 \text{ kg}) = 0.45 \text{ kg}$$

(*b*)   The magnitude of the force F is found by using the mass and acceleration of either object:

$$F = m_1 a_1 = (1 \text{ kg})(5 \text{ m/s}^2) = 5 \text{ N}$$

Exercise A force of 3 N produces an acceleration of 2 m/s^2 on an object of unknown mass. (*a*) What is the mass of the object? (*b*) If the force is increased to 4 N, what is the acceleration? (*Answers*   (*a*) 1.5 kg, (*b*) 2.67 m/s^2)

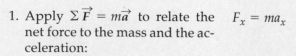

It is found experimentally that two or more forces acting on an object accelerate it as if the object were acted on by a single force equal to the vector sum of the individual forces. That is, forces combine as vectors. Newton's second law is thus

$$\sum \vec{F} = \vec{F}_{\text{net}} = m\vec{a}$$

Example 4-2

You're stranded in space away from your spaceship. Fortunately, you have a propulsion unit that provides a constant force $\vec{F}$ for 3 s. After 3 s you have moved 2.25 m. If your mass is 68 kg, find $\vec{F}$.

Picture the Problem The force acting on you is constant, so your acceleration $\vec{a}$ is also constant. Hence, we use the kinematic equations of Chapter 2 to find $\vec{a}$, and then obtain the force from $\sum \vec{F} = m\vec{a}$. Choose $\vec{F}$ to be along the *x* axis, so that $\vec{F} = F_x \hat{i}$ (Figure 4-1). The component of Newton's second law along the *x* axis is then $F_x = ma_x$.

Figure 4-1

1. Apply $\sum \vec{F} = m\vec{a}$ to relate the net force to the mass and the acceleration:

$$F_x = ma_x$$

2. To find the acceleration, we use Equation 2-14 with $v_0 = 0$:

$$\Delta x = x - x_0 = v_0 t + \frac{1}{2} a_x t^2 = \frac{1}{2} a_x t^2$$

$$a_x = \frac{2\Delta x}{t^2} = \frac{2(2.25 \text{ m})}{(3 \text{ s})^2} = 0.500 \text{ m/s}^2$$

3. Substitute $a_x = 0.500$ m/s^2 and $m = 68$ kg to find the force:

$$F_x = ma_x = (68 \text{ kg})(0.500 \text{ m/s}^2) = 34.0 \text{ N}$$

Example 4-3   *try it yourself*

A particle of mass 0.4 kg is subjected simultaneously to two forces $\vec{F_1} = 2\,\mathrm{N}\,\hat{i} - 4\,\mathrm{N}\,\hat{j}$ and $\vec{F_2} = 2.6\,\mathrm{N}\,\hat{i} + 5\,\mathrm{N}\,\hat{j}$. If the particle is at the origin and starts from rest at $t = 0$, find (a) its position vector $\vec{r}$ and (b) its velocity $\vec{v}$ at $t = 1.6$ s.

Picture the Problem Since $\vec{F_1}$ and $\vec{F_2}$ are constant, the acceleration of the particle is constant. Hence, you can use the kinematic equations of Chapter 2 to determine the particle's position and velocity as functions of time.

Cover the column to the right and try these on your own before looking at the answers.

Steps

Answers

(a) 1. Write the general equation for the position vector $\vec{r}$ as a function of time t for constant acceleration $\vec{a}$ in terms of $\vec{r_0}$, $\vec{v}$, and $\vec{a}$, and substitute $\vec{r_0} = 0$, $\vec{v_0} = 0$.

$\vec{r} = \vec{r_0} + \vec{v_0}t + \frac{1}{2}\vec{a}t^2 = \frac{1}{2}\vec{a}t^2$

2. Use $\Sigma\vec{F} = m\vec{a}$ to write the acceleration $\vec{a}$ in terms of the resultant force $\Sigma\vec{F}$ and the mass m.

$\vec{a} = \dfrac{\Sigma\vec{F}}{m}$

3. Compute $\Sigma\vec{F}$ from the given forces.

$\Sigma\vec{F} = \vec{F_1} + \vec{F_2} = -0.6\,\mathrm{N}\,\hat{i} + 1.0\,\mathrm{N}\,\hat{j}$

4. Find the acceleration vector $\vec{a}$.

$\vec{a} = \dfrac{\Sigma\vec{F}}{m} = -1.5\,\mathrm{m/s^2}\,\hat{i} + 2.5\,\mathrm{m/s^2}\,\hat{j}$

5. Find the position vector $\vec{r}$ for a general time t.

$\vec{r} = \frac{1}{2}\vec{a}t^2 = \frac{1}{2}a_x\,t^2\hat{i} + \frac{1}{2}a_y\,t^2\hat{j}$

$= -0.75\,\mathrm{m/s^2}\,t^2\hat{i} + 1.25\,\mathrm{m/s^2}\,t^2\hat{j}$

6. Find $\vec{r}$ at $t = 1.6$ s.

$\vec{r} = -1.92\,\mathrm{m}\,\hat{i} + 3.20\,\mathrm{m}\,\hat{j}$

(b) Write the velocity vector $\vec{v}$ in terms of the acceleration and time and compute its components for the time $t = 1.6$ s.

$\vec{v} = \vec{a}t = (-1.5\,\mathrm{m/s^2}\,\hat{i} + 2.5\,\mathrm{m/s^2}\,\hat{j})t$

$= -2.4\,\mathrm{m/s}\,\hat{i} + 4.00\,\mathrm{m/s}\,\hat{j}$

4-3   The Force Due to Gravity: Weight

If we drop an object near the earth's surface, it accelerates toward the earth. If we neglect air resistance, all objects have the same acceleration, called the acceleration due to gravity $\vec{g}$ at any given point in space. The force causing this acceleration is the force of gravity on the object, called its weight $\vec{w}$. If $\vec{w}$ is the only force acting on an object, the object is said to be in **free fall**. If its mass is m, Newton's second law defines the weight $\vec{w}$:

$$\vec{w} = m\vec{g}$$ 4-3

Weight

Since $\vec{g}$ is the same for all objects at a given point, the weight of an object must be proportional to its mass. The vector $\vec{g}$ is the force per unit mass exerted by the earth on any object and is called the **gravitational field** of the earth. It is equal to the free-fall acceleration experienced by an object. Near the surface of the earth, g has the value

$$g = 9.81\,\mathrm{N/kg} = 9.81\,\mathrm{m/s^2}$$

Careful measurements show that $\vec{g}$ varies with location. In particular, at points above the surface of the earth, $\vec{g}$ points toward the center of the earth and varies inversely with the square of the distance to the center of the earth. Thus, an object weighs slightly less at very high altitudes than it does at sea level. The gravitational field also varies slightly with latitude because the earth is not exactly spherical but is slightly flattened at the poles. Thus, weight, unlike mass, is *not* an intrinsic property of an object. Although the weight of an object varies from place to place because of changes in *g*, this variation is too small to be noticed in most practical applications on or near the surface of the earth.

An example should help clarify the difference between mass and weight. Consider a bowling ball near the moon. Its weight is the force exerted on it by the moon, but that force is a mere one-sixth of the force exerted on the bowling ball when it is similarly positioned on earth. The ball weighs about one-sixth as much on the moon, and lifting the ball on the moon requires one-sixth the force. However, because the mass of the ball is the same on the moon as on the earth, throwing the ball with some horizontal acceleration requires the same force on the moon as on the earth or in free space.

Though an object's weight may vary from one place to another, at any particular location its weight is proportional to its mass. Thus, we can conveniently compare the masses of two objects at a given location by comparing their weights.

Our awareness of our own weight comes from other forces that balance it. When you sit on a chair, you feel a force exerted by the chair that balances your weight and prevents you from falling to the floor. When you stand on a spring scale, your feet feel the force exerted by the scale. The scale is calibrated to read the force it must exert (by the compression of its springs) to balance your weight. This force is called your **apparent weight.** If there is no force to balance your weight, as in free fall, your apparent weight is zero. This condition, called **weightlessness,** is experienced by astronauts in orbiting satellites. As will be discussed in Chapter 5, when an object travels in a circle, the direction of its velocity vector is constantly changing, and the object is therefore accelerating. A satellite in a circular orbit near the surface of the earth is accelerating toward the earth. The only force acting on the satellite is gravity (its weight), so it is in free fall with the acceleration due to gravity. Astronauts in the satellite are also in free fall. The only force on them is their weight, which produces the acceleration *g*. Since there is no force balancing the force of gravity, the astronauts have zero apparent weight.

Units of Force and Mass

Like the second and the meter, the SI unit of mass, the kilogram, is a fundamental unit. The unit of force, the newton, and the units for other quantities that we will study such as momentum and energy, are derived from the three fundamental units second, meter, and kilogram.

As noted above, the newton is defined as the force that produces an acceleration of 1 m/s^2 when it acts on 1 kg. Then Newton's second law gives

$$1 \, N = (1 \, kg)(1 \, m/s^2) = 1 \, kg \cdot m/s^2 \qquad \text{4-4}$$

A convenient standard unit for mass in atomic and nuclear physics is the unified mass unit (u), which is defined as one-twelfth the mass of the neutral carbon-12 (^{12}C) atom. The unified mass unit is related to the kilogram by

$$1 \, u = 1.660 \, 540 \times 10^{-27} \, kg \qquad \text{4-5}$$

The mass of a hydrogen atom is approximately 1 u.

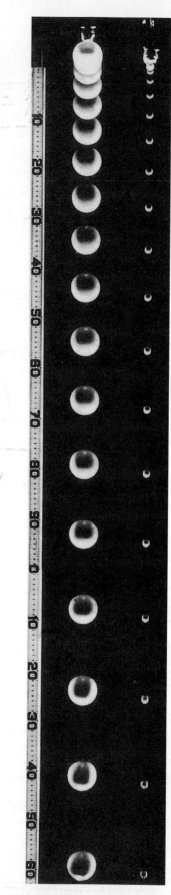

When air resistance can be neglected, objects of different mass fall with the same acceleration due to gravity.

Although we generally use SI units in this book, we need to know another system, the U.S. customary system, still used in the United States, which is based on the foot, the second, and the pound. The U.S. customary system differs from SI in that it uses a unit of force, the pound, as a fundamental unit rather than using a unit of mass. The **pound** was originally defined as the weight of a particular standard object at a particular location. It is now defined as 4.448222 N. Rounding to three places, we have 1 lb ≈ 4.45 N. Since 1 kg weighs 9.81 N, its weight in pounds is

$$9.81 \text{ N} \times \frac{1 \text{ lb}}{4.45 \text{ N}} = 2.20 \text{ lb} \qquad \text{4-6}$$

Weight of 1 kg

The unit of mass in the U.S. customary system is the rarely encountered slug, defined as the mass of an object that weighs 32.2 lb. When working problems in the U.S. customary system, we substitute w/g for mass m, where w is the weight in pounds and g is the acceleration due to gravity in feet per second per second:

$$g = 32.2 \text{ ft/s}^2 \qquad \text{4-7}$$

Example 4-4

The net force acting on a 130-lb student is 25 lb. What is her acceleration?

According to Newton's second law, her acceleration is the force divided by her mass:

$$a = \frac{F}{m} = \frac{F}{w/g} = \frac{25 \text{ lb}}{(130 \text{ lb})/(32.2 \text{ ft/s}^2)}$$

$$= 6.15 \text{ ft/s}^2$$

Exercise What force is needed to give an acceleration of 3 ft/s² to a 5-lb block? (*Answer*   0.466 lb)

4-4 Newton's Third Law

The word *force* is used to describe the interaction between two objects. When two objects interact, they exert forces on each other. Newton's third law states that these forces are equal in magnitude and opposite in direction. If object A exerts a force on object B, object B exerts a force on A that is equal in magnitude and opposite in direction. Thus, forces always occur in pairs. It is common to refer to one force in the pair as an action and the other as a reaction. This terminology is unfortunate because it sounds like one force "reacts" to the other, which is not true. Both forces occur simultaneously. Either can be called the action and the other the reaction. Action and reaction forces can never balance *each other* because they act on *different objects*. In Figure 4-2, a block rests on a table. The force acting downward on the block is the weight $\vec{w}$ due to the attraction of the earth. An equal and opposite force $\vec{w}' = -\vec{w}$ is exerted by the block on the earth. These forces are an action–reaction pair. If they were the only forces present, the block would accelerate downward,

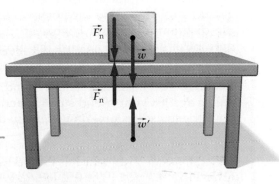

Figure 4-2

because it would have only a single force acting on it (and the earth would accelerate upward toward the block). However, the table exerts an upward force $\vec{F}_n$ on the block that balances the block's weight. The block also exerts a force $\vec{F}_n' = -\vec{F}_n$ downward on the table. The forces $\vec{F}_n$ and $\vec{F}_n'$ are also an action–reaction pair.

Example 4-5

A horse refuses to pull a cart (Figure 4-3). The horse reasons, "According to Newton's third law, whatever force I exert on the cart, the cart will exert an equal and opposite force on me, so the net force will be zero and I will have no chance of accelerating the cart." What is wrong with this reasoning?

Picture the Problem Since we are interested in the motion of the cart, we have enclosed it with a dashed line and have indicated the forces acting on it. The force exerted by the horse is labeled $\vec{T}$. It is exerted by the horse on the harness. (Since the harness is attached to the cart, we are considering it part of the cart.) Other forces acting on the cart are its weight $\vec{w}$, the vertical support force of the ground $\vec{F}_n$, and the horizontal force exerted by the ground, labeled $\vec{f}$ (for friction).

In the idealized diagram, the cart is drawn as a particle with the forces acting on it. The vertical forces $\vec{w}$ and $\vec{F}_n$ sum to zero. (We know this because we know the cart does not accelerate vertically.) The horizontal forces are $\vec{T}$ to the right and $\vec{f}$ to the left. The cart will accelerate if $\vec{T}$ is greater than $\vec{f}$.

Note that the reaction force to $\vec{T}$, which we call $\vec{T}'$, is exerted on the horse, not on the cart. It has no effect on the motion of the cart, but it does affect the motion of the horse. If the horse is to accelerate to the right, there must be a force $\vec{F}$ (to the right) exerted by the ground on the horse's feet that is greater than $\vec{T}'$.

Remark This example illustrates the importance of drawing a simple diagram when solving mechanics problems. Had the horse sketched a simple diagram, he would have seen that he need only push back hard against the ground so that the ground would push him forward.

Exercise As you stand facing a friend, place your palms against your friend's palms and push. Can your friend exert a force on you if you do not exert a force back? Try it.

Exercise True or false: The force exerted by the cart on the horse is equal and opposite to the force exerted by the horse on the cart only when the cart is not accelerating. (*Answer* False! An action–reaction pair of forces describes the interaction of two objects. One force cannot exist without the other. They are always equal and opposite.)

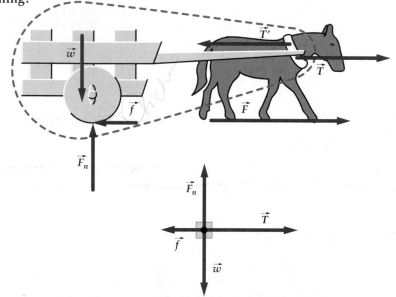

Figure 4-3

4-5   Forces in Nature

The full power of Newton's second law emerges when it is combined with the force laws that describe the interactions of objects. For example, Newton's law for gravitation, which we study in Chapter 11, gives the gravitational force exerted by one object on another in terms of the distance between the objects and the masses of each. This, combined with Newton's second law, enables us to calculate the orbits of planets around the sun, the motion of the moon, and variations with altitude of g, the acceleration due to gravity.

The Fundamental Forces

All the different forces observed in nature can be explained in terms of four basic interactions that occur between elementary particles:

1. The gravitational force
2. The electromagnetic force
3. The strong nuclear force (also called the hadronic force)
4. The weak nuclear force

The everyday forces that we observe between macroscopic objects are due to either the gravitational force or the electromagnetic force.

(a)

(b)

(c)

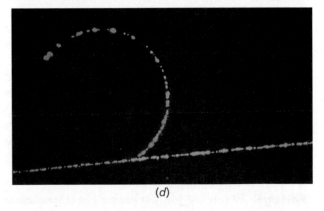

(d)

(a) The gravitational force between the earth and an object near the earth's surface is the weight of the object. The gravitational force exerted by the sun keeps the planets in their orbits. Similarly, the gravitational force exerted by the earth keeps the moon in its nearly circular orbit around the earth. The gravitational forces exerted by the moon and sun on the oceans of the earth cause the tides. Mont-Saint-Michel, France, shown in the photo, is an island when the tide is in. (b) The electromagnetic force includes both the electric and the magnetic forces. A familiar example of the electric force is the attraction between bits of paper and a comb that is electrified after being run through hair. The magnetic force between a magnet and iron arises when electric charges are in motion. The electromagnetic force between charged elementary particles is vastly greater than the gravitational force between them. For example, the electrostatic force of repulsion between two protons is of the order of 10^{36} times the gravitational attraction between them. The lightning bolts shown in the photo are the result of the electromagnetic force. (c) The strong nuclear force occurs between elementary particles called hadrons, which include protons and neutrons. The strong force results from the interaction of quarks, the building blocks of hadrons, and is responsible for holding nuclei together. The magnitude of the strong force decreases rapidly with distance and is negligible beyond a few nuclear diameters. The hydrogen bomb explosion shown in the photo illustrates the strong nuclear force. (d) The weak nuclear force, which also has a short range, occurs between leptons (which include electrons and muons) and between hadrons (which include protons and neutrons). This false-color cloud chamber photograph illustrates the weak interaction between a cosmic ray muon (green) and an electron (red) knocked out of an atom.

Action at a Distance

The fundamental forces of gravity and electromagnetism act between particles that are separated in space. This creates a philosophical problem referred to as **action at a distance.** Newton perceived action at a distance as a flaw in his theory of gravitation but avoided giving any other hypothesis. Today the problem is avoided by introducing the concept of a field, which acts as an intermediary agent. For example, we consider the attraction of the earth by the sun in two steps. The sun creates a condition in space that we call the gravitational field. This field then exerts a force on the earth. Similarly, the earth produces a gravitational field that exerts a force on the sun. Your weight is the force exerted by the gravitational field of the earth on you. When we study electricity and magnetism (Chapters 22–32) we will study electric fields, which are produced by electrical charges, and magnetic fields, which are produced by electrical charges in motion.

Contact Forces

Many forces we encounter are exerted by objects in direct contact. These forces are electromagnetic in origin and are exerted between the molecules of each object.

Solids Consider a book on a table. The weight of the book pulls it downward, pressing it against the molecules in the table's surface, which resist compression and exert a force upward on the book. Such a force, *perpendicular to the surface, is called a **normal force** (one meaning of the word *normal* is "perpendicular"). A supporting surface bends slightly in response to a load, though this is rarely noticeable to the naked eye.

Normal forces can vary over a wide range of magnitude. A table, for instance, will exert an upward force on any object resting on it. As long as the table doesn't break, the normal force will balance the weight of the object. Furthermore, if you press down on the object, the table will exert a support force that counters the extra force, preventing the object from accelerating downward.

Objects in contact can also exert forces on each other that are *parallel to the surfaces in contact. The parallel component of a contact force is called a **frictional force.** We will consider frictional forces in the next chapter.

Springs When a spring is compressed or extended by a small amount Δx, the force it exerts is found experimentally to be

$$F_x = -k\,\Delta x \qquad\qquad 4\text{-}8$$

Hooke's law

where k is the force constant, a measure of the stiffness of the spring (Figure 4-4). The negative sign in Equation 4-8 signifies that when the spring is stretched or compressed, the force it exerts back is in the opposite direction. This relation, known as Hooke's law, turns out to be quite important. An object at rest under the influence of forces that balance is said to be in static equilibrium. If a small displacement results in a net restoring force toward the equilibrium position, the equilibrium is called stable equilibrium. For small displacements, nearly all restoring forces obey Hooke's law.

The molecular force of attraction between atoms in a molecule or solid varies approximately linearly with the change in separation (for small changes); the force varies much like that of a spring. We can therefore use

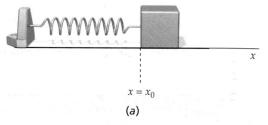

$x = x_0$

(a)

$F_x = -k\,\Delta x$ is negative because Δx is positive.

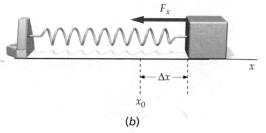

F_x

Δx

x_0

(b)

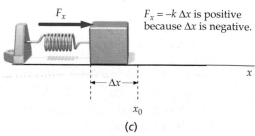

F_x

$F_x = -k\,\Delta x$ is positive because Δx is negative.

Δx

x_0

(c)

Figure 4-4 A horizontal spring. (*a*) When the spring is unstretched, it exerts no force on the block. (*b*) When the spring is stretched so that Δx is positive, it exerts a force of magnitude $k\,\Delta x$ in the negative x direction. (*c*) When the spring is compressed so that Δx is negative, the spring exerts a force of magnitude $k\,\Delta x$ in the positive direction.

two masses on a spring to model a diatomic molecule, or a set of masses connected by springs to model a solid, as shown in Figure 4-5.

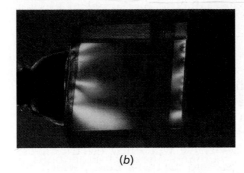

(a)   (b)

Figure 4-5 (*a*) Model of a solid consisting of atoms connected to each other by springs. The springs are very stiff (large force constant) so that when a weight is placed on the solid its deformation is not visible. However, compression such as that produced by the C clamp on the plastic block in (*b*) leads to stress patterns that are visible when viewed with polarized light.

Example 4-6

A 110-kg basketball player hangs on the rim following a slam dunk. When he comes to rest, the rim is bent down a distance of 15 cm. Assume that the rim can be approximated by a spring and calculate the force constant *k*.

Figure 4-6

Picture the Problem Since the acceleration of the player is zero, the net force exerted on him must be zero. The upward force exerted by the rim balances his weight (Figure 4-6). Let $y = 0$ be the original position of the rim and choose down to be positive. Then Δy is positive, the weight mg is positive, and the force exerted by the rim, $-k\,\Delta y$ is negative.

1. Apply $\Sigma \vec{F} = m\vec{a}$ to the player:

$$\Sigma F_y = mg + (-k\,\Delta y) = ma_y = 0$$

2. Solve for *k*:

$$k = \frac{mg}{\Delta y} = \frac{(110\ \text{kg})(9.81\ \text{N/kg})}{0.15\ \text{m}}$$

$$= 7.19 \times 10^3\ \text{N/m}$$

Remark Although a basketball rim doesn't look much like a spring, when the displacement is small, the force it exerts is proportional to the displacement and oppositely directed. Note that we used N/kg for the units of *g* so that kg cancels, giving N/m for the units of *k*. We can use either 9.81 N/kg or 9.81 m/s² for *g*, whichever is more convenient, because 1 N/kg = 1 m/s².

Exercise A 4-kg bunch of bananas is suspended motionless from a spring balance whose force constant is $k = 300$ N/m. By how much is the spring stretched? (*Answer* 13.1 cm)

Exercise A spring of force constant 400 N/m is attached to a 3-kg block that rests on a horizontal air track that renders friction negligible. What extension of the spring is needed to give the block an acceleration of 4 m/s² upon release? (*Answer* 3.0 cm)

Exercise in Dimensional Analysis An object of mass *m* oscillates at the end of an ideal spring of force constant *k*. The time for one complete oscillation is the period *T*. Assuming that *T* depends on *m* and *k*, use dimensional analysis to find the form of the relationship $T = f(m, k)$, ignoring numerical constants. This is most easily found by looking at the units. Note that the units of *k* are N/m = (kg·m/s²)/m = kg/s², and the units of *m* are kg. (*Answer* $T = C\sqrt{m/k}$, where *C* is some dimensionless constant. The correct expression for the period, as we will see in Chapter 14, is $T = 2\pi\sqrt{m/k}$.)

Strings   If we pull on a string, the string stretches slightly and pulls back with an equal but opposite force (unless the string breaks). We can think of a string as a spring with such a large force constant that the extension of the string is negligible. The string is flexible, however, so we cannot exert a force of compression on it. When we push on a string, it merely flexes or bends.

4-6   Problem Solving: Free-Body Diagrams

Imagine a dogsled being pulled across icy ground. The dog in front pulls on a light rope attached to the sled with a force $\vec{F}$ (Figure 4-7*a*). The taut rope then pulls the sled forward. What forces act on the sled? Both the rope and ice touch the sled, so we know the rope and ice exert contact forces on it. We also know that the earth exerts a gravitational force on the sled (the sled's weight). Thus, three forces act on the sled (assuming that friction is negligible):

1. The weight of the sled, $\vec{w}$

2. The contact force $\vec{F}_n$ exerted by the ice (without friction, the contact force is perpendicular to the ice)

3. The contact force $\vec{T}$ exerted by the rope

Figure 4-7 (*a*) A dog pulling a sled. The first step in problem solving is to isolate the system to be analyzed. In this case, a circle isolates the sled from its surroundings. (*b*) The forces acting on the sled of (*a*).

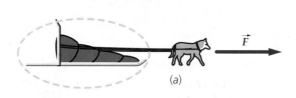

(a)

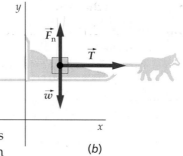

(b)

A diagram that shows schematically all the forces acting on a system, such as Figure 4-7*b*, is called a **free-body diagram.** Note that $\vec{F}_n$ and $\vec{w}$ in the diagram have equal magnitudes. The magnitudes must be equal because the sled doesn't accelerate vertically. Conditions on the motion of an object, such as the requirement that the sled remain on the ice, are called **constraints.**

The *y* component of Newton's second law gives

$$\sum F_y = F_{n,y} + w_y + T_y = ma_y$$

$$F_n - w + 0 = 0$$

or

$$F_n = w$$

The *x* component of Newton's second law gives

$$\sum F_x = F_{n,x} + w_x + T_x = ma_x$$

or

$$a_x = \frac{T}{m}$$

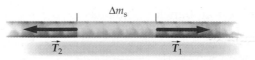

Figure 4-8 Free-body diagram for a segment Δm_s of the rope. Newton's second law applied to this segment gives $T_1 - T_2 = (\Delta m_s)a_x$. If the mass of the segment is negligible, $T_1 = T_2$. The tension T is the magnitude of the force each segment exerts on each adjacent segment. These forces act along the rope, so a light rope connecting two points has a tension that has a constant magnitude throughout.

The tension is constant along the rope, so the force $\vec{F}$ exerted by the dog on the rope equals the force $\vec{T}$ exerted by the rope on the sled (Figure 4-8). Constant tension in a string or rope also holds for a string that passes over a frictionless peg or pulley of negligible mass as long as there are no tangential forces acting on the string between the two points considered.

In this simple example, we found two things: the horizontal acceleration ($a_x = T/m = F/m$), and the vertical force $\vec{F}_n$ exerted by the ice ($F_n = w$). According to Newton's third law, forces always act in pairs. Figure 4-7 shows only those forces that act on the *sled*. Figure 4-9 shows the reaction forces to those in Figure 4-7. These are the gravitational force $\vec{w}'$ exerted by the sled *on the earth*, the force $\vec{F}_n'$ exerted by the sled *on the ice*, and the force $\vec{T}'$ exerted by the sled *on the rope*. Since these forces are not exerted on the sled, they have nothing to do with its motion. Therefore, they are not part of the application of Newton's second law to the motion of the sled.

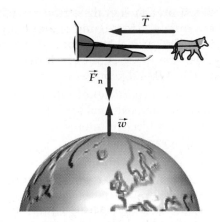

Figure 4-9 The reaction forces corresponding to the three forces shown in Figure 4-7. These forces do *not* act on the sled.

Example 4-7

During your winter break, you enter a dogsled race in which students replace the dogs. Wearing cleats for traction, you begin the race by pulling on a rope attached to the sled with a force of 150 N at 25° with the horizontal. The mass of the sled is 80 kg and there is negligible friction between the sled and ice (Figure 4-10). Find (a) the acceleration of the sled and (b) the normal force $\vec{F}_n$ exerted by the surface on the sled.

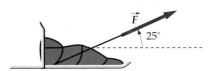

Figure 4-10

Picture the Problem Three forces act on the sled: its weight, $m\vec{g}$, which acts downward; the normal force, $\vec{F}_n$, which acts upward; and the tension in the rope, $\vec{T}$, directed 25° above the horizontal. Since the forces do not lie along a line, we study the system by applying Newton's second law to the x and y directions separately. We choose x to be in the direction of motion, and y to be perpendicular to the ice. Then we draw a free-body diagram for the sled.

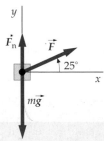

(a)   Apply $\Sigma\vec{F} = m\vec{a}$ to motion along the x axis to determine the acceleration of the sled, a_x:

$$\sum F_x = T\cos\theta = ma_x$$

$$a_x = \frac{T\cos\theta}{m} = \frac{(150\text{ N})(\cos 25°)}{80\text{ kg}} = 1.70\text{ m/s}^2$$

(b)   There is no acceleration in the y direction. Apply $\Sigma\vec{F} = m\vec{a}$ to motion along the y axis to determine F_n:

$$\sum F_y = T\sin\theta + F_n - mg = ma_y = 0$$

$$F_n = mg - T\sin\theta$$

$$= (80\text{ kg})(9.81\text{ N/kg}) - (150\text{ N})(\sin 25°) = 721\text{ N}$$

Remarks Note that only the x component of the tension, $T\cos\theta$, causes the sled to accelerate. Also note that the ice supports less than the full weight of the sled, since part of the weight, $T\sin\theta$, is supported by the rope.

Check the Result If $\theta = 0$, the sled is accelerated by a force T and the ice supports all the weight of the sled. Our results agree, giving $a_x = T/m$ and $F_n = mg$. For $\theta = 90°$, $a_x = 0$ and $F_n = mg - T$, as expected.

Exercise What is the greatest tension that can be applied to the rope without lifting the sled off the surface? (*Answer* $T = 1.86$ kN)

Example 4-7 illustrates a general method for solving problems using Newton's laws:

1. Draw a neat diagram.

2. Isolate the object (particle) of interest, and draw a free-body diagram showing each external force that acts on the object. If there is more than one object of interest in the problem, draw a separate free-body diagram for each.

3. Choose a convenient coordinate system for each object and apply Newton's second law, $\Sigma \vec{F} = m\vec{a}$, in component form. If the direction of the acceleration is known, choose a coordinate axis to be parallel to it. For objects sliding along a surface, choose one coordinate axis parallel to the surface and the other perpendicular to it.

4. Solve for the unknowns in the resulting equations.

5. Check to see whether your results have the correct units and seem reasonable. Substituting extreme values into your solution is a good way to check your work for errors.

Solving problems using Newton's laws

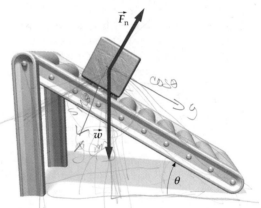

Example 4-8

You unload a moving van by sliding its cargo down a ramp that has rollers (i.e., the ramp is approximately frictionless). The ramp is inclined at an angle θ to the horizontal. For a box of mass m, find both the acceleration of the box as it slides down the ramp and the normal force exerted by the ramp on the box.

Picture the Problem Two forces act on the box, the weight $\vec{w}$ and the normal force $\vec{F}_n$. Since these forces act along different lines, they cannot sum to zero, hence there is a net force on the box causing it to accelerate. The ramp constrains the box to move parallel to its surface, so we choose a coordinate system aligned with the ramp, as shown in Figure 4-11. Then the acceleration has only one nonzero component, a_x.

Figure 4-11

Note that $\vec{w}$ is perpendicular to the horizontal, and the negative y axis is perpendicular to the incline, so the angle between $\vec{w}$ and the negative y axis is the same as the angle θ of the incline.

1. Draw a free-body diagram for the box:

Figure 4-12

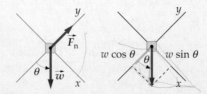

2. Apply $\Sigma \vec{F} = m\vec{a}$ to the box:

$$\Sigma \vec{F} = m\vec{a}$$
$$\vec{F}_n + \vec{w} = m\vec{a}$$

3. The normal force $\vec{F}_n$ is in the y direction:

$$F_{n,x} = 0, \qquad F_{n,y} = F_n$$

4. The weight $\vec{w}$ has both x and y components:

$$w_x = w \sin\theta = mg \sin\theta$$
$$w_y = -w \cos\theta = -mg \cos\theta$$

5. Apply $\Sigma \vec{F} = m\vec{a}$ in component form:

$$\Sigma F_x = ma_x, \qquad 0 + mg \sin\theta = ma_x$$
$$\Sigma F_y = ma_y, \qquad F_n - mg \cos\theta = ma_y = 0$$

6. Solve for a_x and F_n:

$$a_x = g \sin \theta$$

$$F_n = mg \cos \theta$$

Remark The acceleration down the incline is constant and equal to $g \sin \theta$.

Check the Result It is useful to check our results at the extreme values of inclination, $\theta = 0$ and $\theta = 90°$. At $\theta = 0$, the surface is horizontal. The weight has only a y component, which is balanced by the normal force $\vec{F}_n$. The acceleration is zero: $a_x = g \sin 0° = 0$. At the opposite extreme, $\theta = 90°$, the incline is vertical. Then the weight has only an x component along the incline, and the normal force is zero: $F_n = mg \cos 90° = 0$. The acceleration is $a_x = g \sin 90° = g$. That is, the box is in free fall.

Figure 4-13

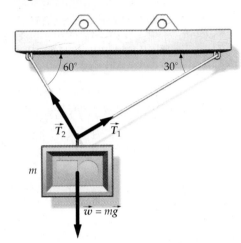

Example 4-9   *try it yourself*

A picture weighing 8 N is supported by two wires with tension $\vec{T}_1$ and $\vec{T}_2$, as shown in Figure 4-13. Find each tension.

Picture the Problem Since the picture does not accelerate, the net force acting on it must be zero. The three forces acting on the picture, its weight $m\vec{g}$, the tension $\vec{T}_1$ in one wire, and the tension $\vec{T}_2$ in the other wire, must therefore sum to zero.

Cover the column to the right and try these on your own before looking at the answers.

Steps

Answers

1. Draw a free-body diagram for the picture. On your diagram show the x and y components of the tensions.

   Figure 4-14

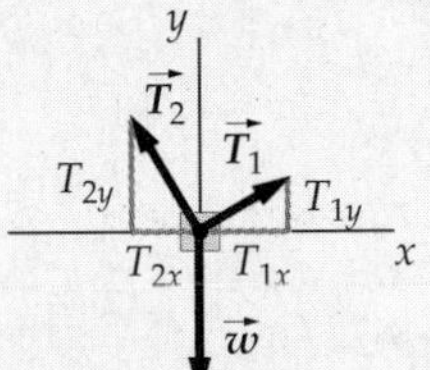

2. Apply $\Sigma \vec{F} = m\vec{a}$ in vector form to the picture.

   $$\sum \vec{F} = \vec{T}_1 + \vec{T}_2 + \vec{w} = m\vec{a} = 0$$

3. Resolve each force into its x and y components. This gives you two equations for the two unknowns T_1 and T_2.

   $$\sum F_x = T_1 \cos 30° - T_2 \cos 60° = 0$$

   $$\sum F_y = T_1 \sin 30° + T_2 \sin 60° - mg = 0$$

4. Solve the x component equation for T_2 in terms of T_1.

   $$T_2 = T_1 \frac{\cos 30°}{\cos 60°} = T_1 \sqrt{3}$$

5. Substitute your result for T_2 from step 3 into the y-component equation and solve for T_1.

   $$T_1 \sin 30° + (T_1 \sqrt{3}) \sin 60° - mg = 0$$

   $$T_1 = \frac{1}{2} mg = 4 \text{ N}$$

6. Use your result for T_1 to find T_2.

   $$T_2 = \sqrt{3} T_1 = \frac{\sqrt{3}}{2} mg = 6.93 \text{ N}$$

Remarks Note that the more vertical of the two wires supports the greater share of the load, as one might expect. Also, we see that $T_1 + T_2 > 8$ N. The "extra" force is due to the wires pulling to the right and left.

Example 4-10

As your jet plane speeds down the runway on takeoff, you decide to determine its acceleration, so you take out your yo-yo and note that when you suspend it, the string makes an angle of 22° with the vertical (Figure 4-15). (*a*) What is the acceleration of the plane? (*b*) If the mass of the yo-yo is 40 g, what is the tension in the string?

Picture the Problem Since the yo-yo accelerates in the horizontal direction, it must be acted on by a net horizontal force. This force is supplied by the horizontal component of the tension $\vec{T}$. The vertical component of $\vec{T}$ balances the weight of the yo-yo. We choose a coordinate system in which the *x* direction is parallel to the acceleration vector $\vec{a}$, and the *y* direction is vertical. Writing Newton's second law for both the *x* and *y* directions gives two equations to determine the two unknowns, *a* and *T*.

Figure 4-15

(*a*)1. Draw a free-body diagram for the yo-yo:

Figure 4-16

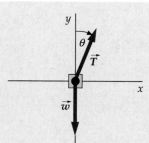

2. Apply $\sum \vec{F} = m\vec{a}$ in component form to the yo-yo:

$$\sum F_x = ma_x, \qquad T_x + w_x = T \sin \theta + 0 = ma_x$$

$$\sum F_y = ma_y, \qquad T_y + w_y = T \cos \theta - mg = ma_y = 0$$

3. Simplify:

$$T \sin \theta = ma_x, \qquad T \cos \theta = mg$$

4. Dividing one of these equations by the other eliminates *T* and allows us to determine *a*:

$$\frac{T \sin \theta}{T \cos \theta} = \frac{ma_x}{mg}$$

or

$$a_x = g \tan \theta = (9.81 \text{ m/s}^2)\tan 22° = 3.96 \text{ m/s}^2$$

(*b*) *T* can be found directly from the *y* component of Newton's second law:

$$T = \frac{mg}{\cos \theta} = \frac{(0.04 \text{ kg})(9.81 \text{ m/s}^2)}{\cos 22°} = 0.423 \text{ N}$$

Remark Notice that *T* is greater than the weight of the yo-yo (*mg* = 0.392 N), since the cord not only keeps the yo-yo from falling but also accelerates it in the horizontal direction. Here we use the units m/s² for *g* because we are calculating acceleration.

Check the Result Note that the *x* equation gives the same value for the tension: $T = ma/\sin \theta = 0.423$ N. At $\theta = 0$, we find that $T = mg$ and $a = 0$.

Exercise For what acceleration *a* would the tension in the string be equal to 3*mg*? What is θ in this case? (*Answers* $a = 27.8 \text{ m/s}^2$, $\theta = 70.5°$)

Our next example applies Newton's laws to objects that are at rest relative to a reference frame that is itself accelerating.

Example 4-11

An 80-kg man stands on a scale fastened to the floor of an elevator. The scale is calibrated in newtons. What does the scale read when (a) the elevator is moving with upward acceleration a; (b) the elevator is moving with downward acceleration a'; (c) the elevator is moving upward at 20 m/s while its speed is decreasing at a rate of 8 m/s^2?

Picture the Problem The scale reading is the magnitude of the normal force $\vec{F}_n$ exerted by the scale on the man (Figure 4-17). Since the man is at rest relative to the elevator, he and the elevator have the same acceleration. Two forces act on the man; the downward force of gravity, $m\vec{g}$, and the upward normal force from the scale, $\vec{F}_n$. The sum of these forces gives the man the observed acceleration. In what follows, we choose upward to be the positive direction.

Figure 4-17

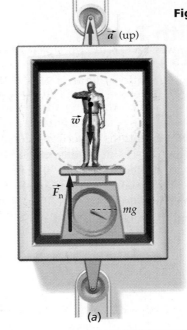
(a)

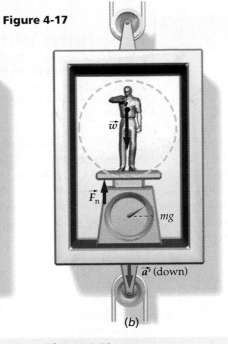
(b)

Figure 4-18

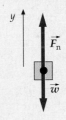

(a) 1. Draw a free-body diagram for the man:

2. Apply $\Sigma \vec{F} = m\vec{a}$ in the y direction:

$$\sum F_y = ma_y$$

3. Solve for F_n. This is the reading on the scale (the man's apparent weight):

$$F_n - mg = ma_y = ma$$
$$F_n = mg + ma$$

(b) 1. Apply $\Sigma \vec{F} = m\vec{a}$ in the y direction for the case in which the elevator accelerates downward with magnitude a':

$$\sum F_y = ma_y$$
$$F_n - mg = ma_y = m(-a')$$

2. Solve for F_n:

$$F_n = mg - ma'$$

(c) 1. Apply $\Sigma \vec{F} = m\vec{a}$ in the y direction. Note that the acceleration of the elevator is downward:

$$\sum F_y = ma_y$$

2. Solve for F_n:

$$F_n - mg = ma_y = (80\ \text{kg})(-8\ \text{m/s}^2)$$
$$F_n = (80\ \text{kg})(-8\ \text{m/s}^2) + (80\ \text{kg})(9.81\ \text{m/s}^2) = 145\ \text{N}$$

Remarks When the elevator accelerates upward, the man's apparent weight is greater than mg by the amount ma. For the man, it is as if gravity were increased from g to $g + a$. When the elevator accelerates downward, the man's apparent weight is less than mg by the amount ma'. He feels lighter, as if gravity were $g - a'$. If $a' = g$, the elevator is in free fall, and the man experiences weightlessness. Note that our conclusions are independent of the speed and direction of motion of the elevator.

Exercise An elevator descending to the ground floor comes to a stop with an acceleration of magnitude 4 m/s^2. If your mass is 70 kg and you are standing on a scale in the elevator, what does the scale read as the elevator is stopping? (*Answer*   967 N)

4-7   Problems With Two or More Objects

In some problems, two or more objects are in contact or are connected by a string or spring. These problems are solved by drawing a free-body diagram for each object and then applying Newton's second law to each object. The resultant equations, together with any equations describing constraints, are solved simultaneously for the unknown forces or accelerations. An example of a constraint would be two objects connected by a string that is always taut. In that case, the objects must have equal speeds and their accelerations must be equal in magnitude. If the objects are in direct contact, the forces they exert on each other must be equal and opposite, as stated in Newton's third law.

Steve

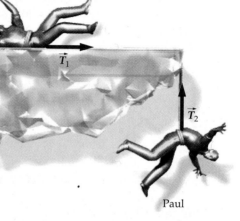

Example 4-12

Paul (mass m_P) accidentally falls off the edge of a cliff. Fortunately he is tied by a long rope to Steve (mass m_S), who has a climbing ax. Before Steve sets his ax to stop them, he slides without friction along the level snow, attached by the rope to Paul (Figure 4-19). Assume there is no friction between the rope and the cliff. Find the acceleration of each person and the tension T in the rope.

Picture the Problem The rope tensions $\vec{T}_1$ and $\vec{T}_2$ have equal magnitudes T because the rope is assumed to be massless, and the cliff is assumed to be frictionless. The rope does not stretch or become slack, so Paul and Steve have the same speed at all times. Their accelerations a_S and a_P must therefore be equal in magnitude (but not in direction). Because Steve has no vertical acceleration, the vertical forces F_n and $m_S g$ must balance. The acceleration of each person is related to the forces acting on him by Newton's second law.

Paul

Figure 4-19

1. Draw free-body diagrams for both Steve and Paul:

Figure 4-20

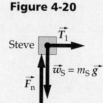

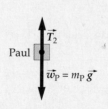

2. Apply $\Sigma \vec{F} = m\vec{a}$ in the horizontal direction to Steve, using $T_1 = T$:

$$\sum F_x = ma_x, \qquad T = m_S a_S$$

3. Apply $\Sigma \vec{F} = m\vec{a}$ to Paul. His acceleration is downward, and the forces acting on him are his weight $m_P g$ downward and $T_2 = T$ upward. Take the downward direction to be positive:

$$\sum F_y = ma_y, \qquad m_P g - T = m_P a_P$$

4. Because they are connected by the length of rope, the accelerations of Paul and Steve must be equal in magnitude. Let $a = a_S = a_P$:

$$T = m_S a, \qquad m_P g - T = m_P a$$

5. We solve these two equations for a and T by first eliminating one of the unknowns. T is eliminated by substituting $m_S a$ for T in the second equation:

$$m_P g - m_S a = m_P a$$

6. Solve for a:

$$a = \frac{m_P}{m_S + m_P} g$$

7. The expression for a can be substituted into the first equation in step 4 to find T:

$$T = m_S a = \frac{m_S m_P}{m_S + m_P} g$$

Remark In step 3, we chose downward to be positive to keep the solution as simple as possible. With this choice, when Steve moves in the positive direction (to the right), Paul also moves in the positive direction (downward). Note that the acceleration a is the same as that for a mass $m = m_S + m_P$ acted on by a force $m_P g$.

Check the Result If m_P is very much greater than m_S, we expect the acceleration to be approximately g and the tension to be approximately zero. Substituting $m_S = 0$ does indeed give $a = g$ and $T = 0$ in this case. If m_P is much less than m_S, we expect the acceleration to be approximately zero and the tension to be $m_P g$. If we neglect m_P in the denominator in steps 6 and 7, we indeed obtain $a \approx (m_P/m_S) g \approx 0$ and $T \approx m_P g$.

Exercise (a) Find the acceleration if the masses are $m_S = 78$ kg and $m_P = 92$ kg. (b) Find the acceleration if these two masses are interchanged. (Answers (a) $a = 0.541g$, (b) $a - 0.159g$)

Example **4-13**	*try it yourself*

While constructing a space station, you push on a box of mass m_1 with a force $\vec{F}$. The box is in direct contact with a second box of mass m_2 (Figure 4-21). (a) What is the acceleration of the boxes? (b) What is the magnitude of the force exerted by one box on the other?

Picture the Problem Let $\vec{F}_{2,1}$ be the contact force exerted by m_2 on m_1 and $\vec{F}_{1,2}$ be the force exerted by m_1 on m_2. These forces are equal and opposite so $\vec{F}_{2,1} = -\vec{F}_{1,2}$ and $F_{2,1} = F_{1,2}$. Apply Newton's second law to each box separately and use the fact that the accelerations a_2 and a_3 are equal.

Figure 4-21

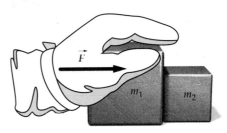

Cover the column to the right and try these on your own before looking at the answers.

Steps

Answers

(a) 1. Draw free-body diagrams for the two boxes.

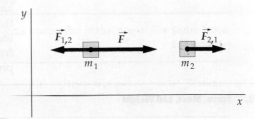

Figure 4-22

2. Apply $\Sigma \vec{F} = m\vec{a}$ to the first box.

$$F - F_{2,1} = m_1 a_1 = m_1 a$$

3. Apply $\Sigma \vec{F} = m\vec{a}$ to the second box.

$$F_{1,2} = m_2 a_2 = m_2 a$$

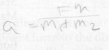

4. Add these equations to eliminate $F_{2,1}$ and $F_{1,2}$ and solve for $a = a_2 = a_3$.

$$a = \frac{F}{m_1 + m_2}$$

(b) Substitute your value for a into the equation in step 3 and solve for $F_{1,2}$.

$$F_{1,2} = \frac{m_2}{m_1 + m_2} F$$

the motions of two objects under the influence of several forces with constraints	Draw a free-body diagram for each object. Choose a coordinate system so that the acceleration is along one axis. Apply $\Sigma \vec{F} = m\vec{a}$ in component form to each object separately. Use the constraints to obtain information relating the magnitudes of the acceleration or the magnitudes of the forces (e.g., constant tension in a string). Solve the equations simultaneously to determine each force and acceleration.
	Examples 4-12, 4-13
3. Apply Newton's laws in the U.S. customary system.	Draw a free-body diagram. Use $m = w/g$ for the mass. **Example 4-4**
4. Find the forces acting on static objects.	Draw a free-body diagram. Choose a coordinate system in which one or more of the forces is along one of the axes. Apply $\Sigma \vec{F} = 0$ in component form, and solve for the desired quantities.
	Example 4-9
5. Apply Newton's laws to an object at rest in an accelerated system (such as a boxcar or elevator).	Draw a free-body diagram. Choose a coordinate system in which the acceleration is along one axis. Apply $\Sigma \vec{F} = m\vec{a}$ in component form, noting that the acceleration is the acceleration of the object in an inertial reference frame. Since the object is at rest in a noninertial frame, $\vec{a}$ is the acceleration of the accelerated frame. Solve for the forces.
	Examples 4-10, 4-11

Problems

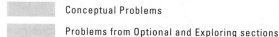

     Conceptual Problems

     Problems from Optional and Exploring sections

In a few problems, you are given more data than you actually need; in a few other problems, you are required to supply data from your general knowledge, outside sources, or informed estimates.

- •   Single-concept, single-step, relatively easy
- ••   Intermediate-level, may require synthesis of concepts
- •••   Challenging, for advanced students

For all problems, use $g = 9.81$ m/s^2 for the acceleration due to gravity and neglect friction and air resistance unless instructed to do otherwise.

Newton's First Law: The Law of Inertia

1   ••   How can you tell if a particular reference frame is an inertial reference frame?

2   ••   Suppose you find that an object in a particular frame has an acceleration $\vec{a}$ when there are no forces acting on it. How can you use this information to find an inertial reference frame?

Force, Mass, and Newton's Second Law

3   •   If an object has no acceleration in an inertial reference frame, can you conclude that no forces are acting on it?

4   •   If only a single force acts on an object, must the object accelerate in an inertial reference frame? Can it ever have zero velocity?

5   •   If an object is acted upon by a single known force, can you tell in which direction the object will move using no other information?

6   •   An object is observed to be moving at constant velocity in an inertial reference frame. It follows that

(a) no forces act on the object.
(b) a constant force acts on the object in the direction of motion.
(c) the net force acting on the object is zero.
(d) the net force acting on the object is equal and opposite to its weight.

7   •   A body moves with constant speed in a straight line in an inertial reference frame. Which of the following statements must be true?

(a) No force acts on the body.
(b) A single constant force acts on the body in the direction of motion.
(c) A single constant force acts on the body in the direction opposite to the motion.
(d) A net force of zero acts on the body.
(e) A constant net force acts on the body in the direction of motion.

8   ••   Figure 4-23 shows the position x versus time t of a particle moving in one dimension. During what time intervals is there a net force acting on the particle? Give the direction (+ or −) of the net force during these time intervals.

Figure 4-23 Problem 8

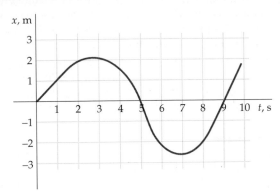

Figure 4-25 Problem 16

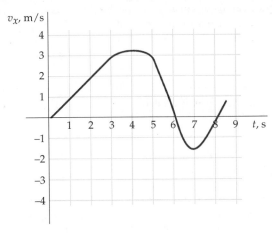

9 • A particle of mass m is traveling at an initial speed $v_0 = 25.0$ m/s. It is brought to rest in a distance of 62.5 m when a net force of 15.0 N acts on it. What is m?

(a) 37.5 kg   (b) 3.00 kg   (c) 1.50 kg
(d) 6.00 kg   (e) 3.75 kg

10 • (a) An object experiences an acceleration of 3 m/s² when a certain force F_0 acts on it. What is its acceleration when the force is doubled? (b) A second object experiences an acceleration of 9 m/s² under the influence of the force F_0. What is the ratio of the masses of the two objects? (c) If the two objects are tied together, what acceleration will the force F_0 produce?

11 • A tugboat tows a ship with a constant force F_1. The increase in the ship's speed in a 10-s interval is 4 km/h. When a second tugboat applies a second constant force F_2 in the same direction, the speed increases by 16 km/h in a 10-s interval. How do the magnitudes of the two forces compare? (Neglect water resistance.)

12 • A force F_0 causes an acceleration of 3 m/s² when it acts on an object of mass m sliding on a frictionless surface. Find the acceleration of the same object in the circumstances shown in Figure 4-24a and b.

Figure 4-24 Problem 12

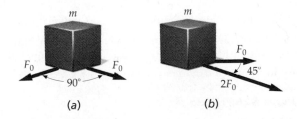

(a)   (b)

13 • A force $\vec{F} = 6\,\text{N}\,\hat{i} - 3\,\text{N}\,\hat{j}$ acts on an object of mass 1.5 kg. Find the acceleration $\vec{a}$. What is the magnitude a?

14 • A single force of 12 N acts on a particle of mass m. The particle starts from rest and travels in a straight line a distance of 18 m in 6 s. Find m.

15 • To drag a 75-kg log along the ground at constant velocity, you have to pull on it with a horizontal force of 250 N. (a) What is the resistive force exerted by the ground? (b) What force must you exert if you want to give the log an acceleration of 2 m/s²?

16 • Figure 4-25 shows a plot of v_x versus t for an object of mass 8 kg moving in a straight line. Make a plot of the net force acting on the object as a function of time.

17 •• A 4-kg object is subjected to two forces, $\vec{F}_1 = 2\,\text{N}\,\hat{i} - 3\,\text{N}\,\hat{j}$ and $\vec{F}_2 = 4\,\text{N}\,\hat{i} - 11\,\text{N}\,\hat{j}$. The object is at rest at the origin at time $t = 0$. (a) What is the object's acceleration? (b) What is its velocity at time $t = 3$ s? (c) Where is the object at time $t = 3$ s?

Weight and Mass

18 • Suppose an object were sent far out in space, away from galaxies, stars, or other bodies. How would its mass change? Its weight?

19 • How would an astronaut in apparent weightlessness be aware of her mass?

20 • Under what circumstances would your apparent weight be greater than your true weight?

21 • On the moon, the acceleration due to gravity is only about 1/6 of that on earth. An astronaut whose weight on earth is 600 N travels to the lunar surface. His mass as measured on the moon will be

(a) 600 kg.   (b) 100 kg.   (c) 61.2 kg.
(d) 9.81 kg.   (e) 360 kg.

22 • Find the weight of a 54-kg girl in (a) newtons and (b) pounds.

23 • Find the mass of a 165-lb man in kilograms.

24 • After watching a space documentary, Lou speculates that there is money to be made by combining the phenomenon of weightlessness in space with the widespread longing for weight loss in the general population. Researching the matter, he learns that the gravitational force on a mass m at a height h above the earth's surface is given by $F - mgR_E^2/(R_E + h)^2$, where R_E is the radius of the earth (about 6370 km) and g is the acceleration due to gravity at the earth's surface. (a) Using this expression, find the weight in newtons and pounds of an 83-kg person at the earth's surface. (b) If this person were weight-conscious and rich, and Lou managed to sell the person a trip to a height of 400 km above the earth's surface, how much weight would the person lose? (c) What is the person's mass at this altitude?

25 •• Caught without a map again, Hayley lands her spacecraft on an unknown planet. Visibility is poor, but she finds someone on a local communications channel and asks for directions to Earth. "You're already on Earth," is the reply, "Wait there and I'll be right over." Hayley is suspicious, however, so she drops a lead ball of mass 76.5 g from the top of her ship, 18 m above the surface of the planet. It takes 2.5 s to reach the ground. (*a*) If Hayley's mass is 68.5 kg, what is her weight on this planet? (*b*) Is she on Earth?

Newton's Third Law

26 • True or false:

(*a*) Action–reaction forces never act on the same object.
(*b*) Action equals reaction only if the objects are not accelerating.

27 • An 80-kg man on ice skates pushes a 40-kg boy also on skates with a force of 100 N. The force exerted by the boy on the man is

(*a*) 200 N.
(*b*) 100 N.
(*c*) 50 N.
(*d*) 40 N.

28 • A boy holds a bird in his hand. The reaction force to the force exerted on the bird by the boy's hand is

(*a*) the force of the earth on the bird.
(*b*) the force of the bird on the earth.
(*c*) the force of the hand on the bird.
(*d*) the force of the bird on the hand.
(*e*) the force of the earth on the hand.

The reaction force to the weight of the bird is

(*a*) the force of the earth on the bird.
(*b*) the force of the bird on the earth.
(*c*) the force of the hand on the bird.
(*d*) the force of the bird on the hand.
(*e*) the force of the earth on the hand.

29 • A baseball player hits a ball with a bat. If the force with which the bat hits the ball is considered the action force, what is the reaction force?

(*a*) The force the bat exerts on the batter's hands.
(*b*) The force on the ball exerted by the glove of the person who catches it.
(*c*) The force the ball exerts on the bat.
(*d*) The force the pitcher exerts on the ball while throwing it.
(*e*) Friction, as the ball rolls to a stop.

30 •• Dean reads in his physics book that when two people pull on the end of a rope in a tug-of-war, the forces exerted by each on the other are equal and opposite, according to Newton's third law. Misunderstanding the law tragically, Dean runs out to challenge Hugo the Large, convinced that the laws of physics guarantee a tie. Hugo lumbers over, picks up the rope, pulls Dean off his feet, and then drags him through a puddle, across the road, and up the steps of the physics building. Use a force diagram to show Dean that, in spite of Newton's third law, it is possible for one side to win a tug-of-war.

31 • A 2.5-kg object hangs at rest from a string attached to the ceiling. (*a*) Draw a diagram showing all forces acting on the object and indicate each reaction force. (*b*) Do the same for each force acting on the string.

32 • A 9-kg box rests on a 12-kg box that rests on a horizontal table. (*a*) Draw a diagram showing all forces acting on the 9-kg box and indicate each reaction force. (*b*) Do the same for all forces acting on the 12-kg box.

Contact Forces

33 • A vertical spring of force constant 600 N/m has one end attached to the ceiling and the other to a 12-kg block resting on a horizontal surface so that the spring exerts an upward force on the block. The spring is stretched by 10 cm. (*a*) What force does the spring exert on the block? (*b*) What is the force that the surface exerts on the block?

34 • A 6-kg box on a frictionless horizontal surface is attached to a horizontal spring with a force constant of 800 N/m. If the spring is stretched 4 cm from its equilibrium length, what is the acceleration of the box?

35 •• The acceleration *a* versus spring length *L* observed when a 0.5-kg mass is pulled along a frictionless table by a single spring is shown in the following table:

L, cm	4	5	6	7	8	9	10	11	12	13	14
a, m/s^2	0	2.0	3.8	5.6	7.4	9.2	11.2	12.8	14.0	14.6	14.6

(*a*) Make a plot of the force exerted by the spring versus length *L*. (*b*) If the spring is extended to 12.5 cm, what force does it exert? (*c*) How much does the spring extend when the mass is suspended from it at rest near sea level, where $g = 9.81$ N/kg?

Problem Solving

36 • A picture is supported by two wires as in Example 4-9. Do you expect the tension in the wire that is more nearly vertical to be greater than or less than the tension in the other wire?

37 • A clothesline is stretched taut between two poles. Then a wet towel is hung at the center of the line. Can the line remain horizontal? Explain.

38 •• Which of the free-body diagrams in Figure 4-26 represents a block sliding down a frictionless inclined surface?

Figure 4-26 Problem 38

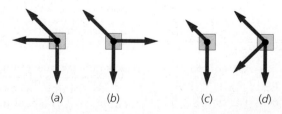

(*a*)   (*b*)   (*c*)   (*d*)

39 • A lamp with a mass $m = 42.6$ kg is hanging from wires as shown in Figure 4-27. The tension T_1 in the vertical handle is

(a) 209 N.
(b) 418 N.
(c) 570 N.
(d) 360 N.
(e) 730 N.

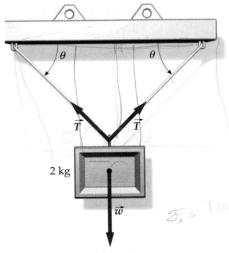

Figure 4-27
Problem 39

---

whose mass is 61.2 kg, realizes that the rope won't hold him unless he slides down with an appropriate acceleration. What must his acceleration be if the rope is not to break and ruin the whole effect?

45 •• A student has to escape from his girlfriend's dormitory through a window that is 15.0 m above the ground. He has a 24-m rope, but it will break when the tension exceeds 360 N, and the student weighs 600 N. The student will be injured if he hits the ground with a speed greater than 8 m/s. (a) Show that he cannot safely slide down the rope. (b) Find a strategy using the rope that will permit the student to reach the ground safely.

46 •• A rifle bullet of mass 9 g starts from rest and exits from the 0.6-m barrel at 1200 m/s. Find the force exerted on the bullet, assuming it to be constant, while the bullet is in the barrel.

---

40 • A 40.0-kg object supported by a vertical rope is initially at rest. The object is then accelerated upward. The tension in the rope needed to give the object an upward speed of 3.50 m/s in 0.700 s is

(a) 590 N.   (b) 390 N.   (c) 200 N.
(d) 980 N.   (e) 720 N.

41 • A hovering helicopter of mass m_h is lowering a truck of mass m_t. If the truck's downward speed is increasing at the rate 0.1g, what is the tension in the supporting cable?

(a) $1.1m_t g$ (b) $m_t g$ (c) $0.9m_t g$
(d) $1.1(m_h + m_t)g$ (e) $0.9(m_h + m_t)g$

42 • A 10-kg object on a frictionless table is subjected to two horizontal forces, $\vec{F}_1$ and $\vec{F}_2$, with magnitudes $F_1 = 20$ N and $F_2 = 30$ N, as shown in Figure 4-28. (a) Find the acceleration $\vec{a}$ of the object. (b) A third force $\vec{F}_3$ is applied so that the object is in static equilibrium. Find $\vec{F}_3$.

47 •• A 2-kg picture is hung by two wires of equal length. Each makes an angle of θ with the horizontal, as shown in Figure 4-30. (a) Find the general equation for tension T, given θ and weight w for the picture. For what angle θ is T the least? The greatest? (b) If $\theta = 30°$, what is the tension in the wires?   **Figure 4-30** Problem 47

Figure 4-28 Problem 42   **Figure 4-29** Problem 43

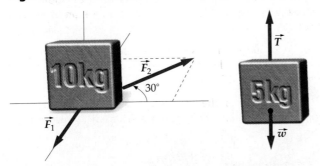

43 • A vertical force $\vec{T}$ is exerted on a 5-kg body near the surface of the earth, as shown in Figure 4-29. Find the acceleration of the body if (a) $T = 5$ N, (b) $T = 10$ N, and (c) $T = 100$ N.

44 •• To compensate for a distinct lack of personality, Herbert relies on the Grand Entrance technique when he attends parties. His latest plan for appearing at a pool party is to arrive by helicopter and then slide down a nylon rope as the helicopter hovers above poolside. However, as the helicopter approaches its destination, the pilot tells Herbert that the rope will break if the tension exceeds 300 N. Herbert,

48 •• A bullet of mass 1.8×10^{-3} kg moving at 500 m/s impacts with a large fixed block of wood and travels 6 cm before coming to rest. Assuming that the deceleration of the bullet is constant, find the force exerted by the wood on the bullet.

49 •• A 1000-kg load is being moved by a crane. Find the tension in the cable that supports the load as (a) it is accelerated upward at 2 m/s², (b) it is lifted at constant speed, and (c) it moves upward with speed decreasing by 2 m/s each second.

50 •• A horse-drawn coach is decelerating at 3.0 m/s² while moving in a straight line. A lamp of mass 0.844 kg is hanging from the ceiling of the coach on a string 0.6 m long. The angle that the string makes with the vertical is

(a) 8.5° toward the front of the coach.
(b) 17° toward the front of the coach.
(c) 17° toward the back of the coach.
(d) 2.5° toward the front of the coach.
(e) 0° or straight down.

51 •• For the systems in equilibrium in Figure 4-31, find the unknown tensions and masses.

Figure 4-31 Problem 51

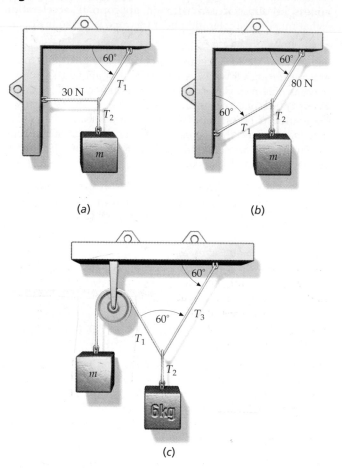

(a)   (b)

(c)

52   •• Your car is stuck in a mudhole. You are alone, but you have a long, strong rope. Having studied physics, you tie the rope tautly to a telephone pole and pull on it sideways, as shown in Figure 4-32. (*a*) Find the force exerted by the rope on the car when the angle θ is 3° and you are pulling with a force of 400 N but the car does not move. (*b*) How strong must the rope be if it takes a force of 600 N to move the car when θ = 4°?

Figure 4-32 Problem 52

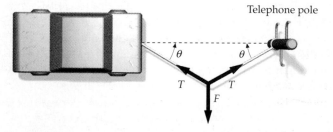

Inclined Planes

53   • A box slides down a frictionless inclined plane. Draw a diagram showing the forces acting on the box. For each force in your diagram, indicate the reaction force.

54   • The system shown in Figure 4-33 is in equilibrium. It follows that the mass *m* is

(*a*) 3.5 kg.
(*b*) 3.5 sin 40° kg.
(*c*) 3.5 tan 40° kg.
(*d*) none of the above.

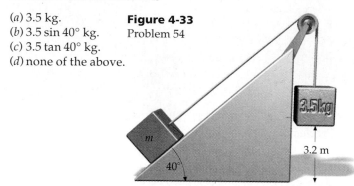

Figure 4-33
Problem 54

55   • In Figure 4-34, the objects are attached to spring balances calibrated in newtons. Give the readings of the balances in each case, assuming that the strings are massless and the incline is frictionless.

Figure 4-34 Problem 55

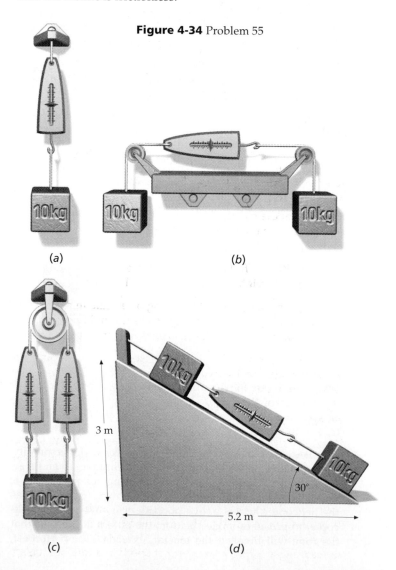

(a)   (b)

(c)   (d)

56 •• A box is held in position by a cable along a frictionless incline (Figure 4-35). (*a*) If $\theta = 60°$ and $m = 50$ kg, find the tension in the cable and the normal force exerted by the incline. (*b*) Find the tension as a function of θ and m, and check your result for $\theta = 0°$ and $\theta = 90°$.

Figure 4-35
Problem 56

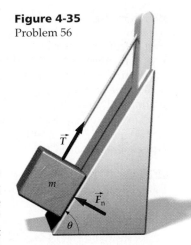

57 •• A horizontal force of 100 N pushes a 12-kg block up a frictionless incline that makes an angle of 25° with the horizontal. (*a*) What is the normal force that the incline exerts on the block? (*b*) What is the acceleration of the block?

58 •• A 65-kg boy weighs himself by standing on a scale mounted on a skateboard that is rolling down an incline, as shown in Figure 4-36. Assume there is no friction so that the force exerted by the incline on the skateboard is perpendicular to the incline. What is the reading on the scale if $\theta = 30°$?

Figure 4-36
Problem 58

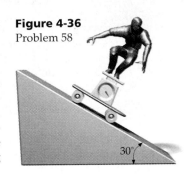

Elevators

59 • An object is suspended from the ceiling of an elevator that is descending at a constant speed of 9.81 m/s. The tension in the string holding the object is

(*a*) equal to the weight of the object.
(*b*) less than the weight of the object but not zero.
(*c*) greater than the weight of the object.
(*d*) zero.

60 • What effect does the velocity of an elevator have on the apparent weight of a person in the elevator?

61 • Suppose you are standing on a scale in a descending elevator as it comes to a stop on the ground floor. Will the scale's report of your weight be high, low, or correct?

62 • A person of weight w is in an elevator going up when the cable suddenly breaks. What is the person's apparent weight immediately after the cable breaks?

(*a*) w (*b*) Greater than w (*c*) Less than w
(*d*) $9.8w$ (*e*) Zero

63 • A person in an elevator is holding a 10-kg block by a cord rated to withstand a tension of 150 N. When the elevator starts up, the cord breaks. What was the minimum acceleration of the elevator?

64 • A 60-kg girl weighs herself by standing on a scale in an elevator. What does the scale read when (*a*) the elevator is descending at a constant rate of 10 m/s; (*b*) the elevator is descending at 10 m/s and gaining speed at a rate of 2 m/s²;

(*c*) the elevator is ascending at 10 m/s but its speed is decreasing by 2 m/s in each second?

65 •• A 2-kg block hangs from a spring balance calibrated in newtons that is attached to the ceiling of an elevator (Figure 4-37). What does the balance read when (*a*) the elevator is moving up with a constant velocity of 30 m/s; (*b*) the elevator is moving down with a constant velocity of 30 m/s; (*c*) the elevator is ascending at 20 m/s and gaining speed at a rate of 10 m/s²?

From $t = 0$ to $t = 2$ s, the elevator moves upward at 10 m/s. Its velocity is then reduced uniformly to zero in the next 2 s, so that it is at rest at $t = 4$ s. Describe the reading on the balance during the interval $0 < t < 4$ s.

Figure 4-37 Problem 65

66 •• A man stands on a scale in an elevator that has an upward acceleration a. The scale reads 960 N. When he picks up a 20-kg box, the scale reads 1200 N. Find the mass of the man, his weight, and the acceleration a.

Two or More Objects

67 • Two boxes of mass m_1 and m_2 connected together by a massless string are accelerated uniformly on a frictionless surface, as shown in Figure 4-38. The ratio of the tensions T_1/T_2 is given by

(*a*) m_1/m_2.   (*b*) m_2/m_1.   (*c*) $(m_1 + m_2)/m_2$.
(*d*) $m_1/(m_1 + m_2)$.   (*e*) $m_2/(m_1 + m_2)$.

Figure 4-38
Problem 67

68 • A box of mass $m_2 = 3.5$ kg rests on a frictionless horizontal shelf and is attached by strings to boxes of masses $m_1 = 1.5$ kg and $m_3 = 2.5$ kg, which hang freely, as shown in Figure 4-39. Both pulleys are frictionless and massless. The system is initially held at rest. After it is released, find (*a*) the acceleration of each of the boxes, and (*b*) the tension in each string.

Figure 4-39
Problem 68

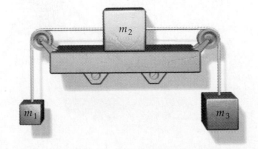

69 •• Two blocks are in contact on a frictionless, horizontal surface. The blocks are accelerated by a horizontal force $\vec{F}$ applied to one of them (Figure 4-40). Find the acceleration and the contact force for (a) general values of F, m_1, and m_2, and (b) for $F = 3.2$ N, $m_1 = 2$ kg, and $m_2 = 6$ kg.

Figure 4-40 Problem 69

70 •• Repeat the previous problem, but with the two blocks interchanged.

71 •• Two 100-kg blocks are dragged along a frictionless surface with a constant acceleration of 1.6 m/s², as shown in Figure 4-41. Each rope has a mass of 1 kg. Find the force F and the tension in the ropes at points A, B, and C.

Figure 4-41
Problem 71

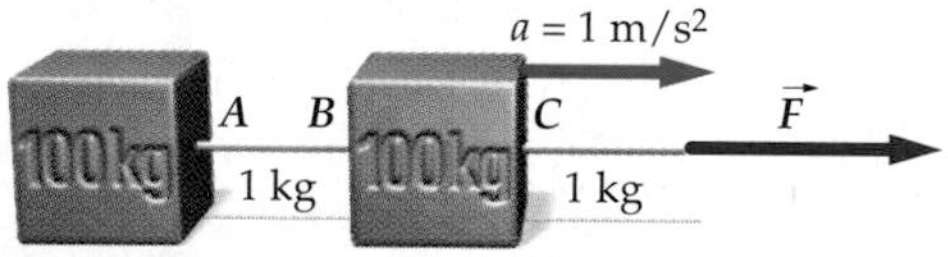

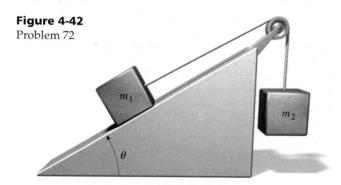

72 •• Two objects are connected by a massless string, as shown in Figure 4-42. The incline and pulley are frictionless. Find the acceleration of the objects and the tension in the string for (a) general values of θ, m_1, and m_2, and (b) $\theta = 30°$ and $m_1 = m_2 = 5$ kg.

Figure 4-42
Problem 72

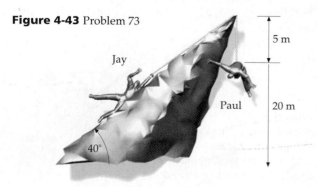

73 •• Two climbers on an icy (frictionless) slope, tied together by a 30-m rope, are in the predicament shown in Figure 4-43. At time $t = 0$, the speed of each is zero, but the top climber, Paul (mass 52 kg), has taken one step too many and his friend Jay (mass 74 kg) has dropped his pick. (a) Find the tension in the rope as Paul falls and his speed just before he hits the ground. (b) If Paul unhooks his rope after hitting the ground, find Jay's speed as he hits the ground.

Figure 4-43 Problem 73

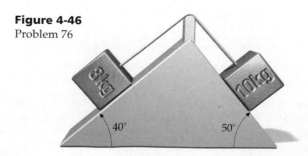

74 • The northwest face of Half Dome, a large rock in Yosemite National Park, makes an angle of $\theta = 7.0°$ with the vertical. Suppose a rock climber lying horizontal on the top is trying to support her unfortunate friend of equal mass who is hanging from a rope over the edge, as shown in Figure 4-44. If the friction is negligible (the top is icy!), at what acceleration will they slide down before the top partner manages to grab someone's hand and stop?

Figure 4-44 Problem 74

75 •• In a stage production of Peter Pan, the 50-kg actor playing Peter has to fly in vertically, and to be in time with the music, he must be lowered a distance of 3.2 m in 2.2 s. Backstage, a smooth surface sloped at 50° supports a counterweight of mass m, as shown in Figure 4-45. Show the calculations that the stage manager must perform to find (a) the mass of the counterweight that must be used, and (b) the tension in the wire.

Figure 4-45
Problem 75

76 •• An 8-kg block and a 10-kg block connected by a rope that passes over a frictionless peg slide on frictionless inclines, as shown in Figure 4-46. (a) Find the acceleration of the blocks and the tension in the rope. (b) The two blocks are replaced by two others of mass m_1 and m_2 such that there is no acceleration. Find whatever information you can about the mass of these two new blocks.

Figure 4-46
Problem 76

77 •• A heavy rope of length 5 m and mass 4 kg lies on a frictionless horizontal table. One end is attached to a 6-kg block. At the other end of the rope, a constant horizontal force of 100 N is applied. (*a*) What is the acceleration of the system? (*b*) Give the tension in the rope as a function of position along the rope.

78 •• A 60-kg housepainter stands on a 15-kg aluminum platform. The platform is attached to a rope that passes through an overhead pulley, which allows the painter to raise herself and the platform (Figure 4-47). (*a*) To accelerate herself and the platform at a rate of 0.8 m/s², with what force must she pull on the rope? (*b*) When her speed reaches 1 m/s, she pulls in such a way that she and the platform go up at a constant speed. What force is she exerting on the rope? (Ignore the mass of the rope.)

Figure 4-47
Problem 78

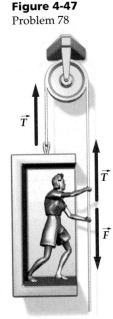

79 ••• Figure 4-48 shows a 20-kg block sliding on a 10-kg block. All surfaces are frictionless. Find the acceleration of each block and the tension in the string that connects the blocks.

Figure 4-48 Problem 79

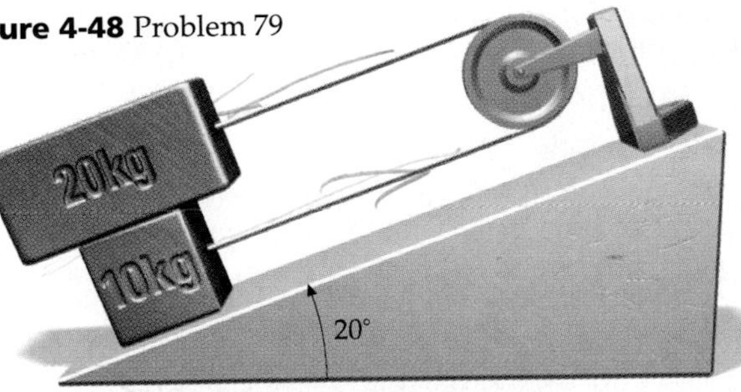

80 ••• A 20-kg block with a pulley attached slides along a frictionless ledge. It is connected by a massless string to a 5-kg block via the arrangement shown in Figure 4-49. Find the acceleration of each block and the tension in the connecting string.

Figure 4-49 Problem 80

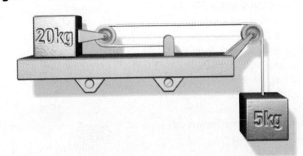

Atwood's Machine

81 •• The apparatus in Figure 4-50 is called an *Atwood's machine* and is used to measure the acceleration due to gravity *g* by measuring the acceleration of the two blocks. Assuming a massless, frictionless pulley and a massless string, show that the magnitude of the acceleration of either body and the tension in the string are

$$a = \frac{m_1 - m_2}{m_1 + m_2} \quad \text{and} \quad T = \frac{2m_1 m_2 g}{m_1 + m_2}$$

Figure 4-50
Problems 81–84

82 •• If one of the masses of the Atwood's machine in Figure 4-50 is 1.2 kg, what should the other mass be so that the displacement of either mass during the first second following release is 0.3 m?

83 •• A small pebble of mass *m* rests on the block of mass m_2 of the Atwood's machine in Figure 4-50. Find the force exerted by the pebble on m_2.

84 •• Find the force exerted by the Atwood's machine on the hanger to which the pulley is attached, as shown in Figure 4-50, while the blocks accelerate. Neglect the mass of the pulley. Check your answer by considering appropriate variations for m_1 and/or m_2.

85 ••• The acceleration of gravity *g* can be determined by measuring the time *t* it takes for a mass m_2 in an Atwood's machine to fall a distance *L*, starting from rest. (*a*) Find an expression for *g* in terms of m_1, m_2, *L*, and *t*. (*b*) Show that if there is a small error in the time measurement *dt*, it will lead to an error in the determination of *g* by an amount *dg* given by $dg/g = -2\, dt/t$. If *L* = 3 m and m_1 is 1 kg, find the value of m_2 such that *g* can be measured with an accuracy of ±5% with a time measurement that is accurate to 0.1 s. Assume that the only significant uncertainty in the measurement is the time of fall.

General Problems

86 • True or false:

(*a*) If there are no forces acting on an object, it will not accelerate.
(*b*) If an object is not accelerating, there must be no forces acting on it.
(*c*) The motion of an object is always in the direction of the resultant force.
(*d*) The mass of an object depends on its location.

87 • A skydiver of weight *w* is descending near the surface of the earth. What is the magnitude of the force exerted by her body *on the earth*?

(*a*) *w* (*b*) Greater than *w* (*c*) Less than *w* (*d*) 9.8*w*
(*e*) 0 (*f*) It depends on the air resistance.

88 • The net force on a moving object is suddenly reduced to zero. As a consequence, the object

(*a*) stops abruptly.
(*b*) stops during a short time interval.
(*c*) changes direction.
(*d*) continues at constant velocity.
(*e*) changes velocity in an unknown manner.

89 • A force of 12 N is applied to an object of mass m. The object moves in a straight line, with its speed increasing by 8 m/s every 2 s. Find m.

90 • A certain force F_1 gives an object an acceleration of 6×10^6 m/s². Another force F_2 gives the same object an acceleration of 15×10^6 m/s². What is the acceleration of the object if (a) the two forces act together on the object in the same direction; (b) the two forces act in opposite directions on the object; (c) the two forces act on the object at 90° to each other?

91 • A certain force applied to a particle of mass m_1 gives it an acceleration of 20 m/s². The same force applied to a particle of mass m_2 gives it an acceleration of 50 m/s². If the two particles are tied together and the same force is applied to the pair, find the acceleration.

92 • A 6-kg object is pulled along a frictionless horizontal surface by a horizontal force of 10 N. (a) If the object is at rest at $t = 0$, how fast is it moving after 3 s? (b) How far does it travel during these 3 s?

93 • If you weigh 125 lb on the earth, what would your weight be in pounds on the moon, where the free-fall acceleration due to gravity is 5.33 ft/s²?

94 • A redheaded woodpecker hits the bark of a tree extremely hard—the speed of its head reaches approximately $v = 3.5$ m/s before impact. If the mass of the bird's head is 0.060 kg, and the average force acting on the head during impact is $F = 6.0$ N, find (a) the acceleration of the head (assuming it is constant); (b) the depth of penetration into the bark; (c) the time t it takes the woodpecker's head to stop.

95 •• A simple accelerometer can be made by suspending a small object from a string attached to a fixed point on an accelerating object—to the ceiling of a passenger car, for example. When there is an acceleration, the object will deflect and the string will make some angle with the vertical. (a) How is the direction in which the suspended object deflects related to the direction of the acceleration? (b) Show that the acceleration a is related to the angle θ that the string makes by $a = g \tan \theta$; (c) Suppose the accelerometer is attached to the ceiling of an automobile that brakes to rest from 50 km/h in a distance of 60 m. What angle will the accelerometer make? Will the object swing forward or backward?

96 •• The mast of a sloop is supported at bow and stern by stainless steel wires, the forestay and backstay, anchored 10 m apart (Figure 4-51). The 12-m-long mast weighs 800 N and stands vertically on the deck of the sloop. The mast is positioned 3.6 m behind where the forestay is attached. The tension in the forestay is 500 N. Find the tension in the backstay and the force that the mast exerts on the deck of the sloop.

Figure 4-51
Problem 96

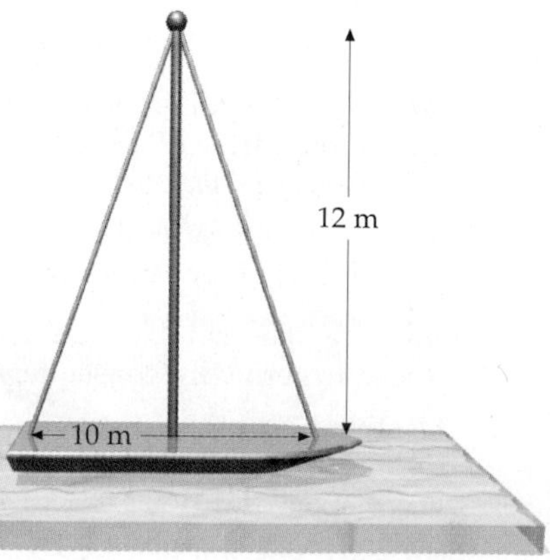

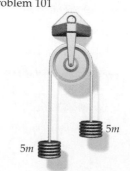

97 •• A block of mass m_1 is pulled along a smooth horizontal surface by a force $\vec{F}$ exerted at the end of a rope that has a much smaller mass m_2, as shown in Figure 4-52. (a) Find the acceleration of the rope and block, assuming them to be one object. (b) What is the net force acting on the rope? (c) Find the tension in the rope at the point where it is attached to the block. (d) The diagram, with the rope perfectly horizontal along its length, is not quite accurate. Correct the diagram, and state how this correction affects your solution.

Figure 4-52
Problem 97

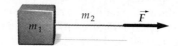

98 •• Joe and Sal are in a rollerbladers' club that is building a ramp to reach new levels of extremeness. The ramp is to be a simple incline, so that after coasting horizontally, a skater will ride up the slope at some angle θ. Sal suggests making the slope as steep as possible to maximize the height that will be reached. Joe whips out a pencil and paper to prove to Sal that, if the surfaces are smooth, the height reached is independent of the angle of the slope. Sal acknowledges that even though Joe is being smug and obnoxious, his argument is sound. Show Joe's proof.

99 •• A car traveling 90 km/h crashes into the rear end of an unoccupied stalled vehicle. Fortunately, the driver is wearing a seat belt. Using reasonable values for the mass of the driver and the stopping distance, estimate the force (assuming it to be constant) exerted on the driver by the seat belt.

100 •• A 2-kg body rests on a frictionless wedge that has an inclination of 60° and an acceleration a to the right such that the mass remains stationary relative to the wedge (Figure 4-53). (a) Find a. (b) What would happen if the wedge were given a greater acceleration?

Figure 4-53
Problem 100

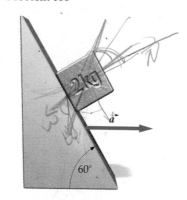

101 •• The masses attached to each side of an Atwood's machine consist of a stack of five washers each of mass m, as shown in Figure 4-54. The tension in the string is T_0. When one of the washers is removed from the left side, the remaining washers accelerate and the tension decreases by 0.3 N. (a) Find m. (b) Find the new tension and the acceleration of each mass when a second washer is removed from the left side.

Figure 4-54
Problem 101

102 •• Consider the Atwood's machine in Figure 4-54. When N washers are transferred from the left side to the right side, the right side drops 47.1 cm in 0.40 s. Find N.

103 •• Blocks of mass m and $2m$ are connected by a string (Figure 4-55). (*a*) If the forces are constant, find the tension in the connecting string. (*b*) If the forces vary with time as $F_1 = Ct$ and $F_2 = 2Ct$, where C is a constant and t is time, find the time t_0 at which the tension in the string is T_0.

Figure 4-55 Problem 103

104 ••• Find the normal force and the tangential force exerted by the road on the wheels of your bicycle (*a*) as you climb an 8% grade at constant speed, (*b*) as you descend the 8% grade at constant speed. (An 8% grade means that the angle of inclination θ is given by tan $\theta = 0.08$).

105 ••• The pulley in an Atwood's machine is given an upward acceleration $\vec{a}$, as shown in Figure 4-56. Find the acceleration of each mass and the tension in the string that connects them.

Figure 4-56 Problem 105

Figure 4-57 Problem 106

106 ••• The pulley in an Atwood's machine has a mass m_p. A force $\vec{F}$ is exerted on the pulley, as shown in Figure 4-57. Find the acceleration of each mass and the tension in the string that connects them.

CHAPTER 5

Applications of Newton's Laws

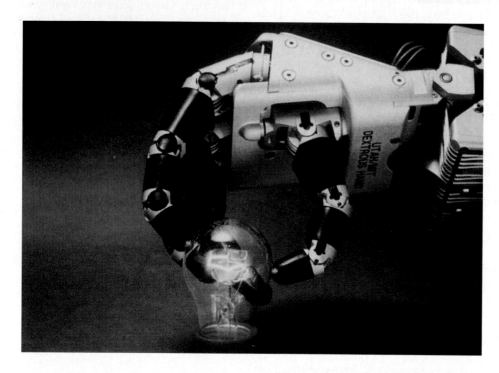

The Utah/MIT Dextrous Hand is a tendon-operated, multidegree-of-freedom dextrous hand that has multichannel touch sensing capability.

In this chapter we give examples of the application of Newton's laws to problems that involve frictional forces and to problems involving circular motion. We will also briefly discuss the motion of an object under the influence of drag forces, which are not constant but depend on the velocity of the object.

5-1 Friction

Static Friction

Friction is a complicated, incompletely understood phenomenon that arises due to the bonding of molecules between two surfaces that are in close contact. This bonding is the same as the molecular bonding that holds an object together. When you apply a small horizontal force to a large box resting on the floor, the box may not move because the force of **static friction**, $\vec{f}_s$, exerted by the floor on the box, balances the force you are applying (Figure 5-1). The force of static fric-

Figure 5-1

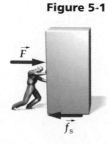

tion, which opposes the applied force, can adjust from zero to some maximum force $f_{s,max}$, depending on how hard you push. You might expect $f_{s,max}$ to be proportional to the area of contact between the two surfaces, but this is not the case. To a good approximation, $f_{s,max}$ is independent of the area of contact and is simply proportional to the normal force exerted by one surface on the other:

$$f_{s,max} = \mu_s F_n \qquad \text{5-1}$$

Definition—Coefficient of static friction

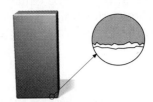

Figure 5-2 The microscopic area of contact between box and floor is only a small fraction of the macroscopic area of the box. This fraction is proportional to the normal force exerted between the surfaces. If the box rests on its side, the macroscopic area is increased, but the force per unit area is decreased by the same factor, so the microscopic area of contact is not changed.

where μ_s is called the **coefficient of static friction,** a dimensionless quantity that depends on the nature of the surfaces in contact (Figure 5-2). If you exert a horizontal force smaller than $f_{s,max}$ on the box, the frictional force will just balance this horizontal force. In general, we can write

$$f_s \leq \mu_s F_n \qquad \text{5-2}$$

Kinetic Friction

If you push the box in Figure 5-1 hard enough, it will slide across the floor. When the box is sliding, molecular bonds are continually being formed and ruptured, and small pieces of the surfaces are being broken off. The result is a force of **kinetic friction,** $\vec{f}_k$ (also called sliding friction) that opposes the motion. To keep the box sliding with constant velocity, you must exert a force on the box that is equal in magnitude and opposite in direction to the force of kinetic friction exerted by the floor.

The **coefficient of kinetic friction,** μ_k, is defined as the ratio of the magnitudes of the kinetic frictional force f_k and the normal force F_n:

$$f_k = \mu_k F_n \qquad \text{5-3}$$

Definition—Coefficient of kinetic friction

where μ_k depends on the nature of the surfaces in contact. Experimentally, it is found that μ_k is less than μ_s, and is approximately constant for speeds ranging from about 1 cm/s to several meters per second, the only situations we will consider. To a good approximation μ_k, like μ_s, is independent of the (macroscopic) area of contact.

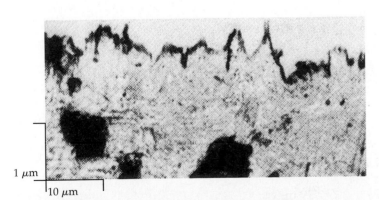

1 μm

10 μm

Magnified section of a polished steel surface showing surface irregularities. The irregularities are about 5×10^{-5} cm high, corresponding to several thousand atomic diameters.

Computer graphic showing gold atoms (bottom) adhering to the fine point of a nickel probe (top) that has been in contact with the gold surface.

Figure 5-3 shows the frictional force exerted on the box by the floor as a function of the applied force. The force of friction balances the applied force until the applied force equals $\mu_s F_n$, at which point the box begins to slide. Then the frictional force is constant and equal to $\mu_k F_n$. Table 5-1 lists some approximate values of μ_s and μ_k for various surfaces.

Figure 5-3

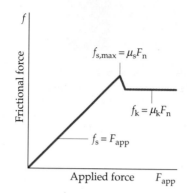

Table 5-1

Approximate Values of Frictional Coefficients

Materials	μ_s	μ_k
Steel on steel	0.7	0.6
Brass on steel	0.5	0.4
Copper on cast iron	1.1	0.3
Glass on glass	0.9	0.4
Teflon on Teflon	0.04	0.04
Teflon on steel	0.04	0.04
Rubber on concrete (dry)	1.0	0.80
Rubber on concrete (wet)	0.30	0.25
Waxed ski on snow (0ªC)	0.10	0.05

Example 5-1

A bartender slides a beer stein of mass 0.45 kg horizontally along the bar with an initial speed of 3.5 m/s. The stein comes to rest near the customer after sliding 2.8 m. Find the coefficient of kinetic friction.

Picture the Problem The force of kinetic friction is the only horizontal force acting on the stein (Figure 5-4). Since the frictional force is constant, the acceleration is constant. We can find a_c from the constant-acceleration equations of Chapter 2 and relate it to μ_k using $\Sigma F_x = ma_x$. Choose the direction of motion of the stein to be positive.

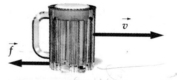

Figure 5-4

1. Draw a free-body diagram for the beer stein:

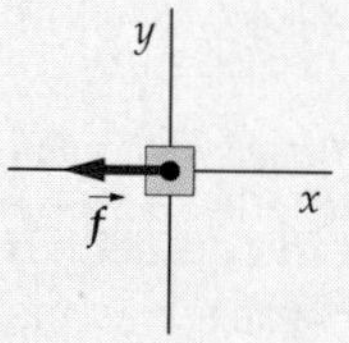

Figure 5-5

2. The coefficient of friction is related to the frictional force:

$$f_k = \mu_k F_n$$

3. Apply $\Sigma \vec{F} = m\vec{a}$ in component form to the beer stein:

$$\sum F_y = ma_y, \qquad F_n - mg = 0$$
$$\sum F_y = ma_y, \qquad -f_k = ma_x$$

4. Substitute mg for F_n to write the frictional force in terms of mg:

$$f_k = \mu_k F_n = \mu_k mg = -ma_x$$
$$\mu_k = \frac{-a_x}{g}$$

5. Relate the constant acceleration to the total distance traveled and the initial velocity using Equation 2–14:

$$v^2 = v_0^2 + 2a_x \Delta x = 0$$
$$a_x = -\frac{v_0^2}{2\,\Delta x} = -\frac{(3.5 \text{ m/s})^2}{2(2.8 \text{ m})} = -2.19 \text{ m/s}^2$$

6. Substitute this value of a_x to calculate μ_k:

$$\mu_k = -\frac{a_x}{g} = -\frac{-2.19 \text{ m/s}^2}{9.81 \text{ m/s}^2} = 0.223$$

Remark The mass m of the beer stein cancels. The greater the mass, the harder it is to stop the stein, but the greater mass is also accompanied by greater friction. The net result is that mass has no effect.

Example 5-2

A block rests on an inclined plane surface. The angle of inclination is increased until it reaches a critical angle θ_c, after which the block begins to slide. Find the coefficient of static friction μ_s.

Picture the Problem The forces acting on the block are its weight mg, the normal force F_n exerted by the plane, and the force of friction f (Figure 5-6). At angles less than the critical angle θ_c, the frictional force balances the component $mg \sin \theta$ down the incline. At the critical angle, $f_s = \mu_s F_n$. If we write $\Sigma \vec{F} = 0$, we can relate μ_s to the angle θ_c. Choose the x axis to be parallel to the plane and the y axis to be perpendicular to the plane.

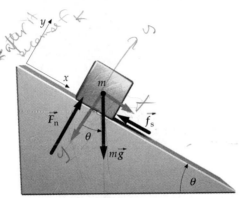

Figure 5-6

1. Draw a free-body diagram for the block:

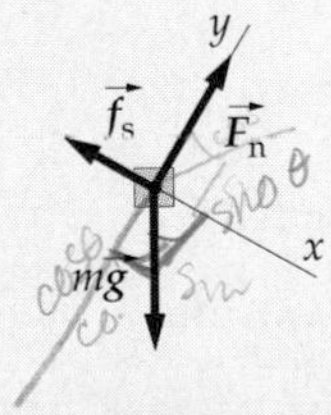

Figure 5-7

2. Apply $\Sigma \vec{F} = m\vec{a}$ in component form to the block:

$$\sum F_y = F_n - mg \cos \theta = ma_y = 0$$

$$\sum F_x = mg \sin \theta - f_s = ma_x = 0$$

3. Substitute $\mu_s F_n$ for f_s in the x equation:

$$mg \sin \theta - \mu_s F_n = 0$$

4. Solve for μ_s using $F_n = mg \cos \theta$ from the y equation:

$$\mu_s = \frac{mg \sin \theta}{F_n} = \frac{mg \sin \theta}{mg \cos \theta} = \tan \theta$$

Exercise The coefficient of static friction between a car's tires and the road on a particular day is 0.7. What is the steepest angle of inclination of the road for which the car can be parked with its wheels locked and not slide down the hill? (*Answer*   35°)

From Example 5-2 we see that the coefficient of static friction is related to the critical angle θ_c at which an object begins to slip by

$$\mu_s = \tan \theta_c \qquad\qquad 5\text{-}4$$

Example 5-3

Two children are pulled on a sled over snow-covered ground. The sled, which is initially at rest, is pulled by a rope that makes an angle of 40° with the horizontal. The children have a combined mass of 45 kg and the sled has a mass of 5 kg. The coefficients of static and kinetic friction are $\mu_s = 0.2$ and $\mu_k = 0.15$. Find the frictional force exerted by the ground on the sled and the acceleration of the children and sled, starting from rest, if the tension in the rope is (a) 100 N and (b) 140 N.

Picture the Problem First we need to find out whether the frictional force is static or kinetic. To do this we compare the maximum frictional force with the horizontal force exerted by the tension in the rope. We choose the coordinate system shown, where $\vec{f}$ is the frictional force, $\vec{F}_n$ is the vertical force exerted by the ground, and $\vec{T}$ is the tension in the rope (Figure 5-8).

Figure 5-8

$F_n = f + T \cos 90°$

(a)1. Draw a free-body diagram for the sled:

Figure 5-9

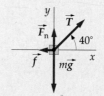

2. The maximum possible force of static friction is related to the magnitude of the normal force F_n:

$$f_{s,max} = \mu_s F_n$$

3. Apply $\Sigma \vec{F} = m\vec{a}$ in component form to the sled to obtain equations for f and F_n:

$$\sum F_x = T_x - f = ma_x$$
$$\sum F_y = F_n + T_y - mg = ma_y$$

4. Compute the horizontal and vertical components of the tension in the rope:

$$T_x = T \cos 40° = (100 \text{ N})(0.766) = 76.6 \text{ N}$$
$$T_y = T \sin 40° = (100 \text{ N})(0.643) = 64.3 \text{ N}$$

5. F_n can be determined by setting a_y equal to zero:

$$F_n + T_y - mg = 0$$
$$F_n = mg - T_y = (50 \text{ kg})(9.81 \text{ m/s}^2) - 64.3 \text{ N}$$
$$= 490 \text{ N} - 64.3 \text{ N} = 426 \text{ N}$$

6. Substitution of this result into step 1 above gives the maximum frictional force:

$$f_{s,max} = \mu_s F_n = 0.2(426 \text{ N}) = 85.2 \text{ N}$$

7. Since the applied horizontal force T_x does not exceed the maximum possible force of static friction, the sled remains at rest. The horizontal component of Newton's second law then gives:

$$T_x - f = ma_x = 0$$
$$f = T_x = 76.6 \text{ N}$$

(b)1. Now we find T_x and T_y for $T = 140$ N:

$$T_x = (140 \text{ N})(\cos 40°) = 107 \text{ N}$$
$$T_y = (140 \text{ N})(\sin 40°) = 90.0 \text{ N}$$

2. Apply $\Sigma \vec{F} = m\vec{a}$ in the y direction:

$$\Sigma F_y = F_n + T_y - mg = ma_y$$

3. Set a_y equal to zero and solve for F_n:

$$F_n + T_y - mg = ma_y = 0$$
$$F_n = mg - T_y = 490 \text{ N} - 90.0 \text{ N} = 400 \text{ N}$$

4. Use this new value of F_n to find the new maximum possible force of static friction:

$$f_{s,max} = \mu_s F_n = 0.2(400 \text{ N}) = 80.0 \text{ N}$$

5. Since this value for T_x is greater than the maximum force of static friction, the sled will slide. The frictional force on the sled will thus be due to kinetic friction:

$$f_k = \mu_k f_n = 0.15(400 \text{ N}) = 60.0 \text{ N}$$

6. Apply $\Sigma \vec{F} = m\vec{a}$ in the x direction to find the acceleration of the sled:

$$\Sigma F_x = T_x - f_k = ma_x$$
$$a_x = \frac{T_x - f_k}{m} = \frac{107 \text{ N} - 60.0 \text{ N}}{50 \text{ kg}} = 0.940 \text{ m/s}^2$$

Remark There are two important points in this example: (1) The normal force is not equal to the weight of the children and the sled because the vertical component of the tension helps lift the sled off the ground. (2) In part (a), the force of static friction is not equal to $\mu_s F_n$; it is less than this maximum possible limiting value.

Example 5-4   *try it yourself*

The mass m_2 in Figure 5-10 has been adjusted so that the block of mass m_1 is on the verge of sliding. (a) If $m_1 = 7$ kg and $m_2 = 5$ kg, what is the coefficient of static friction between the shelf and the block? (b) With a slight nudge, the blocks move with acceleration a. Find a if the coefficient of kinetic friction between the shelf and the block is $\mu_k = 0.54$.

Figure 5-10

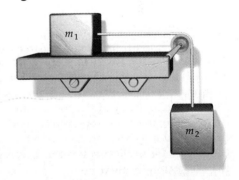

Picture the Problem Apply Newton's second law to each block, using the fact that T has the same magnitude throughout the rope, so $T_1 = T_2$, and that the accelerations have the same magnitude because the rope does not stretch. Choose the positive direction to be rightward for m_1 and downward for m_2.

To find the coefficient of static friction μ_s, as required in part (a), set the force of static friction on m_1 equal to its maximum value $f_{max} = \mu_s g$, and set the acceleration equal to zero.

Cover the column to the right and try these on your own before looking at the answers.

Steps

Answers

(a)1. Draw a free-body diagram for each block:

Figure 5-11

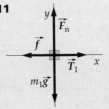

Block 1

Figure 5-12

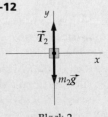

Block 2

2. Apply $\Sigma \vec{F} = m\vec{a}$ in component form to block 1 using $T_1 = T_2 = T$.

$$\sum F_x = ma_{1x}, \qquad T - f = 0$$

$$\sum F_y = ma_{1y}, \qquad F_n - m_1 g = 0$$

3. Apply $\Sigma \vec{F} = m\vec{a}$ to block 2.

$$\sum F_y = ma_{2y}, \qquad m_2 g - T = 0$$

4. Add the two equations containing the tension T to eliminate T, and then solve for f.

$$f = m_2 g$$

5. Relate f to μ_s and the weight of block 1.

$$f = \mu_s F_n = \mu_s m_1 g$$

6. Combine your two equations for f and solve for μ_s.

$$f = m_2 g = \mu_s m_1 g$$

$$\mu_s = \frac{m_2}{m_1} = 0.714$$

(b)1. Apply $\Sigma \vec{F} = m\vec{a}$ to the horizontal motion of block 1 and the vertical motion of block 2. Use $f = \mu_k m_1 g$ for the frictional force on block 1 and note that T and a are the same for both blocks.

$$\sum F_x = ma_{1x}, \qquad T - \mu_k m_1 g = m_1 a$$

$$\sum F_x = ma_{1y}, \qquad m_2 g - T = m_2 a$$

2. Eliminate T from your equations in step 1 of part (b) and solve for a.

$$a = \frac{m_2 - \mu_k m_1}{m_1 + m_2} g = 0.997 \text{ m/s}^2$$

Check the Result Note that $\mu_k = 0$ gives the result derived in Example 4-12.

Exercise What is the tension in the rope when the blocks are sliding? (Answer $T = m_2(g - a) = m_1(a + \mu_k g) = 44.1$ N)

Example 5-5

A runaway baby buggy is sliding without friction across a frozen pond toward a hole in the ice. You race after the buggy on skates. As you grab it, you and the buggy are moving toward the hole at speed v_0. The coefficient of friction between your skates and the ice as you turn out the blades to brake is μ_k. D is the distance to the hole when you reach the buggy, M is the total mass of the buggy, and m is your mass. (a) What is the least value of D such that you stop the buggy before it reaches the hole in the ice? (b) What force do you exert on the buggy?

Picture the Problem Initially, you and the buggy are moving toward the hole with speed v_0, which we take to be in the positive x direction. If F is the magnitude of the force you exert on the buggy, the net force on the buggy is $-F$ and that on you is $F - f$, where $f = \mu_k mg$, the force of kinetic friction. The minimum value of D is that for which your speed is zero just as you reach the hole (Figure 5-13).

Figure 5-13

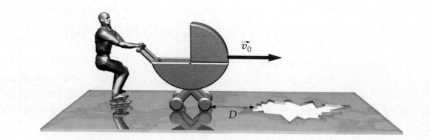

(a) 1. Draw free-body diagrams for you and the
buggy:

Figure 5-14

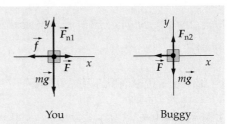

You

Buggy

2. The minimum distance D is related to the
initial speed v_0 and the final speed $v = 0$:

$$v^2 = v_0^2 + 2a_x D = 0$$

$$D = -\frac{v_0^2}{2a_x}$$

3. Apply $\Sigma \vec{F} = m\vec{a}$ to the motion of both you
and the buggy in the x direction to relate the
acceleration of each object to the forces acting upon them:

You $\Sigma F_x = ma_x$, $F - f = F - \mu_k mg = ma_x$

Buggy $\Sigma F_x = Ma_x$, $-F = Ma$

4. Add these two equations to eliminate F, then
solve for a_x:

$$-\mu_k mg = (M + m)a_x$$

The acceleration is negative, as expected:

$$a_x = -\left(\frac{m}{M + m}\right)\mu_k g$$

5. Substitute this result for a_x into the equation
for D in step 2:

$$D = -\frac{v_0^2}{2a_x} = \left(\frac{M + m}{m}\right)\frac{v_0^2}{2\mu_k g}$$

(b) F is found from Newton's second law applied to the buggy:

$$F = -Ma_x = \left(\frac{Mm}{M + m}\right)\mu_k g$$

Remark The minimum value of D is proportional to v_0^2 and inversely proportional to μ_k. Figure 5-15 shows the stopping distance D versus initial
velocity squared for values of M/m equal to 0.1, 0.3, and 1.0, with $\mu_k = 0.5$.
Note that when the mass of the buggy is larger, a greater stopping distance is
required for a given initial velocity. This is somewhat like stopping a car
while the car is pulling a trailer that does not have its own brakes. The mass
of the trailer increases the stopping distance for a given speed.

Figure 5-15

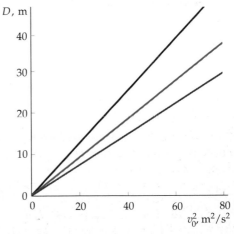

Example 5-6   *try it yourself*

A girl of mass m_g sits on a toboggan of mass m_t, which in turn sits on a frozen
pond assumed to be frictionless. The toboggan is pulled with a horizontal
force $\vec{F}$ (Figure 5-16). The coefficients of static and sliding friction between the
girl and toboggan are μ_s and μ_k. (a) Find the maximum value of F for which
the girl will not slide relative to the toboggan. (b) Find the acceleration of the
toboggan and girl when F is greater than this value.

Picture the Problem The forces acting on the girl are her weight $\vec{w}_g$, friction $\vec{f}$, and the normal force $\vec{F}_{n1}$ exerted by the toboggan. The forces on the
toboggan are its weight $\vec{w}_t$, friction $\vec{f}'$, the applied force $\vec{F}$, and the normal
force $\vec{F}'_{n1}$ exerted by the girl. The forces $\vec{f}'$ and $\vec{f}$ are action–reaction forces,
as are $\vec{F}_{n1}$ and $\vec{F}'_{n1}$. The vertical forces balance in both cases. The only horizontal force on the girl is that of friction. If static friction is great enough, the
girl has the same acceleration as the toboggan. Otherwise, the girl starts to
slide, and then kinetic friction causes her acceleration.

Figure 5-16

Cover the column to the right and try these on your own before looking at the answers.

Steps   **Answers**   **Figure 5-17**

(a) 1. Draw free-body diagrams for each object.

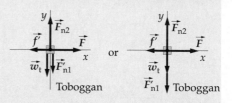

2. Apply $\Sigma \vec{F} = m\vec{a}$ to the horizontal motions of the girl and the toboggan, using $f' = f$ for the frictional force.

$$f = m_g a_1, \qquad F - f = m_t a_2$$

3. Eliminate the frictional force from your equations to obtain an expression for the force F in terms of the masses and the common acceleration a.

$$F = (m_g + m_t)a$$

4. The maximum acceleration of the girl occurs when the force of static friction on her is maximum. Find this maximum frictional force and use it to find her maximum acceleration a_{max}.

$$f_{s,max} = \mu_s F_{n1} = \mu_s m_g g$$
$$a_{max} = \mu_s g$$

5. Use your result for a_{max} to find the maximum force F_{max} for which the girl does not slide.

$$F_{max} = (m_g + m_t)\mu_s g$$

(b) 1. When the girl slides, the horizontal force on her is that of kinetic friction, f_k. Find f_k and the acceleration of the girl, a_g.

$$f_k = \mu_k m_g g, \qquad a_g = \mu_k g$$

2. Write the horizontal component of $\Sigma \vec{F} = m\vec{a}$ for the toboggan.

$$F - f_k = m_t a$$

3. Solve for the acceleration of the toboggan a_t using the value of f_k found previously.

$$a_t = \frac{F - \mu_k m_g g}{m_t}$$

Remarks Two alternative styles for the free-body diagram of the toboggan are shown. In the first, the two downward forces are displaced slightly so they can be seen. In the second, they are drawn head to tail. In Figure 5-18, the accelerations of the toboggan and the girl are shown as functions of the applied force F.

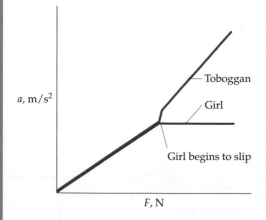

Figure 5-18 The accelerations of the toboggan and the girl as functions of the applied force F. When the friction is static, they move together with the same acceleration. When the girl begins to slip, the frictional force is smaller ($\mu_k m_g g$ rather than $\mu_s m_g g$), so the net force on the toboggan is greater and the slope of a_t versus F is greater. While the girl slips on the toboggan, the net force on the girl is constant ($\mu_k m_g g$), so the acceleration of the girl is constant.

Figure 5-19 Forces acting on a car with front-wheel drive. The normal forces $\vec{F}_n$ are not generally equal on the front and rear tires.

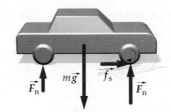

Figure 5-19 shows the forces acting on a car that is just starting to move from rest. The weight of the car is balanced by the normal force F_n exerted on the tires. To start the car moving, the engine delivers power to the axle that makes the wheels rotate (we discuss power in Chapter 6). If the road were perfectly frictionless, the wheels would merely spin. When friction is present, the frictional force exerted by the road on the tires is in the forward direction and provides the force needed to accelerate the car. If the power delivered by the engine is small enough so that the force exerted by the tire surface on the road surface is not too great, the two surfaces do not slip. Then the wheels roll without slipping and the tire tread touching the road is at rest relative to it (Figure 5-20). The friction between the road and the tire is then static friction. The largest frictional force that the tires can exert on the road (and that the road can exert on the tires) is $\mu_s F_n$.

Figure 5-20 When a wheel rolls without slipping, each point on the rim has a velocity of magnitude v relative to the center of the wheel, where v is the speed of the center of the wheel relative to the ground. The velocity of the point on the tire in contact with the ground is zero relative to the ground. In this figure, dashed lines represent velocities relative to the center of the wheel, and solid lines represent velocities relative to the ground.

If the power delivered by the engine is too great, the surfaces in contact will slip and the wheels will spin. Then the force that accelerates the car is the force of kinetic friction, which is less than the force of static friction. If we are stuck on ice or snow, our chances of getting free are better if we use a light touch on the accelerator pedal. Similarly, when braking a car to a stop, the force exerted by the road on the tires may be either static friction or kinetic friction, depending on how the brakes are applied. If the brakes are applied so hard that the wheels lock, the tires will slide along the road and the stopping force will be that of kinetic friction. If the brakes are applied gently, so that no slipping occurs between the tires and the road, the stopping force will be that of static friction. Antilock braking systems in cars allow you to brake hard without locking the wheels to provide maximum friction for stopping.

When an ideal, rigid wheel rolls *at constant speed* along a horizontal road without slipping, no force accelerates it. But because a real tire continually deforms, and the tread and road are continually peeled apart, a small force is needed to maintain the constant velocity. The **coefficient of rolling friction,** μ_r, is the ratio of the force needed to keep a wheel rolling at constant velocity on a level surface to the normal force exerted by the surface on the wheel. Typical values of μ_r are 0.01 to 0.02 for rubber tires on concrete, and 0.001 to 0.002 for steel wheels on steel rails.

Example 5-7

A car is traveling at 30 m/s along a horizontal road. The coefficients of friction between the road and the tires are $\mu_s = 0.5$ and $\mu_k = 0.3$. How far does the car travel before stopping if (a) the car is braked with an antilock braking system so that the wheels do not slip, and (b) the car is braked hard with no antilock braking system so that the wheels lock?

Picture the Problem The force that stops a car when it brakes is the force of friction exerted by the road on the tires (Figure 5-21). Since frictional force exerted by the road is constant, the acceleration is constant, and we can use the constant acceleration equations of Chapter 2 to relate the stopping distance to the acceleration. We then find the acceleration from Newton's second law. We take the direction of motion to be the positive x direction.

Figure 5-21

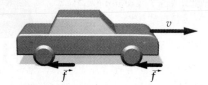

(a)1. Draw a free-body diagram for the car. Treat all four wheels as if they were a single point of contact with the ground. Assume further that the brakes are applied to all four wheels:

Figure 5-22

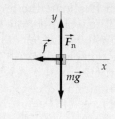

2. Equation 2-14 relates the stopping distance Δx to the initial speed v_0:

$$v^2 = v_0^2 + 2a_x\Delta x = 0$$

$$\Delta x = -\frac{v_0^2}{2a_x}$$

3. Apply $\Sigma\vec{F} = m\vec{a}$ to the car. Since the wheels do not slip, the horizontal force exerted by the road is that of static friction:

$$\sum F_x = ma_x, \qquad -\mu_s F_n = ma_x$$

$$\sum F_y = ma_y, \qquad F_n - mg = m(0) = 0$$

4. Since there is no vertical acceleration, the normal force F_n exerted by the road balances the weight mg of the car. Substitute mg for F_n and solve for a_x:

$$F_n = mg$$

$$a_x = \frac{-\mu_s F_n}{m} = \frac{-\mu_s mg}{m} = -\mu_s g$$

$$= -(0.5)(9.81 \text{ m/s}^2) = -4.90 \text{ m/s}^2$$

5. Substituting these results in the equation for Δx in step 2 gives the stopping distance:

$$\Delta x = -\frac{v_0^2}{2a}$$

$$= -\frac{(30 \text{ m/s})^2}{2(-4.90 \text{ m/s}^2)}$$

$$= 91.8 \text{ m}$$

(b)1. When the wheels lock, the force exerted by the road on the car is that of kinetic friction. Using reasoning similar to that in part (a), we obtain for the acceleration:

$$a_x = -\mu_k g = -(0.3)(9.81 \text{ m/s}^2)$$

$$= -2.94 \text{ m/s}^2$$

2. The stopping distance is then:

$$\Delta x = -\frac{v_0^2}{2a}$$

$$= -\frac{(30 \text{ m/s})^2}{2(-2.94 \text{ m/s}^2)} = 153 \text{ m}$$

Remark The stopping distance is more than 50% greater when the wheels are locked. This is why antilock braking systems were developed for automobiles. Also note that the stopping distance is independent of the car's mass—the stopping distance is the same for a subcompact or a large truck, provided the coefficients of friction are the same.

Exercise What must the coefficient of static friction be between the road and the tires of a four-wheel-drive car if the car is to accelerate from rest to 25 m/s in 8 s? (*Answer*   0.319)

5-2   Circular Motion

Centripetal Acceleration

Newton showed that a particle moving with constant speed v in a circle of radius r has an acceleration of magnitude v^2/r directed radially inward toward the center of the circle. This acceleration, called **centripetal acceleration**, requires a net force directed toward the center of the circle. Figure 5-23 shows a satellite moving in a circular orbit around the earth. At an altitude of 200 km, the gravitational force on the satellite is just slightly less than at the earth's surface. Why doesn't the satellite fall toward the earth? Actually, the satellite does "fall." But because of its horizontal velocity, it continually misses the earth. If the satellite in Figure 5-23 were not accelerating, it would move from point P_1 to P_2 in some time t. Instead, it arrives at point P_2' on its circular orbit. In a sense, the satellite "falls" the distance h shown in Figure 5-23. If t is small, P_2 and P_2' are nearly on a radial line. In that case we can calculate h from the right triangle of sides vt, r, and $r + h$. Since $r + h$ is the hypotenuse of the right triangle, the Pythagorean theorem gives

$$(r + h)^2 = (vt)^2 + r^2$$
$$r^2 + 2hr + h^2 = v^2t^2 + r^2$$

or

$$h(2r + h) = v^2t^2$$

For very short times, h will be much less than r, so we can neglect h compared with $2r$ for the term in parentheses. Then

$$2rh \approx v^2t^2$$

or

$$h \approx \frac{1}{2}\left(\frac{v^2}{r}\right)t^2$$

Comparing this with the constant-acceleration expression $h = \frac{1}{2}at^2$, we see that the magnitude of the acceleration of the satellite is

$$a = \frac{v^2}{r} \qquad\qquad 5\text{-}5$$

Centripetal acceleration

From Figure 5-23 we see that the direction is inward toward the center of the circle. A geometric proof that this result holds in general for circular motion with constant speed is given in Figure 5-24. An algebraic proof is outlined in Problem 108.

The motion of a particle moving in a circle with constant speed is often described in terms of the time required for one complete revolution T, called the **period.** During one period, the particle travels a distance of $2\pi r$ (where r is the radius of the circle), so its speed is related to r and T by

$$v = \frac{2\pi r}{T} \qquad\qquad 5\text{-}6$$

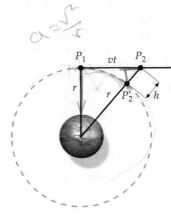

Figure 5-23 The satellite is moving with speed v in a circular orbit of radius r about the earth. If the satellite did not accelerate toward the earth, it would move in a straight line from point P_1 to P_2. Because of its acceleration, it instead falls a distance h. For small time t, $h = \frac{1}{2} (v^2/r) t^2 = \frac{1}{2} at^2$.

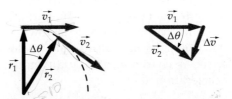

Figure 5-24 Position and velocity vectors for a particle moving in a circle at constant speed. The angle $\Delta\theta$ between $\vec{v}_1$ and $\vec{v}_2$ is the same as that between $\vec{r}_1$ and $\vec{r}_2$ because the position and velocity vectors must both move through equal angles to remain mutually perpendicular. For very small time intervals, the velocity change $\Delta\vec{v}$ is approximately perpendicular to $\vec{v}$ and points inward toward the center of the circle. The magnitude of the acceleration can be found from $\Delta\theta \approx v\,\Delta t/r$ and $\Delta\theta = \Delta v/v$. Then $\Delta v/\Delta t \approx v^2/r$.

Example 5-8

A satellite moves at constant speed in a circular orbit about the center of the earth and near the surface of the earth. If its acceleration is 9.81 m/s^2, find (*a*) its speed and (*b*) the time for one complete revolution.

Picture the Problem Since the satellite orbits near the surface of the earth, we take the radius of the orbit to be the radius of the earth, $r = 6370$ km.

(*a*) Setting the centripetal acceleration v^2/r equal to g yields the speed v:

$$a = \frac{v^2}{r} = g \quad \text{or}$$

$$v = \sqrt{rg} = \sqrt{(6370 \text{ km})(9.81 \text{ m/s}^2)}$$

$$= 7.91 \text{ km/s} = 17{,}700 \text{ mi/h}$$

(*b*) We use Equation 5-6 to get the period T:

$$T = \frac{2\pi r}{v} = \frac{2\pi (6370 \text{ km})}{7.91 \text{ km/s}} = 5060 \text{ s} = 84.3 \text{ min}$$

Remark For satellites in orbit a few hundred kilometers above the earth's surface, the orbital radius r is slightly greater than 6370 km. As a result, the centripetal acceleration is slightly less than 9.81 m/s^2 because of the decrease in the gravitational force with distance from the center of the earth. Many satellites are launched into such orbits, and their periods are roughly 90 min.

Exercise A car rounds a curve of radius 40 m at 48 km/h. What is its centripetal acceleration? (*Answer*   4.44 m/s^2)

A particle moving in a circle with *varying speed* has a component of acceleration tangent to the circle, dv/dt, as well as the radially inward centripetal acceleration, v^2/r. For general motion along a curve, we can treat a portion of the curve as an arc of a circle (Figure 5-25). The particle then has acceleration v^2/r toward the center of curvature, and if the speed is changing, it has tangential acceleration dv/dt.

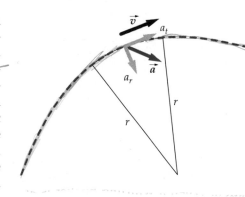

Figure 5-25 A particle moving along an arbitrary curve can be considered to be moving in a circular arc during a small time interval. Its instantaneous acceleration vector has a component $a_r = v^2/r$ toward the center of curvature of the arc and a component $a_t = dv/dt$ that is tangential to the curve.

Centripetal Force

As with any acceleration, there must be a net force in the direction of the acceleration to produce it. For centripetal accelerations, this force is called the **centripetal force.** It is *not* a new kind of force, but merely a name for the force needed for circular motion. The centripetal force may be due to a string, spring, or other contact force such as a normal force or friction; it may be an action-at-a-distance type of force such as a gravitational force, or it may be any combination of these. It is always directed inward, toward the center of the circle of motion.

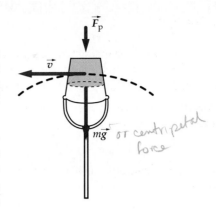

Figure 5-26

Example 5-9

You swing a pail of water in a vertical circle of radius r. The speed of the pail is v_t at the top of the circle. (a) Find the force exerted on the water by the pail at the top of the circle. (b) Find the minimum value of v_t for the water to remain in the pail. (c) Find the force exerted by the pail on the water at the bottom of the circle, where the pail's speed is v_b.

Picture the Problem We apply Newton's second law to find the force exerted by the pail. Two forces act on the water: gravity, $m\vec{g}$, and the force of the pail on the water, $\vec{F}_p$. Since the water moves in a circular path, it has a centripetal acceleration toward the center of the circle (Figure 5-26). We choose up for the positive y direction. Then at the top of the circle the acceleration is downward, $a_{y,\text{top}} = -v_t^2/r$; at the bottom it is upward, $a_{y,\text{bottom}} = +v_b^2/r$.

(a)1. Draw free-body diagrams for the water at the top and bottom of the circle.

Figure 5-27

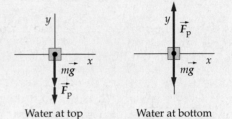

Water at top   Water at bottom

2. Apply $\Sigma\vec{F} = m\vec{a}$ to the water at the top of the circle. Both $\vec{F}_p$ and the weight are toward the center of the circle in the negative y direction:

$$\sum F_y = ma_y = m\left(-\frac{v_t^2}{r}\right)$$

$$-F_p - mg = m\left(-\frac{v_t^2}{r}\right) \quad \text{or} \quad F_p = m\frac{v_t^2}{r} - mg$$

(b) The pail cannot exert an upward force on the water at the top of the circle. The minimum force it can exert is zero. Set $F_p = 0$ and solve for the minimum speed, $v_{t,\text{min}}$:

$$0 = m\frac{v_{t,\text{min}}^2}{r} - mg \quad \text{or} \quad v_{t,\text{min}} = \sqrt{rg}$$

(c) Apply $\Sigma\vec{F} = m\vec{a}$ to the water at the bottom of the circle where $\vec{F}_p$ and the acceleration are upward:

$$\sum F_{by} = ma_y = m\left(+\frac{v^2}{r}\right)$$

$$F_p - mg = m\frac{v_b^2}{r} \quad \text{or} \quad F_p = m\frac{v_b^2}{r} + mg$$

Remark Note that there is no arrow for centripetal force in the free-body diagrams. Centripetal force is not a kind of force exerted by some agent; it is just the name for the resultant force that must point toward the center of the circle to provide the centripetal acceleration. When a whirling bucket is at the top of its circle, both gravity and the contact force of the pail contribute to the necessary centripetal force on the water. When the water is moving at the minimum speed at the top of the circle, its acceleration is $\vec{g}$, the free-fall acceleration due to gravity, and the only force acting on it at this point is its weight, $m\vec{g}$. At the bottom of the circle, F_p must be greater than the weight mg by enough to provide the necessary centripetal force.

Check the Result When $v = 0$ at the bottom, $F_p = mg$.

Exercise Estimate (a) the minimum speed at the top of the circle, and (b) the maximum period of revolution that will keep you from getting wet if you swing a pail of water in a vertical circle at constant speed. (Answers (a) Assuming $r \sim 1$ m, we find $v_{t,\text{min}} \sim 3$ m/s, (b) $T = (2\pi r/v) \sim 2$ s)

exploring

Noninertial Reference Frames, Pseudoforces, and Cyclones

Newton's laws are not valid in noninertial reference frames (frames that are accelerating relative to an inertial reference frame). Imagine standing on a subway train moving at constant speed in a straight line. You are holding onto a vertical pole, and a ball rests on the floor. When the train slows for the next station, the ball immediately accelerates forward until it hits the front of the car. You would also accelerate forward if you didn't hold onto the pole. Since there is no apparent force on the ball, its acceleration violates Newton's second law. Similarly, you feel the force of

the pole pulling you backward even though, relative to the train, you are not accelerating. There seems to be a very real force acting in the forward direction trying to push you, the ball, and everything else toward the front of the car.

For an observer standing in the station, the ball continues to move forward with constant speed (since no net force acts on it) and the train slows (accelerates backward) so the ball hits the front of the car. The backward force exerted by the pole on you causes you to decelerate along with the train. For the observer in the station, who is in an inertial reference frame, Newton's laws are obeyed.

Your frame, which is accelerating relative to the station, is a noninertial reference frame. You can use $\Sigma \vec{F} = m\vec{a}$ in your frame if you introduce a so-called **pseudoforce** that acts on each object, $\vec{F}_p = -m\vec{a}_t$, where m is the mass of the object. No agent exerts this pseudoforce; it is a fictitious force that exists only in your noninertial frame.

Now imagine riding on the rim of a merry-go-round, again holding onto a vertical pole. You have to grip the pole to keep from flying off. It seems like there is an outward force acting on you and everything else in this frame. This pseudoforce is called a centrifugal ("center-fleeing")

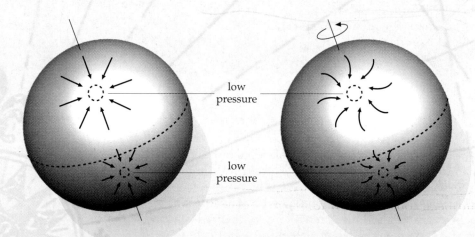

Figure 1 Suppose a low-pressure system develops at some middle latitude in the Northern Hemisphere. Since the pressure is low, air flows toward that region from all directions. Air moving down from the north is like the ball moving from the center of the platform toward the rim; it is deflected sideways to the west. Air moving up from the south is like a ball thrown from the rim toward the center, and the earth's rotation causes the air to move more rapidly eastward than the low-pressure region; thus, it outraces the low and deflects to the east. The net effect of the

deflection of the winds sideways is a counterclockwise circulation pattern established around the low-pressure area, as indicated by the red arrows in the figure. This weather pattern is known as a cyclone; hurricanes are particularly fierce cyclones. Counterclockwise circulation is characteristic of cyclones in the Northern Hemisphere; in the Southern Hemisphere, cyclones rotate in a clockwise direction. On a smaller scale, the Coriolis effect causes water draining in a bathtub to rotate counterclockwise in the Northern Hemisphere and clockwise in the Southern Hemisphere.

Typhoon Pat photographed over the Western Pacific by astronauts in the Space Shuttle *Discovery*. Shown in the photo is the counterclockwise rotation of the typhoon due to the Coriolis effect. Similar phenomena occur elsewhere in the universe with respect to other rotating bodies. An example is the Great Red Spot on Jupiter, a cyclone so vast that the earth could fit comfortably inside. The Great Red Spot has endured for hundreds of years, unlike the relatively short-lived hurricanes on earth.

force. Again, no agent exerts this fictitious force; it exists only in the noninertial rotating frame. From the frame of reference of an observer on the ground, you are hanging onto the pole so that it will exert the centripetal force required to make you move in a circular path.

There is another pseudoforce that occurs in rotating frames that is interesting because of its connection with weather patterns. Suppose you toss a ball from the center of the merry-go-round to a friend who rides at the rim. You throw the ball straight to your friend, but because your friend is moving sideways, the ball misses its target. From your point of view, the ball was deflected sideways away from your friend, as if a sideways force had acted on it. This sideways pseudoforce, which depends on the velocity of an object in a ro-

tating frame, is called the Coriolis force, and the sideways deflection is called the Coriolis effect. The same effect has dramatic consequences for weather on the earth (Figure 1).

Example 5-10   *try it yourself*

A tether ball of mass m is suspended from a rope of length L and travels at constant speed v in a horizontal circle of radius r. The rope makes an angle θ given by $\sin\theta = r/L$, as shown in Figure 5-28. Find (*a*) the tension in the rope, and (*b*) the speed of the ball.

Picture the Problem Two forces act on the ball: its weight, $m\vec{g}$, and the tension in the rope, $\vec{T}$. Because of the ball's circular motion, it has a horizontal acceleration of magnitude v^2/r directed toward the center of the circle. Hence, the vertical component of the tension balances the ball's weight, and the horizontal component is the centripetal force. Choose y to be vertical and x to be directed toward the center of the circle.

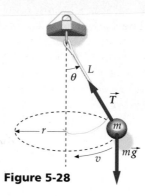

Figure 5-28

Cover the column to the right and try these on your own before looking at the answers.

Steps   **Answers**   **Figure 5-29**

(*a*)1. Draw a free-body diagram for the ball.

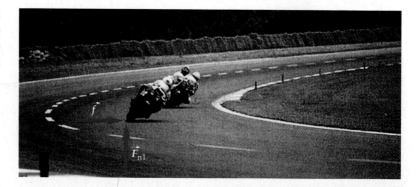

2. Apply $\Sigma\vec{F} = m\vec{a}$ in component form to the ball.

$$\Sigma F_x = T\sin\theta = ma_x = \frac{mv^2}{r}$$

$$\Sigma F_y = T\cos\theta - mg = ma_y = 0$$

3. Solve the y equation for T.

$$T = \frac{mg}{\cos\theta}$$

(*b*) Substitute the result for T in the x equation and solve for v.

$$v = \sqrt{rg\tan\theta}$$

Remarks An object attached to a string and moving in a horizontal circle so that the string makes an angle θ with the vertical is called a *conical pendulum*.

When a car rounds a curve on a horizontal road, the centripetal force is provided by the force of friction exerted by the road on the tires of the car. If the car does not slide radially, the friction is static friction.

As the motorcycle rounds the curve, it is tilted so that the resultant of the normal and frictional forces exerted by the road acts along the plane of the cycle.

Example 5-11

In a skid test, a recent-model BMW 530i was able to travel in a circle of radius 45.7 m in 15.2 s without skidding. (*a*) What was its average speed? (*b*) Assuming v to be constant, what was the centripetal acceleration? (*c*) Again assuming v to be constant, what is the minimum value for the coefficient of static friction?

Picture the Problem Figure 5-30 shows the forces acting on the car. The normal force F_n balances the downward force due to gravity mg. The horizontal force is the force of static friction, which provides the centripetal force. The faster the car travels, the greater the required centripetal force. The average speed can be found from the circumference of the circle and the period T. This average speed puts a lower limit on the maximum value of the coefficient of static friction.

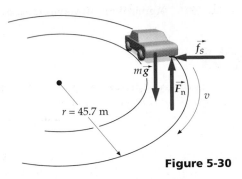

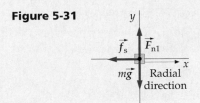

$r = 45.7$ m

Figure 5-30

(a) 1. Draw a free-body diagram for the car:

Figure 5-31

2. The average speed is the total circumference $2\pi r$ divided by the period T:

$$v = \frac{2\pi r}{T} = \frac{2\pi(45.7 \text{ m})}{15.2 \text{ s}} = 18.9 \text{ m/s}$$

(b) Use v to calculate the centripetal acceleration:

$$a_c = \frac{v^2}{r} = \frac{(18.9 \text{ m/s})^2}{45.7 \text{ m}} = 7.82 \text{ m/s}^2$$

(c) 1. Apply $\Sigma \vec{F} = m\vec{a}$ to the vertical and radial motions of the car. Choose the outward radial direction to be positive:

$$\Sigma F_y = ma_y, \qquad F_n - mg = 0$$

$$\Sigma F_r = ma_r, \qquad -f_s = -m\frac{v^2}{r}$$

2. The maximum value for static friction is proportional to the normal force:

$$f_{s,max} = \mu_s F_n = \mu_s mg$$

3. Substituting $f_{s,max}$ into $\Sigma \vec{F} = m\vec{a}$ relates μ_s to the speed:

$$\mu_s mg = m\frac{v^2}{r}$$

$$\mu_s = \frac{v^2}{rg} = \frac{(18.9 \text{ m/s})^2}{(45.7 \text{ m})(9.81 \text{ m/s}^2)} = 0.797$$

Check the Result If μ_s were equal to 1, the inward force would be equal to mg and the centripetal acceleration would be g. Here μ_s is about 0.8 and the centripetal acceleration is about $0.8g$.

Banked Curves

If a curved road is not horizontal but banked, the normal force of the road will have a component directed inward toward the center of the circle that will contribute to the centripetal force. The banking angle can be chosen such that, for a given speed, no friction is needed for a car to handle the curve without sliding.

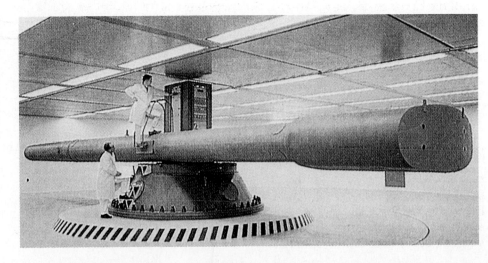

A large centrifuge used for research at Sandia National Laboratories.

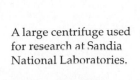

Example 5-12

A curve of radius 30 m is banked at an angle θ. Find θ for which a car can round the curve at 40 km/h even if the road is frictionless.

Figure 5-32

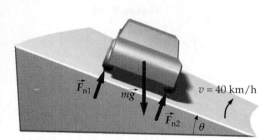

Figure 5-33

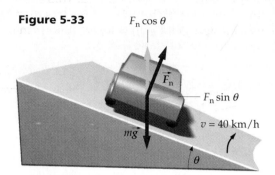

Picture the Problem In this case only two forces act on the car: gravity and the normal force. Since the road is banked, the normal force has a horizontal component that supplies the needed centripetal force. In Figure 5-32, the forces exerted by the road on the car are represented by $\vec{F}_{n1}$ and $\vec{F}_{n2}$. These forces are combined into $\vec{F}_n$ in Figure 5-33. We see that the angle between the normal force F_n and the vertical is θ, the same as the banking angle. We find this angle by applying Newton's second law. The total normal force has the component $F_n \sin \theta$ directed toward the center of the curve, which provides the centripetal acceleration of the car.

1. Draw a free-body diagram for the car. We choose y to be the vertical direction and x to be the horizontal direction toward the center of the circle. We call the total normal force exerted by the road F_n:

Figure 5-34

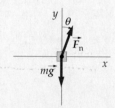

2. Apply $\Sigma \vec{F} = m\vec{a}$ in component form:

$$\sum F_x = ma_x, \quad F_n \sin \theta = ma_x = m\frac{v^2}{r}$$
$$\sum F_y = ma_y, \quad F_n \cos \theta - mg = ma_y = 0$$

3. Solve the y equation for the normal force F_n:

$$F_n = \frac{mg}{\cos \theta}$$

4. Substitute this result for F_n into the x equation:

$$\frac{mg}{\cos \theta}\sin \theta = m\frac{v^2}{r}$$

5. Solve for θ:

$$\tan \theta = \frac{v^2}{rg}$$

6. Substitute $v = 40$ km/h $= 11.1$ m/s, $r = 30$ m, and $g = 9.81$ m/s^2 and find θ:

$$\tan \theta = \frac{(11.1 \text{ m/s})^2}{(30 \text{ m})(9.81 \text{ m/s}^2)} = 0.419$$

$$\theta = 22.7°$$

Remark The banking angle θ depends on v and r, but not the mass m; θ increases with increasing v, and decreases with increasing r. When the banking angle, speed, and radius satisfy $\tan \theta = v^2/rg$, the car rounds the curve smoothly, with no tendency to slide either inward or outward. If the car speed is greater than $rg \tan \theta$, the road will exert a frictional force down the incline (Figure 5-35). This force has an inward horizontal component, which provides the additional centripetal force needed to keep the car from moving outward (sliding up the incline). If the car speed is less than this amount, the road must exert a frictional force up the incline.

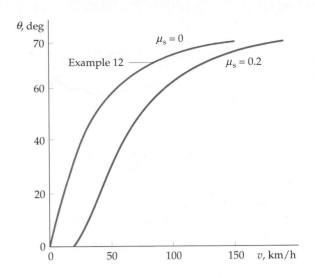

Figure 5-35 The addition of friction allows the car in this example to travel a bit faster around the curve without slipping. The two plots show banking angles θ versus the maximum velocities that can be attained without slipping, both without friction (red; $\mu_s = 0$) and with friction (green; $\mu_s = 0.2$). When the banking angle is zero, the maximum speed through the turn without slipping is zero with no friction, and about 30 km/h with $\mu_s = 0.2$.

5-3   Drag Forces

When an object moves through a fluid such as air or water, the fluid exerts a **drag force** or retarding force that tends to reduce the speed of the object. The drag force depends on the shape of the object, the properties of the fluid, and the speed of the object relative to the fluid. Unlike ordinary friction, the drag force increases as the speed of the object increases. At low speeds, the drag force is approximately proportional to the speed of the object; at higher speeds, it is more nearly proportional to the square of the speed.

Consider an object dropped from rest and falling under the influence of the force of gravity, which we assume to be constant. Now add a drag force of magnitude bv^n, where b is a constant that depends on the shape of the object and the properties of the air, and the exponent n is approximately 1 at low speeds and approximately 2 at higher speeds. We then have a constant downward force mg and an upward force bv^n (Figure 5-36). If we take the downward direction to be positive, we obtain from Newton's second law

$$\sum F_y = mg - bv^n = ma_y \qquad 5\text{-}7$$

At $t = 0$, the instant when the object is dropped, the speed is zero, so the retarding force is zero and the acceleration is g downward. As the speed of the object increases, the drag force increases and the acceleration becomes less than g. Eventually, the speed is great enough for the drag force bv^n to equal the force of gravity mg, at which point the acceleration is zero. The object then continues moving at a constant speed v_t, called its **terminal speed.** Setting the acceleration a in Equation 5-7 equal to zero, we obtain

$$bv_t^n = mg$$

Solving for the terminal speed, we get

$$v_t = \left(\frac{mg}{b}\right)^{1/n} \qquad 5\text{-}8$$

The larger the constant b, the lower the terminal speed. A parachute is designed to maximize b so that the terminal speed will be small. Cars, on the other hand, are designed to minimize b to reduce the effect of wind resistance.

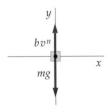

Figure 5-36 A free-body diagram showing forces on an object falling with air resistance.

exploring

Numerical Methods: Euler's Method

If a particle moves under the influence of a *constant* force, its acceleration is constant and we can find its velocity and position from the constant-acceleration formulas in Chapter 2. But consider a particle moving through space where the force on it, and therefore its acceleration, depends on its position and velocity. The velocity and acceleration of the particle at one instant determine its position and velocity at the next instant, which then determines its acceleration at that instant. The actual position, velocity, and acceleration of an object all change continuously with time. We can approximate this by replacing the continuous time variations with small time steps of duration Δt. The simplest approximation is to assume constant acceleration during each step. This approximation is called **Euler's method**. If the time interval is sufficiently short, the change in acceleration during the interval will be small and can be neglected.

Let x_0, v_0, and a_0 be the known position, velocity, and acceleration of a particle at some initial time t_0. If we assume constant acceleration during Δt, velocity at time $t_1 = t_0 + \Delta t$ is given by

$$v_1 = v_0 + a\,\Delta t \qquad\qquad 1$$

Similarly, if we neglect the change in velocity during the time interval, the new position is given by

$$x_1 = x_0 + v_0\Delta t \qquad\qquad 2$$

(Other methods of numerical integration are more accurate but less simple to use. For example, the accuracy is improved if a and v are computed at the midpoint of the interval rather than at the beginning.)

We can use the values v_1 and x_1 to compute the new acceleration a_1 from Newton's second law, and then use a_1 for the next time interval to compute v_2 and x_2.

$$v_2 = v_1 + a_1\Delta t$$

$$x_2 = x_1 + v_1\Delta t$$

In general, the connection between the position and velocity at time t_n and time $t_{n+1} = t_n + \Delta t$ is given by

$$v_{n+1} = v_n + a_n\Delta t \qquad\qquad 3$$

and

$$x_{n+1} = x_n + v_n\Delta t \qquad\qquad 4$$

To find the velocity and position at some time t, we therefore divide the time interval $t - t_0$ into a large number of smaller intervals Δt and apply Equations 3 and 4, beginning at the initial time t_0. This involves a large number of simple, repetitive calculations that are easily done on a computer. The technique of breaking the time interval into small steps and computing the acceleration, velocity, and position at each step using the values from the previous step is called numerical integration.

Drag Forces

To illustrate the use of numerical methods, let us consider a problem in which a sky diver is dropped from rest at some height under the influences of gravity and a drag force that is proportional to the square of the speed. We will find the velocity v and the distance traveled x as functions of time.

The equation describing the motion of an object of mass m dropped from rest is Equation 5-7 with $n = 2$:

$$\sum F_y = mg - bv^n = ma_y$$

The acceleration is thus

$$a = g - \left(\frac{b}{m}\right)v^2 \qquad\qquad 5$$

It is convenient to write the constant b/m in terms of the terminal speed v_t. Setting $a = 0$ in Equation 5 we obtain

$$0 = g - \left(\frac{b}{m}\right)v_t^2$$

$$\frac{b}{m} = \frac{g}{v_t^2}$$

Substituting g/v_t^2 for b/m in Equation 5 gives

$$a = g\left(1 - \frac{v^2}{v_t^2}\right) \qquad\qquad 6$$

To solve Equation 6 numerically, we need numerical values for g and v_t. A reasonable terminal speed for a sky diver is 60 m/s. Using this and $g = 9.81$ m/s^2, we obtain

$$a = 9.81\left(1 - \frac{v^2}{3600}\right) \qquad\qquad 7$$

We have omitted the units in this equation. Since we are using SI units, the unit for v is meters per second, and the unit for x is meters. If we choose $x_0 = 0$ for the initial position, the initial values are $x_0 = 0$, $v_0 = 0$, and $a_0 = g = 9.81$. To find the velocity v and position x after some time, say $t = 20$ s, we divide the time interval $0 < t < 20$ s into many small intervals Δt and apply Equations 3 and 4. We do this by writing a computer program or by using a computer spreadsheet. Figure 1 shows graphs of v versus t and x versus t based on data found using a spreadsheet with $\Delta t = 0.5$ s. At $t = 20$ s, the computed values are $v = 59.97$ m/s and $x = 957.5$ m.

But how accurate are our computations? We can estimate the accuracy by running the program again using a smaller time interval. If we use $\Delta t = 0.25$ s, one-half of the value we originally used, we obtain $v = 59.92$ m/s and $x = 952.0$ m at $t = 20$ s. The difference in v is about 0.1% and that in x is about 0.5%. These are our estimates of the accuracy of the original computations.

Since the difference between the value of a_{av} for some time interval Δt and the value of a_i at the beginning of the interval becomes smaller as the time interval becomes smaller, we might expect that it would be better to use very small time intervals, say $\Delta t = 0.000\ 000\ 001$ s. But there are two reasons for not using very small time intervals. First, the smaller the time interval, the larger the number of calculations required, and the longer the program takes to run. Second, the computer keeps only a fixed number of digits at each step of the calculation, so that at each step there is a round-off error. These round-off errors add up. The larger the number of calculations, the more significant the total round off error becomes. When we first decrease the time interval, the accuracy improves because a_i more nearly approximates a_{av} for the interval. However, as the time interval is decreased further, the round-off errors build up and the accuracy of the computation decreases. A good rule of thumb to follow is to use no more than about 10^4 or 10^5 time intervals for a typical numerical integration.

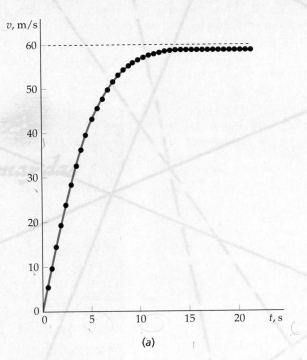

(a)

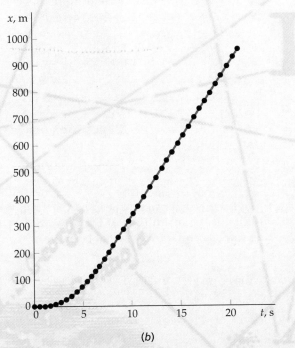

(b)

Figure 1 (a) Graph of v versus t for a sky diver found by numerical integration using $\Delta t = 0.5$ s. The horizontal dashed line is the terminal speed $v_t = 60$ m/s. (b) Graph of x versus t using $\Delta t = 0.5$ s.

The terminal speed of a sky diver before release of the parachute is about 60 m/s, or ≈ 200 km/h. When the parachute is opened, the drag force shoots up to become temporarily greater than the force of gravity, and the sky diver experiences an upward acceleration while falling; that is, the downward speed of the sky diver decreases. As the speed of the sky diver drops, the drag force decreases, until a new terminal speed, about 20 km/h, is reached.

A golf ball and Styrofoam ball falling in air. The air resistance is negligible for the heavier golf ball, which falls with essentially constant acceleration. The Styrofoam ball reaches terminal speed quickly, as indicated by the nearly equal spacing at the bottom.

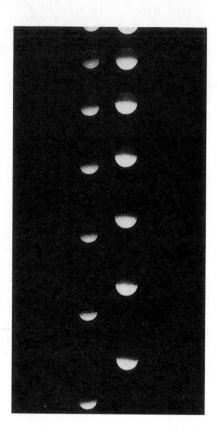

Example 5-13

A sky diver of mass 64 kg reaches a terminal speed of 180 km/h with her arms and legs outspread. (*a*) What is the magnitude of the upward drag force on the sky diver? (*b*) If the drag force is equal to bv^2, what is the value of b?

(*a*)1. Draw a free-body diagram:

Figure 5-37

2. Apply $\Sigma \vec{F} = m\vec{a}$. Since the sky diver is moving with constant velocity, the acceleration is zero:

$$\sum F_y = ma_y$$

$$mg - F_d = 0$$

$$F_d = mg = (64 \text{ kg})(9.81 \text{ N/kg}) = 628 \text{ N}$$

(*b*) To find b we set $F_d = bv^2$:

$$F_d = mg = bv^2$$

$$b = \frac{mg}{v^2} = 0.251 \text{ N·s}^2/\text{m}^2 = 0.251 \text{ kg/m}$$

Summary

1. The acceleration due to circular motion ($a = v^2/r$) is an important *derived* result in kinematics.
2. Friction and drag forces are complex phenomena empirically approximated by simple equations.

Topic	Remarks and Relevant Equations
1. Friction	Two objects in contact exert frictional forces on each other. These forces are parallel to the surfaces of the objects at the points of contact and directed opposite to the direction of sliding or tendency to slide.
Static friction	$$f_s \leq \mu_s F_n \qquad \text{5-2}$$ where F_n is the normal force of contact and μ_s is the coefficient of static friction.
Kinetic friction	$$f_k \leq \mu_k F_n \qquad \text{5-3}$$ where μ_k is the coefficient of kinetic friction. The coefficient of kinetic friction is slightly less than the coefficient of static friction.
2. Circular Motion	When an object moves in a circle with constant speed, it is accelerating, because its velocity is changing in direction. The acceleration is called centripetal acceleration and is directed toward the center of the circle. There must be a net inward force to provide the centripetal acceleration.
Centripetal acceleration	$$a = \frac{v^2}{r} \qquad \text{5-5}$$
Speed and period	$$v = \frac{2\pi r}{T} \qquad \text{5-6}$$
3. General Motion Along a Curve in Space	A particle moving along an arbitrary curve can be considered to be moving in a circular arc during a small time interval. Its instantaneous acceleration vector has a component $a_r = v^2/r$ toward the center of curvature of the arc and a component $a_t = dv/dt$ that is tangential to the curve.
4. Drag Forces (optional)	When an object moves through a fluid, it experiences a drag force that opposes its motion. The drag force increases with increasing speed. If the body is dropped from rest, its speed increases until the drag force equals the force of gravity, after which it moves with a constant speed called its terminal speed. The terminal speed depends on the shape of the body and on the medium through which it falls.

Problem-Solving Guide

1. Begin by drawing a neat diagram that includes the important features of the problem.
2. Isolate the object of interest and draw a free-body diagram that shows how each external force acts on the object. If there is more than one object of interest in the problem, draw a separate free-body diagram for each object.
3. Choose a convenient coordinate system for each object and apply Newton's second law, $\vec{F}_{net} = m\vec{a}$, in component form.
4. Solve the resulting equations for the unknowns.
5. Check to see whether your results are reasonable.

Summary of Worked Examples

Type of Calculation	Procedure and Relevant Examples
1. **Problems With Friction**	Remember that the force of static friction may not be equal to its limiting value of $\mu_s F_n$ and that the normal force between the surfaces of two objects is not necessarily equal to the weight of the upper object. **Example 5-3**
Find the stopping distance of an object.	Apply $\Sigma\vec{F} = m\vec{a}$ in component form to the object. If the object is sliding, the frictional force on it is $\mu_k F_n$, where F_n is the normal force. If the friction is static (as can be the case in braking a car), the maximum frictional force is $\mu_s F_n$. In both cases, the force is constant, so the constant acceleration equations of Chapter 2 can be used. **Examples 5-1, 5-4**
Determine how an object on an inclined plane will move.	Apply $\Sigma\vec{F} = m\vec{a}$ in component form to the object. The frictional force is $f_s \leq \mu_s F_n$ if the object is static and $\mu_k F_n$ if it is sliding, where F_n is the normal force. **Example 5-2**
Determine if friction is static or kinetic and find the force of static friction.	Determine what the frictional force would have to be to prevent sliding. If this value is greater than $\mu_s F_n$, sliding will occur. Apply $\Sigma\vec{F} = m\vec{a}$ in component form to the object. **Examples 5-3, 5-4, 5-10**
Find the acceleration of one or more objects, taking into account various forces, including friction.	Determine if the frictional force is static or kinetic. Apply $\Sigma\vec{F} = m\vec{a}$ in component form to each object. **Examples 5-3, 5-4, 5-5, 5-6**
2. **Circular Motion**	Apply $\Sigma\vec{F} = m\vec{a}$ in component form to the object. When an object is moving in a circle, resolve the forces acting on the object into radial and tangential components. The inward radial component of the resultant force is the centripetal force, which equals mv^2/r. The tangential component of the resultant force equals $m\,dv/dt$.
Find the period of an object undergoing circular motion at constant speed.	Apply $\Sigma\vec{F} = m\vec{a}$ in component form, using v^2/r for the acceleration. Then use $T = 2\pi r/v$. **Example 5-8**
Find the forces on an object in circular motion.	Apply $\Sigma\vec{F} = m\vec{a}$. Resolve the forces into radial and tangential components. Set the net inward radial component equal to $ma = mv^2/r$. **Examples 5-9, 5-10, 5-11**
Find the forces on a car traveling on a banked curve. (optional)	Apply $\Sigma\vec{F} = m\vec{a}$. The normal force has an inward component $F_n \sin\theta$. If there is a frictional force, its radial component is $\pm f\cos\theta$, depending on whether the car is sliding or tending to slide up or down the bank. **Example 5-12**
3. **Drag Force and Terminal Speed** (optional)	Set $F_d = mg$ for the drag force on an object falling at terminal speed. **Example 5-13**

Problems

In a few problems, you are given more data than you actually need; in a few other problems, you are required to supply data from your general knowledge, outside sources, or informed estimates.

- Single-concept, single-step, relatively easy
- •• Intermediate-level, may require synthesis of concepts
- ••• Challenging, for advanced students

Friction

1 • Various objects lie on the floor of a truck moving along a horizontal road. If the truck accelerates, what force acts on the objects to cause them to accelerate?

2 • Any object resting on the floor of a truck will slide if the truck's acceleration is too great. How does the critical acceleration at which a light object slips compare with that at which a much heavier object slips?

3. • True or false:

(a) The force of static friction always equals $\mu_s F_n$.
(b) The force of friction always opposes the motion of an object.
(c) The force of friction always opposes sliding.
(d) The force of kinetic friction always equals $\mu_k F_n$.

4 • A block of mass m rests on a plane inclined at an angle θ with the horizontal. It follows that the coefficient of static friction between the block and plane is

(a) $\mu_s \geq g$.
(b) $\mu_s = \tan \theta$.
(c) $\mu_s \leq \tan \theta$.
(d) $\mu_s \geq \tan \theta$.

5 • A block of mass m is at rest on a plane inclined at an angle of 30° with the horizontal, as in Figure 5-38. Which of the following statements about the force of static friction is true?

(a) $f_s > mg$
(b) $f_s > mg \cos 30°$
(c) $f_s = mg \cos 30°$
(d) $f_s = mg \sin 30°$
(e) None of these statements is true.

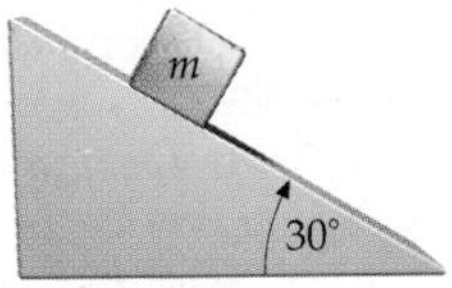

Figure 5-38 Problem 5

6 • A block of mass m slides at constant speed down a plane inclined at an angle θ with the horizontal. It follows that

(a) $\mu_k = mg \sin \theta$.
(b) $\mu_k = \tan \theta$.
(c) $\mu_k = 1 - \cos \theta$.
(d) $\mu_k = \cos \theta - \sin \theta$.

7 • A block of wood is pulled by a horizontal string across a horizontal surface at a constant velocity with a force of 20 N. The coefficient of kinetic friction between the surfaces is 0.3. The force of friction is

(a) impossible to determine without knowing the mass of the block.
(b) impossible to determine without knowing the speed of the block.
(c) 0.3 N.
(d) 6 N.
(e) 20 N.

8 • A 20-N block rests on a horizontal surface. The coefficients of static and kinetic friction between the surface and the block are $\mu_s = 0.8$ and $\mu_k = 0.6$. A horizontal string is attached to the block and a constant tension T is maintained in the string. What is the force of friction acting on the block if (a) $T = 15$ N or (b) $T = 20$ N.

9 • A block of mass m is pulled at a constant velocity across a horizontal surface by a string as in Figure 5-39. The magnitude of the frictional force is

(a) $\mu_k mg$.
(b) $T \cos \theta$.
(c) $\mu_k(T - mg)$.
(d) $\mu_k T \sin \theta$.
(e) $\mu_k(mg + T \sin \theta)$.

Figure 5-39
Problem 9

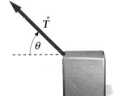

10 • A tired worker pushes with a horizontal force of 500 N on a 100-kg crate resting on a thick pile carpet. The coefficients of static and kinetic friction are 0.6 and 0.4, respectively. Find the frictional force exerted by the surface.

11 • A box weighing 600 N is pushed along a horizontal floor at constant velocity with a force of 250 N parallel to the floor. What is the coefficient of kinetic friction between the box and the floor?

12 • The coefficient of static friction between the tires of a car and a horizontal road is $\mu_s = 0.6$. If the net force on the car is the force of static friction exerted by the road, (a) what is the maximum acceleration of the car when it is braked? (b) What is the least distance in which the car can stop if it is initially traveling at 30 m/s?

13 • The force that accelerates a car along a flat road is the frictional force exerted by the road on the car's tires. (a) Explain why the acceleration can be greater when the wheels do not spin. (b) If a car is to accelerate from 0 to 90 km/h in 12 s at constant acceleration, what is the minimum coefficient of friction needed between the road and tires? Assume that half the weight of the car is supported by the drive wheels.

14 • On the current tour of the rock band Dead Wait, the show opens with a dark stage. Suddenly there is the sound of a large automobile accident. Lead singer Sharika comes sliding to the front of the stage on her knees. Her initial speed is 3 m/s. After sliding 2 m, she comes to rest in a dry ice fog as flash pots explode on either side. What is the coefficient of kinetic friction between Sharika and the stage?

15 • A 5-kg block is held at rest against a vertical wall by a horizontal force of 100 N. (a) What is the frictional force exerted by the wall on the block? (b) What is the minimum horizontal force needed to prevent the block from falling if the coefficient of friction between the wall and the block is $\mu_s = 0.40$?

16 • On a snowy day with the temperature near the freezing point, the coefficient of static friction between a car's tires and an icy road is 0.08. What is the maximum incline that this four-wheel-drive vehicle can climb with zero acceleration?

17 • A 50-kg box that is resting on a level floor must be moved. The coefficient of static friction between the box and the floor is 0.6. One way to move the box is to push down on it at an angle θ with the horizontal. Another method is to pull up on the box at an angle θ with the horizontal. (a) Explain why one method is better than the other. (b) Calculate the force necessary to move the box by each method if $\theta = 30°$ and compare the answer with the results when $\theta = 0°$.

18 • A 3-kg box resting on a horizontal shelf is attached to a 2-kg box by a light string as in Figure 5-40. (a) What is the minimum coefficient of static friction such that the objects remain at rest? (b) If the coefficient of static friction is less than that found in part (a), and the coefficient of kinetic friction between the box and the shelf is 0.3, find the time for the 2-kg mass to fall 2 m to the floor if the system starts from rest.

Figure 5-40
Problem 18

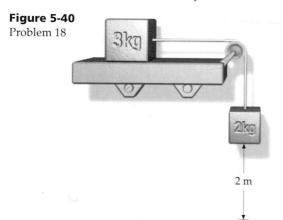

19 •• A block on a horizontal plane is given an initial velocity v. It comes to rest after a displacement d. The coefficient of kinetic friction between the block and the plane is given by

(a) $\mu_k = v^2 d / 2g$.
(b) $\mu_k = v^2 / 2dg$.
(c) $\mu_k = v^2 g / d^2$.
(d) none of the above.

20 •• A block of mass $m_1 = 250$ g is at rest on a plane that makes an angle $\theta = 30°$ above the horizontal (Figure 5-41). The coefficient of kinetic friction between the block and the plane is $\mu_k = 0.100$. The block is attached to a second block of mass $m_2 = 200$ g that hangs freely by a string that passes over a frictionless and massless pulley. When the second block has fallen 30.0 cm, its speed is

(a) 83 cm/s.
(b) 48 cm/s.
(c) 160 cm/s.
(d) 59 cm/s.
(e) 72 cm/s.

Figure 5-41 Problems 20–22

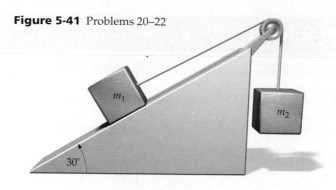

21 •• Returning to Figure 5-41, this time $m_1 = 4$ kg. The coefficient of static friction between the block and the incline is 0.4. (a) Find the range of possible values for m_2 for which the system will be in static equilibrium. (b) What is the frictional force on the 4-kg block if $m_1 = 1$ kg?

22 •• Returning once again to Figure 5-41, this time $m_1 = 4$ kg, $m_2 = 5$ kg, and the coefficient of kinetic friction between the inclined plane and the 4-kg block is $\mu_k = 0.24$. Find the acceleration of the masses and the tension in the cord.

23 •• The coefficient of static friction between the bed of a truck and a box resting on it is 0.30. The truck is traveling at 80 km/h along a horizontal road. What is the least distance in which the truck can stop if the box is not to slide?

24 •• A 4.5-kg mass is given an initial velocity of 14 m/s up an incline that makes an angle of 37° with the horizontal. When its displacement is 8.0 m, its upward velocity has diminished to 5.2 m/s. Find (a) the coefficient of kinetic friction between the mass and plane, (b) the displacement of the mass from its starting point at the time when it momentarily comes to rest, and (c) the speed of the block when it again reaches its initial position.

25 •• An automobile is going up a grade of 15° at a speed of 30 m/s. The coefficient of static friction between the tires and the road is 0.7. (a) What minimum distance does it take to stop the car? (b) What minimum distance would it take if the car were going down the grade?

26 •• A block of mass m slides with initial speed v_0 on a horizontal surface. If the coefficient of kinetic friction between the block and the surface is μ_k, find the distance d that the block moves before coming to rest.

27 •• A rear-wheel-drive car supports 40% of its weight on its two drive wheels and has a coefficient of static friction of 0.7. (a) What is the vehicle's maximum acceleration? (b) What is the shortest possible time in which this car can achieve a speed of 100 km/h? (Assume that the engine has unlimited power.)

28 •• Lou bets an innocent stranger that he can place a 2-kg block against the side of a cart, as in Figure 5-42, and that the block will not fall to the ground, even though Lou will use no hooks, ropes, fasteners, magnets, glue, or adhesives of any kind. When the stranger accepts the bet, Lou begins to push the cart in the direction shown. The coefficient of static friction between the block and the cart is 0.6. (a) Find the minimum acceleration for which Lou will win the bet. (b) What is the magnitude of the frictional force in this case?

(c) Find the force of friction on the block if a is twice the minimum needed for the block not to fall. (d) Show that, for a block of any mass, the block will not fall if the acceleration is $a \geq g/\mu_s$, where μ_s is the coefficient of static friction.

Figure 5-42 Problem 28

29 •• Two blocks attached by a string slide down a $20°$ incline. The lower block has a mass of $m_1 = 0.25$ kg and a coefficient of kinetic friction $\mu_k = 0.2$. For the upper block, $m_2 = 0.8$ kg and $\mu_k = 0.3$. Find (a) the acceleration of the blocks and (b) the tension in the string.

30 •• Two blocks attached by a string are at rest on an inclined surface. The lower block has a mass of $m_1 = 0.2$ kg and a coefficient of static friction $\mu_s = 0.4$. The upper block has a mass $m_2 = 0.1$ kg and $\mu_s = 0.6$. (a) At what angle θ_c do the blocks begin to slide? (b) What is the tension in the string just before sliding begins?

31 •• Two blocks connected by a massless, rigid rod slide on a surface inclined at an angle of $20°$. The lower block has a mass $m_1 = 1.2$ kg, and the upper block's mass is $m_2 = 0.75$ kg. (a) If the coefficients of kinetic friction are $\mu_k = 0.3$ for the lower block and $\mu_k = 0.2$ for the upper block, what is the acceleration of the blocks? (b) Determine the force transmitted by the rod.

32 •• A block of mass m rests on a horizontal surface (Figure 5-43). The block is pulled by a massless rope with a force $\vec{F}$ at an angle θ. The coefficient of static friction is 0.6. The minimum value of the force needed to move the block depends on the angle θ. (a) Discuss qualitatively how you would expect this force to depend on θ. (b) Compute the force for the angles $\theta = 0°, 10°, 20°, 30°, 40°, 50°,$ and $60°$, and make a plot of F versus θ for $mg = 400$ N. From your plot, at what angle is it most efficient to apply the force to move the block?

Figure 5-43 Problem 32

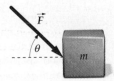

33 •• Answer the same questions as in Problem 32, only this time with a force $\vec{F}$ that pushes down on the block in Figure 5-44 at an angle θ with the horizontal.

Figure 5-44 Problem 33

34 •• A 100-kg mass is pulled along a frictionless surface by a horizontal force $\vec{F}$ such that its acceleration is 6 m/s² (Figure 5-45). A 20-kg mass slides along the top of the 100-kg mass and has an acceleration of 4 m/s². (It thus slides backward relative to the 100-kg mass.) (a) What is the frictional force exerted by the 100-kg mass on the 20-kg mass? (b) What is the net force acting on the 100-kg mass? What is the force F? (c) After the 20-kg mass falls off the 100-kg mass, what is the acceleration of the 100-kg mass? (Assume that the force F does not change.)

Figure 5-45 Problem 34

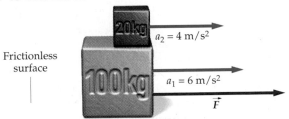

Frictionless surface

35 •• A 60-kg block slides along the top of a 100-kg block with an acceleration of 3 m/s² when a horizontal force $\vec{F}$ of 320 N is applied, as in Figure 5-46. The 100-kg block sits on a horizontal frictionless surface, but there is friction between the two blocks. (a) Find the coefficient of kinetic friction between the blocks. (b) Find the acceleration of the 100-kg block during the time that the 60-kg block remains in contact.

Figure 5-46 Problem 35

36 •• The coefficient of static friction between a rubber tire and the road surface is 0.85. What is the maximum acceleration of a 1000-kg four-wheel-drive truck if the road makes an angle of $12°$ with the horizontal and the truck is (a) climbing, and (b) descending?

37 •• A 2-kg block sits on a 4-kg block that is on a frictionless table (Figure 5-47). The coefficients of friction between the blocks are $\mu_s = 0.3$ and $\mu_k = 0.2$. (a) What is the maximum force F that can be applied to the 4-kg block if the 2-kg block is not to slide? (b) If F is half this value, find the acceleration of each block and the force of friction acting on each block. (c) If F is twice the value found in (a), find the acceleration of each block.

Figure 5-47 Problem 37

38 •• In Figure 5-48, the mass $m_2 = 10$ kg slides on a frictionless shelf. The coefficients of static and kinetic friction between m_2 and $m_1 = 5$ kg are $\mu_s = 0.6$ and $\mu_k = 0.4$. (a) What is the maximum acceleration of m_1? (b) What is the maximum value of m_3 if m_1 moves with m_2 without slipping? (c) If $m_3 = 30$ kg, find the acceleration of each body and the tension in the string.

Figure 5-48 Problem 38

39 ••• A box of mass m rests on a horizontal table. The coefficient of static friction is μ_s. A force $\vec{F}$ is applied at an angle θ as shown in Problem 5-32. (a) Find the force F needed to move the box as a function of angle θ. (b) At the angle θ for which this force is minimum, the slope $dF/d\theta$ of the curve F versus θ is zero. Compute $dF/d\theta$ and show that this derivative is zero at the angle θ that obeys $\tan\theta = \mu_s$. Compare this general result with that obtained in Problem 5-32.

40 ••• A 10-kg block rests on a 5-kg bracket like the one in Figure 5-49. The 5-kg bracket sits on a frictionless surface. The coefficients of friction between the 10-kg block and the bracket on which it rests are $\mu_s = 0.40$ and $\mu_k = 0.30$. (a) What is the maximum force F that can be applied if the 10-kg block is not to slide on the bracket? (b) What is the corresponding acceleration of the 5-kg bracket?

Figure 5-49 Problem 40

41 ••• Lou has set up a kiddie ride at the Winter Ice Fair. He builds a right-angle triangular wedge, which he intends to push along the ice with a child sitting on the hypotenuse. If he pushes too hard, the kid will slide up and over the top, and Lou could be looking at a lawsuit. If he doesn't push hard enough, the kid will slide down the wedge, and the parents will want their money back. If the angle of inclination of the wedge is 40°, what are the minimum and maximum values for the acceleration that Lou must achieve? Use m for the child's mass, and μ_s for the coefficient of static friction between the child and the wedge.

Figure 5-50 Problem 42

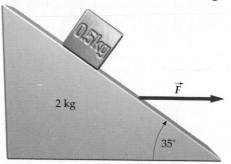

42 ••• A block of mass 0.5 kg rests on the inclined surface of a wedge of mass 2 kg, as in Figure 5-50. The wedge is acted on by a horizontal force $\vec{F}$ and slides on a frictionless surface. (a) If the coefficient of static friction between the wedge and the block is $\mu_s = 0.8$, and the angle of the incline is 35°, find the maximum and minimum values of F for which the block does not slip. (b) Repeat part (a) with $\mu_s = 0.4$.

Circular Motion

43   True or false: An object cannot move in a circle unless there is a net force acting on it.

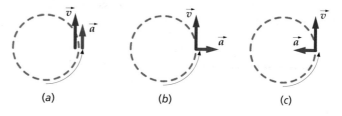

(a)   (b)   (c)

Figure 5-51 Problem 44

44 • An object moves in a circle counterclockwise with constant speed (Figure 5-51). Which figure shows the correct velocity and acceleration vectors?

(d)

45 • A particle is traveling in a vertical circle at constant speed. One can conclude that the _____ is constant.

(a) velocity   (b) acceleration   (c) net force
(d) apparent weight   (e) None of the above.

46 • An object travels with a constant speed v in a circular path of radius r. (a) If v is doubled, how is the acceleration a affected? (b) If r is doubled, how is a affected? (c) Why is it impossible for an object to travel around a perfectly sharp angular turn?

47 • A boy whirls a ball on a string in a horizontal circle of radius 0.8 m. How many revolutions per minute must the ball make if the magnitude of its centripetal acceleration is to be the same as the free-fall acceleration due to gravity g?

48 • A 0.20-kg stone attached to a 0.8-m long string is rotated in the horizontal plane. The string makes an angle of 20° with the horizontal. Determine the speed of the stone.

49 • A 0.75-kg stone attached to a string is whirled in a horizontal circle of radius 35 cm as in the conical pendulum of Example 5-10. The string makes an angle of 30° with the vertical. (*a*) Find the speed of the stone. (*b*) Find the tension in the string.

50 •• A stone with a mass $m = 95$ g is being whirled in a horizontal circle on the end of a string that is 85 cm long. The length of time required for the stone to make one complete revolution is 1.22 s. The angle that the string makes with the horizontal is ———— .

(*a*) 52°   (*b*) 46°   (*c*) 26°   (*d*) 23°   (*e*) 3°

51 •• A pilot of mass 50 kg comes out of a vertical dive in a circular arc such that her upward acceleration is 8.5*g*. (*a*) What is the magnitude of the force exerted by the airplane seat on the pilot at the bottom of the arc? (*b*) If the speed of the plane is 345 km/h, what is the radius of the circular arc?

52 •• A 65-kg airplane pilot pulls out of a dive by following the arc of a circle whose radius is 300 m. At the bottom of the circle, where her speed is 180 km/h, (*a*) what are the direction and magnitude of her acceleration? (*b*) What is the net force acting on her at the bottom of the circle? (*c*) What is the force exerted on the pilot by the airplane seat?

53 •• Mass m_1 moves with speed *v* in a circular path of radius *R* on a frictionless horizontal table (Figure 5-52). It is attached to a string that passes through a frictionless hole in the center of the table. A second mass m_2 is attached to the other end of the string. Derive an expression for *R* in terms of m_1, m_2, and *v*.

Figure 5-52 Problem 53

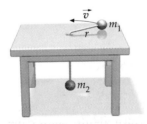

54 •• In Figure 5-53 particles are shown traveling counterclockwise in circles of radius 5 m. The acceleration vectors are indicated at three specific times. Find the values of *v* and *dv/dt* for each of these times.

Figure 5-53 Problem 54

55 •• A block of mass m_1 is attached to a cord of length L_1, which is fixed at one end. The block moves in a horizontal circle on a frictionless table. A second block of mass m_2 is attached to the first by a cord of length L_2 and also moves in a circle, as shown in Figure 5-54. If the period of the motion is *T*, find the tension in each cord.

Figure 5-54 Problem 55

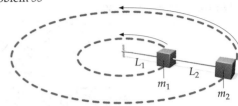

56 •• A particle moves with constant speed in a circle of radius 4 cm. It takes 8 s to make a complete trip. Draw the path of the particle to scale, and indicate the particle's position at 1-s intervals. Draw displacement vectors for each interval. These vectors also indicate the directions for the average-velocity vectors for each interval. Find graphically the change in the average velocity $\Delta \vec{v}$ for two consecutive 1-s intervals. Compare $\Delta \vec{v}/\Delta t$, measured in this way, with the instantaneous acceleration computed from $a = v^2/r$.

57 •• A man swings his child in a circle of radius 0.75 m, as shown in the photo. If the mass of the child is 25 kg and the child makes one revolution in 1.5 s, what are the magnitude and direction of the force that must be exerted by the man on the child? (Assume the child to be a point particle.)

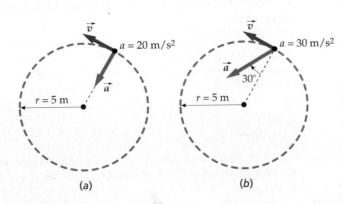

(*a*)   (*b*)   (*c*)

58 •• The string of a conical pendulum is 50 cm long and the mass of the bob is 0.25 kg. Find the angle between the string and the horizontal when the tension in the string is six times the weight of the bob. Under those conditions, what is the period of the pendulum?

59 •• Frustrated with his inability to make a living through honest channels, Lou sets up a deceptive weight-loss scam. The trick is to make insecure customers believe that they can "think those extra pounds away" if they will only take a ride in a van that Lou claims to be "specially equipped to enhance mental-mass fluidity." The customer sits on a platform scale in the back of the van, and Lou drives off at a constant speed of 14 m/s. Lou then asks the customer to "think heavy" as he drives through the bottom of a dip in the road having a radius of curvature of 80 m. Sure enough, the scale's reading increases, until Lou says, "Now think light," and drives over the crest of a hill having a radius of curvature of 100 m. If the scale reads 800 N when the van is on level ground, what is the range of readings for the trip described here?

60 •• A 100-g disk sits on a horizontally rotating turntable. The turntable makes one revolution each second. The disk is located 10 cm from the axis of rotation of the turntable. (a) What is the frictional force acting on the disk? (b) The disk will slide off the turntable if it is located at a radius larger than 16 cm from the axis of rotation. What is the coefficient of static friction?

61 •• A tether ball of mass 0.25 kg is attached to a vertical pole by a cord 1.2 m long. Assume the cord attaches to the center of the ball. If the cord makes an angle of 20° with the vertical, then (a) what is the tension in the cord? (b) What is the speed of the ball?

62 •• An object on the equator has an acceleration toward the center of the earth because of the earth's rotation and an acceleration toward the sun because of the earth's motion along its orbit. Calculate the magnitudes of both accelerations, and express them as fractions of the free-fall acceleration due to gravity g.

63 •• A small bead with a mass of 100 g slides along a semicircular wire with a radius of 10 cm that rotates about a vertical axis at a rate of 2 revolutions per second, as in Figure 5-55. Find the values of θ for which the bead will remain stationary relative to the rotating wire.

64 ••• Consider a bead of mass m that is free to move on a thin, circular wire of radius r. The bead is given an initial speed v_0, and there is a coefficient of kinetic friction μ_k. The experiment is performed in a

spacecraft drifting in space. Find the speed of the bead at any subsequent time t.

65 ••• Revisiting the previous problem, (a) find the centripetal acceleration of the bead. (b) Find the tangential acceleration of the bead. (c) What is the magnitude of the resultant acceleration?

Loop-the-Loop

66 • A block is sliding on a frictionless surface along a loop-the-loop, as in Figure 5-56. The block is moving fast enough that it never loses contact with the track. Match the points along the track to the appropriate free-body diagrams (Figure 5-57).

Figure 5-55
Problem 63

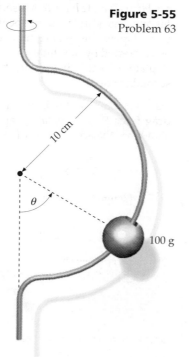

10 cm

θ

100 g

Figure 5-56
Problem 66

D

A   C

B

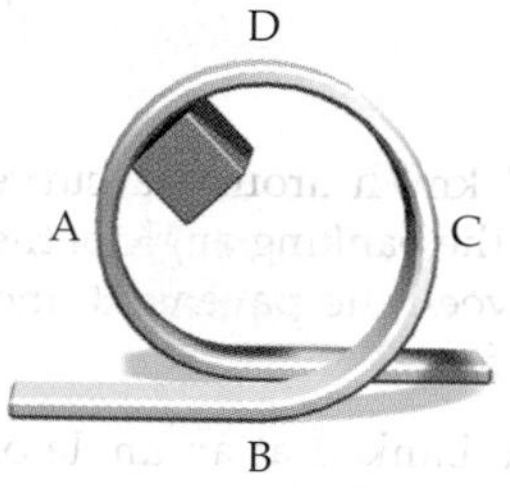

Figure 5-57
Problem 66

Point A
Point B
Point C
Point D

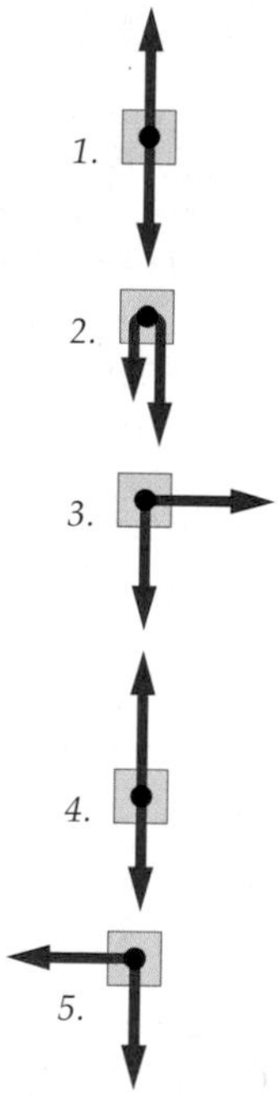

1.

2.

3.

4.

5.

67 • A person rides a loop-the-loop at an amusement park. The cart circles the track at a constant speed. At the top of the loop, the normal force exerted by the seat equals the person's weight, mg. At the bottom of the loop, the force exerted by the seat will be _____ .

(a) 0
(b) mg
(c) 2mg
(d) 3mg
(e) greater than mg, but the exact value cannot be calculated from the information given

68 • The radius of curvature of a loop-the-loop roller coaster is 12.0 m. At the top of the loop, the force that the seat exerts on a passenger of mass m is 0.4mg. Find the speed of the roller coaster at the top of the loop.

Unbanked Curves

69 • Realizing that he has left the gas stove on, Aaron races for his car to drive home. He lives at the other end of a long, unbanked curve in the highway, and he knows that when he is traveling alone in his car at 40 km/h, he can just make it around the curve without skidding. He yells at his friends, "Get in the car! With greater mass, I can take the curve at a higher speed!" Carl says, "No, that will make you skid at an even *lower* speed." Bonita says, "The mass doesn't matter. Just get going!" Who is right?

70 • A car speeds along the curved exit ramp of a free-way. The radius of the curve is 80 m. A 70-kg passenger holds the arm rest of the car door with a 220-N force to keep from sliding across the front seat of the car. (Assume that the exit ramp is not banked and ignore friction with the car seat.) What is the car's speed?

(a) 16 m/s
(b) 57 m/s
(c) 18 m/s
(d) 50 m/s
(e) 28 m/s

71 ••• Suppose you ride a bicycle on a horizontal surface in a circle with a radius of 20 m. The resultant force exerted by the road on the bicycle (normal force plus frictional force) makes an angle of 15° with the vertical. (a) What is your speed? (b) If the frictional force is half its maximum value, what is the coefficient of static friction?

Banked Curves (optional)

72 • A 750-kg car travels at 90 km/h around a curve with a radius of 160-m. What should the banking angle of the curve be so that the only force between the pavement and tires of the car is the normal reaction force?

73 •• A curve of radius 150 m is banked at an angle of 10°. An 800-kg car negotiates the curve at 85 km/h without skidding. Find (a) the normal force on the tires exerted by the pavement, (b) the frictional force exerted by the pavement on the tires of the car, and (c) the minimum coefficient of static friction between the pavement and tires.

74 •• On another occasion, the car in the previous problem negotiates the curve at 38 km/h. Find (a) the normal force exerted on the tires by the pavement, and (b) the frictional force exerted on the tires by the pavement.

75 ••• A civil engineer is asked to design a curved section of roadway that meets the following conditions: With ice on the road, when the coefficient of static friction between the road and rubber is 0.08, a car at rest must not slide into the ditch and a car traveling less than 60 km/h must not skid to the outside of the curve. What is the minimum radius of curvature of the curve and at what angle should the road be banked?

76 ••• A curve of radius 30 m is banked so that a 950-kg car traveling 40 km/h can round it even if the road is so icy that the coefficient of static friction is approximately zero. Find the range of speeds at which a car can travel around this curve without skidding if the coefficient of static friction between the road and the tires is 0.3.

Drag Forces (optional)

77 • How would you expect the value of b for air resistance to depend on the density of air?

78 • True or false: The terminal speed of an object depends on its shape.

79 • As a skydiver falls through the air, her terminal speed

(a) depends on her mass.
(b) depends on her orientation as she falls.
(c) equals her weight.
(d) depends on the density of the air.
(e) depends on all of the above.

80 • What are the dimensions and SI units of the constant b in the retarding force bv^n if (a) $n = 1$, and (b) $n = 2$?

81 • A small pollution particle settles toward the earth in still air with a terminal speed of 0.3 mm/s. The particle has a mass of 10^{-10} g and a retarding force of the form bv. What is the value of b?

82 • A Ping-Pong ball has a mass of 2.3 g and a terminal speed of 9 m/s. The retarding force is of the form bv^2. What is the value of b?

83 • A sky diver of mass 60 kg can slow herself to a constant speed of 90 km/h by adjusting her form. (a) What is the magnitude of the upward drag force on the sky diver? (b) If the drag force is equal to bv^2, what is the value of b?

84 • Newton showed that the air resistance on a falling object with a circular cross section should be approximately $\frac{1}{2}\rho\pi r^2 v^2$, where $\rho = 1.2$ kg/m³, the density of air. Find the terminal speed for a 56-kg sky diver, assuming that his cross-sectional area is equivalent to that of a disk of radius 0.30 m.

85 •• An 800-kg car rolls down a very long 6° grade. The drag force for motion of the car has the form $F_d = 100$ N + $(1.2$ N·s²/m²)v^2. What is the terminal velocity for the car rolling down this grade?

86 •• While claims of hailstones the size of golf balls may be a slight exaggeration, hailstones are often substantially larger than raindrops. Estimate the terminal velocity of a raindrop and a large hailstone. (See Problem 84.)

87 •• (a) A parachute creates enough air resistance to keep the downward speed of an 80-kg sky diver to a constant 6.0 m/s. Assuming that the force of air resistance is given by $f = bv^2$, calculate b for this case. (b) A sky diver free-falls until his speed is 60 m/s before opening his parachute. If the parachute opens instantaneously, calculate the initial upward force exerted by the chute on the sky diver moving at 60 m/s. Explain why it is important that the parachute takes a few seconds to open.

88 ••• An object falls under the influence of gravity and a drag force $F_d = -bv$. (a) By applying Newton's second law, show that the acceleration of the object can be written

$$a = \frac{dv}{dt} = g - \frac{b}{m}v$$

(b) Rearrange this equation to obtain

$$\frac{dv}{v - v_t} = -\frac{g}{v_t}dt$$

where $v_t = mg/b$.
(c) Integrate this equation to obtain the exact solution

$$v = \frac{mg}{b}\left(1 - e^{-bt/m}\right) = v_t\left(1 - e^{-gt/v_t}\right)$$

(d) Plot v versus t for $v_t = 60$ m/s.

89 ••• Small spherical particles experience a viscous drag force given by Stokes' law: $F_d = 6\pi\eta r v$, where r is the radius of the particle, v is its speed, and η is the viscosity of the fluid medium. (*a*) Estimate the terminal speed of a spherical pollution particle of radius 10^{-5} m and density of 2000 kg/m^3. (*b*) Assuming that the air is still and that η is 1.8×10^{-5} N·s/m^2, estimate the time it takes for such a particle to fall from a height of 100 m.

90 ••• An air sample containing pollution particles of the size and density given in Problem 89 is captured in a test tube 8.0 cm long. The test tube is then placed in a centrifuge with the midpoint of the test tube 12 cm from the center of the centrifuge. The centrifuge spins at 800 revolutions per minute. Estimate the time required for nearly all of the pollution particles to sediment at the end of the test tube and compare this to the time required for a pollution particle to fall 8.0 cm under the action of gravity and subject to the viscous drag of air.

General Problems

91 • The mass of the moon is about 1% that of the earth. The centripetal force that keeps the moon in its orbit around the earth

(*a*) is much smaller than the gravitational force exerted on the moon by the earth.

(*b*) depends on the phase of the moon.

(*c*) is much greater than the gravitational force exerted on the moon by the earth.

(*d*) is the same as the gravitational force exerted on the moon by the earth.

(*e*) I cannot answer; we haven't studied Newton's law of gravity yet.

92 • True or false: Centripetal force is one of the four fundamental forces.

93 • On an icy winter day, the coefficient of friction between the tires of a car and a roadway might be reduced to one-half of its value on a dry day. As a result, the maximum speed at which a curve of radius R can be safely negotiated is

(*a*) the same as on a dry day.

(*b*) reduced to 71% of its value on a dry day.

(*c*) reduced to 50% of its value on a dry day.

(*d*) reduced to 37% of its value on a dry day.

(*e*) reduced by an unknown amount depending on the car's mass.

94 • A 4.5-kg block slides down an inclined plane that makes an angle of 28° with the horizontal. Starting from rest, the block slides a distance of 2.4 m in 5.2 s. Find the coefficient of kinetic friction between the block and plane.

95 • A model airplane of mass 0.4 kg is attached to a horizontal string and flies in a horizontal circle of radius 5.7 m. (The weight of the plane is balanced by the upward "lift" force of the air on the wings of the plane.) The plane makes 1.2 revolutions every 4 s. (*a*) Find the speed v of the plane. (*b*) Find the tension in the string.

96 •• Show with a force diagram how a motorcycle can travel in a circle on the inside vertical wall of a hollow cylinder. Assume reasonable parameters (coefficient of friction, radius of the circle, mass of the motorcycle, or whatever is required), and calculate the minimum speed needed.

97 •• An 800-N box rests on a plane inclined at 30° to the horizontal. A physics student finds that she can prevent the box from sliding if she pushes on it with a force of at least 200 N parallel to the surface. (*a*) What is the coefficient of static friction between the box and the surface? (*b*) What is the greatest force that can be applied to the box parallel to the incline before the box slides up the incline?

98 •• The position of a particle is given by the vector $\vec{r} = -10$ m cos $\omega t \hat{i} + 10$ m sin $\omega t \hat{j}$, where $\omega = 2$ s^{-1}. (*a*) Show that the path of the particle is a circle. (*b*) What is the radius of the circle? (*c*) Does the particle move clockwise or counterclockwise around the circle? (*d*) What is the speed of the particle? (*e*) What is the time for one complete revolution?

99 •• A crate of books is to be put on a truck with the help of some planks sloping up at 30°. The mass of the crate is 100 kg, and the coefficient of sliding friction between it and the planks is 0.5. You and your friends push *horizontally* with a force $\vec{F}$. Once the crate has started to move, how large must F be to keep the crate moving at constant speed?

100 •• Brother Bernard is a very large dog with a taste for tobogganing. Ernie gives him a ride down Idiots' Hill—so named because it is a steep slope that levels out at the bottom for 10 m, and then drops into a river. When they reach the level ground at the bottom, their speed is 40 km/h, and Ernie, sitting in front, starts to dig in his heels to make the toboggan stop. He knows, however, that if he brakes too hard, he will be mashed by Brother Bernard. If the coefficient of static friction between the dog and the toboggan is 0.8, what is the minimum stopping distance that will keep Brother Bernard off Ernie's back?

101 •• An object with a mass of 5.5 kg is allowed to slide from rest down an inclined plane. The plane makes an angle of 30° with the horizontal and is 72 m long. The coefficient of kinetic friction between the plane and the object is 0.35. The speed of the object at the bottom of the plane is

(*a*) 5.3 m/s.

(*b*) 15 m/s.

(*c*) 24 m/s.

(*d*) 17 m/s.

(*e*) 11 m/s.

102 •• A brick slides down an inclined plank at constant speed when the plank is inclined at an angle θ_0. If the angle is increased to θ_1, the block accelerates down the plank with acceleration a. The coefficient of kinetic friction is the same in both cases. Given θ_0 and θ_1, calculate a.

103 •• One morning, Lou was in a particularly deep and peaceful slumber. Unfortunately, he had spent the night in the back of a dump truck, and Barry, the driver, was keen to go off to work and start dumping things. Rather than risk a ruckus with Lou, Barry simply raised the back of the truck, and when it reached an angle of 30°, Lou slid down the 4-m incline in 2 s, plopped onto a pile of sand, rolled over, and continued to sleep. Calculate the coefficients of static and kinetic friction between Lou and the truck.

104 •• In a carnival ride, the passenger sits on a seat in a compartment that rotates with constant speed in a vertical circle of radius $r = 5$ m. The heads of the seated passengers always point toward the axis of rotation. (a) If the carnival ride completes one full circle in 2 s, find the acceleration of the passenger. (b) Find the slowest rate of rotation (in other words, the longest time T to complete one full circle) if the seat belt is to exert no force on the passenger at the top of the ride.

105 •• A flat-topped toy cart moves on frictionless wheels, pulled by a rope under tension T. The mass of the cart is m_1. A load of mass m_2 rests on top of the cart with a coefficient of static friction μ_s. The cart is pulled up a ramp that is inclined at angle θ above the horizontal. The rope is parallel to the ramp. What is the maximum tension T that can be applied without making the load slip?

106 •• A sled weighing 200 N rests on a 15° incline, held in place by static friction (Figure 5-58). The coefficient of static friction is 0.5. (a) What is the magnitude of the normal force on the sled? (b) What is the magnitude of the static friction on the sled? (c) The sled is now pulled up the incline at constant speed by a child. The child weighs 500 N and pulls on the rope with a constant force of 100 N. The rope makes an angle of 30° with the incline and has negligible weight. What is the magnitude of the kinetic friction force on the sled? (d) What is the coefficient of kinetic friction between the sled and the incline? (e) What is the magnitude of the force exerted on the child by the incline?

Figure 5-58 Problem 106

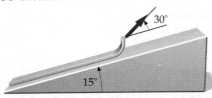

107 •• A child slides down a slide inclined at 30° in time t_1. The coefficient of kinetic friction between her and the slide is μ_k. She finds that if she sits on a small cart with frictionless wheels, she slides down the same slide in time $t_1/2$. Find μ_k.

108 •• The position of a particle of mass $m = 0.8$ kg as a function of time is

$$\vec{r} = x\hat{i} + y\hat{j} = R \sin \omega t \hat{i} + R \cos \omega t \hat{j}$$

where $R = 4.0$ m, and $\omega = 2\pi$ s^{-1}. (a) Show that the path of this particle is a circle of radius R with its center at the origin. (b) Compute the velocity vector. Show that $v_x/v_y = -y/x$. (c) Compute the acceleration vector and show that it is in the radial direction and has the magnitude v^2/r. (d) Find the magnitude and direction of the net force acting on the particle.

109 •• In an amusement-park ride, riders stand with their backs against the wall of a spinning vertical cylinder. The floor falls away and the riders are held up by friction. If the radius of the cylinder is 4 m, find the minimum number of revolutions per minute necessary to prevent the riders from

dropping when the coefficient of static friction between a rider and the wall is 0.4.

110 •• Some bootleggers race from the police down a road that has a sharp, level curve with a radius of 30 m. As they go around the curve, the bootleggers squirt oil on the road behind them, reducing the coefficient of static friction from 0.7 to 0.2. When taking this curve, what is the maximum safe speed of (a) the bootleggers' car, and (b) the police car?

111 •• A mass m_1 on a horizontal shelf is attached by a thin string that passes over a frictionless peg to a 2.5-kg mass m_2 that hangs over the side of the shelf 1.5 m above the ground (Figure 5-59). The system is released from rest at $t = 0$ and the 2.5-kg mass strikes the ground at $t = 0.82$ s. The system is now placed in its initial position and a 1.2-kg mass is placed on top of the block of mass m_1. Released from rest, the 2.5-kg mass now strikes the ground 1.3 s later. Determine the mass m_1 and the coefficient of kinetic friction between m_1 and the shelf.

Figure 5-59 Problem 111

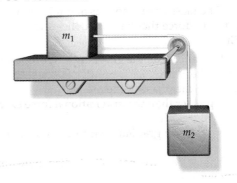

112 ••• (a) Show that a point on the surface of the earth at latitude θ has an acceleration relative to a reference frame not rotating with the earth with a magnitude of $3.37 \cos \theta$ cm/s^2. What is the direction of this acceleration? (b) Discuss the effect of this acceleration on the apparent weight of an object near the surface of the earth. (c) The free-fall acceleration of an object at sea level measured *relative to the earth's surface* is 9.78 m/s^2 at the equator and 9.81 m/s^2 at latitude $\theta = 45°$. What are the values of the gravitational field g at these points?

CHAPTER 6

Work and Energy

Work and energy are important concepts in physics as well as in our everyday life. In physics, a force does **work** when it acts on an object that moves through a distance, and there is a component of the force along the line of motion. For a constant force in one dimension, the work done equals the force times the distance. (This differs somewhat from the everyday use of the word work. When you study hard for an exam, the only work you do as the term is understood in physics is in moving your pencil or turning the pages of your book.)

The concept of **energy** is closely associated with that of work. When work is done by one system on another, energy is transferred between the two systems. For example, when you do work pushing a swing, chemical energy in your body is transferred to the swing and appears as kinetic energy of motion or gravitational potential energy of the earth–swing system. There are many forms of energy. Kinetic energy is associated with the motion of an object. Potential energy is associated with the configuration of a system, such as the separation distance between some object and the earth. Thermal energy is associated with the random motion of the molecules within a system and is closely connected with the temperature of the system.

The action of the pole vaulter shown here demonstrates several kinds of energy. First the vaulter transforms the internal chemical energy of his body into kinetic energy as he runs. Some of this kinetic energy is then converted into elastic potential energy, represented by the deformation of the pole. The rest of the vaulter's kinetic energy is eventually converted into gravitational potential energy, which in turn is converted into kinetic energy as the vaulter drops. Mechanical energy is finally converted into thermal energy when the athlete drops onto the mat.

6-1 Work and Kinetic Energy

Motion in One Dimension With Constant Forces

The work W done by a constant force $\vec{F}$ whose point of application moves through a distance Δx is defined to be

$$W = F \cos\theta\, \Delta x = F_x \Delta x \qquad \text{6-1}$$

Definition—Work by a constant force

where θ is the angle between $\vec{F}$ and the x axis, and Δx is the displacement of the force as shown in Figure 6-1.

Work is a scalar quantity that is positive if Δx and F_x have the same signs and negative if they have opposite signs. The dimensions of work are those of force times distance. The SI unit of work and energy is the **joule** (J), which equals the product of a newton and a meter*:

$$1\,\text{J} = 1\,\text{N}\cdot\text{m} \qquad\qquad\qquad 6\text{-}2$$

A convenient unit of work and energy in atomic and nuclear physics is the electron volt (eV):

$$1\,\text{eV} = 1.6 \times 10^{-19}\,\text{J} \qquad\qquad\qquad 6\text{-}3$$

Commonly used multiples are keV (1000 eV) and MeV (10^6 eV). The work required to remove an electron from an atom is of the order of several eV, whereas the work needed to remove a proton or neutron from an atomic nucleus is of the order of several MeV.

Exercise   A force of 12 N is exerted on a box at an angle of $\theta = 20°$, as in Figure 6-1. How much work is done by the force as the box moves along the table a distance of 3 m? (*Answer*   33.8 J)

When there are several forces that do work, the total work is found by computing the work done by each force and summing:

$$W_{\text{total}} = F_{1x}\Delta x_1 + F_{2x}\Delta x_2 + F_{3x}\Delta x_3 + \cdots$$

When the forces do work on a *particle*, the displacement of the force Δx_i is the same for each force and is equal to the displacement of the particle Δx:

$$W_{\text{total}} = F_{1x}\Delta x + F_{2x}\Delta x + F_{3x}\Delta x + \cdots$$
$$= (F_{1x} + F_{2x} + F_{3x})\Delta x$$
$$= F_{\text{net }x}\Delta x \qquad\qquad\qquad 6\text{-}4$$

Thus for a particle, the total work can be found by summing all the forces to find the net force and then computing the work done by the net force.

The Work–Kinetic Energy Theorem

There is an important relation between the total work done on a particle and the initial and final speeds of the particle. If F_x is the net force acting on a particle, Newton's second law gives

$$F_x = ma_x$$

Since the work done by the net force equals the total work done on the particle,

$$W_{\text{total}} = F_x\Delta x = ma_x\,\Delta x$$

For a constant force, the acceleration is constant, and we can relate the distance the particle moves to its initial speed v_i and final speed v_f by using the constant-acceleration formula (Equation 2-15):

$$v_f^2 = v_i^2 + 2a_x\Delta x$$

Substituting $\frac{1}{2}(v_f^2 - v_i^2)$ for $a_x\,\Delta x$ yields

$$W_{\text{total}} = \frac{1}{2}mv_f^2 - \frac{1}{2}mv_i^2 \qquad\qquad\qquad 6\text{-}5$$

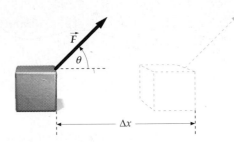

Figure 6-1 When a constant force $\vec{F}$ moves through a distance Δx, the work done is $F \cos\theta\,\Delta x = F_x\,\Delta x$.

---

* In the U.S. customary system, the unit of work is the foot-pound: 1 ft-lb = 1.356 J.

The quantity $\frac{1}{2}mv^2$ is a scalar quantity called the **kinetic energy** K of the particle:

$$K = \frac{1}{2}mv^2$$

6-6

Definition—Kinetic energy

The quantity on the right side of Equation 6-5 is the change in the kinetic energy of the particle. Thus,

The *total* work done on a particle is equal to the *change* in its kinetic energy:

$$W_{total} = \Delta K = \frac{1}{2}mv_f^2 - \frac{1}{2}mv_i^2$$

6-7

Work–kinetic energy theorem

This result is known as the **work–kinetic energy theorem.** It holds whether the net force is constant or variable, as we will see in the next section.

Exercise   A girl of mass 50 kg is running at 3.5 m/s. What is her kinetic energy? (*Answer*   306 J)

Example 6-1

A truck of mass 3000 kg is to be loaded onto a ship by a crane that exerts an upward force of 31 kN on the truck. This force, which is just strong enough to get the truck started upward, is applied over a distance of 2 m. Find (*a*) the work done by the crane, (*b*) the work done by gravity, and (*c*) the upward speed of the truck after 2 m.

Picture the Problem   The applied force is in the direction of motion, so the work it does is positive. On the other hand, gravity is opposite in direction to the motion, so the work done by gravity is negative (Figure 6-2). The final speed of the truck can be obtained from its final kinetic energy, which equals the total work done on the truck because it starts from rest. The total work is the sum of the results for (*a*) and (*b*).

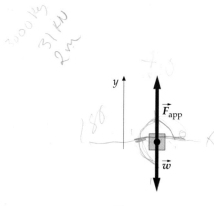

Figure 6-2

(*a*)   Calculate the work done by the applied force:

$$W_{app} = F_{app} \cos 0° \, \Delta y = (31 \text{ kN})(1)(2 \text{ m}) = 62 \text{ kJ}$$

(*b*)   Calculate the work done by gravity:

$$W_g = mg \cos 180° \, \Delta y$$

$$= (3000 \text{ kg})(9.81 \text{ N/kg})(-1)(2 \text{ m}) = -59 \text{ kJ}$$

(*c*)1. The final speed is related to the final kinetic energy:

$$K_f = \frac{1}{2}mv_f^2$$

$$v_f = \sqrt{\frac{2K_f}{m}}$$

2. Apply the work–kinetic energy theorem, with $v_i = 0$:

$$W_{total} = \Delta K = K_f - K_i = K_f$$

3. The total work is the sum of the applied work and the work done by gravity:

$$W_{total} = W_{app} + W_g = 62 \text{ kJ} - 59 \text{ kJ} = 3.0 \text{ kJ}$$

4. Substitute $K_f = 3.0$ J to obtain the final speed of the truck:

$$v_f = \sqrt{\frac{2K_f}{m}} = \sqrt{\frac{2(3.0 \text{ kJ})}{3000 \text{ kg}}} = 1.4 \text{ m/s}$$

Remark We treat each force separately when calculating the work done. We could also find the total work by first adding the forces to obtain the net force, then applying $W_{total} = F_{net\,x}\,\Delta x$. In either case, the work–kinetic energy theorem applies only to the total work. We could have also found the speed using Newton's second law.

Exercise Find the final speed of the truck if the same upward force were applied for 2 m after it was already moving upward at 1 m/s. (*Answer* 1.73 m/s. Note that the answer is *not* 1.4 m/s + 1 m/s. Why not?)

Example 6-2

In a television tube, an electron is accelerated from rest to a kinetic energy of 2.5 keV over a distance of 80 cm. (The force that accelerates the electron is an electric force due to the electric field in the tube.) Find the force on the electron, assuming it to be constant and in the direction of motion.

Picture the Problem Since the electron starts from rest, the work done equals the final kinetic energy. To find the force in newtons, we must convert the energy from keV to joules.

1. Set the work done to be equal to the change in $W = F\,\Delta x = \Delta K = K_f - K_i = K_f = 2.5\text{ keV}$
   kinetic energy:

2. Solve for F and convert the energy to joules: $F = \dfrac{W}{\Delta x} = \dfrac{2.5\text{ keV}}{0.8\text{ m}} \times \dfrac{1.6 \times 10^{-19}\text{ J}}{1\text{ eV}} = 5.0 \times 10^{-16}\text{ N}$

Remark When we discuss electricity we will see that the work done per charge is called the potential difference and is measured in volts. Thus, 1 eV is the energy acquired or lost by a particle of charge e (an electron or proton, for example) when its potential difference changes by 1 V.

Example 6-3

Your professor enters the dogsled race during winter break. To get started, he pulls his sled (total mass 80 kg) with a force of 180 N at 20° to the horizontal. Find (*a*) the work he does, and (*b*) the final speed of the sled after it moves $\Delta x = 5$ m, assuming that it starts from rest and there is no friction.

Figure 6-3

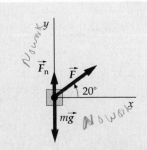

Picture the Problem The work done by the professor is $F_x\,\Delta x$ since there is no motion in the y direction. This is also the total work done on the sled because the other forces, mg and F_n, have no x components (Figure 6-3). The final speed of the sled is found by applying the work–kinetic energy theorem with $v_i = 0$.

(*a*) The work done by the professor is $F_x\,\Delta x$. This is also the total work done on the sled:

$$W = F_x\Delta x = (F\cos 20°)\Delta x$$
$$= (180\text{ N})(\cos 20°)(5\text{ m}) = 846\text{ J} = W_{total}$$

(*b*) Apply the work–kinetic energy theorem and solve for the final speed:

$$W_{total} = \frac{1}{2}mv_f^2 - \frac{1}{2}mv_i^2 = \frac{1}{2}mv_f^2$$
$$v_f = \sqrt{\frac{2W_{total}}{m}} = \sqrt{\frac{2(846\text{ J})}{80\text{ kg}}} = 4.60\text{ m/s}$$

Remark We do not need to work out the units. If we have a correct equation, and all quantities are in SI units, the result will be in the correct SI units. However, as a check on the equation, we can show that $1 \text{ J/kg} = 1 \text{ m}^2/\text{s}^2$. We have $1 \text{ J/kg} = 1 \text{ N·m /kg} = (1 \text{ kg·m/s}^2) \text{ m·kg} = 1 \text{ m}^2/\text{s}^2$.

Exercise What force did your professor exert if the sled starts with a speed of 2 m/s, and its final speed is 4.5 m/s after he pulls it through a distance of 5 m? (*Answer* 138 N)

What if you hold a weight in a fixed position? You are expending energy, but are you doing work? According to the definition of work, you are not do-ing work *on the weight* because the weight does not move (Figure 6-4). But your muscles are continually contracting and relaxing as you hold the weight. Molecular assemblies in your muscle *do* move, and work is done. In the process, internal chemical energy in your body is converted to thermal energy (Figure 6-5).

Figure 6-4 The man standing on a ledge does not do work on the weight when holding it at a fixed position. The same task could be accomplished by tying the rope to a fixed point.

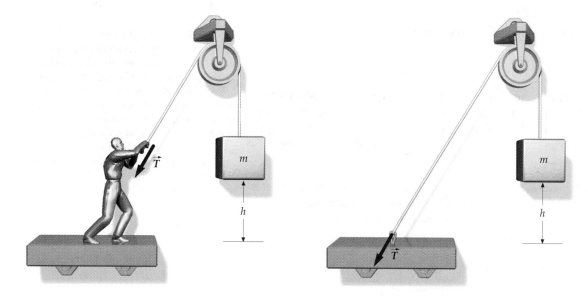

In working muscle, fuel molecules such as sugar drive the motion of molecular "machines."

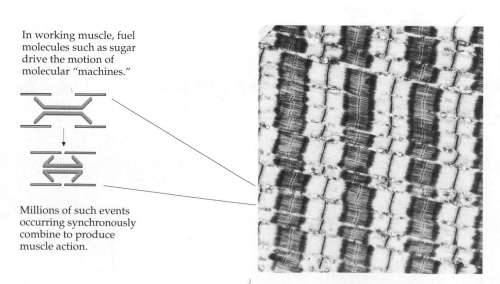

Millions of such events occurring synchronously combine to produce muscle action.

Figure 6-5 Muscle work. While the man holding the weight in Figure 6-4 may be doing no work on the weight, his body is putting out work on the molecular level, as structures within the muscle slide over each other during muscular extension and contraction.

Work Done by a Variable Force

In Figure 6-6, we plot a constant force F_x as a function of position x. The work done on a particle whose displacement is Δx is represented by the area under the force-versus-position curve, indicated by the shading in Figure 6-6.

Many forces vary with distance. For example, a spring exerts a force proportional to the distance it is stretched or compressed. And the gravitational force the earth exerts on a spaceship varies inversely with the square of the distance between the two bodies. We can approximate a variable force by a series of constant forces (Figure 6-7). The work done by a variable force is then

$$W = \lim_{\Delta x_i \to 0} \sum_i F_x \Delta x_i = \text{area under the } F_x\text{-versus-}x \text{ curve} \qquad 6\text{-}8$$

This limit is the integral of F_x over x. So the work done by a variable force F_x acting on a particle as it moves from x_1 to x_2 is

$$W = \int_{x_1}^{x_2} F_x\, dx = \text{area under the } F_x\text{-versus-}x \text{ curve} \qquad 6\text{-}9$$

Definition—Work by a variable force

For each rectangular area, the force is constant, so the work done equals the change in the ~~kinetic energy over that interval~~. The total work done is the sum of the areas over all intervals, which equals the change in kinetic energy over the complete interval. Thus, $W_{\text{total}} = \Delta K$ holds for variable forces as well as for constant forces.

> **Exercise in Dimensional Analysis**   A spring is characterized by its force constant k, which has dimensions N/m. How does the work required to stretch a spring by an amount x_0 depend on k and x_0? (*Answer*   Since work has dimensions of N·m, the work must depend on k and x_0 in the combination kx_0^2. We will see in Example 6-5 that the actual expression is $W = \frac{1}{2}kx_0^2$. The factor $\frac{1}{2}$ arises because the force varies from 0 to a maximum value of kx_0, and has the average value $\frac{1}{2}kx_0$.)

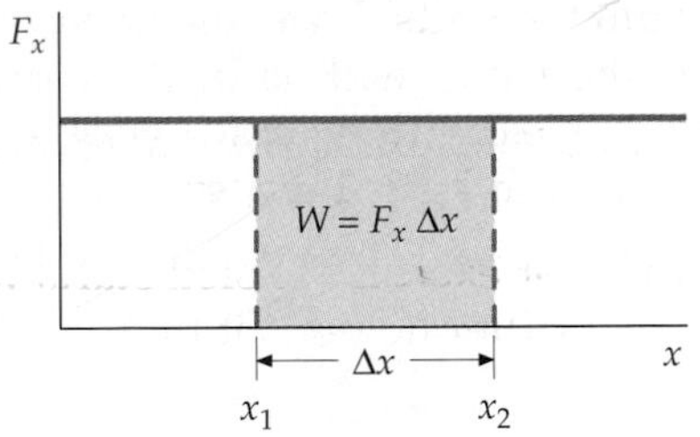

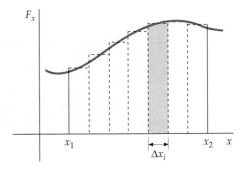

Figure 6-6 The work done by a constant force is represented graphically as the area under the F_x-versus-x curve.

Figure 6-7 A variable force can be approximated by a series of constant forces over small intervals. The work done by the constant force in each interval is the area of the rectangle beneath the force curve. The sum of these rectangular areas is the sum of the work done by the set of constant forces that approximates the varying force. In the limit of infinitesimally small Δx_i, the sum of the areas of the rectangles equals the area under the complete curve.

Example 6-4

A force F_x varies with x as shown in Figure 6-8. Find the work done by the force on a particle as the particle moves from $x = 0$ to $x = 6$ m.

1. We find the work done by calculating the area under the F_x-versus-x curve:

$$W = A$$

2. This area is the sum of the two areas shown:

$$W = A = A_1 + A_2$$

$$= (5\text{ N})(4\text{ m}) + \frac{1}{2}(5\text{ N})(2\text{ m})$$

$$= 20\text{ J} + 5\text{ J} = 25\text{ J}$$

Figure 6-8

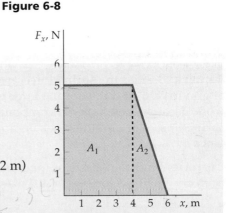

Exercise The force shown is the only force that acts on a particle of mass 3 kg. If the particle starts from rest at $x = 0$, how fast is it moving when it reaches $x = 6$ m? (*Answer*   4.08 m/s)

Example 6-5

A 4-kg block on a frictionless table is attached to a horizontal spring that obeys Hooke's law and exerts a force $\vec{F} = -kx\hat{i}$, where $k = 400$ N/m and x is measured from the equilibrium position of the block. The spring is originally compressed with the block at $x_1 = -5$ cm (Figure 6-9). Find (a) the work done by the spring on the block as the block moves from $x_1 = -5$ cm to its equilibrium position $x_2 = 0$, and (b) the speed of the block at $x_2 = 0$.

Figure 6-9

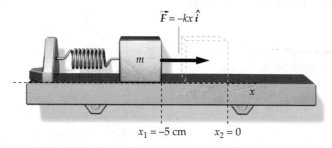

$\vec{F} = -kx\hat{i}$

m

x

$x_1 = -5$ cm $x_2 = 0$

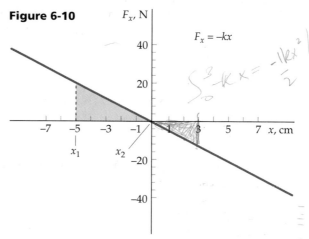

Figure 6-10

Picture the Problem The work done on the block as it moves from x_1 to $x_2 = 0$ equals the area under the F_x-versus-x curve between these limits (the shaded area in Figure 6-10), which can be calculated by integrating the force over the distance. The work done equals the change in kinetic energy, which is just the final kinetic energy since the initial kinetic energy is zero. The speed of the block at $x = 0$ is found from the kinetic energy of the block.

(a) The work W done by the spring on the block is the integral of $F_x\,dx$ from $x_1 = -5$ cm to $x_2 = 0$:

$$W = \int_{x_1}^{x_2} F_x\,dx = \int_{x_1}^{0} -kx\,dx = -k\int_{x_1}^{0} x\,dx = -\frac{1}{2}kx^2\Big|_{x_1}^{0}$$

$$= \frac{1}{2}kx_1^2 = \frac{1}{2}(400 \text{ N/m})(0.05 \text{ m})^2 = 0.500 \text{ J}$$

(b) Apply the work–kinetic energy theorem with $v_1 = 0$ and solve for v_2:

$$W = \frac{1}{2}mv_2^2 - \frac{1}{2}mv_1^2 = \frac{1}{2}mv_2^2$$

$$v_2 = \sqrt{\frac{2W}{m}} = \sqrt{\frac{2(0.500 \text{ J})}{4 \text{ kg}}} = 0.50 \text{ m/s}$$

Remark Besides the spring force, two other forces act on the block; the force of gravity, $m\vec{g}$, and the normal force of the table, $\vec{F}_n$. These latter forces do no work because they have no component in the direction of motion. Only the spring does work on the block because the force it exerts acts through a distance Δx.

Exercise Find the speed of the block when it reaches $x = 3$ cm if it starts from $x = 0$ with velocity $v_x = 0.5$ m/s. (*Answer*   0.4 m/s)

Note that we could not have solved Example 6-5 by finding the acceleration and then using the constant-acceleration equations. The force exerted by the spring on the block, $F_x = -kx$, varies with position, so the acceleration also varies.

6-2   Work and Energy in Three Dimensions

Figure 6-11 shows a particle of mass m acted on by a force $\vec{F}$ as it moves along a curve in space. Consider a small displacement Δs, where s is the distance measured along the curve. $\vec{F}$ has components F_s parallel to and $F_\perp$ perpendicular to the displacement. The component $F_\perp$ provides the centripetal force needed for the particle to round the curve, but since it is perpendicular to the motion, it does not contribute to the work done on the particle by $\vec{F}$, which is

$$\Delta W = F_s \Delta s$$

To find the work done as the particle moves along the curve from point 1 to point 2, we compute $F_s \, \Delta s$ for each element of the path and sum. In the limit of smaller and smaller displacement elements, this sum becomes an integral:

$$W = \int_{s_1}^{s_2} F_s \, ds$$

From Newton's second law,

$$F_s = m \frac{dv}{dt}$$

If we think of the speed as a function of the distance s, we can apply the chain rule for derivatives:

$$\frac{dv}{dt} = \frac{dv}{ds}\frac{ds}{dt} = v \frac{dv}{ds}$$

where we have used $ds/dt = v$, the speed. The work done by the net force is then

$$W_{\text{total}} = \int_{s_1}^{s_2} F_s \, ds = \int_{s_1}^{s_2} m \frac{dv}{dt} \, ds = \int_{s_1}^{s_2} mv \frac{dv}{ds} \, ds = \int_{v_1}^{v_2} mv \, dv$$

or

$$W_{\text{total}} = \int_{s_1}^{s_2} F_s \, ds = \frac{1}{2} mv_2^2 - \frac{1}{2} mv_1^2 \qquad \text{6-10}$$

Work–kinetic energy theorem in three dimensions

Equation 6-10, along with its one-dimensional counterpart, Equation 6-7, follows directly from the definition of work and from Newton's second law of motion.

The Dot Product

The component F_s in Figure 6-11 is related to the angle ϕ between $\vec{F}$ and $\Delta\vec{s}$ by $F_s = F \cos \phi$, so the work done by $\vec{F}$ for a displacement $\Delta\vec{s}$ is

$$\Delta W = F_s \Delta s = (F \cos \phi) \, \Delta s$$

This combination of two vectors and the cosine of the angle between them is called the **dot product** (or **scalar product**) of the vectors. The dot product of two general vectors $\vec{A}$ and $\vec{B}$ is written $\vec{A} \cdot \vec{B}$ and is defined by

$$\vec{A} \cdot \vec{B} = AB \cos \phi \qquad \text{6-11}$$

Definition—Dot product

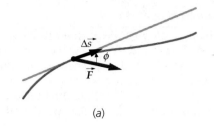

(a)

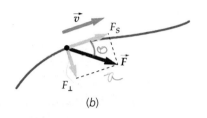

(b)

Figure 6-11 (*a*) A particle moving along an arbitrary curve in space. (*b*) The perpendicular component of the force $F_\perp$ changes the direction of the particle's motion but not its speed. The tangential component F_s changes the particle's speed but not its direction. F_s equals the mass m times the tangential acceleration dv/dt. Only this component does work.

Table 6-1

Properties of Dot Products

If	then
$\vec{A}$ and $\vec{B}$ are perpendicular,	$\vec{A}\cdot\vec{B} = 0$ (since $\phi = 90°$, $\cos 90° = 0$)
$\vec{A}$ and $\vec{B}$ are parallel,	$\vec{A}\cdot\vec{B} = AB$ (since $\phi = 0°$, $\cos 0° = 1$)
$\vec{A}\cdot\vec{B} = 0$,	Either $\vec{A} = 0$ or $\vec{B} = 0$ or $\vec{A}$ and $\vec{B}$ are perpendicular
Furthermore,	
$\vec{A}\cdot\vec{A} = A^2$	Since $\vec{A}$ is parallel to itself
$\vec{A}\cdot\vec{B} = \vec{B}\cdot\vec{A}$	Commutative rule of multiplication
$(\vec{A} + \vec{B})\cdot\vec{C} = \vec{A}\cdot\vec{C} + \vec{B}\cdot\vec{C}$	Distributive rule of multiplication

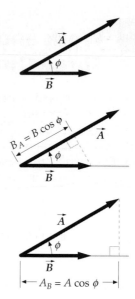

where ϕ is the angle between $\vec{A}$ and $\vec{B}$. The dot product $\vec{A}\cdot\vec{B}$ can be thought of as A times the component of $\vec{B}$ in the direction of $\vec{A}$ (that is, A times $B \cos \phi$), or as B times the component of $\vec{A}$ in the direction of $\vec{B}$ (that is, B times $A \cos \phi$). Figure 6-12 shows a geometric representation of the dot product $\vec{A}\cdot\vec{B}$. Properties of the dot product are summarized in Table 6-1. We can use unit vectors to write the dot product in terms of the rectangular components of the two vectors:

$$\vec{A}\cdot\vec{B} = (A_x\hat{i} + A_y\hat{j} + A_z\hat{k})\cdot(B_x\hat{i} + B_y\hat{j} + B_z\hat{k})$$

Since the unit vectors $\hat{i}$, $\hat{j}$, and $\hat{k}$ are mutually perpendicular, $\hat{i}\cdot\hat{j} = \hat{i}\cdot\hat{k} = \hat{j}\cdot\hat{k} = 0$. So the cross terms like $A_x\hat{i}\cdot B_y\hat{j}$ are zero. In addition, the dot product of a unit vector with itself is 1; $\hat{i}\cdot\hat{i} = \hat{j}\cdot\hat{j} = \hat{k}\cdot\hat{k} = 1$, so a term like $A_x\hat{i}\cdot B_x\hat{i}$ equals A_xB_x. The result is

$$\vec{A}\cdot\vec{B} = A_xB_x + A_yB_y + A_zB_z \qquad \text{6-12}$$

The component of a vector along some axis can be written as the dot product of the vector and the unit vector along that axis. For example, the component A_x is found from

$$\vec{A}\cdot\hat{i} = (A_x\hat{i} + A_y\hat{j} + A_z\hat{k})\cdot\hat{i} = A_x \qquad \text{6-13}$$

Figure 6-12 The dot product $\vec{A}\cdot\vec{B}$ is the product of A and the projection of $\vec{B}$ on $\vec{A}$ or the product of B and the projection of $\vec{A}$ on $\vec{B}$.

Example 6-6

(a) Find the angle between the vectors $\vec{A} = (3 \text{ m})\hat{i} + (2 \text{ m})\hat{j}$ and $\vec{B} = (4 \text{ m})\hat{i} - (3 \text{ m})\hat{j}$ (Figure 6-13). **(b)** Find the component of $\vec{A}$ in the direction of $\vec{B}$.

Picture the Problem We find the angle ϕ from the definition of the dot product. The component of $\vec{A}$ in the direction of $\vec{B}$ is found from the dot product of $\vec{A}$ with the unit vector $\vec{B}/B$.

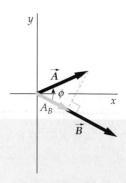

Figure 6-13

(a)1. Write the dot product of $\vec{A}$ and $\vec{B}$ in terms of A, B, and $\cos \phi$ and solve for $\cos \phi$:

$$\vec{A}\cdot\vec{B} = AB \cos \phi$$

$$\cos \phi = \frac{\vec{A}\cdot\vec{B}}{AB}$$

2. Find $\vec{A}\cdot\vec{B}$ from their components:

$$\vec{A}\cdot\vec{B} = A_xB_x + A_yB_y$$

$$= (3 \text{ m})(4 \text{ m}) + (2 \text{ m})(-3 \text{ m})$$

$$= 12 \text{ m}^2 - 6 \text{ m}^2 = 6 \text{ m}^2$$

3. The magnitudes of the vectors are obtained from the dot product of the vector with itself:

$$\vec{A}\cdot\vec{A} = A^2 = A_x^2 + A_y^2$$
$$= (2\text{ m})^2 + (3\text{ m})^2 = 13\text{ m}^2$$
$$A = \sqrt{13}\text{ m}$$

and

$$\vec{B}\cdot\vec{B} = B^2 = B_x^2 + B_y^2$$
$$= (4\text{ m})^2 + (-3\text{ m})^2 = 25\text{ m}^2$$
$$B = 5\text{ m}$$

4. Substitute these values into the equation in step 1 for $\cos\phi$ to find ϕ:

$$\cos\phi = \frac{\vec{A}\cdot\vec{B}}{AB} = \frac{6\text{ m}^2}{(\sqrt{13}\text{ m})(5\text{ m})} = 0.333$$
$$\phi = 70.6°$$

(b) The component of $\vec{A}$ along $\vec{B}$ is the dot product of $\vec{A}$ with the unit vector $\vec{B}/B$:

$$A_B = \vec{A}\cdot\frac{\vec{B}}{B} = \frac{\vec{A}\cdot\vec{B}}{B} = \frac{6\text{ m}^2}{5\text{ m}} = 1.2\text{ m}$$

Check the Result The component of A along B is $A\cos\phi = (\sqrt{13}\text{ m})\cos 70.6° = 1.2$ m.

Exercise (a) Find $\vec{A}\cdot\vec{B}$ for $\vec{A} = (3\text{ m})\hat{i} + (4\text{ m})\hat{j}$ and $\vec{B} = (2\text{ m})\hat{i} + (8\text{ m})\hat{j}$. (b) Find A, B, and the angle between $\vec{A}$ and $\vec{B}$ for these vectors. (*Answers* (a) 38 m^2, (b) $A = 5$ m, $B = 8.25$ m, $\phi = 23°$)

In dot-product notation, the work dW done by a force $\vec{F}$ on a particle undergoing a displacement $d\vec{s}$ is

$$dW = F\cos\phi\,ds = \vec{F}\cdot d\vec{s} \qquad 6\text{-}14$$

and the work done on the particle as it moves from point 1 to point 2 is

$$W = \int_{s_1}^{s_2}\vec{F}\cdot d\vec{s} \qquad 6\text{-}15$$

The general definition of work

When several forces $\vec{F}_i$ act on a particle whose displacement is $d\vec{s}$, the total work is

$$dW_{\text{total}} - \vec{F}_1\cdot d\vec{s} + \vec{F}_2\cdot d\vec{s} + \cdots = \left(\sum_i\vec{F}_i\right)\cdot d\vec{s} \qquad 6\text{-}16$$

Example 6-7 *try it yourself*

A particle is given a displacement $\Delta\vec{s} = 2\text{ m }\hat{i} - 5\text{ m }\hat{j}$ along a straight line. During the displacement, a constant force $\vec{F} = 3\text{ N }\hat{i} + 4\text{ N }\hat{j}$ acts on the particle (Figure 6-14). Find (a) the work done by the force, and (b) the component of the force in the direction of the displacement.

Picture the Problem The work W is found by computing $W = \vec{F}\cdot\Delta\vec{s} = F_x\Delta x + F_y\Delta y + F_z\Delta z$. Since $\vec{F}\cdot\Delta\vec{s} = F\cos\phi|\Delta\vec{s}|$, we can find the component of $\vec{F}$ in the direction of the displacement from

$$F\cos\phi = \frac{(\vec{F}\cdot\Delta\vec{s})}{|\Delta\vec{s}|} = \frac{W}{|\Delta\vec{s}|}$$

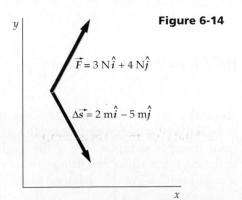

Figure 6-14

$\vec{F} = 3\text{ N}\hat{i} + 4\text{ N}\hat{j}$

$\Delta\vec{s} = 2\text{ m}\hat{i} - 5\text{ m}\hat{j}$

Cover the column to the right and try these on your own before looking at the answers.

Steps   **Answers**

(a)   Compute the work done W. $W = -14 \, \text{N} \cdot \text{m}$

(b) 1. Compute $\Delta \vec{s} \cdot \Delta \vec{s}$ and use your result to find the distance $|\Delta \vec{s}|$. $|\Delta \vec{s}| = \sqrt{29} \, \text{m}$

   2. Compute $F \cos \phi = W / |\Delta \vec{s}|$. $F \cos \phi = -2.60 \, \text{N}$

Remark The component of the force in the direction of the displacement is negative, so the work done is negative.

Exercise Find the magnitude of $\vec{F}$, and the angle ϕ between $\vec{F}$ and $\Delta \vec{s}$. (*Answer* $F = 5 \, \text{N}$, $\phi = 121°$)

Example 6-8

You ski downhill on waxed skis that are nearly frictionless. (a) What work is done on you as you ski a distance s down the hill? (b) What is your speed on reaching the bottom of the run? Assume the length of the ski run is s, its angle of incline is θ, and your mass is m. The height of the hill is then h = s sin θ.

Figure 6-15a   **Figure 6-15b**

Picture the Problem We assume that you are a particle. Two forces act on you: gravity, $m\vec{g}$, and the normal force exerted by the hill, $\vec{F}_n$ (Figure 6-15a). Only gravity does work on you, because the normal force is perpendicular to the hill, and hence has no component in the direction of your motion. The work–kinetic energy theorem with $v_i = 0$ gives the final speed v.

   Figure 6-15b shows a free-body diagram for you on skies. The net force is $mg \sin \theta$, which is the component of the weight in the direction of the displacement Δs.

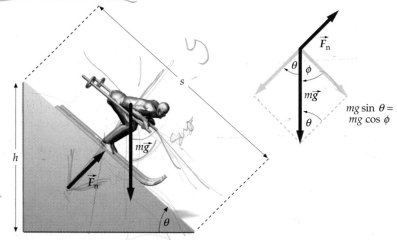

$mg \sin \theta = mg \cos \phi$

(a) 1. The work done by gravity as you traverse the slope is $m\vec{g} \cdot \vec{s}$: $W = m\vec{g} \cdot \vec{s} = mgs \cos \phi = mgs \sin \theta$

   2. From Figure 6-15a, the angle θ is related to h and s: $\sin \theta = \dfrac{h}{s}$

   3. Substitute h for $s \sin \theta$: $W = mgh$

(b)   Apply the work–kinetic energy theorem to find the final speed v: $W = mgh = \dfrac{1}{2} mv^2 - 0$ or $v = \sqrt{2gh}$

Remarks $mg \sin \theta = mg \cos \phi$ is the component of the weight in the direction of the displacement. This is the component that does work on you. The final speed is independent of the angle θ, and the same as if the skier had dropped vertically a height h with acceleration g. If θ were smaller, the skier would travel a greater distance to drop the same vertical distance h, but the component of the force of gravity in the direction of motion would be less. The two effects cancel, and the work done by gravity is *mgh* independent of the angle of the slope. Figure 6-16 shows that for a hill of arbitrary shape, the work done by the earth on the skier is *mgh*.

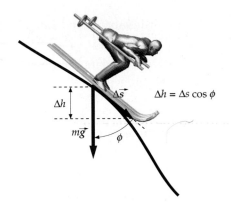

Figure 6-16 Skier skiing down a hill of arbitrary shape. The work done by the earth during a displacement $\Delta\vec{s}$ is $m\vec{g}\cdot\Delta\vec{s} = mg\,\Delta s\cos\phi = mg\,\Delta h$, where Δh is the vertical distance dropped. The total work done by the earth when the skier skis down a vertical distance h is $W = \int_0^s m\vec{g}\cdot d\vec{s} = mg\int_0^s \cos\phi\,ds = mg\int_0^h dh = mgh$, independent of the shape of the hill.

$\Delta h = \Delta s\cos\phi$

6-3   Power

The **power** P supplied by a force is the rate at which the force does work. Consider a particle moving with instantaneous velocity $\vec{v}$. In a short time interval dt, the particle has displacement $d\vec{s} = \vec{v}\,dt$. The work done by a force $\vec{F}$ acting on the particle during this time interval is

$$dW = \vec{F}\cdot d\vec{s} = \vec{F}\cdot\vec{v}\,dt$$

The power delivered to the particle is then

$$P = \frac{dW}{dt} = \vec{F}\cdot\vec{v}$$

6-17

Definition—Power

The SI unit of power, one joule per second, is called a watt (W):

$$1\text{ W} = 1\text{ J/s}$$

Note the difference between power and work. Two motors that lift a given load a given distance do the same amount of work, but the one that does it in the least time supplies more power. Gas and electric companies charge for energy, not power, usually by the kilowatt-hour (kW·h). A kilowatt-hour of energy is

$$1\text{ kW·h} = (10^3\text{ W})(3600\text{ s}) = 3.6\times10^6\text{ W·s} = 3.6\text{ MJ}$$

In the U.S. customary system, the unit of energy is the foot-pound and the unit of power is the foot-pound per second. A commonly used multiple of this unit, called a horsepower (hp), is defined as

$$1\text{ hp} = 550\text{ ft·lb/s} = 746\text{ W}$$

Example 6-9

A small motor is used to operate a lift that raises a load of bricks weighing 800 N to a height of 10 m in 20 s. What is the minimum power the motor must produce?

Picture the Problem Assuming that the bricks are lifted without acceleration, the upward force exerted by the motor is equal to the weight of the bricks, $F = 800$ N. The speed of the bricks is $v = 10\text{ m}/(20\text{ s}) = 0.5\text{ m/s}$.

Figure 6-17

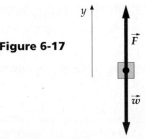

The power is the product of the speed v and the component of force in the direction of motion, which is simply F in this case (Figure 6-17):

$$P = Fv = (800\text{ N})(0.5\text{ m/s})$$
$$= 400\text{ N·m/s} = 400\text{ J/s} = 400\text{ W}$$

Remark This minimum power output of 400 W is a little more than $\frac{1}{2}$ horse-power.

Exercise (a) Find the total work done by the force. (b) Calculate the power by dividing the total work by the total time. (*Answers*   (a) 8000 J, (b) 400 W)

Consider a net force F_x acting on a particle in one dimension. The rate at which this force does work is

$$P = F_x v_x$$

Substituting $F_x = ma_x$ we have

$$P = F_x v_x = ma_x v_x \qquad\qquad 6\text{-}18$$

or

$$a_x = \frac{P}{mv_x} \qquad\qquad 6\text{-}19$$

Thus, for a constant power P, the acceleration varies inversely as the speed. A familiar example is the difficulty in passing an automobile at high speeds. For a given power, the acceleration at high speeds is smaller than at lower speeds. Alternatively, it takes more power to give the same acceleration to an automobile moving at 80 km/h than one moving at 60 km/h.

If we write $a_x = dv_x/dt$ in Equation 6-18, then

$$P = ma_x v_x = mv_x \frac{dv_x}{dt} = \frac{d}{dt}\left(\frac{1}{2}mv_x^2\right) = \frac{dK}{dt}$$

Assuming that P is constant, and integrating over some time interval, we get

$$P\,\Delta t = \Delta K \quad \text{(constant power)} \qquad\qquad 6\text{-}20$$

So the time it takes an automobile or airplane at constant power to accelerate from one speed to another speed is proportional to the change in the kinetic energy.

Example 6-10

A new Cadillac can accelerate from 0 to 96 km/h in 6.5 s. How quickly would you expect it to be able to accelerate from 80 km/h to 112 km/h?

Picture the Problem According to Equation 6-20, the time is related to the power and change in kinetic energy $\Delta t = \Delta K/P$. If we assume constant power, the time required is proportional to the change in kinetic energy. Since we only need to calculate ratios, we do not need to convert the units. Let m be the mass of the Cadillac.

1. The time Δt_1 for a change in kinetic energy ΔK_1 is:

$$\Delta t_1 = \frac{\Delta K_1}{P}$$

2. If Δt_2 is the time needed for a change in kinetic energy ΔK_2, the times Δt_1 and Δt_2 are related by:

$$\frac{\Delta t_2}{\Delta t_1} = \frac{\Delta K_2}{\Delta K_1} = \frac{\frac{1}{2}mv_{2f}^2 - \frac{1}{2}mv_{2i}^2}{\frac{1}{2}mv_{1f}^2 - \frac{1}{2}mv_{1i}^2} = \frac{v_{2f}^2 - v_{2i}^2}{v_{1f}^2 - v_{1i}^2}$$

3. Substitute the given values for the speeds:

$$\frac{\Delta t_2}{\Delta t_1} = \frac{v_{2f}^2 - v_{2i}^2}{v_{1f}^2 - v_{1i}^2} = \frac{(112 \text{ km/h})^2 - (80 \text{ km/h})^2}{(96 \text{ km/h})^2 - 0} = 0.667$$

4. Solve for Δt_2:

$$\Delta t_2 = 0.667\,\Delta t_1 = (0.667)(6.5 \text{ s}) = 4.33 \text{ s}$$

Remark The actual time to accelerate from 80 km/h to 112 km/h as measured in tests was 4.0 s. Air resistance, which we have neglected, is greater at higher speeds, but automobile engines actually have somewhat greater power at higher speeds.

Exercise A car accelerates from 0 to 40 km/h in T seconds. If the power output of the car is constant, how long does it take for the car to accelerate from 40 km/h to 80 km/h? (*Answer* $3T$ seconds; see Figure 6-18.)

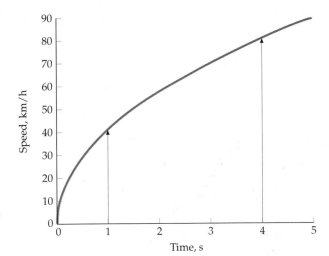

Figure 6-18 The arrows indicate the times at which the speed is 40 km/h and 80 km/h.

Example 6-11	*try it yourself*

A truck of mass m is accelerated from rest at $t = 0$ with constant power P along a level road. (*a*) **Find the speed of the truck as a function of time.** (*b*) **Show that if $x = 0$ at time $t = 0$, the position function $x(t)$ is given by**

$$x = \sqrt{\frac{8P}{9m}}\, t^{3/2}$$

Picture the Problem You can calculate the velocity function by integrating the acceleration $a = dv/dt = P/mv$, as given by Equation 6-19. The position function $x(t)$ can then be obtained by integrating the velocity.

Cover the column to the right and try these on your own before looking at the answers.

Steps   **Answers**

(*a*)1. Show that Equation 6-19 can be written as $v\, dv = \dfrac{P}{m}\, dt$
        $v\, dv = $ (constant) dt.

2. Integrate to obtain v^2 using the fact that $\dfrac{v^2}{2} = \dfrac{P}{m}\, t$
        $v_0 = 0$.

3. Solve for v. $v = \left(\dfrac{2P}{m}\right)^{1/2} t^{1/2}$

(*b*)1. Set $v = dx/dt$ and solve for dx. $dx = \left(\dfrac{2P}{m}\right)^{1/2} t^{1/2}\, dt$

2. Integrate to obtain $x(t)$. $x = \displaystyle\int dx = \int \left(\dfrac{2P}{m}\right)^{1/2} t^{1/2}\, dt = \left(\dfrac{8P}{9m}\right)^{1/2} t^{3/2}$

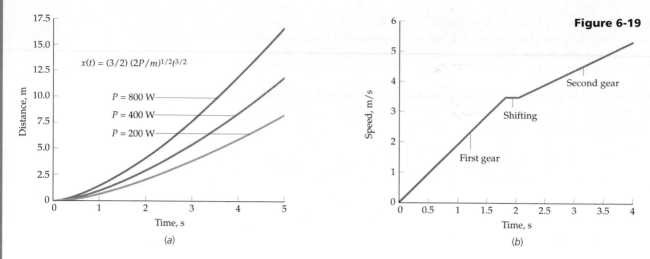

$$x(t) = (3/2)(2P/m)^{1/2}t^{3/2}$$

P = 800 W
P = 400 W
P = 200 W

(a)

Figure 6-19

Second gear

Shifting

First gear

(b)

Remark Figure 6-19 shows two graphical views of power. Figure 6-19a shows distance versus time for various constant powers for a truck of mass $m = 1600$ kg. Figure 6-19b shows a vehicle's speed increasing approximately linearly with time up to a maximum speed, and then climbing with a lesser slope after gears are shifted at approximately 2 s.

6-4   Potential Energy

The total work done on a particle equals the change in its kinetic energy. But we are frequently interested in the work done on a *system* of two or more particles.* Often, the work done by external forces on a *system* does not increase the kinetic energy *of the system*, but instead is stored as **potential energy.**

Consider lifting a barbell of mass m to a height h. The work you do on the barbell is mgh. The kinetic energy of the barbell does not increase because the earth does negative work $-mgh$, so the total work on the barbell is zero. Now consider the barbell and the planet Earth (but not you) to be a *system* of particles. The external forces on the earth–barbell system are the gravitational attraction you exert on the earth, w, the force your feet exert on the earth, $w + mg$, and the force mg exerted by your hands on the barbell (Figure 6-20). (We can neglect the gravitational force you exert on the barbell.) The barbell moves, but the earth doesn't, so the only external force exerted on the system that does work is the force you exert on the barbell. The total work done on the earth–barbell system by forces *external* to the system is mgh. This work is stored as potential energy, which is associated with the configuration of the earth–barbell system.

Consider another system consisting of a dart and spring in a toy dart gun. You compress the spring by pushing the dart into the gun. The work you do on this system is stored as potential energy in the dart–spring system. Its configuration has been changed because the spring has been compressed. Figure 6-21 shows a schematic description of such a system. The spring is compressed by the two forces $\vec{F}_1$ and $\vec{F}_2$, which are equal and opposite. Note that even though each force does (positive) work on the two-mass system, the net external force on the system is zero.

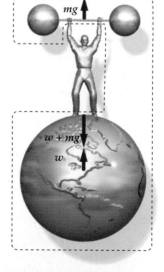

Figure 6-20 A system consisting of a barbell and the earth. When you lift the barbell, you do work on this system.

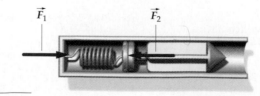

$\vec{F}_1$ $\vec{F}_2$

Figure 6-21 Potential energy of a dart gun.

---

* Systems of particles are discussed more thoroughly in Chapter 8.

Conservative Forces

When you ride a ski lift to the top of a hill of height h, the work done by the lift on you is mgh and that done by gravity is $-mgh$. When you ski down the hill to the bottom, the work done by gravity is $+mgh$ independent of the shape of the hill. The total work done by gravity on you during the round trip is zero independent of the path you take. The force of gravity exerted by the earth on you is called a **conservative force.**

> A force is conservative if the total work it does on a particle is zero when the particle moves around any closed path returning to its initial position.

Definition—Conservative force

From Figure 6-22 we see that this definition implies the following:

> The work done by a conservative force on a particle is independent of the path taken as the particle moves from one point to another.

Alternative definition—Conservative force

Now consider you and the earth to be a *two-particle system*. When a ski lift raises you to the top of the hill, it does work mgh on the system. This work is stored as potential energy of the system. When you ski down the hill, this potential energy is converted to kinetic energy of motion.

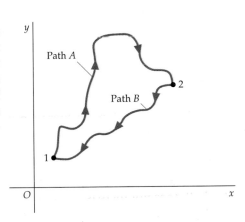

Figure 6-22 Two paths in space connecting the points 1 and 2. If the work done by a conservative force along path A from 1 to 2 is W, then the work done on the return trip along path B must be $-W$ because the round-trip work is zero. When traversing path B from 1 to 2, the force is the same at each point, but the displacement is opposite that when going from 2 to 1. Then the work done along path B from 1 to 2 must also be W. It follows that the work done as a particle going from point 1 to 2 is the same along any path connecting the two points.

Potential-Energy Functions

Since the work done by a conservative force on a particle does not depend on the path, it can depend only on the endpoints 1 and 2. We can use this property to define the **potential-energy function** U that is associated with a conservative force. Note that when the skier skis down the hill, the work done by gravity *decreases* the potential energy of the system. In general, we define the potential energy function such that the work done by a conservative force equals the decrease in the potential-energy function:

$$W = \int \vec{F} \cdot d\vec{s} = -\Delta U$$

or

$$\Delta U = U_2 - U_1 = -\int_{s_1}^{s_2} \vec{F} \cdot d\vec{s} \qquad \text{6-21a}$$

Definition—Potential-energy function

For infinitesimal displacement, we have

$$dU = -\vec{F} \cdot d\vec{s} \qquad \text{6-21b}$$

Gravitational Potential Energy Near the Earth's Surface   We can calculate the potential-energy function associated with the gravitational force near the surface of the earth from Equation 6-21b. For the force $\vec{F} = -mg\hat{j}$, we have

$$dU = -\vec{F} \cdot d\vec{s} = -(-mg\hat{j}) \cdot (dx\hat{i} + dy\hat{j} + dz\hat{k}) = +mg\,dy$$

Integrating, we obtain

$$U = \int mg\,dy = mgy + U_0$$

$$U = U_0 + mgy$$

6-22

Gravitational potential energy near the earth's surface

where U_0, the arbitrary constant of integration, is the value of the potential energy at $y = 0$. Since only a change in the potential energy is defined, the actual value of U is not important. We are free to choose U to be zero at any convenient reference point. For example, if the gravitational potential energy of the earth–skier system is chosen to be zero when the skier is at the bottom of the hill, its value when the skier is at a height h above that level is mgh. Or we could choose the potential energy to be zero when the skier is at sea level, in which case its value at any other point would be mgy, where y is measured from sea level.

Exercise   A 55-kg girl stands on a ledge that is 8 m above the ground. What is the potential energy U of the girl–earth system if (*a*) U is chosen to be zero on the ground and (*b*) U is chosen to be zero 4 m above the ground and (*c*) U is chosen to be zero 10 m above the ground? (*Answers*   (*a*) 4.32 kJ, (*b*) 2.16 kJ, (*c*) −1.08 kJ)

Example 6-12

A bottle of mass 0.350 kg falls from rest from a shelf that is 1.75 m above the floor. Find the original potential energy of the bottle–earth system relative to the floor, and the kinetic energy of the bottle just before it hits the floor.

Picture the Problem   We choose the potential energy of the bottle–earth system to be zero when the bottle is on the floor. The work done by the earth on the bottle as it falls equals the change in its kinetic energy.

1. The original potential energy U at $y = 1.75$ m is:
$$U = mgy = (0.350\text{ kg})(9.81\text{ N/kg})(1.75\text{ m})$$
$$= 6.01\text{ J}$$

2. Set the work done to be equal to the change in kinetic energy. The total work on the bottle is the work done by the earth:
$$\Delta K = W_{\text{total}} = mgy = 6.01\text{ J}$$

3. Since the original kinetic energy is zero, the final kinetic energy equals the change in kinetic energy:
$$K = \Delta K = 6.01\text{ J}$$

Remark   In this example, the potential energy lost by the bottle–earth system is converted entirely to kinetic energy of the bottle as it falls. Note that in step 1 we used the definition 1 J = 1 N·m.

Potential Energy of a Spring   Another example of a conservative force is that of a stretched spring. Suppose we pull a block attached to a spring from a position $x = 0$ (equilibrium) to x_1 (Figure 6-23). The spring does negative work because its force is opposite the direction of motion. If we then release the block, the spring does positive work as it accelerates the block toward its initial position. The total work done by the spring when the block reaches its initial position is zero independent of how far we stretched the spring (assuming we did not stretch the spring so far that it was damaged). The force exerted by the spring is therefore a conservative force. We can calculate the potential-energy function associated with this force from Equation 6-21b:

$$dU = -\vec{F} \cdot d\vec{s} = -F_x \, dx = -(-kx)dx = +kx \, dx$$

Then

$$U = \int kx \, dx = \frac{1}{2}kx^2 + U_0$$

where U_0 is the potential energy when $x = 0$, that is, when the spring is unstretched. Choosing U_0 to be zero gives

$$U = \frac{1}{2}kx^2 \qquad\qquad\qquad 6\text{-}23$$

Potential energy of a spring

When we pull the block from $x = 0$ to x_1, we must exert an applied force $F_{\text{app}} = +kx$ to balance the spring force. The work we do is

$$W_{\text{app}} = \int_0^{x_1} kx \, dx = \frac{1}{2}kx_1^2$$

This work is stored as potential energy in the spring–block system.

$F_{\text{app}} = kx$

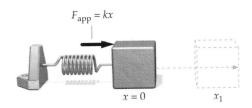

$x = 0$ x_1

Figure 6-23 To stretch the spring, a force $F_{\text{app}} = +kx$ must be applied to the block.

Example 6-13

Find the total potential energy of the basketball player hanging on the rim in Example 4-6 (Figure 6-24). Assume the player can be described as a point mass of 110 kg at 2 m above the floor and the force constant of the rim is 7.2 kN/m. The rim is displaced a distance $s = 15$ cm.

Picture the Problem The potential energy consists of gravitational potential energy, $U_\text{g} = mgy$, and energy stored in the displaced rim, whose potential energy is assumed to be the same as if it were a spring: $U_\text{s} = \frac{1}{2}ks^2$. Choose $y = 0$ at the floor for the gravitational potential energy.

$s = 15 \text{ cm} = 0.15 \text{ m}$

Figure 6-24

The total potential energy is the sum of gravitational potential energy and potential energy of the rim (see Figure 6-25):

$$U = U_g + U_s = mgy + \frac{1}{2}ks^2$$

$$= (110 \text{ kg})(9.81 \text{ N/kg})(2 \text{ m}) + \frac{1}{2}(7.2 \text{ kN/m})(0.15 \text{ m})^2$$

$$= 2158 \text{ J} + 81.0 \text{ J} = 2239 \text{ J}$$

Remark Nearly all of the potential energy is gravitational in this case, because even though the force constant of the "spring" is very large, the displacement is very small.

Exercise A 3-kg block is hung vertically from a spring with a force constant of 600 N/m. (*a*) By how much is the spring stretched when the block is in equilibrium? (*b*) How much potential energy is stored in the spring–block system? (*Answers* (*a*) 4.9 cm, (*b*) 0.72 J)

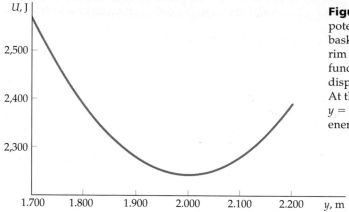

Figure 6-25 Total potential energy of the basketball player plus rim $U = U_s + U_g$ as a function of the vertical displacement of the player. At the equilibrium position $y = 2.00$ m, the potential energy is at a minimum.

Potential energy is associated with the configuration of a *system of particles*, but we sometimes have systems such as the earth–skier system, in which only one particle moves (the earth's motion is negligible). For brevity, then, we sometimes refer to the potential energy of the earth–skier system as simply the potential energy of the skier.

Nonconservative Forces

Not all forces are conservative. An example of a nonconservative force is kinetic friction. Suppose you push a box around some closed path on a rough table so that the box ends up at its original position. The force of kinetic friction is always opposite the direction of motion, so the work it does is always negative, and the total round-trip work it does cannot be zero. Another example of a nonconservative force is a force applied by a human agent. The work that you do in pushing a box around a closed path on a rough table is not generally zero. It depends on how great a force you decide to exert on the box. Thus, neither the force you exert nor the force of kinetic friction is conservative, and no potential-energy function can be defined for either.

Potential Energy and Equilibrium in One Dimension

For a general conservative force in one dimension, $\vec{F} = F_x\hat{i}$, Equation 6-21*b* is

$$dU = -\vec{F}\cdot d\vec{s} = -F_x \, dx$$

The force is therefore the negative derivative of the potential-energy function:

$$F_x = -\frac{dU}{dx} \qquad\qquad 6\text{-}24$$

We can illustrate this general relation for a block–spring system by differentiating the function $U = \frac{1}{2}kx^2$. We obtain

$$F_x = -\frac{dU}{dx} = -\frac{d}{dx}\left(\frac{1}{2}kx^2\right) = -kx$$

Figure 6-26 shows a plot of $U = \frac{1}{2}kx^2$ versus x for a block and spring. The derivative of this function is represented graphically as the slope of the line tangent to the curve. The force is thus equal to the negative of the slope of the curve. At $x = 0$, the force $F_x = -dU/dx$ is zero and the block is in equilibrium.

A particle is in equilibrium if the net force acting on it is zero.

Condition for equilibrium

When x is positive in Figure 6-26, the slope is positive and the force F_x is negative. When x is negative the slope is negative and the force F_x is positive. In either case, the force is in the direction that will accelerate the block toward lower potential energy. If the block is displaced slightly from $x = 0$, the force is directed back toward $x = 0$. The equilibrium at $x = 0$ is thus **stable equilibrium**.

In stable equilibrium, a small displacement results in a restoring force that accelerates the particle back toward its equilibrium position.

Figure 6-27 shows a potential-energy curve with a maximum rather than a minimum at the equilibrium point $x = 0$. Such a curve could represent the potential energy of a skier at the top of a hill. For this curve, when x is positive, the slope is negative and the force F_x is positive, and when x is negative, the slope is positive and the force F_x is negative. Again, the force is in the direction that will accelerate the particle toward lower potential energy, but this time the force is away from the equilibrium position. The maximum at $x = 0$ in Figure 6-27 is a point of **unstable equilibrium**.

In unstable equilibrium, a small displacement results in a force that accelerates the particle away from its equilibrium position.

Figure 6-28 shows a potential-energy curve that is flat in the region near $x = 0$. No force acts on a particle at $x = 0$, and hence the particle is at equilibrium; furthermore, there will be no resulting force if the particle is displaced slightly in either direction. This is an example of **neutral equilibrium.**

In neutral equilibrium, a small displacement results in zero force and the particle remains in equilibrium.

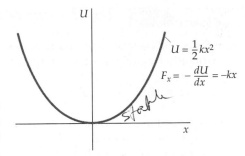

Figure 6-26 Plot of the potential-energy function U versus x for an object on a spring. A minimum in a potential energy curve is a point of stable equilibrium. Displacement in either direction results in a force directed toward the equilibrium position.

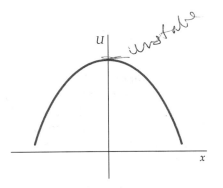

Figure 6-27 A particle at $x = 0$ on this potential-energy curve will be in unstable equilibrium because a displacement in either direction results in a force directed away from the equilibrium position.

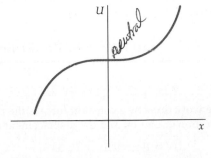

Figure 6-28 Neutral equilibrium. The force $Fx = -dU/dx$ is zero at $x = 0$ and at neighboring points, so displacement away from $x = 0$ results in no force, and the system remains in equilibrium.

Example 6-14   *try it yourself*

The force between two atoms in a diatomic molecule can be represented approximately by the potential-energy function

$$U = U_0\left[\left(\frac{a}{x}\right)^{12} - 2\left(\frac{a}{x}\right)^{6}\right]$$

where U_0 and a are constants (Figure 6-29). (*a*) At what value of x is the potential energy zero? (*b*) Find the force F_x. (*c*) At what value of x is the potential energy a minimum? Show that $U_{min} = -U_0$.

Picture the Problem The force is the negative derivative of the potential-energy function. The potential energy has its minimum value when its slope is zero; that is, when the force is zero.

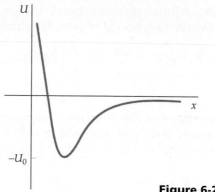

Figure 6-29

Cover the column to the right and try these on your own before looking at the answers.

Steps	Answers
(*a*) Set $U = 0$ and solve for x.	$x = \dfrac{a}{\sqrt[6]{2}}$
(*b*) Compute $F_x = -dU/dx$	$F_x = \dfrac{12U_0}{a}\left[\left(\dfrac{a}{x}\right)^{13} - \left(\dfrac{a}{x}\right)^{7}\right]$
(*c*) 1. Set F_x equal to zero and solve for x.	$x = a$
2. Use your result to find U_{min}.	$U_{min} = -U_0$

Remark This potential-energy function is generally known as the "Lennard-Jones" or "6–12" potential. The minimum occurs at $x = a$, which is the average spacing between atoms in such a molecule. The lowest energy of the molecule is slightly greater than the minimum $-U_0$, so the energy needed to separate the atoms is slightly less than U_0.

Summary

1. Work, kinetic energy, potential energy, and power are important derived dynamic quantities.

2. The work–kinetic energy theorem is an important relation derived from Newton's laws applied to a particle.

3. The dot product of vectors is a mathematical definition that is useful throughout physics.

Topic	Remarks and Relevant Equations
1. Work	
Constant force	The work done by a constant force is the product of the component of the force in the direction of motion and the displacement of the force:

$$W = F \cos\theta\, \Delta x = F_x\, \Delta x \qquad \text{6-1}$$

Variable force	$$W = \int_{x_1}^{x_2} F_x\, dx = \text{area under the } F_x\text{-versus-}x \text{ curve}$$	6-9

Force in three dimensions	$$W = \int_1^2 \vec{F} \cdot d\vec{s}$$	6-14

Units	The SI unit of work and energy is the joule (J): $$1\,\text{J} = 1\,\text{N} \cdot \text{m}$$	6-2

2. Kinetic Energy	$$K = \frac{1}{2} m v^2$$	6-6

3. Work–Kinetic Energy Theorem	$$W_{\text{total}} = \Delta K = \frac{1}{2} m v_f^2 - \frac{1}{2} m v_i^2$$	6-7

4. Dot Product	$$\vec{A} \cdot \vec{B} = AB \cos \phi$$ where ϕ is the angle between the vectors.	6-11

In terms of components	$$\vec{A} \cdot \vec{B} = A_x B_x + A_y B_y + A_z B_z$$	6-12

Vector component	$$\vec{A} \cdot \hat{i} = A_x$$	6-13

5. Power	$$P = \frac{dW}{dt} = \vec{F} \cdot \vec{v}$$	6-17

6. Conservative Force	A force is conservative if the total work it does on a particle is zero when the particle moves along any path that returns it to its initial position. The work done by a conservative force on a particle is independent of the path taken by the particle as it moves from one point to another.	

7. Potential Energy	The potential energy of a system is the energy associated with the configuration of the system. The change in the potential energy of a system is defined as the negative of the work done by conservative forces acting on the system.	

| Definition | $$\Delta U = U_2 - U_1 = -W = -\int_1^2 \vec{F} \cdot d\vec{s}$$ | 6-21a |
| | $$dU = -\vec{F} \cdot d\vec{s}$$ | 6-21b |

Gravitational	$$U = U_0 + mgy$$	6-22

Spring	$$U = \frac{1}{2} k x^2$$	6-23

Conservative force	In one dimension, a conservative force equals the negative derivative of the potential-energy function associated with it: $$F_x = -\frac{dU}{dx}$$	6-24

Potential-energy curve	At a minimum on the curve of the potential-energy function versus the displacement, the force is zero and the system is in stable equilibrium. At a maximum, the force is zero and the system is in unstable equilibrium. A conservative force always tends to accelerate a particle toward a position of lower potential energy.	

Problem-Solving Guide

1. Begin by drawing a neat diagram that includes the important features of the problem.
2. The work–kinetic energy theorem relates the initial and final speeds of a particle to the total work done on the particle.

Summary of Worked Examples

Type of Calculation	Procedure and Relevant Examples

1. Work

Find the work done by a constant force.	$W = F_x \Delta x$	**Examples 6-1, 6-2, 6-3**
Find the work done by a force that varies with position.	The work is given by $W = \int_{x_1}^{x_2} F \, dx$. This integral equals the area under the F-versus-x curve. For a spring, $$W = \int_{x_1}^{x_2} -kx \, dx = -\frac{1}{2}k\left(x_2^2 - x_1^2\right).$$	**Examples 6-4,6-5**
Find the work done by a constant force when $\vec{F}$ and $\Delta\vec{s}$ are given in terms of unit vectors.	Compute $W = \vec{F} \cdot \Delta\vec{s}$.	**Example 6-7**
Find the work done by gravity.	The work is $+mgh$ if the object moves downward and $-mgh$ if it moves upward, independent of the path.	**Example 6-8**

2. Work–Kinetic Energy Theorem

Find the final speed of an object.	The final speed is found from the kinetic energy, which is obtained from the work–kinetic energy theorem. $W_{\text{total}} = \Delta K$.	**Examples 6-1, 6-2, 6-3, 6-5, 6-8, 6-12**

3. Find the angle between two vectors or the component of one vector along another.

	Find the dot product $\vec{A} \cdot \vec{B} = AB \cos \phi$. Then divide by the magnitudes obtained from $A = \sqrt{\vec{A} \cdot \vec{A}}$ and $B = \sqrt{\vec{B} \cdot \vec{B}}$. The component of $\vec{B}$ along $\vec{A}$ is $B_A = \vec{B} \cdot (\vec{A}/A) = B \cos \phi$. **Examples 6-6, 6-7**

4. Power

Find the power supplied by a force.	The instantaneous power is $P = \vec{F} \cdot \vec{v}$.	**Example 6-9**
Find the time needed to accelerate from one speed to another at constant power.	Find the change in kinetic energy and use $P \, \Delta t = \Delta K$.	**Example 6-10**
Find v and x at a given time t.	If $x_0 = 0$ and $v_0 = 0$, use the results of Example 6-2. Otherwise, derive expressions for v from $a = dv/dt = P/mv$ and $dx/dt = v$.	**Example 6-11**

5. Potential Energy

Find the gravitational potential energy of an object.	Use $U_g = mgy$.	**Example 6-12**
Find the potential energy of a stretched spring.	Use $U_s = \frac{1}{2}kx^2$.	**Example 6-13**
Find the force from the potential energy function.	Compute $F_x = -dU/dx$.	**Example 6-14**

Problems

In a few problems, you are given more data than you actually need; in a few other problems, you are required to supply data from your general knowledge, outside sources, or informed estimates.

- • Single-concept, single-step, relatively easy
- •• Intermediate-level, may require synthesis of concepts
- ••• Challenging, for advanced students

Take g = 9.81 N/kg = 9.81 m/s² and neglect friction in all problems unless otherwise stated.

Work and Kinetic Energy

1 • True or false:

(a) Only the net force acting on an object can do work.
(b) No work is done on a particle that remains at rest.
(c) A force that is always perpendicular to the velocity of a particle never does work on the particle.

2 • A heavy box is to be moved from the top of one table to the top of another table of the same height on the other side of the room. Is work required to do this?

3 • To get out of bed in the morning, do you have to do work?

4 • By what factor does the kinetic energy of a car change when its speed is doubled?

5 • An object moves in a circle at constant speed. Does the force that accounts for its acceleration do work on it? Explain.

6 • An object initially has kinetic energy K. The object then moves in the opposite direction with three times its initial speed. What is the kinetic energy now?

(a) K
(b) $3K$
(c) $-3K$
(d) $9K$
(e) $-9K$

7 • A 15-g bullet has a speed of 1.2 km/s. (a) What is its kinetic energy in joules? (b) What is its kinetic energy if its speed is halved? (c) What is its kinetic energy if its speed is doubled?

8 • Find the kinetic energy in joules of (a) a 0.145-kg baseball moving with a speed of 45 m/s and (b) a 60-kg jogger running at a steady pace of 9 min/mi.

9 • A 6-kg box is raised from rest a distance of 3 m by a vertical force of 80 N. Find (a) the work done by the force, (b) the work done by gravity, and (c) the final kinetic energy of the box.

10 • A constant force of 80 N acts on a box of mass of 5.0 kg that is moving in the direction of the applied force with a speed of 20 m/s. Three seconds later the box is moving with a speed of 68 m/s. Determine the work done by this force.

11 •• You run a race with your girlfriend. At first you each have the same kinetic energy, but you find that she is beating you. When you increase your speed by 25%, you are running at the same speed she is. If your mass is 85 kg, what is her mass?

Work Done by a Variable Force

12 • How does the work required to stretch a spring 2 cm from its natural length compare with that required to stretch it 1 cm from its natural length?

13 •• A 3-kg particle is moving with a speed of 2 m/s when it is at $x = 0$. It is subjected to a single force F_x that varies with position as shown in Figure 6-30. (a) What is the kinetic energy of the particle when it is at $x = 0$? (b) How much work is done by the force as the particle moves from $x = 0$ to $x = 4$ m? (c) What is the speed of the particle when it is at $x = 4$ m?

Figure 6-30
Problem 13

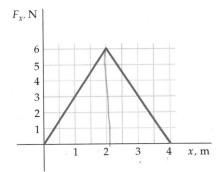

14 •• A 4-kg particle is initially at rest at $x = 0$. It is subjected to a single force F_x that varies with position as shown in Figure 6-31. Find the work done by the force as the particle moves (a) from $x = 0$ to $x = 3$ m, and (b) from $x = 3$ m to $x = 6$ m. Find the kinetic energy of the particle when it is at (c) $x = 3$ m and (d) $x = 6$ m.

Figure 6-31
Problem 14

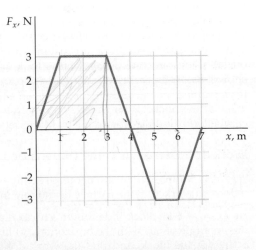

15 •• A force F_x acts on a particle. The force is related to the position of the particle by the formula $F_x = Cx^3$, where C is a constant. Find the work done by this force on the particle when the particle moves from $x = 1.5$ m to $x = 3$ m.

16 •• Lou's latest invention, aimed at urban dog owners, is the X-R-Leash. It is made of a rubber-like material that exerts a force $F_x = -kx - ax^2$ when it is stretched a distance x, where k and a are constants. The ad claims, "You'll never go back to your old dog leash after you've had the thrill of an X-R-Leash experience. And you'll see a new look of respect in the eyes of your proud pooch." Find the work done on a dog by the leash if the person remains stationary and the dog bounds off, stretching the X-R-Leash from $x = 0$ to $x = x_0$.

17 •• A 3-kg object is moving with a speed of 2.40 m/s in the x direction when it passes the origin. It is acted on by a single force F_x that varies with x as shown in Figure 6-32. (a) What is the work done by the force from $x = 0$ to $x = 2$ m? (b) What is the kinetic energy of the object at $x = 2$ m? (c) What is the speed of the object at $x = 2$ m? (d) What is the work done on the object from $x = 0$ to $x = 4$ m? (e) What is the speed of the object at $x = 4$ m?

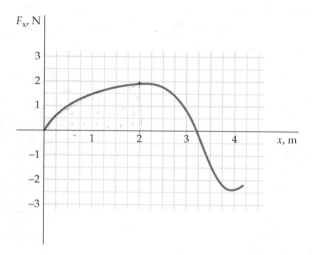

Figure 6-32 Problem 17

18 •• Near Margaret's cabin is a 20-m water tower that attracts many birds during the summer months. During a hot spell last year, the tower went dry, and Margaret had to have her water hauled in. She got lonesome without the birds visiting, so she decided to carry some water up the tower to attract them back. Her bucket has a mass of 10 kg and holds 30 kg of water when it is full. However, the bucket has a hole, and as Margaret climbed at a constant speed, water leaked out at a constant rate. Several birds took advantage of the shower below, but when she got to the top, only 10 kg of water remained for the birdbath. (a) Write an expression for the mass of the bucket plus water as a function of the height y climbed. (b) Find the work done by Margaret on the bucket.

Work and Energy in Three Dimensions

19 • Suppose there is a net force acting on a particle but it does no work. Can the particle be moving in a straight line?

20 • A 6-kg block slides down a frictionless incline making an angle of 60° with the horizontal. (a) List all the forces acting on the block, and find the work done by each force when the block slides 2 m (measured along the incline). (b) What is the total work done on the block? (c) What is the speed of the block after it has slid 1.5 m if it starts from rest?

(d) What is its speed after 1.5 m if it starts with an initial speed of 2 m/s?

21 • An 85-kg cart is deposited on a 1.5-m platform after being rolled up an incline formed by a plank of length L that has been laid from the lower level to the top of the platform. (Assume that the rolling is equivalent to sliding without friction.) (a) Find the force parallel to the incline needed to push the cart up without acceleration for $L = 3, 4$, and 5 m. (b) Calculate directly from Equation 6-15 the work needed to push the cart up the incline for each value of L. (c) Since the work found in (b) is the same for each value of L, what advantage, if any, is there in choosing one length over another?

22 • A 2-kg object attached to a horizontal string moves with a speed of 2.5 m/s in a circle of radius 3 m on a frictionless horizontal surface. (a) Find the tension in the string. (b) List the forces acting on the object, and find the work done by each force during one revolution.

Dot Products

23 • What is the angle between the vectors $\vec{A}$ and $\vec{B}$ if $\vec{A} \cdot \vec{B} = -AB$?

24 • Two vectors $\vec{A}$ and $\vec{B}$ have magnitudes of 6 m and make an angle of 60° with each other. Find $\vec{A} \cdot \vec{B}$.

25 • Find $\vec{A} \cdot \vec{B}$ for the following vectors: (a) $\vec{A} = 3\hat{i} - 6\hat{j}, \vec{B} = -4\hat{i} + 2\hat{j}$; (b) $\vec{A} - 5\hat{i} + 5\hat{j}, \vec{B} = 2\hat{i} - 4\hat{j}$; and (c) $\vec{A} = 6\hat{i} + 4\hat{j}, \vec{B} = 4\hat{i} - 6\hat{j}$.

26 • Find the angles between the vectors $\vec{A}$ and $\vec{B}$ in Problem 25.

27 • A 2-kg object is given a displacement $\Delta\vec{s} = 3$ m $\hat{i} + 3$ m $\hat{j} - 2$ m $\hat{k}$ along a straight line. During the displacement, a constant force $\vec{F} = 2$ N $\hat{i} - 1$ N $\hat{j} + 1$ N $\hat{k}$ acts on the object. (a) Find the work done by $\vec{F}$ for this displacement. (b) Find the component of $\vec{F}$ in the direction of the displacement.

28 •• (a) Find the unit vector that is parallel to the vector $\vec{A} = A_x\hat{i} + A_y\hat{j} + A_z\hat{k}$. (b) Find the component of the vector $\vec{A} = 2\hat{i} - \hat{j} - \hat{k}$ in the direction of the vector $\vec{B} = 3\hat{i} + 4\hat{j}$.

29 •• When a particle moves in a circle with constant speed, the magnitudes of its position vector and velocity vectors are constant. (a) Differentiate $\vec{r} \cdot \vec{r} = r^2 = $ constant with respect to time to show that $\vec{v} \cdot \vec{r} = 0$ and therefore $\vec{v} \perp \vec{r}$. (b) Differentiate $\vec{v} \cdot \vec{v} = v^2 = $ constant with respect to time to show that $\vec{a} \cdot \vec{v} = 0$ and therefore $\vec{a} \perp \vec{v}$. What do the results of (a) and (b) imply about the direction of $\vec{a}$? (c) Differentiate $\vec{v} \cdot \vec{r} = 0$ with respect to time and show that $\vec{a} \cdot \vec{r} + v^2 = 0$ and therefore $a_r = -v^2/r$.

30 •• Vectors $\vec{A}$, $\vec{B}$, and $\vec{C}$ form a triangle as shown in Figure 6-33. The angle between $\vec{A}$ and $\vec{B}$ is θ, and the vectors are related by $\vec{C} = \vec{A} - \vec{B}$. Compute $\vec{C} \cdot \vec{C}$ in terms of A, B, and θ, and derive the law of cosines, $C^2 = A^2 + B^2 - 2AB \cos\theta$.

Figure 6-33 Problem 30

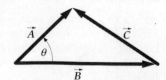

Power

31 • The dimension of power is _____.

(a) $[M][L]^2[T]^2$
(b) $[M][L]^2/[T]$
(c) $[M][L]^2/[T]^2$
(d) $[M][L]^2/[T]^3$

32 • True or false: A kilowatt-hour is a unit of power.

33 • The engine of a car operates at constant power. The ratio of acceleration of the car at a speed of 60 km/h to that at 30 km/h (neglecting air resistance) is _____.

(a) $\frac{1}{2}$ (b) $1/\sqrt{2}$ (c) $\sqrt{2}$ (d) 2

34 •• A car starts from rest and travels at constant acceleration. Which of the following statements are true?

(a) The power delivered by the engine is constant.
(b) The power delivered by the engine increases as the car gains speed.
(c) The power delivered by the engine decreases as the car gains speed.
(d) Both (b) and (c) are correct.

35 •• Force A does 5 J of work in 10 s. Force B does 3 J of work in 5 s. Which force delivers greater power?

36 • A 5-kg box is lifted by a force equal to the weight of the box. The box moves upward at a constant velocity of 2 m/s. (a) What is the power input of the force? (b) How much work is done by the force in 4 s?

37 • Fluffy has just caught a mouse, and decides that the only decent thing to do is to bring it to the bedroom so that his human roommate can admire it when she wakes up. A constant horizontal force of 3 N is enough to drag the mouse across the rug at a constant speed v. If Fluffy's force does work at the rate of 6 W, (a) what is her speed, v? (b) How much work does Fluffy do in 4 s?

38 • A single force of 5 N acts in the x direction on an 8-kg object. (a) If the object starts from rest at $x = 0$ at time $t = 0$, find its velocity v as a function of time. (b) Write an expression for the power input as a function of time. (c) What is the power input of the force at time $t = 3$ s?

39 • Find the power input of a force $\vec{F}$ acting on a particle that moves with a velocity $\vec{v}$ for (a) $\vec{F} = 4\,\text{N}\,\hat{\imath} + 3\,\text{N}\,\hat{k}$, $\vec{v} = 6\,\text{m/s}\,\hat{\imath}$; (b) $\vec{F} = 6\,\text{N}\,\hat{\imath} - 5\,\text{N}\,\hat{\jmath}$, $\vec{v} = -5\,\text{m/s}\,\hat{\imath} + 4\,\text{m/s}\,\hat{\jmath}$; and (c) $\vec{F} = 3\,\text{N}\,\hat{\imath} + 6\,\text{N}\,\hat{\jmath}$, $\vec{v} = 2\,\text{m/s}\,\hat{\imath} + 3\,\text{m/s}\,\hat{\jmath}$.

40 •• A particle of mass m moves from rest at $t = 0$ under the influence of a single force of magnitude F. Show that the power delivered by the force at time t is $P = F^2t/m$.

41 •• At a speed of 20 km/h, a 1200-kg car accelerates at 3 m/s² using 20 kW of power. How much power must be expended to accelerate the car at 2 m/s² at a speed of 40 km/h?

42 •• A car manufacturer claims that his car can accelerate from rest to 100 km/h in 8 s. The car's mass is 800 kg. (a) Assuming that this performance is achieved at constant power, determine the power developed by the car's engine. (b) What is the car's speed after 4 s? (Neglect friction and air resistance.)

43 •• Show that the position of the truck in Example 6-11 is related to its speed by $x = (m/3P)v^3$.

44 •• A 700-kg car accelerates from rest under constant power. At the end of 8.0 s, its speed is 90 km/h and it is located 133 m from its starting point. If the car continues to accelerate using the same power, what will its speed be at the end of 10 s, and how far will the car be from the starting point?

45 •• A 4.0-kg object initially at rest at $x = 0$ is accelerated at constant power of 8.0 W. At $t = 9.0$ s, it is at $x = 36.0$ m. Find its speed at $t = 6.0$ s and its position at that instant.

46 •• A 700-kg car accelerates from rest under constant power at $t = 0$. At $t = 9$ s it is 117.7 m from its starting point and its acceleration is then 1.09 m/s². Find the power expended by the car's engine, neglecting frictional losses.

Potential Energy

47 • Two knowledge seekers decide to ascend a mountain. Sal chooses a short, steep trail, while Joe, who weighs the same as Sal, goes up via a long, gently sloped trail. At the top, they get into an argument about who gained more potential energy. Which of the following is true?

(a) Sal gains more gravitational potential energy than Joe.
(b) Sal gains less gravitational potential energy than Joe.
(c) Sal gains the same gravitational potential energy as Joe.
(d) To compare energies, we must know the height of the mountain.
(e) To compare energies, we must know the length of the two trails.

48 • The gravitational potential energy of an object changes by -6 J. It follows that the work done by the gravitational force on this object is

(a) -6 J and the elevation of the object is increased.
(b) -6 J and the elevation of the object is decreased.
(c) $+6$ J and the elevation of the object is increased.
(d) $+6$ J and the elevation of the object is decreased.

49 • A woman runs up a flight of stairs. The gain in her gravitational potential energy is U. If she runs up the same stairs with twice the speed, what will be her gain in potential energy?

(a) U
(b) $2U$
(c) $U/2$
(d) $4U$
(e) $U/4$

50 • Which of the following statements is true?

(a) The kinetic and potential energies of an object must always be positive quantities.
(b) The kinetic and potential energies of an object must always be negative quantities.
(c) Kinetic energy can be negative, but potential energy cannot.
(d) Potential energy can be negative, but kinetic energy cannot.
(e) None of the preceding statements is true.

51 • A block slides a certain distance down an incline. The work done by gravity is W. What is the work done by gravity if this block slides the same distance up the incline?

(a) W
(b) Zero
(c) −W
(d) Gravity can't do work; some other force does work.
(e) Cannot be determined unless given the distance traveled.

52 • True or false:

(a) Only conservative forces can do work.
(b) If only conservative forces act, the kinetic energy of a particle does not change.
(c) The work done by a conservative force equals the decrease in the potential energy associated with that force.

53 • When you climb a mountain, is the work done on you by gravity different if you take a short, steep trail instead of a long, gentle trail? If not, why do you find one trail easier?

54 • Which of the following forces are conservative and which are nonconservative?

(a) the frictional force exerted on a sliding box
(b) the force exerted by a linear spring that obeys Hooke's law
(c) the force of gravity
(d) the wind resistance on a moving car

55 • An 80-kg man climbs up a 6-m high flight of stairs. What is the increase in gravitational potential energy?

56 • One of the highlights of Sharika's concert is her daredevil swan dive into the audience from a height of 2 m above the crowd's outstretched hands. If her mass is 60 kg, and the time of her dive is defined as $t = 0$, (a) what is her initial potential energy relative to $U = 0$ at the position of the crowd's hands? (b) From Newton's laws, find the distance she has fallen and her speed at $t = 0.20$ s. (c) Find her potential and kinetic energy at $t = 0.40$ s. (d) Find her kinetic energy and speed just as she reaches the hands of the crowd in the mosh pit.

57 • Water flows over Victoria Falls, which is 128 m high, at an average rate of 1.4×10^6 kg/s. If half the potential energy of this water were converted into electric energy, how much power would be produced by these falls?

58 • A 2-kg box slides down a long, frictionless incline of angle 30°. It starts from rest at time $t = 0$ at the top of the incline at a height of 20 m above the ground. (a) What is the original potential energy of the box relative to the ground? (b) From Newton's laws, find the distance the box travels in 1 s and its speed at $t = 1$ s. (c) Find the potential energy and the kinetic energy of the box at $t = 1$ s. (d) Find the kinetic energy and the speed of the box just as it reaches the bottom of the incline.

59 • A force $F_x = 6$ N is constant. (a) Find the potential-energy function U associated with this force for an arbitrary reference position x_0 at which $U = 0$. (b) Find U such that $U = 0$ at $x = 4$ m. (c) Find U such that $U = 14$ J at $x = 6$ m.

60 • A spring has a force constant of $k = 10^4$ N/m. How far must it be stretched for its potential energy to be (a) 50 J and (b) 100 J?

61 •• A simple Atwood's machine uses two masses, m_1 and m_2 (Figure 6-34). Starting from rest, the speed of the two masses is 4.0 m/s at the end of 3.0 s. At that instant, the kinetic energy of the system is 80 J and each mass has moved a distance of 6.0 m. Determine the values of m_1 and m_2.

Figure 6-34 Problem 61

62 •• A straight rod of negligible mass is mounted on a frictionless pivot as in Figure 6-35. Masses m_1 and m_2 are suspended at distances l_1 and l_2. (a) Write an expression for the gravitational potential energy of the masses as a function of the angle θ made by the rod and the horizontal. (b) For what angle θ is the potential energy a minimum? Is the statement "systems tend to move toward a configuration of minimum potential energy" consistent with your result? (c) Show that if $m_1 l_1 = m_2 l_2$, the potential energy is the same for all values of θ. (When this holds, the system will balance at any angle θ. This result is known as *Archimedes' law of the lever*.)

Figure 6-35 Problem 62

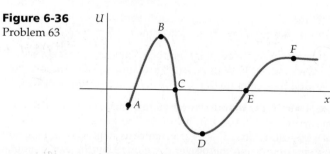

Force, Potential Energy, and Equilibrium

63 •• Figure 6-36 shows the plot of a potential-energy function U versus x. (a) At each point indicated, state whether the force F_x is positive, negative, or zero. (b) At which point does the force have the greatest magnitude? (c) Identify any equilibrium points, and state whether the equilibrium is stable or unstable.

Figure 6-36 Problem 63

64 • (a) Find the force F_x associated with the potential-energy function $U = Ax^4$, where A is a constant. (b) At what point(s) is the force zero?

65 •• A potential-energy function is given by $U = C/x$, where C is a positive constant. (a) Find the force F_x as a function of x. (b) Is this force directed toward the origin or away from it? (c) Does the potential energy increase or decrease as x increases? (d) Answer parts (b) and (c) where C is a negative constant.

66 •• On the potential-energy curve for U versus y shown in Figure 6-37, the segments AB and CD are straight lines. Sketch a plot of the force F_y versus y.

Figure 6-37 U, J
Problem 66

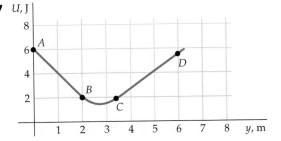

67 •• The force acting on an object is given by $F_x = a/x^2$. Determine the potential energy of the object as a function of x.

68 •• The potential energy of an object is given by $U(x) = 3x^2 - 2x^3$, where U is in joules and x is in meters. (a) Determine the force acting on this object. (b) At what positions is this object in equilibrium? (c) Which of these equilibrium positions are stable and which are unstable?

69 •• During a Dead Wait concert, Sharika and Chico, each of mass M, are attached to the ends of a light rope that is hung over two frictionless pegs, as shown in Figure 6-38. A large gong of mass m is attached to the middle of the rope, between the pegs, and Sharika and Chico beat it madly in lieu of the usual guitar solo. (a) Find the potential energy of the system as a function of the distance y shown in the figure. (b) Find the value of y for which the potential energy function of the system is a minimum. (c) Find the equilibrium distance y_0 using the potential energy function. (d) Check your answer by applying Newton's laws to the gong.

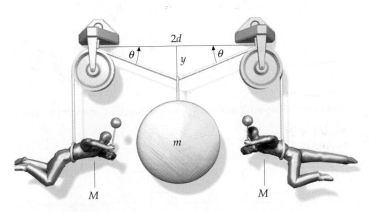

Figure 6-38 Problem 69

70 ••• The potential energy of an object is given by $U(x) = 8x^2 - x^4$, where U is in joules and x is in meters. (a) Determine the force acting on this object. (b) At what positions is this object in equilibrium? (c) Which of these equilibrium positions are stable and which are unstable?

71 ••• The force acting on an object is given by $F(x) = x^3 - 4x$. Locate the positions of unstable and stable equilibrium and show that at these points $U(x)$ is a local maximum or minimum, respectively.

72 ••• The potential energy of a 4-kg object is given by $U = 3x^2 - x^3$ for $x \le 3$ m, and $U = 0$ for $x \ge 3$ m, where U is in joules and x is in meters. (a) At what positions is this object in equilibrium? (b) Sketch a plot of U versus x. (c) Discuss the stability of the equilibrium for the values of x found in (a). (d) If the total energy of the particle is 12 J, what is its speed at $x = 2$ m?

73 ••• A force is given by $F_x = Ax^{-3}$, where $A = 8$ N·m³. (a) For positive values of x, does the potential energy associated with this force increase or decrease with increasing x? (You can determine the answer to this question by imagining what happens to a particle that is placed at rest at some point x and is then released.) (b) Find the potential-energy function U associated with this force such that U approaches zero as x approaches infinity. (c) Sketch U versus x.

General Problems

74 • True or false:

(a) Only the net force acting on an object can do work.
(b) Work is the area under the force-versus-time curve.

75 • Negative work by an applied force implies that

(a) the kinetic energy of the object increases.
(b) the applied force is variable.
(c) the applied force is perpendicular to the displacement.
(d) the applied force has a component that is opposite to the displacement.
(e) nothing; there is no such thing as negative work.

76 •• A movie crew is in the Badlands when their car overheats. After they stop to let it cool down, an argument breaks out. They agree that they must go easy on the engine, but they disagree about when the engine works the hardest, and therefore about how they should drive for the rest of the trip. Carolyn claims that the work done by the car in accelerating from 0 to 20 km/h is less than that required to accelerate from 20 to 30 km/h, meaning they should drive more slowly. Ted says no, the work done between 0 and 20 km/h is more than the work done between 20 and 30 km/h. Ernie says it all depends on the mass of the car, and Bloop says it all depends on how long you take to change from one speed to another. Who is right?

77 • Figure 6–39 shows two pulleys arranged to help lift a heavy load. A rope runs around two massless, frictionless pulleys and the weight $\vec{w}$ hangs from one pulley. You exert a force of magnitude F on the free end of the cord. (a) If the weight is to move up a distance h, through what distance must the force move? (b) How much work is

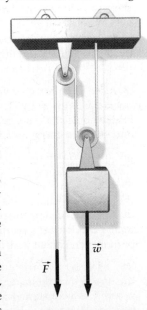

Figure 6-39 Problem 77

done on the weight? (c) How much work do you do? (This is an example of a simple machine in which a small force F_1 moves through a large distance x_1 to exert a large force F_2 (= w) through a smaller distance $x_2 = h$.)

78 • In February 1995, a total of 54.3 billion kW-h of electrical energy was generated by nuclear power plants in the United States. At the same time, the population of the United States was about 255 million people. If the average American has a mass of 60 kg, and if the entire energy output of all nuclear power plants was diverted to supplying energy for a single giant elevator, estimate the height h at which the entire population of the country could be lifted by the elevator. In your calculations, assume that 25% of the energy goes into lifting the people; assume also that g is constant over the entire height h.

79 • One of the most powerful cranes in the world, operating in Switzerland, can slowly raise a load of $M = 6000$ tonne to a height of $h = 12.0$ m (1 tonne = 1000 kg). (a) How much work is done by the crane? (b) If it takes 1.00 min to lift the load at constant velocity to this height, find the power developed by the crane.

80 • In Australia, there used to be a ski lift of length 5.6 km. It took about 60 min for a gondola to travel all the way up. If there were 12 gondolas going up at once, each of mass 550 kg, and the angle of ascent was 30°, estimate the power P of the engine needed to operate the ski lift.

81 • A 2.4-kg object attached to a horizontal string moves with constant speed in a circle of radius R on a frictionless horizontal surface. The kinetic energy of the object is 90 J and the tension in the string is 360 N. Find R.

82 • How high must an 800-kg Ford Escort be lifted to gain an amount of potential energy equal to the kinetic energy it has when it is moving at 100 km/h?

83 • The movie crew arrives in the Badlands ready to shoot a scene. The script calls for a car to crash into a vertical rock face at 100 km/h. Unfortunately, the car won't start, and there is no mechanic in sight. They are about to skulk back to the studio to face the producer's wrath when the cameraman gets an idea. They use a crane to lift the car by its rear end and then drop it, filming at an angle that makes the car appear to be traveling horizontally. How high should the 800-kg car be lifted so that it reaches a speed of 100 km/h in the fall?

84 •• The force acting on a particle that is moving along the x axis is given by $F_x = -ax^2$, where a is a constant. Calculate the potential-energy function U relative to $U = 0$ at $x = 0$, and sketch a graph of U versus x.

85 •• Water from behind a dam flows through a large turbine at a rate of 1.5×10^6 kg/min. The turbine is located 50 m below the surface of the reservoir, and the water leaves the turbine with a speed of 5 m/s. (a) Neglecting any energy dissipation, what is the power output of the turbine? (b) How many U.S. citizens would be supplied with energy by this dam if each citizen uses 3×10^{11} J of energy per year?

86 •• A force acts on a cart of mass m in such a way that the speed v of the cart increases with distance x as $v = Cx$, where C is a constant. (a) Find the force acting on the cart as a function of position. (b) What is the work done by the force in moving the cart from $x = 0$ to $x = x_1$?

87 •• A force $\vec{F} = (2$ N $/m^2)x^2\,\hat{i}$ is applied to a particle. Find the work done on the particle as it moves a total distance of 5 m (a) parallel to the y axis from point (2 m, 2 m) to point (2 m, 7 m) and (b) in a straight line from (2 m, 2 m) to (5 m, 6 m).

88 •• A particle of mass m moves along the x axis. Its position varies with time according to $x = 2t^3 - 4t^2$ where x is in meters and t is in seconds. Find (a) the velocity and acceleration of the particle at any time t; (b) the power delivered to the particle at any time t; and (c) the work done by the force from $t = 0$ to $t = t_1$.

89 •• A 3-kg particle starts from rest at $x = 0$ and moves under the influence of a single force $F_x = 6 + 4x - 3x^2$ where F_x is in newtons and x is in meters. (a) Find the work done by the force as the particle moves from $x = 0$ to $x = 3$ m. (b) Find the power delivered to the particle when it is at $x = 3$ m.

90 •• The initial kinetic energy imparted to a 20-g bullet is 1200 J. Neglecting air resistance, find the range of this projectile when it is fired at an angle such that the range equals the maximum height attained.

91 •• A force F_x acting on a particle is shown as a function of x in Figure 6-40. (a) From the graph, calculate the work done by the force when the particle moves from $x = 0$ to the following values of x: -4, -3, -2, -1, 0, 1, 2, 3, and 4 m. (b) Plot the potential energy U versus x for the range of values of x from -4 m to $+4$ m, assuming that $U = 0$ at $x = 0$.

Figure 6-40 Problem 91

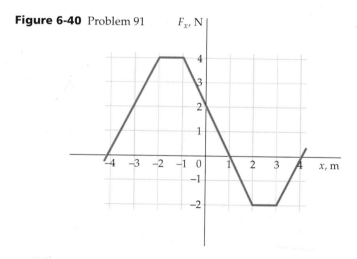

92 •• Repeat Problem 91 for the force F_x shown in Figure 6-41.

Figure 6-41 Problem 92

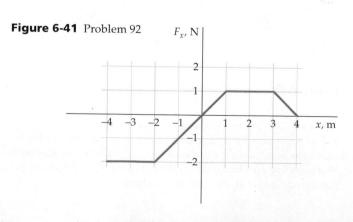

93 •• A rope of length L and mass per unit length of u lies coiled on the floor. (*a*) What force F is required to hold one end of the rope a distance $y < L$ above the floor as shown in Figure 6-42? (*b*) Find the work required to lift one end of the rope from the floor to a height $l < L$ by integrating $F \, dy$ from $y = 0$ to $y = l$.

Figure 6-42 Problem 93

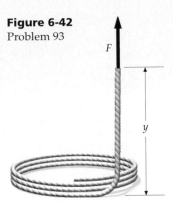

94 ••• A box of mass M is at the bottom of a frictionless inclined plane (Figure 6-43). The box is attached to a string that pulls with a constant tension T. (*a*) Find the work done by the tension T when the box has moved a distance x along the plane. (*b*) Find the speed of the box as a function of x and θ. (*c*) Determine the power produced by the tension in the string as a function of x and θ.

Figure 6-43 Problem 94

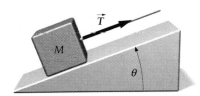

95 ••• A force in the xy plane is given by $\vec{F} = (F_0/r)(y\hat{i} - x\hat{j})$, where F_0 is a constant and $r = \sqrt{x^2 + y^2}$. (*a*) Show that the magnitude of this force is F_0 and that its direction is perpendicular to $\vec{r} = x\hat{i} + y\hat{j}$. (*b*) Find the work done by this force on a particle that moves in a circle of radius 5 m centered at the origin. Is this force conservative?

96 ••• A theoretical formula for the potential energy associated with the nuclear force between two protons, two neutrons, or a neutron and a proton is the *Yukawa potential*:

$$U = -U_0 \left(\frac{a}{x} \right) e^{-x/a}$$

where U_0 and a are constants. (*a*) Sketch U versus x using $U_0 = 4$ pJ (a picojoule, pJ, is 1×10^{-12} J) and $a = 2.5$ fm (a femtometer, fm, is 1×10^{-15} m). (*b*) Find the force F_x. (*c*) Compare the magnitude of the force at the separation $x = 2a$ to that at $x = a$. (*d*) Compare the magnitude of the force at the separation $x = 5a$ to that at $x = a$.

CHAPTER

Conservation of Energy

The waterfall in this 1961 lithograph by the Swiss artist M. C. Escher violates the law of conservation of energy. As the water falls, part of its potential energy is converted into the kinetic energy of the waterwheel. How then does the water get back to the top of the waterfall?

The potential energy of a system is defined in such a way that the work done by an internal conservative force on the system equals the *decrease* in potential energy. If the conservative force is the only force that does work, the work it does also equals the *increase* in kinetic energy. Since the decrease in potential energy equals the increase in kinetic energy, the sum of potential and kinetic energy, or the total mechanical energy, does not change. This is known as the law of conservation of mechanical energy. It follows from Newton's laws, and presents a useful alternative to Newton's laws for solving many problems in mechanics. The use of conservation of energy is limited, however, because there are usually nonconservative forces present, such as friction. When friction is present, the mechanical energy of the system decreases.

Since mechanical energy is often not conserved, the importance of energy was not realized until the nineteenth century, when it was discovered that the disappearance of macroscopic mechanical energy is always accompanied by the appearance of some other kind of energy, often thermal energy, which is usually indicated by an increase in temperature. We now know that, on the microscopic scale, this thermal energy consists of the kinetic and potential energies of the molecules in the system.

There are other forms of energy, such as the internal chemical energy in your body, the energy of sound, and electromagnetic energy. Whenever the energy of a system changes, we can account for the change by the appearance or disappearance of energy somewhere else. This experimental observation is the law of conservation of energy, one of the most fundamental and important laws in all of science. Although energy changes from one form to another, it is never created or destroyed.

We begin by considering systems in which mechanical energy is conserved. We then extend the discussion to include thermal and chemical energy, and we develop specific methods to deal with dissipative systems, wherein kinetic friction converts mechanical energy to thermal energy. After discussing Einstein's famous relation between mass and energy, we conclude by considering the quantization of energy, the surprising result, discovered in the first quarter of the twentieth century, that energy changes in a system are not continuous, but occur in lumps or quanta. Although the quantum of energy is so small that the incremental quality of energy goes unnoticed in the macroscopic world, the quantization of energy has profound consequences for microscopic systems such as atoms and molecules.

7-1   The Conservation of Mechanical Energy

Consider a system for which the only forces acting are internal, conservative forces. The skier–earth system discussed in Chapter 6 is such a system. The total work done on each particle in the system equals the increase in the kinetic energy of that particle, so the total work done by all the forces equals the increase in the total kinetic energy of the system:

$$W_{total} = \sum \Delta K_i = \Delta K \qquad\qquad 7\text{-}1$$

Since each internal force is conservative, the work it does decreases the potential energy associated with that force. So the total work done by all the internal forces equals the total decrease in potential energy of the system:

$$W_{total} = \sum - \Delta U_i = -\Delta U \qquad\qquad 7\text{-}2$$

Thus, $\Delta K = -\Delta U$ or

$$\Delta K + \Delta U = \Delta(K + U) = 0 \qquad\qquad 7\text{-}3$$

The sum of the kinetic energy K and the potential energy U of a system is called the **total mechanical energy** E_{mech}:

$$E_{mech} = K + U \qquad\qquad 7\text{-}4$$

Definition—Total mechanical energy

We have just shown that when only internal, conservative forces do work on a system of two or more particles, the total mechanical energy of the system does not change:

$$\Delta(K + U) = \Delta E_{mech} = 0$$

$$E_{mech} = K + U = \text{constant} \qquad\qquad 7\text{-}5$$

Conservation of mechanical energy

This is the **law of conservation of mechanical energy** and is the origin of the expression "conservative force."

If $E_i = K_i + U_i$ is the initial mechanical energy of the system and $E_f = K_f + U_f$ is the final mechanical energy, conservation of mechanical energy implies that

$$E_f = E_i$$

$$K_f + U_f = K_i + U_i \qquad \qquad \text{7-6}$$

Conservation of mechanical energy

Many mechanics problems can be solved by setting the final mechanical energy of a system equal to its initial mechanical energy.

Multiflash photograph of a simple pendulum. As the bob descends, gravitational potential energy is converted into kinetic energy, and the speed increases as indicated by the increased spacing of the recorded positions. The speed decreases as the bob moves up, and the kinetic energy is changed into potential energy.

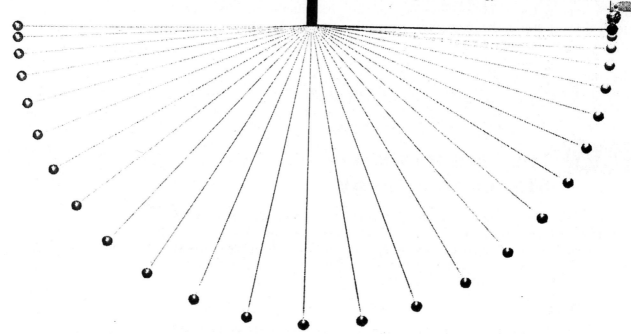

Applications

Consider a downhill skier who starts at rest from a height h above the bottom of a hill, which we assume to be frictionless. What is the skier's speed at a height y above the bottom of the hill? The mechanical energy of the earth–skier system is conserved because the only force doing work is the internal, conservative force of gravity. If we choose $U = 0$ at the bottom of the hill, the original potential energy is mgh. This is also the total mechanical energy because the initial kinetic energy is zero. Thus,

$$E_i = K_i + U_i = 0 + mgh = mgh$$

At the height y, the potential energy is mgy and the speed is v. Hence,

$$E_f = K_f + U_f = \frac{1}{2}mv^2 + mgy$$

Setting $E_f = E_i$ we find

$$\frac{1}{2}mv^2 + mgy = mgh$$

or

$$v = \sqrt{2g(h - y)}$$

The speed of the skier is the same as if she had undergone free fall through a distance $h - y$.

Example 7-1

Standing near the edge of the roof of a 12-m high building, you kick a ball with an initial speed of $v_i = 16$ m/s at an angle of 60° above the horizontal. Neglecting air resistance, find (a) how high above the building the ball rises, and (b) its speed just before it hits the ground.

Picture the Problem Since gravity is the only force that does work on the ball–earth system, mechanical energy is conserved. At the top of its flight, the ball is moving horizontally with its initial horizontal velocity $v_{top} = v_i \cos 60°$. We choose $U = 0$ at the top of the building (Figure 7-1).

Figure 7-1

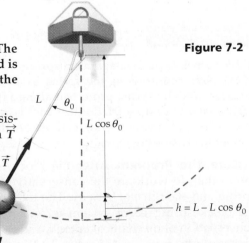

(a)1. Conservation of mechanical energy relates the height h to the initial velocity v_i and the velocity at the top of its flight v_{top}:

$$E_{top} = E_i$$

$$\frac{1}{2} m v_{top}^2 + mgh = \frac{1}{2} m v_i^2$$

2. Solve for h:

$$h = \frac{v_i^2 - v_{top}^2}{2g}$$

3. The velocity at the top of its flight equals its initial horizontal velocity:

$$v_{top} = v_i \cos \theta = (16 \text{ m/s}) \cos 60° = 8 \text{ m/s}$$

4. Substitute this value for v_t and $v_i = 16$ m/s and solve for h:

$$h = \frac{v_i^2 - v_{top}^2}{2g} = \frac{(16 \text{ m/s})^2 - (8 \text{ m/s})^2}{2(9.81 \text{ m/s}^2)} = 9.79 \text{ m}$$

(b)1. If v_f is the speed of the ball just before it hits the ground, its energy is:

$$E_f = \frac{1}{2} m v_f^2 + mgy$$

2. Apply conservation of mechanical energy:

$$\frac{1}{2} m v_f^2 + mgy = \frac{1}{2} m v_i^2$$

3. Solve for v_f, and set $y = -12$ m to find the final velocity:

$$v_f = \sqrt{v_i^2 - 2gy}$$

$$= \sqrt{16 \text{ (m/s)}^2 - 2(9.81 \text{ m/s}^2)(-12 \text{ m})} = 22.2 \text{ m/s}$$

Example 7-2

A pendulum consists of a bob of mass m attached to a string of length L. The bob is pulled aside so that the string makes an angle θ_0 with the vertical and is released from rest. Find expressions for (a) the speed v at the bottom of the swing, and (b) the tension in the string at that time.

Figure 7-2

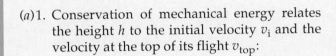

Picture the Problem The two forces acting on the bob (neglecting air resistance) are the force of gravity $m\vec{g}$, which is conservative, and the tension $\vec{T}$ (Figure 7-2). Since $\vec{T}$ is perpendicular to the motion, it does no work. Hence, the mechanical energy of the bob–earth system is conserved. To find the speed of the bob, equate the initial and final energies. The tension in the string is found from Newton's second law. We choose $U = 0$ at the bottom of the swing. The initial height h above the bottom is related to the initial angle θ_0 by $h = L - L \cos \theta_0$.

(*a*)1. Apply conservation of mechanical energy:

$$E_f = E_i$$

2. The initial energy E_i of the system is its potential energy:

$$E_i = K_i + U_i = 0 + mgh$$

3. At the bottom, the energy E_f is all kinetic:

$$E_f = K_f + U_f = \frac{1}{2}mv^2 + 0 = \frac{1}{2}mv^2$$

4. Conservation of mechanical energy thus relates the speed v to the height h:

$$\frac{1}{2}mv^2 = mgh$$

5. Solve for the speed v:

$$v = \sqrt{2gh}$$

6. To express speed in terms of the initial angle θ_0, we need to relate h to θ_0. This is done in Figure 7-2:

$$h = L - L\cos\theta_0 = L(1 - \cos\theta_0)$$

7. Substitute this value for h to express the speed at the bottom in terms of θ_0:

$$v = \sqrt{2gh} = \sqrt{2gL(1 - \cos\theta_0)}$$

(*b*)1. The forces on the bob are its weight $m\vec{g}$ down and $\vec{T}$, which is up when the bob is at the bottom of the circle. Choose up to be the positive y direction, and apply $\Sigma F_y = ma_y$:

$$T - mg = ma$$

2. At the bottom, the bob has a centripetal acceleration v^2/L toward the center of the circle, which is upward at this point:

$$a = \frac{v^2}{L} = \frac{2gL(1 - \cos\theta_0)}{L} = 2g(1 - \cos\theta_0)$$

3. Substitute this value of a into the equation in (*b*)1 to find T:

$$T = mg + ma = mg + 2mg(1 - \cos\theta_0)$$
$$= mg(3 - 2\cos\theta_0)$$

Remarks The tension at the bottom is greater than the weight of the bob because the bob is accelerating upward. If the bob is released at $\theta_0 = 90°$, the tension at the bottom is $3mg$. The speed of the bob can also be found using Newton's laws (see Problem 97), but the solution is difficult because the acceleration tangential to the curve varies with the angle θ and therefore with time, so the constant-acceleration formulas do not apply. Finally, step 4 in part (*a*) shows that the speed at the bottom is the same as if the bob had dropped in free fall from a height h.

Figure 7-3

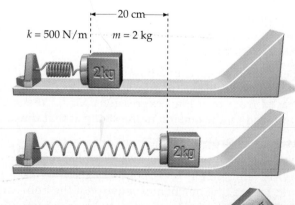

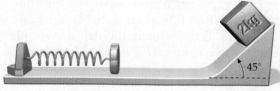

| Example **7-3** | *try it yourself* |

A 2-kg block is pushed against a spring that has a force constant of 500 N/m, compressing the spring by 20 cm. The block is then released, and the spring projects it along a frictionless horizontal surface and then up a frictionless incline of angle 45° as shown in Figure 7-3. How far up the incline does the block travel before momentarily coming to rest?

Picture the Problem After the block is released, the only forces that do work are the conservative forces exerted by the spring and the force of gravity. The total mechanical energy of the block–spring–earth system is conserved. Find h from the conservation of mechanical energy, and then find the distance s up the incline from $\sin 45° = h/s$.

Cover the column to the right and try these on your own before looking at the answers.

Steps	Answers

1. Write the initial mechanical energy in terms of the compression distance x.

$E_i = \frac{1}{2}kx^2$

2. Write the final mechanical energy in terms of the height h.

$E_f = mgh$

3. Apply conservation of mechanical energy, and solve for h.

$mgh = \frac{1}{2}kx^2$

$h = \frac{kx^2}{2mg} = 0.51\ \text{m}$

4. Find the distance s from $h = s \sin \theta$.

$s = 0.721\ \text{m}$

Remark In this problem, the initial potential energy in the spring is converted first into kinetic energy and then into gravitational potential energy.

Exercise Find the speed of the block just after it leaves the spring. (*Answer*   3.16 m/s)

$v = 2 \dfrac{\pi r}{T\ (\text{period})}$

Example 7-4

A spring with a force constant of k hangs vertically. A block of mass m is attached to the unstretched spring and allowed to fall from rest. Find an expression for the maximum distance the block falls before it begins moving upward.

Picture the Problem As the block drops, its speed first increases, then reaches some maximum value, and then decreases until it is again zero when the block is at its lowest point (Figure 7-4). Only conservative forces are present, so we apply the conservation of mechanical energy to the earth–spring–block system. The initial and final positions of the block are shown. Choose the gravitational potential energy of the block to be zero at the original position $y = 0$. The initial potential energy of the spring is zero because the spring is unstretched at this position. Since the block is at rest at this point, the total mechanical energy is zero. Let d be the distance the block falls.

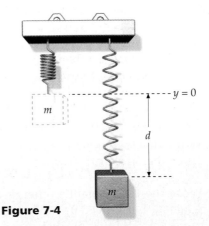

Figure 7-4

1. At a general point y, the total mechanical energy is the sum of the kinetic energy K, the gravitational potential energy U_g, which is equal to $-mgy$, and the spring potential energy U_s, which is equal to $\frac{1}{2}ky^2$:

$E = K + U_g + U_s = \frac{1}{2}mv^2 - mgy + \frac{1}{2}ky^2$

2. Apply conservation of mechanical energy:

$E = E_i = 0$

$\frac{1}{2}mv^2 - mgy + \frac{1}{2}ky^2 = 0$

3. Set $v = 0$ and solve for y. There are two solutions. One gives the initial position and the other is the one we want:

$$0 - mgy + \frac{1}{2}ky^2 = 0$$

$$y = 0 \qquad \text{(initial position)}$$

$$y = d = \frac{2mg}{k}$$

Remarks Gravitational potential energy is converted into the kinetic energy of the block plus the potential energy of the spring. At the lowest point, where the block is momentarily at rest, the gain in potential energy of the spring equals the loss in gravitational potential energy of the block. In bungee jumping from a bridge, you are attached to an elastic cord of length L less than the height of the bridge. When you have dropped a distance L, the cord is unstretched and you have kinetic energy equal to mgL. You therefore fall farther than $2mg/k$ (assuming the cord acts like a spring of constant k) before you come back up. If mechanical energy were perfectly conserved, you would come up and strike the bridge from which you had leapt. Instead, some mechanical energy is dissipated because of nonelastic properties of the cord.

| **Example 7-5** | *try it yourself* | **Figure 7-5** |

Two blocks are attached to a light string that passes over a massless, frictionless pulley. The two blocks have masses m_1 and m_2 and are initially at rest. Find the speed of either block when the heavier one falls a distance h.

Picture the Problem Mechanical energy is conserved. The net work done by the string tension is zero; it does positive work lifting the lighter object and an equal amount of negative work as the heavier object moves downward the same distance in the opposite direction (Figure 7-5). If we choose $U_i = 0$ when the blocks are at rest, the total energy is zero. Because the string does not stretch, both blocks move with the same speed v.

Cover the column to the right and try these on your own before looking at the answers.

Steps   **Answers**

1. Write the total kinetic energy of the system when the blocks are moving with speed v.

$$K = \frac{1}{2}m_1v^2 + \frac{1}{2}m_2v^2$$

2. Write the total potential energy of the system when m_1 has moved up a distance h and m_2 has moved down the same distance.

$$U = m_1gh - m_2gh$$

3. Add U and K to obtain the total energy E.

$$E = K + U = \frac{1}{2}m_1v^2 + \frac{1}{2}m_2v^2 + m_1gh - m_2gh$$

4. Apply conservation of mechanical energy.

$$E = E_i = 0$$

$$\frac{1}{2}m_1v^2 + \frac{1}{2}m_2v^2 + m_1gh - m_2gh = 0$$

5. Solve for v.

$$v = \sqrt{\frac{2(m_2 - m_1)}{m_1 + m_2}gh}$$

Remarks This device, called an *Atwood's machine*, is analyzed in terms of forces in Problems 81–85 in Chapter 4. If a pulley has mass, it has kinetic energy when it rotates. Our pulley is massless, so we can neglect its energy of rotation. We consider the more complicated problem of a pulley with mass in Chapter 9.

Since all the forces are constant, the acceleration of the blocks is constant. From the constant-acceleration equation $v^2 = a\,\Delta x$, we see that the acceleration is given by $a = [(m_2 - m_1)/(m_1 + m_2)]g$, so $g = [(m_1 + m_2)/(m_2 - m_1)]a$. If m_1 and m_2 are not too different, the acceleration of either object is a small fraction of g. It was easily measured with the rather crude timing devices available in the eighteenth century, whereas a direct measurement of g was difficult if not impossible.

Exercise What is the magnitude of the acceleration of either block if the masses are $m_1 = 3$ kg and $m_2 = 5$ kg? (*Answer* $a = 0.25g = 2.45$ m/s^2)

We've seen that the law of conservation of mechanical energy can be used as an alternative to Newton's laws for solving certain problems in mechanics. When we are not interested in the time t, the conservation of mechanical energy is often much easier to use than Newton's second law (Figure 7-6). Since the conservation of mechanical energy was derived from Newton's laws, any problem that can be solved using it can also be solved directly from Newton's laws, though often with much more difficulty.

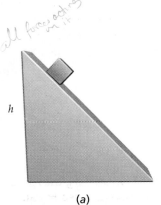

(a)

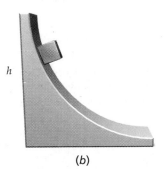

(b)

Figure 7-6 (*a*) One can easily find the speed of a block sliding down a frictionless incline of constant slope by applying Newton's second law or by using conservation of mechanical energy. However, if the incline is frictionless but not of constant slope, as in (*b*), the problem can still be solved easily using conservation of mechanical energy, whereas it can be solved using Newton's second law only if the slope of the incline is known at each point, and then the calculation is quite tedious.

7-2   The Conservation of Energy

In the macroscopic world, nonconservative forces are always present to some extent, the most common being frictional forces, which decrease the mechanical energy of a system. However, the decrease in mechanical energy is found to be equal to the increase in thermal energy produced by the frictional forces. Another type of nonconservative force is that involved in the deformations of objects. When you bend a coat hanger back and forth, you do work on the coat hanger, but the work you do does not appear as mechanical energy. Instead, the coat hanger becomes warm. The work done in deforming the hanger is dissipated as thermal energy. Similarly, when a ball of putty is dropped to the floor, it warms as it deforms on impact, and the original potential energy appears as thermal energy. If thermal energy is added to mechanical energy, the total energy is conserved even when there are frictional forces or forces of deformation.

A third type of nonconservative force is associated with chemical reactions. When we include systems in which chemical reactions take place, the sum of mechanical energy plus thermal energy is not conserved. For example, suppose that you begin running from rest. Originally you have no kinetic energy. When you begin to run, internal chemical energy in your muscle is converted to kinetic energy of your body, and thermal energy is produced. It is possible to identify and measure the chemical energy that is used. In this case, the sum of mechanical, thermal, and chemical energy is conserved.

Even when thermal energy and chemical energy are included, the total energy of the system does not always remain constant. The energy of a system can change because of some form of radiation, such as sound waves or electromagnetic waves. However, *the increase or decrease in the total energy of a system can always be accounted for by the appearance or disappearance of energy somewhere else.* This experimental result is known as the **law of conservation of energy**. Let E_{sys} be the total energy of a given system, E_{in} be the energy that enters the system, and E_{out} be the energy that leaves the system. The law of conservation of energy then states:

$$E_{in} - E_{out} = \Delta E_{sys} \qquad\qquad 7\text{-}7$$

Law of conservation of energy

Alternatively,

> The total energy of the universe is constant. Energy can be converted from one form to another, or transmitted from one region to another, but energy can never be created or destroyed.

Law of conservation of energy

The total energy E of many systems familiar from everyday life can be accounted for completely by mechanical energy E_{mech}, thermal energy E_{therm}, and chemical energy E_{chem}. To be comprehensive and include other possible forms of energy, such as electromagnetic or nuclear energy, we include E_{other}, and write generally

$$E_{sys} = E_{mech} + E_{therm} + E_{chem} + E_{other} \qquad\qquad 7\text{-}8$$

The Work–Energy Theorem

A common way to transfer energy into or out of a system is to do work on the system from the outside. If this is the only source of energy transferred,* the law of conservation of energy becomes

$$W_{ext} = \Delta E_{sys} \qquad\qquad 7\text{-}9$$

Work–energy theorem

where W_{ext} is the work done on the system by external forces, and ΔE_{sys} is the change in the system's total energy. This work–energy theorem for systems, which we will call simply the work–energy theorem, is a powerful tool for studying a wide variety of systems. Note that if the system is just a single particle, its energy can only be kinetic, so Equation 7-9 is equivalent to the work–kinetic energy theorem studied in Chapter 6.

---

*Energy can also be transferred when heat is exchanged between a system and its surroundings. Exchanges of heat energy, which occur when there is a temperature difference between a system and its surroundings, are discussed in Chapter 19.

Example 7-6

A ball of putty of mass m is released from rest from a height h and falls to the hard floor (Plop!). Discuss the application of the law of conservation of energy to (a) the system consisting of the ball alone, and (b) the system consisting of the earth and ball.

Picture the Problem   Two forces act on the ball: gravity and the force of the floor. Since the floor does not move, the force it exerts does no work. There are no chemical or other energy changes, so we can neglect E_{chem} and E_{other}. If we neglect the sound energy radiated when the ball hits the floor, the only energy transferred to or from the ball is the work done by gravity, so we can use the work–energy theorem.

(a)1. Write the work–energy theorem:

$$W_{ext} = \Delta E_{sys} = \Delta E_{mech} + \Delta E_{therm}$$

2. The two external forces on the system are gravity and the force exerted by the floor. The floor does not move and therefore does no work. The only work done on the ball is by gravity:

$$W_{ext} = mgh$$

3. Since the ball alone is our system, its mechanical energy is entirely kinetic, which is zero both initially and finally. Thus, the change in mechanical energy is zero:

$$\Delta E_{mech} = 0$$

4. Substitute mgh for W_{ext} and 0 for ΔE_{mech} in step 1:

$$W_{ext} = \Delta E_{therm} = mgh$$

(b)1. There are no external forces acting on this system (the force of gravity and the force of the floor are now internal to the system), so there is no external work done:

$$W_{ext} = 0$$

2. Write the work–energy theorem with $W_{ext} = 0$:

$$W_{ext} = \Delta E_{sys} = \Delta E_{mech} + \Delta E_{therm} = 0$$
$$\Delta E_{therm} = -\Delta E_{mech}$$

3. The original mechanical energy of the ball–earth system is the original gravitational potential energy, and the final mechanical energy is zero:

$$E_i = mgh$$
$$E_f = 0$$

4. The change in mechanical energy of the ball–earth system is thus:

$$\Delta E_{mech} = E_f - E_i = 0 - mgh = -mgh$$

5. The work–energy theorem thus gives the same result found in (a):

$$\Delta E_{therm} = -\Delta E_{mech} = mgh$$

Remarks   In (a), energy is transferred to the ball by the work done on it by gravity. This energy appears as the kinetic energy of the ball before it hits the floor and as thermal energy after. The ball warms slightly and the energy is eventually transferred to the surroundings as heat. In (b), the original potential energy of the ball–earth system is converted to kinetic energy of the ball just before it hits and then into thermal energy.

(a)

(b)

(c)

(a) In this power plant in Kansas, energy stored in the fossil fuel coal (the black mound at lower right) is released by burning the coal to produce steam; the steam is then used to drive turbines to produce electricity. The excess heat is dissipated by cooling towers. (b) The potential energy of the water at the top of Niagara Falls is used to produce electrical energy. (c) This wind farm at Altamont Pass in California uses hundreds of windmills to convert wind energy into electrical energy.

Problems Involving Kinetic Friction

Kinetic frictional forces exerted by one surface on another when the surfaces slide across each other decrease the total mechanical energy of a system and increase the thermal energy. Consider a block that begins with initial velocity v_i and slides on a rough table until it stops (Figure 7-7). We choose the block and table to be our system. Then $\Delta E_{chem} = \Delta E_{other} = 0$ and no external work is done on this system. The work–energy theorem gives

$$0 = \Delta E_{mech} + \Delta E_{therm}$$

The mechanical energy lost is the initial kinetic energy of the block

$$\Delta E_{mech} = -\frac{1}{2} mv_i^2 \qquad\qquad 7\text{-}10$$

We can relate the loss in mechanical energy to frictional force. If f is the magnitude of the frictional force, Newton's second law gives

$$-f = ma$$

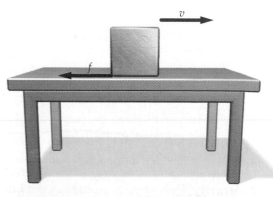

Figure 7-7 A block sliding on a rough table. The force of friction reduces the mechanical energy of the block–table system.

Multiplying both sides of this equation by Δs, we find

$$-f\,\Delta s = ma\,\Delta s = m\left(\frac{1}{2}v_f^2 - \frac{1}{2}v_i^2\right) = -\frac{1}{2}mv_i^2 \qquad \text{7-11}$$

*The work done by kinetic friction is examined in detail in "Work and Heat Transfer in the Presence of Sliding Friction" by B. A. Shewood and W. H. Bernard, *American Journal of Physics*, **52**, 1001 (1984).

where we have used the constant-acceleration formula $2a\,\Delta s = v_f^2 - v_i^2$, and $v_f = 0$. Comparing Equations 7-10 and 7-11 we find

$$f\,\Delta s = -\Delta E_{\text{mech}} \qquad \text{7-12}$$

Note that the quantity $-f\,\Delta s$ is *not* the work done by friction on the sliding block, because the displacement of the frictional force is not, in general, equal to the displacement of the block. However, it can be shown that $f\,\Delta s$ does equal the increase in thermal energy due to the dissipation of mechanical energy on the surfaces as they slide across one another.* Thus,

$$f\,\Delta s = \Delta E_{\text{therm}} \qquad \text{7-13}$$

Energy dissipated by friction

Substituting this result into the work–energy theorem (with $E_{\text{chem}} = E_{\text{other}} = 0$), we obtain

$$W_{\text{ext}} = \Delta E_{\text{mech}} + \Delta E_{\text{therm}} = \Delta E_{\text{mech}} + f\,\Delta s \qquad \text{7-14}$$

Work–energy theorem for problems with friction

When there is no external work done on the system, the energy dissipated by friction equals the decrease in mechanical energy:

$$\Delta E_{\text{therm}} = f\,\Delta s = -\Delta E_{\text{mech}} \qquad (W_{\text{ext}} = 0) \qquad \text{7-15}$$

Figure 7-8

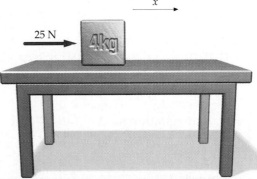

Example 7-7

A horizontal force of 25 N is applied to a 4-kg block, which is initially at rest on a horizontal table. The coefficient of kinetic friction μ_k between the block and table is 0.35. Find (*a*) the external work done on the block–table system, (*b*) the energy dissipated by friction, (*c*) the kinetic energy of the block after it has been pushed 3 m, and (*d*) the speed of the block after it has been pushed 3 m.

Picture the Problem We choose the block plus table as our system (Figure 7-8). The speed of the block is found from its final kinetic energy, which we find using the work–energy theorem with $\Delta E_{\text{chem}} = 0$ and $\Delta E_{\text{therm}} = f\,\Delta s$. The mechanical energy is increased by the external work and decreased by the energy dissipated by friction.

(*a*)   The external work done is the product of the external force and the distance traveled:

$$W_{\text{ext}} = F_{\text{ext}}\,\Delta x = (25\text{ N})(3\text{ m}) = 75\text{ J}$$

ask
shouldn't

(*b*)   The energy dissipated by friction is $f\,\Delta x$:

$$\Delta E_{\text{therm}} = f\,\Delta x = \mu_k mg\,\Delta x$$

$$= (0.35)(4\text{ kg})(9.81\text{ N/kg})(3\text{ m}) = 41.2\text{ J}$$

(*c*)1. Apply the work–energy theorem to find the final kinetic energy:

$$W_{\text{ext}} = \Delta E_{\text{mech}} + f\,\Delta x$$

2. Since the initial kinetic energy is zero and there is no change in the potential energy, the final kinetic energy equals the change in mechanical energy:

$$\Delta E_{\text{mech}} = \Delta K = K_f - K_i = K_f$$

3. Substitute this result into the work–energy theorem:

$$W_{ext} = K_f + f\,\Delta x$$

$$K_f = W_{ext} - f\,\Delta x = 75\text{ J} - 41.2\text{ J} = 33.8\text{ J}$$

(d)1. The final speed of the block is related to its kinetic energy:

$$K_f = \frac{1}{2}mv^2$$

2. Solve for the final speed of the block:

$$v = \sqrt{\frac{2K_f}{m}} = \sqrt{\frac{2(33.8\text{ J})}{4\text{ kg}}} = 4.11\text{ m/s}$$

Example 7-8   *try it yourself*

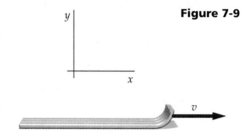

Figure 7-9

A 5-kg toboggan is sliding with an initial speed of 4 m/s. If the coefficient of friction between the toboggan and the snow is 0.14, how far will the toboggan go before coming to rest?

Picture the Problem We choose the toboggan and snow as our system (Figure 7-9). Then $W_{ext} = 0$, and the work–energy theorem implies that the energy dissipated by friction equals the change in mechanical energy.

Cover the column to the right and try these on your own before looking at the answers.

Steps	Answers
1. Write the work–energy theorem with $W_{ext} = 0$:	$W_{ext} = \Delta E_{mech} + f\,\Delta x = 0$
2. Solve for Δx.	$\Delta x = -\dfrac{\Delta E_{mech}}{f}$
3. Write the change in mechanical energy in terms of the initial speed of the toboggan.	$\Delta E_{mech} = -\dfrac{1}{2}mv^2 \quad \Delta K$
4. Write the frictional force f in terms of the coefficient of friction and the weight of the toboggan.	$f = \mu_k mg$
5. Substitute your results for f and ΔE_{mech} into your equation for Δx in step 1 and calculate Δx.	$\Delta x = -\dfrac{-\frac{1}{2}mv^2}{\mu_k\,mg} = 5.82\text{ m}$

Remark Figure 7-10 shows stopping distance versus initial speed of the toboggan for three different values of the coefficient of kinetic friction.

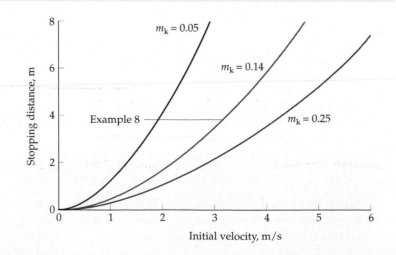

Figure 7-10

Example 7-9

A child of mass 40 kg goes down a rough slide inclined at 30°. The coefficient of kinetic friction between the child and the slide is $\mu_k = 0.2$. If the child starts from rest at the top of the slide, a height 4 m above the bottom, how fast is she traveling when she reaches the bottom?

Picture the Problem As the child slides down, some of her potential energy is converted into kinetic energy and some into thermal energy because of friction. We choose the child–slide–earth as our system (Figure 7-11). Then $W_{ext} = 0$, and the work–energy theorem implies that the energy dissipated by friction equals the change in mechanical energy. We choose $y = 0$ at the bottom of the slide so the final potential energy of the child is zero.

Figure 7-11

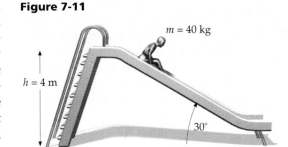

$m = 40$ kg

$h = 4$ m

30°

1. The speed at the bottom is related to the final kinetic energy:

$$v = \sqrt{\frac{2K_f}{m}}$$

2. The final kinetic energy equals the initial potential energy minus the energy dissipated by friction:

$$K_f = mgh - f\,\Delta s$$

3. The frictional force is related to the normal force F_n:

$$f = \mu_k F_n = \mu_k mg \cos 30°$$

4. The total distance traveled is related to the height h:

$$\Delta s = \frac{h}{\sin 30°}$$

5. Substitute these results to calculate the final kinetic energy:

$$K_f = mgh - f\,\Delta s$$

$$= mgh - \mu_k mg \cos 30° \frac{h}{\sin 30°}$$

$$= mgh(1 - \mu_k \cot 30°)$$

6. Use this final kinetic energy to find v:

$$v = \sqrt{\frac{2K_f}{m}} = \sqrt{2gh(1 - \mu_k \cot 30°)}$$

$$= \sqrt{2(9.81 \text{ m/s}^2)(4 \text{ m})[1 - 0.2\,(1.73)]} = 7.16 \text{ m/s}$$

Remark The energy dissipated by friction decreases the total mechanical energy of the child–slide–earth system from mgh to its final value $mgh(1 - \cot 30°)$, which equals the kinetic energy at the bottom. Note that the result is independent of the mass of the child.

Exercise For the earth–child–slide system, calculate (a) the initial mechanical energy, (b) the final mechanical energy, and (c) the energy dissipated by friction. (*Answers*   (a) 1570 J, (b) 1026 J, (c) 544 J)

Example **7-10**	*try it yourself*

A 4-kg block hangs by a light string that passes over a massless, frictionless pulley and is connected to a 6-kg block that rests on a rough shelf. The coefficient of kinetic friction is $\mu_k = 0.2$. The 6-kg block is pushed against a spring to which it is not attached. The spring has a force constant of 180 N/m, and it is compressed 30 cm. Find the speed of the blocks after the spring is released and the 4-kg block has fallen a distance of 40 cm.

Picture the Problem   The speed of the blocks is obtained from their final kinetic energy. Consider the system to be the earth, the shelf, the spring, and the two blocks $m_1 = 6$ kg and $m_2 = 4$ kg (Figure 7-12). Then $W_{ext} = 0$ and the work–energy theorem implies that the energy dissipated by friction equals the change in mechanical energy. Choose the initial gravitational potential energy to be zero.

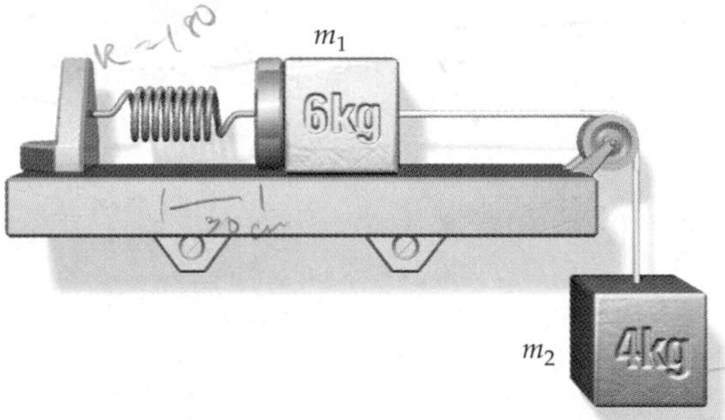

Figure 7-12

Cover the column to the right and try these on your own before looking at the answers.

Steps

Answers

1. Write expressions for the initial mechanical energy and the final mechanical energy when each block has moved a distance Δs.

$$E_i = \frac{1}{2} kx^2$$

$$E_f = K_f + U_f = \frac{1}{2}(m_1 + m_2)v^2 - m_2 g\,\Delta s$$

2. Write an expression for the energy dissipated by friction in terms of the coefficient of friction and Δs.

$$f\,\Delta s = \mu_k m_1 g\,\Delta s$$

3. Set the energy dissipated by friction equal to the loss in mechanical energy of the system.

$$\mu_k m_1 g\,\Delta s = \frac{1}{2} kx^2 - \left[\frac{1}{2}(m_1 + m_2)v^2 - m_2 g\,\Delta s\right]$$

4. Solve your equation for v^2 and substitute the numerical values.

$$v^2 = \frac{kx^2 + 2m_2 g\,\Delta s - 2\mu_k m_1 g\,\Delta s}{m_1 + m_2} = 3.82 \text{ m}^2/\text{s}^2$$

5. Solve for v.

$$v = 1.95 \text{ m/s}$$

Remarks   This solution assumes that the string remains taut at all times. This will be true if the acceleration of m_1 is less than g, that is, if the net force on m_1 is less than $m_1 g$. Initially, the force exerted by the spring on m_1 has the magnitude $k\,\Delta x_1 = (180 \text{ N/m})(0.3 \text{ m}) = 54$ N, which is less than $m_1 g = (6 \text{ kg})(9.81 \text{ N/kg}) = 58.9$ N. Since the spring force decreases as block m_1 moves forward, and the frictional force decreases the net force, the acceleration of the 6-kg block will always be less than g, and the string will remain taut.

Systems With Chemical Energy

Sometimes a system's internal chemical energy is converted into mechanical energy and thermal energy with no work being done by an outside agent. For example, to walk forward, you push back on the floor and the floor pushes forward on you with the force of static friction. This force accelerates you, but it does *not* do work. The displacement of the point of application of the force is zero (assuming your shoes do not slip on the floor), therefore no work is done and no energy is transferred from the floor to your body. The kinetic energy of your body comes from the conversion of chemical energy in your body derived from the food you eat. We consider a similar case in the next example.

This pizza contains about 16 megajoules of energy, approximately the same as the energy in a gallon (3.78 L) of gasoline.

Example 7-11

A man of mass m walks with a small constant speed up a flight of stairs to a height h. Discuss the application of energy conservation to the system consisting of the man alone.

Picture the Problem Two forces act on the man: gravity and the force of the stairs. Since the stairs do not move, they do no work. In this case we *cannot* neglect changes in chemical energy.

1. Write the work–energy theorem:

$$W_{ext} = \Delta E_{sys} = \Delta E_{mech} + \Delta E_{therm} + \Delta E_{chem}$$

2. The only work done on the man is done by gravity. This work is negative because the force is in the opposite direction of the displacement:

$$W_{ext} = -mgh$$

3. Since the man alone is our system, his mechanical energy is entirely kinetic, which is zero both initially and finally:

$$\Delta E_{mech} = 0$$

4. Substitute these results into the work–energy theorem:

$$-mgh = \Delta E_{therm} + \Delta E_{chem}$$

Remark If there were no change in thermal energy, the chemical energy of the man would decrease by mgh. Because the body is relatively inefficient, the amount of chemical energy converted in the man's body will be considerably greater than mgh. The amount of energy above mgh appears as thermal energy, which is eventually transferred from the man to his surroundings as heat.

Exercise Discuss the energy conservation for the system of man plus earth. (*Answer* For this system no external work is done, so the total energy, which now includes potential energy, is conserved. The change in mechanical energy is mgh, so the work–energy theorem gives $0 = mgh + \Delta E_{therm} + \Delta E_{chem}$.)

Example 7-12

A 1000-kg car travels at a constant speed of 100 km/h = 28 m/s = 62 mi/h up a 10% grade. (A 10% grade means that the road rises 1 m for each 10 m of horizontal distance—that is, the angle of inclination θ is given by $\tan \theta = 0.1$ [Figure 7-13].) What is the minimum power that must be delivered by the car's engine? (Neglect rolling friction and air drag.)

Picture the Problem The power delivered by the car's engine is the rate of decrease of its chemical energy. Some of it goes into increasing the potential energy of the car as it climbs the hill, and some goes into an increase in thermal energy, which is expelled as exhaust. For a 10% grade, $\tan \theta = 0.10$ is given, and $\sin \theta \approx \tan \theta$ because the angle is small (Figure 7-13). For the car–earth system, $W_{ext} = 0$, so the total energy is conserved.

Figure 7-13

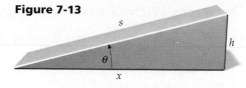

$$\tan \theta = h/x \sim \sin \theta = h/s$$

$$P = \frac{dw}{dt} = F \cdot v$$

1. The power input by the engine is the rate of decrease of its chemical energy:

$$P = -\frac{dE_{chem}}{dt}$$

2. The chemical energy change is found from the work–energy theorem:

$$W_{ext} = \Delta E_{mech} + \Delta E_{therm} + \Delta E_{chem} = 0$$

$$\Delta E_{chem} = -\Delta E_{mech} - \Delta E_{therm}$$

3. Convert the changes to time derivatives:

$$P = -\frac{dE_{chem}}{dt} = \frac{dE_{mech}}{dt} + \frac{dE_{therm}}{dt}$$

4. Since the speed $v = ds/dt$ is constant, the rate of change of the mechanical energy is just the rate of change of potential energy:

$$\frac{dE_{mech}}{dt} = \frac{dU}{dt} = \frac{d(mgh)}{dt} = mg\frac{dh}{dt}$$

5. From Figure 7-13 we can see that when the car travels a distance s along the road, it climbs a height h, which is related to s by:

$$h = s \sin \theta$$

6. We can use the approximation $\tan \theta \approx \sin \theta$ because the angle is small:

$$h = s \sin \theta \approx s \tan \theta = 0.1s$$

7. We can now relate the rate of change of mechanical energy to the speed:

$$\frac{dE_{mech}}{dt} = mg\frac{dh}{dt} = 0.1\, mg\frac{ds}{dt} = 0.1\, mgv \quad \overset{F \cdot v}{}$$

8. Substitute these results into the equation for power in step 3:

$$P = \frac{dE_{mech}}{dt} + \frac{dE_{therm}}{dt}$$

$$= 0.1mgv + \frac{dE_{therm}}{dt}$$

$$= (0.1)(1000\ \text{kg})(9.81\ \text{N/kg})(28\ \text{m/s}) + \frac{dE_{therm}}{dt}$$

$$= 27.5\ \text{kW} + \frac{dE_{therm}}{dt}$$

9. The minimum power occurs when $dE_{therm}/dt = 0$:

$$P_{min} = 27.5\ \text{kW}$$

Remarks The actual power needed by a car is considerably greater than our result because cars are typically only about 15% efficient. About 85% of the power generated by a car's engine goes to internal thermal energy that is expelled as heat exhaust plus thermal energy created by rolling friction and wind resistance.

Remark Figure 7-14 shows the speed of a car versus power for various incline angles. The speed is larger at a fixed power for smaller inclines. The blue curve shows the approximate effect of a term in the required power proportional to v (from rolling friction) and v^3 (from air drag).

Figure 7-14

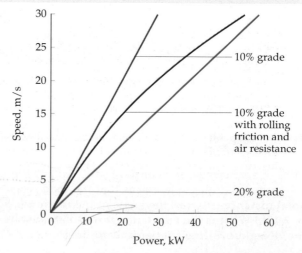

*e*xploring

Transducers

Devices that convert one form of energy to another are called transducers. Those shown here convert nonelectrical energy to electrical energy.

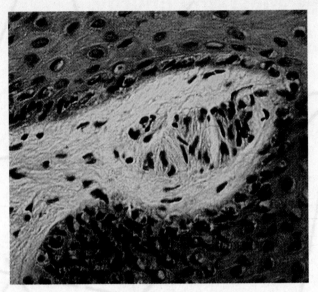

Our sense of touch arises from encapsulated nerve fibers called Meissner's corpuscles, shown above, that are located directly under the outer layer of skin. When skin overlying them is touched, the corpuscles are deformed, triggering electrical impulses in the nerve fibers. A stronger touch produces greater deformation and increases the frequency of impulses. The system is in some respects a biological counterpart to a strain gauge; in both transducers, the energy of mechanical stress is used to modulate changes in electrical conduction.

The strain gauge shown above consists of a grid of very fine wires or foils of a substance such as carbon that changes its electrical resistance when mechanically stressed. The wire is bonded to a thin insulating backing, which is attached by adhesive to an object. Stresses that distort the object deform the attached strain gauge as well. The degree of deformation is measured by the change in resistance of the gauge. If a fixed voltage is applied across the ends of the gauge wire, a varying resistance will produce variations in the current.

A microphone converts sound energy to electrical energy. In the kind shown here, a copper ring is attached to a thin plastic membrane. Sound waves hitting the membrane cause it and the ring to vibrate. The ring is mounted in the field of a permanent magnet. Motion back and forth across the magnetic field, caused by the vibration, induces an alternating current in the ring. This current causes a secondary alternating magnetic field to arise, which in turn creates a secondary alternating current, this time in a wire coil (connected to the output leads) positioned behind the ring. These processes are discussed more fully in Chapters 28 (Section 28-2), 29 (Section 29-2), and 30 (Section 30-7).

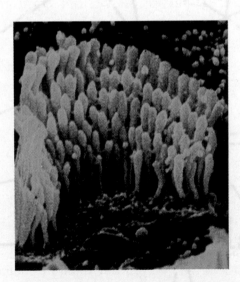

Sound waves transmitted to the spiral chamber of the inner ear cause the fluid there to vibrate. Sensory receptor cells (above, magnified 3500 times) are attached to the walls of the chamber. The receptor cells, stimulated by the vibrating fluid, cause neurons connected to their base to transmit electrical impulses. The impulses, traveling along a chain of neurons, eventually register in the brain as the sensation "sound." This system for converting sound energy to electical energy is a biological counterpart to a microphone.

7-3   Mass and Energy

In 1905, Albert Einstein published his special theory of relativity, a result of which is the famous equation

$$E_0 = mc^2 \quad \text{rest energy}$$ 7-16

where $c = 3 \times 10^8$ m/s is the speed of light in vacuum. We will study this theory in some detail in Chapter 39.

According to Equation 7-16, a particle or system of mass m has "rest" energy mc^2. This energy is intrinsic to the particle. Consider the positron, a particle emitted in a nuclear process called beta decay. Positrons and electrons have identical masses, but equal and opposite electrical charge. When a positron encounters an electron in matter, electron–positron annihilation occurs, a process in which the electron and positron disappear and their energy appears as electromagnetic radiation. If the two particles are initially at rest, the energy of the electromagnetic radiation equals the rest energy of the electron plus that of the positron.

Energies in atomic and nuclear physics are usually expressed in units of electron volts (eV) or mega-electron-volts (1 MeV = 10^6 eV). A convenient unit for the masses of atomic particles is eV/c^2 or MeV/c^2. Table 7-1 lists the rest energies (and therefore the masses) of some elementary particles and light nuclei. The total rest energy of a positron plus electron is 2(0.511 MeV), which is the radiation energy emitted upon annihilation.

The quasar 3C 273 is shown imaged via X-ray energy. The X-ray energy emitted by this quasar is more than a million times that emitted by the entire Milky Way galaxy. The mechanism that generates this enormous energy is not known. It is conjectured to be caused by the annihilation of vast amounts of matter and antimatter.

Table 7-1

Rest Energies of Some Elementary Particles and Light Nuclei

Particle		Rest Energy (MeV)	
Electron	e^-	0.5110	
Positron	e^+	0.5110	
Proton	p	938.280	
Neutron	n	939.573	
Deuteron	d	1875.628	
Triton	t	2808.944	
Alpha particle	α	3727.409	

The rest energy of a *system* can consist of the potential energy of the system, or other internal energies of the system in addition to the intrinsic rest energies of the particles in the system. If the system at rest absorbs energy, ΔE, its rest energy increases, and its mass increases by

$$\Delta M = \frac{\Delta E}{c^2}$$ 7-17

Consider two 1-kg blocks connected by a spring of force constant k. If we stretch the spring a distance A, the potential energy of the system increases by $\Delta U = \frac{1}{2}kA^2$. According to Equation 7-17, the mass of the system has also increased by $\Delta M = \Delta U/c^2$. Because c is such a large number, this increase in mass cannot be observed in macroscopic systems. For example, suppose $k = 800$ N/m, and $A = 10$ cm $= 0.1$ m. The potential energy of the spring system is then $\frac{1}{2}kA^2 = \frac{1}{2}(800$ N/m$)(0.1$ m$)^2 = 4$ J. The increase in mass of the system is

$$\Delta M = \frac{\Delta U}{c^2} = \frac{4 \text{ J}}{(3 \times 10^8 \text{ m/s})^2} = 4.44 \times 10^{-17} \text{ kg}$$

The relative mass increase $\Delta M/M \approx 2 \times 10^{-17}$ is much too small to be observed.

Nuclear Energy

In nuclear reactions, the energy changes are often an appreciable fraction of the rest energy of the system. Consider the deuteron, which is the nucleus of deuterium, an isotope of hydrogen called heavy hydrogen. The deuteron consists of a proton and neutron bound together. From Table 7-1, we see that the mass of the proton is 938.28 MeV/c^2 and the mass of the neutron is 939.57 MeV/c^2. The sum of these two masses is 1877.85 MeV/c^2. But the mass of the deuteron is 1875.63 MeV/c^2, which is less than the sum of the masses of the proton and neutron by 2.22 MeV/c^2. Note that this mass difference is about 0.12%, much greater than the uncertainties inherent in the measurement of these masses, and very much greater than the unobservable 10^{-17} relative mass increase discussed above for a macroscopic system. So where do we find the missing mass of 2.22 MeV/c^2?

Deuterons can be produced by letting neutrons from a reactor collide with protons. When a neutron is captured to form a deuteron, 2.22 MeV of energy is released, usually in the form of electromagnetic radiation. Thus, the mass of the proton–neutron system decreases by 2.22 MeV/c^2 when the particles combine to form a deuteron. Similarly, to break up a deuteron into its constituent parts, a proton plus a neutron, 2.22 MeV of energy must be put into the system. The energy needed to break up a nucleus into its constituent parts is called the **binding energy** of the nucleus. Deuterons can be broken up by bombarding them with energetic particles or electromagnetic radiation possessing energy of at least 2.22 MeV. If the energy is greater, 2.22 MeV is converted to the excess mass of the neutron and proton over that of the deuteron, and the rest appears as kinetic energy of the outgoing particles.

The deuteron is an example of a bound system. Its rest energy is less than the rest energy of its parts, so energy must be put into the system to break it apart. If the rest energy of a system is greater than the rest energy of its parts, the system is unbound. An example is uranium-236, which breaks apart or **fissions** into two smaller nuclei.* The sum of the masses of the resultant parts is less than the mass of the original nucleus. Thus the mass of the system decreases, and energy is released.

In nuclear fusion, two very light nuclei, such as a deuteron and a triton (nucleus of the hydrogen isotope tritium), fuse together. The mass of the resultant nucleus is less than that of the original parts and again energy is released. In a chemical reaction that produces energy, such as coal burning, the mass decrease is of the order of 1 eV/c^2 per atom. This is more than a million times smaller than the mass changes in nuclear reactions, and is not readily observable.

---

* Uranium-236, written ^{236}U, is made in a nuclear reactor when the stable isotope uranium-235 absorbs a neutron.

Example 7-13

A hydrogen atom consisting of a proton and an electron has a binding energy of 13.6 eV. By what percentage is the mass of the proton plus the electron greater than that of the hydrogen atom?

1. The percentage difference between the mass of the hydrogen atom and the masses of its parts is the binding energy E_b divided by $m_e + m_p$:

$$\% \text{ difference} = \frac{E_b}{m_e + m_p} = \frac{13.6 \text{ eV}/c^2}{m_e + m_p}$$

2. Obtain the rest masses of the proton and electron from Table 7-1:

$$m_p = 938.28 \text{ MeV}/c^2, \qquad m_e = 0.511 \text{ MeV}/c^2$$

3. Add to find the sum of these masses:

$$m_p + m_e = 938.79 \text{ MeV}/c^2$$

4. The rest mass of the hydrogen atom is less than this by 13.6 eV/c^2. The percentage difference is:

$$\% \text{ difference} = \frac{13.6 \text{ eV}/c^2}{938.79 \times 10^6 \text{ eV}/c^2} = 1.45 \times 10^{-8}$$

$$= 1.45 \times 10^{-6} \%$$

Remark This mass difference is too small to be measured directly. However, binding energies can be accurately measured, so the mass difference can be found from $E_b = \Delta m/c^2$.

Example 7-14   *try it yourself*

In a typical nuclear fusion reaction, a tritium nucleus (^{3}H) and a deuterium nucleus (^{2}H) fuse together to form a helium nucleus (^{4}He) plus a neutron (Figure 7-15). The reaction is written ^{2}H + ^{3}H → ^{4}He + n. If the initial kinetic energy of the particles is negligible, how much energy is released in this fusion reaction?

Picture the Problem Since energy is released, the total rest energy of the initial particles must be greater than that of the final particles. This difference equals the energy released.

Figure 7-15

Cover the column to the right and try these on your own before looking at the answers.

Steps

Answers

1. Write down the rest energies of ^{2}H and ^{3}H from Table 7-1 and add to find the total initial rest energy.

E_0 (initial) = 1875.628 MeV + 2808.944 MeV

= 4684.572 MeV

2. Do the same for ^{4}He and n to find the final rest energy.

E_0 (final) = 3727.409 MeV + 939.573 MeV

= 4666.982 MeV

3. Find the energy released from $E_{\text{released}} = E_0$ (initial) − E_0 (final).

E_{released} = 4684.572 MeV − 4666.982 MeV

= 17.59 MeV ≈ 17.6 MeV

Remarks This and other fusion reactions occur in the sun. The energy that is released bathes the earth and is ultimately responsible for all life on the planet. The energy constantly pouring out from the sun is matched by a continuous decrease in the sun's rest mass.

Newtonian Mechanics and Relativity

When the speed of a particle approaches the speed of light, Newton's second law breaks down, and we must modify Newtonian mechanics according to Einstein's theory of relativity.* The criterion for the validity of Newtonian mechanics can also be stated in terms of the energy of a particle. In Newtonian mechanics, the kinetic energy of a particle moving with speed v is

$$K = \frac{1}{2}mv^2 = \frac{1}{2}mc^2\frac{v^2}{c^2} = \frac{1}{2}E_0\frac{v^2}{c^2}$$

where $E_0 = mc^2$ is the rest energy of the particle. Then

$$\frac{v}{c} = \sqrt{\frac{2K}{E_0}}$$

Newtonian mechanics is valid if the speed of the particle is much less than the speed of light, or, alternatively, if the kinetic energy of the particle is much less than its rest energy.

7-4   Quantization of Energy

When energy is put into a system that remains at rest, the internal energy of the system increases. It would seem that we could choose to put any amount of energy into a system. However, this is not true for microscopic systems such as atoms or molecules. The internal energy of a microscopic system can increase only by discrete increments.

If we have two blocks attached to a spring and we pull the blocks apart, we do work on the block–spring system, and its potential energy increases. If we then release the blocks, they oscillate back and forth. The energy of oscillation E, which is the kinetic energy of motion of the blocks plus the potential energy due to the stretching of the spring, equals the original potential energy. In time, the energy of the system decreases because of various damping effects such as friction and air resistance. As close as we can measure, the energy decreases continuously. Eventually all the energy is dissipated and the energy of oscillation is zero.

Now consider a diatomic molecule such as molecular oxygen, O_2. The force of attraction between the two oxygen atoms varies approximately linearly with the change in separation (for small changes) much like that of a spring. If a diatomic molecule is set oscillating with some energy E, the energy decreases with time as the molecule radiates, or interacts with its surroundings, but the decrease is *not continuous*. The energy decreases in finite steps, and the lowest energy state, called the ground state, is not zero. The vibrational energy of a diatomic molecule is said to be **quantized**; that is, the molecule can possess energies only in certain amounts, known as quanta.

When blocks on a spring or diatomic molecules oscillate, the time for one oscillation is called the period T. The reciprocal of the period is the frequency of oscillation $f = 1/T$. We will see in Chapter 12 that the period and frequency of an oscillator do not depend on the energy of oscillation. As the en-

---

* Einstein published two very different theories of relativity. His special theory of relativity, which applies to our discussion here, was published in 1905 and applies to particles moving at speeds near the speed of light. Einstein's general theory of relativity, published in 1916, deals with gravity.

ergy decreases, the frequency remains the same. Figure 7-16 shows an **energy-level diagram** for an oscillator. The allowed energies are approximately equally spaced, and are given by*

$$E_n = \left(n + \frac{1}{2}\right)hf, \qquad n = 0, 1, 2, 3, \ldots \qquad \text{7-18}$$

where f is the frequency of oscillation and h is a fundamental constant of nature called Planck's constant[†]:

$$h = 6.626 \times 10^{-34}\,\text{J·s} \qquad \text{7-19}$$

The integer n in Equation 7-18 is called a **quantum number.** The lowest possible energy is the **ground-state energy** $E_0 = \frac{1}{2}hf$.

Microscopic systems often gain or lose energy by absorbing or emitting electromagnetic radiation. By conservation of energy, if E_i and E_f are the initial and final energies of a system, the energy of the radiation emitted or absorbed is

$$E_{\text{rad}} = E_i - E_f$$

Since the system energies E_i and E_f are quantized, the radiated energy is also quantized. Historically, the quantization of electromagnetic radiation, as proposed by Max Planck and Albert Einstein, was the first "discovery" of energy quantization. The quantum of radiation energy is called a **photon**. The energy of a photon is given by

$$E_{\text{photon}} = hf \qquad \text{7-20}$$

where f is the frequency of the electromagnetic radiation. Electromagnetic radiation includes light, microwaves, radio waves, television waves, X rays, and gamma rays. These differ from one another only in their range of frequencies and thereby in the range in energy of their photons.

As far as we know, all systems exhibit energy quantization. For macroscopic systems, the steps between energy levels are so small as to be unobservable. For example, typical oscillation frequencies for two blocks on a spring are 1 to 10 times per second. If $f = 10$ oscillations per second, the spacing between allowed levels is $hf = (6.626 \times 10^{-34}\,\text{J·s}) \times (10/\text{s}) \approx 6 \times 10^{-33}$ J. Since the energy of a macroscopic system is of the order of joules, a quantum step of 10^{-33} J is too small to be noticed. To put it another way, if the energy of a system is 1 J, the value of n is of the order of 10^{32} and changes of one or two quantum units will not be observable.

For a diatomic molecule, a typical frequency of vibration is 10^{14} vibrations per second, and a typical energy is 10^{-19} J. The spacing between allowed levels is then $E_{n+1} - E_n = hf \approx (6.63 \times 10^{-34}\,\text{J·s})(10^{14}\,\text{s}) \approx 6 \times 10^{-20}$ J. Thus, changes in the energy of oscillation are on the same order as the energy of the molecule, and quantization is definitely noticeable.

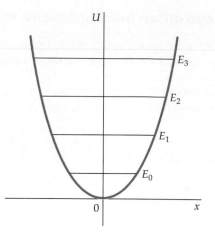

Figure 7-16 Energy-level diagram for an oscillator.

---

* A diatomic molecule can also have rotational energy. The rotational energy is quantized, as expected, but the energy levels are not equally spaced, and the lowest possible energy is zero. We study rotational energy in Chapters 9 and 10.

† In 1900 the German physicist Max Planck had introduced this constant as a calculational device to explain discrepancies between theory and experiment on the spectrum of blackbody radiation. The significance of Planck's constant was not appreciated by Planck or anyone else until Einstein postulated in 1905 that the energy of electromagnetic radiation is not continuous, but occurs in packets of size hf, where f is the frequency of the radiation and h is the constant discovered by Planck.

Summary

1. The conservation of mechanical energy is an important relation derived from Newton's laws for conservative forces. It is useful in solving many problems.

2. The work–energy theorem and the conservation of energy are fundamental laws of nature that have applications in all areas of physics.

3. Einstein's equation $E_0 = mc^2$ is a fundamental relation between mass and energy.

4. Quantization is a fundamental property of the energy in bound systems.

Topic	Remarks and Relevant Equations
1. Mechanical Energy	The sum of the kinetic and potential energy of a system is called the total mechanical energy
	$$E_{mech} = K + U \qquad \text{7-4}$$
Conservation of mechanical energy	If no external forces do work on a system, and the internal forces are all conservative, the total mechanical energy of the system remains constant
	$$E_{mech} = K + U = \text{constant} \qquad \text{7-5}$$
	$$K_f + U_f = K_i + U_i \qquad \text{7-6}$$
2. Total Energy of a System	The energy of a system consists of mechanical energy E_{mech}, thermal energy E_{therm}, chemical energy E_{chem}, and other types of energy E_{other}, such as sound radiation and electromagnetic radiation.
	$$E_{sys} = E_{mech} + E_{therm} + E_{chem} + E_{other} \qquad \text{7-8}$$
3. Conservation of Energy	
Universe	The total energy of the universe is constant. Energy can be converted from one form to another, or transmitted from one region to another, but energy can never be created or destroyed.
System	The energy of a system can be changed by various means such as work done on the system, heat transfer, and emission or absorption of radiation. The increase or decrease in the energy of the system can always be accounted for by the appearance or disappearance of some kind of energy somewhere else.
	$$E_{in} - E_{out} = \Delta E_{sys} \qquad \text{7-7}$$
Work–energy theorem	$$W_{ext} = \Delta E_{sys} = \Delta E_{mech} + \Delta E_{therm} + \Delta E_{chem} + \Delta E_{other} \qquad \text{7-9}$$
4. Energy Dissipated by Friction	For a system that involves a pair of sliding surfaces, the total energy dissipated by friction on both surfaces equals the increase in thermal energy of the system and is given by
	$$f \, \Delta s = \Delta E_{therm} \qquad \text{7-13}$$
	where Δs is the displacement of one surface relative to the other. If there is no external work done, the increase in thermal energy equals the decrease in mechanical energy of the system:
	$$\Delta E_{therm} = f \, \Delta s = -\Delta E_{mech} \qquad \text{7-14}$$
5. Problem Solving	The conservation of mechanical energy and the work–energy theorem can be used as an alternative to Newton's laws to solve mechanics problems that require the determination of the speed of a particle as a function of its position.

6.	Mass and Energy	A particle with mass m has an intrinsic rest energy E_0 given by

$$E_0 = mc^2 \qquad \textbf{7-15}$$

where $c = 3 \times 10^8$ m/s is the speed of light in vacuum.

A system with mass M also has a rest energy $E_0 = Mc^2$. If a system gains or loses internal energy ΔE, it simultaneously gains or loses mass $\Delta M = \Delta E/c^2$.

7.	Binding Energy	The energy required to separate a system into its constituent parts is called its binding energy. The binding energy is ΔMc^2, where ΔM is the increase in mass of the parts over the mass of the system.

8.	Newtonian Mechanics and the Theory of Relativity	When the speed of a particle approaches the speed of light c, or the kinetic energy of the particle approaches its rest energy, Newtonian mechanics breaks down and must be replaced by Einstein's special theory of relativity.

9.	Energy Quantization	The internal energy of a microscopic system is found to have only a discrete set of possible values. For a system oscillating with frequency f, the allowed energy values are separated by an amount hf, where h is Planck's constant:

$$h = 6.626 \times 10^{-34} \, \text{J·s} \qquad \textbf{7-19}$$

Photons   Microscopic systems often exchange energy with their surroundings by emitting or absorbing electromagnetic radiation, which is also quantized. The quantum of radiation energy is called the photon:

$$E_{\text{photon}} = hf \qquad \textbf{7-20}$$

where f is the frequency of the electromagnetic radiation.

Problem-Solving Guide

1. Begin by drawing a neat diagram with a suitable coordinate system.
2. Choose your system and indicate any external forces that act on it. Include in your system objects for which we have a potential-energy function, such as springs, and the earth. If there is sliding friction, make sure your system includes both surfaces.
3. Determine whether mechanical energy is conserved. If it is, write expressions for the initial and final mechanical energy and set them equal to each other. Choose a convenient point for the zero of potential energy.
4. If there is sliding friction, equate $f\Delta s$ to ΔE_{therm}.
5. Apply the work–energy theorem.

Summary of Worked Examples

Type of Calculation	Procedure and Relevant Examples
1. Mechanical Energy Conserved	
Find the speed of an object that is falling in the earth's gravitational field.	Use conservation of mechanical energy to find the final kinetic energy. The gravitational potential energy is mgy, where y is measured from an arbitrary point. **Examples 7-1, 7-2**
Find the distance traveled or the final velocity of an object attached to a spring moving in the earth's gravitational field.	Use conservation of mechanical energy. The potential energy of a spring that is stretched or compressed by an amount x is $\frac{1}{2}kx^2$. **Examples 7-3, 7-4**

Find the final energy of two objects connected by a string.	Use conservation of mechanical energy for the two-object system. The work done by the string tension is internal to the system. **Example 7-5**
2. Analyzing Energy Changes When Thermal or Chemical Energies Are Involved	Choose your system and apply conservation of energy as expressed in the work–energy theorem. **Examples 7-6, 7-7, 7-8**
Find the power needed to climb a hill.	Use conservation of energy as expressed in the work–energy theorem. Convert each term to a time rate of change. **Example 7-12**
3. Problems With Sliding Friction	If W_{ext} is zero, the energy dissipated by friction equals the decrease in mechanical energy. If there is work done by external forces, use conservation of energy as expressed in the work–energy theorem with $\Delta E_{therm} = f\Delta s$. **Examples 7-8, 7-9, 7-11**
4. Find the Energy Released in a Nuclear Reaction	Compute the initial and final rest energies. Then $E_{released} = E_0(\text{initial}) - E_0(\text{final})$. **Example 7-14**

Problems

Conceptual Problems

Problems from Optional and Exploring sections

In a few problems, you are given more data than you actually need; in a few other problems, you are required to supply data from your general knowledge, outside sources, or informed estimates.

- Single-concept, single-step, relatively easy
- Intermediate-level, may require synthesis of concepts
- Challenging, for advanced students

Take $g = 9.81$ *N/kg* $= 9.81$ *m/s^2 and neglect friction in all problems unless otherwise stated.*

The Conservation of Mechanical Energy

1 •• What are the advantages and disadvantages of using the conservation of mechanical energy rather than Newton's laws to solve problems?

2 •• Two objects of unequal mass are connected by a massless cord passing over a frictionless peg. After the objects are released from rest, which of the following statements are true? (U = gravitational potential energy, K = kinetic energy of the system.)

(a) $\Delta U < 0$ and $\Delta K > 0$
(b) $\Delta U = 0$ and $\Delta K > 0$
(c) $\Delta U < 0$ and $\Delta K = 0$
(d) $\Delta U = 0$ and $\Delta K = 0$
(e) $\Delta U > 0$ and $\Delta K < 0$

3 •• Two stones are thrown with the same initial speed at the same instant from the roof of a building. One stone is thrown at an angle of 30° above the horizontal; the other is thrown horizontally. (Neglect air resistance.) Which statement below is true?

(a) The stones strike the ground at the same time and with equal speeds.
(b) The stones strike the ground at the same time with different speeds.
(c) The stones strike the ground at different times with equal speeds.
(d) The stones strike the ground at different times with different speeds.

4 • A block of mass m is pushed up against a spring, compressing it a distance x, and the block is then released. The spring projects the block along a frictionless horizontal surface, giving the block a speed v. The same spring projects a second block of mass $4m$, giving it a speed $3v$. What distance was the spring compressed in the second case?

5 • A woman on a bicycle traveling at 10 m/s on a horizontal road stops pedaling as she starts up a hill inclined at 3.0° to the horizontal. Ignoring friction forces, how far up the hill will she travel before stopping?

(a) 5.1 m   (b) 30 m
(c) 97 m   (d) 10.2 m
(e) The answer depends on the mass of the woman.

6 • A pendulum of length L with a bob of mass m is pulled aside until the bob is a distance $L/4$ above its equilibrium position. The bob is then released. Find the speed of the bob as it passes the equilibrium position.

7 • When she hosts a garden party, Julie likes to launch bagels to her guests with a spring device that she has devised. She places one of her 200-g bagels against a horizontal spring mounted on her gazebo. The force constant of the spring is 300 N/m, and she compresses it 9 cm. (a) Find the work done by Julie and the spring when Julie launches a bagel. (b) If the released bagel leaves the spring at the spring's equilibrium position, find the speed of the bagel at that point. (c) If the bagel launcher is 2.2 m above the grass, what is Julie's horizontal range firing 200-g bagels?

8 • A 3-kg block slides along a frictionless horizontal surface with a speed of 7 m/s (Figure 7-17). After sliding a distance of 2 m, the block makes a smooth transition to a frictionless ramp inclined at an angle of 40° to the horizontal.

Figure 7-17
Problem 8

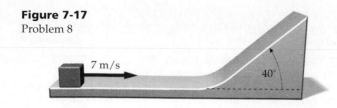

7 m/s   40°

How far up the ramp does the block slide before coming momentarily to rest?

9   • The 3-kg object in Figure 7-18 is released from rest at a height of 5 m on a curved frictionless ramp. At the foot of the ramp is a spring of force constant $k = 400$ N/m. The object slides down the ramp and into the spring, compressing it a distance x before coming momentarily to rest. (a) Find x. (b) What happens to the object after it comes to rest?

Figure 7-18 Problem 9

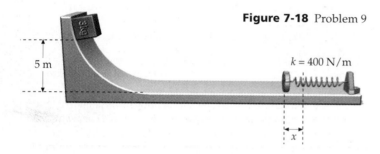

3kg

5 m $k = 400$ N/m

x

10   • A vertical spring compressed a distance x sits on a concrete floor. When a block of mass m_1 is placed on the spring and the spring is released, the block is projected upward to a height h. If a block of mass $m_2 = 2m_1$ is placed on the spring and the spring is again compressed a distance x and released, to what height will the block rise?

(a) $h/4$ (b) $h/2$ (c) $h/\sqrt{2}$ (d) h

11   • If the spring in Problem 10 is compressed an amount $2x$ when the block of mass m_2 is placed on it, to what height will the block rise when the spring is released?

(a) $2h$ (b) $\sqrt{2}\,h$ (c) h (d) $h/\sqrt{2}$

12   • A 15-g ball is shot from a spring gun whose spring has a force constant of 600 N/m. The spring can be compressed 5 cm. How high will the ball go if the gun is aimed vertically?

13   • A stone is projected horizontally with a speed of 20 m/s from a bridge 16 m above the surface of the water. What is the speed of the stone as it strikes the water?

14   • At a dock, a crane lifts a 4000-kg container 30 m, swings it out over the deck of a freighter, and lowers the container into the hold of the freighter, which is 8 m below the level of the dock. How much work is done by the crane? (Neglect friction losses.)

15   • A 16-kg child on a playground swing moves with a speed of 3.4 m/s when the 6-m-long swing is at its lowest point. What is the angle that the swing makes with the vertical when the child is at the highest point?

16   •• In 1983, Jacqueline De Creed, driving a 1967 Ford Mustang, made a jump of 71 m, taking off from a ramp inclined at 30° with the horizontal. If the mass of the car and driver was about 900 kg, find the kinetic energy K and potential energy U of De Creed's vehicle at the top point of her flight.

17   •• The system in Figure 7-19 is initially at rest when the lower string is cut. Find the speed of the objects when they are at the same height.

18   •• While traveling in the far north, one of your companions gets snow blindness, and you have to lead him along by the elbow. Looking back, you see your other companion, Sandy, fall and slide along the frictionless surface of the frozen river valley shown in Figure 7-20. If point Q is 4.5 m higher than point P, and your hapless companion fell at point P with a velocity v_0 down the slope, describe his motion to your snow-blind friend if (a) $v_0 = 2$ m/s and (b) $v_0 = 5$ m/s. (c) What is the minimum initial speed required for the fall to carry your partner past point Q?

Figure 7-19 Problem 17

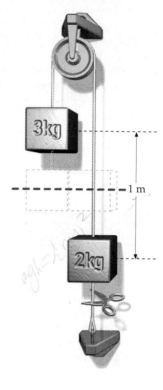

3kg

1 m

2kg

Figure 7-20 Problem 18

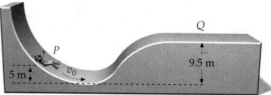

Q

P

v_0

5 m

9.5 m

19   •• A block rests on an inclined plane as in Figure 7-21. A spring to which it is attached via a pulley is being pulled downward with gradually increasing force. The value of μ_s is known. Find the potential energy U of the spring at the moment when the block begins to move.

Figure 7-21 Problem 19

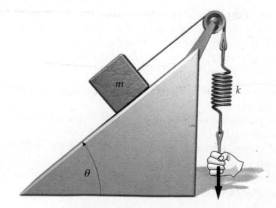

m

k

θ

20 •• Sandy is sliding helplessly across the frictionless ice with her climbing rope trailing behind (Figure 7-22). Racing after her, you get hold of her rope just as she goes over the edge of a cliff. You manage to grab a tree branch in time to keep from going over yourself. Let $U = 0$ for the position of Sandy dangling in midair at the other end of the rope. Snap! The branch to which you are clinging breaks. (*a*) Write an expression for the total mechanical energy of this two-body system after Sandy has fallen a distance y. (*b*) There is another tree branch 2 m closer to the cliff edge than the first. What is your speed as you reach it?

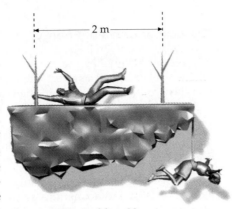

Figure 7-22 Problem 20

21 •• A 2.4-kg block is dropped from a height of 5.0 m onto a spring of spring constant 3955 N/m. When the block is momentarily at rest, the spring has compressed by 25 cm. Find the speed of the block when the compression of the spring is 15.0 cm.

22 •• Red is a girl of mass m who is taking a picnic lunch to her grandmother. She ties a rope of length R to a tree branch over a creek and starts to swing from rest at point A, which is a distance $R/2$ lower than the branch (Figure 7-23). What is the minimum breaking tension for the rope if it is not to break and drop Red into the creek?

Figure 7-23 Problem 22

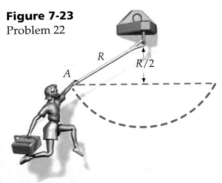

23 •• A ball at the end of a string moves in a vertical circle with constant energy E. What is the difference between the tension at the bottom of the circle and the tension at the top?

24 •• A roller coaster car of mass 1500 kg starts a distance $H = 23$ m above the bottom of a loop 15 m in diameter (Figure 7-24). If friction is negligible, the downward force of the rails on the car when it is upside down at the top of the loop is _____.

(*a*) 4.6×10^4 N   (*b*) 3.1×10^4 N   (*c*) 1.7×10^4 N
(*d*) 980 N   (*e*) 1.6×10^3 N

25 •• A stone is thrown upward at an angle of 53° above the horizontal. Its maximum height during the trajectory is 24 m. What was the stone's initial speed?

26 •• A baseball of mass 0.17 kg is thrown from the roof of a building 12 m above the ground. Its initial velocity is 30 m/s at an angle of 40° above the horizontal. (*a*) What is the maximum height of the ball? (*b*) What is the work done by gravity as the ball moves from the roof to its maximum height? (*c*) What is the speed of the ball as it strikes the ground?

27 •• An 80-cm-long pendulum with a 0.6-kg bob is released from rest at initial angle θ_0 with the vertical. At the bottom of the swing, the speed of the bob is 2.8 m/s. (*a*) What was the initial angle of the pendulum? (*b*) What angle does the pendulum make with the vertical when the speed of the bob is 1.4 m/s?

28 •• The Royal Gorge bridge over the Arkansas River is about $L = 310$ m high. A bungee jumper of mass 60 kg has an elastic cord of length $d = 50$ m attached to her feet. Assume that the cord acts like a spring of force constant k. The jumper leaps, barely touches the water, and after numerous ups and downs comes to rest at a height h above the water. (*a*) Find h. (*b*) Find the maximum speed of the jumper.

29 •• A pendulum consists of a 2-kg bob attached to a light string of length 3 m. The bob is struck horizontally so that it has an initial horizontal velocity of 4.5 m/s. For the point at which the string makes an angle of 30° with the vertical, what is (*a*) the speed? (*b*) the potential energy? (*c*) the tension in the string? (*d*) What is the angle of the string with the vertical when the bob reaches its greatest height?

30 •• Lou is trying to kill mice by swinging a clock of mass m attached to one end of a light (massless) stick 1.4 m in length hanging on a nail in the wall (Figure 7-25). The clock end of the stick is free to rotate around its other end in a vertical circle. Lou raises the clock until the stick is horizontal, and when mice peek their heads out from the hole to their den, he gives it an initial downward velocity v. The clock misses a mouse and continues on its circular path with just enough energy to complete the circle and bonk Lou on the

Figure 7-25 Problem 30

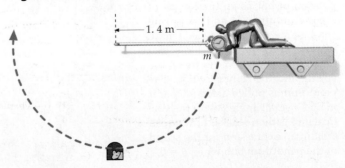

Figure 7-24 Problem 24

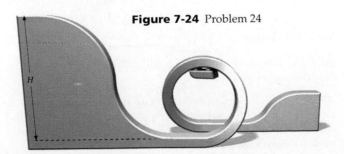

back of his head, to the sound of cheering mice. (*a*) What was the value of *v*? (*b*) What was the clock's speed at the bottom of its swing?

31 •• A pendulum consists of a string of length *L* and a bob of mass *m*. The string is brought to a horizontal position and the bob is given the minimum initial speed enabling the pendulum to make a full turn in the vertical plane. (*a*) What is the maximum kinetic energy *K* of the bob? (*b*) What is the tension in the string when the kinetic energy is maximum?

32 •• A child whose weight is 360 N swings out over a pool of water using a rope attached to the branch of a tree at the edge of the pool. The branch is 12 m above ground level and the surface of the pool is 1.8 m below ground level. The child holds onto the rope at a point 10.6 m from the branch and moves back until the angle between the rope and the vertical is 23°. When the rope is in the vertical position, the child lets go and drops into the pool. Find the speed of the child at the surface of the pool.

33 •• Walking by a pond, you find a rope attached to a tree limb 5.2 m off the ground. You decide to use the rope to swing out over the pond. The rope is a bit frayed but supports your weight. You estimate that the rope might break if the tension is 80 N greater than your weight. You grab the rope at a point 4.6 m from the limb and move back to swing out over the pond. (*a*) What is the maximum safe initial angle between the rope and the vertical so that it will not break during the swing? (*b*) If you begin at this maximum angle, and the surface of the pond is 1.2 m below the level of the ground, with what speed will you enter the water if you let go of the rope when the rope is vertical?

Figure 7-26 Problem 34

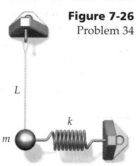

34 •• A pendulum of length *L* has a bob of mass *m* attached to a light string, which is attached to a spring of force constant *k*. With the pendulum in the position shown in Figure 7-26, the spring is at its unstretched length. If the bob is now pulled aside so that the string makes a *small* angle θ with the vertical, what is the speed of the bob after release as it passes through the equilibrium position?

Figure 7-27 Problem 35

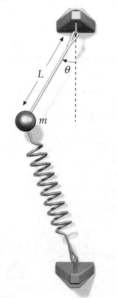

35 ••• A pendulum is suspended from the ceiling and attached to a spring fixed at the opposite end directly below the pendulum support (Figure 7-27). The mass of the pendulum bob is *m*, the length of the pendulum is *L*, and the spring constant is *k*. The unstretched length of the spring is *L*/2 and the distance between the bottom of the spring and the ceiling is 1.5*L*. The pendulum is pulled aside so that it makes a small angle θ with the vertical and is then released from rest. Obtain an expression for the speed of the pendulum bob when θ = 0.

The Conservation of Energy

36 • True or false:
(*a*) The total energy of a system cannot change.
(*b*) When you jump into the air, the floor does work on you, increasing your potential energy.

37 • A man stands on roller skates next to a rigid wall. To get started, he pushes off against the wall. Discuss the energy changes pertinent to this situation.

38 • Discuss the energy changes involved when a car starts from rest and accelerates so that the car's wheels do not slip. What external force accelerates the car? Does this force do work?

39 • A body falling through the atmosphere (air resistance is present) gains 20 J of kinetic energy. The amount of gravitational potential energy that is lost is
(*a*) 20 J.   (*b*) more than 20 J.   (*c*) less than 20 J.
(*d*) impossible to tell without knowing the mass of the body.
(*e*) impossible to tell without knowing how far the body falls.

40 • Assume that you can expend energy at a constant rate of 250 W. Estimate how fast you can run up four flights of stairs, with each flight 3.5 m high.

41 • A 70-kg skater pushes off the wall of a skating rink, acquiring a speed of 4 m/s. (*a*) How much work is done on the skater? (*b*) What is the change in the mechanical energy of the skater? (*c*) Discuss the conservation of energy as applied to the skater.

42 • In a volcanic eruption, 4 km³ of mountain with a density of 1600 kg/m³ was lifted an average height of 500 m. (*a*) How much energy in joules was released in this eruption? (*b*) The energy released by thermonuclear bombs is measured in megatons of TNT, where 1 megaton of TNT = 4.2 × 10¹⁵ J. Convert your answer for (*a*) to megatons of TNT.

43 •• An 80-kg physics student climbs a 120-m hill. (*a*) What is the increase in the gravitational potential energy of the student? (*b*) Where does this energy come from? (*c*) The student's body is 20% efficient; that is, for every 20 J that are converted to mechanical energy, 100 J of internal energy are expended, with 80 J going into thermal energy. How much chemical energy is expended by the student during the climb?

44 •• In 1993, Carl Fentham of Great Britain raised a full keg of beer (mass 62 kg) to a height of about 2 m 676 times in 6 h. Assuming that work was done only as the keg was going up, estimate how many such kegs of beer he would have to drink to reimburse his energy expenditure. (1 liter of beer is approximately 1 kg and provides about 1.5 MJ of energy; in your calculations, neglect the mass of the empty keg.)

Kinetic Friction

45 • Discuss the energy considerations when you pull a box along a rough road.

46 • A 2000-kg car moving at an initial speed of 25 m/s along a horizontal road skids to a stop in 60 m. (*a*) Find the energy dissipated by friction. (*b*) Find the coefficient of kinetic friction between the tires and the road.

47 • An 8-kg sled is initially at rest on a horizontal road. The coefficient of kinetic friction between the sled and the road is 0.4. The sled is pulled a distance of 3 m by a force of 40 N applied to the sled at an angle of 30° to the horizontal. (a) Find the work done by the applied force. (b) Find the energy dissipated by friction. (c) Find the change in the kinetic energy of the sled. (d) Find the speed of the sled after it has traveled 3 m.

48 • Returning to Problem 8, suppose the surfaces described are not frictionless and that the coefficient of kinetic friction between the block and the surfaces is 0.30. Find (a) the speed of the block when it reaches the ramp, and (b) the distance that the block slides up the ramp before coming momentarily to rest. (Neglect the energy dissipated along the transition curve.)

49 • The 2-kg block in Figure 7-28 slides down a frictionless curved ramp, starting from rest at a height of 3 m. The block then slides 9 m on a rough horizontal surface before coming to rest. (a) What is the speed of the block at the bottom of the ramp? (b) What is the energy dissipated by friction? (c) What is the coefficient of friction between the block and the horizontal surface?

Figure 7-28 Problem 49

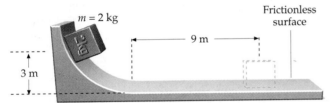

m = 2 kg

Frictionless surface

9 m

3 m

50 •• A 20-kg girl slides down a playground slide that is 3.2 m high. When she reaches the bottom of the slide, her speed is 1.3 m/s. (a) How much energy was dissipated by friction? (b) If the slide is inclined at 20°, what is the coefficient of friction between the girl and the slide?

Figure 7-29 Problem 51

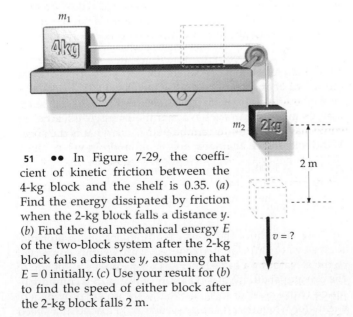

m_1

4kg

m_2 2kg

2 m

$v = ?$

51 •• In Figure 7-29, the coefficient of kinetic friction between the 4-kg block and the shelf is 0.35. (a) Find the energy dissipated by friction when the 2-kg block falls a distance y. (b) Find the total mechanical energy E of the two-block system after the 2-kg block falls a distance y, assuming that E = 0 initially. (c) Use your result for (b) to find the speed of either block after the 2-kg block falls 2 m.

52 •• Nils Lied, an Australian meteorologist, once played golf on the ice in Antarctica and drove a ball a horizontal distance of 2400 m. For a rough estimate, let us assume that the ball took off at θ = 45°, flew a horizontal distance of 200 m without air resistance, and then slid on the ice without bouncing, its velocity being equal to the horizontal component of the initial velocity. Estimate the coefficient of kinetic friction μ_k between the ice and ball.

53 •• A particle of mass m moves in a horizontal circle of radius r on a rough table. It is attached to a horizontal string fixed at the center of the circle. The speed of the particle is initially v_0. After completing one full trip around the circle, the speed of the particle is $\frac{1}{2}v_0$. (a) Find the energy dissipated by friction during that one revolution in terms of m, v_0, and r. (b) What is the coefficient of kinetic friction? (c) How many more revolutions will the particle make before coming to rest?

54 •• In 1987, British skier Graham Wilkie achieved a speed of v = 211 km/h going downhill. Assuming that he reached the maximum speed at the end of the hill and then continued on the horizontal surface, find the maximum distance d he *could have* covered on the horizontal surface. Take the coefficient of kinetic friction μ_k to be constant throughout the run; neglect air resistance. Assume the hill is 225 m high with a constant slope of 30° with the horizontal.

55 •• During a move, Kate and Lou have to push Kate's 80-kg stove up a rough loading ramp, pitched at an angle of 10°, to get it into a truck. They push it along the horizontal floor to pick up speed and give it one last push at the bottom of the ramp, hoping for the best. Unfortunately, the stove stops short and then slides down the ramp, sending them leaping to the side. (a) If the stove has a speed of 3.0 m/s at the bottom of the ramp, and a speed of 0.8 m/s when it is 2 m up the ramp, what is the maximum height reached by the stove? (b) What is the stove's speed when it passes the 2-m spot again? (c) What is the energy dissipated by friction during the complete round trip back to the bottom of the ramp?

56 •• A 2.4-kg box has an initial velocity of 3.8 m/s upward along a rough plane inclined at 37° to the horizontal. The coefficient of kinetic friction between the box and plane is 0.30. How far up the incline does the box travel? What is its speed when it passes its starting point on its way down the incline?

57 ••• A block of mass m rests on a rough plane inclined at θ with the horizontal (Figure 7-30). The block is attached to a spring of constant k near the top of the plane. The coefficients of static and kinetic friction between the block and plane are μ_s and μ_k, respectively. The spring is slowly pulled upward along the plane until the block starts to move. (a) Obtain an expression for the extension d of the spring the

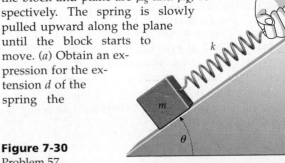

k

m

θ

Figure 7-30
Problem 57

instant the block moves. (b) Determine the value of μ_k such that the block comes to rest just as the spring is in its unstressed condition, i.e., neither extended nor compressed.

Mass and Energy

58 • How much rest mass is consumed in the core of a nuclear-fueled electric generating plant in producing (a) one joule of thermal energy? (b) enough energy to keep a 100-W light bulb burning for 10 years?

59 • (a) Calculate the rest energy in 1 g of dirt. (b) If you could convert this energy into electrical energy and sell it for 10 cents per kilowatt-hour, how much money would you get? (c) If you could power a 100-W light bulb with this energy, for how long could you keep the bulb lit?

60 • A muon has a rest energy of 105.7 MeV. Calculate its rest mass in kilograms.

61 • For the fusion reaction in Example 7-14, calculate the number of reactions per second that are necessary to generate 1 kW of power.

62 • How much energy is needed to remove one neutron from ^{4}He, leaving ^{3}He plus a neutron? (The rest energy of ^{3}He is 2808.41 MeV.)

63 • A free neutron at rest decays into a proton plus an electron:

$$n \rightarrow p + e$$

Use Table 7-1 to calculate the energy released in this reaction.

64 •• In one nuclear fusion reaction, two ^{2}H nuclei combine to produce ^{4}He. (a) How much energy is released in this reaction? (b) How many such reactions must take place per second to produce 1 kW of power?

65 •• A large nuclear power plant produces 3000 MW of power by nuclear fission, which converts matter into energy. (a) How many kilograms of matter does the plant consume in one year? (b) In a coal-burning power plant, each kilogram of coal releases 31 MJ of energy when burned. How many kilograms of coal are needed each year for a 3000-MW plant?

General Problems

66 •• A block of mass m, starting from rest, is pulled by a string up a frictionless inclined plane that makes an angle θ with the horizontal. The tension in the string is T and the string is parallel to the plane. After traveling a distance L, the speed of the block is v. The work done by the tension T is

(a) $mgL \sin \theta$ (b) $mgL \cos \theta + \frac{1}{2} mv^2$ (c) $mgL \sin \theta + \frac{1}{2} mv^2$
(d) $mgL \cos \theta$ (e) $TL \cos \theta$

67 •• A block of mass m slides with constant velocity v down a plane inclined at θ with the horizontal. During the time interval Δt, what is the magnitude of the energy dissipated by friction?

(a) $mgv \Delta t \tan \theta$ (b) $mgv \Delta t \sin \theta$ (c) $\frac{1}{2} mv^3 \Delta t$
(d) The answer cannot be determined without knowing the coefficient of kinetic friction.

68 •• Assume that on applying the brakes a constant frictional force acts on the wheels of a car. If that is so, it follows that

(a) the distance the car travels before coming to rest is proportional to the speed of the car before the brakes are applied.
(b) the car's kinetic energy diminishes at a constant rate.
(c) the kinetic energy of the car is inversely proportional to the time that has elapsed since the application of the brakes.
(d) none of the above apply.

69 • Our bodies convert internal chemical energy into work and heat at the rate of about 100 W, which is called our metabolic rate. (a) How much internal chemical energy do we use in 24 h? (b) The energy comes from the food that we eat and is usually measured in kilocalories, where 1 kcal = 4.184 kJ. How many kilocalories of food energy must we ingest per day if our metabolic rate is 100 W?

70 • A 3.5-kg box rests on a horizontal frictionless surface in contact with a spring of spring constant 6800 N/m. The spring is fixed at its other end and is initially at its uncompressed length. A constant horizontal force of 70 N is applied to the box so that the spring compresses. Determine the distance the spring is compressed when the box is momentarily at rest.

71 • The average energy per unit time per unit area that reaches the upper atmosphere of the earth from the sun, called the solar constant, is 1.35 kW/m^2. Because of absorption and reflection by the atmosphere, about 1 kW/m^2 reaches the surface of the earth on a clear day. How much energy is collected in 8 h of daylight by a solar panel 1 m by 2 m on a rotating mount that is always perpendicular to the sun's rays.

72 • When the jet-powered car *Spirit of America* went out of control during a test drive at Bonneville Salt Flats, Utah, it left skid marks about 9.5 km long. (a) If the car was moving initially at a speed of $v = 708$ km/h, estimate the coefficient of kinetic friction μ_k. (b) What was the kinetic energy K of the car at time $t = 60$ s after the brakes were applied? Take the mass of the car to be 1250 kg.

73 •• A T-bar tow is required to pull 80 skiers up a 600-m slope inclined at 15° above horizontal at a speed of 2.5 m/s. The coefficient of kinetic friction is 0.06. Find the motor power required if the mass of the average skier is 75 kg.

74 •• A 2-kg box is projected with an initial speed of 3 m/s up a rough plane inclined at 60° above horizontal. The coefficient of kinetic friction is 0.3. (a) List all the forces acting on the box. (b) How far up the plane does the box slide before it stops momentarily? (c) What is the energy dissipated by friction as the box slides up the plane? (d) What is the speed of the box when it again reaches its initial position?

75 •• A 1200-kg elevator driven by an electric motor can safely carry a maximum load of 800 kg. What is the power provided by the motor when the elevator ascends with a full load at a speed of 2.3 m/s?

76 •• To reduce the power requirement of elevator motors, elevators are counterbalanced with weights connected to the elevator by a cable that runs over a pulley at the top of the elevator shaft. If the elevator in Problem 75 is counterbalanced with a mass of 1500 kg, what is the power provided by the motor when the elevator ascends fully loaded at a speed of 2.3 m/s? How much power is provided by the motor when the elevator ascends without a load at 2.3 m/s?

77 •• The spring constant of a toy dart gun is 5000 N/m. To cock the gun the spring is compressed 3 cm. The 7-g dart, fired straight upward, reaches a maximum height of 24 m. Determine the energy dissipated by air friction during the dart's ascent. Estimate the speed of the projectile when it returns to its starting point.

78 •• A 0.050-kg dart is fired vertically from a spring gun that has a spring constant of 4000 N/m. Prior to release, the spring was compressed by 10.7 cm. When the dart is 6.8 m above the gun, its upward speed is 28 m/s. Determine the maximum height reached by the dart.

79 •• In a volcanic eruption, a 2-kg piece of porous volcanic rock is thrown vertically upward with an initial speed of 40 m/s. It travels upward a distance of 50 m before it begins to fall back to the earth. (a) What is the initial kinetic energy of the rock? (b) What is the increase in thermal energy due to air friction during ascent ? (c) If the increase in thermal energy due to air friction on the way down is 70% of that on the way up, what is the speed of the rock when it returns to its initial position?

80 •• A block of mass m starts from rest at a height h and slides down a frictionless plane inclined at θ with the horizontal as shown in Figure 7-31. After sliding a distance L, the block strikes a spring of force constant k. Find the compression of the spring when the block is momentarily at rest.

Figure 7-31 Problem 80

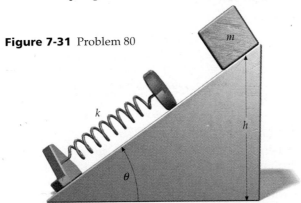

81 •• A car of mass 1500 kg traveling at 24 m/s is at the foot of a hill that rises 120 m in 2.0 km. At the top of the hill, the speed of the car is 10 m/s. Find the average power delivered by the car's engine, neglecting any frictional losses.

82 •• In a new ski jump event, a loop is installed at the end of the ramp as shown in Figure 7-32. This problem addresses the physical requirement such an event would place on the skiers. Neglect friction.

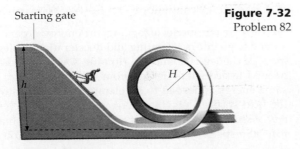

Starting gate

Figure 7-32 Problem 82

(a) Along the track of the ramp and the loop, where do the legs of the skier have to support the maximum weight?
(b) If the loop has a radius R, where should the starting gate (indicated by h) be placed so that the maximum force on the skier's legs is 4 times the skier's body weight?
(c) With the starting gate at the height found in part (b), will the skier be able to make it completely around the loop? Why or why not?
(d) What is the minimum height h such that the skier can make it around the loop? What is the maximum force on the skier's legs for this height?

83 •• A mass m is suspended from the ceiling by a spring and is free to move vertically in the y direction as indicated in Figure 7-33. We are given that the potential energy as a function of position is $U = \frac{1}{2}ky^2 - mgy$.

(a) Sketch U as a function of y. What value of y corresponds to the *unstretched* condition of the spring, y_0?
(b) From the given expression for U, find the net downward force acting on m at any position y.
(c) The mass is released from rest at y = 0; if there is no friction, what is the maximum value y_{max} that will be reached by the mass? Indicate y_{max} on your sketch.
(d) Now consider the effect of friction. The mass ultimately settles down into an equilibrium position y_{eq}. Find this point on your sketch.
(e) Find the amount of thermal energy produced by friction from the start of the operation to the final equilibrium.

Figure 7-33
Problem 83

84 •• A spring-loaded gun is cocked by compressing a short, strong spring by a distance d. The gun fires a signal flare of mass m directly upward. The flare has speed v_0 as it leaves the spring and is observed to rise to a maximum height h above the point where it leaves the spring. After it leaves the spring, effects of drag force by the air on the packet are significant. (Express answers in terms of m, v_0, d, h, and g, the acceleration due to gravity.) (a) How much work is done on the spring in the course of the compression? (b) What is the value of the spring constant k? (c) How much mechanical energy is converted to thermal energy because of the drag force of the air on the flare between the time of firing and the time at which maximum elevation is reached?

85 •• A roller-coaster car having a total mass (including passengers) of 500 kg travels freely along the winding frictionless track in Figure 7-34. Points A, E, and G are horizontal straight sections, all at the same height of 10 m above ground. Point C is at a height of 10 m above the ground on a section sloped at an angle of 30°. Point B is at the top of a hill, while point D is at ground level at the bottom of a valley. The radius of curvature at each of these points is 20 m. Point F is at the middle of a banked horizontal curve of radius of curvature of 30 m, and at the same height of 10 m above the ground as points A, E, and G. At point A the speed of the car is 12 m/s.

(a) If the car is just barely able to make it over the hill at point B, what is the height of that point above the ground?

(b) If the car is just barely able to make it over the hill at point B, what is the magnitude of the total force exerted on the car by the track at that point?

(c) What is the acceleration of the car at point C?

(d) What are the magnitude and direction of the total force exerted on the car by the track at point D?

(e) What are the magnitude and direction of the total force exerted on the car by the track at point F?

(f) At point G, a constant braking force is applied to the car, bringing the car to a halt in a distance of 25 m. What is the braking force?

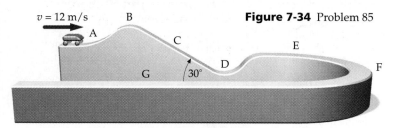

Figure 7-34 Problem 85

$v = 12$ m/s

86 • An elevator (mass $M = 2000$ kg) is moving downward at $v_0 = 1.5$ m/s. A braking system prevents the downward speed from increasing. (a) At what rate (in J/s) is the braking system converting mechanical energy to thermal energy? (b) While the elevator is moving downward at $v_0 = 1.5$ m/s, the braking system fails and the elevator is in free fall for a distance $d = 5$ m before hitting the top of a large safety spring with force constant $k = 1.5 \times 10^4$ N/m. After the elevator cage hits the top of the spring, we want to know the distance Δy that the spring is compressed before the cage is brought to rest. Write an algebraic expression for the value of Δy in terms of the known quantities M, v_0, g, k, and d, and substitute the given values to find Δy.

87 • To measure the force of friction on a moving car, engineers turn off the engine and allow the car to coast down hills of known steepness. The engineers collect the following data:

1. On a 2.87° hill, the car can coast at a steady 20 m/s.
2. On a 5.74° hill, the steady coasting speed is 30 m/s.

The total mass of the car is 1000 kg.

(a) What is the force of friction at 20 m/s (F_{20}) and at 30 m/s (F_{30})?

(b) How much useful power must the engine deliver to drive the car on a level road at steady speeds of 20 m/s (P_{20}) and 30 m/s (P_{30})?

(c) At full throttle, the engine delivers 40 kW. What is the angle of the steepest incline up which the car can maintain a steady 20 m/s?

(d) Assume that the engine delivers the same total useful work from each liter of gas, no matter what speed. At 20 m/s on a level road, the car goes 12.7 km/L. How many kilometers per liter does it get if it goes 30 m/s instead?

88 •• A 50,000-kg barge is pulled along a canal at a constant speed of 3 km/h by a heavy tractor. The towrope makes an angle of 18° with the velocity vector of the barge. The tension in the towrope is 1200 N. If the towrope breaks, how far will the barge move before coming to rest? Assume that the drag force between the barge and water is independent of velocity.

89 •• A 2-kg block is released 4 m from a massless spring with a force constant $k = 100$ N/m that is fixed along a frictionless plane inclined at 30°, as shown in Figure 7-35. (a) Find the maximum compression of the spring. (b) If the plane is rough rather than frictionless, and the coefficient of kinetic friction between the plane and the block is 0.2, find the maximum compression. (c) For the rough plane, how far up the incline will the block travel after leaving the spring?

Figure 7-35 Problem 89

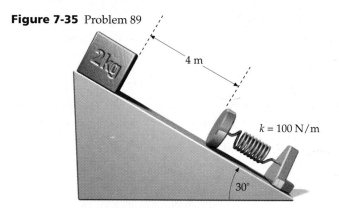

$k = 100$ N/m

90 •• A train with a total mass of 2×10^6 kg rises 707 m in a travel distance of 62 km at an average speed of 15.0 km/h. If the frictional force is 0.8% of the weight, find (a) the kinetic energy of the train, (b) the total change in its potential energy, (c) the energy dissipated by kinetic friction, and (d) the power output of the train's engines.

91 •• While driving, one expects to spend more energy accelerating than driving at a constant speed. (a) Neglecting friction, calculate the energy required to give a 1200-kg car a speed of 50 km/h. (b) If friction results in a retarding force of 300 N at a speed of 50 km/h, what is the energy needed to move the car a distance of 300 m at a constant speed of 50 km/h? (c) Assuming that the energy losses due to friction in part (a) are 75% of those found in part (b), estimate the ratio of the energy consumption for the two cases considered.

92 •• In one model of jogging, the energy expended is assumed to go into accelerating and decelerating the legs. If the mass of the leg is m and the running speed is v, the energy needed to accelerate the leg from rest to v is $\frac{1}{2}mv^2$, and the same energy is needed to decelerate the leg back to rest for the next stride. Thus, the energy required for each stride is mv^2. Assume that the mass of a man's leg is 10 kg and that he runs at a speed of 3 m/s with 1 m between one footfall and the next. Therefore, the energy he must provide to his legs in each second is $3 \times mv^2$. Calculate the rate of the man's energy expenditure using this model and assuming that his muscles have an efficiency of 25%.

93 •• On July 31, 1994, Sergei Bubka pole-vaulted over a height of 6.14 m. If his body was momentarily at rest at the top of the leap, and all of the energy required to raise his body derived from his kinetic energy just prior to planting his pole, how fast was he moving just before takeoff? Neglect the mass of the pole. If he could maintain that speed for a 100-m sprint, how fast would he cover that distance? Since the world record for the 100-m dash is just over 9.8 s, what do you conclude about world-class pole-vaulters?

94 •• A 5-kg block is held against a spring of force constant 20 N/cm, compressing it 3 cm. The block is released and the spring extends, pushing the block along a rough horizontal surface. The coefficient of friction between the surface and the block is 0.2. (a) Find the work done on the block by the spring as it extends from its compressed position to its equilibrium position. (b) Find the energy dissipated by friction while the block moves the 3 cm to the equilibrium position of the spring. (c) What is the speed of the block when the spring is at its equilibrium position? (d) If the block is not attached to the spring, how far will it slide along the rough surface before coming to rest?

95 •• A pendulum of length L has a bob of mass m. It is released from some angle θ_1. The string hits a peg at a distance x directly below the pivot as in Figure 7-36, effectively shortening the length of the pendulum. Find the maximum angle θ_2 between the string and the vertical when the bob is to the right of the peg.

Figure 7-36 Problem 95

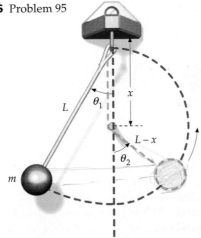

96 ••• A block of mass m is dropped onto the top of a vertical spring whose force constant is k. If the block is released from a height h above the top of the spring, (a) what is the maximum kinetic energy of the block? (b) What is the maximum compression of the spring? (c) At what compression is the block's kinetic energy half its maximum value?

97 ••• The bob of a pendulum of length L is pulled aside so the string makes an angle θ_0 with the vertical, and the bob is then released. In Example 7-2, the conservation of energy was used to obtain the speed of the bob at the bottom of its swing. In this problem, you are to obtain the same result using Newton's second law. (a) Show that the tangential component of Newton's second law gives $dv/dt = -g \sin \theta$, where v is the speed and θ is the angle made by the string and the vertical. (b) Show that v can be written $v = L \, d\theta/dt$. (c) Use this result and the chain rule for derivatives to obtain

$$\frac{dv}{dt} = \frac{dv}{d\theta}\frac{d\theta}{dt} = \frac{dv}{d\theta}\frac{v}{L}$$

(d) Combine the results of (a) and (c) to obtain

$$v \, dv = -gL \sin \theta \, d\theta$$

(e) Integrate the left side of the equation in part (d) from $v = 0$ to the final speed v and the right side from $\theta = \theta_0$ to $\theta = 0$, and show that the result is equivalent to $v = \sqrt{2gh}$, where h is the original height of the bob above the bottom.

CHAPTER 8

Systems of Particles and Conservation of Momentum

This break shot illustrates the transfer of momentum in two dimensions from the white cue ball to the other balls on a pool table.

We've discussed Newton's laws in terms of the motion of point particles, but many applications concern extended objects—cars, rockets, people. We will justify these applications by showing that there is one point of a system, the **center of mass,** that moves as if all the mass of the system were concentrated at that point, and all the external forces acting on the system were acting exclusively on that point. The motion of any object or system of particles can be described in terms of the motion of the center of mass (which may be thought of as the bulk motion of the system) plus the motion of individual particles in the system relative to the center of mass.

The mass of a particle times its velocity is called the **momentum** of the particle. The momentum of a system is the sum of the momenta of the individual particles in the system. When the net external force acting on a system is zero, the system's total momentum remains constant. Momentum in an isolated system is a conserved quantity, just like energy. We will use conservation of momentum to analyze collisions between billiard balls, cars, and subatomic particles, and we will apply it to the decay of radioactive nuclei.

8-1   The Center of Mass

We first consider a simple system of two particles in one dimension. If two point masses, m_1 and m_2, have coordinates x_1 and x_2 on the x axis, then the center-of-mass coordinate x_{cm} is defined by

$$Mx_{cm} = m_1x_1 + m_2x_2 \qquad 8\text{-}1$$

where $M = m_1 + m_2$ is the total mass of the system. In the case of just two particles, the center of mass lies at some point on the line between the particles; if the particles have equal masses, the center of mass is midway between them (Figure 8-1).

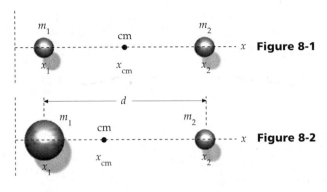

Figure 8-1

Figure 8-2

If the particles are of unequal mass, the center of mass is closer to the more massive particle (Figure 8-2).

If we choose the position of m_1 to be the origin, x_2 is the distance d between the particles (Figure 8-3) and the center of mass is given by

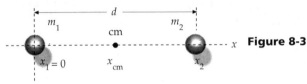

Figure 8-3

$$Mx_{cm} = m_1x_1 + m_2x_2 = m_1(0) + m_2d$$

$$x_{cm} = \frac{m_2}{M}d = \frac{m_2}{m_1 + m_2}d \qquad 8\text{-}2$$

Exercise   A 4-kg mass is at the origin and a 2-kg mass is at $x = 6$ cm. Find x_{cm}. (*Answer* $x_{cm} = 2$ cm)

We can generalize from two particles in one dimension to a system of many particles in three dimensions. For N particles,

$$Mx_{cm} = m_1x_1 + m_2x_2 + m_3x_3 + \cdots + m_Nx_N = \sum_i m_ix_i \qquad 8\text{-}3a$$

where again $M = \Sigma m_i$ is the total mass of the system. Similarly,

$$My_{cm} = \sum_i m_iy_i \quad \text{and} \quad Mz_{cm} = \sum_i m_iz_i \qquad 8\text{-}3b$$

In vector notation, $\vec{r}_i = x_i\hat{i} + y_i\hat{j} + z_i\hat{k}$ is the position vector of the ith particle. The position vector of the **center of mass,** $\vec{r}_{cm}$, is defined by

$$M\vec{r}_{cm} = \sum_i m_i\vec{r}_i \qquad 8\text{-}4$$

Definition—Center of mass, system of particles

where $\vec{r}_{cm} = x_{cm}\hat{i} + y_{cm}\hat{j} + z_{cm}\hat{k}$.
   To find the center of mass of a continuous object, we replace the sum in Equation 8-4 with an integral:

$$M\vec{r}_{cm} = \int \vec{r}\, dm \qquad 8\text{-}5$$

Definition—Center of mass, continuous object

where dm is an element of mass located at position $\vec{r}$, as shown in Figure 8-4. Examples showing how to calculate the center of mass using integration are given in Section 8-2. For regularly shaped objects, we can use symmetry to find the center of mass. For example, the center of mass of a uniform cylinder or sphere is at the geometric center of the object.

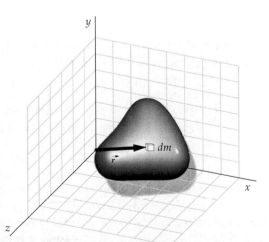

Figure 8-4 Mass element dm located at $\vec{r}$ for finding the center of mass by integration.

Example 8-1

Find the center of mass of a water molecule.

Picture the Problem A water molecule consists of an oxygen atom and two hydrogen atoms (Figure 8-5). Oxygen has a mass of 16 unified mass units (u) and each hydrogen has a mass of 1 u. The hydrogen atoms are each at an average distance of 9.6 nm (9.6×10^{-9} m) from the oxygen atom, and are separated from one another by an angle of 104.5°. The calculation is simplified if we place the origin at the location of the oxygen atom, with the x axis bisecting the angle between the hydrogen atoms. Then, given the symmetries of the molecule, the center of mass will be on the x axis, and the line from the oxygen atom to each hydrogen atom will make an angle of 52.2°.

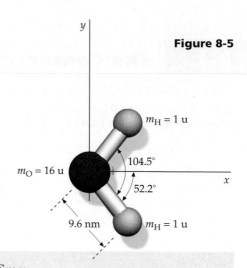

Figure 8-5

1. The location of the center of mass is given by its coordinates, x_{cm} and y_{cm}:

$$x_{cm} = \frac{\Sigma m_i x_i}{M}, \qquad y_{cm} = \frac{\Sigma m_i y_i}{M}$$

2. By symmetry, the center of mass is on the x axis:

$$y_{cm} = 0$$

3. Write out the expressions for x_{cm} explicitly:

$$x_{cm} = \frac{m_H x_{H1} + m_H x_{H2} + m_O x_O}{m_H + m_H + m_O}$$

4. We have chosen the origin to be the location of oxygen, so the x coordinate of oxygen is zero. The x coordinates of the hydrogen atoms are calculated from the 52.2° angle each hydrogen atom makes with the x axis:

$$x_O = 0$$

$$x_{H1} = x_{H2} = 9.6 \text{ nm cos } 52.2° = 5.9 \text{ nm}$$

5. Substituting the x coordinates and the mass values into step 3 gives x_{cm}:

$$x_{cm} = \frac{(1 \text{ u})5.9 \text{ nm} + (1 \text{ u})5.9 \text{ nm} + (16 \text{ u})0}{1 \text{ u} + 1 \text{ u} + 16 \text{ u}}$$

$$= 0.66 \text{ nm}$$

Example 8-1 can also be solved by first finding the center of mass of just the two hydrogen atoms. For a system of three particles, Equation 8-4 is

$$M\vec{r}_{cm} = m_1\vec{r}_1 + m_2\vec{r}_2 + m_3\vec{r}_3$$

The first two terms on the right side of this equation are related to the center of mass of the first two particles $\vec{r}'_{cm}$:

$$m_1\vec{r}_1 + m_2\vec{r}_2 = (m_1 + m_2)\vec{r}'_{cm}$$

The center of mass of the three-particle system can then be written

$$M\vec{r}_{cm} = (m_1 + m_2)\vec{r}'_{cm} + m_3\vec{r}_3$$

So we can first find the center of mass for two of the particles, the hydrogen atoms, for example, and then replace them with a single particle of total mass $m_1 + m_2$ at that center of mass (Figure 8-6). The same technique enables us to

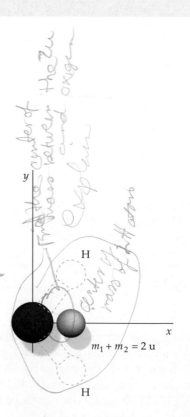

Figure 8-6 Example 8-1 with the two H atoms replaced by a single particle of mass $m_1 + m_2 = 2$ u on the x axis at the center of mass of the original atoms. The center of mass then falls between the oxygen atom at the origin and the calculated center of mass of the two hydrogen atoms.

calculate centers of mass for more complex systems, such as two uniform sticks (Figure 8-7). The
center of mass of each stick separately is at the
center of the stick. The center of mass of the system is found by treating each stick as a point particle at its individual center of mass.

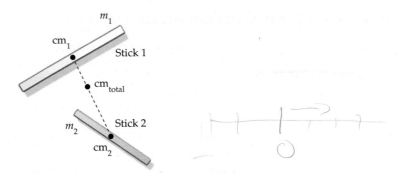

Figure 8-7

try it yourself

Find the center of mass of the uniform sheet of plywood in Figure 8-8.

Picture the Problem The sheet can be divided into two symmetrical parts.
The center of mass of each part is at its geometric center. Let m_1 be the mass
of part 1 and m_2 be the mass of part 2. The total mass is $M = m_1 + m_2$. The
masses are proportional to the areas.

Figure 8-8

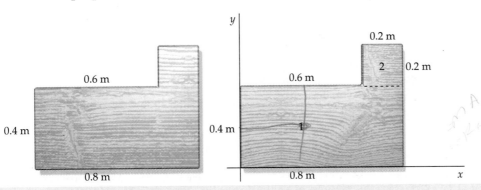

Cover the column to the right and try these on your own before looking at the answers.

Steps

Answers

1. Write the x and y coordinates of the center of
   mass in terms of m_1 and m_2.

   $mx_{cm} = m_1 x_{cm,1} + m_2 x_{cm,2}$

   $my_{cm} = m_1 y_{cm,1} + m_2 y_{cm,2}$

2. The mass of each part is proportional to its area.
   Calculate the areas A_1 and A_2 and the ratio
   A_1/A_2.

   $A_1 = 0.32 \text{ m}^2$, $A_2 = 0.04 \text{ m}^2$, $\dfrac{A_1}{A_2} = 8$

3. Express the masses m_1 and $M = m_1 + m_2$ in
   terms of the smallest mass m_2.

   $m_1 = 8m_2$, $M = 9m_2$

4. Write the x and y coordinates of the center of
   mass coordinates for each part by inspection of
   the figure.

   $x_1 = 0.4 \text{ m}$, $y_1 = 0.2 \text{ m}$

   $x_2 = 0.7 \text{ m}$, $y_2 = 0.5 \text{ m}$

5. Substitute these results to calculate x_{cm} and y_{cm}.

   $x_{cm} = 0.433 \text{ m}$, $y_{cm} = 0.233 \text{ m}$

Remark The center of mass is very near the center of mass of part 1 because $m_1 = 8m_2$.

Gravitational Potential Energy of a System

The gravitational potential energy of a system of particles in a uniform gravitational field is the same as if all the mass were concentrated at the center of mass. Let h_i be the height of the ith particle in a system above some reference level. The gravitational potential energy of the system is

$$U = \sum_i m_i g h_i = g \sum_i m_i h_i$$

But, by definition of the center of mass, the height of the center of mass is given by

$$M h_{cm} = \sum_i m_i h_i$$

so

$$U = M g h_{cm} \qquad\qquad 8\text{-}6$$

We can use this result to locate the center of mass of an object experimentally. For example, two objects connected by a light rod will balance if the pivot is at the center of mass (Figure 8-9). If we pivot the system at any other point, the system will rotate until the potential energy is at a minimum, which occurs when the center of mass is at its lowest possible point directly below the pivot (Figure 8-10).

If we suspend any irregular object from a pivot, the object will hang so that its center of mass lies somewhere on the vertical line drawn directly downward from the pivot. Now suspend the object from another point and note where the vertical line now passes across the object. The center of mass will lie at the intersection of the two lines (Figure 8-11).

Figure 8-9

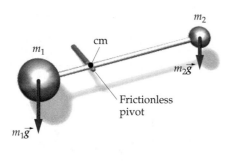

Figure 8-10

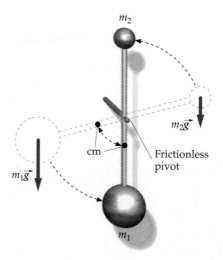

Figure 8-11 The center of mass of an irregular object can be found by suspending it from two points.

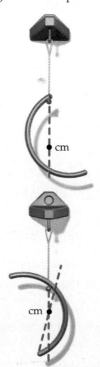

8-2 Finding the Center of Mass by Integration

In this section we illustrate finding the center of mass by integration (Equation 8-5):

$$M\vec{r}_{cm} = \int \vec{r}\,dm$$

Uniform Stick

This problem, whose answer we can guess from symmetry considerations, illustrates the technique for setting up the integration. We first choose a coordinate system with the x axis along the stick and one end of the stick at the origin (Figure 8-12). Let the mass per unit length of the stick be λ. Since the stick is uniform, $\lambda = M/L$. In Figure 8-9, we have indicated a mass element dm of length dx at a distance x from the origin. The mass of an element of length dx is

$$dm = M\frac{dx}{L} = \frac{M}{L}\,dx = \lambda\,dx$$

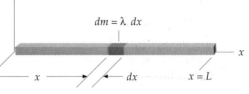

$dm = \lambda\,dx$

Figure 8-12

Equation 8-5 thus gives

$$Mx_{cm} = \int x\,dm = \int_0^L x\lambda\,dx = \frac{\lambda x^2}{2}\Big|_0^L$$

Using $\lambda = M/L$, we obtain the expected result:

$$x_{cm} = \frac{\lambda L^2}{2M} = \frac{M}{L}\left(\frac{L^2}{2M}\right) = \frac{1}{2}L$$

Semicircular Hoop

The calculation for determining the center of mass of a semicircular hoop is simplest with the origin at the center of curvature and the y axis on the hoop's line of symmetry (Figure 8-13). Then $x_{cm} = 0$ because of symmetry. However, $y_{cm} > 0$, since all of the mass is at positive values of y. In Figure 8-13, we indicate a mass element of length $ds = R\,d\theta$. Since the total length of the hoop is πR, the mass per unit length is $\lambda = M/\pi R$, where M is the total mass. The mass of the element is thus

$$dm = \lambda\,ds = \lambda R\,d\theta$$

The y coordinate of the mass element is related to the angle θ by $y = R\sin\theta$. The angle θ varies from 0 to π. We thus have

$$My_{cm} = \int y\,dm = \int y\lambda\,ds = \int y\lambda R\,d\theta$$

$$= \int_0^\pi (R\sin\theta)\lambda R\,d\theta = R^2\lambda\int_0^\pi \sin\theta\,d\theta$$

$$- R^2\lambda(-\cos\theta)\Big|_0^\pi = 2R^2\lambda$$

Using $\lambda = M/\pi R$, we have

$$My_{cm} = 2R^2\frac{M}{\pi R}, \qquad y_{cm} = \frac{2R}{\pi}$$

In this case, the center of mass is not within the body of the object.

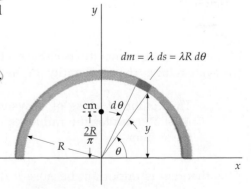

$dm = \lambda\,ds = \lambda R\,d\theta$

Figure 8-13 Geometry for calculating the center of mass of a semicircular hoop by integration. The center of mass lies on the y axis.

8-3 Motion of the Center of Mass

Figure 8-14 is a multiflash photograph of a baton thrown into the air. Although the motion of the baton is complicated, the motion of the center of mass is simple. While the baton is in the air, the center of mass follows a parabolic path, the same path that would be followed by a point particle. We will show in general that the acceleration of the center of mass of a system of particles equals the net external force acting on the system divided by the total mass of the system. For the baton thrown into the air, the acceleration of the center of mass is $\vec{g}$ downward.

To find the acceleration of the center of mass, we first find its velocity by differentiating Equation 8-4 with respect to time:

$$M\frac{d\vec{r}_{cm}}{dt} = m_1\frac{d\vec{r}_1}{dt} + m_2\frac{d\vec{r}_2}{dt} + \cdots = \sum_i m_i\frac{d\vec{r}_i}{dt}$$

or

$$M\vec{v}_{cm} = m_1\vec{v}_1 + m_2\vec{v}_2 + \cdots = \sum_i m_i\vec{v}_i \qquad 8\text{-}7$$

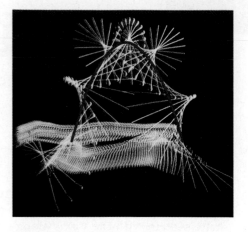

Figure 8-14 Multiflash photo of a baton thrown into the air. The center of mass follows the same simple parabolic path it would if it were a single point particle.

We differentiate again to obtain the acceleration of the center of mass:

$$M\vec{a}_{cm} = m_1\vec{a}_1 + m_2\vec{a}_2 + \cdots = \sum_i m_i\vec{a}_i \qquad 8\text{-}8$$

According to Newton's second law, we can replace the quantity $m_i\vec{a}_i$ with $\vec{F}_i$, the net force acting on the ith particle. Forces acting on a particle fall into two categories: *internal* forces due to interactions with other particles within the system, and *external* forces due to agents outside the system:

$$\vec{F}_i = m_i\vec{a}_i = \vec{F}_{i,\text{int}} + \vec{F}_{i,\text{ext}}$$

Substituting this into Equation 8-8 gives

$$M\vec{a}_{cm} = \sum_i \vec{F}_{i,\text{int}} + \sum_i \vec{F}_{i,\text{ext}} \qquad 8\text{-}9$$

According to Newton's third law, for each internal force acting on one particle, there is an equal but opposite force acting on another particle. The internal forces thus occur in pairs of equal and opposite forces. When we sum over all the particles in the system, the internal forces cancel, $\Sigma\vec{F}_{i,\text{int}} = 0$, leaving only the external forces. Equation 8-9 then becomes

$$\vec{F}_{\text{net,ext}} = \sum_i \vec{F}_{i,\text{ext}} = M\vec{a}_{cm} \qquad 8\text{-}10$$

Newton's second law for a system

That is, the net external force acting on the system equals the total mass M of the system times the acceleration of the center of mass $\vec{a}_{cm}$. Thus,

> The center of mass of a system moves like a particle of mass $M = \Sigma m_i$ under the influence of the net external force acting on the system.

This theorem is important because it describes the motion of the center of mass for any system of particles: The center of mass behaves just like a single point particle acted on by the external forces. The motions of the individual particles of the system are usually much more complex and are not described by Equation 8-10. The baton thrown into the air in Figure 8-14 is an example. The only external force acting is gravity, so the center of mass of the baton moves in a simple parabolic path, as would a point particle. (The rotation of the baton about its center of mass is not described by Equation 8-10.)

Exercise A cylinder rests on a sheet of paper on a table. You pull the paper to the right, causing the cylinder to roll backward relative to the paper (Figure 8-15). How does the cylinder's center of mass move? (*Answer* It accelerates to the right, because the net external force acting on the cylinder is that of friction to the right. Try it. The cylinder may *appear* to accelerate to the left, because it rolls backward on the paper. But mark its original position on the *table*: While the cylinder is on the paper, the motion of the center of mass is to the right.)

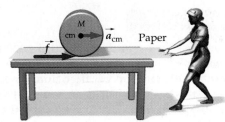

Figure 8-15

Example 8-3

A projectile is fired into the air over level ground with an initial velocity of 24.5 m/s at 36.9° to the horizontal. At its highest point, it explodes into two fragments of equal mass. One fragment falls straight down to the ground. Where does the other fragment land?

Figure 8-16

Picture the Problem Since the only *external* force acting on the system is gravity, the center of mass, which is midway between the fragments, continues on its parabolic path as if there had been no explosion (Figure 8-16). It lands at $x = R$, where R is the range. The fragment that falls straight down lands at a point $x_1 = 0.5R$. The other fragment must then land at $x_2 = 1.5R$. Let m be the mass of each fragment.

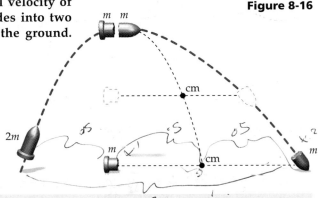

1. The landing positions x_1 and x_2 of the fragments are related to the final position of the center of mass by:

$$mx_1 + mx_2 = (2m)x_{cm}$$

2. Set $x_{cm} = R$, $x_1 = 0.5R$, and solve for x_2:

$$x_2 = 2x_{cm} - x_1 = 2R - 0.5R = 1.5R$$

3. Find the range for the given initial velocity:

$$R = \frac{v_0^2}{g}\sin 2\theta = \frac{(24.5 \text{ m/s})^2}{9.81 \text{ m/s}^2}\sin(73.8°) = 58.8 \text{ m}$$

4. Substitute this value of R to find x_2:

$$x_2 = 1.5R = 88.2 \text{ m}$$

Remarks Figure 8-17 plots the height versus distance for exploding projectiles when the first fragment has a horizontal velocity of half the initial horizontal velocity. As in the original example, in which the first fragment falls straight down, the center of mass follows a normal parabolic trajectory.

Exercise If one of the fragments lands back at the initial position of the projectile, where does the other one land? (*Answer* 2R)

Remarks If both fragments have no vertical component of velocity after the explosion, they land at the same time. If one fragment is moving downward after the explosion, the other fragment will have an upward component of velocity. The downward-moving fragment will then hit the ground first, and since the ground exerts a force on it before the other fragment lands, our analysis breaks down because there is an unbalanced force on the system.

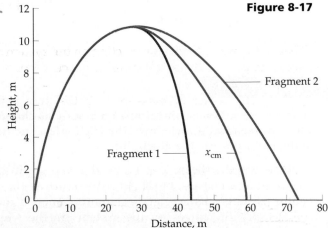

Figure 8-17

A special case of a system's center of mass in motion is the system with zero net external force acting on it. Here, $\vec{a}_{cm} = 0$, so the center of mass is at rest or moves with constant velocity. The internal forces and motion may be complex, but the behavior of the center of mass is simple.

Example 8-4

You (mass 80 kg) and Bubba (mass 120 kg) are in a rowboat (mass 60 kg) on a calm lake. You are at the center of the boat, rowing, and he is at the back, 2 m from the center. You get tired and stop rowing. Bubba offers to row, and after the boat comes to rest, you change places. How far does the boat move? (Neglect any horizontal force exerted by the water.)

Picture the Problem Since there are no external forces in the horizontal direction, the center of mass does not move. This determines the distance the boat must move (Figure 8-18). Choose the origin at the center of the boat. Then $x_{\text{Bubba}} = 2$ m, $x_{\text{you}} = 0$, and $x_{\text{boat}} = 0$. After you and Bubba switch places, the center of mass will have a different x value because the origin moves. The difference in the x values for the center of mass is the distance the origin moves.

Figure 8-18

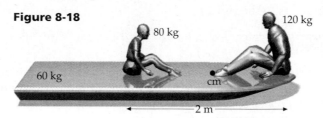

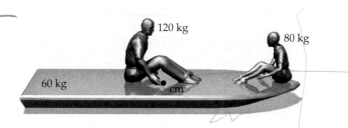

1. Find the initial x value of the center of mass:

$$x_{\text{cm}} = \frac{\Sigma m_i x_i}{M}$$

$$= \frac{(80 \text{ kg})(0) + (60 \text{ kg})(0) + (120 \text{ kg})(2 \text{ m})}{80 \text{ kg} + 60 \text{ kg} + 120 \text{ kg}} = 0.923 \text{ m}$$

2. Now compute the new coordinate x'_{cm}:

$$x'_{\text{cm}} = \frac{(120 \text{ kg})(0) + (60 \text{ kg})(0) + (80 \text{ kg})(2 \text{ m})}{80 \text{ kg} + 60 \text{ kg} + 120 \text{ kg}} = 0.615 \text{ m}$$

3. The difference $x_{\text{cm}} - x'_{\text{cm}}$ is the distance that the boat has moved:

$$\Delta x = x_{\text{cm}} - x'_{\text{cm}} = 0.923 \text{ m} - 0.615 \text{ m} = 0.308 \text{ m}$$

Remark Choosing the origin at the center of the boat rather than at one end simplifies the calculation of the center of mass position because two of the masses are then located at $x = 0$.

Example 8-5

A wedge of mass m_2 sits at rest on a scale as shown in Figure 8-19. A small block of mass m_1 slides down the frictionless incline of the wedge. Find the scale reading while the block slides.

Picture the Problem We choose the wedge plus block to be the system. Since the block accelerates down the wedge, the center of mass has acceleration components to the right and downward. The forces on the system are the weights of the block and wedge, the force F_x exerted by the scale on the wedge to the right, and the normal force F_n exerted upward by the scale. The scale reading equals the magnitude of F_n.

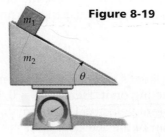

Figure 8-19

1. Draw a free-body diagram for the wedge–block system:

Figure 8-20

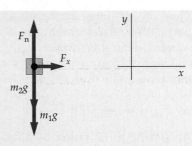

2. Write the vertical component of Newton's second law for the system and solve for F_n:

$$F_n - m_1g - m_2g = Ma_{cm,y} = (m_1 + m_2)a_{cm,y}$$

$$F_n = (m_1 + m_2)g + (m_1 + m_2)a_{cm,y}$$

3. Write the vertical component of the acceleration of the center of mass in terms of the acceleration of the block:

$$Ma_{cm,y} = m_1a_{1y} + m_2a_{2y} = m_1a_{1y}$$

$$a_{cm,y} = \frac{m_1}{m_1 + m_2}a_{1y}$$

4. From Example 4-8, a block sliding down a stationary incline has acceleration $g \sin\theta$ down the incline. Find the y component of this acceleration and use it to find $a_{cm,y}$:

$$a_{1y} = -a_1 \cos\theta = -g \sin^2\theta$$

$$a_{cm,y} = \frac{m_1}{m_1 + m_2}a_{1y} = -\frac{m_1}{m_1 + m_2}g \sin^2\theta$$

5. Substitute this value for $a_{cm,y}$ and calculate F_n:

$$F_n = (m_1 + m_2)g + (m_1 + m_2)a_{cm,y}$$

$$= (m_1 + m_2)g - m_1g \sin^2\theta$$

Exercise Find the force F_x exerted on the wedge by the scale (*Answer* $F_x = m_1g \sin\theta \cos\theta$)

8-4   Conservation of Momentum

A particle's **momentum** $\vec{p}$ is defined as the product of its mass and velocity:

$$\vec{p} = m\vec{v} \qquad\qquad \text{8-11}$$

Definition—Momentum of a particle

Momentum is a vector quantity that may be thought of as a measurement of the effort needed to bring a particle to rest.* For example, a heavy truck has more momentum than a light car traveling at the same speed. It takes a greater force to stop the truck in a given time than it does to stop the car.

Newton's second law can be written in terms of the momentum of a particle. Differentiating Equation 8-11 with respect to time, we obtain

$$\frac{d\vec{p}}{dt} = \frac{d(m\vec{v})}{dt} = m\frac{d\vec{v}}{dt} = m\vec{a}$$

Then substituting the force $\vec{F}_{net}$ for $m\vec{a}$, we get

$$\vec{F}_{net} = \frac{d\vec{p}}{dt} \qquad\qquad \text{8-12}$$

---

* The quantity $m\vec{v}$ is sometimes referred to as the *linear momentum* of a particle to distinguish it from the *angular momentum*, which is discussed in Chapter 10.

Thus the net force acting on a particle equals the time rate of change of the particle's linear momentum. Newton's original statement of his second law was in fact in this form.

The total momentum $\vec{P}$ of a system of many particles is the sum of the momenta of the individual particles:

$$\vec{P} = \sum_i m_i \vec{v}_i = \sum_i \vec{p}_i$$

According to Equation 8-7, $\sum m_i \vec{v}_i$ equals the total mass M times the velocity of the center of mass:

$$\vec{P} = \sum_i m_i \vec{v}_i = M\vec{v}_{cm} \qquad\qquad\qquad \text{8-13}$$

Total momentum of a system

Differentiating this equation with respect to time, we obtain

$$\frac{d\vec{P}}{dt} = M\frac{d\vec{v}_{cm}}{dt} = M\vec{a}_{cm}$$

But according to Newton's second law (Equation 8-10), $M\vec{a}_{cm}$ equals the net external force acting on the system. Thus,

$$\sum_i \vec{F}_{ext} = \vec{F}_{net,ext} = \frac{d\vec{P}}{dt} \qquad\qquad\qquad \text{8-14}$$

When the net external force acting on a system of particles is zero, the rate of change of the total momentum is zero, and the total momentum of the system remains constant:

$$\vec{P} = \sum_i m_i \vec{v}_i = M\vec{v}_{cm} = \text{constant} \qquad (\vec{F}_{net,ext} = 0) \qquad \text{8-15}$$

Conservation of momentum

This result is known as the **law of conservation of momentum:**

> If the net external force on a system is zero, the total momentum of the system remains constant.

This law is one of the most important in physics. It is more widely applicable than the law of conservation of mechanical energy, because internal forces exerted by one particle in a system on another are often not conservative. Thus, these internal forces can change the total mechanical energy of the system, though they have no effect on the system's total momentum. We see from Equation 8-15 that if the total momentum is constant, the velocity of the center of mass of the system is constant.

Example 8-6

During repair of the Hubble Space Telescope, an astronaut replaces two solar panels whose frames are bent. Pushing the detached panels away into space, she is propelled in the opposite direction. The astronaut's mass is 60 kg and the panel's mass is 80 kg. The astronaut is at rest relative to her spaceship when she shoves away the panel, and she shoves it at 0.3 m/s relative to the spaceship. What is her subsequent velocity relative to the space ship? (During this operation the astronaut is tethered to the ship; for our calculation, assume that the tether remains slack.)

Picture the Problem The velocity of the astronaut can be found from the velocity of the panel using conservation of momentum. Choose the direction of motion of the panel to be positive.

1. Apply conservation of momentum to find the velocity of the astronaut. Since the total momentum is initially zero, it remains zero:

$$p_p + p_a = m_p v_p + m_a v_a = 0$$

2. Solve for the astronaut's velocity:

$$v_a = -\frac{m_p}{m_a} v_p = -\frac{80 \text{ kg}}{60 \text{ kg}} (0.3 \text{ m/s}) = -0.4 \text{ m/s}$$

Remark Although momentum is conserved, the mechanical energy of this system increased because chemical energy of the astronaut was converted to kinetic energy.

Exercise Find the final kinetic energy of the astronaut–panel system. (*Answer* 8.4 J)

(handwritten) $KE_{total} = \frac{1}{2}mv_p^2 + \frac{1}{2}mv_i^2$

(handwritten) $\frac{1}{2}(80)(.3)^2 + \frac{1}{2}(60)(.4)^2$

Figure 8-21

Example 8-7

A runaway 14,000-kg railroad car is rolling at 4 m/s toward a switchyard. A sudden downpour fills the open-topped car with 2000 kg of rainwater. After the rainstorm, how long does it take the car to cover the 500-m distance to the switchyard? Assume that the rain comes straight down and that slowing due to friction is negligible.

$m_w = 2000 \text{ kg}$

$m_c = 14{,}000 \text{ kg}$

$v_1 = 4 \text{ ms}$

Picture the Problem We find the travel time that we seek from the distance traveled and the speed of the car. Consider the car and the water falling into the car as our system (Figure 8-21). No horizontal external forces act on this system, so the horizontal component of the momentum of the system is conserved. The final speed of the rain-filled car is found from its final momentum, which equals the car's initial momentum. The water initially has no horizontal momentum. Let m_c and m_w be the masses of the car and water, respectively.

1. The time from the end of the storm until the car reaches the yard is the distance d to the yard divided by the car's final speed v_f:

$$\Delta t = \frac{d}{v_f} = \frac{500 \text{ m}}{v_f}$$

2. Apply conservation of momentum to relate the final speed v_f to the initial speed v_i:

$$m_c v_i + m_w(0) = (m_c + m_w) v_f$$

3. Solve for v_f:

$$v_f = \frac{m_c v_i}{m_c + m_w} = \frac{(14{,}000 \text{ kg})(4 \text{ m/s})}{14{,}000 \text{ kg} + 2000 \text{ kg}} = 3.5 \text{ m/s}$$

4. Substitute the result for v_f into step 1:

$$\Delta t = \frac{500 \text{ m}}{v_f} = \frac{500 \text{ m}}{3.5 \text{ m/s}} = 143 \text{ s}$$

Remark Mechanical energy of the system is converted to thermal energy. Let K_w be the kinetic energy of the rainwater just as it hits the car. The initial mechanical energy is $K_w + \frac{1}{2}m_c v_i^2 = K_w + \frac{1}{2}(14{,}000 \text{ kg})(4 \text{ m/s})^2 = K_w + 112 \text{ kJ}$, whereas the final energy is $\frac{1}{2}(m_c + m_w)v_f^2 = \frac{1}{2}(16{,}000 \text{ kg})(3.5 \text{ m/s})^2 = 98 \text{ kJ}$.

Exercise Suppose that there is a small hole in the bottom of the car so that the water leaks out at 10 kg/s. Assume that the car is full when the rain stops. How long does it take the car to cover the 500 m? (*Answer* 143 s. The water leaking out does not impart any momentum to the rest of the system. If the ground were frictionless and nonporous, all of the water initially in the car would arrive at the switchyard along with the car.)

Example 8-8

A 40-kg skateboarder on a 3-kg board is training with two 5-kg weights. Beginning from rest, she throws the weights horizontally one at a time from her board. The velocity of each weight is 7 m/s relative to her and the board after it is thrown. How fast is she propelled in the opposite direction after throwing the second weight? Assume that the board rolls without friction.

Figure 8-22

$v_1 - 7$ m/s

5 kg

40 kg

v_2

3 kg

x

Picture the Problem No external horizontal forces act on the system, so the horizontal component of momentum is conserved. We need to find the velocity of the skateboarder after throwing each weight (Figure 8-22). Choose the direction of her motion to be the positive direction. If v_1 is her velocity relative to the ground after throwing the first weight, the weight travels at $v_1 - 7$ m/s relative to the ground. After finding v_1 from conservation of momentum, we use it to find v_2.

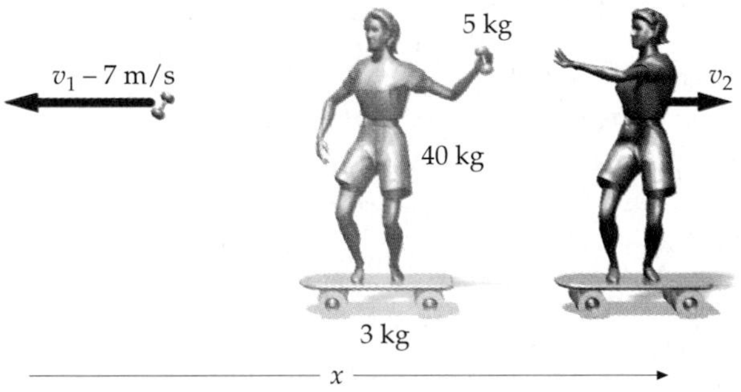

1. Her final velocity v_f is related to her final momentum:

$$p_f = mv_f$$

2. Apply conservation of momentum to relate the final momentum p_f to the initial momentum p_i:

$$p_i = p_f$$

3. The initial momentum p_i is zero. Let v_1 and p_1 be her velocity and momentum after throwing the first weight. The momentum p_1 is that of the skateboard plus girl (43 kg), and one weight (5 kg) with velocity v_1, plus the other weight (5 kg) with velocity $v_1 - 7$ m/s:

$$0 = (48 \text{ kg})v_1 + (5 \text{ kg})(v_1 - 7 \text{ m/s})$$

$$v_1 = \frac{35 \text{ kg·m/s}}{53 \text{ kg}} = 0.660 \text{ m/s}$$

4. When the second weight is thrown, the initial momentum of the girl, skateboard, and weight is $(48 \text{ kg})v_1$. Apply conservation of momentum and solve for v_2:

$$(48 \text{ kg})v_1 = (43 \text{ kg})v_2 + (5 \text{ kg})(v_2 - 7 \text{ m/s})$$

$$v_2 = \frac{(48 \text{ kg})v_1 + 35 \text{ kg·m/s}}{48 \text{ kg}} = 1.73 \text{ m/s}$$

Remark This example illustrates the principle of the rocket; a rocket moves forward by throwing its fuel out the back in the form of exhaust gases.

Exercise How fast is the skateboarder moving if, starting from rest, she throws both weights together, and the weights have velocity 7 m/s relative to her and the board *after they are thrown*? (*Answer* 1.32 m/s)

Example **8-9**   *try it yourself*

A thorium-227 nucleus at rest decays into a radium-223 nucleus (mass 223 u) by emitting an α particle (mass 4 u) (Figure 8-23). The kinetic energy of the α particle is found to be 6.00 MeV. What is the kinetic energy of the recoiling radium nucleus?

Picture the Problem Since the thorium nucleus before decay is at rest, its total momentum is zero. You can therefore relate the velocity of the radium nucleus to that of the α particle using conservation of momentum.

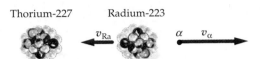

Thorium-227   Radium-223

Figure 8-23

Cover the column to the right and try these on your own before looking at the answers.

Steps	*Answers*
1. Write the kinetic energy of the radium nucleus K_{Ra} in terms of its mass m_{Ra} and speed v_{Ra}.	$K_{\text{Ra}} = \dfrac{1}{2} m_{\text{Ra}} v_{\text{Ra}}^2$
2. Use conservation of momentum to relate v_{Ra} to the speed of the α particle v_α.	$m_\alpha v_\alpha = m_{\text{Ra}} v_{\text{Ra}}$
3. Substitute your expression for v_α into the kinetic energy expression in step 1.	$K_{\text{Ra}} = \dfrac{1}{2} m_{\text{Ra}} \times \left(\dfrac{m_\alpha v_\alpha}{m_{\text{Ra}}} \right)^2$
4. Factor $K_{\text{Ra}} = \frac{1}{2} m_\alpha v_\alpha^2$ from your expression in step 3.	$K_{\text{Ra}} = \dfrac{m_\alpha}{m_{\text{Ra}}} \left(\dfrac{1}{2} m_\alpha v_\alpha^2 \right)$
5. Substitute the given values to calculate K_{Ra}.	$K_{\text{Ra}} = 0.107 \text{ MeV}$

Remark In this process, rest energy of the thorium nucleus is converted into kinetic energy of the α particle plus radium nucleus. The mass of the thorium nucleus is greater than that of the α particle plus radium nucleus by about 6.1 MeV/c^2.

8-5   Kinetic Energy of a System

Although the total momentum of a system of particles must be constant if the net external force on the system is zero, the total mechanical energy of the system can change. As we saw in the examples of the previous section, internal forces that cannot change the total momentum may be nonconservative and thus change the total mechanical energy of the system. There is an important theorem concerning the kinetic energy of a system of particles that allows us to treat the energy of complex systems more easily and gives us insight into energy changes within a system:

The kinetic energy of a system of particles can be written as the sum of two terms: (1) the kinetic energy associated with the motion of the center of mass, $\frac{1}{2} M v_{\text{cm}}^2$, where M is the total mass of the system; and (2) the kinetic energy associated with the motion of the particles of the system relative to the center of mass, $\Sigma \frac{1}{2} m_i u_i^2$, where $\vec{u}_i$ is the velocity of the ith particle relative to the center of mass.

The kinetic energy of a system of particles is the sum of the kinetic energies of the individual particles:

$$K = \sum_i K_i = \sum_i \frac{1}{2} m_i v_i^2 = \sum_i \frac{1}{2} m_i (\vec{v}_i \cdot \vec{v}_i)$$

The velocity of each particle can be written as the sum of the velocity of the center of mass, $\vec{v}_{cm}$, and the velocity of the particle relative to the center of mass, $\vec{u}_i$:

$$\vec{v}_i = \vec{v}_{cm} + \vec{u}_i \qquad \text{8-16}$$

Then

$$K = \sum_i \frac{1}{2} m_i(\vec{v}_i \cdot \vec{v}_i) = \sum_i \frac{1}{2} m_i(\vec{v}_{cm} + \vec{u}_i) \cdot (\vec{v}_{cm} + \vec{u}_i)$$

$$= \sum_i \frac{1}{2} m_i v_{cm}^2 + \sum_i \frac{1}{2} m_i u_i^2 + \vec{v}_{cm} \cdot \sum_i m_i \vec{u}_i$$

where in the last term we have removed $\vec{v}_{cm}$ from the sum because it is the same for each particle; that is, it refers to the system and not to any particular particle. The quantity $\Sigma m_i \vec{u}_i$ is the total momentum of the system *relative to the center of mass*. This quantity is necessarily zero. Relative to the center of mass, the velocity of the center of mass, $\vec{u}_{cm}$, is zero, and the total momentum $M\vec{u}_{cm}$ is also zero. Then

$$K = \sum_i \frac{1}{2} m_i v_{cm}^2 + \sum_i \frac{1}{2} m_i u_i^2 = \frac{1}{2} M v_{cm}^2 + K_{rel} \qquad \text{8-17}$$

Kinetic energy of a system of particles

where M is the total mass and K_{rel} is the kinetic energy of the particles relative to the center of mass. When there are no external forces, v_{cm} is constant and the kinetic energy associated with bulk motion ($\frac{1}{2} M v_{cm}^2$) does not change. Only the relative kinetic energy can change in an isolated system.

8-6 Collisions

In a collision, two objects approach and interact strongly for a very short time. During the brief time of collision, any external forces are much smaller than the forces of interaction between the objects. Thus, the only important forces acting on the two-object system are the interaction forces, which are equal and opposite, so the total momentum of the system remains unchanged. The collision time is usually so small that the displacement of the objects during the collision can be neglected. Before and after the collision, the interaction of the two objects is small compared with the interaction during the collision. Examples are a cue ball hitting an object ball, a baseball being hit by a bat, a dart colliding with a dart board, and a comet swinging around the sun.

When the total kinetic energy of the two objects is the same after the collision as before, the collision is called an **elastic collision**; otherwise, it is called an **inelastic collision**. An extreme case is the **perfectly inelastic collision**, in which all of the kinetic energy relative to the center of mass is converted to thermal or internal energy of the system, and the two objects stick together after the collision.

Impulse and Average Force

Figure 8-24 shows the time variation of the magnitude of a typical force exerted by one object on another during a collision. During the collision time $\Delta t = t_f - t_i$, the force is large. For other times, the force is negligibly small.

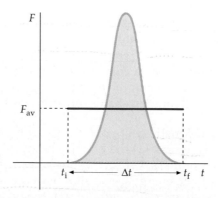

Figure 8-24 Typical time variation of force during a collision. The area under the F-versus-t curve is the magnitude of the impulse, I. F_{av} is the average force for time interval Δt. The rectangular area $F_{av} \Delta t$ is the same as the area under the F-versus-t curve.

The **impulse** $\vec{I}$ of the force is a vector defined as

$$\vec{I} = \int_{t_i}^{t_f} \vec{F}\,dt \qquad\qquad 8\text{-}18$$

Definition—Impulse

The magnitude of the impulse of the force is the area under its *F*-versus-*t* curve. The unit of impulse is N·s. If $\vec{F}$ is the net force acting on a particle, it is related to the rate of change of momentum of the particle by Newton's second law, $\vec{F} = d\vec{p}/dt$. Then the impulse of the net force equals the total change in momentum during the time interval:

$$\vec{I}_{\text{net}} = \int_{t_i}^{t_f} \vec{F}_{\text{net}}\,dt = \int_{t_i}^{t_f} \frac{d\vec{p}}{dt}\,dt = \vec{p}_f - \vec{p}_i = \Delta\vec{p} \qquad\qquad 8\text{-}19$$

The **average force** for the interval $\Delta t = t_f - t_i$ is defined as

$$\vec{F}_{\text{av}} = \frac{1}{\Delta t}\int_{t_i}^{t_f} \vec{F}\,dt = \frac{\vec{I}}{\Delta t} \qquad\qquad 8\text{-}20$$

Definition—Average force

The average force is the constant force that gives the same impulse as the actual force in the time interval Δt, as shown by the rectangle in Figure 8-24. The average force can be calculated from the change in momentum if a collision time is known. This time is often estimated using the distance traveled by one of the objects during the collision.

Example 8-10

With an expert karate blow, you shatter a concrete block. Consider your fist to have a mass 0.70 kg, to be moving 5.0 m/s as it strikes the block, and to stop within 6 mm of the point of contact. (*a*) What impulse does the block exert on your fist? (*b*) What is the approximate collision time and the average force the block exerts on your fist?

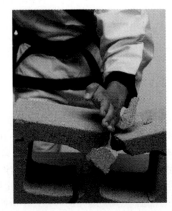

Picture the Problem The impulse equals the change in momentum $\Delta\vec{p}$. We find $\Delta\vec{p}$ from the mass and velocity of the fist. The time of collision for part (*b*) comes from the given distance $\Delta y = 6$ mm and the average velocity during the collision, which we can estimate by assuming constant acceleration. We will choose upward as the positive direction.

(*a*)1. Set the impulse equal to the change in momentum: $\qquad \vec{I} = \Delta\vec{p} = \vec{p}_f - \vec{p}_i$

2. The initial momentum is that of the fist just before it hits the block with speed *v*, and the final momentum is zero:

$$\vec{p}_i = -(0.7\ \text{kg})(5.0\ \text{m/s})\,\hat{j} = -3.5\ \text{kg·m/s}\,\hat{j}$$
$$\vec{p}_f = 0$$

3. Find the impulse exerted by the block on the fist:

$$\vec{I} = \vec{p}_f - \vec{p}_i = 0 - (-3.5\ \text{kg·m/s}\,\hat{j}) = 3.5\ \text{kg·m/s}\,\hat{j}$$

(*b*)1. The collision time is the distance moved divided by the average speed:

$$\Delta t = \frac{\Delta y}{v_{\text{av}}}$$

2. Assuming constant acceleration, $v_{\text{av}} = \frac{1}{2}v$. Since we have chosen upward to be positive, both Δy and v_{av} are negative. Calculate Δt:

$$\Delta t = \frac{\Delta y}{\frac{1}{2}v} = \frac{-0.006\ \text{m}}{-2.5\ \text{m/s}} = 0.0024\ \text{s} = 2.4\ \text{ms}$$

3. The average force is the impulse divided by the collision time. It is upward, as expected:

$$\vec{F}_{\text{av}} = \frac{\vec{I}}{\Delta t} = \frac{3.5\ \text{N·s}\,\hat{j}}{0.0024\ \text{s}} = 1.46\ \text{kN}\,\hat{j}$$

Remark The average force is large—about 116 times the weight of the fist.

Example 8-11 *try it yourself*

A car equipped with an 80-kg crash-test dummy drives into a wall at 25 m/s (about 56 mi/h). Estimate the force that the seat belt exerts on the dummy upon impact.

Picture the Problem Assume that the car and dummy travel about 1 m as the car comes to rest, and that the acceleration is constant during the crash (Figure 8-25). This means that the average speed of the car during the collision is one-half the initial speed, or $v_{av} = 12.5$ m/s. To find the force, calculate the impulse I, then divide by the collision time Δt.

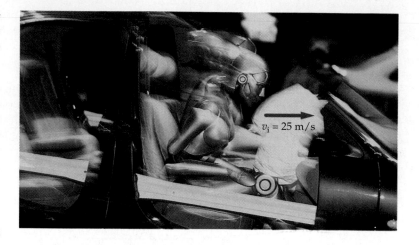

$v_i = 25$ m/s

Figure 8-25

Cover the column to the right and try these on your own before looking at the answers.

Steps

Answers

1. Find the dummy's initial momentum.

$mv = 2000$ kg·m/s

2. Set the impulse equal to the change in momentum to find the magnitude of the impulse exerted by the seat belt on the dummy.

$I = 2000$ N·s

3. Estimate the collision time using $\Delta x = 1$ m and $v_{av} = 12.5$ m/s.

$\Delta t = 0.08$ s

4. Compute the average force.

$F_{av} = 25,000$ N

Remark The average acceleration is $a_{av} = \Delta v/\Delta t = 313$ m/s², or roughly 32 times the acceleration due to gravity. A large acceleration means a large force, as step 4 of the example reveals. 25,000 N (about 5600 lb) is clearly enough to cause serious injuries. An air bag increases the stopping distance somewhat, which helps to prevent injury. The air bag also allows the force to be distributed over a much larger area.

Remark (*a*) Figure 8-26 shows the average force exerted by the seat belt on the dummy as a function of the stopping distance x. With no seat belt or air bag, you either fly through the windshield or are stopped in a fraction of a meter by the dashboard or steering wheel. (*b*) The plot shows the force as a function of the initial velocity for three stopping distances: 2 m, 1.5 m, and 1 m.

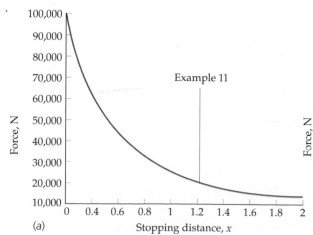

(*a*)

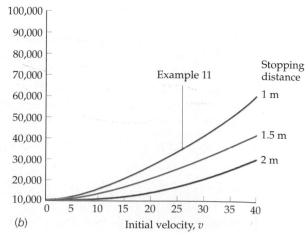

(*b*)

Figure 8-26

Example 8-12

You strike a golf ball with a club. What are reasonable estimates for (a) the impulse I, (b) the collision time Δt, and (c) the average force F_{av}? A typical golf ball has a mass $m = 45$ g and a radius $r = 2$ cm (Figure 8-27). For a typical drive, the range is roughly $R = 160$ m (about 175 yards).

Picture the Problem The impulse equals the change in momentum of the ball, which is mv_0. We estimate the initial speed v_0 from the range. We estimate the collision time from the distance traveled Δx and the average speed $\frac{1}{2}v_0$, assuming constant acceleration. We use $\Delta x = 2$ cm, the radius of the ball. The average force is then obtained from the impulse I and collision time Δt.

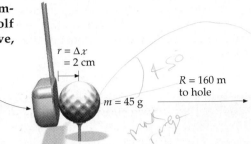

Figure 8-27

(a)1. Set the impulse equal to the change in momentum of the ball:

$$I = mv_0$$

2. The initial speed is related to the range R, which is given by Equation 2-22:

$$R = \frac{v_0^2}{g} \sin 2\theta_0$$

3. Take $\theta_0 = 45°$ corresponding to maximum range ($\sin 2\theta_0 = 1$), and calculate the initial speed:

$$v_0 = \sqrt{Rg} = \sqrt{(160 \text{ m})(9.81 \text{ m/s}^2)} = 40 \text{ m/s}$$

4. Use this value of v_0 to calculate the magnitude of the impulse:

$$I = mv_0 = (0.045 \text{ kg})(40 \text{ m/s})$$
$$= 1.8 \text{ kg·m/s} = 1.8 \text{ N·s}$$

(b) Calculate the collision time Δt using $x = 2$ cm, and $v_{av} = \frac{1}{2}v_0$:

$$\Delta t = \frac{\Delta x}{v_{av}} = \frac{\Delta x}{\frac{1}{2}v_0} = \frac{0.02 \text{ m}}{20 \text{ m/s}} = 0.001 \text{ s}$$

(c) Use the calculated values of I and Δt to find the magnitude of the average force:

$$f_{av} = \frac{I}{\Delta t} = \frac{1.8 \text{ N·s}}{0.001 \text{ s}} = 1800 \text{ N}$$

Remark Again we see very large forces exerted during a collision. Here the force exerted on the golf ball by the club is roughly 4000 times the weight of the ball, giving it a brief acceleration of $4000g$. In comparison, the force of friction exerted on the ball during the collision is negligible.

Collisions in One Dimension

Consider an object of mass m_1 with initial velocity v_{1i} approaching a second object of mass m_2 that is moving in the same direction with initial velocity v_{2i}. If $v_{2i} < v_{1i}$, the objects collide. Let v_{1f} and v_{2f} be their final velocities after the collision. (The velocities can be positive or negative, depending on whether the objects are moving to the right or left.) Conservation of momentum gives one relation between the two unknown velocities v_{1f} and v_{2f}:

$$m_1 v_{1f} + m_2 v_{2f} = m_1 v_{1i} + m_2 v_{2i} \qquad \text{8-21}$$

To determine v_{1f} and v_{2f}, we must have a second relation. That second relation, which we shall now develop, depends on the type of collision.

Perfectly Inelastic Collisions in One Dimension In perfectly inelastic collisions, the particles stick together after the collision. The second relation between the final velocities is that they are equal to each other and to the velocity of the center of mass:

$$v_{1f} = v_{2f} = v_{cm}$$

This result combined with conservation of momentum gives

$$(m_1 + m_2)v_{cm} = m_1 v_{1i} + m_2 v_{2i} \qquad\qquad 8\text{-}22$$

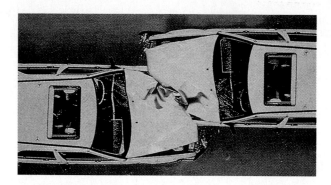

Perfectly inelastic collision of two cars.

Example 8-13

An astronaut of mass 60 kg is on a space walk to repair a communications satellite. Suddenly she needs to consult her physics book. You happen to have it with you, so you throw it to her with speed 4 m/s relative to your spacecraft. She is at rest relative to the spacecraft just before catching the 3.0-kg book (Figure 8-28). Find (*a*) her velocity just after she catches the book, (*b*) the initial and final mechanical energy of the book–astronaut system, and (*c*) the impulse exerted by the book on the astronaut.

Picture the Problem (*a*) The final velocity of the book and astronaut is the velocity of the center of mass. We find this using conservation of momentum, as expressed in Equation 8-21. The initial and final kinetic energies are calculated from the initial and final velocities. Since the book and astronaut move with the same final velocity, the collision is perfectly inelastic. (*b*) The mechanical energies of the book and astronaut are calculated directly from their masses and speeds. (*c*) The impulse exerted by the book on the astronaut equals the change in momentum of the astronaut.

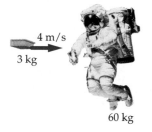

4 m/s

3 kg

60 kg   63 kg

Figure 8-28

(*a*)1. Use conservation of momentum to relate the final velocity of the system, v_{cm}, to the initial velocities:

$$m_b v_b + m_a v_a = (m_b + m_a)v_{cm}$$
$$(3.0\ \text{kg})(4\ \text{m/s}) + (60\ \text{kg})(0) = (63\ \text{kg})v_{cm}$$

2. Solve for v_{cm}:

$$v_{cm} = \frac{(3.0\ \text{kg})(4\ \text{m/s})}{63\ \text{kg}} = 0.19\ \text{m/s}$$

(*b*)1. The initial mechanical energy of the book–astronaut system is the kinetic energy of the book:

$$E_1 = K_b = \frac{1}{2}m_b v_b = \frac{1}{2}(3.0\ \text{kg})(4\ \text{m/s})^2 = 24\ \text{J}$$

2. The final mechanical energy is the kinetic energy of the book and astronaut moving together:

$$E_f = K_f = \frac{1}{2}(m_b + m_a)v_{cm}^2$$

$$= \frac{1}{2}(63\ \text{kg})(0.19\ \text{m/s})^2 = 1.38\ \text{J}$$

(c)   Set the impulse exerted on the astronaut equal to the change in momentum of the astronaut:

$$I = \Delta p_{ast} = m_{ast}\Delta v_{ast} = (60 \text{ kg})(0.19 \text{ m/s})$$
$$= 11.4 \text{ kg·m/s} = 11.4 \text{ N·s}$$

Remark Most of the initial mechanical energy in this collision is lost by conversion to thermal energy. The impulse exerted by the book on the astronaut is equal and opposite to that exerted by the astronaut on the book, so the total change in momentum is zero.

It is useful to express the kinetic energy K of a particle in terms of its momentum p. For a mass m moving with speed v, we have

$$K = \frac{1}{2}mv^2 = \frac{(mv)^2}{2m}$$

Since $\vec{p} = m\vec{v}$,

$$K = \frac{p^2}{2m} \qquad\qquad 8\text{-}23$$

We can apply this to a perfectly inelastic collision where one object is initially at rest. The momentum of the system is that of the incoming object:

$$P = m_1 v_{1i}$$

The initial kinetic energy is

$$K_i = \frac{P^2}{2m_1} \qquad\qquad 8\text{-}24$$

After colliding, the objects move together as a single mass $m_1 + m_2$ with v_{cm}. Momentum is conserved, so the final momentum equals P. The final kinetic energy is then

$$K_f = \frac{P^2}{2(m_1 + m_2)} \qquad\qquad 8\text{-}25$$

Comparing Equations 8-24 and 8-25, we see that the final energy is less than the initial energy.

Example 8-14

In a feat of public marksmanship, you fire a bullet into a hanging target (Figure 8-29). The target, with bullet embedded, swings upward. Noting the height reached at the top of the swing, you immediately inform the crowd of the bullet's speed. For arbitrary masses m_1 and m_2, and height h, how would you calculate the speed?

Picture the Problem The initial speed of the bullet, v_{1i}, is related to the speed of the bullet–block system, v_f, just after the inelastic collision by conservation of momentum. The speed v_f is related to the height h by conservation of mechanical energy. Let m_1 be the mass of the bullet and m_2 be the mass of the target.

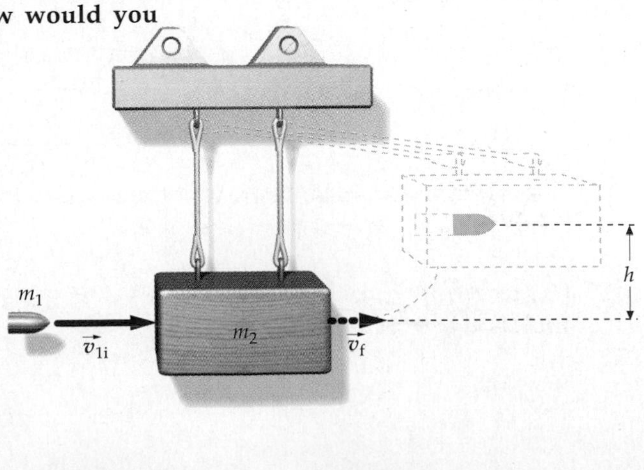

m_1

Figure 8-29

$\vec{v}_{1i}$ m_2 $\vec{v}_f$ h

1. Use conservation of momentum during the collision to find v_{1i} in terms of v_f:

$$m_1 v_{1i} + m_2(0) = (m_1 + m_2)v_f$$

$$v_{1i} = \frac{m_1 + m_2}{m_1} v_f$$

2. Use conservation of mechanical energy after the collision to find v_f in terms of the height h:

$$\frac{1}{2}(m_1 + m_2)v_f^2 = (m_1 + m_2)gh$$

$$v_f = \sqrt{2gh}$$

3. Substituting v_f into the equation in step 1, we can solve for v_{1i}:

$$v_{1i} = \frac{m_1 + m_2}{m_1} v_f = \frac{m_1 + m_2}{m_1} \sqrt{2gh}$$

Remark In this problem, as in all collision problems, we assume that the time of the collision is so short that the displacement of the block during the collision is negligible. Devices such as the one pictured are called *ballistic pendulums*.

Exercise If the mass of the bullet is 12 g, the mass of the block on the ballistic pendulum is 2 kg, and the final height is 10.4 cm, what speed did you announce to the crowd? (*Answer*   240 m/s)

Exercise A 2000-kg car moving 25 m/s runs head-on into a 1500-kg car initially at rest. If the collision is perfectly inelastic, find (*a*) each car's speed after the collision, and (*b*) the ratio of the system's final kinetic energy to its initial kinetic energy. (*Answers*   (*a*) 14.3 m/s, (*b*) 0.57)

Example 8-15   *try it yourself*

You repeat your feat of Example 8-14, this time with an empty box as a target. The bullet strikes the target and passes through it completely. A laser ranging device indicates that the bullet emerged with half its initial velocity. Hearing this, you correctly report how high the target must have swung. How high did it swing?

Picture the Problem The height h is related to the box's speed after colliding, v_2, by conservation of mechanical energy (Figure 8-30). This speed can be determined using conservation of momentum.

Figure 8-30

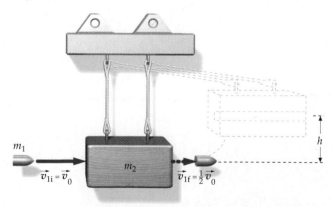

Cover the column to the right and try these on your own before looking at the answers.

Steps   *Answers*

1. Use conservation of mechanical energy to relate the final $m_2 gh = \dfrac{1}{2} m_2 v_2^2$
   height h to the speed v_2 of the box after the collision.

2. Use conservation of momentum to write an equation relating $m_2 v_2 + m_1 \left(\dfrac{1}{2} v_0 \right) = m_1 v_0$
   the speed v_2 of the box to v_0.

3. Solve for v_2. $v_2 = \dfrac{m_1}{2m_2} v_0$

4. Substitute this value of v_2 into your equation in step 1 for h. $h = \dfrac{v_2^2}{2g} = \dfrac{m_1^2 v_0^2}{8 m_2^2 g}$

Remark Inelastic collisions also occur in microscopic systems. For example, when an electron is scattered by an atom, the atom is sometimes excited to a higher internal energy state. As a result, the total kinetic energy of the atom and the electron is lower after the collision than it was before.

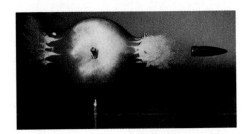

A bullet traveling 850 m/s collides inelastically with an apple, which moments later disintegrates completely. Exposure time is less than a millionth of a second.

Elastic Collisions in One Dimension   For elastic collisions, the initial and final kinetic energies are equal:

$$\frac{1}{2} m_1 v_{1f}^2 + \frac{1}{2} m_2 v_{2f}^2 = \frac{1}{2} m_1 v_{1i}^2 + \frac{1}{2} m_2 v_{2i}^2 \qquad \text{8-26}$$

This, together with conservation of momentum (Equation 8-21), is sufficient to determine the final velocities of the two objects. However, the quadratic nature of Equation 8-26 often complicates the solution of an elastic collision problem. Such problems can be treated more easily if we express the relative velocity of the two particles after the collision in terms of the relative velocity before the collision. Rearranging Equation 8-26 gives

$$m_2 (v_{2f}^2 - v_{2i}^2) = m_1 (v_{1i}^2 - v_{1f}^2)$$

or

$$m_2 (v_{2f} - v_{2i})(v_{2f} + v_{2i}) = m_1 (v_{1i} - v_{1f})(v_{1i} + v_{1f}) \qquad \text{8-27}$$

From conservation of momentum, we know that

$$m_1 v_{1f} + m_2 v_{2f} = m_1 v_{1i} + m_2 v_{2i}$$

so that

$$m_2 (v_{2f} - v_{2i}) = m_1 (v_{1i} - v_{1f}) \qquad \text{8-28}$$

Then dividing Equation 8-27 by Equation 8-28, we get

$$v_{2f} + v_{2i} = v_{1i} + v_{1f}$$

or

$$v_{2f} - v_{1f} = -(v_{2i} - v_{1i}) \qquad \text{8-29}$$

Relative velocities in an elastic collision

Figure 8-31 Approach and recession in an elastic collision.

If two objects are to collide, $v_{2i} - v_{1i}$ must be negative (Figure 8-31), making their **speed of approach** $-(v_{2i} - v_{1i})$. After colliding, the objects' **speed of recession** is $v_{2f} - v_{1f}$, which is positive. Equation 8-29 states

> In elastic collisions, the speed of recession equals the speed of approach.

Solving elastic-collision problems is usually easier using Equation 8-29 than Equation 8-26. But remember, Equation 8-29 depends on conservation of mechanical energy, so it applies only to *elastic* collisions.

Example 8-16

A 4-kg block moving right at 6 m/s collides elastically with a 2-kg block moving right at 3 m/s (Figure 8-32). Find their final velocities.

Picture the Problem Conservation of momentum (Equation 8-18) and conservation of energy (expressed as a reversal of relative velocities, Equation 8-24) give two equations for the two unknown final velocities. Let subscript 1 denote the 4-kg block, and subscript 2 denote the 2-kg block.

Figure 8-32

1. Apply conservation of momentum:

$$p_i = m_1 v_{1i} + m_2 v_{2i} = 30 \text{ kg·m/s}$$
$$p_f = m_1 v_{1f} + m_2 v_{2f} = p_i$$

2. Substituting numerical values in step 1 relates v_{1f} and v_{2f}:

$$4v_{1f} + 2v_{2f} = 30 \text{ m/s}$$

3. Calculate the velocity of approach:

$$v_{2i} - v_{1i} = -3 \text{ m/s}$$

4. Use conservation of mechanical energy to set the velocity of recession equal to the negative of the velocity of approach:

$$v_{2f} - v_{1f} = -(v_{2i} - v_{1i}) = 3 \text{ m/s}$$

5. With the two relations for two unknowns in steps 2 and 4, we solve for the final velocities:

$$v_{1f} = 4 \text{ m/s} \quad \text{and} \quad v_{2f} = 7 \text{ m/s}$$

Check the Result As a check, we calculate the initial and final kinetic energies: $K_i = \frac{1}{2}(4 \text{ kg})(6 \text{ m/s})^2 + \frac{1}{2}(2 \text{ kg})(3 \text{ m/s})^2 = 72 \text{ J} + 9 \text{ J} = 81 \text{ J}$; $K_f = \frac{1}{2}(4 \text{ kg})(4 \text{ m/s})^2 + \frac{1}{2}(2 \text{ kg})(7 \text{ m/s})^2 = 32 \text{ J} + 49 \text{ J} = 81 \text{ J} = K_i$.

Example 8-17

A neutron of mass m_1 and speed v_{1i} collides elastically with a carbon nucleus of mass m_2 at rest (Figure 8-33). (a) What are the final velocities of both particles? (b) What fraction of its initial energy does the neutron lose?

Figure 8-33

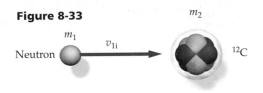

Picture the Problem Conservation of momentum and conservation of energy allow us to find the final velocities. Since the initial energy of the carbon nucleus is zero, its final energy equals the energy lost by the neutron.

(a)1. Use conservation of momentum to obtain one relation for the final velocities:

$$m_1 v_{1i} = m_1 v_{1f} + m_2 v_{2f}$$

2. Use conservation of mechanical energy to set the velocity of recession equal to the negative of the velocity of approach:

$$v_{2f} - v_{1f} = -(v_{2i} - v_{1i}) = v_{1i}$$

3. To eliminate v_{2f}, solve step 2 for v_{2f}, and substitute the answer in step 1:

$$v_{2f} = v_{1i} + v_{1f}$$
$$m_1 v_{1i} = m_1 v_{1f} + m_2(v_{1i} + v_{1f})$$

4. Solve for v_{1f} (Note that v_{1f} is negative because the mass of the carbon nucleus m_2 is greater than the mass of the neutron.):

$$v_{1f} = \frac{m_1 - m_2}{m_1 + m_2} v_{1i}$$

5. Use v_{1f} to find v_{2f}:

$$v_{2f} = v_{1i} + v_{1f} = \frac{2m_1}{m_1 + m_2} v_{1i}$$

(b)1. The energy lost by the neutron is the final energy of the carbon nucleus:

$$-\Delta K_n = K_{2f} = \frac{1}{2} m_2 v_{2f}^2 = \frac{2m_2 m_1 v_{1i}^2}{(m_1 + m_2)^2}$$

$$= \frac{4m_1 m_2}{(m_1 + m_2)^2}\left(\frac{1}{2}\right) m_2 v_{1i}^2$$

2. Divide $-\Delta K_n$ by K_n to find the fraction of the initial energy of the neutron lost:

$$f = \frac{-\Delta K_n}{K_n} = \frac{4m_1 m_2}{(m_1 + m_2)^2}$$

Remark An important application of energy transfer in elastic collisions is the slowing down of neutrons in a nuclear reactor. High-energy neutrons are emitted in the fission of a uranium nucleus. If these neutrons are to cause another uranium nucleus to fission, their energy must be reduced, that is, they must be slowed down or "moderated." One mechanism for slowing down neutrons is the elastic scattering of the neutrons with the nuclei in the reactor. The fractional energy loss $f = -\Delta K_n/K_n$ depends on the ratio of the mass of the moderator nucleus to that of the neutron, as shown in Figure 8-34. For uranium, $m_2 \approx 235 m_1$ and $f \approx 0.017 = 1.7\%$. For carbon, $m_2 \approx 12 m_1$ and $f = 0.28 = 28\%$; for hydrogen, $m_1 \approx m_1$ and $f = 1 = 100\%$. A moderator such as graphite or water is added to a reactor to slow down the neutrons so that they can be captured by uranium nuclei.

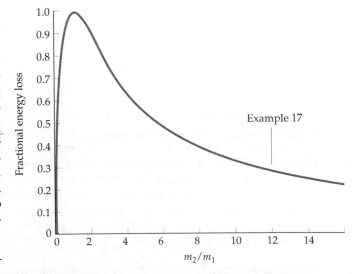

Figure 8-34 Fractional energy loss as a function of the ratio of the two masses. The maximum energy loss occurs when $m_1 = m_2$.

Exercise A 2-kg box moving at 3 m/s makes an elastic collision with a stationary 4-kg box. (a) What is the original mechanical energy? (b) How much energy is transferred to the 4-kg box? (*Answers* (a) 9 J, (b) 8 J)

The results of Example 8-17 for the final velocities of an incoming particle colliding with a second particle initially at rest are worth noting. The final velocity of the incoming particle, v_{1f}, and that of the originally stationary particle, v_{2f}, are related to the initial velocity of the incoming particle by

$$v_{1f} = \frac{m_1 - m_2}{m_1 + m_2} v_{1i}$$
8-30a

and

$$v_{2f} = \frac{2m_1}{m_1 + m_2} v_{1i}$$
8-30b

When a very massive object (say a bowling ball) collides with a light stationary object (say a Ping-Pong ball), the massive object is essentially unaffected. Before the collision, the relative velocity of approach is v_{1i}. If the massive object continues with a velocity that is essentially v_{1i} after the collision, the velocity of the smaller object must be $2v_{1i}$ so that the speed of recession is equal to the speed of approach. This result also follows from Equations 8-30a and 8-30b if we take m_2 to be much smaller than m_1, in which case $v_{1f} \approx v_{1i}$ and $v_{2f} \approx 2v_{1i}$, as expected. An example of such a collision, shown in Figure 8-35, is that between a golf ball and a golf club (whose mass is augmented by the mass of the golfer swinging the club).

(left) A baseball and bat collide. The ball briefly deforms due to the large force the bat exerts during contact. Rebounding, the ball springs back to its original shape, converting elastic potential energy of deformation into kinetic energy.

Figure 8-35 Multiflash photograph of a golfer hitting a ball. The ball travels at approximately twice the speed of the club, as can be seen by comparing the distances Δs_c traveled by the club and Δs_b traveled by the ball between flashes.

The Coefficient of Restitution   Most collisions lie somewhere between the extreme cases of elastic, in which the relative velocities are reversed, and perfectly inelastic, in which there is no relative velocity after the collision. The **coefficient of restitution**, e, is a measure of the elasticity of a collision. It is defined as the ratio of the relative speed of recession to the relative speed of approach:

$$e = \frac{|v_{2f} - v_{1f}|}{|v_{2i} - v_{1i}|} = \frac{v_{rec}}{v_{app}}$$
8-31

Definition—Coefficient of restitution

For an elastic collision, $e = 1$; for a perfectly inelastic collision, $e = 0$.

Collisions in Three Dimensions

Perfectly Inelastic Collisions in Three Dimensions   For collisions in three dimensions, the total initial momentum is the sum of the initial momentum vectors of each object involved in the collision. For perfectly inelastic collisions, the objects stick together, and since their final momentum equals the initial momentum, they move off in the direction of the resultant total momentum with velocity $\vec{v}_{cm}$ given by

$$\vec{v}_{cm} = \frac{\vec{P}}{m_1 + m_2}$$
8-32

where $\vec{P} = \vec{p}_1 + \vec{p}_2$ is the total momentum of the system. Since $\vec{P}$ is in the plane formed by $\vec{p}_1$ and $\vec{p}_2$, the collision takes place in this plane.

optional

Example 8-18 _try it yourself_

A small car of mass 1.2 Mg (1.2×10^3 kg) traveling east at 60 km/h collides at an intersection with a truck of mass 3 Mg traveling north at 40 km/h, as shown. The car and truck stick together. Find the velocity of the wreckage just after the collision.

Picture the Problem Choose our coordinate system so that initially the car is traveling in the x direction and the truck is traveling in the y direction (Figure 8-36). Then write the momentum of each object in vector form, and use conservation of momentum.

Figure 8-36

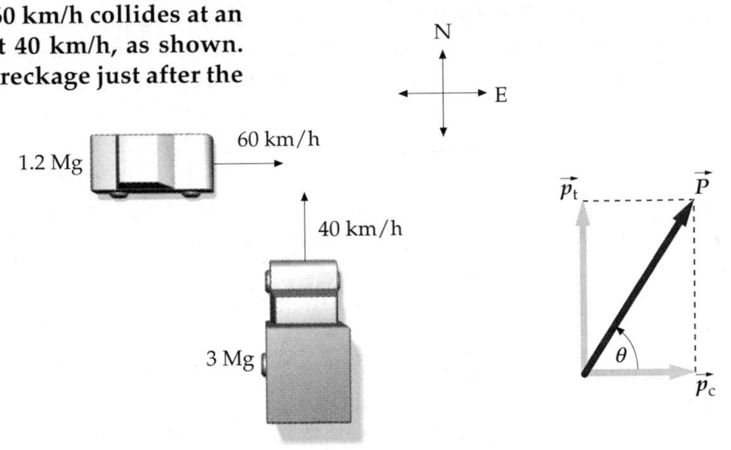

1.2 Mg   60 km/h

40 km/h

3 Mg

$\vec{p}_t$ $\vec{P}$

θ

$\vec{p}_c$

N

E

Cover the column to the right and try these on your own before looking at the answers.

Steps

Answers

1. Write the initial momentum vectors for the car and truck, $\vec{p}_c$ and $\vec{p}_t$, using the unit vectors $\hat{i}$ and $\hat{j}$.

$\vec{p}_c = (72\ \text{Mg·km/h})\,\hat{i}, \qquad \vec{p}_t = (120\ \text{Mg·km/h})\,\hat{j}$

2. Add the vectors in step 1 to obtain the total initial momentum $\vec{P}$, which is also the final momentum.

$\vec{P} = (72\ \text{Mg·km/h})\,\hat{i} + (120\ \text{Mg·km/h})\,\hat{j}$

3. Divide by the total mass M to find the velocity of the center of mass, which is the final velocity of the wreckage.

$\vec{v}_{cm} = \vec{P}/M = (17.1\ \text{km/h})\,\hat{i} + (28.6\ \text{km/h})\,\hat{j}$

4. Find the magnitude of the final velocity.

$v_{cm} = 33.3\ \text{km/h}$

5. Find the direction of the final velocity from $\tan\theta = P_y/P_x$.

$\theta = 59°$

Elastic Collisions in Three Dimensions   Elastic collisions in three dimensions are more complicated than those we have covered previously. Figure 8-37 shows an off-center collision between an object of mass m_1 moving with velocity $\vec{v}_{1i}$ parallel to the x axis toward an object of mass m_2 that is initially at rest at the origin. The distance b between the centers measured perpendicular to the direction of $\vec{v}_{1i}$ is called the **impact parameter**. After the collision, object 1 moves off with velocity $\vec{v}_{1f}$, making an angle θ_1 with its initial velocity, and object 2 moves with velocity $\vec{v}_{2f}$, making an angle θ_2 with $\vec{v}_{1f}$. Conservation of momentum gives

$$\vec{P} = m_1\vec{v}_{1i} = m_1\vec{v}_{1f} + m_2\vec{v}_{2f}$$

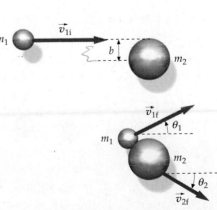

Figure 8-37 Off-center collision. The final velocities depend on the impact parameter b and on the type of force exerted by one object on the other.

optional

We can see from this equation that the vector $\vec{v}_{2f}$ must lie in the plane formed by $\vec{v}_{1i}$ and $\vec{v}_{1f}$, which we will take to be the xy plane. Assuming that we know the initial velocity $\vec{v}_{1i}$, we have four unknowns: the x and y components of both final velocities; or alternatively, the two final speeds and the two angles of deflection. The x and y components of the conservation-of-momentum equation give us two of the needed relations among these quantities. Conservation of energy gives a third relation. To find the four unknowns, we need another relation. The fourth relation depends on the impact parameter b and on the type of interacting force exerted by the bodies on each other. In practice, the fourth relation is often found experimentally, by measuring the angle of deflection or the angle of recoil. Such a measurement can then give us information about the type of interacting force between the bodies.

We omit further discussion of elastic collisions in three dimensions except for the interesting special case of the off-center elastic collision of two objects *of equal mass* when one is initially at rest (Figure 8-38a). If $\vec{v}_{1i}$ and $\vec{v}_{1f}$ are the initial and final velocities of object 1 and $\vec{v}_{2f}$ is the final velocity of object 2, conservation of momentum gives

$$m\vec{v}_{1i} = m\vec{v}_{1f} + m\vec{v}_{2f}$$

or

$$\vec{v}_{1i} = \vec{v}_{1f} + \vec{v}_{2f}$$

These vectors form the triangle shown in Figure 8-38b. Since energy is conserved in the collision,

$$\frac{1}{2}mv_{1i}^2 = \frac{1}{2}mv_{1f}^2 + \frac{1}{2}mv_{2f}^2$$

or

$$v_{1i}^2 = v_{1f}^2 + v_{2f}^2 \qquad \text{8-33}$$

Equation 8-33 is the Pythagorean theorem for a right triangle formed by the vectors $\vec{v}_{1f}$, $\vec{v}_{2f}$, and $\vec{v}_{1i}$ with the hypotenuse of the triangle being $\vec{v}_{1i}$. So for this special case, the final velocity vectors $\vec{v}_{1f}$ and $\vec{v}_{2f}$ are perpendicular to each other, as shown in Figure 8-38b.

Figure 8-38 (*a*) Off-center elastic collision of two spheres of equal mass when one sphere is initially at rest. After the collision, the spheres move off at right angles to each other. (*b*) The velocity vectors for this collision form a right triangle.

Before collision

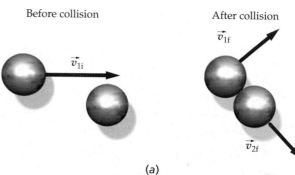

After collision

(a)

(b)

Multiflash photograph of an off-center elastic collision of two balls of equal mass. The dotted ball, entering from the left, strikes the striped ball, which is initially at rest. The final velocities of the two balls are perpendicular to each other.

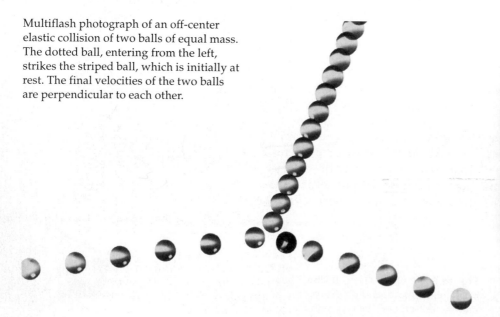

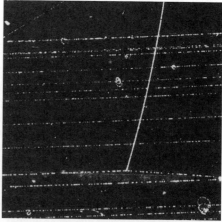

Proton–proton collision in a liquid-hydrogen bubble chamber. A proton entering from the left interacts with a stationary proton. The two then move off at right angles. The slight curvature of the tracks is due to a magnetic field.

*e*xploring

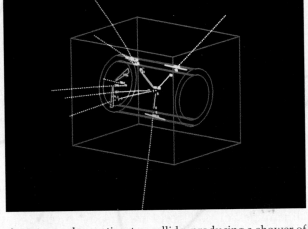

Collisions on Extreme Scales

A proton and an antiproton collide, producing a shower of other particles, including the rare Z particle. The electrically neutral Z leaves no track, but quickly decays into other particles that are charged and do leave tracks. Here the Z decays into an electron (pink track) and a positron (white track). The energy of the electron and positron is measured by the curvature of their tracks in a magnetic field. The total energy of these decay particles equals $m_Z c^2$, where m_Z is the predicted mass of the Z particle.

Tracks of a complicated spray of particles emitted when a neutrino (entering from the left) collides with a proton in the Big European Bubble Chamber at CERN. Neutrinos were first discovered when it was observed that the decay of a neutron into a proton and electron appeared to violate conservation of energy and momentum.

Computer simulation of the collision of two galaxies. Each step represents 100 million years.

The collision of the large spiral galaxy NGC 5194 (bottom), which has a mass of about 10^{11} solar masses, and a much younger galaxy, NGC 5195, which is about one-third as massive.

8-7   The Center-of-Mass Reference Frame

When the net external force on a system is zero, the velocity of the center of mass is constant. It is often convenient to choose a coordinate system with the origin at the center of mass. Then, relative to the original coordinate system, this coordinate system moves with a constant velocity $\vec{v}_{cm}$. The frame of reference attached to the center of mass is called the **center-of-mass reference frame.** If a particle has velocity $\vec{v}$ in the original reference frame, its velocity relative to the center of mass is $\vec{u}_1 = \vec{v}_1 - \vec{v}_{cm}$. In the center-of-mass frame, the velocity of the center of mass is zero. Since the total momentum of a system equals the total mass times the velocity of the center of mass, the total momentum is also zero in the center-of-mass frame. Thus, the center-of-mass reference frame is also called the **zero-momentum reference frame.**

The mathematics of collisions are greatly simplified when considered within the center-of-mass reference frame. The momenta of the two incoming objects are equal and opposite. After a perfectly inelastic collision, the objects remain at rest. All of the original energy is lost to thermal energy. A perfectly elastic collision in one dimension reverses the velocity of each object but does not change the magnitude of v (you will derive this in Problem 93).

Consider a simple two-particle system in a reference frame in which one particle of mass m_1 is moving with a velocity $\vec{v}_1$ and a second particle of mass m_2 is moving with a velocity $\vec{v}_2$ (Figure 8-39). In this frame, the velocity of the center of mass is

$$\vec{v}_{cm} = \frac{m_1\vec{v}_1 + m_2\vec{v}_2}{m_1 + m_2}$$

We can transform the velocities of the two particles to their velocities in the center-of-mass reference frame by subtracting $\vec{v}_{cm}$. The velocities of the particles in the center-of-mass frame are $\vec{u}_1$ and $\vec{u}_2$, given by

$$\vec{u}_1 = \vec{v}_1 - \vec{v}_{cm} \qquad\qquad 8\text{-}34a$$

and

$$\vec{u}_2 = \vec{v}_2 - \vec{v}_{cm} \qquad\qquad 8\text{-}34b$$

Since the total momentum is zero in the center-of-mass frame, the particles have equal and opposite momenta in this frame.

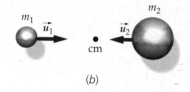

Original reference frame

(a)

Center-of-mass reference frame

(b)

Figure 8-39 (a) Two particles moving in a general reference frame in which the center of mass has a velocity $\vec{v}_{cm}$. (b) In the center-of-mass reference frame, the center of mass is at rest and the particles have equal and opposite momenta. The velocities in the two frames are related by $\vec{u}_1 = \vec{v}_1 - \vec{v}_{cm}$ and $\vec{u}_2 = \vec{v}_2 - \vec{v}_{cm}$.

Example 8-19

Find the final velocities for the elastic collision in Example 8-16 (in which a 4-kg block moving right at 6 m/s collides elastically with a 2-kg block moving right at 3 m/s) by transforming their velocities to the center-of-mass reference frame.

Picture the Problem We transform to the center-of-mass frame by first finding v_{cm} and subtracting it from each velocity. We then solve the collision by reversing the velocities and transforming back to the original frame.

1. Calculate the velocity of the center of mass, v_{cm}:

$$v_{cm} = \frac{m_1 v_{1i} + m_2 v_{2i}}{m_1 + m_2}$$

$$= \frac{(4\text{ kg})(6\text{ m/s}) + (2\text{ kg})(3\text{ m/s})}{4\text{ kg} + 2\text{ kg}} = 5\text{ m/s}$$

Figure 8-40

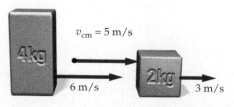

Initial conditions

$v_{cm} = 5$ m/s

Transform to the center-of-mass
frame by subtracting v_{cm}

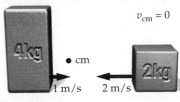

2. Transform the initial velocities to the center-of-mass reference frame by subtracting v_{cm} from the initial velocities:

$u_{1i} = v_{1i} - v_{cm} = 6 \text{ m/s} - 5 \text{ m/s} = 1 \text{ m/s}$

$u_{2i} = v_{2i} - v_{cm} = 3 \text{ m/s} - 5 \text{ m/s} = -2 \text{ m/s}$

Figure 8-41

Solve collision

3. Solve the collision in the center-of-mass reference frame by reversing the velocity of each object:

$u_{1f} = -u_{1i} = -1 \text{ m/s}$

$u_{2f} = -u_{2i} = +2 \text{ m/s}$

Figure 8-42

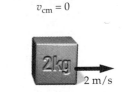

Transform back to the original frame by adding v_{cm}

4. To find the final velocities in the original frame, add v_{cm} to each final velocity:

$v_{1f} = u_{1f} + v_{cm} = -1 \text{ m/s} + 5 \text{ m/s} = 4 \text{ m/s}$

$v_{2f} = u_{2f} + v_{cm} = 2 \text{ m/s} + 5 \text{ m/s} = 7 \text{ m/s}$

Figure 8-43

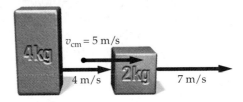

Remark This is the same result found in Example 8-16.

Exercise Show that the total momentum both before and after the collision is zero in the center-of-mass reference frame. (*Answer*   Before the collision: $P_i = (4 \text{ kg})(1 \text{ m/s}) + (2 \text{ kg})(-2 \text{ m/s}) = 0$; after the collision: $P_f = (4 \text{ kg}) \times (-1 \text{ m/s}) + (2 \text{ kg})(2 \text{ m/s}) = 0$)

8-8   Rocket Propulsion

Rocket propulsion is a striking example of the conservation of momentum in action. The mathematical description of rocket propulsion can become quite complex because the mass of the rocket changes continuously as it burns fuel and expels exhaust gas. The easiest approach is to compute the change in the momentum of the total system (including the exhaust gas) for some time interval and use Newton's law in the form $F_{ext} = dP/dt$, where F_{ext} is the net force acting on the rocket.

Consider a rocket moving with speed v relative to the earth (Figure 8-44). If the fuel is burned at a constant rate, $R = |dm/dt|$, the rocket's mass at time t is

$$m = m_0 - Rt \qquad \text{8-35}$$

where m_0 is the initial mass of the rocket. The momentum of the system at time t is

$$P_i = mv$$

At a later time $t + \Delta t$, the rocket has expelled gas of mass $R\,\Delta t$. If the gas is exhausted at a speed u_{ex} *relative to the rocket*, the velocity of the gas relative to the earth is $v - u_{ex}$. The rocket then has a mass $m - R\,\Delta t$ and is moving at a speed $v + \Delta v$ (Figure 8-45).

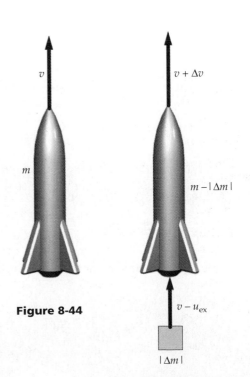

Figure 8-44

Figure 8-45

The momentum of the system at $t + \Delta t$ is

$$P_f = (m - R\,\Delta t)(v + \Delta v) + R\,\Delta t(v - u_{ex})$$

$$= mv + m\,\Delta v - v\,R\,\Delta t - R\,\Delta t\,\Delta v + v\,R\,\Delta t - u_{ex}\,R\,\Delta t$$

$$\approx mv + m\,\Delta v - u_{ex}\,R\,\Delta t$$

where we have dropped the term $R\,\Delta t\,\Delta v$, which is the product of two very small quantities, and therefore negligible compared with the others. The change in momentum is

$$\Delta P = P_f - P_i = m\,\Delta v - u_{ex}\,R\,\Delta t$$

and

$$\frac{\Delta P}{\Delta t} = m\frac{\Delta v}{\Delta t} - u_{ex}R \qquad\qquad 8\text{-}36$$

As Δt approaches zero, $\Delta v/\Delta t$ approaches the derivative dv/dt, which is the acceleration. For a rocket moving upward near the surface of the earth, $F_{ext} = -mg$. Setting $dP/dt = F_{ext} = -mg$ gives us the **rocket equation**:

$$m\frac{dv}{dt} = Ru_{ex} + F_{ext} = Ru_{ex} - mg \qquad\qquad 8\text{-}37$$

Rocket equation

or

$$\frac{dv}{dt} = \frac{Ru_{ex}}{m} - g = \frac{Ru_{ex}}{m_0 - Rt} - g \qquad\qquad 8\text{-}38$$

The quantity Ru_{ex} is the force exerted on the rocket by the exhausting fuel. This is called the **thrust**:

$$F_{th} = Ru_{ex} = \left|\frac{dm}{dt}\right|u_{ex} \qquad\qquad 8\text{-}39$$

Definition—Rocket thrust

Equation 8-38 is solved by integrating both sides with respect to time. For a rocket starting at rest at $t = 0$, the result is

$$v = -u_{ex}\ln\!\left(\frac{m_0 - Rt}{m_0}\right) - gt \qquad\qquad 8\text{-}40$$

as can be verified by taking the time derivative of v. The **payload** of a rocket is the final mass, m_f, after all the fuel has been burned. The **burn time** t_b is given by $m_f = m_0 - Rt_b$, or

$$t_b = \frac{m_0 - m_f}{R} \qquad\qquad 8\text{-}41$$

Thus, a rocket starting at rest with mass m_0, and payload of m_f, attains a final speed

$$v_f = -u_{ex}\ln\frac{m_f}{m_0} - gt_b \qquad\qquad 8\text{-}42$$

Final speed of rocket

assuming the acceleration of gravity to be constant.

Example 8-20	*try it yourself*

The Saturn V rocket used in the Apollo moon-landing program had an initial mass m_0 of 2.85×10^6 kg, a payload of 27%, a burn rate R of 13.84×10^3 kg/s, and a thrust F_{th} of 34×10^6 N. Find (a) the exhaust speed, (b) the burn time t_b, (c) the acceleration at liftoff, (d) the acceleration at burnout t_b, and (e) the final speed of the rocket.

Picture the Problem (a) The exhaust speed can be found from the thrust and burn rate. (b) To find the burn time, you need to find the total mass of fuel burned, which is the initial mass minus the payload. (c) The acceleration is found from Equation 8-38. (d) The final speed is given by Equation 8-42.

Cover the column to the right and try these on your own before looking at the answers.

Steps	**Answers**
(a) Calculate u_{ex} from the given thrust and burn rate.	$u_{ex} = 2.46$ km/s
(b)1. Calculate the final mass m_f of the rocket.	$m_f = (0.27)m_0 = 7.70 \times 10^5$ kg
2. Use your result to calculate the burn time t_b.	$t_b = \dfrac{m_0 - m_f}{R} = 150$ s
(c) Calculate dv/dt for $m = m_0$ and for $m = m_f$.	Initially, $dv/dt = 2.14$ m·s² finally, $dv/dt = 34.4$ m/s²
(d) Calculate the final speed from Equation 8-42.	$v_f = 1.75$ km/s

Remarks The initial acceleration is small—only $0.21g$. At burnout, the rocket's acceleration has increased to 3.5 g. The speed of the rocket at burnout, after two and a half minutes of burning, is roughly 6300 km/h (3900 mi/h).

Summary

1. The conservation of momentum for an isolated system is a fundamental law of nature that has applications in all areas of physics.

Topic	Remarks and Relevant Equations	
1. Center of Mass		
Position for multiple objects	$M\vec{r}_{cm} = m_1\vec{r}_1 + m_2\vec{r}_2 + \cdots = \sum_i m_i\vec{r}_i$	8-4
Position for continuous objects	$M\vec{r}_{cm} = \int \vec{r}\,dm$	8-5
Motion of center of mass	$\vec{F}_{net,ext} = \sum_i \vec{F}_{i,ext} = M\vec{a}_{cm}$	8-10

2.   Momentum

Definition for a particle	$$\vec{p} = m\vec{v}$$	8-11
Kinetic energy of a particle in terms of momentum	$$K = \frac{p^2}{2m}$$	8-23
Definition for a system of particles	$$\vec{P} = \sum_i m_i\vec{v}_i = M\vec{v}_{cm}$$	8-13
Newton's second law for systems	$$\sum_i \vec{F}_{ext} = \vec{F}_{net,ext} = \frac{d\vec{P}}{dt}$$	8-14
Law of conservation of momentum	If the net external force acting on a system is zero, the total momentum of the system is conserved.	

3.   Energy of a System

Kinetic energy	The kinetic energy of a system of particles can be written as the sum of two terms: (1) the kinetic energy associated with the motion of the center of mass, $\frac{1}{2}Mv_{cm}^2$, and (2) the kinetic energy associated with the motion of the particles of the system relative to the center of mass, $\Sigma\frac{1}{2}m_i u_i^2$, where $\vec{u}_i$ is the velocity of the ith particle relative to the center of mass.	

$$K = \sum_i \frac{1}{2}m_i v_{cm}^2 + \sum_i \frac{1}{2}m_i u_i^2 = \frac{1}{2}Mv_{cm}^2 + K_{rel} \qquad \text{8-17}$$

Gravitational potential energy	$$U = Mgh_{cm}$$	8-6

4.   Collisions

Impulse	The impulse of a force is defined as the integral of the force over the time interval during which the force acts. It equals the total change in momentum of the particle.	

$$\vec{I} = \int_{t_i}^{t_f} \vec{F}\,dt = \Delta\vec{p} \qquad \text{8-19}$$

Average force	$$\vec{F}_{av} = \frac{1}{\Delta t}\int_{t_i}^{t_f} \vec{F}\,dt = \frac{\vec{I}}{\Delta t}$$	8-20
Elastic collisions	An elastic collision is one in which the total kinetic energy of the two objects is the same before and after the collision.	
Relative speeds of approach and recession	The relative speed of recession of the objects after an elastic collision in one dimension equals the relative speed of approach before the collision.	

$$v_{2f} - v_{1f} = -(v_{2i} - v_{1i}) \qquad \text{8-29}$$

Perfectly inelastic collisions	In a perfectly inelastic collision, the objects stick together and move with the velocity of the center of mass.	
Coefficient of restitution (optional)	The coefficient of restitution e is a measure of the elasticity of a collision and is defined as the ratio of the relative speed of recession to the relative speed of approach:	

$$e = \frac{|v_{2f} - v_{1f}|}{|v_{2i} - v_{1i}|} = \frac{v_{rec}}{v_{app}} \qquad \text{8-31}$$

For an elastic collision, $e = 1$; for a perfectly inelastic collision, $e = 0$.

5.   Center-of-Mass Reference Frame
(optional)

The center-of-mass reference frame is one that moves with the velocity of the center of mass. In this frame, the total momentum of a system is zero. In perfectly inelastic collisions, the particles remain at rest within the center-of-mass reference frame after the collision. In elastic collisions in one dimension, the velocity of each particle is reversed.

Transforming into the center-of-mass reference frame

The velocity of a particle in the center-of-mass frame, $\vec{u}_i$, is related to the velocity in the original frame, $\vec{v}_i$, by

$$\vec{u}_i = \vec{v}_i - \vec{v}_{cm} \qquad \text{8-34}$$

6.   Rocket Propulsion (optional)

Rockets achieve thrust by burning fuel and exhausting the resulting gases. The force exerted by the exhaust gases on the rocket propels the rocket forward.

Rocket equation

$$m\frac{dv}{dt} = Ru_{ex} + F_{ext} = Ru_{ex} - mg \qquad \text{8-37}$$

where u_{ex} is the exhaust speed and R is the thrust.

Thrust

$$F_{th} = Ru_{ex} = \left|\frac{dm}{dt}\right|u_{ex} \qquad \text{8-39}$$

Final speed

$$v_f = -u_{ex}\ln\frac{m_f}{m_0} - gt_b \qquad \text{8-42}$$

Problem-Solving Guide

1.   Begin by drawing a neat diagram with a suitable coordinate system.

2.   If possible, choose a system for which the net external force is zero, and then apply conservation of momentum.

3.   In collision problems, choose a system that includes both objects. Any external forces will generally be negligible compared to the collision forces during the collision. Apply conservation of momentum. If the collision is elastic, apply conservation of mechanical energy by setting the velocity of recession equal to the velocity of approach.

Summary of Worked Examples

Type of Calculation	Procedure and Relevant Examples
1. Find the Center of Mass	Draw a coordinate system and show the locations of all the masses in the system. A thoughtful choice for the coordinate system can greatly simplify the calculation. For symmetric systems, the center of mass is at the geometric center.
System of particles	Determine x_{cm}, y_{cm}, and z_{cm} separately using $$Mx_{cm} = \sum_i m_i x_i, \text{ etc.} \qquad \textbf{Example 8-1}$$
Continuous system	The center of mass of a uniform symmetric object is at its geometric center. Divide nonsymmetric objects into sets of symmetric objects if possible. In general, x_{cm} can be found from $x_{cm} = \int x\, dm$. **Example 8-2**
2. Find the Forces Acting on a System	Use $\Sigma\vec{F}_{i,ext} = M\vec{a}_{cm}$. Calculate a_{cm} from $M\vec{a}_{cm} = \Sigma m_i\vec{a}_i$ **Example 8-5**
3. Problems With Variable Mass	Choose a system that includes all of the masses and apply conservation of momentum to the directions where there is no external force. **Examples 8-3, 8-7, 8-8**

4.	**Collisions**	Sketch the system before and after the collision. Indicate masses and velocities in your sketch. Apply conservation of momentum.
	Find the impulse	Set the impulse equal to the change in momentum. **Examples 8-10, 8-11, 8-12, 8-13**
	Perfectly inelastic	Use conservation of momentum, $m_1\vec{v}_1 + m_2\vec{v}_2 = (m_1 + m_2)\vec{v}_{cm}$. The final velocity equals the velocity of the center of mass. The loss in mechanical energy can be found by writing the kinetic energy as $K = P^2/2m$, and using the fact that the momentum P is the same before and after the collision. **Examples 8-7, 8-13, 8-14, 8-15, 8-18**
	Elastic, one dimension	Use conservation of momentum and conservation of energy as expressed by the fact that the relative speed of recession equals the relative speed of approach. **Examples 8-16, 8-17**
	In center-of-mass reference frame (optional)	Find the velocity of each particle in the center-of-mass reference frame by subtracting the velocity of the center of mass in the original frame. In the center-of-mass frame, the velocity of each particle is reversed by the collision. The final velocities in the original frame can then be found by adding $\vec{v}_{cm}$ to each velocity. **Example 8-19**
5.	**Estimate Time of Collision, Impulse, Acceleration, and Average Force** (optional)	Use a reasonable guess for the distance traveled during the collision, Δs, and assume constant acceleration to find v_{av}. Then the time of collision is $\Delta t = \Delta s/v_{av}$, the acceleration is $a = \Delta v/\Delta t$, the impulse is $\vec{I} = \Delta\vec{p}$, and the average force is $\vec{F}_{av} = \vec{I}/\Delta t = \Delta\vec{p}/\Delta t$. **Examples 8-10, 8-11, 8-12**
6.	**Rocket Propulsion** (optional)	
	Find the final speed of the rocket	Use

$$v_f = -u_{ex}\ln\frac{m_f}{m_0} - gt_b$$

where u_{ex} is the exhaust speed, m_f is the mass of the rocket without fuel, m_0 is the initial mass with fuel, and t_b is the burn time.   **Example 8-20**

Problems

In a few problems, you are given more data than you actually need; in a few other problems, you are required to supply data from your general knowledge, outside sources, or informed estimates.

- • Single-concept, single-step, relatively easy
- •• Intermediate-level, may require synthesis of concepts
- ••• Challenging, for advanced students

Take g = 9.81 N/kg = 9.81 m/s² and neglect friction in all problems unless otherwise stated.

The Center of Mass

1 • Give an example of a three-dimensional object that has no mass at its center of mass.

2 • Three point masses of 2 kg each are located on the x axis at the origin, $x = 0.20$ m, and $x = 0.50$ m. Find the center of mass of the system.

3 • A 24-kg child is 20 m from an 86-kg adult. Where is the center of mass of this system?

4 • Three objects of 2 kg each are located in the xy plane at points (10 cm, 0), (0, 10 cm), and (10 cm, 10 cm). Find the location of the center of mass.

5 • Find the center of mass x_{cm} of the three masses in Figure 8-46.

Figure 8-46
Problem 5

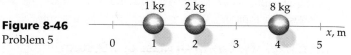

6 • Alley Oop's club-ax consists of a symmetrical 8-kg stone attached to the end of a uniform 2.5-kg stick that is 98 cm long. The dimensions of the club-ax are shown in Figure 8-47. How far is the center of mass from the handle end of the club-ax?

Figure 8-47 Problem 6

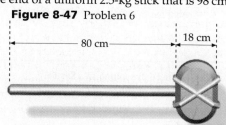

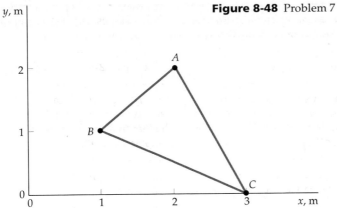

Figure 8-48 Problem 7

7 • Three balls A, B, and C, with masses of 3 kg, 1 kg, and 1 kg, respectively, are connected by massless rods. The balls are located as in Figure 8-48. What are the coordinates of the center of mass?

8 • By symmetry, locate the center of mass of an equilateral triangle of side length a located with one vertex on the y axis and the others at $(-a/2, 0)$ and $(+a/2, 0)$.

9 •• The uniform sheet of plywood in Figure 8-49 has a mass of 20 kg. Find its center of mass.

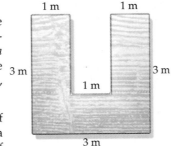

Figure 8-49 Problem 9

Finding the Center of Mass by Integration (optional)

10 •• Show that the center of mass of a uniform semicircular disk of radius R is at a point $(4/3\pi)$ R from the center of the circle.

11 •• A baseball bat of length L has a peculiar linear density (mass per unit length) given by $\lambda = \lambda_0(1 + x^2/L^2)$. Find the x coordinate of the center of mass in terms of L.

12 ••• Find the center of mass of a homogeneous solid hemisphere of radius R and mass M.

13 ••• Find the center of mass of a thin hemispherical shell.

14 ••• A sheet of metal is cut in the shape of a parabola. The edge of the sheet is given by the expression $y = ax^2$, and y ranges from $y = 0$ to $y = b$. Find the center of mass in terms of a and b.

Motion of the Center of Mass of a System

15 • On the night before your physics exam, you hear a banging on your door, and in walks Kelly. She says, "There's a big problem here. According to Newtonian physics, only external forces can cause the center of mass of a system to accelerate. But a car accelerates because of its own engine, so obviously Newton was wrong." She crosses her arms in a way that suggests that she is not going anywhere until she gets a satisfactory explanation. How can you explain Kelly's error to her in order to rescue Newton and get back to your studying?

16 • Two pucks of mass m_1 and m_2 lie unconnected on a frictionless table. A horizontal force F_1 is exerted on m_1 only. What is the magnitude of the acceleration of the center of mass of the pucks?

(a) F_1/m_1
(b) $F_1/(m_1 + m_2)$
(c) F_1/m_2
(d) $(m_1 + m_2) F_1/m_1 m_2$

17 • The two pucks in Problem 16 are lying on a frictionless table and connected by a spring of force constant k. A horizontal force F_1 is again exerted only on m_1 along the spring away from m_2. What is the magnitude of the acceleration of the center of mass?

(a) F_1/m_1
(b) $F_1/(m_1 + m_2)$
(c) $(F_1 + k\,\Delta x)/m_1 m_2$, where Δx is the amount the spring is stretched
(d) $(m_1 + m_2)F_1/m_1 m_2$

18 • Two 3-kg masses have velocities $\vec{v}_1 = 2\,\text{m/s}\,\hat{i} + 3\,\text{m/s}\,\hat{j}$ and $\vec{v}_2 = 4\,\text{m/s}\,\hat{i} - 6\,\text{m/s}\,\hat{j}$. Find the velocity of the center of mass for the system.

19 • A 1500-kg car is moving westward with a speed of 20 m/s, and a 3000-kg truck is traveling east with a speed of 16 m/s. Find the velocity of the center of mass of the system.

20 • A force $\vec{F} = 12\,\text{N}\,\hat{i}$ is applied to the 3-kg ball in Problem 7. What is the acceleration of the center of mass?

21 •• A block of mass m is attached to a string and suspended inside a hollow box of mass M. The box rests on a scale that measures the system's weight. (a) If the string breaks, does the reading on the scale change? Explain your reasoning. (b) Assume that the string breaks and mass m falls with constant acceleration g. Find the acceleration of the center of mass, giving both direction and magnitude. (c) Using the result from (b), determine the reading on the scale while m is in free fall.

22 •• A vertical spring of force constant k is attached at the bottom to a platform of mass m_p, and at the top to a massless cup, as in Figure 8-50. The platform rests on a scale. A ball of mass m_b is placed in the cup. What is the reading on the scale when (a) the spring is compressed an amount $d = mg/k$? (b) the ball comes to rest momentarily with the spring compressed? (c) the ball again comes to rest in its original position?

Figure 8-50 Problem 22

23 •• In the Atwood's machine in Figure 8-51, the string passes over a fixed, frictionless cylinder of mass m_c. (a) Find the acceleration of the center of mass of the two-block-and-cylinder system. (b) Use Newton's second law for systems to find the force F exerted by the support. (c) Find the tension in the string connecting the blocks and show that $F = m_c g + 2T$.

24 •• Repeat Problems 22a and 22b with the ball dropped into the cup from a height h above the cup.

Figure 8-51 Problem 23

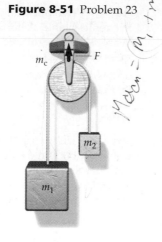

The Conservation of Momentum

25 • True or false:

(a) The momentum of a heavy object is greater than that of a light object moving at the same speed.
(b) The momentum of a system may be conserved even when mechanical energy is not.
(c) The velocity of the center of mass of a system equals the total momentum of the system divided by its total mass.

26 • How is the recoil of a rifle related to momentum conservation?

27 • A man is stranded in the middle of an ice rink that is perfectly frictionless. How can he get to the edge?

28 • A girl jumps from a boat to a dock. Why does she have to jump with more energy than she would need if she were jumping the same distance from one dock to another?

29 •• Much early research in rocket motion was done by Robert Goddard, physics professor at Clark College in Worcester, Massachusetts. A quotation from a 1921 editorial in the *New York Times* illustrates the public acceptance of his work: "That Professor Goddard with his 'chair' at Clark College and the countenance of the Smithsonian Institution does not know the relation between action and reaction, and the need to have something better than a vacuum against which to react—to say that would be absurd. Of course, he only seems to lack the knowledge ladled out daily in high schools." The belief that a rocket needs something to push against was a prevalent misconception before rockets in space were commonplace. Explain why that belief is wrong.

30 •• Liz, Jay, and Tara discover that sinister chemicals are leaking at a steady rate from a hole in the bottom of a railway car. To collect evidence of a potential environmental mishap, they videotape the car as it rolls without friction at an initial speed v_0. Tara claims that careful analysis of the videotape will show that the car's speed is increasing, because it is losing mass as it drains. The increase in speed will help to prove that the leak is occurring. Liz says no, that with a loss of mass, the car's speed will be decreasing. Jay says the speed will remain the same. (a) Who is right? (b) What forces are exerted on the system of the car plus chemical cargo?

31 • A girl of mass 55 kg jumps off the bow of a 75-kg canoe that is initially at rest. If her velocity is 2.5 m/s to the right, what is the velocity of the canoe after she jumps?

32 • Two masses of 5 kg and 10 kg are connected by a compressed spring and rest on a frictionless table. After the spring is released, the smaller mass has a velocity of 8 m/s to the left. What is the velocity of the larger mass?

33 • Figure 8-52 shows the behavior of a projectile just after it has broken up into three pieces. What was the speed of the projectile the instant before it broke up?

(a) v_3
(b) $v_3/3$
(c) $v_3/4$
(d) $4v_3$
(e) $(v_1 + v_2 + v_3)/4$

Figure 8-52 Problem 33

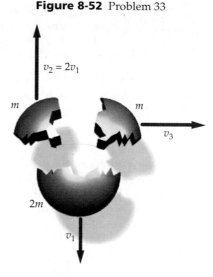

34 • A shell of mass m and speed v explodes into two identical fragments. If the shell was moving horizontally with respect to the earth, and one of the fragments is subsequently moving vertically with the speed v, find the velocity $\vec{v}'$ of the other fragment.

35 •• In a circus act, Marcello (mass 70.0 kg) is shot from a cannon with a muzzle velocity of 24.0 m/s at an angle of 30° above horizontal. His partner, Tina (mass 50.0 kg), stands on an elevated platform located at the top of his trajectory. He grabs her as he flies by and the two fly off together. They land in a net at the same elevation as the cannon a horizontal distance x away. Find x.

36 •• A block and a handgun loaded with one bullet are firmly affixed to opposite ends of a massless cart that rests on a level frictionless air table (Figure 8-53). The mass of the handgun is m_g, the mass of the block is m_{bk}, and the mass of the bullet is m_{bt}. The gun is aimed so that when fired, the bullet will go into the block. When the bullet leaves the barrel of the handgun, it has a velocity v_b as measured by an observer at rest with the table. Take the fall of the bullet to be negligible and its penetration into the block to be small. (a) What is the velocity of the cart immediately after the bullet leaves the gun barrel? (b) What is the velocity of the cart immediately after the bullet comes to rest in the block? (c) How far has the block moved from its initial position at the moment when the bullet comes to rest in the block?

Figure 8-53 Problem 36

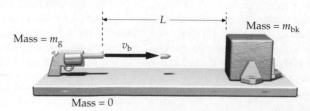

37 •• A small object of mass m slides down a wedge of mass $2m$ and exits smoothly onto a frictionless table. The wedge is initially at rest on the table. If the object is initially at rest at a height h above the table, find the velocity of the wedge when the object leaves it.

Kinetic Energy of a System

38 • Describe how a basketball is moving when (a) its total kinetic energy is just the energy of motion of its center of mass, and (b) its total kinetic energy is the energy of its motion relative to its center of mass.

39 • Two bowling balls are moving with the same velocity, but one just slides down the alley, whereas the other rolls down the alley. Which ball has more energy?

40 • A 3-kg block is traveling to the right at 5 m/s, and a second 3-kg block is traveling to the left at 2 m/s. (a) Find the total kinetic energy of the two blocks in this reference frame. (b) Find the velocity of the center of mass of the two-body system. (c) Find the velocities of the two blocks relative to the center of mass. (d) Find the kinetic energy of the motion of the blocks relative to the center of mass. (e) Show that your answer for part (a) is greater than your answer for part (d) by an amount equal to the kinetic energy of the center of mass.

41 • Repeat Problem 40 with the second, 3-kg block replaced by a block having a mass of 5 kg and moving to the right at 3 m/s.

Impulse and Average Force

42 • Explain why a safety net can save the life of a circus performer.

43 • How might you estimate the collision time of a baseball and bat?

44 • Why does a wine glass survive a fall onto a carpet but not onto a concrete floor?

45 • A soccer ball of mass 0.43 kg leaves the foot of the kicker with an initial speed of 25 m/s. (a) What is the impulse imparted to the ball by the kicker? (b) If the foot of the kicker is in contact with the ball for 0.008 s, what is the average force exerted by the foot on the ball?

46 • A 0.3-kg brick is dropped from a height of 8 m. It hits the ground and comes to rest. (a) What is the impulse exerted by the ground on the brick? (b) If it takes 0.0013 s from the time the brick first touches the ground until it comes to rest, what is the average force exerted by the ground on the brick?

47 • A meteorite of mass 30.8 tonne (1 tonne = 1000 kg) is exhibited in the Hayden Planetarium in New York. Suppose the kinetic energy of the meteorite as it hit the ground was 617 MJ. Find the impulse I experienced by the meteorite up to the time its kinetic energy was halved (which took about 3.0 s). Find also the average force F exerted on the meteorite during this time interval.

48 •• When a 0.15-kg baseball is hit, its velocity changes from +20 m/s to −20 m/s. (a) What is the magnitude of the impulse delivered by the bat to the ball? (b) If the baseball is in contact with the bat for 1.3 ms, what is the average force exerted by the bat on the ball?

49 •• A 300-g handball moving with a speed of 5.0 m/s strikes the wall at an angle of 40° and then bounces off with the same speed at the same angle. It is in contact with the wall for 2 ms. What is the average force exerted by the ball on the wall?

50 •• A 2000-kg car traveling at 90 km/h crashes into a concrete wall that does not give at all. (a) Estimate the time of the collision, assuming that the center of the car travels halfway to the wall with constant deceleration. (Use any reasonable length for the car.) (b) Estimate the average force exerted by the wall on the car.

51 •• You throw a 150-g ball to a height of 40 m. (a) Use a reasonable value for the distance the ball moves while it is in your hand to calculate the average force exerted by your hand and the time the ball is in your hand while you throw it. (b) Is it reasonable to neglect the weight of the ball while it is being thrown?

52 •• A handball of mass 300 g is thrown straight against a wall with a speed of 8 m/s. It rebounds with the same speed. (a) What impulse is delivered to the wall? (b) If the ball is in contact with the wall for 0.003 s, what average force is exerted on the wall by the ball? (c) The ball is caught by a player who brings it to rest. In the process, her hand moves back 0.5 m. What is the impulse received by the player? (d) What average force was exerted on the player by the ball?

53 ••• The great limestone caverns were formed by dripping water. (a) If water droplets of 0.03 mL fall from a height of 5 m at a rate of 10 per minute, what is the average force exerted on the limestone floor by the droplets of water? (b) Compare this force to the weight of a water droplet.

54 ••• A favorite game at picnics is the egg toss. Two people toss a raw egg back and forth as they move farther apart. If the force required to break the egg's shell is about 5 N and the mass of the egg is 50 g, estimate the maximum separation distance for the egg throwers. Make whatever assumptions seem reasonable.

Collisions in One Dimension

55 • True or false:

(a) In any perfectly inelastic collision, all the kinetic energy of the bodies is lost.

(b) In a head-on elastic collision, the relative speed of recession after the collision equals the relative speed of approach before the collision.

(c) Kinetic energy is conserved in an elastic collision.

56 •• Under what conditions can all the initial kinetic energy of colliding bodies be lost in a collision?

57 •• Consider a perfectly inelastic collision of two objects of equal mass. (a) Is the loss of kinetic energy greater if the two objects have oppositely directed velocities of equal magnitude $v/2$, or if one of the two objects is initially at rest and the other has an initial velocity of v? (b) In which situation is the percentage loss in kinetic energy the greatest?

58 •• A mass m_1 traveling with a speed v makes a head-on elastic collision with a stationary mass m_2. In which scenario will the energy imparted to m_2 be greatest?

(a) $m_2 \ll m_1$
(b) $m_2 = m_1$
(c) $m_2 \gg m_1$
(d) None of the above.

59 • Joe and Sal decide that little Ronny is well-behaved enough to sit at the table with the family for Thanksgiving dinner. They are wrong. Ronny throws a 150-g handful of mashed potatoes horizontally with a speed of 5 m/s. It strikes a 1.2-kg gravy boat that is initially at rest on the frictionless table. If the potatoes stick to the gravy boat, what is the speed of the combined system as it slides down the table toward Grandpa?

60 • A 2000-kg car traveling to the right at 30 m/s is chasing a second car of the same mass that is traveling to the right at 10 m/s. (a) If the two cars collide and stick together, what is their speed just after the collision? (b) What fraction of the initial kinetic energy of the cars is lost during this collision? Where does it go?

61 • An 85-kg running back moving at 7 m/s makes a perfectly inelastic collision with a 105-kg linebacker who is initially at rest. What is the speed of the players just after their collision?

62 • A 5.0-kg object with a speed of 4.0 m/s collides head-on with a 10-kg object moving toward it with a speed of 3.0 m/s. The 10-kg object stops dead after the collision. (a) What is the final speed of the 5-kg object? (b) Is the collision elastic?

63 • A ball of mass m moves with speed v to the right toward a much heavier bat that is moving to the left with speed v. Find the speed of the ball after it makes an elastic collision with the bat.

64 •• During the Great Muffin Wars of '98, students from rival residences became familiar with the characteristics of various muffins. Mushy Pumpkin Surprise, for example, was good for temporarily blinding an attacker, while Mrs. O'Brien's Bran Muffins, having the density of lacrosse balls, were used more sparingly, and mainly as a deterrent. According to the rules, all muffins must have a mass of 0.3 kg. During one of the more memorable battles, a muffin moving to the right at 5 m/s collides with a muffin moving to the left at 2 m/s. Find the final velocities if (a) it is a perfectly inelastic collision of two pumpkin muffins, and (b) it is an elastic collision of two bran muffins.

65 •• Repeat Problem 64 with a second (illegal) muffin having a mass of 0.5 kg and moving to the right at 3 m/s.

66 •• A proton of mass m undergoes a head-on elastic collision with a stationary carbon nucleus of mass $12m$. The speed of the proton is 300 m/s. (a) Find the velocity of the center of mass of the system. (b) Find the velocity of the proton after the collision.

67 •• A 3-kg block moving at 4 m/s makes an elastic collision with a stationary block of mass 2 kg. Use conservation of momentum and the fact that the relative velocity of recession equals the relative velocity of approach to find the veloc-

ity of each block after the collision. Check your answer by calculating the initial and final kinetic energies of each block.

68 •• Night after night, Lucy is tormented by nocturnal wailing from the house next door. One day she seizes a revolver, stalks to the neighbors' window with a crazed look in her eye, takes aim, and fires a 10-g bullet into her target: a 1.2-kg saxophone that rests on a frictionless surface. The bullet passes right through and emerges on the other side with a speed of 100 m/s, and the saxophone is given a speed of 4 m/s. Find the initial speed of the bullet, and the amount of energy dissipated in its trip through the saxophone.

69 •• A block of mass $m_1 = 2$ kg slides along a frictionless table with a speed of 10 m/s. Directly in front of it, and moving in the same direction with a speed of 3 m/s, is a block of mass $m_2 = 5$ kg. A massless spring with spring constant $k = 1120$ N/m is attached to the second block as in Figure 8-54. (a) Before m_1 runs into the spring, what is the velocity of the center of mass of the system? (b) After the collision, the spring is compressed by a maximum amount Δx. What is the value of Δx? (c) The blocks will eventually separate again. What are the final velocities of the two blocks measured in the reference frame of the table?

Figure 8-54 Problem 69

70 •• A bullet of mass m is fired vertically from below into a block of wood of mass M that is initially at rest, supported by a thin sheet of paper. The bullet blasts through the block, which rises to a height of H above its initial position before falling back down. The bullet continues rising to height h. (a) Express the upward velocity of the bullet and the block immediately after the bullet exits the block in terms of h and H. (b) Use conservation of momentum to express the speed of the bullet before it enters the block of wood in terms of given parameters. (c) Obtain expressions for the mechanical energies of the system before and after the inelastic collision. (d) Express the energy dissipated in the block of wood in terms of $m, h, M,$ and H.

71 •• A proton of mass m is moving with initial speed v_0 toward an α particle of mass $4m$, which is initially at rest. Because both particles carry positive electrical charge, they repel each other. Find the speed v' of the α particle (a) when the distance between the two particles is least, and (b) when the two particles are far apart.

Ballistic Pendulums

72 •• A 16-g bullet is fired into the bob of a ballistic pendulum of mass 1.5 kg. When the bob is at its maximum height, the strings make an angle of 60° with the vertical. The length of the pendulum is 2.3 m. Find the speed of the bullet.

73 •• A bullet of mass m_1 is fired with a speed v into the bob of a ballistic pendulum of mass m_2. The bob is attached to a very light rod of length L that is pivoted at the other end. The bullet is stopped in the bob. Find the minimum v such that the bob will swing through a complete circle.

74 •• A bullet of mass m_1 is fired with a speed v into the bob of a ballistic pendulum of mass m_2. Find the maximum height h attained by the bob if the bullet passes through the bob and emerges with a speed $v/2$.

Exploding Objects and Radioactive Decay

75 •• A 3-kg bomb slides along a frictionless horizontal plane in the x direction at 6 m/s. It explodes into two pieces, one of mass 2 kg and the other of mass 1 kg. The 1-kg piece moves along the horizontal plane in the y direction at 4 m/s. (a) Find the velocity of the 2-kg piece. (b) What is the velocity of the center of mass after the explosion?

76 •• The beryllium isotope ^{4}Be is unstable and decays into two α particles (helium nuclei of mass $m = 6.68 \times 10^{-27}$ kg) with the release of 1.5×10^{-14} J of energy. Determine the velocities of the two α particles that arise from the decay of a ^{4}Be nucleus at rest.

77 •• The light isotope of lithium, ^{5}Li, is unstable and breaks up spontaneously into a proton (hydrogen nucleus) and an α particle (helium nucleus). In this process, a total energy of 3.15×10^{-13} J is released, appearing as the kinetic energy of the two reaction products. Determine the velocities of the proton and α particle that arise from the decay of a ^{5}Li nucleus at rest. (*Note:* The masses of the proton and alpha particle are $m_p = 1.67 \times 10^{-27}$ kg and $m_\alpha = 4m_p = 6.68 \times 10^{-27}$ kg.)

78 •• Jay and Dave decide that the best way to protest the opening of a new incinerator is to launch a stink bomb into the middle of the ceremony. They calculate that a 6-kg projectile launched with an initial speed of 40 m/s at an angle of 30° will do the trick. The bomb will explode on impact, no one will get hurt, but everyone will stink. Perfect. However, at the top of its flight, the bomb explodes into two fragments, each having a horizontal trajectory. To top it off—this really isn't their day—the 2-kg fragment lands right at the feet of Dave and Jay. (a) Where does the 4-kg fragment land? (b) Find the energy of the explosion by comparing the kinetic energy of the projectiles just before and just after the explosion.

79 •• A projectile of mass $m = 3$ kg is fired with initial speed of 120 m/s at an angle of 30° with the horizontal. At the top of its trajectory, the projectile explodes into two fragments of masses 1 kg and 2 kg. The 2-kg fragment lands on the ground directly below the point of explosion 3.6 s after the explosion. (a) Determine the velocity of the 1-kg fragment immediately after the explosion. (b) Find the distance between the point of firing and the point at which the 1-kg fragment strikes the ground. (c) Determine the energy released in the explosion.

80 ••• The boron isotope ^{9}B is unstable and disintegrates into a proton and two α particles. The total energy released as kinetic energy of the decay products is 4.4×10^{-14} J. In one such event, with the ^{9}B nucleus at rest prior to decay, the velocity of the proton is measured to be 6.0×10^6 m/s. If the two α particles have equal energies, find the magnitude and the direction of their velocities with respect to that of the proton.

The Coefficient of Restitution (optional)

81 • The coefficient of restitution for steel on steel is measured by dropping a steel ball onto a steel plate that is rigidly attached to the earth. If the ball is dropped from a height of 3 m and rebounds to a height of 2.5 m, what is the coefficient of restitution?

82 • According to the official rules of racquetball, a ball acceptable for tournament play must bounce to a height of between 173 and 183 cm when dropped from a height of 254 cm at room temperature. What is the acceptable range of values for the coefficient of restitution for the racquetball–floor system?

83 • A ball bounces to 80% of its original height. (a) What fraction of its mechanical energy is lost each time it bounces? (b) What is the coefficient of restitution of the ball–floor system?

84 •• A 2-kg object moving at 6 m/s collides with a 4-kg object that is initially at rest. After the collision, the 2-kg object moves backward at 1 m/s. (a) Find the velocity of the 4-kg object after the collision. (b) Find the energy lost in the collision. (c) What is the coefficient of restitution for this collision?

85 •• A 2-kg block moving to the right with speed 5 m/s collides with a 3-kg block that is moving in the same direction at 2 m/s, as in Figure 8-55. After the collision, the 3-kg block moves at 4.2 m/s. Find (a) the velocity of the 2-kg block after the collision, and (b) the coefficient of restitution for the collision.

Figure 8-55 Problem 85

Collisions in Three Dimensions (optional)

86 • In a pool game, the cue ball, which has an initial speed of 5 m/s, makes an elastic collision with the eight ball, which is initially at rest. After the collision, the eight ball moves at an angle of 30° with the original direction of the cue ball. (a) Find the direction of motion of the cue ball after the collision. (b) Find the speed of each ball. Assume that the balls have equal mass.

87 •• An object of mass $M_1 = m$ collides with velocity $v_0\hat{i}$ into an object of mass $M_2 = 2m$ with velocity $\frac{1}{2}v_0\hat{j}$. Following the collision, the mass m_2 has a velocity $\frac{1}{4}v_0\hat{i}$. (a) Determine the velocity of the mass m_1 after the collision. (b) Was this an elastic collision? If not, express the energy change in terms of m and v_0.

88 •• A puck of mass 0.5 kg approaches a second, similar puck that is stationary on frictionless ice. The initial speed of the moving puck is 2 m/s. After the collision, one puck leaves with a speed v_1 at 30° to the original line of motion; the

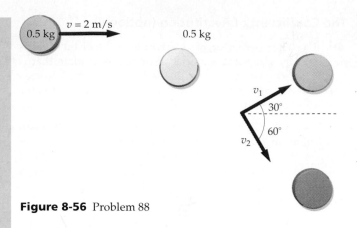

Figure 8-56 Problem 88

second puck leaves with speed v_2 at 60°, as in Figure 8-56. (a) Calculate v_1 and v_2. (b) Was the collision elastic?

89 •• Figure 8-57 shows the result of a collision between two objects of unequal mass. (a) Find the speed v_2 of the larger mass after the collision and the angle θ_2. (b) Show that the collision is elastic.

Figure 8-57 Problem 89

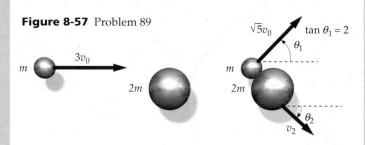

90 •• A ball moving at 10 m/s makes an off-center elastic collision with another ball of equal mass that is initially at rest. The incoming ball is deflected at an angle of 30° from its original direction of motion. Find the velocity of each ball after the collision.

91 ••• A particle has an initial speed v_0. It collides with a second particle that is at rest and is deflected through an angle ϕ. Its speed after the collision is v. The second particle recoils. Its velocity makes an angle θ with the initial direction of the first particle. (a) Show that

$$\tan \theta = \frac{v \sin \phi}{v_0 - v \cos \phi}$$

(b) Do you have to assume that the collision is either elastic or inelastic to get the result in part (a)?

The Center-of-Mass Reference Frame (optional)

92 • Describe a perfectly inelastic collision as viewed in the center-of-mass reference frame.

93 •• A particle with momentum p_1 in one dimension makes an elastic collision with a second particle of momentum $p_2 = -p_1$ in the center-of-mass reference frame. After the

collision its momentum is p_1'. Write the total initial and final energies in terms of p_1 and p_1' and show that $p_1' = \pm p_1$. If $p_1' = -p_1$, the particle is merely turned around by the collision and leaves with the speed it had initially. What is the significance of the plus sign in your solution?

94 •• A 3-kg block is traveling to the right at 5 m/s, and a 1-kg block is traveling to the left at 3 m/s. (a) Find the velocity v_{cm} of the center of mass. (b) Subtract v_{cm} from the velocity of each block to find the velocity of each block in the center-of-mass reference frame. (c) After they make an elastic collision, the velocity of each block is reversed in this frame. Find the velocity of each block after an elastic collision. (d) Transform back into the original frame by adding v_{cm} to the velocity of each block. (e) Check your result by finding the initial and final kinetic energies of the blocks in the original frame.

95 •• Repeat Problem 94 with a second block having a mass of 5 kg and moving to the right at 3 m/s.

Rocket Propulsion (optional)

96 •• A rocket burns fuel at a rate of 200 kg/s and exhausts the gas at a relative speed of 6 km/s. Find the thrust of the rocket.

97 •• The payload of a rocket is 5% of its total mass, the rest being fuel. If the rocket starts from rest and moves with no external forces acting on it, what is its final velocity if the exhaust velocity of its gas is 5 km/s?

98 •• A rocket moves in free space with no external forces acting on it. It starts from rest and has an exhaust speed of 3 km/s. Find the final velocity if the payload is (a) 20%, (b) 10%, (c) 1%.

99 •• A rocket has an initial mass of 30,000 kg, of which 20% is the payload. It burns fuel at a rate of 200 kg/s and exhausts its gas at a relative speed of 1.8 km/s. Find (a) the thrust of the rocket, (b) the time until burnout, and (c) its final speed assuming it moves upward near the surface of the earth where the gravitational field g is constant.

General Problems

100 • Why can friction and the force of gravity usually be neglected in collision problems?

101 • The condition necessary for the conservation of momentum of a given system is that

(a) energy is conserved.
(b) one object is at rest.
(c) no external force acts.
(d) internal forces equal external forces.
(e) the net external force is zero.

102 • As a pendulum bob swings back and forth, is the momentum of the bob conserved? Explain why or why not.

103 • A model-train car of mass 250 g traveling with a speed of 0.50 m/s links up with another car of mass 400 g that is initially at rest. What is the speed of the cars immediately after they have linked together? Find the initial and final kinetic energies.

104 • (a) Find the total kinetic energy of the two model-train cars of Problem 103 before they couple. (b) Find the initial velocities of the two cars relative to the center of mass of the system, and use them to calculate the initial kinetic energy of the system relative to the center of mass. (c) Find the kinetic energy of the center of mass. (d) Compare your answers for (b) and (c) with that for (a).

105 • A 4-kg fish is swimming at 1.5 m/s to the right. He swallows a 1.2-kg fish swimming toward him at 3 m/s. Neglecting water resistance, what is the velocity of the larger fish immediately after his lunch?

106 • A 3-kg block moves at 6 m/s to the right while a 6-kg block moves at 3 m/s to the right. Find (a) the total kinetic energy of the two-block system, (b) the velocity of the center of mass, (c) the center-of-mass kinetic energy, and (d) the kinetic energy relative to the center of mass.

107 • A 1500-kg car traveling north at 70 km/h collides at an intersection with a 2000-kg car traveling west at 55 km/h. The two cars stick together. (a) What is the total momentum of the system before the collision? (b) Find the magnitude and direction of the velocity of the wreckage just after the collision.

108 • The great white shark can have a mass as great as 3000 kg. Suppose such a shark is cruising the ocean when it spots a meal below it: a 200.0-kg fish swimming horizontally at 8.00 m/s. The shark rushes vertically downward at 3.00 m/s and swallows the prey at once. At what angle to the vertical θ will the shark be moving immediately after the snack? What is the final speed of the shark? (Neglect any drag effects of the water.)

109 • Repeat Problem 106 for a 3-kg block moving at 6 m/s to the right and a 6-kg block moving at 3 m/s to the left.

110 • Repeat Problem 106 for a 3-kg block moving at 10 m/s to the right and a 6-kg block moving at 1 m/s to the right.

111 •• A 60-kg woman stands on the back of a 6-m-long, 120-kg raft that is floating at rest in still water with no friction. The raft is 0.5 m from a fixed pier, as in Figure 8-58.

(a) The woman walks to the front of the raft and stops. How far is the raft from the pier now?
(b) While the woman walks, she maintains a constant speed of 3 m/s relative to the raft. Find the total kinetic energy of the system (woman plus raft), and compare with the kinetic energy if the woman walked at 3 m/s on a raft *tied to the pier*.

Figure 8-58 Problem 111

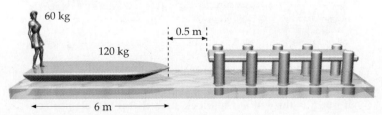

Figure 8-59 Problem 112

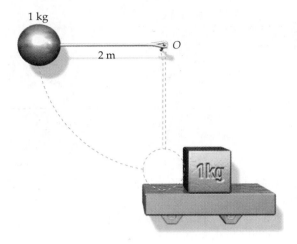

(c) Where does this energy come from, and where does it go when the woman stops at the front of the raft?
(d) On land, the woman can put a lead shot 6 m. She stands at the back of the raft, aims forward, and puts the shot so that just after it leaves her hand, it has the same velocity *relative to her* as it does when she throws it from the ground. Where does the shot land?

112 •• A 1-kg steel ball and a 2-m cord of negligible mass make up a simple pendulum that can pivot without friction about the point O, as in Figure 8-59. This pendulum is released from rest in a horizontal position and when the ball is at its lowest point it strikes a 1-kg block sitting at rest on a rough shelf. Assume that the collision is perfectly elastic and take the coefficient of friction between the block and shelf to be 0.1. (a) What is the velocity of the block just after impact? (b) How far does the block move before coming to rest?

113 •• In World War I, the most awesome weapons of war were huge cannons mounted on railcars. Figure 8-60 shows such a cannon, mounted so that it will project a shell at an angle of 30°. With the car initially at rest, the cannon fires a 200-kg projectile at 125 m/s. Now consider a system composed of a cannon, shell, and railcar, all rolling on the track without frictional losses. (a) Will the total vector momentum of that system be the same (i.e., "conserved") before and after the shell is fired? Explain your answer in a few words. (b) If the mass of the railcar plus cannon is 5000 kg, what will be the recoil velocity of the car along the track after the firing? (c) The shell is observed to rise to a maximum height of 180 m as it moves through its trajectory. At this point, its speed is 80 m/s. On the basis of this information, calculate the amount of thermal energy produced by air friction on the shell on its way from firing to this maximum height.

Figure 8-60 Problem 113

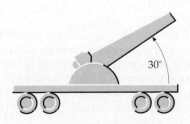

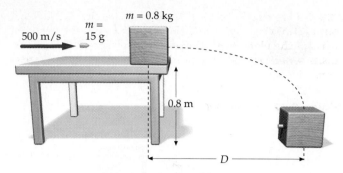

Figure 8-61 Problem 114

114 •• A 15-g bullet traveling at 500 m/s strikes an 0.8-kg block of wood that is balanced on a table edge 0.8 m above the ground (Figure 8-61). If the bullet buries itself in the block, find the distance D at which the block hits the floor.

115 •• In hand-pumped railcar races, a speed of 32 km/h has been achieved by teams of four. A car of mass 350 kg is moving at that speed toward a river when Carlos, the chief pumper, notices that the bridge ahead is out. All four people (of mass 75 kg each) jump simultaneously backward off the car with a velocity that has a horizontal component of 4 m/s *relative to the car after jumping*. The car proceeds off the bank and falls in the water a distance 25.0 m off the bank. (*a*) Estimate the time of the fall of the railcar. (*b*) What happens to the team of pumpers?

116 •• A constant force $\vec{F} = 12 \text{ N} \vec{i}$ is applied to the 8-kg mass of Problem 5 at $t = 0$. (*a*) What is the velocity of the center of mass of the three-particle system at $t = 5$ s? (*b*) What is the location of the center of mass at $t = 5$ s?

117 •• Two particles of mass m and $4m$ are moving in a vacuum at right angles as in Figure 8-62. A force $\vec{F}$ acts on both particles for a time T. As a result, the velocity of the particle m is $4v$ in its original direction. Find the new velocity $\vec{v}'$ of the particle of mass $4m$.

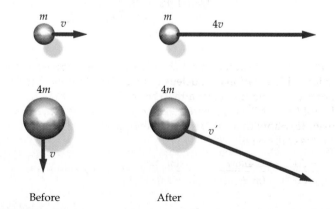

Before   After

Figure 8-62 Problem 117

118 •• An open railroad car of mass 20,000 kg is rolling without friction at 5 m/s along a level track when it starts to rain. After the car has collected 2000 kg of water, the rain stops. (*a*) What is the car's velocity? (*b*) As the car is rolling along, the water begins leaking out of a hole in the bottom at a rate of 5 kg/s. What is the velocity after half the water has

leaked out? (*c*) What is the velocity after all the water has leaked out?

119 •• In the "slingshot effect," the transfer of energy in an elastic collision is used to boost the energy of a space probe so that it can escape from the solar system. Figure 8-63 shows a space probe moving at 10.4 km/s (relative to the sun) toward Saturn, which is moving at 9.6 km/s (relative to the sun) toward the probe. Because of the gravitational attraction between Saturn and the probe, the probe swings around Saturn and heads back in the opposite direction with speed v_f. (*a*) Assuming this collision to be a one-dimensional elastic collision with the mass of Saturn much greater than that of the probe, find v_f. (*b*) By what factor is the kinetic energy of the probe increased? Where does this energy come from?

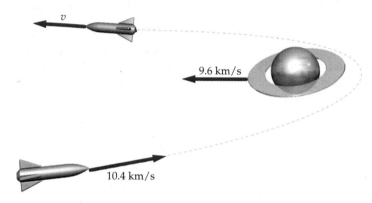

Figure 8-63 Problem 119

120 •• You (mass 80 kg) and your friend (mass unknown) are in a rowboat (mass 60 kg) on a calm lake. You are at the center of the boat rowing and she is at the back, 2 m from the center. You get tired and stop rowing. She offers to row and after the boat comes to rest, you change places. You notice that after changing places the boat has moved 20 cm relative to a fixed log. What is your friend's mass?

121 •• A small car of mass 800 kg is parked behind a small truck of mass 1600 kg on a level road (Figure 8-64). The brakes of both the car and the truck are off so that they are free to roll with negligible friction. A man sitting on the tailgate of the truck shoves the car away by exerting a constant force on the car with his feet. The car accelerates at 1.2 m/s². (*a*) What is the acceleration of the truck? (*b*) What is the magnitude of the force exerted on either the truck or the car?

Figure 8-64 Problem 121

122 •• A 13-kg block is at rest on a level floor. A 400-g glob of putty is thrown at the block such that it travels horizontally, hits the block, and sticks to it. The block and putty slide 15 cm along the floor. If the coefficient of sliding friction is 0.4, what is the initial speed of the putty?

123 •• A careless driver rear-ends a car that is halted at a stop sign. Just before impact, the driver slams on his brakes, locking the wheels. The driver of the struck car also has his foot solidly on the brake pedal, locking his brakes. The mass of the struck car is 900 kg, and that of the initially moving vehicle is 1200 kg. On collision, the bumpers of the two cars mesh. Police determine from the skid marks that after the collision the two cars moved 0.76 m together. Tests revealed that the coefficient of sliding friction between the tires and pavement was 0.92. The driver of the moving car claims that he was traveling at less than 15 km/h as he approached the intersection. Is he telling the truth?

124 •• A pendulum consists of a 0.4-kg bob attached to a string of length 1.6 m. A block of mass M rests on a horizontal frictionless surface (Figure 8-65). The pendulum is released from rest at an angle of 53° with the vertical and the bob collides elastically with the block. Following the collision, the maximum angle of the pendulum with the vertical is 5.73°. Determine the mass M.

Figure 8-65 Problem 124

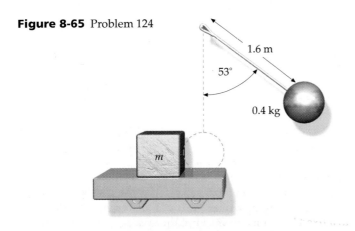

1.6 m

53°

0.4 kg

m

125 •• Initially, mass $m = 1.0$ kg and mass M are both at rest on a frictionless inclined plane (Figure 8-66). Mass M rests against a spring that has a spring constant of 11,000 N/m. The distance along the plane between m and M is 4.0 m. Mass m is released, makes an elastic collision with mass M, and rebounds a distance of 2.56 m back up the inclined plane. Mass M comes to rest momentarily 4.0 cm from its initial position. Find the mass M.

Figure 8-66 Problem 125

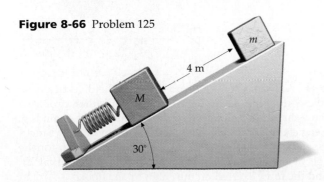

4 m

m

M

30°

126 •• A circular plate of radius r has a circular hole cut out of it having radius $r/2$ (Figure 8-67). Find the center of mass of the plate. *Hint:* The hole can be represented by two disks superimposed, one of mass m and the other of mass $-m$.

Figure 8-67 Problem 126

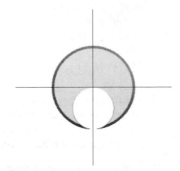

127 •• Using the hint from Problem 126, find the center of mass of a solid sphere of radius r that has a spherical cavity of radius $r/2$, as in Figure 8-68.

Figure 8-68 Problem 127

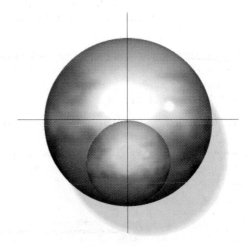

128 •• A neutron of mass m makes an elastic head-on collision with a stationary nucleus of mass M. (a) Show that the energy of the nucleus after the collision is $K_{nucleus} = [4mM/(m + M)^2]K_n$, where K_n is the initial energy of the neutron. (b) Show that the fraction of energy lost by the neutron in this collision is

$$\frac{-\Delta K_n}{K_n} = \frac{4mM}{(m + M)^2} = \frac{4(m/M)}{(1 + m/M)^2}$$

129 •• The mass of a carbon nucleus is approximately 12 times the mass of a neutron. (a) Use the results of Problem 128 to show that after N head-on collisions of a neutron with carbon nuclei at rest, the energy of the neutron is approximately $0.716^N E_0$, where E_0 is its original energy. Neutrons emitted in the fission of a uranium nucleus have an energy of about 2 MeV. For such a neutron to cause the fission of another uranium nucleus in a reactor, its energy must be reduced to about 0.02 eV. (b) How many head-on collisions are needed to reduce the energy of a neutron from 2 MeV to 0.02 eV, assuming elastic head-on collisions with stationary carbon nuclei?

130 •• On average, a neutron loses 63% of its energy in an elastic collision with a hydrogen atom and 11% of its energy in an elastic collision with a carbon atom. The numbers are lower than the ones we have been using in earlier problems because most collisions are not head-on. Calculate the number of collisions, on average, needed to reduce the energy of a neutron from 2 MeV to 0.02 eV (a desirable outcome for reasons explained in Problem 129) if the neutron collides with (a) hydrogen atoms and (b) carbon atoms.

131 •• A rope of length L and mass M lies coiled on a table. Starting at $t = 0$, one end of the rope is lifted from the table with a force F such that it moves with a constant velocity v. (a) Find the height of the center of mass of the rope as a function of time. (b) Differentiate your result in (a) twice to find the acceleration of the center of mass. (c) Assuming that the force exerted by the table equals the weight of the rope still there, find the force F you exert on the top of the rope.

132 •• A tennis ball of mass m_t is held a small distance above a basketball of mass m_b. Both are dropped from a height h above the floor. (Take h to be the distance to the center of the basketball.) The basketball collides elastically with the floor. Find the speed v_t of the tennis ball after it then collides elastically with the basketball. Calculate the height reached by the tennis ball if $m_b = 0.480$ kg, $m_t = 0.060$ kg, and $h = 2$ m. (*Caution:* If you try this experimentally, get out of the way of the tennis ball!)

133 •• Repeat Problem 24 if the cup has a mass m_c and the ball collides with it inelastically.

134 •• Two astronauts at rest face each other in space. One, with mass m_1, throws a ball of mass m_b to the other, whose mass is m_2. She catches the ball and throws it back to the first astronaut. If they each throw the ball with a speed of v relative to themselves, how fast are they moving after each has made one throw and one catch?

135 •• The ratio of the mass of the earth to the mass of the moon is $M_e/m_m = 81.3$. The radius of the earth is about 6370 km, and the distance from the earth to the moon is about 384,000 km. (a) Locate the center of mass of the earth–moon system relative to the surface of the earth. (b) What external forces act on the earth–moon system? (c) In what direction is the acceleration of the center of mass of this system? (d) Assume that the center of mass of this system moves in a circular orbit around the sun. How far must the center of the earth move in the radial direction (toward or away from the sun) during the 14 days between the time the moon is farthest from the sun (full moon) and the time it is closest to the sun (new moon)?

136 •• You wish to enlarge a skating surface so you stand on the ice at one end and aim a hose horizontally to spray water on the schoolyard pavement. Water leaves the hose at 2.4 kg/s with speed 30 m/s. If your mass is 75 kg, what is your recoil acceleration? (Neglect friction and the mass of the hose.)

137 •• A neutron at rest decays into a proton plus an electron. The conservation of momentum implies that the electron and proton should have equal and opposite momentum. However, experimentally they do not. This apparent nonconservation of momentum led Wolfgang Pauli to suggest in 1931 that there was a third, unseen particle emitted in the decay. This particle is called a neutrino, and it was finally observed directly in 1957. Suppose that the electron has momentum $p = 4.65 \times 10^{-22}$ kg·m/s along the negative x direction and the proton ($m = 1.67 \times 10^{-27}$ kg) moves with speed 2.93×10^5 m/s at an angle 17.9° above the x axis. Find the momentum of the neutrino. (The kinetic energy of the electron is comparable to its rest energy, so its energy and momentum are related relativistically rather than classically. However, the rest energy of the proton is large compared with its kinetic energy so the classical relation $E = \frac{1}{2}mv^2 = p^2/2m$ is valid.)

138 ••• A stream of glass beads, each with a mass of 0.5 g, comes out of a horizontal tube at a rate of 100 per second (Figure 8-69). The beads fall a distance of 0.5 m to a balance pan and bounce back to their original height. How much mass must be placed in the other pan of the balance to keep the pointer at zero?

Figure 8-69 Problem 138

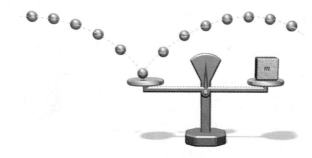

139 ••• A dumbbell consisting of two balls of mass m connected by a massless rod of length L rests on a frictionless floor against a frictionless wall until it begins to slide down the wall as in Figure 8-70. Find the speed v of the bottom ball at the moment when it equals the speed of the top one.

Figure 8-70 Problem 139

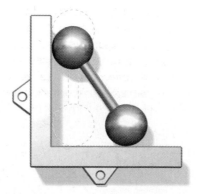

140 ••• A chain of length L and mass m is held vertically so that the bottom link just touches the floor. It is then dropped. (a) Show that the acceleration of the top end of the chain is g. (b) If the chain is moving downward with speed v at time t, and speed $v + \Delta v$ at time $t + \Delta t$, find an expression for the change in momentum of the chain during the interval Δt. (c) Find the force exerted on the chain by the floor.

9

Rotation

Star tracks in a time exposure of the night sky.

Rotational motion is all around us from molecules to galaxies. The earth rotates about its axis. Wheels, gears, propellers, motors, the drive shaft in a car, a CD in its player, a pirouetting ice skater, all rotate. Our study of rotation is simplified by analogies between linear motion and rotational motion. In this chapter, we consider rotation about an axis that is fixed in space, or one that is moving parallel to itself as in a rolling ball. More general examples of rotational motion are discussed in Chapter 10.

9-1   Angular Velocity and Angular Acceleration

Imagine a disk rotating about a fixed axis perpendicular to the disk and through its center (Figure 9-1). Points near the rim move faster than points near the axis. But when a point near the rim moves through a complete circle, so does any other point on the disk. As the disk rotates through a given angle, *all* points on the disk rotate through the same angle. The angle through which a disk rotates is a characteristic of the disk as a whole, as is the rate at which the angle changes. As the disk turns, the distance between any two

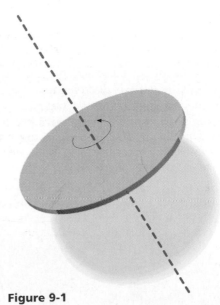

Figure 9-1

particles that make up the disk remains fixed. Such a system is called a **rigid body.**

Consider a typical particle in the disk (Figure 9-2). Let r_i be the distance from the center of the disk to the ith particle, and θ_i be the angle measured counterclockwise from a fixed reference line in space to a line from the center to the particle. As the disk rotates through an angle $d\theta$, the particle moves through a circular arc of length

$$ds_i = r_i|d\theta| \qquad\qquad 9\text{-}1$$

where $d\theta$ is measured in radians. The distance ds_i varies from particle to particle, but the angle $d\theta$, called the **angular displacement,** is the same for all particles of the disk. For one complete revolution, the arc length Δs_i is $2\pi r$ and the angular displacement $\Delta\theta$ is

$$\Delta\theta = \frac{2\pi r_i}{r_i} = 2\pi\,\text{rad} = 360° = 1\,\text{rev}$$

The time rate of change of the angle, $d\theta/dt$, is the same for all particles of the disk, and is called the **angular velocity** ω of the disk:

$$\omega = \frac{d\theta}{dt} \qquad\qquad 9\text{-}2$$

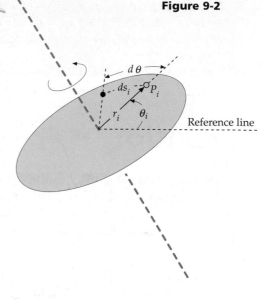

Figure 9-2

Definition—Angular velocity

For counterclockwise rotation, θ increases, so ω is positive. For clockwise rotation, θ decreases and ω is negative. The units of ω are radians per second. Since radians are dimensionless, the dimensions of angular velocity are those of reciprocal time (T^{-1}). The magnitude of the angular velocity is called the **angular speed.** We often use revolutions per minute (rev/min) to describe rotation. To convert between revolutions, radians, and degrees, we use

$$1\,\text{rev} = 2\pi\,\text{rad} = 360°$$

Exercise A CD-ROM disc is rotating at 3000 revolutions per minute. What is its angular speed in radians per second? (*Answer* 314 rad/s)

The time rate of change of angular velocity is called the **angular acceleration** α:

$$\alpha = \frac{d\omega}{dt} = \frac{d^2\theta}{dt^2} \qquad\qquad 9\text{-}3$$

Definition—Angular acceleration

The units of α are radians per second per second (rad/s^2). If ω is increasing, α is positive; if ω is decreasing, α is negative.

The linear velocity of a particle on the disk is tangent to the circular path of the particle and has magnitude $v_{it} = ds_i/dt$. We can relate this *tangential velocity* to the angular velocity of the disk using Equations 9-1 and 9-2:

$$v_{it} = \frac{ds_i}{dt} = \frac{r_i\,d\theta}{dt} = r_i\omega \qquad\qquad 9\text{-}4$$

Similarly, the tangential acceleration of a particle on the disk is

$$a_{it} = \frac{dv_{it}}{dt} = r_i\frac{d\omega}{dt}$$

so

$$a_{it} = r_i\alpha \qquad\qquad 9\text{-}5$$

A tiny device called a wobble motor. This motor, which has a diameter of the order of a millimeter, has achieved angular speeds in excess of 120,000 rev/min. The edge of a dime is visible in the background.

Each particle of the disk also has a radial acceleration, the centripetal acceleration, which points inward along the radial line, and has the magnitude

$$a_{ic} = \frac{v_{it}^2}{r_i} = \frac{(r_i\omega)^2}{r_i} = r_i\omega^2$$ 9-6

Exercise A point on the rim of a compact disc is 6.0 cm from the axis of rotation. Find the tangential speed v_t, tangential acceleration a_t, and centripetal acceleration a_c of the point when the disc is rotating at a constant angular speed of 300 rev/min. (*Answers* $v_t = 188$ cm/s, $a_t = 0$, $a_c = 5.92 \times 10^3$ cm/s^2)

handwritten: w $w = (31.41)^2 \cdot 6 = 5.92 \times 10^3$

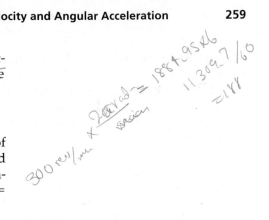

handwritten (right margin):
$300 \text{ rev}/\text{min} \times 2\pi \text{ rad} = 1884.95 \times 6$
s min
$\frac{11309.7}{60}$
$= 188$

The angular displacement θ, angular velocity ω, and angular acceleration α are analogous to the linear displacement x, linear velocity v, and linear acceleration a in one-dimensional motion. If the angular acceleration α is constant, we can integrate Equation 9-3 to find ω:

$$\omega = \omega_0 + \alpha t$$ 9-7

where the constant of integration ω_0 is the initial angular velocity. This is the rotational analog of $v = v_0 + at$. Integrating again, we obtain

$$\theta = \theta_0 + \omega_0 t + \frac{1}{2}\alpha t^2$$ 9-8

which is the rotational analog of $x = x_0 + v_0 t + \frac{1}{2}at^2$ with θ replacing x, ω replacing v, and α replacing a. Similarly, by eliminating t from Equations 9-7 and 9-8, we get

$$\omega^2 = \omega_0^2 + 2\alpha(\theta - \theta_0)$$ 9-9

which is the rotational analog of $v^2 = v_0^2 + 2a(x - x_0)$. The equations for constant angular acceleration have the same form as those for constant linear acceleration.

Example 9-1

A compact disc rotates from rest to 500 rev/min in 5.5 s. (*a*) What is its angular acceleration, assuming that it is constant? (*b*) How many revolutions does it make in 5.5 s? (*c*) How far does a point on the rim 6 cm from the center travel during the 5.5 s it takes to get to 500 rev/min?

Picture the Problem Part (*a*) is analogous to the linear problem of finding the acceleration given the final velocity. To find α in rad/s^2 we need to convert ω to rad/s. Part (*b*) is analogous to finding the distance traveled in a given time.

handwritten (right margin):
$\omega = \omega_0 \text{ rotat}$ $\alpha(5.5 \text{ sec})$
$500 = \alpha(5.5 \text{ sec})$
$\alpha = \frac{500}{5.5 \text{ sec}}$

(*a*)1. The angular acceleration is related to the initial and final angular velocities: $\omega = \omega_0 + \alpha t = 0 + \alpha t$

2. Solve for α: $\alpha = \dfrac{\omega}{t} = \dfrac{500 \text{ rev/min}}{5.5 \text{ s}} \times \dfrac{2\pi \text{ rad}}{1 \text{ rev}} \times \dfrac{1 \text{ min}}{60 \text{ s}} = 9.52 \text{ rad/s}^2$

(*b*)1. The angular displacement is related to the time by Equation 9-8: $\theta - \theta_0 = \omega_0 t + \dfrac{1}{2}\alpha t^2$

$= 0 + \dfrac{1}{2}(9.52 \text{ rad/s}^2)(5.5 \text{ s})^2 = 144 \text{ rad}$

2. Convert radians to revolutions: $144 \text{ rad} \times \dfrac{1 \text{ rev}}{2\pi \text{ rad}} = 22.9 \text{ rev}$

(c)   The distance traveled Δs is r times the angular displacement: $\Delta s = r\,\Delta\theta = (6\text{ cm})(144\text{ rad}) = 8.64\text{ m}$

Check the Result The average angular velocity in revolutions per minute is 250 rev/min. In 5.5 s, the compact disc rotates $(250\text{ rev}/60\text{ s})(5.5\text{ s}) = 22.9$ rev.

Remarks A compact disc is scanned by a laser that begins at the inner radius of about 2.4 cm and moves out to the edge at 6.0 cm. As the laser moves outward, the angular velocity of the disc decreases from 500 rev/min to 200 rev/min so that the linear (tangential) velocity of the disc at the point where the laser beam strikes remains constant.

Exercise Convert 500 rev/min to rad/s. (*Answer* 500 rev/min = 52.4 rad/s)

Exercise Check the result of part (*b*) in the example using $\omega^2 = \omega_0^2 + 2\alpha(\theta - \theta_0)$.

Exercise Find the linear speed of a point on the disc at (*a*) $r = 2.4$ cm when the disc rotates at 500 rev/min, and (*b*) $r = 6.0$ cm when the disc rotates at 200 rev/min. (*Answers* (*a*) 126 cm/s, (*b*) 126 cm/s)

9-2   Torque, Moment of Inertia, and Newton's Second Law for Rotation

To set a top spinning, you twist it. In Figure 9-3, a disk is set spinning by the forces $\vec{F}_1$ and $\vec{F}_2$ exerted at the edges of the disk. The points at which these forces are applied is important. The same forces applied so that their lines of action pass through the center of the disk, as in Figure 9-4, will not spin the disk. Figure 9-5 shows a single force $\vec{F}_i$ acting on the ith particle of a disk. The perpendicular distance between the line of action of a force and the axis of rotation is called the **lever arm** ℓ of the force. A force times its lever arm is the magnitude of the **torque** τ_i:

$$\tau_i = F_i\ell$$ 9-10

Torque can be thought of as a twist, just as a force is a push or a pull. It is the torque that affects the angular velocity of the object.

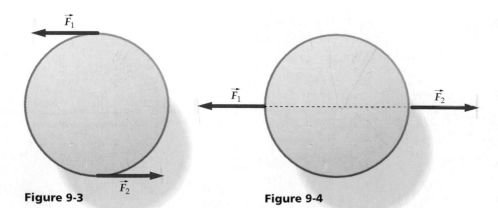

Figure 9-3

Figure 9-4

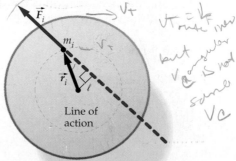

Figure 9-5 The force $\vec{F}_i$ produces a torque $F_i\ell$ about the center.

The lever arm of the force in Figure 9-5 is $\ell = |r_i \sin\phi|$, where ϕ is the angle between $\vec{F}_i$ and the position vector $\vec{r}_i$ from the center of the disk to the point of application of the force. The torque exerted by this force is

$$\tau_i = F_i r_i \sin\phi \qquad 9\text{-}11$$

The torque is taken to be positive if it tends to rotate the disk counterclockwise, and negative if it tends to rotate the disk clockwise. In Figure 9-6, $\vec{F}_i$ is resolved into two components, $F_{ir} = F_i \cos\phi$ along the radial line $\vec{r}_i$ and $F_{it} = F_i \sin\phi$ perpendicular to the radial line. The radial component has no effect on the rotation of the disk. The torque exerted by $\vec{F}_i$ can be written in terms of F_{it}. From Equation 9-10,

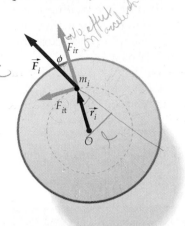

Figure 9-6

$$\tau_i = F_i \ell = F_i r_i \sin\phi = F_{it} r_i \qquad 9\text{-}12$$

We now show that a rigid body's angular acceleration is proportional to the net torque acting on it. Let $\vec{F}_i$ be the net external force acting on the ith particle. The tangential acceleration of the ith particle is, by Newton's second law,

$$F_{it} = m_i a_{it} = m_i r_i \alpha \qquad 9\text{-}13$$

where we have used $a_{it} = r_i \alpha$ (Equation 9-5). Multiply each side by r_i:

$$r_i F_{it} = m_i r_i^2 \alpha \qquad 9\text{-}14$$

The left side of Equation 9-14 is the torque τ_i exerted by the force $\vec{F}_i$ about the pivot O. So

$$\tau_i = m_i r_i^2 \alpha \qquad 9\text{-}15$$

Summing over all the particles in the object gives

$$\sum_i \tau_i = \sum_i m_i r_i^2 \alpha \qquad 9\text{-}16$$

$\sum \tau_i$ is the net torque acting on the object. For a rigid body, the angular acceleration is the same for all the particles of the object and can therefore be taken out of the sum. The quantity $\sum m_i r_i^2$ is called the **moment of inertia I.** For a continuous object, the sum is replaced by an integral.

$$I = \sum_i m_i r_i^2 \qquad \text{(system of particles)} \qquad 9\text{-}17$$

$$I = \int r^2 dm \qquad \text{(continuous object)} \qquad 9\text{-}18$$

Definition—Moment of inertia

In Equation 9-17, r_i is the distance of the ith particle from the axis of rotation, and in Equation 9-18, r is the distance of the mass element dm from the axis of rotation. For a disk with the origin at the center on the axis, this is the same as the distance to the origin.

Writing I for the moment of inertia, Equation 9-16 becomes

$$\sum_i \tau_i = I\alpha \qquad 9\text{-}19$$

In Chapter 8, we saw that the net force acting on a system of particles is equal to the net *external* force acting on the system because the internal forces (those exerted by the particles within the system on one another) cancel in pairs. The treatment of internal torques exerted by the particles within a system on one another leads to a similar result, that is, the net torque acting on a system equals the net *external* torque acting on the system. (We discuss this further in Chapter 10.) We can thus write Equation 9-19 as

$$\tau_{net,ext} = \sum_i \tau_{i,ext} = I\alpha \qquad \text{9-20}$$

Newton's second law for rotation

This is the rotational analog of Newton's second law for linear motion, $\sum \vec{F} = m\vec{a}$.

The torque exerted by a wrench on a nut is proportional to the force and to the lever arm. Charlie could exert a greater torque with the same force if he held the wrenches nearer their ends.

9-3   Calculating the Moment of Inertia

The moment of inertia is a measure of the resistance of an object to changes in its rotational motion. It is the rotational analog of mass. The moment of inertia depends on the distribution of mass within the object relative to the axis of rotation. The farther the mass from the axis, the greater the moment of inertia. Thus, unlike the mass of an object, which is a property of the object itself, the moment of inertia of an object also depends on the location of the axis of rotation.

Systems of Particles

For systems consisting of discrete particles, we can compute the moment of inertia about a given axis directly from Equation 9-17.

Axis of rotation

Example 9-2

Four particles of mass m are connected by massless rods to form a rectangle of sides $2a$ and $2b$ as shown. The system rotates about an axis in the plane of the figure through the center (Figure 9-7). Find the moment of inertia about this axis.

Picture the Problem Since we are given that the objects are particles, we use Equation 9-17. In that equation, r_i is the perpendicular distance from the particle of mass m_i to the axis (line) of rotation.

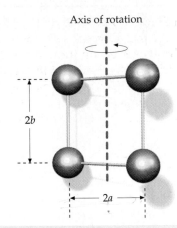

Figure 9-7

1. Apply the definition of moment of inertia for discrete particles (Equation 9-17):

$$I = \sum_i m_i r_i^2 = m_1 r_1^2 + m_2 r_2^2 + m_3 r_3^2 + m_4 r_4^2$$

2. The masses m_i and the distances r_i are given:

$$m_1 = m_2 = m_3 = m_4 = m$$

$$r_1 = r_2 = r_3 = r_4 = a$$

3. Substitution gives the moment of inertia:

$$I = ma^2 + ma^2 + ma^2 + ma^2 = 4ma^2$$

Remark Notice that I is independent of the length b, which has no effect on how far the masses are from the axis of rotation. This example and exercise illustrate the fact that the moment of inertia depends on the location of the axis of rotation. In Figure 9-8, the moment of inertia, represented by the dimensionless quantity I/ma^2, is plotted versus the distance from the two left particles to the axis of rotation. Note that the moment of inertia is a minimum when the axis is directly in the middle.

Exercise Find the moment of inertia of this system for rotation about an axis parallel to the first axis but passing through two of the particles as shown in Figure 9-9. (*Answer* $I = 8ma^2$)

Figure 9-9

Axis of rotation

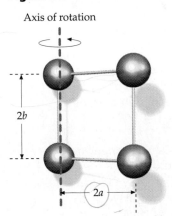

Figure 9-8

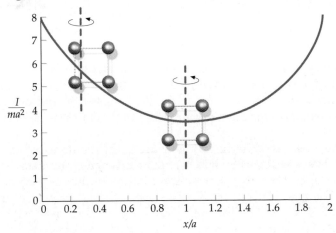

Continuous Objects

To calculate the moment of inertia for continuous objects, we use Equation 9-18 where r is the distance of the mass element dm from the axis of rotation. Table 9-1 lists the moments of inertia of various uniform objects. We now give several more examples of calculating I for continuous objects.

Table 9-1

Moments of Inertia of Uniform Bodies of Various Shapes

Cylindrical shell about axis

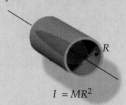

$$I = MR^2$$

Solid cylinder about axis

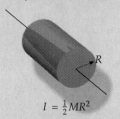

$$I = \tfrac{1}{2} MR^2$$

Hollow cylinder about axis

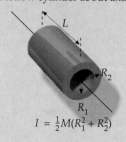

$$I = \tfrac{1}{2} M(R_1^2 + R_2^2)$$

Cylindrical shell about diameter through center

$$I = \tfrac{1}{2} MR^2 + R \tfrac{1}{12} ML^2$$

Solid cylinder about diameter through center

$$I = \tfrac{1}{4} MR^2 + \tfrac{1}{12} ML^2$$

Thin rod about perpendicular line through center

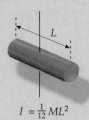

$$I = \tfrac{1}{12} ML^2$$

Thin rod about perpendicular line through one end

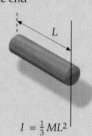

$$I = \tfrac{1}{3} ML^2$$

Thin spherical shell about diameter

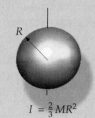

$$I = \tfrac{2}{3} MR^2$$

Solid sphere about diameter

$$I = \tfrac{2}{5} MR^2$$

Solid rectangular parallelpiped about axis through center perpendicular to face

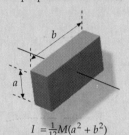

$$I = \tfrac{1}{12} M(a^2 + b^2)$$

Example 9-3

Figure 9-10

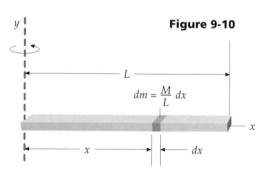

Find the moment of inertia of a uniform stick of length L and mass M about an axis perpendicular to the stick and through one end (Figure 9-10). Assume that the stick has negligible thickness.

Picture the Problem Let the stick lie along the x axis with its end at the origin. To calculate I_y about the y axis, we choose a mass element dm at a distance x from the axis. Since the total mass M is uniformly distributed along the length L, the mass per unit length (linear mass density) is $\lambda = M/L$.

1. The moment of inertia is given by the integral:

$$I = \int_0^L x^2\, dm$$

2. Write dm in terms of the mass density λ and dx:

$$dm = \lambda\, dx = \frac{M}{L}\, dx$$

3. Substitute and perform the integration:

$$I_y = \int_0^L x^2\, dm = \int_0^L x^2 \frac{M}{L}\, dx = \frac{M}{L} \int_0^L x^2\, dx$$

$$= \frac{M}{L} \frac{1}{3} x^3 \Big|_0^L = \frac{M}{L} \frac{L^3}{3} = \frac{1}{3} ML^2$$

Remark The moment of inertia about the z axis is also $\tfrac{1}{3} ML^2$ and that about the x axis is zero, assuming that all of the mass is right on the x axis.

Hoop About a Perpendicular Axis Through Its Center Assume that a hoop has mass M and radius R (Figure 9-11). The axis of rotation is the axis of the hoop, which is perpendicular to the plane of the hoop. All the mass is at a distance $r = R$, and the moment of inertia is

$$I = \int r^2\, dm = R^2 \int dm = MR^2$$

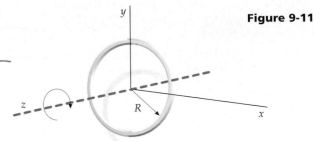

Figure 9-11

Uniform Disk About a Perpendicular Axis Through Its Center For the case of a uniform disk, we expect that I will be smaller than MR^2 since the mass is uniformly distributed from $r = 0$ to $r = R$ rather than being concentrated at $r = R$ as it is in a hoop. In Figure 9-12, each mass element is a hoop of radius r and thickness dr. The moment of inertia of any given mass element is $r^2\, dm$. Since the area of each element is $dA = 2\pi r\, dr$, the mass of each element is

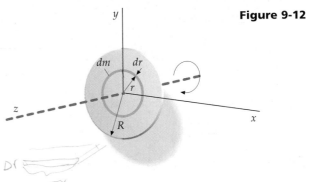

Figure 9-12

$$dm = \frac{M}{A}\, dA = \frac{M}{A}\, 2\pi r\, dr$$

where $A = \pi R^2$ is the area of the disk. We thus have

$$I = \int r^2\, dm = \int_0^R r^2 \frac{M}{A}\, 2\pi r\, dr$$

$$= \frac{2\pi M}{\pi R^2} \int_0^R r^3\, dr = \frac{2M}{R^2}\frac{R^4}{4} = \frac{1}{2} MR^2$$

Uniform Cylinder About Its Axis We consider a cylinder to be a set of disks, each with mass dm and moment of inertia $\frac{1}{2} dm\, R^2$ (Figure 9-13). The moment of inertia of the complete cylinder is then

$$I = \int \frac{1}{2}\, dm\, R^2 = \frac{1}{2} R^2 \int dm = \frac{1}{2} MR^2$$

where M is the total mass of the cylinder.

Figure 9-13

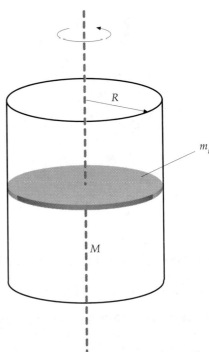

The Parallel-Axis Theorem

We can often simplify the calculation of moments of inertia for various bodies by using the **parallel-axis theorem,** which relates the moment of inertia about an axis through the center of mass of an object to the moment of inertia about a second, parallel axis (Figure 9-14). Let I_{cm} be the moment of inertia about an axis through the center of mass of an object of total mass M, and let I be that about a parallel axis a distance h away. The parallel-axis theorem states that

$$I = I_{cm} + Mh^2 \qquad\qquad\qquad 9\text{-}21$$

Parallel-axis theorem

Example 9-2 and the exercise following it illustrated a special case of this theorem with $h = a$, $M = 4m$, and $I_{cm} = 4ma^2$. A proof of the parallel-axis theorem is given at the end of this section.

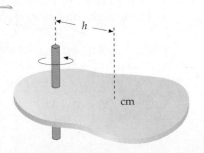

Figure 9-14 An object rotating about an axis parallel to an axis through the center of mass and a distance h from it.

Example 9-4	*try it yourself*

Figure 9-15

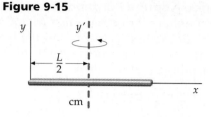

Find the moment of inertia of a stick of uniform density about the y' axis through the center of mass (Figure 9-15).

Picture the Problem Here you know $I = \frac{1}{3}ML^2$ about one end and wish to find I_{cm}. Use the parallel-axis theorem with $h = \frac{1}{2}L$.

Cover the column to the right and try these on your own before looking at the answers.

Steps   **Answers**

1. Apply the parallel-axis theorem to write I about $I = Mh^2 + I_{cm}$
   the end in terms of I_{cm}.

2. Substitute $I = \frac{1}{3}ML^2$ about the end and solve for $I_{cm} = \dfrac{1}{12}ML^2$
   I_{cm}.

Remark The moment of inertia is least when an object is rotated about its center of mass.

Proof of the Parallel-Axis Theorem

We can prove the parallel-axis theorem using the result developed in Chapter 8 that the kinetic energy of a system of particles is the sum of the kinetic energy of the motion of the center of mass plus the kinetic energy of the motion relative to the center of mass:

$$K = \frac{1}{2}MV_{cm}^2 + K_{rel} \qquad\qquad 9\text{-}22$$

Consider a rigid object rotating with an angular velocity ω about an axis a distance h from a parallel axis through the center of mass as shown in Figure 9-16. When the body rotates through an angle $d\theta$ measured about the axis of rotation, it rotates through the same angle $d\theta$ measured about any other parallel axis. The motion of the object relative to the center of mass is thus a rotation about the center-of-mass axis with the same angular velocity ω. The kinetic energy of this relative motion is

Figure 9-16

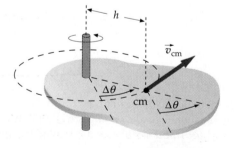

$$K_{rel} = \frac{1}{2}I_{cm}\omega^2$$

The velocity of the center of mass relative to any point on the axis of rotation is $v_{cm} = h\omega$. The kinetic energy of the motion of the center of mass is thus

$$\frac{1}{2}Mv_{cm}^2 = \frac{1}{2}M(h\omega)^2 = \frac{1}{2}M\omega^2h^2$$

The relative energy is $\frac{1}{2}I\omega^2$. Equation 9-21 then becomes

$$K = \frac{1}{2}M\omega^2h^2 + \frac{1}{2}I_{cm}\omega^2 = \frac{1}{2}(Mh^2 + I_{cm})\omega^2 = \frac{1}{2}I\omega^2$$

with

$$I = Mh^2 + I_{cm}$$

which is the parallel-axis theorem.

9-4   Applications of Newton's Second Law for Rotation

In this section, we give several applications of Newton's second law for rotation as expressed in Equation 9-20.

Example 9-5

To get some exercise without going anywhere, you set your bike on a stand so that the rear wheel is free to turn (Figure 9-17). As you pedal, the chain applies a force of 18 N to the sprocket at a distance of $r_s = 7$ cm from the axle of the wheel. Consider the wheel to be a hoop ($I = MR^2$) of radius $R = 35$ cm and mass 2.4 kg. What is the angular velocity of the wheel after 5 s?

Picture the Problem The angular velocity is found from the angular acceleration, which is found from Newton's second law for rotation. Since the forces are constant, the torques are constant and the constant angular acceleration equations apply. Note that $\vec{F}$ acts in the direction of the chain, so the line of force is tangent to the wheel and the lever arm is the radius r_s of the sprocket.

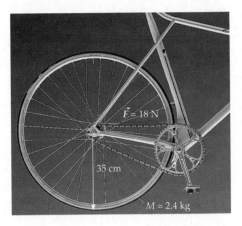

Figure 9-17

1. The angular velocity is related to the angular acceleration and the time:

$$\omega = \omega_0 + \alpha t = 0 + \alpha t$$

2. Apply Newton's second law for rotational motion, $\Sigma \tau_{i,\text{ext}} = I\alpha$, to relate α to the net torque and the moment of inertia:

$$\alpha = \frac{\tau_{\text{net}}}{I}$$

3. The only torque acting on the system is the applied force F with lever arm r_s:

$$\tau_{\text{net}} = Fr_s$$

4. Substitute this value for the torque and $I = MR^2$ for the moment of inertia:

$$\alpha = \frac{\tau_{\text{net}}}{I} = \frac{Fr_s}{MR^2}$$

5. Substitute the given values to obtain α:

$$\alpha = \frac{Fr_s}{MR^2} = \frac{(18\text{ N})(0.07\text{ m})}{(2.4\text{ kg})(0.35\text{ m})^2} = 4.29\text{ rad/s}^2$$

6. Use this value of α to find the angular velocity after 5 s:

$$\omega = \alpha t = (4.29\text{ rad/s}^2)5\text{ s} = 21.4\text{ rad/s}$$

Rotation Under Nonslip Conditions

There are many physical situations in which a string is wrapped around a rotating cylinder. If the string doesn't slip, its linear velocity must equal the tangential velocity of the rim of the cylinder:

$$v_t = R\omega \qquad\qquad\qquad 9\text{-}23$$

Nonslip condition for v and ω

The situation expressed in Equation 9-23 is called a nonslip condition. If we differentiate it with respect to time, we can relate the tangential acceleration of the wheel to the linear acceleration of the chain:

$$a_t = R\alpha \qquad\qquad\qquad 9\text{-}24$$

Nonslip condition for a and α

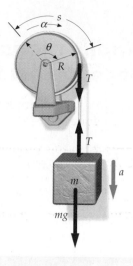

Figure 9-18

Example 9-6

An object of mass m is tied to a light string wound around a wheel that has a moment of inertia I and radius R. The wheel bearing is frictionless, and the string does not slip on the rim. Find the tension in the string and the acceleration of the object.

Picture the Problem In this system, the object descends with a constant downward acceleration a, while the wheel turns with a constant angular acceleration α (Figure 9-18). Because the string unwinds from the wheel without slipping, $a = R\alpha$. We apply Newton's law for rotation to the wheel to determine α, and apply Newton's second law to the object to obtain a. Since the object moves downward and the wheel rotates clockwise, we take these directions to be positive.

1. The only force that exerts a torque on the wheel is the tension T, which has lever arm R. Apply Newton's second law for rotational motion, $\Sigma\tau_{i,\text{ext}} = I\alpha$ to relate T and the angular acceleration α:

$$TR = I\alpha$$

2. Draw a free-body diagram for the suspended object (Figure 9-19), and apply $\Sigma\vec{F} = m\vec{a}$ to relate T to the linear acceleration a:

$$mg - T = ma$$

Figure 9-19

3. We have two equations for three unknowns, T, a, and α. A third equation is the nonslip condition relating a and α:

$$a = R\alpha$$

4. We now have three equations to determine T, a, and α. Use $a = R\alpha$ in the equation in step 1 to eliminate α, and solve for a:

$$TR = I\alpha = I\frac{a}{R} \quad \text{or} \quad a = \frac{TR^2}{I}$$

5. Substitute this result for a into the linear Newton's second-law equation, and solve for T:

$$mg - T = m\frac{TR^2}{I}$$

$$T = \frac{mg}{1 + mR^2/I} = \frac{I}{I + mR^2}mg$$

6. Substituting this value for T in step 4 yields a:

$$a = \frac{mR^2}{I + mR^2}g$$

Check the Result Let's check a couple extreme limits. If $I = 0$, the object should fall freely, and the string should be slack; our results give $T = 0$, $a = g$. What happens if $I \to \infty$? For $I \gg mR^2$, our equations give $T \approx mg$, and $a \approx 0$.

Remark We see that the tension and acceleration depend on the dimensionless quantity I/mR^2. A plot of the acceleration in units of g (a/g), versus I/mR^2 is shown in Figure 9-20.

Figure 9-20

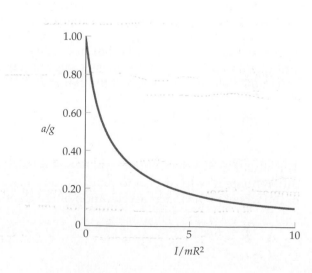

Example 9-7   *try it yourself*

Two blocks are connected by a string that passes over a pulley of radius R and moment of inertia I. The block of mass m_1 slides on a frictionless, horizontal surface; the block of mass m_2 is suspended from the string (Figure 9-21). Find the acceleration a of the blocks and the tensions T_1 and T_2 assuming that the string does not slip on the pulley.

Picture the Problem In this problem, the tensions T_1 and T_2 are not equal because there is friction between the string and the pulley. (Otherwise, the pulley would not turn.) Note that T_2 exerts a clockwise torque and T_1 exerts a counterclockwise torque on the pulley. Use Newton's second law for each block and $\tau = I\alpha$ for the pulley, then relate α and a by $a = R\alpha$.

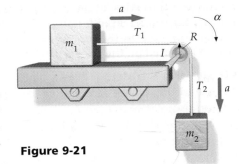

Figure 9-21

Cover the column to the right and try these on your own before looking at the answers.

Steps	**Answers**

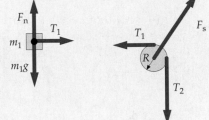

Figure 9-22

1. Draw a free-body diagram for each block and for the pulley (Figure 9-22). Note that the pulley does not accelerate, so the support must exert a force on the axle F_s that balances the forces exerted by the string.

2. Apply $\Sigma \vec{F} = m\vec{a}$ to each block.

$$T_1 = m_1 a, \qquad m_2 g - T_2 = m_2 a$$

3. Add the two equations in step 2 and rearrange to get an equation for $T_2 - T_1$.

$$T_2 - T_1 = m_2 g - (m_1 + m_2)a$$

4. Apply Newton's second law for rotational motion, $\Sigma \tau_{i,\text{ext}} = I\alpha$, to the pulley and obtain another equation for $T_2 - T_1$. Use the nonslip condition to eliminate α.

$$(T_2 - T_1)R = I\alpha, \qquad T_2 - T_1 = \left(\frac{I}{R^2}\right)a$$

5. Set the equations for $T_2 - T_1$ in steps 3 and 4 equal and solve for a in terms of the masses, I, and R^2.

$$a = \frac{m_2}{m_1 + m_2 + I/R^2}g$$

6. Substitute your result for a into each of the equations in step 2 and solve for T_1 and T_2.

$$T_1 = \frac{m_1}{m_1 + m_2 + I/R^2}m_2 g$$

$$T_2 = \frac{(m_1 + I/R^2)}{m_1 + m_2 + I/R^2}m_2 g$$

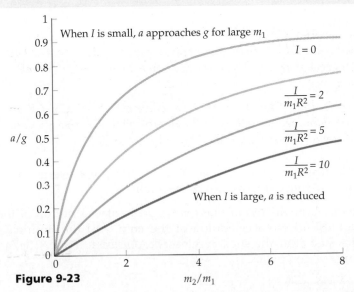

Check the Result If $I = 0$, $T_1 = T_2$, and the acceleration is $a = m_2 g/(m_1 + m_2)$, as expected. If I is very large, $I/R^2 \gg (m_1 + m_2)$, then $T_1 \approx 0$, $T_2 \approx m_2 g$, and $a \approx 0$.

Remarks In general, the acceleration can be written in dimensionless form

$$\frac{a}{g} = \frac{m_2/m_1}{1 + m_2/m_1 + I/m_1 R^2}$$

We see from this expression that $I/m_1 R^2$ is a convenient dimensionless parameter to characterize the moment of inertia for this problem. A plot of a/g versus m_2/m_1 for various values of $I/m_1 R^2$ is shown in Figure 9-23. As m_2/m_1 becomes large, a/g approaches 1. The approach is more rapid for small $I/m_1 R^2$.

When I is small, a approaches g for large m_1

$I = 0$

$\dfrac{I}{m_1 R^2} = 2$

$\dfrac{I}{m_1 R^2} = 5$

$\dfrac{I}{m_1 R^2} = 10$

When I is large, a is reduced

a/g

m_2/m_1

Figure 9-23

Example 9-8

A uniform thin stick of length L and mass M is pivoted at one end. It is held horizontal and released (Figure 9-24). Assume the pivot is frictionless. Find (a) the angular acceleration of the stick immediately after it is released, and (b) the force F_0 exerted on the stick by the pivot at this time.

Picture the Problem The angular acceleration is found from $\tau = I\alpha$, where τ is the torque on the rod relative to the pivot exerted by gravity. Since the rod has an angular acceleration, its center of mass has a tangential acceleration $a_{cm} = \frac{1}{2}L\alpha$. The initial centripetal acceleration of the rod is zero because its velocity is zero just after release. The force exerted by the pivot is found by applying Newton's second law to the rod. Since the acceleration is downward and the weight is downward, F_0 must be vertical. Assume that it is upward and take the positive direction to be downward.

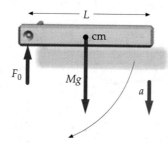

Figure 9-24

(a)1. Write Newton's second law for rotation:

$$\sum \tau_{i,\text{ext}} = I\alpha$$

2. Compute the torque about the end of the stick:

$$\tau = Mg\frac{L}{2}$$

3. Find the moment of inertia about the end of the stick from Table 9-1:

$$I = \frac{1}{3}ML^2$$

4. Substitute these values to compute α:

$$\alpha = \frac{\tau}{I} = \frac{MgL/2}{ML^2/3} = \frac{3}{2}\frac{g}{L}$$

(b)1. Write $\sum \vec{F} = m\vec{a}$ for the stick:

$$Mg - F_0 = Ma_{cm}$$

2. Relate a_{cm} to α:

$$a_{cm} = r\alpha = \frac{L}{2}\alpha = \frac{L}{2}\frac{3}{2}\frac{g}{L} = \frac{3}{4}g$$

3. Substitute a_{cm} into Newton's second law and solve for F_0:

$$F_0 = Mg - Ma_{cm} = Mg - M\left(\frac{3}{4}g\right) = \frac{1}{4}Mg$$

Remark Just after the stick is released, the pivot exerts an upward force equal to one-fourth the weight of the stick.

Exercise: A small coin of mass $m \ll M$ is placed on top of the stick at its center. Find (a) the acceleration of the coin, and (b) the force it exerts on the stick just after the stick is released. (*Answers* (a) $a = 3g/4$ downward, (b) $f = mg/4$ downward)

9-5   Rotational Kinetic Energy

The kinetic energy of a rotating object is the sum of the kinetic energies of the individual particles in the object. The kinetic energy of a mass element m_i is

$$K = \frac{1}{2} m_i v_i^2$$

Summing over all the elements and using $v_i = r_i \omega$ gives

$$K_{rot} = \sum_i \frac{1}{2} m_i v_i^2 = \sum_i \frac{1}{2} m_i (r_i \omega)^2 = \frac{1}{2} \left(\sum_i m_i r_i^2 \right) \omega^2$$

The term in the second set of parentheses is the moment of inertia I relative to the axis of rotation. The kinetic energy is thus

$$K_{rot} = \frac{1}{2} I \omega^2 \qquad\qquad 9\text{-}25$$

Kinetic energy of rotation

Equation 9-25 is the rotational analog of $K = \frac{1}{2} mv^2$ for linear motion.

The Crab Pulsar is one of the fastest-rotating neutron stars known, but it is slowing down. It appears to blink on (left) and off (right) like the rotating lamp in a lighthouse, at the fast rate of about 30 times per second, but the period is increasing by about 10^{-5} s per year. The loss in rotational energy, which is equivalent to the power output of 100,000 suns, appears as light emitted by electrons accelerated in the magnetic field of the pulsar.

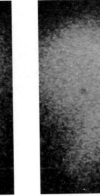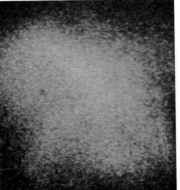

Example 9-9

A flywheel used for storing energy consists of a uniform disk of mass 1.5 × 10⁵ kg and radius 2.2 m that rotates at 3000 rev/min about its center of mass. Find its kinetic energy.

Picture the Problem The kinetic energy is calculated directly from $K = \frac{1}{2} I \omega^2$. To obtain K in joules, we must express the angular velocity in radians per second.

1. The kinetic energy of rotation is: $\qquad K_{rot} = \frac{1}{2} I \omega^2$

2. Calculate the moment of inertia of the disk:

$$I = \frac{1}{2} mR^2 = \frac{1}{2}(1.5 \times 10^5 \text{ kg})(2.2 \text{ m})^2$$
$$= 3.63 \times 10^5 \text{ kg·m}^2$$

3. Convert ω to rad/s:

$$\omega = \frac{3000 \text{ rev}}{60 \text{ s}} \times \frac{2\pi \text{ rad}}{1 \text{ rev}} = 314 \text{ rad/s}$$

4. Substitute these values to find the kinetic energy:

$$K_{rot} = \frac{1}{2} I \omega^2 = \frac{1}{2}(3.63 \times 10^5 \text{ kg·m}^2)(314 \text{ rad/s})^2$$
$$= 1.79 \times 10^{10} \text{ J}$$

Remark To use $K = \frac{1}{2} I \omega^2$, we must express ω in radians per second. Since a radian is dimensionless, the units of step 4 are kg·m²/s² = J. This energy is about 5000 kW·h.

Example 9-10

The stick of Example 9-8 is again released from rest when it is horizontal. Assuming the pivot to be frictionless, find (a) the angular velocity of the stick when it reaches its vertical position, and (b) the force exerted by the pivot at this time. (c) What initial angular velocity is needed for the stick to reach a vertical position at the top of its swing?

Figure 9-25

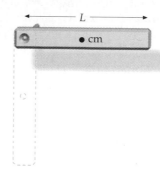

Picture the Problem (a) As the stick swings down, its potential energy decreases, and its kinetic energy of rotation about the pivot point increases (Figure 9-25). As it swings up, its kinetic energy decreases and its potential energy increases. Since the pivot is frictionless, we use conservation of mechanical energy. The angular velocity of the stick is then found from its rotational kinetic energy. Choose $U = 0$ initially. Then use $U = -MgL/2$ when the stick hangs vertically. (b) When the stick is vertical, there is no torque on it, so the stick has no angular acceleration, and the center of mass has no tangential acceleration. But the center of mass has a tangential velocity, so it has a centripetal acceleration toward the pivot. We apply $\Sigma \vec{F}_{ext} = m\vec{a}_{cm}$ to the stick to find the force exerted by the pivot. (c) We find the initial angular velocity from conservation of mechanical energy.

(a)1. The angular velocity of the stick is related to its kinetic energy of rotation K:	$K_f = \dfrac{1}{2} I\omega_f^2$
2. Apply conservation of mechanical energy with $E_f = K_f - MgL/2$, and $E_i = 0$:	$E_f = E_i$ $\dfrac{1}{2} I\omega_f^2 - Mg\dfrac{L}{2} = 0$
3. Solve for ω_f:	$\omega_f = \sqrt{\dfrac{MgL}{I}} = \sqrt{\dfrac{MgL}{\frac{1}{3}ML^2}} = \sqrt{\dfrac{3g}{L}}$
(b)1. Let $\vec{F}_p$ be the force exerted by the pivot when the stick is vertical. Apply $\Sigma \vec{F}_{ext} = m\vec{a}_{cm}$, taking the upward direction to be positive:	$\Sigma \vec{F}_{ext} = m\vec{a}_{cm}$ $F_p - Mg = Ma_{cm}$
2. a_{cm} is the centripetal acceleration:	$a_{cm} = \dfrac{v_{cm}^2}{r} = \dfrac{(\frac{1}{2}L\omega_f)^2}{\frac{1}{2}L} = \dfrac{L}{2}\omega_f^2 = \dfrac{L}{2}\dfrac{3g}{L} = \dfrac{3}{2}g$
3. Substitute and calculate F_p:	$F_p = Mg + Ma_{cm} = M\left(g + \dfrac{3}{2}g\right) = \dfrac{5}{2}Mg$
(c)1. The initial angular velocity ω_i is related to the initial kinetic energy:	$K_i = \dfrac{1}{2} I\omega_i^2 = \dfrac{1}{2}\left(\dfrac{1}{3}ML^2\right)\omega_i^2 = \dfrac{1}{6}ML^2\omega_i^2$
2. Apply conservation of mechanical energy with $K_f = 0$ and $U_i = 0$ to relate the initial kinetic energy to the final position:	$K_f + U_f = K_i + U_i$ $0 + U_f = K_i + 0$ $Mg\dfrac{L}{2} = K_i = \dfrac{1}{6}ML^2\omega_i^2$
3. Solve for the initial angular velocity:	$\omega_i = \sqrt{\dfrac{3g}{L}}$

Remarks We could not easily have used Newton's laws to solve this problem because the acceleration is not constant. The angular velocity ω
versus angle θ is shown in Figure 9-26 for a stick
of length 1 m and various values of the initial angular velocity ω_i. For no initial angular velocity
(bottom curve), θ varies from 0 to π (180°). For
any initial velocity, ω is maximum at $\theta = \pi/2$, corresponding to the stick being vertical and below
the pivot. If the initial angular velocity is great
enough, the stick will swing completely around.
Then ω oscillates, as shown in the top curve.

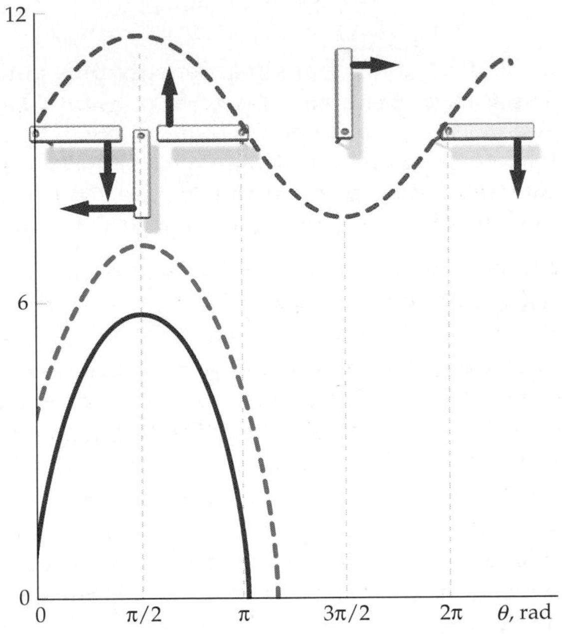

Figure 9-26

| Example **9-11** | *try it yourself* |

The drum of a winch has mass M and radius R. A cable wound around the
drum suspends a load of mass m. The entire cable has a length L and density
(mass per unit length) λ, with a total mass $m_c = L\lambda$. The load begins to fall toward the ground, unwinding cable as it goes. How fast is the load moving after
it has fallen a distance d?

Picture the Problem As the load falls, mechanical energy is conserved.
Choose the initial potential energy to be zero. Then the total mechanical energy is zero. When the load has fallen a distance d, its potential energy is
$-mgd$ (Figure 9-27). In addition, the center of mass of the hanging cable
(mass λd) has dropped a distance $d/2$, so the potential energy of the cable is
$-(\lambda d)g(d/2)$. When the load is moving at speed v, the drum is rotating at angular speed $\omega = v/R$. Since the hanging part of the cable moves with speed v
and the cable does not stretch or become slack, the entire cable must move at
speed v. We find v from the conservation of mechanical energy. Assume that
the drum is a uniform cylinder of moment of inertia $\frac{1}{2}MR^2$.

Mass = M

Figure 9-27

Cover the column to the right and try these on your own before looking at the answers.

Steps

Answers

1. Apply conservation of mechanical energy.

$E_f = E_i = 0$

$K_f + U_f = 0$

2. Write an expression for the total potential energy of the load
and cable when the load has fallen a distance d.

$U = -mgd - (\lambda d)g\dfrac{d}{2} = -mgd - \dfrac{m_c g d^2}{2L}$

3. Express the kinetic energy of the winch in terms of its moment of inertia I and angular speed ω.

$K_w = \dfrac{1}{2}I\omega^2$

4. Use the nonslip condition and $I = \frac{1}{2}MR^2$ to express the kinetic energy of the winch in terms of M and v.

$K_w = \dfrac{1}{2}I\omega^2 = \dfrac{1}{4}Mv^2$

5. Write an expression for the kinetic energy of the cable and of
the load.

$K_c + K_{load} = \dfrac{1}{2}m_c v^2 + \dfrac{1}{2}mv^2$

6. Find the total final kinetic energy plus potential energy and set it equal to zero.

$$\frac{1}{4}Mv^2 + \frac{1}{2}m_cv^2 + \frac{1}{2}mv^2 - mgd - \frac{m_cgd^2}{2L} = 0$$

7. Solve for v.

$$v = \sqrt{\frac{4mgd + 2m_cgd^2/L}{M + 2m + 2m_c}}$$

Power

When you spin an object you do work on it, increasing its kinetic energy. Consider a force F_i acting on the ith particle of a rotating object. As the object turns through an angle $d\theta$, the ith particle moves a distance $ds_i = r_i\,d\theta$, and the force does work:

$$dW_i = F_{it}\,ds_i = F_{it}r_i\,d\theta = \tau_i\,d\theta$$

where τ_i is the torque exerted by the force F_i. In general, the work done by a torque τ when an object turns through a small angle $d\theta$ is

$$dW = \tau\,d\theta \qquad\qquad \text{9-26}$$

The rate at which the torque does work is the power input of the torque:

$$P = \frac{dW}{dt} = \tau\frac{d\theta}{dt}$$

or

$$P = \tau\omega \qquad\qquad \text{9-27}$$

Power

The Archimedes screw is a device for lifting water. The rotational work done by the torque exerted at the handle is converted into increased potential energy of the water.

Equations 9-26 and 9-27 are the rotational analogs of $dW = F_s\,ds$ and $P = F_sv_s$.

Table 9-2 compares rotational and linear motion. (Angular momentum, which appears in the last two table entries, is discussed in Chapter 10.)

Table 9-2

Analogs in Rotational and Linear Motion

Rotational Motion		Linear Motion	
Angular displacement	$\Delta\theta$	Displacement	Δx
Angular velocity	$\omega = \dfrac{d\theta}{dt}$	Velocity	$v = \dfrac{dx}{dt}$
Angular acceleration	$\alpha = \dfrac{d\omega}{dt} = \dfrac{d^2\theta}{dt^2}$	Acceleration	$a = \dfrac{dx}{dt} = \dfrac{d^2x}{dt^2}$
Constant angular acceleration equations	$\omega = \omega_0 + \alpha t$ $\Delta\theta = \omega_{av}\,\Delta t$ $\omega_{av} = \frac{1}{2}(\omega_0 + \omega)$ $\theta = \theta_0 + \omega_0 t + \frac{1}{2}\alpha t^2$ $\omega^2 = \omega_0^2 + 2\alpha\,\Delta\theta$	Constant acceleration equations	$v = v_0 + at$ $\Delta x = v_{av}\,\Delta t$ $v_{av} = \frac{1}{2}(v_0 + v)$ $x = x_0 + v_0 + \frac{1}{2}at^2$ $v^2 = v_0^2 + 2a\,\Delta x$
Torque	τ	Force	F
Moment of inertia	I	Mass	m
Work	$dW = \tau\,d\theta$	Work	$dW = F_s\,ds$
Kinetic energy	$K = \frac{1}{2}I\omega^2$	Kinetic energy	$K = \frac{1}{2}mv^2$
Power	$P = \tau\omega$	Power	$P = Fv$
Angular momentum	$L = I\omega$	Momentum	$p = mv$
Newton's second law	$\tau_{net} = I\alpha = \dfrac{dL}{dt}$	Newton's second law	$F_{net} = ma = \dfrac{dp}{dt}$

Example 9-12	*try it yourself*

A Dodge Neon delivers 175 N·m of torque at 5000 rev/min. Find the power output of the car at that engine speed.

Picture the Problem The power equals the product of the torque and angular velocity, which are given. You must express ω in rad/s to obtain the power in watts.

Cover the column to the right and try these on your own before looking at the answers.

Steps	Answers
1. Write the power in terms of τ and ω.	$P = \tau\omega$
2. Convert ω to rad/s.	$\omega = 523$ rad/s
3. Calculate the power.	$P = 91.5$ kW

Remark This power output is about 123 horsepower.

9-6   Rolling Objects

Rolling Without Slipping

Consider a ball of radius R rolling without slipping along a plane surface. As the ball turns through the angle ϕ (Figure 9-28), the point of contact between the ball and the plane moves a distance s that is related to ϕ by

$$s = R\,\phi \qquad\qquad 9\text{-}28$$

Nonslip condition for displacement

Since the ball's center of mass lies directly over the point of contact, it also moves through s. The velocity of the center of mass is therefore

$$v_{cm} = \frac{ds}{dt} = R\frac{d\phi}{dt}$$

or

$$v_{cm} = R\omega \qquad\qquad 9\text{-}29$$

Nonslip condition for velocity

Differentiating each side again gives

$$a_{cm} = R\alpha \qquad\qquad 9\text{-}30$$

Nonslip condition for acceleration

These conditions are the same as the nonslip conditions for a string wrapped around a pulley or peg.

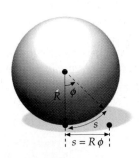

Figure 9-28

When a ball rotates with angular velocity ω, the top and bottom of the ball move with speed $v = R\omega$ relative to the center of the ball (Figure 9-29a). When the ball rolls with speed v without slipping, the top of the ball moves with speed $2v$ and the bottom of the ball in contact with the surface is instantaneously at rest (Figure 9-29b). If a frictional force is exerted by the surface on the ball, it is static friction and no energy is dissipated.

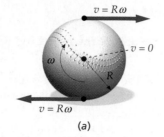 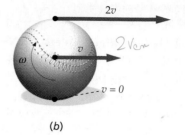

(a)   (b)

We saw in Chapter 7 that the kinetic energy of a system can be written as the sum of the kinetic energy of motion of the center of mass plus the kinetic energy relative to the center of mass. For a rolling object, the relative kinetic energy is $\frac{1}{2}I_{cm}\omega^2$. Thus, the kinetic energy of a rolling object is

Figure 9-29 (a) Rotation without translation. The top of the ball moves to the right with a speed $v = R\omega$ relative to the center, which is at rest. The bottom moves to the left with the same speed relative to the center. (b) Rolling without slipping. If the center moves with speed v, the top moves with speed $2v$ and the bottom of the ball is momentarily at rest.

$$K = \frac{1}{2}I_{cm}\omega^2 + \frac{1}{2}Mv_{cm}^2 \qquad 9\text{-}31$$

> Kinetic energy of a rolling object

Example 9-13   *try it yourself*

A bowling ball of radius 11 cm and mass $m = 7.2$ kg is rolling without slipping on a horizontal ball return at 2 m/s. It then rolls without slipping up a hill to a height h before momentarily coming to rest. Find h.

Picture the Problem Mechanical energy is conserved. The initial kinetic energy, which is the translational kinetic energy of the center of mass, $\frac{1}{2}mv_{cm}^2$, plus the kinetic energy of rotation about the center of mass, $\frac{1}{2}I_{cm}\omega^2$, is converted to potential energy mgh. Since the sphere rolls without slipping, the linear and angular speeds are related by $v_{cm} = R\omega$.

Cover the column to the right and try these on your own before looking at the answers.

Steps	Answers
1. Apply conservation of mechanical energy with $U_i = 0$ and $K_f = 0$.	$E_f = E_i$ $U_f = K_i$
2. Write the total initial kinetic energy K_i in terms of the speed v_{cm} and angular speed ω.	$K_i = \frac{1}{2}mv_{cm}^2 + \frac{1}{2}I_{cm}\omega^2$
3. Substitute $\omega = R/v_{cm}$ and $I_{cm} = \frac{2}{5}mR^2$ and solve for K_i in terms of the mass m and v_{cm}.	$K_i = \frac{1}{2}mv_{cm}^2 + \frac{1}{2}\left(\frac{2}{5}mR^2\right)\left(\frac{v_{cm}}{R}\right)^2 = \frac{7}{10}mv_{cm}^2$
4. Set this initial kinetic energy equal to the final potential energy mgh.	$\frac{7}{10}mv_{cm}^2 = mgh$
5. Solve for h.	$h = \frac{7v_{cm}^2}{10g} = 0.285$ m $= 28.5$ cm

Remark The height is independent of the mass or radius of the ball.

Exercise Find the total mechanical energy of the ball. (*Answer*   84 J)

Example 9-14

A cue stick hits a cue ball horizontally a distance x above the center of the ball (Figure 9-30). Find the value of x for which the cue ball will roll without slipping from the beginning. Express your answer in terms of the radius R of the ball.

Picture the Problem If the stick hits at the level of the ball's center, the ball initially translates with no rotation. If the stick hits below the center, the ball initially has back spin. At a certain value of x, the ball has just the right forward spin and forward acceleration to satisfy the nonslip condition. The value of x determines the torque exerted on the ball, and hence its angular acceleration α. The linear acceleration a is F/m independent of x. For the ball to roll without slipping from the start, we find α and a, then set $a = R\alpha$ (nonslip condition) to find x. The weight and normal force act through the center of mass and thus exert no torque about it. The frictional force is much smaller than the collision force of the stick and can be neglected.

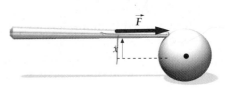

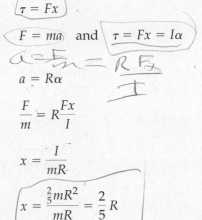

Figure 9-30

1. The torque about the center of the ball equals F times x:

$$\tau = Fx$$

2. Apply Newton's second law $\Sigma \vec{F} = m\vec{a}$ and Newton's second law for rotational motion about the center of the ball, $\Sigma \tau = I\alpha$:

$$F = ma \quad \text{and} \quad \tau = Fx = I\alpha$$
$$a = \frac{F}{m} = \frac{R F_x}{I}$$

3. The nonslip condition relates a and α:

$$a = R\alpha$$

4. Express a and α in terms of the force F from step 2:

$$\frac{F}{m} = R\frac{Fx}{I}$$

5. Solve for x:

$$x = \frac{I}{mR}$$

6. For a sphere, $I = \frac{2}{5}mR^2$:

$$x = \frac{\frac{2}{5}mR^2}{mR} = \frac{2}{5}R$$

Remark If the ball is struck at a point higher than $2R/5$ or lower than $2R/5$ from the center, it will have topspin or backspin and slip. Rolling with slipping is discussed in the next subsection.

When an object rolls down an incline, its center of mass is accelerated. The analysis of such a problem is simplified by an important theorem concerning the center of mass:

> If the torques are computed relative to the center of mass, Newton's second law for rotation holds for rotation about an axis through the center of mass, no matter how the center of mass is moving.

$$\tau_{\text{net,cm}} = I_{\text{cm}}\alpha \qquad \qquad 9\text{-}32$$

This is the same as Equation 9-19 except that here the torque is computed relative to the center of mass rather than relative to some fixed point, and the moment of inertia is computed about an axis through the center of mass. It is often convenient to compute torques about the center of mass. When the center of mass is accelerating (a ball rolling down an incline, for example), its reference frame is a noninertial one, where we would not necessarily expect Newton's second law for rotation to hold. Nevertheless, it does.*

---

*A proof is given in most intermediate-level mechanics books—for example, G. R. Fowles, *Analytical Mechanics*, Holt, Rinehart & Winston, New York, 1993.

Example 9-15

A uniform solid ball of mass m and radius R rolls without slipping down a plane inclined at an angle θ. Find the acceleration of the center of mass.

Picture the Problem From Newton's second law, the acceleration of the center of mass equals the net force divided by the mass. The forces acting are the weight $m\vec{g}$ downward, the normal force $\vec{F}_n$ that balances the normal component of the weight, and the force of friction $\vec{f}$ acting up the incline (Figure 9-31). As the object accelerates down the incline, the angular velocity of rotation must increase to maintain the nonslip condition. We apply Newton's second law for rotation about a horizontal axis through the center of mass to find α, which is related to the acceleration by the nonslip condition. The only torque about the cm is due to $\vec{f}$. (Both $m\vec{g}$ and $\vec{F}_n$ act through the center of mass.) Choose the positive direction to be down the incline:

Figure 9-31

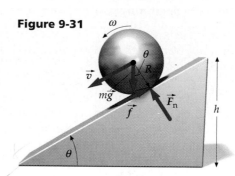

1. Apply $\Sigma\vec{F} = m\vec{a}$ along the incline:

$$mg\sin\theta - f = ma_{cm}$$

2. Apply $\Sigma\tau = I_{cm}\alpha$:

$$fR = I_{cm}\alpha$$

3. Use the nonslip condition to eliminate α and solve for f:

$$fR = I_{cm}\frac{a_{cm}}{R}$$

$$f = \frac{I_{cm}}{R^2}a_{cm}$$

4. Substitute this result for f and solve for a_{cm}:

$$mg\sin\theta - \frac{I_{cm}}{R^2}a_{cm} = ma_{cm}$$

$$a_{cm} = \frac{1}{1 + I_{cm}/mR^2}g\sin\theta$$

5. Substitute $I_{cm} = \frac{2}{5}mR^2$ for a sphere:

$$a_{cm} = \frac{1}{1 + 2/5}g\sin\theta = \frac{5}{7}g\sin\theta$$

Remarks Since the ball rolls without slipping, the friction is static friction. Note that the result is independent of the coefficient of static friction as long as it is great enough so the ball does not slip.

The results of steps 3 and 4 in the above example apply equally to any rolling object:

$$f = \frac{I_{cm}}{R^2}a_{cm} \qquad\qquad 9\text{-}33$$

$$a_{cm} = \frac{1}{1 + I_{cm}/mR^2}g\sin\theta \qquad\qquad 9\text{-}34$$

For a cylinder, $I_{cm} = \frac{1}{2}mR^2$, and the acceleration is $\frac{2}{3}g\sin\theta$. For a hoop, $I_{cm} = mR^2$, and the acceleration is $\frac{1}{2}g\sin\theta$. The linear acceleration of any object rolling down an incline is less than $g\sin\theta$ because of the frictional force directed up the incline. Note that these accelerations are independent of both the mass and radius of the objects. If we release a sphere, a cylinder, and a hoop at the top of an incline, and if they all roll without slipping, the sphere will reach the bottom first because it has the greatest acceleration. The cylinder will be second and the hoop third (Figure 9-32). If any object could slide down the incline without friction, it would reach the bottom before any of the rolling objects.

Since the friction is static, it does no work, and there is no dissipation of mechanical energy. We can therefore use the conservation of mechanical energy to find the speed of an object rolling without slipping down an incline. At the top of the incline, the total energy is the potential energy mgh. At the bottom, the total energy is kinetic energy. Conservation of mechanical energy therefore gives

$$\frac{1}{2}mv_{cm}^2 + \frac{1}{2}I_{cm}\omega^2 = mgh$$

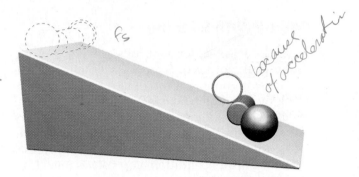

Figure 9-32 A sphere, a cylinder, and a hoop are started together from rest at the top of an incline. The sphere reaches the bottom first, followed by the cylinder and then the hoop.

We can use the nonslip condition to eliminate either v_{cm} or ω. Substituting $\omega = v_{cm}/R$, we obtain

$$\frac{1}{2}mv_{cm}^2 + \frac{1}{2}I_{cm}\left(\frac{v_{cm}}{R}\right)^2 = mgh$$

$$v_{cm}^2 = \frac{1}{1 + I_{cm}/mR^2}2gh \qquad\qquad 9\text{-}35$$

For a cylinder, with $I_{cm} = \frac{1}{2}mR^2$, we obtain $v_{cm} = \sqrt{\frac{4}{3}gh}$. Note that the speed is independent of the mass and radius of the cylinder, and is less than $\sqrt{2gh}$, the speed of an object sliding with no friction down the incline.

We can find the force of static friction f for an object rolling down an incline from Equations 9-33 and 9-34:

$$f = \frac{I_{cm}}{R^2}a_{cm} = \frac{I_{cm}}{R^2}\frac{1}{1 + I_{cm}/R^2}g\sin\theta$$

or

$$f = \frac{1}{1 + mR^2/I_{cm}}mg\sin\theta \qquad\qquad 9\text{-}36$$

For a cylinder, for example, $I_{cm} = \frac{1}{2}mR^2$, so $f = \frac{1}{3}mg\sin\theta$. Note that we have found f without considering the coefficient of static friction, μ_s. For an object rolling down an incline without slipping, f is less than its maximum value; that is,

$$f \le \mu_s F_n = \mu_s mg\cos\theta$$

Then for a cylinder,

$$f = \frac{1}{3}mg\sin\theta \le \mu_s mg\cos\theta$$

or

$$\boxed{\tan\theta \le 3\mu_s} \qquad (\text{cylinder}) \qquad\qquad 9\text{-}37$$

If the tangent of the angle of incline is greater than $3\mu_s$, the cylinder will slip as it moves down the incline.

Exercise A cylinder rolls down a plane inclined at $\theta = 50°$. What is the minimum value of the coefficient of static friction for which the cylinder will roll without slipping? (*Answer* 0.40)

Exercise For a hoop rolling down an incline, (*a*) what is the force of friction, and (*b*) what is the maximum value of $\tan\theta$ for which the hoop will roll without slipping? (*Answers* (*a*) $f = \frac{1}{2}mg\sin\theta$; (*b*) $\tan\theta \le 2\mu_s$)

Rolling With Slipping

When an object slides as it rolls, the nonslip condition does not hold. Suppose a bowling ball is thrown with no initial rotation. As the ball slides along the bowling lane, kinetic friction reduces its linear velocity (Figure 9-33). The frictional force also causes the ball to start rotating. The linear velocity decreases and the angular velocity increases until the nonslip condition $v_{cm} = R\omega$ is met. Then the ball rolls without slipping.

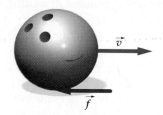

Figure 9-33 A bowling ball moving with no initial rotation. The frictional force $\vec{f}$ exerted by the floor reduces the speed v and increases the angular speed ω until $v = R\omega$.

Another example of rolling with slipping is a ball with topspin, such as a cue ball struck at a point greater than $2R/5$ above the center (see Example 9-14). Then the frictional force increases v and reduces ω until the nonslip condition is met (Figure 9-34).

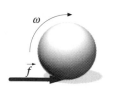

Figure 9-34 Ball with topspin. The frictional force accelerates the ball in the direction of motion.

Example 9-16

A bowling ball of mass M and radius R is thrown such that the instant it touches the floor it is moving horizontally with speed $v_0 = 5$ m/s and is not rotating. The coefficient of kinetic friction between the ball and the floor is $\mu_k = 0.08$. Find (a) the time the ball slides before the nonslip condition is met, and (b) the distance the ball slides before it rolls without slipping.

Picture the Problem We calculate v_{cm} and ω as functions of time, set $v_{cm} = R\omega$, and solve for t. The linear and angular accelerations are found from $\Sigma \vec{F} = m\vec{a}$, and $\tau = I\alpha$. Let the direction of motion be positive.

(a)1. The net force on the ball is the force of kinetic friction, f_k, which acts in the negative direction. Apply $\Sigma \vec{F} = m\vec{a}$:

$$f_k = -\mu_k Mg = Ma_{cm} \quad \text{or}$$
$$a_{cm} = -\mu_k g$$

2. The linear velocity is related to the acceleration and the time:

$$v_{cm} = v_0 + a_{cm}t = v_0 - \mu_k gt$$

3. The torque about the center of the ball is the frictional force times the lever arm R. Apply $\tau = I\alpha$ with $I = \frac{2}{5}MR^2$:

$$\tau = \mu_k MgR = I_{cm}\alpha \quad \text{or}$$
$$\alpha = \frac{\mu_k MgR}{I_{cm}} = \frac{\mu_k MgR}{\frac{2}{5}MR^2} = \frac{5}{2}\left(\frac{\mu_k g}{R}\right)$$

4. The angular velocity is related to the angular acceleration and the time:

$$\omega = \omega_0 + \alpha t = 0 + \alpha t = \frac{5}{2}\left(\frac{\mu_k g}{R}\right)t$$

5. Solve for time t_1 at which $v_{cm} = R\omega$:

$$v_{cm} = v_0 - \mu_k gt_1 = R\omega = \frac{5}{2}\mu_k gt_1$$

$$t_1 = \frac{2v_0}{7\mu_k g} = \frac{2(5 \text{ m/s})}{7(0.08)(9.81 \text{ m/s}^2)} = 1.82 \text{ s}$$

(b)1. The distance traveled in time t_1 is:

$$\Delta x = v_0 t_1 + \frac{1}{2}a_{cm}t_1^2$$

$$= v_0 \frac{2v_0}{7\mu_k g} + \frac{1}{2}(-\mu_k g)\left(\frac{2v_0}{7\mu_k g}\right)^2 = \frac{12}{49}\frac{v_0^2}{\mu_k g}$$

2. Substitute the given values:

$$\Delta x = \frac{12}{49}\frac{v_0^2}{\mu_k g} = \frac{12}{49}\frac{(5 \text{ m/s})^2}{(0.08)(9.81 \text{ m/s}^2)} = 7.80 \text{ m}$$

Exercise Find the speed of the bowling ball when it begins to roll without slipping. (*Answer* $v_{cm} = \frac{5}{7}v_0$. This result is independent of the coefficient of kinetic friction. The rolling speed is $\frac{5}{7}v_0$ whether friction is large or small. The total mechanical energy lost is thus independent of μ_k. The time and distance are sensitive to the value of μ_k, however.)

Exercise Find the total kinetic energy of the ball after it begins to roll without slipping. (*Answer* $K = \frac{5}{14}mv_0^2$)

Remarks In a well-maintained bowling alley, the lanes are lightly oiled and very slick, so that the ball slides over a great distance, which gives the bowler added control. Figure 9-35 shows the sliding distance Δx versus initial velocity for the bowling ball for $\mu_k = 0.08$, as in this example (slippery floor), and for $\mu_k = 0.24$ (sticky floor).

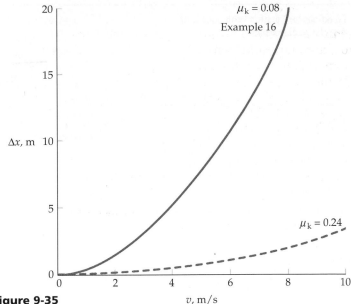

Figure 9-35

Summary

1. Angular displacement, angular velocity, and angular acceleration are fundamental defined quantities in rotational kinematics.

2. Torque and moment of inertia are important derived dynamic concepts. Torque is a measure of the ability of a force to cause an object to start or stop rotating. Moment of inertia is the measure of an object's inertial resistance to angular accelerations. The moment of inertia depends on the distribution of the mass relative to the rotation axis.

3. The parallel-axis theorem, which follows from the definition of the moment of inertia, often simplifies the calculation of I.

4. Newton's second law for rotation, $\Sigma\tau_{i,\text{ext}} = I\alpha$, is derived from Newton's second law and the definitions of τ, I, and α. It is an important relation for problems involving the rotation of a rigid object about an axis of fixed direction.

Topic	Remarks and Relevant Equations	
1. Angular Velocity and Angular Acceleration		
Angular velocity	$\omega = \dfrac{d\theta}{dt}$ (Definition)	9-2
Angular acceleration	$\alpha = \dfrac{d\omega}{dt} = \dfrac{d^2\theta}{dt^2}$ (Definition)	9-3
Tangential speed	$v_{it} = r_i\omega$	9-4
Tangential acceleration	$a_{it} = r_i\alpha$	9-5
Centripetal acceleration	$a_{ic} = \dfrac{v_i^2}{r_i} = r_i\omega^2$	9-6

2. **Equations for Rotation With Constant Angular Acceleration**	$\omega = \omega_0 + \alpha t$	9-7

$$\theta = \theta_0 + \omega_0 t + \frac{1}{2}\alpha t^2$$

9-8

$$\omega^2 = \omega_0^2 + 2\alpha(\theta - \theta_0)$$

9-9

3. **Torque**

The magnitude of the torque exerted by a force on an object is defined as the product of the force and the lever arm.

$$\tau = F\ell \qquad \text{(Definition)}$$

9-10

4. **Moment of Inertia**

System of particles

$$I = \sum m_i r_i^2 \qquad \text{(Definition)}$$

9-17

Continuous object

$$I = \int r^2\, dm \qquad \text{(Definition)}$$

9-18

Parallel-axis theorem

The moment of inertia about an axis a distance h from a parallel axis through the center of mass is

$$I = I_{cm} + Mh^2$$

9-21

where I_{cm} is the moment of inertia about the axis through the center of mass and M is the total mass of the object.

5. **Newton's Second Law for Rotation**

$$\tau_{net,ext} = \sum_i \tau_{i,ext} = I\alpha$$

9-20

If the torques are computed relative to the center of mass, Newton's second law for rotation holds for rotation about an axis through the center of mass, no matter how the center of mass is moving.

6. **Nonslip Conditions**

When a string that is wrapped around a pulley or disk does not slip, the linear and angular quantities are related by

$$v = R\omega$$

9-19

$$a = R\alpha$$

9-20

7. **Energy**

Kinetic energy of rotation

$$K = \frac{1}{2}I\omega^2$$

9-25

K for rotation plus translation

$$K = \frac{1}{2} I_{cm}\omega^2 + \frac{1}{2} Mv_{cm}^2$$

9-31

Power

$$P = \tau\omega$$

9-27

8. **Rolling Objects**

Rolling without slipping

$$v_{cm} = R\omega$$

9-29

Rolling with slipping (optional)

When an object rolls and slips, $v_{cm} \neq R\omega$. Kinetic friction exerts a force that tends to change v_{cm}, and also exerts a torque that changes ω until $v_{cm} = R\omega$ and rolling without slipping sets in.

Problem-Solving Guide

1. Begin by drawing a neat diagram including the important features of the problem. Draw a free-body diagram for each object showing the forces acting, the chosen coordinate system, and appropriate axes of rotation.

2. Write $\Sigma \vec{F} = m\vec{a}$ for each translating object and $\Sigma \tau_i = I\alpha$ for each rotating object, and relate the translational and rotational velocities and accelerations by nonslip conditions whenever applicable.

3. If there is no energy dissipation, conservation of mechanical energy provides a useful method to find final velocities or angular velocities.

Summary of Worked Examples

Type of Calculation	Procedure and Relevant Examples
1. Kinematics	
Find $\Delta\theta$, ω, or α for rotation due to a constant torque	Use $\tau = I\alpha$ and constant angular acceleration equations. **Examples 9-1, 9-5**
2. Moment of Inertia	
Calculate I for a system of particles	Use $I = \Sigma m_i r_i^2$ **Example 9-2**
Calculate I for a continuous object	Use $I = \int r^2 \, dm$ **Example 9-3**
Calculate I about an axis parallel to one through the center of mass	Use the parallel axis theorem $I = I_{cm} + Mh^2$ where h is the distance between the axes **Example 9-4**
3. Finding Forces and Accelerations	
Find the acceleration of an object hanging from a string wrapped around a disk or pulley	Use $\Sigma \vec{F} = m\vec{a}$ for the object and $\Sigma \tau_i = I\alpha$ for the disk or pulley, and the nonslip condition $a = R\alpha$. **Examples 9-6, 9-7**
Find the acceleration of an object rotating about a pivot and the forces exerted by the pivot	Apply $\Sigma \tau = I\alpha$ to find α, then use it to find a_{cm}. Apply $\Sigma \vec{F} = m\vec{a}$ to find the forces. **Example 9-8**
4. Energy and Power	
Find the kinetic energy of a rotating object	Use $K = \frac{1}{2}I\omega^2$. Remember that ω must be in rad/s. **Example 9-9**
Find the power of a motor	Use $P = \tau\omega$. Remember that ω must be in rad/s. **Example 9-12**
Find the final speed or angular speed of a system for which there is no energy dissipation	Use conservation of mechanical energy with $K = \frac{1}{2}I\omega^2$ for rotating the object. Remember that ω must be in rad/s. **Examples 9-15, 9-11**
5. Rolling Objects	
Hit a cue ball, or throw a ball such that it rolls without slipping	The torque and force must be such that the acceleration and angular acceleration are related by $a = R\alpha$. **Example 9-14**
Find the acceleration of an object rolling down an incline	Apply $\Sigma \tau_i = I\alpha$ about the center of mass, and $\Sigma \vec{F} = m\vec{a}_{cm}$ to eliminate the frictional force, and use the nonslip condition $a_{cm} = R\alpha$ to write α in terms of a_{cm}. **Example 9-15**

| Find the kinetic energy of a rolling object | Add $\frac{1}{2}Mv_{cm}^2$ and $\frac{1}{2}I\omega^2$ and use the nonslip condition $v_{cm} = R\omega$. | **Example** 9-13 |

| Rolling and sliding objects (optional) | While sliding, $v_{cm} \neq R\omega$. Kinetic friction exerts a force that accelerates the center of mass and also a torque that produces angular acceleration. Then v_{cm} and ω change until $v_{cm} = R\omega$, at which time rolling without slipping begins. | **Example 9-16** |

Problems

▢ Conceptual Problems

▢ Problems from Optional and Exploring sections

In a few problems, you are given more data than you actually need; in a few other problems, you are required to supply data from your general knowledge, outside sources, or informed estimates.

•   Single-concept, single-step, relatively easy
•• Intermediate-level, may require synthesis of concepts
••• Challenging, for advanced students

Take $g = 9.81$ N/kg $= 9.81$ m/s^2 and neglect friction in all problems unless otherwise stated. Assume that all objects are points unless otherwise indicated.

Angular Velocity and Angular Acceleration

1 • Two points are on a disk turning at constant angular velocity, one point on the rim and the other halfway between the rim and the axis. Which point moves the greater distance in a given time? Which turns through the greater angle? Which has the greater speed? The greater angular velocity? The greater tangential acceleration? The greater angular acceleration? The greater centripetal acceleration?

2 • True or false:

(a) Angular velocity and linear velocity have the same dimensions.
(b) All parts of a rotating wheel must have the same angular velocity.
(c) All parts of a rotating wheel must have the same angular acceleration.

3 •• Starting from rest, a disk takes 10 revolutions to reach an angular velocity ω. At constant angular acceleration, how many additional revolutions are required to reach an angular velocity 2ω?

(a) 10 rev   (b) 20 rev   (c) 30 rev   (d) 40 rev   (e) 50 rev

4 • A particle moves in a circle of radius 90 m with a constant speed of 25 m/s. (a) What is its angular velocity in radians per second about the center of the circle? (b) How many revolutions does it make in 30 s?

5 • A wheel starts from rest with a constant angular acceleration of 2.6 rad/s^2 and rolls for 6 s. At the end of that time, (a) what is its angular velocity? (b) Through what angle has the wheel turned? (c) How many revolutions has it made? (d) What is the speed and acceleration of a point 0.3 m from the axis of rotation?

6 • When a turntable rotating at $33\frac{1}{3}$ rev/min is shut off, it comes to rest in 26 s. Assuming constant angular acceleration, find (a) the angular acceleration, (b) the average angular velocity of the turntable, and (c) the number of revolutions it makes before stopping.

7 • A disk of radius 12 cm, initially at rest, begins rotating about its axis with a constant angular acceleration of 8 rad/s^2. At $t = 5$ s, what are (a) the angular velocity of the disk, and (b) the tangential acceleration a_t and the centripetal acceleration a_c of a point on the edge of the disk?

8 • Radio announcers who still play vinyl records have to be careful when cueing up live recordings. While studio albums have blank spaces between the songs, live albums have audiences cheering. If the volume levels are left up when the turntable is turned on, it sounds as though the audience has suddenly burst through the wall. If a turntable begins at rest and rotates through 10° in 0.5 s, how long must an announcer wait before the record reaches the required angular speed of $33\frac{1}{3}$ rev/min? Assume constant angular acceleration.

9 • A Ferris wheel of radius 12 m rotates once in 27 s. (a) What is its angular velocity in radians per second? (b) What is the linear speed of a passenger? (c) What is the centripetal acceleration of a passenger?

10 • A cyclist accelerates from rest. After 8 s, the wheels have made 3 rev. (a) What is the angular acceleration of the wheels? (b) What is the angular velocity of the wheels after 8 s?

11 • What is the angular velocity of the earth in rad/s as it rotates about its axis?

12 • A wheel rotates through 5.0 rad in 2.8 s as it is brought to rest with constant angular acceleration. The initial angular velocity of the wheel before the braking began was

(a) 0.6 rad/s.   (b) 0.9 rad/s.   (c) 1.8 rad/s.
(d) 3.6 rad/s.   (e) 7.2 rad/s.

13 • A circular space station of radius 5.10 km is a long way from any star. Its rotational speed is controllable to some degree, and so the apparent gravity changes according to the tastes of those who are making the decisions. Dave the Earthling puts in a request for artificial gravity of 9.8 m/s^2 at the circumference. His secret agenda is to give the Earthlings a home-gravity advantage in the upcoming interstellar basketball tournament. Dave's request would require an angular speed of

(a) 4.4×10^{-2} rad/s.　　(b) 7.0×10^{-3} rad/s.
(c) 0.28 rad/s.　　　　　　　(d) -0.22 rad/s.
(e) 1300 rad/s.

14 • A bicycle has wheels of 1.2 m diameter. The bicyclist accelerates from rest with constant acceleration to 24 km/h in 14.0 s. What is the angular acceleration of the wheels?

15 •• The tape in a standard VHS videotape cassette has a length $L = 246$ m; the tape plays for $t = 2.0$ h (Figure 9-36). As the tape starts, the full reel has an outer radius of about $R = 45$ mm, and an inner radius of about $r = 12$ mm. At some point during the play, both reels have the same angular speed. Calculate this angular speed in rad/s and in rev/min.

Figure 9-36 Problem 15

45 mm
12 mm

Torque, Moment of Inertia, and Newton's Second Law for Rotation

16 • The dimension of torque is the same as that of

(a) impulse.　　　　　　(b) energy.
(c) momentum.　　　　　(d) none of the above.

17 • The moment of inertia of an object of mass M

(a) is an intrinsic property of the object.
(b) depends on the choice of axis of rotation.
(c) is proportional to M regardless of the choice of axis.
(d) Both (b) and (c) are correct.

18 • Can an object continue to rotate in the absence of torque?

19 • Does an applied net torque always increase the angular speed of an object?

20 • True or false:

(a) If the angular velocity of an object is zero at some instant, the net torque on the object must be zero at that instant.
(b) The moment of inertia of an object depends on the location of the axis of rotation.
(c) The moment of inertia of an object depends on the angular velocity of the object.

21 • A disk is free to rotate about an axis. A force applied a distance d from the axis causes an angular acceleration α. What angular acceleration is produced if the same force is applied a distance $2d$ from the axis?

(a) α　　(b) 2α　　(c) $\alpha/2$　　(d) 4α　　(e) $\alpha/4$

22 • A disk-shaped grindstone of mass 1.7 kg and radius 8 cm is spinning at 730 rev/min. After the power is shut off, a woman continues to sharpen her ax by holding it against the grindstone for 9 s until the grindstone stops rotating. (a) What is the angular acceleration of the grindstone? (b) What is the torque exerted by the ax on the grindstone? (Assume constant angular acceleration and a lack of other frictional torques.)

23 • A 2.5-kg cylinder of radius 11 cm is initially at rest. A rope of negligible mass is wrapped around it and pulled with a force of 17 N. Find (a) the torque exerted by the rope, (b) the angular acceleration of the cylinder, and (c) the angular velocity of the cylinder at $t = 5$ s.

24 •• A wheel mounted on an axis that is not frictionless is initially at rest. A constant external torque of 50 N·m is applied to the wheel for 20 s, giving the wheel an angular velocity of 600 rev/min. The external torque is then removed, and the wheel comes to rest 120 s later. Find (a) the moment of inertia of the wheel, and (b) the frictional torque, which is assumed to be constant.

25 •• A pendulum consisting of a string of length L attached to a bob of mass m swings in a vertical plane. When the string is at an angle θ to the vertical, (a) what is the tangential component of acceleration of the bob? (b) What is the torque exerted about the pivot point? (c) Show that $\tau = I\alpha$ with $a_t = L\alpha$ gives the same tangential acceleration found in part (a).

26 ••• A uniform rod of mass M and length L is pivoted at one end and hangs as in Figure 9-37 so that it is free to rotate without friction about its pivot. It is struck by a horizontal force F_0 for a short time Δt at a distance x below the pivot as shown. (a) Show that the speed of the center of mass of the rod just after being struck is given by $v_0 = 3F_0 x \, \Delta t/2ML$. (b) Find the force delivered by the pivot, and show that this force is zero if $x = 2L/3$. (Note: The point $x = 2L/3$ is called the *center of percussion* of the rod.)

Figure 9-37
Problem 26

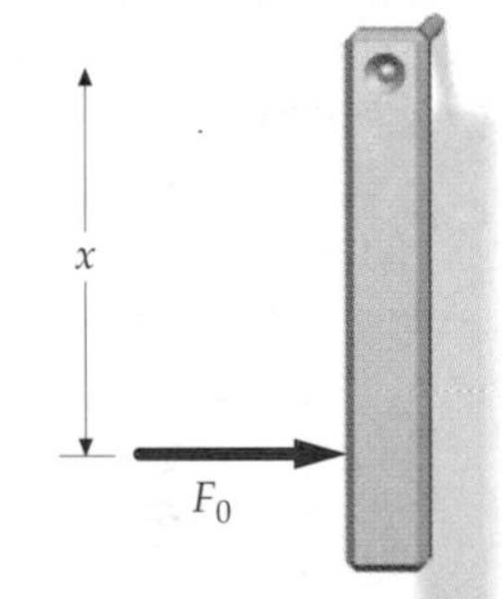

x

F_0

27 ••• A uniform horizontal disk of mass M and radius R is rotating about its vertical axis with an angular velocity ω. When it is placed on a horizontal surface, the coefficient of kinetic friction between the disk and surface is μ_k. (a) Find the torque $d\tau$ exerted by the force of friction on a circular element of radius r and width dr. (b) Find the total torque exerted by friction on the disk. (c) Find the time required to bring the disk to a halt.

Calculating the Moment of Inertia

28 • The moment of inertia of an object about an axis that does not pass through its center of mass is _____ the moment of inertia about a parallel axis through its center of mass.

(a) always less than　　　　(b) sometimes less than
(c) sometimes equal to　　　(d) always greater than

29 • A tennis ball has a mass of 57 g and a diameter of 7 cm. Find the moment of inertia about its diameter. Assume that the ball is a thin spherical shell.

30 • Four par-
ticles at the corners
of a square with a
side length $L = 2$ m
are connected by
massless rods (Fig-
ure 9-38). The parti-
cle masses are $m_1 =$
$m_3 = 3$ kg and $m_2 =$
$m_4 = 4$ kg. Find the
moment of inertia of
the system about the
z axis.

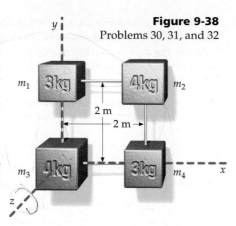

Figure 9-38
Problems 30, 31, and 32

m_1 3kg   4kg m_2

2 m
2 m→

m_3 4kg   3kg m_4 x

31 • Use the parallel-axis theorem and your results for
Problem 30 to find the moment of inertia of the four-particle
system in Figure 9-38 about an axis that is perpendicular to
the plane of the masses and passes through the center of
mass of the system. Check your result by direct computation.

32 • For the four-particle system of Figure 9-38, (a) find
the moment of inertia I_x about the x axis, which passes
through m_3 and m_4, and (b) find I_y about the y axis, which
passes through m_1 and m_3.

33 • Use the parallel-axis theorem to find the moment of inertia
of a solid sphere of mass M and radius R about an axis that is tangent
to the sphere (Figure 9-39).

Figure 9-39
Problem 33

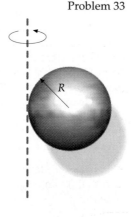

R

34 •• A wagon wheel 1.0 m in
diameter consists of a thin rim
having a mass of 8 kg and six
spokes each having a mass of
1.2 kg. Determine the moment of
inertia of the wagon wheel for rotation about its axis.

35 •• Two point masses m_1
and m_2 are separated by a massless rod of length L. (a) Write an
expression for the moment of inertia about an axis perpendicular to the rod and passing through it at a distance x from
mass m_1. (b) Calculate dI/dx and show that I is at a minimum
when the axis passes through the center of mass of the system.

36 •• A uniform rectangular plate has mass m and sides a
and b. (a) Show by integration that the moment of inertia of
the plate about an axis that is perpendicular to the plate and
passes through one corner is $\frac{1}{3}m(a^2 + b^2)$. (b) What is the moment of inertia about an axis that is perpendicular to the
plate and passes through its center of mass?

37 •• Tracey and Corey are doing intensive research on
theoretical baton twirling. Each is using "The Beast" as a
model baton: two uniform spheres, each of mass 500 g and
radius 5 cm, mounted at the ends of a 30-cm uniform rod of
mass 60 g (Figure 9-40). Tracey and Corey want to calculate
the moment of inertia of The Beast about an axis perpendicular to the rod and passing through its center. Corey uses the
approximation that the two spheres can be treated as point
particles that are 20 cm from the axis of rotation, and that the
mass of the rod is negligible. Tracey, however, makes her cal-

culations without approximations. (a) Compare the two results. (b) If the spheres retained the same mass but were hollow, would the rotational inertia increase or decrease? Justify
your choice with a sentence or two. It is not necessary to calculate the new value of I.

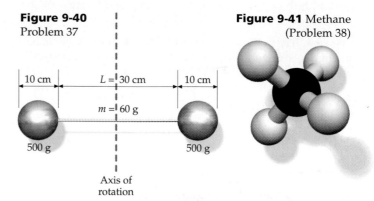

Figure 9-40
Problem 37

10 cm $L = 30$ cm   10 cm

$m = 60$ g

500 g   500 g

Axis of
rotation

Figure 9-41 Methane
(Problem 38)

38 •• The methane molecule (CH_4) has four hydrogen
atoms located at the vertices of a regular tetrahedron of side
length 1.4 nm, with the carbon atom at the center of the tetrahedron (Figure 9-41). Find the moment of inertia of this molecule for rotation about an axis that passes through the carbon atom and one of the hydrogen atoms.

39 ••• A hollow cylinder has mass m, an outside radius
R_2, and an inside radius R_1. Show that its moment of inertia
about its symmetry axis is given by $I = \frac{1}{2}m(R_2^2 + R_1^2)$.

40 ••• Show that the moment of inertia of a spherical shell
of radius R and mass m is $\frac{2}{3}mR^2$. This can be done by direct
integration or, more easily, by finding the increase in the moment of inertia of a solid sphere when its radius changes. To
do this, first show that the moment of inertia of a solid sphere
of density ρ is $I = \frac{8}{15}\pi\rho R^5$. Then compute the change dI in I for
a change dR, and use the fact that the mass of this shell is $m =$
$4\pi R^2\rho\, dR$.

41 ••• The density of the earth is not quite uniform. It
varies with the distance r from the center of the earth as $\rho =$
$C(1.22 - r/R)$, where R is the radius of the earth and C is a
constant. (a) Find C in terms of the total mass M and the radius R. (b) Find the moment of inertia of the earth. (See Problem 40.)

42 ••• Use integration to determine the moment of inertia
of a right circular homogeneous cone of height H, base radius
R, and mass density ρ about its symmetry axis.

43 ••• Use integration to determine the moment of inertia
of a hollow, thin-walled, right circular cone of mass M, height
H, and base radius R about its symmetry axis.

44 ••• Use integration to determine the moment of inertia
of a thin uniform disk of mass M and radius R for rotation
about a diameter. Check your answer by referring to Table
9-1.

45 ••• Use integration to determine the moment of inertia
of a thin circular hoop of radius R and mass M for rotation
about a diameter. Check your answer by referring to Table
9-1.

46 ••• A roadside ice-cream stand uses rotating cones to catch the eyes of travelers. Each cone rotates about an axis perpendicular to its axis of symmetry and passing through its apex. The sizes of the cones vary, and the owner wonders if it would be more energy-efficient to use several smaller cones or a few big ones. To answer this, he must calculate the moment of inertia of a homogeneous right circular cone of height H, base radius R, and mass density ρ. What is the result?

Rotational Kinetic Energy

47 • A constant torque acts on a merry-go-round. The power input of the torque is

(a) constant.
(b) proportional to the angular speed of the merry-go-round.
(c) zero.
(d) none of the above.

48 • The particles in Figure 9-42 are connected by a very light rod whose moment of inertia can be neglected. They rotate about the y axis with angular velocity $\omega = 2$ rad/s. (a) Find the speed of each particle, and use it to calculate the kinetic energy of this system directly from $\Sigma \frac{1}{2}m_i v_i^2$. (b) Find the moment of inertia about the y axis, and calculate the kinetic energy from $K = \frac{1}{2}I\omega^2$.

Figure 9-42
Problem 48

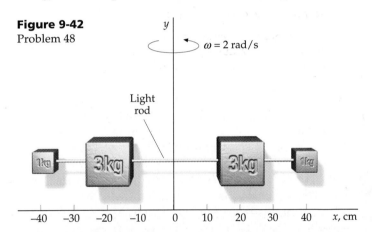

49 • Four 2-kg particles are located at the corners of a rectangle of sides 3 m and 2 m as shown in Figure 9-43. (a) Find the moment of inertia of this system about the z axis. (b) The system is set rotating about this axis with a kinetic energy of 124 J. Find the number of revolutions the system makes per minute.

Figure 9-43
Problem 49

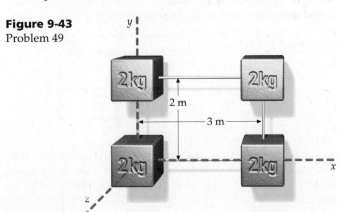

50 • A solid ball of mass 1.4 kg and diameter 15 cm is rotating about its diameter at 70 rev/min. (a) What is its kinetic energy? (b) If an additional 2 J of energy are supplied to the rotational energy, what is the new angular speed of the ball?

51 • An engine develops 400 N·m of torque at 3700 rev/min. Find the power developed by the engine.

52 •• Two point masses m_1 and m_2 are connected by a massless rod of length L to form a dumbbell that rotates about its center of mass with angular velocity ω. Show that the ratio of kinetic energies of the masses is $K_1/K_2 = m_2/m_1$.

53 •• Calculate the kinetic energy of rotation of the earth, and compare it with the kinetic energy of motion of the earth's center of mass about the sun. Assume the earth to be a homogeneous sphere of mass 6.0×10^{24} kg and radius 6.4×10^6 m. The radius of the earth's orbit is 1.5×10^{11} m.

54 •• A 2000-kg block is lifted at a constant speed of 8 cm/s by a steel cable that passes over a massless pulley to a motor-driven winch (Figure 9-44). The radius of the winch drum is 30 cm. (a) What force must be exerted by the cable? (b) What torque does the cable exert on the winch drum? (c) What is the angular velocity of the winch drum? (d) What power must be developed by the motor to drive the winch drum?

Figure 9-44 Problem 54

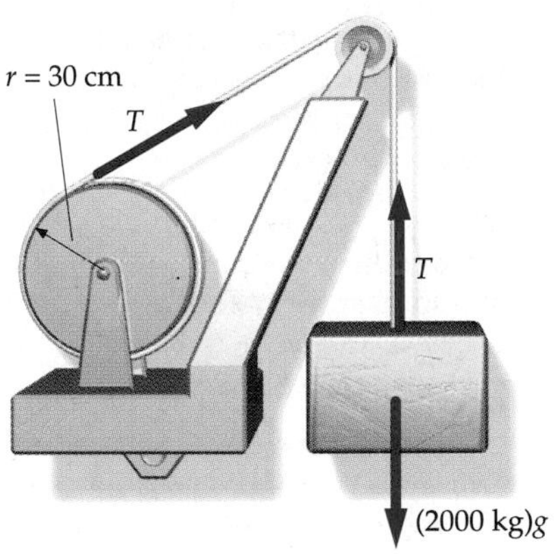

55 •• A uniform disk of mass M and radius R is pivoted such that it can rotate freely about a horizontal axis through its center and perpendicular to the plane of the disk. A small particle of mass m is attached to the rim of the disk at the top, directly above the pivot. The system is given a gentle start, and the disk begins to rotate. (a) What is the angular velocity of the disk when the particle is at its lowest point? (b) At this point, what force must be exerted on the particle by the disk to keep it on the disk?

56 •• A ring 1.5 m in diameter is pivoted at one point on its circumference so that it is free to rotate about a horizontal axis. Initially, the line joining the support and center is horizontal. (a) If released from rest, what is its maximum angular velocity? (b) What must its initial angular velocity be if it is to just make a complete revolution?

57 •• You set out to design a car that uses the energy stored in a flywheel consisting of a uniform 100-kg cylinder of radius R. The flywheel must deliver an average of 2 MJ of mechanical energy per kilometer, with a maximum angular velocity of 400 rev/s. Find the least value of R such that the car can travel 300 km without the flywheel having to be recharged.

58 •• A ladder that is 8.6 m long and has mass 60 kg is placed in a nearly vertical position against the wall of a building. You stand on a rung with your center of mass at the top of the ladder. Assume that your mass is 80 kg. As you lean back slightly, the ladder begins to rotate about its base away from the wall. Is it better to quickly step off the ladder and drop to the ground or to hold onto the ladder and step off just before the top end hits the ground?

59 ••• Consider the situation in Problem 58 with a ladder of length L and mass M. Find the ratio of your speed as you hit the ground if you hang on to the ladder to your speed if you immediately step off as a function of the mass ratio M/m, where m is your mass.

Pulleys, Yo-Yos, and Hanging Things

60 •• A 4-kg block resting on a frictionless horizontal ledge is attached to a string that passes over a pulley and is attached to a hanging 2-kg block (Figure 9-45). The pulley is a uniform disk of radius 8 cm and mass 0.6 kg. (a) Find the speed of the 2-kg block after it falls from rest a distance of 2.5 m. (b) What is the angular velocity of the pulley at this time?

Figure 9-45
Problems 60–63

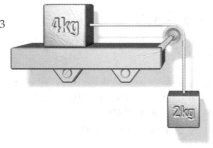

61 •• For the system in Problem 60, find the linear acceleration of each block and the tension in the string.

62 •• Work Problem 60 for the case in which the coefficient of friction between the ledge and the 4-kg block is 0.25.

63 •• Work Problem 61 for the case in which the coefficient of friction between the ledge and the 4-kg block is 0.25.

64 •• In 1993, a giant yo-yo of mass 400 kg and measuring about 1.5 m in radius was dropped from a crane 57 m high. Assuming that the axle of the yo-yo had a radius of $r = 0.1$ m, find the velocity of the descent v at the end of the fall.

65 •• A 1200-kg car is being unloaded by a winch. At the moment shown in Figure 9-46, the gearbox shaft of the winch breaks, and the car falls from rest. During the car's fall, there

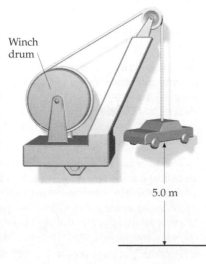

Figure 9-46 Problem 65

is no slipping between the (massless) rope, the pulley, and the winch drum. The moment of inertia of the winch drum is 320 kg·m^2 and that of the pulley is 4 kg·m^2. The radius of the winch drum is 0.80 m and that of the pulley is 0.30 m. Find the speed of the car as it hits the water.

66 •• The system in Figure 9-47 is released from rest. The 30-kg block is 2 m above the ledge. The pulley is a uniform disk with a radius of 10 cm and mass of 5 kg. Find (a) the speed of the 30-kg block just before it hits the ledge, (b) the angular speed of the pulley at that time, (c) the tensions in the strings, and (d) the time it takes for the 30-kg block to reach the ledge. Assume that the string does not slip on the pulley.

Figure 9-47
Problem 66

Figure 9-48 Problem 67

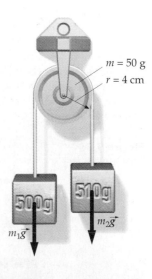

67 •• A uniform sphere of mass M and radius R is free to rotate about a horizontal axis through its center. A string is wrapped around the sphere and is attached to an object of mass m as shown in Figure 9-48. Find (a) the acceleration of the object, and (b) the tension in the string.

68 •• An Atwood's machine has two objects of mass $m_1 = 500$ g and $m_2 = 510$ g, connected by a string of negligible mass that passes over a frictionless pulley (Figure 9-49). The pulley is a uniform disk with a mass of 50 g and a radius of 4 cm. The string does not slip on the pulley. (a) Find the acceleration of the objects. (b) What is the tension in the string supporting m_1? In the string supporting m_2? By how much do they differ? (c) What would your answers have been if you had neglected the mass of the pulley?

Figure 9-49 Problem 68

69 •• Two objects are attached to ropes that are attached to wheels on a common axle as shown in Figure 9-50. The total moment of inertia of the two wheels is 40 kg·m^2. The radii of the wheels are $R_1 = 1.2$ m and $R_2 = 0.4$ m. (a) If $m_1 = 24$ kg, find m_2 such that there is no angular acceleration of the wheels. (b) If 12 kg is gently added to the top of m_1, find the angular acceleration of the wheels and the tensions in the ropes.

Figure 9-50 Problem 69

Figure 9-51
Problems 70 and 71

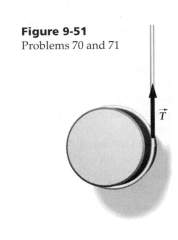

$\vec{T}$

70 •• A uniform cylinder of mass M and radius R has a string wrapped around it. The string is held fixed, and the cylinder falls vertically as shown in Figure 9-51. (a) Show that the acceleration of the cylinder is downward with a magnitude $a = 2g/3$. (b) Find the tension in the string.

71 •• The cylinder in Figure 9-51 is held by a hand that is accelerated upward so that the center of mass of the cylinder does not move. Find (a) the tension in the string, (b) the angular acceleration of the cylinder, and (c) the acceleration of the hand.

72 •• A 0.1-kg yo-yo consists of two solid disks of radius 10 cm joined together by a massless rod of radius 1 cm and a string wrapped around the rod. One end of the string is held fixed and is under constant tension T as the yo-yo is released. Find the acceleration of the yo-yo and the tension T.

73 •• A uniform cylinder of mass m_1 and radius R is pivoted on frictionless bearings. A massless string wrapped around the cylinder connects to a mass m_2, which is on a frictionless incline of angle θ as shown in Figure 9-52. The system is released from rest with m_2 a height h above the bottom of the incline. (a) What is the acceleration of m_2? (b) What is the tension in the string? (c) What is the total energy of the system when m_2 is at height h? (d) What is the total energy when m_2 is at the bottom of the incline and has a speed v? (e) What is the speed v? (f) Evaluate your answers for the extreme cases of $\theta = 0°$, $\theta = 90°$, and $m_1 = 0$.

Figure 9-52 Problem 73

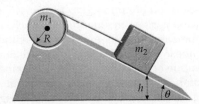

74 •• A device for measuring the moment of inertia of an object is shown in Figure 9-53. A circular platform has a concentric drum of radius 10 cm about which a string is wound. The string passes over a frictionless pulley to a weight of mass M. The weight is released from rest, and the time for it to drop a distance D is measured. The system is then rewound, the object placed on the platform, and the system again released from rest. The time required for the weight to drop the same distance D then provides the data needed to calculate I. With $M = 2.5$ kg, and $D = 1.8$ m, the time is 4.2 s. (a) Find the combined moment of inertia of the platform, drum, shaft, and pulley. (b) With an object placed on the platform, the time is 6.8 s for $D = 1.8$ m. Find I of that object about the axis of the platform.

Figure 9-53
Problem 74

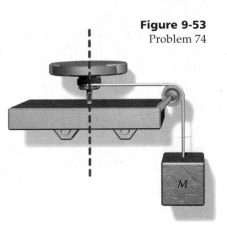

Objects Rolling Without Slipping

75 • True or false: When an object rolls without slipping, friction does no work on the object.

76 • A wheel of radius R is rolling without slipping. The velocity of the point on the rim that is in contact with the surface, relative to the surface, is

(a) equal to $R\omega$ in the direction of motion of the center of mass.
(b) equal to $R\omega$ opposite the direction of motion of the center of mass.
(c) zero.
(d) equal to the velocity of the center of mass and in the same direction.
(e) equal to the velocity of the center of mass but in the opposite direction.

77 •• A solid cylinder and a solid sphere have equal masses. Both roll without slipping on a horizontal surface. If their kinetic energies are the same, then

(a) the translational speed of the cylinder is greater than that of the sphere.
(b) the translational speed of the cylinder is less than that of the sphere.
(c) the translational speeds of the two objects are the same.
(d) (a), (b), or (c) could be correct depending on the radii of the objects.

78 •• Starting from rest at the same time, a coin and a ring roll down an incline without slipping. Which of the following is true?

(a) The ring reaches the bottom first.
(b) The coin reaches the bottom first.
(c) The coin and ring arrive at the bottom simultaneously.
(d) The race to the bottom depends on their relative masses.
(e) The race to the bottom depends on their relative diameters.

79 •• For a hoop of mass M and radius R that is rolling without slipping, which is larger, its translational kinetic energy or its rotational kinetic energy?

(a) Translational kinetic energy is larger.
(b) Rotational kinetic energy is larger.
(c) Both are the same size.
(d) The answer depends on the radius.
(e) The answer depends on the mass.

80 •• For a disk of mass M and radius R that is rolling without slipping, which is larger, its translational kinetic energy or its rotational kinetic energy?

(a) Translational kinetic energy is larger.
(b) Rotational kinetic energy is larger.
(c) Both are the same size.
(d) The answer depends on the radius.
(e) The answer depends on the mass.

81 •• A ball rolls without slipping along a horizontal plane. Show that the frictional force acting on the ball must be zero. *Hint:* Consider a possible direction for the action of the frictional force and what effects such a force would have on the velocity of the center of mass and on the angular velocity.

82 • A homogeneous solid cylinder rolls without slipping on a horizontal surface. The total kinetic energy is K. The kinetic energy due to rotation about its center of mass is

(a) $\frac{1}{2}K$.   (b) $\frac{1}{3}K$.   (c) $\frac{4}{7}K$.
(d) none of the above.

83 • A homogeneous cylinder of radius 18 cm and mass 60 kg is rolling without slipping along a horizontal floor at 5 m/s. How much work is needed to stop the cylinder?

84 • Find the percentages of the total kinetic energy associated with rotation and translation, respectively, for an object that is rolling without slipping if the object is (a) a uniform sphere, (b) a uniform cylinder, or (c) a hoop.

85 • A hoop of radius 0.40 m and mass 0.6 kg is rolling without slipping at a speed of 15 m/s toward an incline of slope 30°. How far up the incline will the hoop roll, assuming that it rolls without slipping?

86 • A ball rolls without slipping down an incline of angle θ. The coefficient of static friction is μ_s. Find (a) the acceleration of the ball, (b) the force of friction, and (c) the maximum angle of the incline for which the ball will roll without slipping.

87 •• An empty can of total mass $3M$ is rolling without slipping. If its mass is distributed as in Figure 9-54, what is the value of the ratio of kinetic energy of translation to the kinetic energy of rotation about its center of mass?

Figure 9-54 Problem 87

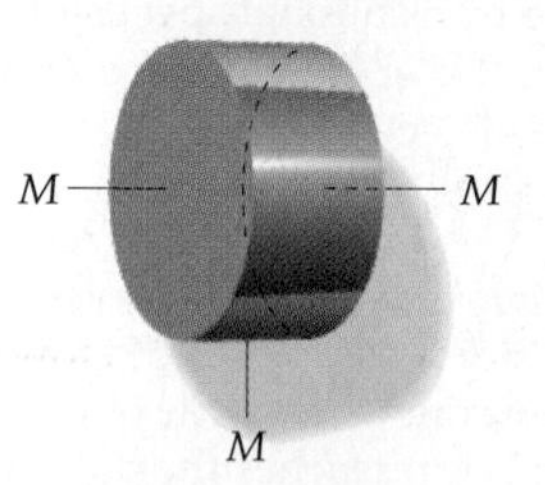

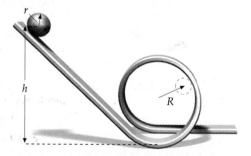

M — — — M

M

88 •• A bicycle of mass 14 kg has 1.2-m-diameter wheels, each of mass 3 kg. The mass of the rider is 38 kg. Estimate the

fraction of the total kinetic energy of bicycle and rider associated with rotation of the wheels.

89 •• A hollow sphere and uniform sphere of the same mass m and radius R roll down an inclined plane from the same height H without slipping (Figure 9-55). Each is moving horizontally as it leaves the ramp. When the spheres hit the ground, the range of the hollow sphere is L. Find the range L' of the uniform sphere.

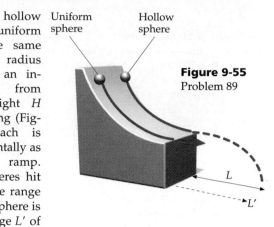

Uniform sphere

Hollow sphere

Figure 9-55
Problem 89

L

L'

90 •• A hollow cylinder and a uniform cylinder are rolling horizontally without slipping. The speed of the hollow cylinder is v. The cylinders encounter an inclined plane that they climb without slipping. If the maximum height they reach is the same, find the initial speed v' of the uniform cylinder.

91 •• A hollow, thin-walled cylinder and a solid sphere start from rest and roll without slipping down an inclined plane of length 3 m. The cylinder arrives at the bottom of the plane 2.4 s after the sphere. Determine the angle between the inclined plane and the horizontal.

92 •• A uniform solid sphere of radius r starts from rest at a height h and rolls without slipping along the loop-the-loop track of radius R as shown in Figure 9-56. (a) What is the smallest value of h for which the sphere will not leave the track at the top of the loop? (b) What would h have to be if, instead of rolling, the ball slides without friction?

Figure 9-56
Problem 92

r

h

R

93 ••• A wheel has a thin 3.0-kg rim and four spokes each of mass 1.2 kg. Find the kinetic energy of the wheel when it rolls at 6.0 m/s on a horizontal surface.

94 ••• Two uniform 20-kg disks of radius 30 cm are connected by a short rod of radius 2 cm and mass 1 kg. When the rod is placed on a plane inclined at 30°, such that the disks hang over the sides, the assembly rolls without slipping. Find (a) the linear acceleration of the system, and (b) the angular acceleration of the system. (c) Find the kinetic energy of translation of the system after it has rolled 2 m down the incline starting from rest. (d) Find the kinetic energy of rotation of the system at the same point.

95 ••• A wheel of radius R rolls without slipping at a speed V. The coordinates of the center of the wheel are X, Y. (a) Show that the x and y coordinates of point P in Figure 9-57 are $X + r_0 \cos \theta$ and $R + r_0 \sin \theta$, respectively. (b) Show that the total velocity v of point P has the components $v_x = V + (r_0 V \sin \theta)/R$ and $v_y = -(r_0 V \cos \theta)/R$. (c) Show that at the instant that $X = 0$, v and r are perpendicular to each other by calculating $\vec{v} \cdot \vec{r}$. (d) Show that $v = r\omega$, where $\omega = V/R$ is the angular velocity of the wheel. These results demonstrate that, in the case of rolling without slipping, the motion is the same as if the rolling object were instantaneously rotating about the point of contact with an angular speed $w = V/R$.

Figure 9-57
Problem 95

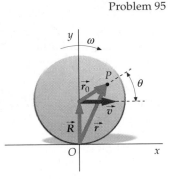

96 ••• A uniform cylinder of mass M and radius R rests on a block of mass m, which in turn is at rest on a horizontal, frictionless table (Figure 9-58). If a horizontal force $\vec{F}$ is applied to the block, it accelerates and the cylinder rolls without slipping. Find the acceleration of the block.

Figure 9-58
Problems 96, 97, and 98

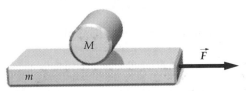

97 ••• (a) Find the angular acceleration of the cylinder in Problem 96. Is the cylinder rotating clockwise or counterclockwise? (b) What is the cylinder's linear acceleration relative to the table? Let the direction of $\vec{F}$ be the positive direction. (c) What is the linear acceleration of the cylinder relative to the block?

98 ••• If the force in Problem 96 acts over a distance d, find (a) the kinetic energy of the block, and (b) the kinetic energy of the cylinder. (c) Show that the total kinetic energy is equal to the work done on the system.

99 ••• A marble of radius 1 cm rolls from rest without slipping from the top of a large sphere of radius 80 cm, which is held fixed (Figure 9-59). Find the angle from the top of the sphere to the point where the marble breaks contact with the sphere.

Figure 9-59 Problem 99

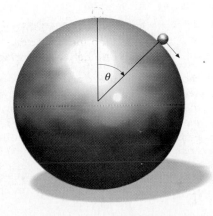

Rolling With Slipping (optional)

100 • True or false: When a sphere rolls and slips on a rough surface, mechanical energy is dissipated.

101 • A cue ball is hit very near the top so that it starts to move with topspin. As it slides, the force of friction
(a) increases v_{cm}.
(b) decreases v_{cm}.
(c) has no effect on v_{cm}.

102 •• A bowling ball of mass M and radius R is thrown such that at the instant it touches the floor it is moving horizontally with a speed v_0 and is not rotating. It slides for a time t_1 a distance s_1 before it begins to roll without slipping. (a) If μ_k is the coefficient of sliding friction between the ball and the floor, find s_1, t_1, and the final speed v_1 of the ball. (b) Find the ratio of the final mechanical energy to the initial mechanical energy of the ball. (c) Evaluate these quantities for $v_0 = 8$ m/s and $\mu_k = 0.06$.

103 •• A cue ball of radius r is initially at rest on a horizontal pool table (Figure 9-60). It is struck by a horizontal cue stick that delivers a force of magnitude P_0 for a very short time Δt. The stick strikes the ball at a point h above the ball's point of contact with the table. (a) Show that the ball's initial angular velocity ω_0 is related to the initial linear velocity of its center of mass v_0 by $\omega_0 = 5v_0(h - r)/2r^2$.

Figure 9-60 Problem 103

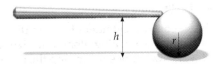

104 ••• A uniform spherical ball is set rotating about a horizontal axis with an angular speed ω_0 and is placed on the floor. If the coefficient of sliding friction between the ball and the floor is μ_k, find the speed of the center of mass of the ball when it begins to roll without slipping.

105 ••• A uniform solid ball resting on a horizontal surface has a mass of 20 g and a radius of 5 cm. A sharp force is applied to the ball in the horizontal direction 9 cm above the horizontal surface. The force increases linearly from 0 to a peak value of 40,000 N in 10^{-4} s and then decreases linearly to 0 in 10^{-4} s. (a) What is the velocity of the ball after impact? (b) What is the angular velocity of the ball after impact? (c) What is the velocity of the ball when it begins to roll without sliding? (d) For how long does the ball slide on the surface? Assume that $\mu_k = 0.5$.

106 ••• A 0.3-kg billiard ball of radius 3 cm is given a sharp blow by a cue stick. The applied force is horizontal and passes through the center of the ball. The initial velocity of the ball is 4 m/s. The coefficient of kinetic friction is 0.6. (a) For how many seconds does the ball slide before it begins to roll without slipping? (b) How far does it slide? (c) What is its velocity once it begins rolling without slipping?

107 ••• A billiard ball initially at rest is given a sharp blow by a cue stick. The force is horizontal and is applied at a distance $2R/3$ below the centerline, as shown in Figure 9-61. The initial speed of the ball is v_0, and the coefficient of kinetic friction is μ_k. (a) What is the initial angular speed ω_0? (b) What is the speed of the ball once it begins to roll without slipping? (c) What is the initial kinetic energy of the ball? (d) What is the frictional work done as it slides on the table?

Figure 9-61 Problem 107

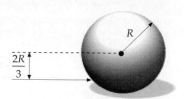

108 ••• A bowling ball of radius R is given an initial velocity v_0 down the lane and a forward spin $\omega_0 = 3v_0/R$. The coefficient of kinetic friction is μ_k. (a) What is the speed of the ball when it begins to roll without slipping? (b) For how long does the ball slide before it begins to roll without slipping? (c) What distance does the ball slide down the lane before it begins rolling without slipping?

109 ••• A solid cylinder of mass M resting on its side on a horizontal surface is given a sharp blow by a cue stick. The applied force is horizontal and passes through the center of the cylinder so that the cylinder begins translating with initial velocity v_0. The coefficient of sliding friction between the cylinder and surface is μ_k. (a) What is the translational velocity of the cylinder when it is rolling without slipping? (b) How far does the cylinder travel before it rolls without slipping? (c) What fraction of its initial mechanical energy is dissipated in friction?

General Problems

110 • The torque exerted on an orbiting communications satellite by the gravitational pull of the earth is

(a) directed toward the earth.
(b) directed parallel to the earth's axis and toward the north pole.
(c) directed parallel to the earth's axis and toward the south pole.
(d) directed toward the satellite.
(e) zero.

111 • The moon rotates as it revolves around the earth so that we always see the same side. Use this fact to find the angular velocity (in rad/s) of the moon about its axis. (The period of revolution of the moon about the earth is 27.3 days.)

112 • Find the moment of inertia of a hoop about an axis perpendicular to the plane of the hoop and through its edge.

113 •• The radius of a park merry-go-round is 2.2 m. To start it rotating, you wrap a rope around it and pull with a force of 260 N for 12 s. During this time, the merry-go-round makes one complete rotation. (a) Find the angular acceleration of the merry-go-round. (b) What torque is exerted by the rope on the merry-go-round? (c) What is the moment of inertia of the merry-go-round?

114 •• A uniform disk of radius 0.12 m and mass 5 kg is pivoted such that it rotates freely about its central axis (Figure 9-62). A string wrapped around the disk is pulled with a force of 20 N. (a) What is the torque exerted on the disk? (b) What is the angular acceleration of the disk? (c) If the disk starts from rest, what is its angular velocity after 5 s? (d) What is its kinetic energy after 5 s? (e) What is the total angle

Figure 9-62 Problem 114

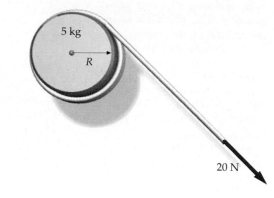

θ that the disk turns through in 5 s? (f) Show that the work done by the torque $\tau \Delta\theta$ equals the kinetic energy.

115 •• A 0.25-kg rod of length 80 cm is suspended by a frictionless pivot at one end. It is held horizontal and released. Immediately after it is released, what is (a) the acceleration of the center of the rod, and (b) the initial acceleration of a point on the end of the rod? (c) Find the linear velocity of the center of mass of the rod when it is vertical.

116 •• A uniform rod of length $3L$ is pivoted as shown in Figure 9-63 and held in a horizontal position. What is the initial angular acceleration α of the rod upon release?

117 •• A uniform rod of length L and mass m is pivoted at the middle as shown in Figure 9-64. It has a load of mass $2m$ attached to one of the ends. If the system is released from a horizontal position, what is the maximum velocity of the load?

Figure 9-63 Problem 116   **Figure 9-64** Problem 117

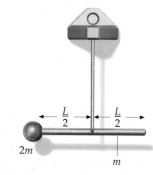

118 •• A marble of mass M and radius R rolls without slipping down the track on the left from a height h_1 as shown in Figure 9-65. The marble then goes up the *frictionless* track on the right to a height h_2. Find h_2.

Figure 9-65 Problem 118

119 •• A uniform disk with a mass of 120 kg and a radius of 1.4 m rotates initially with an angular speed of 1100 rev/min. (a) A constant tangential force is applied at a radial distance of 0.6 m. What work must this force do to stop the wheel? (*b*) If the wheel is brought to rest in 2.5 min, what torque does the force produce? What is the magnitude of the force? (*c*) How many revolutions does the wheel make in these 2.5 min?

120 •• A park merry-go-round consists of a 240-kg circular wooden platform 4.00 m in diameter. Four children running alongside push tangentially along the platform's circumference until, starting from rest, the merry-go-round reaches a steady speed of one complete revolution every 2.8 s. (*a*) If each child exerts a force of 26 N, how far does each child run? (*b*) What is the angular acceleration of the merry-go-round? (*c*) How much work does each child do? (*d*) What is the kinetic energy of the merry-go-round?

121 •• A hoop of mass 1.5 kg and radius 65 cm has a string wrapped around its circumference and lies flat on a horizontal frictionless table. The string is pulled with a force of 5 N. (*a*) How far does the center of the hoop travel in 3 s? (*b*) What is the angular velocity of the hoop about its center of mass after 3 s?

122 •• A vertical grinding wheel is a uniform disk of mass 60 kg and radius 45 cm. It has a handle of radius 65 cm of negligible mass. A 25-kg load is attached to the handle when it is in the horizontal position. Neglecting friction, find (*a*) the initial angular acceleration of the wheel, and (*b*) the maximum angular velocity of the wheel.

123 •• In this problem, you are to derive the perpendicular-axis theorem for planar objects, which relates the moments of inertia about two perpendicular axes in the plane of Figure 9-66 to the moment of inertia about a third axis that is perpendicular to the plane of figure. Consider the mass element *dm* for the figure shown in the *xy* plane. (*a*) Write an expression for the moment of inertia of the figure about the *z* axis in terms of *dm* and *r*. (*b*) Relate the distance *r* of *dm* to the distances *x* and *y*, and thus show that $I_z = I_y + I_x$. (*c*) Apply your result to find the moment of inertia of a uniform disk of radius *R* about a diameter of the disk.

Figure 9-66
Problem 123

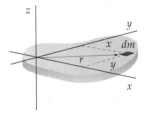

124 •• A uniform disk of radius *R* and mass *M* is pivoted about a horizontal axis parallel to its symmetry axis and passing through its edge such that it can swing freely in a vertical plane (Figure 9-67). It is released from rest with its center of mass at the same height as the pivot. (*a*) What is the angular

Figure 9-67 Problem 124

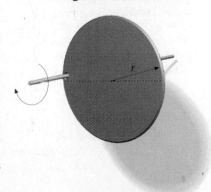

velocity of the disk when its center of mass is directly below the pivot? (*b*) What force is exerted by the pivot at this time?

125 •• A spool of mass *M* rests on an inclined plane at a distance *D* from the bottom. The ends of the spool have radius *R*, the center has radius *r*, and the moment of inertia of the spool about its axis is *I*. A long string of negligible mass is wound many times around the center of the spool. The other end of the string is fastened to a hook at the top of the inclined plane such that the string always pulls parallel to the slope as shown in Figure 9-68. (*a*) Suppose that initially the slope is so icy that there is *no* friction. How does the spool move as it slips down the slope? Use energy considerations to determine the speed of the center of mass of the spool when it reaches the bottom of the slope. Give your answer in terms of *M*, *I*, *r*, *R*, *g*, *D*, and θ. (*b*) Now suppose that the ice is gone and that when the spool is set up in the same way, there is enough friction to keep it from slipping on the slope. What is the direction and magnitude of the friction force in this case?

Figure 9-68 Problem 125

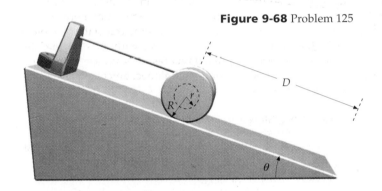

126 •• Ian has suggested another improvement for the game of hockey. Instead of the usual two-minute penalty, he would like to see an offender placed in a barrel at mid-ice and then spun in a circle by the other team. When the offender is silly with dizziness, he is put back into the game. Assume that a penalized player in a barrel approximates a uniform, 100-kg cylinder of radius 0.60 m, and that the ice is smooth (Figure 9-69). Ropes are wound around the barrel, so that pulling them causes rotation. If two players simultaneously pull the ropes with forces of 40 N and 60 N for 6 s, describe the motion of the barrel. Give its acceleration, velocity, and the position of its center of mass as functions of time.

Figure 9-69 Problem 126

127 •• A solid metal rod 1.5 m long is free to rotate without friction about a fixed, horizontal axis perpendicular to the rod and passing through one end. The other end is held in a horizontal position. Small coins of mass *m* are placed on the rod 25 cm, 50 cm, 75 cm, 1 m, 1.25 m, and 1.5 m from the bearing. If the free end is now released, calculate the initial force exerted on each coin by the rod. Assume that the mass of the coins may be neglected in comparison to the mass of the rod.

128 •• A thin rod of length L and mass M is supported in a horizontal position by two strings, one attached to each end as shown in Figure 9-70. If one string is cut, the rod begins to rotate about the point where it connects to the other string (point A in the figure). (*a*) Find the initial acceleration of the center of mass of the rod. (*b*) Show that the initial tension in the string is $mg/4$ and that the initial angular acceleration of the rod about an axis through the point A is $3g/2L$. (*c*) At what distance from point A is the initial linear acceleration equal to g?

Figure 9-70 Problem 128

129 •• Figure 9-71 shows a hollow cylinder of length 1.8 m, mass 0.8 kg, and radius 0.2 m. The cylinder is free to rotate about a vertical axis that passes through its center and is perpendicular to the cylinder's axis. Inside the cylinder are two masses of 0.2 kg each, attached to springs of spring constant k and unstretched lengths 0.4 m. The inside walls of the cylinder are frictionless. (*a*) Determine the value of the spring constant if the masses are located 0.8 m from the center of the cylinder when the cylinder rotates at 24 rad/s. (*b*) How much work was needed to bring the system from $\omega = 0$ to $\omega = 24$ rad/s?

Figure 9-71 Problems 129 and 130

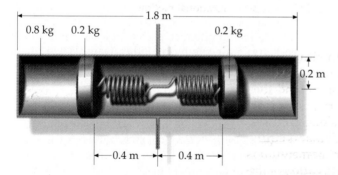

130 •• Suppose that for the system described in Problem 129, the spring constants are each $k = 60$ N/m. The system starts from rest and slowly accelerates until the masses are 0.8 m from the center of the cylinder. How much work was done in the process?

131 •• A string is wrapped around a uniform cylinder of radius R and mass M that rests on a horizontal frictionless surface. The string is pulled horizontally from the top with force F. (*a*) Show that the angular acceleration of the cylinder is twice that needed for rolling without slipping, so that the bottom point on the cylinder slides backward against the table. (*b*) Find the magnitude and direction of the frictional force between the table and cylinder needed for the cylinder to roll without slipping. What is the acceleration of the cylinder in this case?

132 •• Figure 9-72 shows a solid cylinder of mass M and radius R to which a hollow cylinder of radius r is attached. A string is wound about the hollow cylinder. The solid cylinder rests on a horizontal surface. The coefficient of static friction between the cylinder and surface is μ_s. If a light tension is applied to the string in the vertical direction, the cylinder will roll to the left; if the tension is applied with the string horizontally, the cylinder rolls to the right. Find the angle of the string with the horizontal that will allow the cylinder to remain stationary when a small tension is applied to the string.

Figure 9-72 Problem 132

133 ••• A heavy, uniform cylinder has a mass m and a radius R (Figure 9-73). It is accelerated by a force $\vec{T}$, which is applied through a rope wound around a light drum of radius r that is attached to the cylinder. The coefficient of static friction is sufficient for the cylinder to roll without slipping. (*a*) Find the frictional force. (*b*) Find the acceleration a of the center of the cylinder. (*c*) Is it possible to choose r so that a is greater than T/m? How? (*d*) What is the direction of the frictional force in the circumstances of part (*c*)?

Figure 9-73 Problem 133

134 ••• A uniform stick of length L and mass M is hinged at one end. It is released from rest at an angle θ_0 with the vertical. Show that when the angle with the vertical is θ, the hinge exerts a force F_r along the stick and a force F_t perpendicular to the stick given by $F_r = \frac{1}{2}Mg(5\cos\theta - 3\cos\theta_0)$ and $F_t = \frac{1}{4}Mg\sin\theta$.

Conservation of Angular Momentum

Waterspouts off the Grand Bahama Islands offer a stunning visualization of rotational motion.

In this chapter, we extend our study of rotational motion to situations in which the direction of the axis of rotation may change. We begin with an examination of the vector properties of angular velocity and torque and then we introduce the concept of angular momentum, which is the rotational analog of linear momentum. We then show that the net torque acting on a system equals the rate of change of its angular momentum, a result that is equivalent to Newton's second law for rotational motion. Angular momentum is therefore conserved in systems with zero net torque. Like conservation of linear momentum, conservation of angular momentum is a fundamental law of nature, applying even in the atomic domain where Newtonian mechanics fails.

10-1 The Vector Nature of Rotation

To indicate the direction of rotation about a *fixed* axis, we assigned plus and minus signs to the angular velocity vector, just as we used them to indicate the direction of the velocity vector in one-dimensional motion. When the direction of the axis of rotation is not fixed in space, the vector nature of angular velocity becomes important. Consider, for example, the rotating disk in Figure 10-1. We describe the direction of rotation by giving the direction of

Figure 10-1 A disk rotating about an axis through its center and perpendicular to its plane.

the axis of rotation. (By symmetry, all directions in the plane of the disk are equivalent.) We therefore choose the angular velocity vector $\vec{\omega}$ to be along the axis of rotation, and we determine the direction of $\vec{\omega}$ by a convention known as the **right-hand rule**, which is illustrated in Figure 10-2. Thus, if the rotation is counterclockwise, as in Figure 10-1, $\vec{\omega}$ is outward; if it is clockwise, $\vec{\omega}$ is inward.

We apply similar considerations to the torque. Figure 10-3 shows a force $\vec{F}$ acting on a particle at some position $\vec{r}$ relative to the origin O. The torque $\vec{\tau}$ exerted by this force relative to the origin O is defined as a vector that is perpendicular to the plane formed by $\vec{F}$ and $\vec{r}$, and has magnitude $Fr \sin \phi$, where ϕ is the angle between $\vec{F}$ and $\vec{r}$. If $\vec{F}$ and $\vec{r}$ are in the xy plane, as in Figure 10-3, the torque is along the z axis. If $\vec{F}$ is applied to the rim of a disk of radius r, as shown in Figure 10-4, the torque has the magnitude Fr, and is along the axis of rotation as shown.

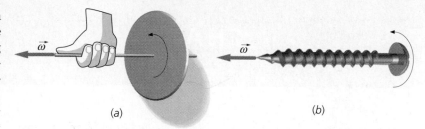

Figure 10-2 The right-hand rule for determining the direction of the angular velocity $\vec{\omega}$. (*a*) When the fingers of the right hand curl in the direction of rotation, the thumb points in the direction of $\vec{\omega}$. (*b*) Looked at another way, the direction of $\vec{\omega}$ is that of the advance of a rotating right-hand screw.

Figure 10-3 Force $\vec{F}$ acting on a particle at position $\vec{r}$.

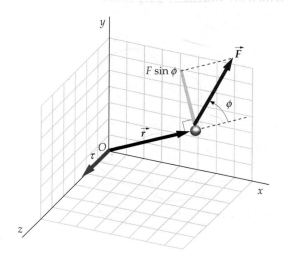

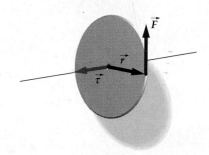

Figure 10-4 The force $\vec{F}$, which is tangent to the disk, exerts a torque along the axis of rotation.

The Cross Product

Torque is expressed mathematically as the **cross product** (or **vector product**) of $\vec{r}$ and $\vec{F}$:

$$\vec{\tau} = \vec{r} \times \vec{F} \qquad \text{10-1}$$

The cross product of two vectors $\vec{A}$ and $\vec{B}$ is defined to be a vector $\vec{C} = \vec{A} \times \vec{B}$ whose magnitude equals the area of the parallelogram formed by the two vectors (Figure 10-5). The vector $\vec{C}$ is perpendicular to the plane containing $\vec{A}$ and $\vec{B}$ in the direction given by the right-hand rule as $\vec{A}$ is rotated into $\vec{B}$ through the smaller angle between these vectors (Figure 10-6). If ϕ is the angle between the two vectors and $\hat{n}$ is a unit vector that is perpendicular to each vector in the direction described, the cross product of $\vec{A}$ and $\vec{B}$ is

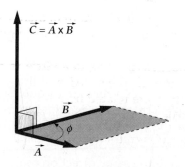

Figure 10-5 The cross product $\vec{A} \times \vec{B}$ is a vector $\vec{C}$ that is perpendicular to both $\vec{A}$ and $\vec{B}$ and has a magnitude $AB \sin \phi$, which equals the area of the parallelogram shown.

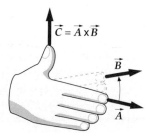

Figure 10-6 The direction of $\vec{A} \times \vec{B}$ is given by the right-hand rule when $\vec{A}$ is rotated into $\vec{B}$ through the angle ϕ.

$$\vec{A} \times \vec{B} = AB \sin \phi \, \hat{n} \qquad \text{10-2}$$

Definition—Cross product

If $\vec{A}$ and $\vec{B}$ are parallel, $\vec{A} \times \vec{B}$ is zero. It follows from the definition of the cross product that

$$\vec{A} \times \vec{A} = 0 \qquad \qquad \text{10-3}$$

and

$$\vec{A} \times \vec{B} = -\vec{B} \times \vec{A} \qquad \qquad \text{10-4}$$

Note that the order in which two vectors are multiplied in a cross product is significant. Below are some properties of the cross product of two vectors:

1. The cross product obeys a distributive law under addition:

$$\vec{A} \times (\vec{B} + \vec{C}) = \vec{A} \times \vec{B} + \vec{A} \times \vec{C} \qquad \qquad \text{10-5}$$

2. If $\vec{A}$ and $\vec{B}$ are functions of some variable such as t, the derivative of $\vec{A} \times \vec{B}$ follows the usual product rule for derivatives:

$$\frac{d}{dt}(\vec{A} \times \vec{B}) = \vec{A} \times \frac{d\vec{B}}{dt} + \frac{d\vec{A}}{dt} \times \vec{B} \qquad \qquad \text{10-6}$$

3. The unit vectors $\hat{i}$, $\hat{j}$, and $\hat{k}$ (Figure 10-7), which are mutually perpendicular, have cross products given by

$$\hat{i} \times \hat{j} = \hat{k}, \qquad \hat{j} \times \hat{k} = \hat{i}, \qquad \hat{k} \times \hat{i} = \hat{j} \qquad \qquad \text{10-7a}$$

Furthermore,

$$\hat{i} \times \hat{i} = \hat{j} \times \hat{j} = \hat{k} \times \hat{k} = 0 \qquad \qquad \text{10-7b}$$

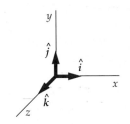

Figure 10-7 The unit vectors $\hat{i}$, $\hat{j}$, and $\hat{k}$ are mutually perpendicular and have magnitude 1.

10-2 Angular Momentum

Figure 10-8 shows a particle of mass m moving with a velocity $\vec{v}$ at a position $\vec{r}$ relative to the origin O. The linear momentum of the particle is $\vec{p} = m\vec{v}$. The **angular momentum** $\vec{L}$ of the particle relative to the origin O is defined to be the cross product of $\vec{r}$ and $\vec{p}$:

$$\vec{L} = \vec{r} \times \vec{p} \qquad \qquad \text{10-8}$$

Definition—Angular momentum of a particle

If $\vec{r}$ and $\vec{p}$ are in the xy plane, as in Figure 10-8, then $\vec{L}$ is along the z axis and is given by $\vec{L} = \vec{r} \times \vec{p} = mvr \sin \phi \, \hat{k}$. Like torque, angular momentum is defined *relative to a point in space*.

Figure 10-9 shows a particle moving in a circle in the xy plane with the center of the circle at the origin. The speed v of the particle and the magnitude of its angular velocity ω are related by $v = r\omega$. The angular momentum of the particle relative to the center of the circle is

$$\vec{L} = \vec{r} \times \vec{p} = \vec{r} \times m\vec{v} = rmv \sin 90° \, \hat{k} = rmv\hat{k} = mr^2\omega\hat{k} = mr^2\vec{\omega}$$

The angular momentum is in the same direction as the angular velocity.

Since mr^2 is the moment of inertia for a single particle about the z axis, we have

$$\vec{L} = mr^2\vec{\omega} = I\vec{\omega}$$

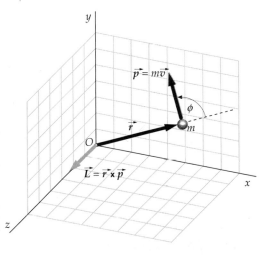

Figure 10-8 A particle with a momentum $\vec{p}$ at position $\vec{r}$ relative to the origin O.

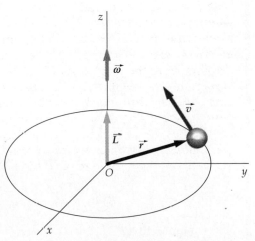

Figure 10-9 A particle moving in a circle has angular momentum relative to the center of the circle $\vec{L} = \vec{r} \times \vec{p} = I\vec{\omega}$.

This result does not hold for a general point. Figure 10-10 shows the angular-momentum vector $\vec{L}'$ for the same particle moving in the same circle but with $\vec{L}'$ computed relative to a point on the z axis that is not at the center of the circle. In this case, the angular momentum is not parallel to the angular velocity $\vec{\omega}$, which is along the z axis.

In Figure 10-11, we add a second particle of equal mass moving in the same circle. The angular-momentum vectors $\vec{L}_1'$ and $\vec{L}_2'$ are shown relative to the same point as in Figure 10-10. The total angular momentum $\vec{L}_1' + \vec{L}_2'$ of the two-particle system is again parallel to the angular velocity $\vec{\omega}$. In this case, the axis of rotation, the z axis, passes through the center of mass of the two-particle system, and the mass distribution is symmetric about this axis. Such an axis is called a **symmetry axis**. For any system of particles that rotates about a symmetry axis, the total angular momentum (which is the sum of the angular momenta of the individual particles) is parallel to the angular velocity and is given by

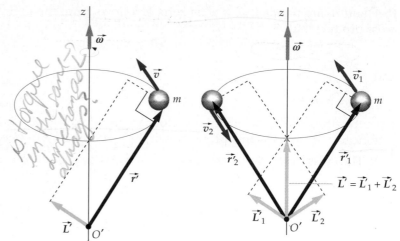

Figure 10-10

Figure 10-11

$$\vec{L} = I\vec{\omega} \qquad\qquad\qquad 10\text{-}9$$

Angular momentum of a system rotating about a symmetry axis

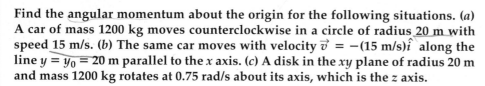

Example 10-1

Find the angular momentum about the origin for the following situations. (*a*) A car of mass 1200 kg moves counterclockwise in a circle of radius 20 m with speed 15 m/s. (*b*) The same car moves with velocity $\vec{v} = -(15 \text{ m/s})\hat{i}$ along the line $y = y_0 = 20$ m parallel to the x axis. (*c*) A disk in the xy plane of radius 20 m and mass 1200 kg rotates at 0.75 rad/s about its axis, which is the z axis.

Picture the Problem

(*a*) $\vec{r}$ and $\vec{p}$ are perpendicular, and $\vec{r} \times \vec{p}$ is in the z direction:

$$\vec{L} = \vec{r} \times \vec{p} = rmv\hat{k} = (20 \text{ m})(1200 \text{ kg})(15 \text{ m/s})\hat{k}$$
$$= 3.6 \times 10^5 \text{ kg·m}^2/\text{s } \hat{k}$$

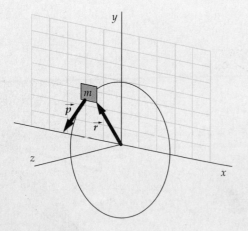

Figure 10-12

(b) 1. For the same car moving to the left along the line $y = y_0 = 20$ m, we express $\vec{r}$ and $\vec{p}$ in terms of unit vectors:

$$\vec{r} = x\hat{i} + y\hat{j} = x\hat{i} + y_0\hat{j}$$

$$\vec{p} = m\vec{v} = -p\hat{i}$$

Figure 10-13

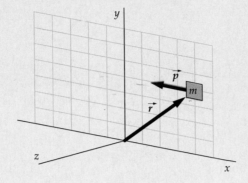

2. Now compute $\vec{r} \times \vec{p}$:

$$\vec{L} = \vec{r} \times \vec{p} = (x\hat{i} + y_0\hat{j}) \times (-p\hat{i})$$

$$= -xp(\hat{i} \times \hat{i}) - y_0 p(\hat{j} \times \hat{i})$$

$$= -xp(0) - y_0 p(-\hat{k}) = y_0 p\hat{k}$$

$$= (20 \text{ m})(1200 \text{ kg})(15 \text{ m/s})\hat{k}$$

$$= 3.6 \times 10^5 \text{ kg}\cdot\text{m}^2/\text{s } \hat{k}$$

(c)   Use $\vec{L} = I\vec{\omega}$:

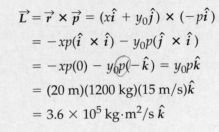

$$\vec{L} = I\vec{\omega} = I\omega\hat{k} = \left(\frac{1}{2}mR^2\right)\omega\hat{k}$$

$$= \frac{1}{2}(1200 \text{ kg})(20 \text{ m})^2(0.75 \text{ rad/s})\hat{k}$$

$$= 1.8 \times 10^5 \text{ kg}\cdot\text{m}^2/\text{s}\hat{k}$$

Figure 10-14

Remarks The angular momentum of the car moving in a circle in (a) is the same as that of the car moving along a straight line in (b). In (c), the velocity of a point on the rim is $v = R\omega = (20 \text{ m})(0.75 \text{ rad/s}) = 15 \text{ m/s}$, the same as the velocity of the car in parts (a) and (b). The moment of inertia of a 1200-kg disk of radius 20 m is less than that of a 1200-kg car at 20 m from the axis because much of the mass of the disk is closer to the axis of rotation.

Figure 10-15 shows a disk rotating about an axis parallel to its symmetry axis a distance h away. The angular momentum $\vec{L}_i$ of the ith particle on the disk relative to point O' is in the z direction, parallel to $\vec{\omega}$, and given by

$$\vec{L}_i = \vec{r}_i \times m_i\vec{v}_i = m_i v_i r_i\hat{k} = m_i r_i^2 \omega\hat{k}$$

Summing over all the particles yields

$$\vec{L} = \sum \vec{L}_i = \sum m_i r_i^2 \omega\hat{k} = I\omega\hat{k} = I\vec{\omega}$$

where I' is the moment of inertia of the disk about the axis through O'. Equation 10-9 thus also holds for rotation about an axis parallel to a symmetry axis. I' and I are related by the parallel-axis theorem (Equation 9-20):

$$I' = I_{\text{cm}} + Mh^2$$

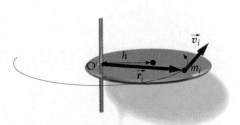

Figure 10-15 A disk rotating about an axis through point O' a distance h from the parallel symmetry axis.

Then

$$\vec{L} = I\omega\hat{k} = (I_{cm} + Mh^2)\omega\hat{k} = I_{cm}\omega\hat{k} + Mh^2\omega\hat{k}$$

or

$$\vec{L} = \vec{L}_{cm} + Mv_{cm}h\hat{k}$$

where $\vec{L}_{cm}$ is the angular momentum about the center of mass, $v_{cm} = h\omega$ is the velocity of the center of mass, and $Mv_{cm}h\hat{k}$ is the angular momentum relative to O' of a particle of mass M moving with speed v_{cm}. This result, obtained for rotation about a fixed axis, holds in general:

> The angular momentum about any point O' is the angular momentum about the center of mass plus the angular momentum associated with the motion of the center of mass about O'.

The angular momentum of an object about its center of mass is called its **spin angular momentum,** whereas the angular momentum associated with the motion of the center of mass is called its **orbital angular momentum.**

$$\vec{L} = \vec{L}_{orbit} + \vec{L}_{spin} = \vec{r}_{cm} \times M\vec{v}_{cm} + \sum_i \vec{r}_i' \times m_i\vec{u}_i \qquad \text{10-10}$$

where $\vec{u}_i$ is the velocity of the ith particle relative to the center of mass. The earth has spin angular momentum due to its rotation about its axis and orbital angular momentum relative to the sun due to its revolving around the sun (Figure 10-16). The total angular momentum of the earth relative to the sun is the vector sum of its spin and orbital angular momenta.

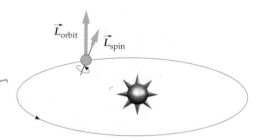

Figure 10-16 Spin angular momentum and orbital angular momentum of the earth.

10-3 Torque and Angular Momentum

We will now show that Newton's second law implies that the rate of change of the angular momentum of a particle equals the net torque acting on the particle. For a number of forces acting on a particle, the net torque relative to the origin O is the sum of the torques due to each force:

$$\vec{\tau}_{net} = \vec{r} \times \vec{F}_1 + \vec{r} \times \vec{F}_2 + \cdots = \vec{r} \times \sum_i \vec{F}_i = \vec{r} \times \vec{F}_{net}$$

According to Newton's second law, the net force equals the rate of change of the linear momentum $d\vec{p}/dt$. Thus,

$$\vec{\tau}_{net} = \vec{r} \times \vec{F}_{net} = \vec{r} \times \frac{d\vec{p}}{dt} \qquad \text{10-11}$$

We now compare this with the rate of change of the angular momentum. We can compute $d\vec{L}/dt$ using the product rule for derivatives:

$$\frac{d\vec{L}}{dt} = \frac{d}{dt}(\vec{r} \times \vec{p}) = \frac{d\vec{r}}{dt} \times \vec{p} + \vec{r} \times \frac{d\vec{p}}{dt}$$

The first term on the right of this equation is zero because

$$\frac{d\vec{r}}{dt} \times \vec{p} = \vec{v} \times m\vec{v} = 0$$

Thus

$$\frac{d\vec{L}}{dt} = \vec{r} \times \frac{d\vec{p}}{dt}$$ 10-12

Comparing this result with Equation 10-11 gives

$$\vec{\tau}_{net} = \frac{d\vec{L}}{dt}$$ 10-13

The net torque acting on a system of particles is the sum of the individual torques. The generalization of Equation 10-13 to a system of particles is then

$$\sum_i \vec{\tau} = \sum_i \frac{d\vec{L}_i}{dt} = \frac{d}{dt} \sum_i \vec{L}_i = \frac{d\vec{L}}{dt}$$

In this equation, the sum of the torques may include internal as well as external torques. We show in Section 10-4 that the sum of the internal torques must be zero. Therefore,

$$\sum_i \vec{\tau}_{i,ext} = \frac{d\vec{L}}{dt}$$ 10-14

The net external torque acting on a system equals the rate of change of the angular momentum of the system.

Newton's second law for rotation

Equation 10-14 is the rotational analog of $\vec{F}_{net,ext} = d\vec{p}/dt$ for linear motion. It holds for a general system of particles, rotating about any axis, whether or not the moment of inertia is constant. For a rigid body rotating about a fixed axis, the moment of inertia is constant and Equation 10-14 becomes

$$\sum_i \vec{\tau}_{i,ext} = \frac{d\vec{L}}{dt} = \frac{d(I\vec{\omega})}{dt} = I\frac{d\vec{\omega}}{dt} = I\vec{\alpha}$$ 10-15

where $\vec{\alpha} = d\vec{\omega}/dt$ is the angular acceleration vector. Equation 10-15 is the same as Equation 9-20.

The direction of rotation is changed by this bevel gear in a diesel engine.

Example 10-2

An Atwood's machine has two blocks of mass m_1 and m_2 ($m_1 > m_2$), connected by a string of negligible mass that passes over a pulley with frictionless bearings (Figure 10-17). The pulley is a uniform disk of mass M and radius R. The string does not slip on the pulley. Apply Equation 10-14 to the system consisting of both blocks plus pulley to find the angular acceleration of the pulley and the acceleration of the blocks.

Picture the Problem Let the pulley and blocks be in the xy plane with the z axis out of the page. We compute the torques and angular momenta about the center of the pulley. Since m_1 is greater than m_2, the disk will rotate counterclockwise corresponding to $\vec{\omega}$ out of the page in the positive z direction.

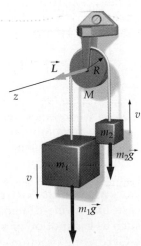

Figure 10-17

The weight of m_1 exerts a torque $m_1 g R$ out of the page, and the weight of m_2 exerts a torque $m_2 g R$ into the page. Since m_1 moves down and m_2 moves up, they both have angular momentum about the center of the pulley directed outward. Since the torque, angular-velocity, and angular-momentum vectors are all along the z axis, we can forget about their vector nature and treat the problem like a one-dimensional problem with positive assigned to counter-clockwise motion and negative to clockwise motion. The speed v of the blocks is related to the angular speed of the pulley ω by the nonslip condition $v = R\omega$.

1. Use $\sum \vec{\tau}_{i,\text{ext}} = d\vec{L}/dt$:

$$\sum \vec{\tau}_{i,\text{ext}} = \frac{d\vec{L}}{dt}$$

2. The total angular momentum about the center of the pulley equals the angular momentum of the pulley plus the angular momentum of the blocks, each of which is in the positive z direction:

$$L_z = L_p + L_1 + L_2 = I\omega + m_1 vR + m_2 vR$$

3. The weight $m_1 \vec{g}$ exerts a torque in the positive z direction, whereas $m_2 \vec{g}$ exerts a torque in the negative z direction. The lever arm for each force is R. The net torque is:

$$\tau_{z,\text{net}} = m_1 g R - m_2 g R$$

4. Substitute these results into Newton's second law for rotation in step 1:

$$m_1 g R - m_2 g R = \frac{dL_z}{dt} = \frac{d}{dt}(I\omega + m_1 vR + m_2 vR)$$
$$= I\alpha + (m_1 + m_2)Ra$$

5. Relate I to M and R, and use the nonslip condition to relate α to a:

$$m_1 g R - m_2 g R = \frac{1}{2}MR^2 \frac{a}{R} + (m_1 + m_2)Ra$$

6. Solve for a:

$$a = \frac{m_1 - m_2}{\frac{1}{2}M + m_1 + m_2} g$$

Remarks This problem could be solved by writing the tensions T_1 on the left and T_2 on the right and using $\tau = I\alpha$ for the pulley and $\Sigma \vec{F} = m\vec{a}$ for each block. If you do not need to know the tensions, using angular momentum is easier. Note that the linear momenta of the blocks are in opposite directions, but their angular momenta are in the same direction.

There are many problems in which the forces, position vectors, and veloci-ties all lie in a plane, so that the torques, angular velocities, and angular-momentum vectors are all along the axis of rotation that remains fixed in space. In such cases, we can assign positive and negative values to counter-clockwise or clockwise rotations, as we did in Example 10-2, and treat the case like a one-dimensional problem. However, there are other situations, such as the motion of a gyroscope (discussed in the "Exploring . . . Motion of a Gyroscope" section) where the vector natures of torque, angular velocity, and angular momentum are important.

exploring

Motion of a Gyroscope

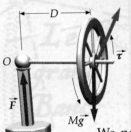

Figure 1

A gyroscope is a common example of motion in which the axis of rotation changes direction. Figure 1 shows a gyroscope that is free to turn on its axle. The axle is pivoted at a point a distance D from the center of the wheel, and is free to turn in any direction. We now give a qualitative understanding of the complex motion of such a system by using Newton's second law for rotation,

$$\vec{\tau}_{\text{net}} = \frac{d\vec{L}}{dt} \quad \text{or} \quad d\vec{L} = \vec{\tau}_{\text{net}}\,dt \qquad 1$$

along with the relations

$$\vec{\tau}_{\text{net}} = \vec{r} \times M\vec{g} \qquad 2$$

and

$$\vec{L} = I_s\vec{\omega}_s \qquad 3$$

where I_s and $\vec{\omega}_s$ are the moment of inertia and angular velocity of the wheel about its spin axis. All we really need to remember in order to describe the motion of a gyroscope is that the *change* in angular momentum of the wheel must be in the direction of the net torque acting on it.

Suppose the axle is held horizontally and then released. If the wheel isn't spinning, it simply falls, rotating about a horizontal axis through O and perpendicular to $\vec{r}$. The torque is horizontal, into the page. For this case, the *change* in angular momentum equals the angular momentum itself, which, in our example, is just $\vec{r} \times M\vec{v}_{\text{cm}}$. However, if the wheel *is* spinning and has a large angular momentum along its axle, it does not fall when the axle is released. If it were to fall, the axle would point downward, resulting in a large component of angular momentum in the downward direction. But there is no torque in the downward direction; the torque is horizontal. What actually happens is that the axle moves horizontally (into the paper in Figure 1). The wheel must move this way so that the *change* in angular momentum is in the direction of the net torque. This is illustrated

in Figure 2*a*, where we see a large angular momentum along the axis of the wheel and a change in angular momentum $d\vec{L}$ in the direction of the torque. This motion, which is always surprising when first encountered, is called **precession**. We can calculate the angular velocity of precession. In a small time interval dt, the change in the angular momentum has a magnitude dL:

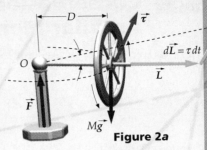

Figure 2a

$$dL = \tau\,dt = MgD\,dt$$

where MgD is the magnitude of the torque about the pivot point. From Figure 2*b*, the angle $d\phi$ through which the axle moves is

$$d\phi = \frac{dL}{L} = \frac{MgD\,dt}{L}$$

Figure 2b

The angular velocity of the precession is thus

$$\omega_P = \frac{d\phi}{dt} = \frac{MgD}{L} = \frac{MgD}{I_s\omega_s} \qquad 4$$

If the angular momentum due to the spin of the wheel is large, the precession can be very slow.

If you perform this experiment, you will notice a small up-and-down oscillation of the axle. This motion is called **nutation**. Suppose the wheel is simply released at the beginning. The bulk motion of the wheel as it precesses results in a component of angular momentum in the upward direction. Since there is no torque in that direction, the axle must initially dip down slightly to make the vertical component of angular momentum zero. Close analysis shows that it first dips down, then overshoots and moves upward, and then continues oscillating (nutating). The nutation can be avoided by giving the wheel a slight push in the direction of precession when released. This provides the upward torque needed for the upward angular momentum. If it is pushed too hard, the axle will move upward initially and then nutate.

Professor Carpenter demonstrates the precession of a gyroscope.

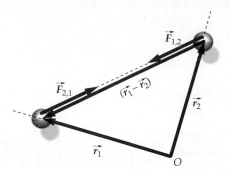

10-4 Conservation of Angular Momentum

When the net external torque acting on a system is zero, we have

$$\frac{d\vec{L}}{dt} = 0 = \vec{\tau}_{ext}$$

or

$$\vec{L} = \text{constant} \qquad \qquad 10\text{-}16$$

Figure 10-18 If the internal forces exerted by one particle on another are along the line joining the particles, the net torque exerted by the forces is zero about any point O.

Equation 10-16 is a statement of the **law of conservation of angular momentum**:

> If the net external torque acting on a system is zero, the total angular momentum of the system is constant.

Conservation of angular momentum

This is the rotational analog of the law of conservation of linear momentum. If a system is isolated from its surroundings, so that there are no external forces or torques acting on it, three quantities are conserved: energy, linear momentum, and angular momentum. The law of conservation of angular momentum is a fundamental law of nature. Even on the microscopic scale of atomic and nuclear physics, where Newtonian mechanics does not hold, the angular momentum of an isolated system is found to be constant over time.

The experimental result that angular momentum is conserved in the absence of a net external torque implies that the internal torques must sum to zero. This fact is suggested by Newton's third law. Consider the two particles shown in Figure 10-18: Let $\vec{F}_{1,2}$ be the force exerted by particle 1 on particle 2, and $\vec{F}_{2,1}$ be that exerted by particle 2 on particle 1. By Newton's third law, $\vec{F}_{2,1} = -\vec{F}_{1,2}$. The sum of the torques exerted by these forces about the origin O is

$$\vec{\tau}_1 + \vec{\tau}_2 = \vec{r}_1 \times \vec{F}_{2,1} + \vec{r}_2 \times \vec{F}_{1,2}$$

$$= \vec{r}_1 \times \vec{F}_{2,1} + \vec{r}_2 \times (-\vec{F}_{1,2}) = (\vec{r}_1 - \vec{r}_2) \times \vec{F}_{2,1}$$

The vector $\vec{r}_1 - \vec{r}_2$ is along the line joining the two particles. If $\vec{F}_{2,1}$ acts parallel to the line joining m_1 and m_2, $\vec{F}_{2,1}$ and $\vec{r}_1 - \vec{r}_2$ are either parallel or antiparallel and

$$(\vec{r}_1 - \vec{r}_2) \times \vec{F}_{2,1} = 0$$

If this is true for all the internal forces, the internal torques cancel in pairs.

There are many examples of the conservation of angular momentum in everyday life. Figures 10-19 and 10-20 illustrate angular momentum conservation in diving and ice skating.

Figure 10-19 Multiflash photograph of a diver. The diver's center of mass moves along a parabolic path after he leaves the board. The angular momentum is provided by the initial external torque due to the force of the board, which does not pass through the diver's center of mass if he leans forward as he jumps. If the diver wanted to undergo one or more somersaults in the air, he would draw in his arms and legs, decreasing his moment of inertia to increase his angular velocity.

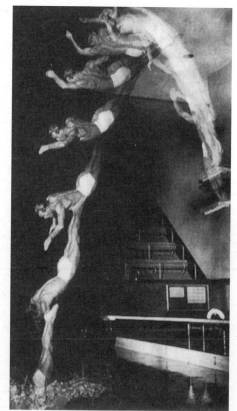

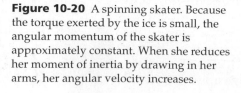

Figure 10-20 A spinning skater. Because the torque exerted by the ice is small, the angular momentum of the skater is approximately constant. When she reduces her moment of inertia by drawing in her arms, her angular velocity increases.

Example 10-3

A disk is rotating with an initial angular speed ω_i about a frictionless shaft through its symmetry axis as shown in Figure 10-21. Its moment of inertia about this axis is I_1. It drops onto another disk of moment of inertia I_2 that is initially at rest on the same shaft. Because of surface friction, the two disks eventually attain a common angular speed ω_f. Find ω_f.

Figure 10-21

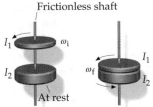

Picture the Problem We find the final angular speed from the final angular momentum, which is equal to the initial angular momentum because there are no external torques acting on the system. Note that we do *not* use conservation of mechanical energy. The angular speed of the upper disk is reduced while that of the lower disk is increased by the force of kinetic friction that acts between the surfaces. We therefore expect that the total mechanical energy is decreased.

1. The final angular speed is related to the initial angular speed by conservation of angular momentum:

$$L_f = L_i$$
$$(I_1 + I_2)\omega_f = I_1\omega_i$$

2. Solve for the final angular speed:

$$\omega_f = \frac{I_1}{I_1 + I_2}\omega_i$$

Check the Result If $I_2 \ll I_1$, the collision should have little effect on disk 1. Our results agree, and give $\omega_f \to \omega_i$. If $I_2 \gg I_1$, then disk 1 should slow to a stop without causing disk 2 to move. Our results give $\omega_f \to 0$, as expected.

In the collision of the two disks in Example 10-3, mechanical energy is not conserved. We can see this by writing the energy in terms of the angular momentum. An object rotating with an angular velocity ω has kinetic energy

$$K = \frac{1}{2}I\omega^2 = \frac{(I\omega)^2}{2I}$$

Using $L = I\omega$, we get

$$K = \frac{L^2}{2I} \qquad\qquad 10\text{-}17$$

Compare this result with that for linear motion, $K = p^2/2m$ (Equation 8-23). The initial kinetic energy in Example 10-3 is

$$K_i = \frac{L_i^2}{2I_1}$$

and the final kinetic energy is

$$K_f = \frac{L_f^2}{2(I_1 + I_2)}$$

Since $L_f = L_i$, the final kinetic energy is less than the initial kinetic energy by the factor $I_1/(I_1 + I_2)$. This interaction of the disks is analogous to a one-dimensional perfectly inelastic collision of two objects.

The rotating plates in the transmission of a truck make inelastic collisions when engaged.

Example 10-4

A merry-go-round of radius 2 m and moment of inertia 500 kg·m² is rotating about a frictionless pivot, making one revolution every 5 s. A child of mass 25 kg originally standing at the center walks out to the rim (Figure 10-22). Find the new angular speed of the merry-go-round.

Picture the Problem The new angular speed ω_f is related to the final angular momentum of the system, which is the sum of the angular momentum of the child, L_c, and the angular momentum of the merry-go-round, L_m. No external torques act on the system, so the final angular momentum equals the initial angular momentum. When the child (considered as a particle of mass m) is a distance r from the center, her moment of inertia is $I_c = mr^2$. Initially, $r = 0$ and $I_{ci} = 0$. At the rim, $I_{cf} = mR^2$.

Figure 10-22

1. The final angular velocity is related to the initial angular velocity by conservation of angular momentum:

$$\vec{L}_f = \vec{L}_i$$

$$I_{sys,f}\omega_f = I_{sys,i}\omega_i$$

2. The moment of inertia of the system is the moment of inertia of the merry-go-round plus that of the child:

$$I_{sys} = I_m + I_c = I_m + mr^2$$

3. Initially $r = 0$, finally $r = R$. Substitute initial and final expressions for I_{sys}:

$$(I_m + mR^2)\omega_f = I_m\omega_i$$

4. Solve for ω_f:

$$\omega_f = \frac{I_m}{mR^2 + I_m}\omega_i = \frac{500\ \text{kg·m}^2}{(25\ \text{kg})(2\ \text{m})^2 + 500\ \text{kg·m}^2}\omega_i$$

$$= \frac{5}{6}\omega_i = \frac{5}{6}\left(\frac{1\ \text{rev}}{5\ \text{s}}\right) = \frac{1\ \text{rev}}{6\ \text{s}}$$

When the child reaches the rim, the merry-go-round rotates once every 6 s.

Remarks To calculate the value of L_i or L_f, or to find the kinetic energy of the system, it is necessary to convert the angular speed to radians per second.

When the child is at the center of the merry-go-round, she is at rest. As she walks outward, she begins to move in a circle. The force that accelerates the child is the friction between her shoes and the merry-go-round. This force exerts a torque on the child, increasing her angular momentum. The child exerts an equal and opposite frictional force on the merry-go-round. The torque associated with this force decreases the angular momentum of the merry-go-round.

From Equation 10-17 we can see that the kinetic energy of the child and merry-go-round system decreases since the angular momentum is constant, but the moment of inertia increases as the child walks toward the rim. At each step, the child makes an inelastic collision with a part of the merry-go-round that is farther out and therefore moving faster than she is. Mechanical energy is lost in each of these inelastic collisions. If the child walks inward, the moment of inertia of the child–merry-go-round system decreases, hence the total kinetic energy of the system must increase. This energy comes from the child's internal chemical energy.

| Example **10-5** | *try it yourself* |

The same child as in Example 10-4 runs with an initial speed 2.5 m/s along a path tangential to the rim of the merry-go-round, which is initially at rest, and then jumps on (Figure 10-23). Find the final angular velocity of the child and the merry-go-round together.

Picture the Problem Once the child's feet leave the ground, no external torques act on the child–merry-go-round system, hence the total angular momentum of the system is conserved. The mass of the child is $m = 25$ kg, her initial speed is $v = 2.5$ m/s, and the radius of the merry-go-round is $R = 2.0$ m. The initial angular speed of the merry-go-round is $w_i = 0$.

Figure 10-23

Cover the column to the right and try these on your own before looking at the answers.

Steps

Answers

1. Write an expression for the initial angular momentum of the running child relative to the center of the merry-go-round.

$L_i = mvR$

2. Write an expression for the total final angular momentum of the child–merry-go-round system in terms of the final angular velocity ω_f.

$L_f = (mr^2 + I_m)\omega_f$

3. Set your expressions in 1 and 2 equal and solve for ω_f.

$$\omega_f = \frac{mvR}{mR^2 + I_m} = 0.208 \text{ rad/s}$$

Exercise Calculate the initial and final kinetic energies of the child–merry-go-round system. (*Answer* $K_i = 78.1$ J, $K_f = 13.0$ J)

The Hubble Space Telescope is aimed by regulating the spin rates of 45-kg flywheels arranged off-axis from each other and spinning at up to 3000 rpm. Software-controlled changes in the spin rates create angular momentum that causes the satellite to slew into new positions. This aiming mechanism can achieve and hold a target to within 0.005 arcsec—equivalent to holding a flashlight beam in Los Angeles on a dime in San Francisco.

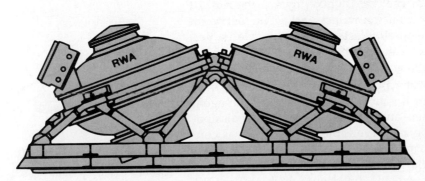

Example 10-6

A particle of mass m moves with speed v_0 in a circle of radius r_0 on a friction-less tabletop. The particle is attached to a string that passes through a hole in the table, as in Figure 10-24. The string is slowly pulled downward so that the particle moves in a smaller circle of radius r_f. (a) Find the final velocity in terms of r_0, v_0, and r_f. (b) Find the tension when the particle is moving in a circle of radius r in terms of m,r, and the angular momentum $L_0 = mv_0r_0$. (c) Calculate the work done on the particle by the tension T by integrating $\vec{T} \cdot d\vec{r}$ from r_0 to r_f. Express your answer in terms of r and L_0.

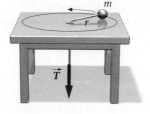

Figure 10-24

Picture the Problem   The speed of the particle is related to its angular momentum. Since the net force acting on the particle is the tension in the string T, which is always directed toward the hole, the torque about the hole is zero. Thus the angular momentum remains constant, $L = mvr = L_0 = mv_0r_0$.

(a) 1. Conservation of angular momentum relates the final speed to the initial speed and the initial and final radii:

$$mv_0r_0 = mv_fr_f$$

2. Solve for v_f:

$$v_f = \frac{r_0}{r_f}v_0$$

(b) 1. Apply $\Sigma \vec{F} = m\vec{a}$ to relate T to v and r:

$$T = m\frac{v^2}{r}$$

2. Apply conservation of angular momentum to obtain a second relation between v and r:

$$mvr = mv_0r_0 = L_0$$

3. Eliminate v and solve for T:

$$T = m\frac{v^2}{r} = m\frac{(L_0/mr)^2}{r} = \frac{L_0^2}{mr^3}$$

(c) 1. Write $dW = \vec{T} \cdot d\vec{r} = T_r\,dr$ with $T_r = -L_0^2/mr^3$ from part (b):

$$dW = \vec{T} \cdot d\vec{r} = T_r\,dr = -\frac{L_0^2}{mr^3}dr = -\frac{L_0^2}{m}r^{-3}\,dr$$

2. Integrate from r_0 to r_f:

$$W = \int_{r_0}^{r_f} T_r\,dr = -\frac{L_0^2}{m}\int_{r_0}^{r_f} r^{-3}\,dr = -\frac{L_0^2}{m}\frac{r^{-2}}{-2}\Big|_{r_0}^{r_f}$$

$$= \frac{L_0^2}{2m}(r_f^{-2} - r_0^{-2})$$

Check the Result   Note that work must be done to pull the string downward. Since r_f is less than r_0, the work is positive. This work is converted into an increased kinetic energy. We can calculate the change in kinetic energy of the particle directly. Using $K = L^2/2I$, with $L_i = L_f$, and $I = mr^2$, the change in kinetic energy is $K_f - K_i = (L_0^2/2mr_f^2) - (L_0^2/2mr_0^2) = (L_0^2/2m)(r_f^{-2} - r_0^{-2})$, which is the same as found by direct integration.

In Figure 10-25, a puck on a frictionless plane is given an initial speed v_0. The puck is attached to a string that wraps around a vertical post. This situation looks similar to Example 10-6, but it is not the same. There is no agent that can do work on the puck, nor is there any mechanism for energy dissipation. Thus, mechanical energy must be conserved. Since $K = L^2/2I$ is constant and I decreases as r_0 decreases, L must also decrease. Note that the tension does not act toward the center of the post. The tension produces a torque about the center of the post in the downward direction, which reduces the angular momentum of the puck, which is in the upward direction.

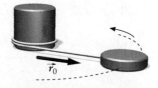

Figure 10-25   A puck sliding around a post on a frictionless table. As the puck slides, the string wraps around the post.

| **Example 10-7** | *try it yourself* |

A thin stick of mass M and length d is attached to a pivot at the top. A piece of clay of mass m and speed v hits the stick a distance x from the pivot and sticks to it (Figure 10-26). Find the ratio of the final energy to the initial energy.

Picture the Problem   The collision is inelastic, so we do not expect mechanical energy to be conserved. During the collision, the pivot exerts a large force on the stick, so linear momentum is also not conserved. However, there are no external torques about the pivot point on the clay–stick system, so angular momentum is conserved. The kinetic energy after the inelastic collision can be written in terms of the angular momentum L_f and the moment of inertia I' of the combined clay–stick system. Conservation of angular momentum allows you to relate L_f to the mass m and velocity v of the clay.

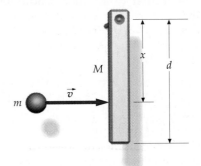

Figure 10-26

Cover the column to the right and try these on your own before looking at the answers.

Steps

Answers

1. Write the kinetic energy after the collision in terms of the angular momentum L_f and the moment of inertia I' of the combined stick–clay system.

$$E_f = \frac{L_f^2}{2I'}$$

2. Apply conservation of angular momentum to relate L_f to m, v, and x.

$$L_f = L_i = mvx$$

3. Write I' in terms of m, x, M, and d.

$$I' = mx^2 + \frac{1}{3}Md^2$$

4. Substitute these expressions for L_f and I' into your equation for E_f.

$$E_f = \frac{L_f^2}{2I'} = \frac{(mvx)^2}{2(mx^2 + \frac{1}{3}Md^2)} = \frac{3}{2}\frac{m^2x^2v^2}{3mx^2 + Md^2}$$

5. Divide the energy after the collision by the initial energy of the clay.

$$\frac{E_f}{E_i} = \frac{\frac{3}{2}m^2x^2v^2/(3mx^2 + Md^2)}{\frac{1}{2}mv^2} = \frac{3mx^2}{3mx^2 + Md^2}$$

Remark   This example is the rotational analog of the ballistic pendulum discussed in Example 8-13. In that example, we used conservation of linear momentum to find the energy of the pendulum after the collision.

| **10-5** | **Quantization of Angular Momentum** |

Angular momentum plays an important role in the description of atoms, molecules, nuclei, and elementary particles. Like energy, angular momentum is **quantized,** that is, changes in angular momentum occur only in discrete amounts.

   The angular momentum of a particle due to its motion is its orbital angular momentum. The magnitude of the orbital angular momentum of a particle can have only the values

$$L = \sqrt{\ell(\ell + 1)}\,\hbar, \qquad \ell = 0, 1, 2, \ldots$$

where $\hbar$ (read "h-bar") is the **fundamental unit of angular momentum,** which is related to another fundamental constant of nature, Planck's constant h:

$$\hbar = \frac{h}{2\pi} = 1.05 \times 10^{-34} \, \text{J·s} \tag{10-18}$$

The component of orbital angular momentum along any line in space is also quantized and can have only the values $\pm m\hbar$ where m is an integer that is less than or equal to ℓ. For example, if $\ell = 2$, m can be 2, 1, or 0.

Because the quantum of angular momentum, $\hbar$, is so small, the quantization of angular momentum is not noticed in the macroscopic world. Consider a particle of mass 1 g ($= 10^{-3}$ kg) moving in a circle of radius 1 cm with a period of 1 s. Its orbital angular momentum is

$$L = mvr = mr^2\omega = mr^2\frac{2\pi}{T} = (10^{-3}\,\text{kg})(10^{-2}\,\text{m})^2\frac{2\pi}{1\,\text{s}} = 6.28 \times 10^{-7}\,\text{J·s}$$

If we divide by $\hbar$, we obtain

$$\frac{L}{\hbar} = \frac{6.28 \times 10^{-7}\,\text{J·s}}{1.05 \times 10^{-34}\,\text{J·s}} = 6 \times 10^{27}$$

The angular momentum of this macroscopic system is equal to 6×10^{27} units of the fundamental unit of angular momentum. Even if we could measure L to one part in a billion, we would never notice the quantization of this macroscopic angular momentum.

The quantization of orbital angular momentum leads to the quantization of rotational energy. Consider a molecule rotating about its center of mass with angular momentum L (Figure 10-27). Let I be its moment of inertia. Its kinetic energy is

$$K = \frac{L^2}{2I} \tag{10-19}$$

But L^2 is quantized to the values $L^2 = \ell(\ell + 1)\hbar^2$ with $\ell = 0, 1, 2, \dots$. Thus the kinetic energy is quantized to the values K_ℓ given by

$$K_\ell = \frac{L^2}{2I} = \frac{\ell(\ell + 1)\hbar^2}{2I} = \ell(\ell + 1)E_{0\text{r}} \tag{10-20a}$$

where

$$E_{0\text{r}} = \frac{\hbar^2}{2I} \tag{10-20b}$$

Figure 10-28 shows an energy-level diagram for a rotating molecule with constant moment of inertia I. Note that, unlike the energy levels for a vibrating system (Section 7-4), the rotational energy levels are not equally spaced, and the lowest level is zero.

Stable matter contains just three kinds of particles: electrons, protons, and neutrons. In addition to its orbital angular momentum, each of these particles also has an intrinsic angular momentum called its **spin**. The spin angular momentum of a particle, like its mass and electric charge, is a fundamental property of the particle that cannot be changed. It does not have anything to do with the particle's motion. The magnitude of the spin angular momentum vector for these particles is $s = \sqrt{\frac{1}{2}(\frac{1}{2} + 1)}\,\hbar$ and the component along any line in space can have just two values, $+\frac{1}{2}\hbar$ and $-\frac{1}{2}\hbar$. Such particles are called "spin-one-half" particles.*

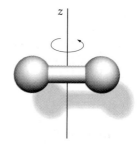

Figure 10-27 Model of a rigid diatomic molecule rotating about the z axis.

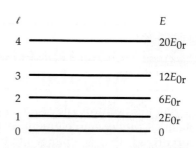

Figure 10-28 Energy level diagram for a rotating molecule.

---

* Electrons, protons, and neutrons and other spin-$\frac{1}{2}$ particles are also called fermions. There are other particles called bosons, such as a photons and α particles, that have zero spin or integral spin.

The picture of an electron as a spinning ball that orbits the nucleus in an atom (like the spinning earth orbiting the sun) is a useful visualization but is not entirely accurate. Spin is a quantum mechanical property, analogous but not identical to "spin" in the macroscopic sense. Whereas the angular momentum of a spinning ball can be increased or decreased, the spin of the electron is a fixed property that does not change. Furthermore, as far as we know, an electron is a point particle that has no size, and hence can be compared only figuratively with a spinning ball.

Summary

1. Angular momentum is an important derived dynamic quantity in macroscopic physics. In microscopic physics, an elementary particle has orbital angular momentum due to its motion, and spin angular momentum, which is an intrinsic, fundamental property of the particle.
2. Conservation of angular momentum is a fundamental law of nature.
3. Quantization of angular momentum is a fundamental law of nature.

Topic	Remarks and Relevant Equations
1. Vector Nature of Rotation	When the axis of rotation changes direction in space, the vector nature of rotational quantities is important.
Angular velocity $\vec{\omega}$	The direction of the angular velocity $\vec{\omega}$ is along the axis of rotation in the sense given by the right-hand rule.
Torque $\vec{\tau}$	$\vec{\tau} = \vec{r} \times \vec{F}$ (10-1)
2. Vector Product	$\vec{A} \times \vec{B} = AB \sin\phi\, \hat{n}$ (10-2) where ϕ is the angle between the vectors and $\hat{n}$ is a unit vector perpendicular to the plane of $\vec{A}$ and $\vec{B}$ in the sense given by the right-hand rule as $\vec{A}$ is rotated into $\vec{B}$.
Properties	$\vec{A} \times \vec{B} = -\vec{B} \times \vec{A}$ (10-4) $\frac{d}{dt}(\vec{A} \times \vec{B}) = \vec{A} \times \frac{d\vec{B}}{dt} + \frac{d\vec{A}}{dt} \times \vec{B}$ (10-6) $\hat{i} \times \hat{j} = \hat{k}, \quad \hat{j} \times \hat{k} = \hat{i}, \quad \hat{k} \times \hat{i} = \hat{j}$ (10-7a) $\hat{i} \times \hat{i} = \hat{j} \times \hat{j} = \hat{k} \times \hat{k} = 0$ (10-7b)
3. Angular Momentum	
For a particle	$\vec{L} = \vec{r} \times \vec{p}$ (10-8)
For a system rotating about a symmetry axis	$\vec{L} = I\vec{\omega}$ (10-9)
For a system rotating and translating	The angular momentum about any point O' is the angular momentum about the center of mass (spin angular momentum) plus the angular momentum associated with center of mass motion about O' (orbital angular momentum). $\vec{L} = \vec{L}_{\text{orbit}} + \vec{L}_{\text{spin}} = \vec{r}_{\text{cm}} \times M\vec{v}_{\text{cm}} + \sum_i \vec{r}_i' \times m_i\vec{u}_i$ (10-10)

Newton's second law for rotation	$\vec{\tau}_{net,ext} = \dfrac{d\vec{L}}{dt}$	**10-13**
Kinetic energy of a rotating object	$K = \dfrac{L^2}{2I}$	**10-16**
Conservation of angular momentum	If the net external torque is zero, the angular momentum of the system is conserved.	
Quantization of angular momentum	The magnitude of the orbital angular momentum of a particle can have only the values $$L = \sqrt{\ell(\ell + 1)}\,\hbar, \qquad \ell = 0, 1, 2, \ldots$$ where	
Fundamental unit of angular momentum	$$\hbar = \frac{h}{2\pi} = 1.05 \times 10^{-34}\,\text{J·s}$$ is the fundamental unit of angular momentum, and h is Planck's constant.	**10-17**
Quantization of any component of orbital angular momentum	The component of orbital angular momentum along any line in space is also quantized and can have only the values $\pm m\hbar$, where m is an integer that is less than or equal to ℓ.	
Spin	Electrons, protons, and neutrons have an intrinsic angular momentum called spin. The magnitude of the spin angular momentum vector for these particles is $$s = \sqrt{\tfrac{1}{2}\left(\tfrac{1}{2} + 1\right)}\,\hbar$$ and the component along any line in space can have just two values, $+\tfrac{1}{2}\hbar$ and $-\tfrac{1}{2}\hbar$.	

Problem-Solving Guide

1. Begin by drawing a neat diagram that includes the important features of the problem.
2. If there is a net torque acting on the system, write an expression for the total angular momentum of the system and apply Newton's second law in the form $\vec{\tau}_{net,ext} = d\vec{L}/dt$.
3. If the net external torque acting on the system is zero, use conservation of angular momentum to relate the final angular velocity to the initial angular velocity.

Summary of Worked Examples

Type of Calculation	Procedure and Relevant Examples
1. Calculate the Angular Momentum	For a particle, use the definition $\vec{L} = \vec{r} \times \vec{p}$. For a system rotating about a symmetry axis or an axis parallel to a symmetry axis, use $\vec{L} = I\vec{\omega}$. **Examples 10-1, 10-2**
2. Applying Newton's Second Law	
Find the angular acceleration of a pulley with masses hanging from it.	Write an expression for the angular momentum of the pulley and each object in the system. Use the nonslip condition to relate v and ω, or a and α. Apply $\Sigma\vec{\tau}_{i,ext} = d\vec{L}/dt$. **Example 10-2**
3. Conservation of Angular Momentum	
Find the final angular speed for a system in which I changes (inelastic collision of rotating disks, walking or jumping on merry-go-round).	Apply conservation of angular momentum to the system. The initial and final energies can be most easily compared if you write $K = L^2/2I$, since L is constant. **Examples 10-3, 10-4, 10-5**

Find the work done by a force that exerts no torque.	Apply conservation of angular momentum to find the final velocity or angular velocity, then use the work–energy theorem.	**Example 10-6**
Find the energy lost in an inelastic collision when there are no external torques acting.	Apply conservation of angular momentum to find the final velocity or angular velocity.	**Example 10-7**

Problems

In a few problems, you are given more data than you actually need; in a few other problems, you are required to supply data from your general knowledge, outside sources, or informed estimates.

Conceptual Problems

Problems from Optional and Exploring sections

● Single-concept, single-step, relatively easy
●● Intermediate-level, may require synthesis of concepts
●●● Challenging, for advanced students

The Vector Nature of Rotation

1 ● True or false:

(a) If two vectors are parallel, their cross product must be zero.
(b) When a disk rotates about its symmetry axis, $\vec{\omega}$ is along the axis.
(c) The torque exerted by a force is always perpendicular to the force.

2 ● Two vectors $\vec{A}$ and $\vec{B}$ have equal magnitude. Their cross product has the greatest magnitude if $\vec{A}$ and $\vec{B}$ are

(a) parallel.
(b) equal.
(c) perpendicular.
(d) antiparallel.
(e) at an angle of 45° to each other.

3 ● A force of magnitude F is applied horizontally in the negative x direction to the rim of a disk of radius R as shown in Figure 10-29. Write $\vec{F}$ and $\vec{r}$ in terms of the unit vectors $\hat{i}, \hat{j}$, and $\hat{k}$, and compute the torque produced by the force about the origin at the center of the disk.

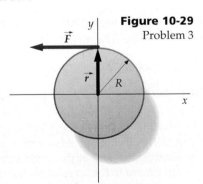

Figure 10-29
Problem 3

4 ● Compute the torque about the origin for the force $\vec{F} = -mg\hat{j}$ acting on a particle at $\vec{r} = x\hat{i} + y\hat{j}$, and show that this torque is independent of the y coordinate.

5 ● Find $\vec{A} \times \vec{B}$ for (a) $\vec{A} = 4\hat{i}$ and $\vec{B} = 6\hat{i} + 6\hat{j}$, (b) $\vec{A} = 4\hat{i}$ and $\vec{B} = 6\hat{i} + 6\hat{k}$, and (c) $\vec{A} = 2\hat{i} + 3\hat{j}$ and $\vec{B} = -3\hat{i} + 2\hat{j}$.

6 ● Under what conditions is the magnitude of $\vec{A} \times \vec{B}$ equal to $\vec{A} \cdot \vec{B}$?

7 ● A particle moves in a circle of radius $\vec{r}$ with an angular velocity $\vec{\omega}$. (a) Show that its velocity is $\vec{v} = \vec{\omega} \times \vec{r}$. (b) Show that its centripetal acceleration is $\vec{a}_c = \vec{\omega} \times \vec{v} = \vec{\omega} \times (\vec{\omega} \times \vec{r})$.

8 ●● If $\vec{A} = 4\hat{i}$, $B_z = 0$, $|\vec{B}| = 5$, and $\vec{A} \times \vec{B} = 12\hat{k}$, determine $\vec{B}$.

9 ● If $\vec{A} = 3\hat{j}$, $\vec{A} \times \vec{B} = 9\hat{i}$, and $\vec{A} \cdot \vec{B} = 12$, find $\vec{B}$.

Angular Momentum

10 ● What is the angle between a particle's linear momentum $\vec{p}$ and its angular momentum $\vec{L}$?

11 ● A particle of mass m is moving with speed v along a line that passes through point P. What is the angular momentum of the particle about point P?

(a) mv
(b) zero
(c) It changes sign as the particle passes through point P.
(d) It depends on the distance of point P from the origin of coordinates.

12 ●● A particle travels in a circular path. (a) If its linear momentum p is doubled, how is its angular momentum affected? (b) If the radius of the circle is doubled but the speed is unchanged, how is the angular momentum of the particle affected?

13 ●● A particle moves along a straight line at constant speed. How does its angular momentum about any point vary over time?

14 ● A particle moving at constant velocity has zero angular momentum about a particular point. Show that the particle either has passed through that point or will pass through it.

15 ● A 2-kg particle moves at a constant speed of 3.5 m/s around a circle of radius 4 m. (a) What is its angular momentum about the center of the circle? (b) What is its moment of inertia about an axis through the center of the circle and perpendicular to the plane of the motion? (*c*) What is the angular speed of the particle?

16   • A 2-kg particle moves at constant speed of 4.5 m/s along a straight line. (*a*) What is the magnitude of its angular momentum about a point 6 m from the line? (*b*) Describe qualitatively how its angular speed about that point varies with time.

17   •• A particle is traveling with a constant velocity $\vec{v}$ along a line that is a distance *b* from the origin *O* (Figure 10-30). Let *dA* be the area swept out by the position vector from *O* to the particle in time *dt*. Show that *dA/dt* is constant in time and equal to $\frac{1}{2}L/m$, where *L* is the angular momentum of the particle about the origin.

Figure 10-30 Problem 17

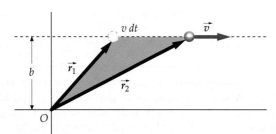

18   •• A 15-g coin of diameter 1.5 cm is spinning at 10 rev/s about a vertical diameter at a fixed point on a tabletop. (*a*) What is the angular momentum of the coin about its center of mass? (*b*) What is its angular momentum about a point on the table 10 cm from the coin? If the coin spins about a vertical diameter at 10 rev/s while its center of mass travels in a straight line across the tabletop at 5 cm/s, (*c*) what is the angular momentum of the coin about a point on the line of motion? (*d*) What is the angular momentum of the coin about a point 10 cm from the line of motion? (There are two answers to this question. Explain why and give both.)

19   •• Two particles of masses m_1 and m_2 are located at $\vec{r}_1$ and $\vec{r}_2$ relative to some origin *O* as in Figure 10-31. They exert equal and opposite forces on each other. Calculate the resultant torque exerted by these internal forces about the origin *O* and show that it is zero if the forces $\vec{F}_1$ and $\vec{F}_2$ lie along the line joining the particles.

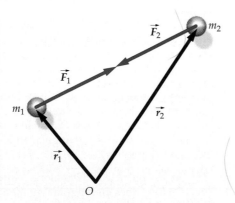

Figure 10-31 Problem 19

Torque and Angular Momentum

20   • True or false:

(*a*) The rate of change of a system's angular momentum is always parallel to the net external torque.

(*b*) If the net torque on a body is zero, the angular momentum must be zero.

21   • A 1.8-kg particle moves in a circle of radius 3.4 m. The magnitude of its angular momentum relative to the center of the circle depends on time according to $L = (4\,\text{N·m})t$. (*a*) Find the magnitude of the torque acting on the particle. (*b*) Find the angular speed of the particle as a function of time.

22   •• A uniform cylinder of mass 90 kg and radius 0.4 m is mounted so that it turns without friction on its fixed symmetry axis. It is rotated by a drive belt that wraps around its perimeter and exerts a constant torque. At time *t* = 0, its angular velocity is zero. At time *t* = 25 s, its angular velocity is 500 rev/min. (*a*) What is its angular momentum at *t* = 25 s? (*b*) At what rate is the angular momentum increasing? (*c*) What is the torque acting on the cylinder? (*d*) What is the magnitude of the force acting on the rim of the cylinder?

23   •• In Figure 10-32, the incline is frictionless and the string passes through the center of mass of each block. The pulley has a moment of inertia *I* and a radius *R*. (*a*) Find the net torque acting on the system (the two masses, string, and pulley) about the center of the pulley. (*b*) Write an expression for the total angular momentum of the system about the center of the pulley when the masses are moving with a speed *v*. (*c*) Find the acceleration of the masses from your results for parts (*a*) and (*b*) by setting the net torque equal to the rate of change of the angular momentum of the system.

Figure 10-32 Problem 23

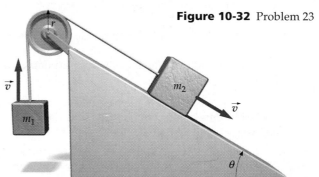

24   •• From her elevated DJ booth at a dance club, Caroline is lowering a 2-kg speaker using a 0.6-kg disk of radius 8 cm as a pulley (Figure 10-33). The speaker wire runs straight up from the speaker, over the pulley, and then horizontally across the table. She attaches the wire to the 4-kg amplifier on her tabletop, and then turns to get the other speaker. The table, however, is nearly frictionless, and the whole system begins to move when she lets go. (*a*) What is the net torque about the center of the pulley? (*b*) What is the total angular momentum of the system 3.5 s after release? (*c*) What is the angular momentum of the pulley at this time? (*d*) What is the ratio of the angular momentum of each piece of equipment to the angular momentum of the pulley?

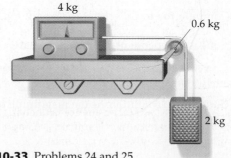

Figure 10-33 Problems 24 and 25

25 •• Work Problem 24 for the case in which the coefficient of friction between the table and the 4-kg amplifier is 0.25.

26 •• Figure 10-34 shows the rear view of a spaceship that is rotating about its longitudinal axis at 6 rev/min. The occupants wish to stop this rotation. They have small jets mounted tangentially, at a distance $R = 3$ m from the axis, as indicated, and can eject 10 g/s of gas from each jet with a nozzle velocity of 800 m/s. For how long must they turn on these jets to stop the rotation? The rotational inertia of the ship around its axis (assumed to be constant) is 4000 kg·m².

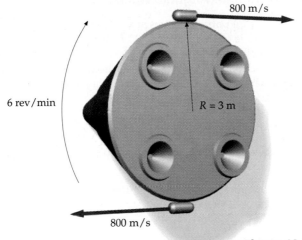

800 m/s

6 rev/min

$R = 3$ m

800 m/s

Figure 10-34
Problem 26

Conservation of Angular Momentum

27 • True or false: If the net torque on a rotating system is zero, the angular velocity of the system cannot change.

28 • Folk wisdom says that a cat always lands on its feet. If a cat starts falling with its feet up, how can it land on its feet without violating the law of conservation of angular momentum?

29 • If the angular momentum of a system is constant, which of the following statements must be true?

(a) No torque acts on any part of the system.
(b) A constant torque acts on each part of the system.
(c) Zero net torque acts on each part of the system.
(d) A constant external torque acts on the system.
(e) Zero net torque acts on the system.

30 • Two identical cylindrical disks have a common axis. Initially, one of the disks is spinning. When the two disks are brought into contact they stick together. Which of the following statements is true?

(a) The total kinetic energy and the total angular momentum are unchanged from their initial values.
(b) Both the total kinetic energy and the total angular momentum are reduced to half of their original values.
(c) The total angular momentum is unchanged, but the total kinetic energy is reduced to half its original value.
(d) The total angular momentum is reduced to half of its original value, but the total kinetic energy is unchanged.
(e) The total angular momentum is unchanged, and the total kinetic energy is reduced to one-quarter of its original value.

31 •• In Example 10-4, does force exerted by the merry-go-round on the child do work?

32 •• Is it easier to crawl radially outward or radially inward on a rotating merry-go-round? Why?

33 •• A block sliding on a frictionless table is attached to a string that passes through a hole in the table. Initially, the block is sliding with speed v_0 in a circle of radius r_0. A student under the table pulls slowly on the string. What happens as the block spirals inward? Give supporting arguments for your choice.

(a) Its energy and angular momentum are conserved.
(b) Its angular momentum is conserved, and its energy increases.
(c) Its angular momentum is conserved, and its energy decreases.
(d) Its energy is conserved, and its angular momentum increases.
(e) Its energy is conserved, and its angular momentum decreases.

34 • A planet moves in an elliptical orbit about the sun with the sun at one focus of the ellipse as in Figure 10-35. (a) What is the torque produced by the gravitational force of attraction of the sun for the planet? (b) At position A, the planet is a distance r_1 from the sun and is moving with a speed v_1 perpendicular to the line from the sun to the planet. At position B, it is at distance r_2 and is moving with speed v_2, again perpendicular to the line from the sun to the planet. What is the ratio of v_1 to v_2 in terms of r_1 and r_2?

Figure 10-35
Problem 34

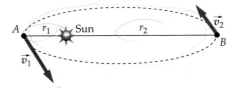

A r_1 Sun r_2 $\vec{v_2}$ B

$\vec{v_1}$

35 • Under gravitational collapse (all forces on various pieces are inward toward the center), the radius of a spinning spherical star of uniform density shrinks by a factor of 2, with the resulting increased density remaining uniform throughout as the star shrinks. What will be the ratio of the final angular speed ω_2 to the initial angular speed ω_1?

(a) 2   (b) 0.5   (c) 4
(d) 0.25   (e) 1.0

36 •• A man stands on a frictionless platform that is rotating with an angular speed of 1.5 rev/s. His arms are outstretched, and he holds a heavy weight in each hand. The moment of inertia of the man, the extended weights, and the platform is 6 kg·m². When the man pulls the weights inward toward his body, the moment of inertia decreases to 1.8 kg·m². (a) What is the resulting angular speed of the platform? (b) What is the change in kinetic energy of the system? (c) Where did this increase in energy come from?

37 •• A small blob of putty of mass m falls from the ceiling and lands on the outer rim of a turntable of radius R and moment of inertia I_0 that is rotating freely with angular speed ω_i about its vertical fixed symmetry axis. (a) What is the postcollision angular speed of the turntable plus putty?

(b) After several turns, the blob flies off the edge of the turntable. What is the angular speed of the turntable after the blob flies off?

38 •• Two disks of identical mass but different radii (r and 2r) are spinning on frictionless bearings at the same angular speed ω_0 but in opposite directions (Figure 10-36). The two disks are brought slowly together. The resulting frictional force between the surfaces eventually brings them to a common angular velocity. What is the magnitude of that final angular velocity in terms of ω_0?

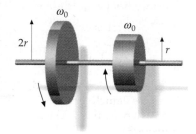

Figure 10-36 Problem 38

39 •• A block of mass m sliding on a frictionless table is attached to a string that passes through a hole in the table. Initially, the block is sliding with speed v_0 in a circle of radius r_0. Find (a) the angular momentum of the block, (b) the kinetic energy of the block, and (c) the tension in the string. A student under the table now pulls slowly on the string. How much work is required to reduce the radius of the circle from r_0 to $r_0/2$?

40 •• At the beginning of each term, a physics professor named Dr. Zeus shows the class his expectations of them through a demonstration that he calls "Lesson #1." He stands at the center of a turntable that can rotate without friction. He then takes a 2-kg globe of the earth and swings it around his head at the end of a 0.8-m chain. The world revolves around him every 3 s, and the professor and the platform have a moment of inertia of 0.5 kg·m². (a) What is the angular speed of the professor? (b) What is the total kinetic energy of the globe, professor, and platform?

41 •• The sun's radius is 6.96×10^8 m, and it rotates with a period of 25.3 d. Estimate the new period of rotation of the sun if it collapses with no loss of mass to become a neutron star of radius 5 km.

42 •• Arriving at the baggage claim area in a small airport, Alan (mass m) discovers a large turntable (radius R and moment of inertia I) that is spinning out of control. Not wanting to pass up an opportunity for magnificence, Alan leaps onto the edge of the turntable, which continues to spin freely with an angular speed of 7.5 rad/s. He struggles on his hands and knees to the center, and then rises up into a pose that resembles a hood ornament and spins like a figure skater in finale. Security is notified, but passengers applaud. Assume that $mR^2 = 2.8I$, and that Alan has a moment of inertia of $I/10$ in his final pose. What is his final angular speed if friction is neglected?

43 •• A 0.2-kg point mass moving on a frictionless horizontal surface is attached to a rubber band whose other end is fixed at point P. The rubber band exerts a force $F = bx$ toward P, where x is the length of the rubber band and b is an unknown coefficient. The mass moves along the dotted line in Figure 10-37. When it passes point A, its velocity is 4 m/s directed as shown. The distance AP is 0.6 m and BP is 1.0 m. (a) Find the velocity of the mass at points B and C. (b) Find b.

Figure 10-37
Problem 43

Quantization of Angular Momentum

44 • A 2-g particle moves at a constant speed of 3 mm/s around a circle of radius 4 mm. (a) Find the magnitude of the angular momentum of the particle. (b) If $L = \sqrt{\ell(\ell + 1)}\,\hbar$, find the value of $\ell(\ell + 1)$ and the approximate value of ℓ. (c) Explain why the quantization of angular momentum is not noticed in macroscopic physics.

45 • The z component of the spin of an electron is $\frac{1}{2}\hbar$, but the magnitude of the spin vector is $\sqrt{0.75}\,\hbar$. What is the angle between the electron's spin angular momentum vector and the z axis?

46 •• Show that the energy difference between one rotational state and the next higher state is proportional to $\ell + 1$ (see Equation 10-20a).

47 •• In the HBr molecule, the mass of the bromine nucleus is 80 times that of the hydrogen nucleus (a single proton); consequently, in calculating the rotational motion of the molecule, one may, to a good approximation, assume that the Br nucleus remains stationary as the H atom (mass 1.67×10^{-27} kg) revolves around it. The separation between the H atom and bromine nucleus is 0.144 nm. Calculate (a) the moment of inertia of the HBr molecule about the bromine nucleus, and (b) the rotational energies for $\ell = 1$, $\ell = 2$, and $\ell = 3$.

48 •• The equilibrium separation between the nuclei of the nitrogen molecule is 0.11 nm. The mass of each nitrogen nucleus is 14 u, where u $= 1.66 \times 10^{-27}$ kg. We wish to calculate the energies of the three lowest angular momentum states of the nitrogen molecule. (a) Approximate the nitrogen molecule as a rigid dumbbell of two equal point masses, and calculate the moment of inertia about its center of mass. (b) Find the rotational energy levels using the relation $E_\ell = \ell(\ell + 1)\hbar^2/2I$.

Collision Problems

49 •• A 16.0-kg, 2.4-m-long rod is supported on a knife edge at its midpoint. A 3.2-kg ball of clay is dropped from rest from a height of 1.2 m and makes a perfectly inelastic collision with the rod 0.9 m from the point of support (Figure 10-38). Find the angular momentum of the rod-and-clay system immediately after the inelastic collision.

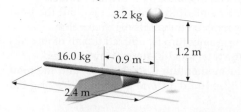

Figure 10-38
Problem 49

50 •• Figure 10-39 shows a thin bar of length L and mass M, and a small blob of putty of mass m. The system is supported on a frictionless horizontal surface. The putty moves to the right with velocity v, strikes the bar at a distance d from the center of the bar, and sticks to the bar at the point of contact. Obtain expressions for the velocity of the system's center of mass and for the angular velocity of the system about its center of mass.

Figure 10-39
Problems 50 and 51

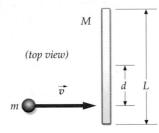

51 •• In Problem 50, replace the blob of putty with a small hard sphere of negligible size that collides elastically with the bar. Find d such that the sphere is at rest after the collision.

52 •• Figure 10-40 shows a uniform rod of length L and mass M pivoted at the top. The rod, which is initially at rest, is struck by a particle of mass m at a point $d = 0.8L$ below the pivot. Assume that the collision is perfectly inelastic. What must be the speed v of the particle so that the maximum angle between the rod and the vertical is 90°?

Figure 10-40
Problems 52 and 53

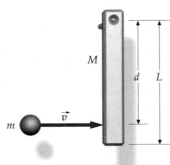

53 •• If, for the system in Problem 52, $L = 1.2$ m, $M = 0.8$ kg, and $m = 0.3$ kg, and the maximum angle between the rod and the vertical is 60°, find the speed of the particle before impact.

54 •• A projectile of mass m_p is traveling at a constant velocity $\vec{v}_0$ toward a stationary disk of mass M and radius R that is free to rotate about a pivot through its axis O (Figure 10-41). Before impact, the projectile is traveling along a line displaced a distance b below the axis. The projectile strikes the disk and sticks to point B. Treat the projectile as a point mass. (a) Before impact, what is the total angular momentum L_0 of the projectile and disk about the O axis? (b) What is the angular speed ω of the disk and projectile system just after the impact? (c) What is the kinetic energy of the disk and projectile system after impact? (d) How much mechanical energy is lost in this collision?

Figure 10-41 Problem 54

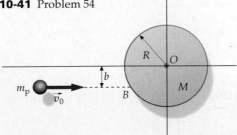

55 •• A uniform rod of length L_1 and mass $M = 0.75$ kg is supported by a hinge at one end and is free to rotate in the vertical plane (Figure 10-42). The rod is released from rest in the position shown. A particle of mass $m = 0.5$ kg is supported by a thin string of length L_2 from the hinge. The particle sticks to the rod on contact. What should be the ratio L_2 / L_1 so that $\theta_{max} = 60°$ after the collision?

56 •• Returning to Figure 10-42, this time set $L_1 = 1.2$ m, $M = 2.0$ kg, and $L_2 = 0.8$ m. After the inelastic collision, $\theta_{max} = 37°$. Find m. How much energy is dissipated in this inelastic collision?

Figure 10-42
Problems 55–58

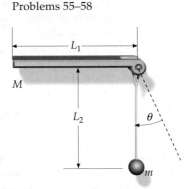

57 ••• Suppose that in Figure 10-42, $m = 0.4$ kg, $M = 0.75$ kg, $L_1 = 1.2$ m, and $L_2 = 0.8$ m. What minimum initial angular velocity must be imparted to the rod so that the system will revolve completely about the hinge following the inelastic collision? How much energy is then dissipated in the inelastic collision?

58 ••• Repeat Problem 56 if the collision between the rod and particle is elastic.

Exploring ... Motion of a Gyroscope

59 • True or false:

(a) Nutation and precession are the same phenomenon.
(b) The direction of precession is the direction of the net torque.
(c) When the gyroscope is not spinning, $\vec{\tau} = d\vec{L}/dt$ does not hold.

60 •• The angular momentum vector for a spinning wheel lies along its axle and is pointed east. To make this vector point south, it is necessary to exert a force on the east end of the axle in which direction?

(a) Up
(b) Down
(c) North
(d) South
(e) East

61 •• A man is walking north carrying a suitcase that contains a spinning gyroscope mounted on an axle attached to the front and back of the case. The angular velocity of the gyroscope points north. The man now begins to turn to walk east. As a result, the front end of the suitcase will

(a) resist his attempt to turn and will try to remain pointed north.
(b) fight his attempt to turn and will pull to the west.
(c) rise upward.
(d) dip downward.
(e) cause no effect whatsoever.

62 •• The angular momentum of the propeller of a small airplane points forward. (a) As the plane takes off, the nose lifts up and the airplane tends to veer to one side. To which side does it veer and why? (*b*) If the plane is flying horizontally and suddenly turns to the right, does the nose of the plane tend to move up or down? Why?

63 •• A car is powered by the energy stored in a single flywheel with an angular momentum $\vec{L}$. Discuss the problems that would arise for various orientations of $\vec{L}$ and various maneuvers of the car. For example, what would happen if $\vec{L}$ points vertically upward and the car travels over a hilltop or through a valley? What would happen if $\vec{L}$ points forward or to one side and the car attempts to turn to the left or right? In each case that you examine, consider the direction of the torque exerted on the car by the road.

64 •• A bicycle wheel of radius 28 cm is mounted at the middle of an axle 50 cm long. The tire and rim weigh 30 N. The wheel is spun at 12 rev/s, and the axle is then placed in a horizontal position with one end resting on a pivot. (*a*) What is the angular momentum due to the spinning of the wheel? (Treat the wheel as a hoop.) (*b*) What is the angular velocity of precession? (*c*) How long does it take for the axle to swing through 360° around the pivot? (*d*) What is the angular momentum associated with the motion of the center of mass, that is, due to the precession? In what direction is this angular momentum?

65 •• A uniform disk of mass 2.5 kg and radius 6.4 cm is mounted in the center of a 10-cm axle and spun at 700 rev/min. The axle is then placed in a horizontal position with one end resting on a pivot. The other end is given an initial horizontal velocity such that the precession is smooth with no nutation. (*a*) What is the angular velocity of precession? (*b*) What is the speed of the center of mass during the precession? (*c*) What are the magnitude and direction of the acceleration of the center of mass? (*d*) What are the vertical and horizontal components of the force exerted by the pivot?

General Problems

66 • An object of mass M is rotating about a fixed axis with angular momentum L. Its moment of inertia about this axis is I. What is its kinetic energy?

(*a*) $IL^2/2$
(*b*) $L^2/2I$
(*c*) $ML^2/2$
(*d*) $IL^2/2M$

67 • Explain why a helicopter with just one main rotor has a second smaller rotor mounted on a horizontal axis at the rear as in Figure 10-43. Describe the resultant motion of the helicopter if this rear rotor fails during flight.

68 •• A woman sits on a spinning piano stool with her arms folded. When she extends her arms out to the side, her kinetic energy

(*a*) increases.
(*b*) decreases.
(*c*) remains the same.

69 •• In tetherball, a ball is attached to a string that is attached to a pole. When the ball is hit, the string wraps around the pole and the ball spirals inward. Neglecting air resistance, what happens as the ball swings around the pole? Give supporting arguments for your choice.

(*a*) The mechanical energy and angular momentum of the ball are conserved.
(*b*) The angular momentum of the ball is conserved, but the mechanical energy of the ball increases.
(*c*) The angular momentum of the ball is conserved, and the mechanical energy of the ball decreases.
(*d*) The mechanical energy of the ball is conserved and the angular momentum of the ball increases.
(*e*) The mechanical energy of the ball is conserved and the angular momentum of the ball decreases.

70 •• A uniform rod of mass M and length L lies on a horizontal frictionless table. A piece of putty of mass $m = M/4$ moves along a line perpendicular to the rod, strikes the rod near its end, and sticks to the rod. Describe qualitatively the subsequent motion of the rod and putty.

71 • A particle of mass 3 kg moves with velocity $\vec{v} = 3 \text{ m/s } \hat{i}$ along the line $z = 0$, $y = 5.3$ m. (*a*) Find the angular momentum $\vec{L}$ relative to the origin when the particle is at $x = 12$ m, $y = 5.3$ m. (*b*) A force $\vec{F} = -3 \text{ N } \hat{i}$ is applied to the particle. Find the torque relative to the origin due to this force.

72 • The position vector of a particle of mass 3 kg is given by $\vec{r} = 4\hat{i} + 3t^2\hat{j}$, where $\vec{r}$ is in meters and t is in seconds. Determine the angular momentum and the torque acting on the particle about the origin.

73 •• An ice skater starts her pirouette with arms outstretched, rotating at 1.5 rev/s. Estimate her rotational speed (in revolutions per second) when she brings her arms flat against her body.

74 •• Two ice skaters hold hands and rotate, making one revolution in 2.5 s. Their masses are 55 kg and 85 kg, and they are separated by 1.7 m. Find (*a*) the angular momentum of the system about their center of mass, and (*b*) the total kinetic energy of the system.

Figure 10-43 Problem 67

75 •• A 2-kg ball attached to a string of length 1.5 m moves in a horizontal circle as a conical pendulum (Figure 10-44). The string makes an angle $\theta = 30°$ with the vertical. (a) Show that the angular momentum of the ball about the point of support P has a horizontal component toward the center of the circle as well as a vertical component, and find these components. (b) Find the magnitude of $d\vec{L}/dt$, and show that it equals the magnitude of the torque exerted by gravity about the point of support.

Figure 10-44 Problem 75

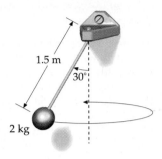

1.5 m

30°

2 kg

76 •• A mass m on a horizontal, frictionless surface is attached to a string that wraps around a vertical cylindrical post so that when it is set into motion it follows a path that spirals inward. (a) Is the angular momentum of the mass conserved? (b) Is the energy of the mass conserved? (c) If the speed of the mass is v_0 when the length of the string is r, what is its speed when the unwrapped length has shortened to $r/2$?

77 •• Figure 10-45 shows a hollow cylindrical tube of mass M, length L, and moment of inertia $ML^2/10$. Inside the cylinder are two masses m, separated a distance ℓ and tied to a central post by a thin string. The system can rotate about a vertical axis through the center of the cylinder. With the system rotating at ω, the strings holding the masses suddenly break. When the masses reach the end of the cylinder, they stick. Obtain expressions for the final angular velocity and the initial and final energies of the system. Assume that the inside walls of the cylinder are frictionless.

Figure 10-45 Problems 77–81

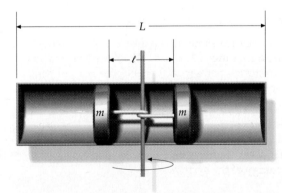

L

ℓ

m   m

78 •• Repeat Problem 77, this time adding friction between the masses and walls of the cylinder. However, the coefficient of friction is not enough to prevent the masses from reaching the ends of the cylinder. Can the final energy of the system be determined without knowing the coefficient of kinetic friction?

79 •• Suppose that in Figure 10-45, $\ell = 0.6$ m, $L = 2.0$ m, $M = 0.8$ kg, and $m = 0.4$ kg. The system rotates at ω such that the tension in the string is 108 N just before it breaks. Determine the initial and final angular velocities and initial and final energies of the system. Assume that the inside walls of the cylinder are frictionless.

80 •• For Problem 77, determine the radial velocity of each mass just before it reaches the end of the cylinder.

81 •• Given the numerical values of Problem 79, suppose the coefficient of friction between the masses and the walls of the cylinder is such that the masses cease sliding 0.2 m from the ends of the cylinder. Determine the initial and final angular velocities of the system and the energy dissipated in friction.

82 •• Kepler's second law states: *The radius vector from the sun to a planet sweeps out equal areas in equal times.* Show that this law follows directly from the law of conservation of angular momentum and the fact that the force of gravitational attraction between a planet and the sun acts along the line joining the two celestial objects.

83 •• Figure 10-46 shows a hollow cylinder of length 1.8 m, mass 0.8 kg, and radius 0.2 m that is free to rotate about a vertical axis through its center and perpendicular to the cylinder's axis. Inside the cylinder are two thin disks of 0.2 kg each, attached to springs of spring constant k and unstretched lengths 0.4 m. The system is brought to a rotational speed of 8 rad/s with the springs clamped so they do not stretch. The springs are then suddenly unclamped. When the disks have stopped their radial motion due to friction between the disks and the wall, they come to rest 0.6 m from the central axis. What is the angular velocity of the cylinder when the disks have stopped their radial motion? How much energy was dissipated in friction between the disks and cylinder wall?

Figure 10-46 Problem 83

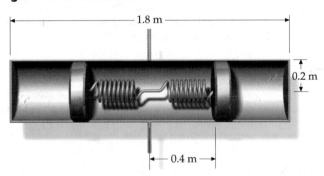

1.8 m

0.2 m

0.4 m

84 •• (a) Assuming the earth to be a homogeneous sphere of radius r and mass m, show that the period T of the earth's rotation about its axis is related to its radius by $T = (4\pi m/5L)r^2$, where L is the angular momentum of the earth due to its rotation. (b) Suppose that the radius r changes by a very small amount Δr due to some internal effect such as thermal expansion. Show that the fractional change in the period ΔT is given approximately by $\Delta T/T = 2 \Delta r/r$. Hint: Use the differentials dr and dT to approximate the changes in these quantities. (c) By how many kilometers would the earth need to expand for the period to change by $\frac{1}{4}d/y$ so that leap years would no longer be necessary?

85 •• The polar ice caps contain about 2.3×10^{19} kg of ice. This mass contributes negligibly to the moment of inertia of the earth because it is located at the poles, close to the axis of rotation. Estimate the change in the length of the day that would be expected if the polar ice caps were to melt and the water were distributed uniformly over the surface of the earth. (The moment of inertia of a spherical shell of mass m and radius r is $\frac{2}{3}mr^2$.)

86 ••• Figure 10-47 shows a hollow cylinder of mass $M = 1.2$ kg and length $L = 1.6$ m that is free to rotate about a vertical axis through its center. Inside the cylinder are two disks, each of mass 0.4 kg that are tied to a central post by a thin string and separated by a distance $\ell = 0.8$ m. The string breaks if the tension exceeds 100 N. Starting from rest, a torque is applied to the system until the string breaks. Assuming the disks are point masses and the radius of the cylinder is negligible, find the amount of work done up to that instant. Suppose that at that instant, the applied torque is removed, and that the walls of the cylinder are frictionless. Obtain an expression for the angular velocity of the system as a function of x for $x < L/2$, where x is the distance between each mass and the central post.

87 ••• For the system of Problem 86, find the angular velocity of the system just before and just after the point masses pass the ends of the cylinder.

88 ••• Repeat Problem 86 with the radius of the hollow cylinder as 0.4 m and the masses treated as thin disks rather than point masses.

89 ••• Figure 10-48 shows a pulley in the shape of a uniform disk with a heavy rope hanging over it. The circumference of the pulley is 1.2 m and its mass is 2.2 kg. The rope is 8.0 m long and its mass is 4.8 kg. At the instant shown in the figure, the system is at rest and the difference in height of the two ends of the rope is 0.6 m. (a) What is the angular velocity of the pulley when the difference in height between the two ends of the rope is 7.2 m? (b) Obtain an expression for the angular momentum of the system as a function of time while neither end of the rope is above the center of the pulley. There is no slippage between rope and pulley.

Figure 10-48 Problem 89

0.6 m

Figure 10-47 Problems 86–88

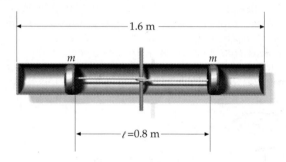

1.6 m

m m

$\ell = 0.8$ m

CHAPTER 11

Gravity

A mechanical model of the solar system, called an orrery, in the collection of Historical Scientific Instruments at Harvard University.

Gravity is the weakest of the four basic forces. It is negligible in the interactions of elementary particles, and thus plays no role in molecules, atoms, and nuclei. The gravitational attraction between objects of ordinary size, such as the gravitational force exerted by a building on a car, is too small to be noticed. Yet when we consider very large objects, such as moons, planets, and stars, gravity is of primary importance. The gravitational force exerted by the earth on us and on the objects around us is a fundamental part of our experience. It is gravity that binds us to the earth and keeps the earth and the other planets on course within the solar system. The gravitational force plays an important role in the life history of stars and in the behavior of galaxies. On the largest of all scales, it is gravity that controls the evolution of the universe.

11-1 Kepler's Laws

The nighttime sky with its myriad stars and shining planets has always fascinated humankind. Toward the end of the sixteenth century, the astronomer Tycho Brahe studied the motions of the planets and made observations that were considerably more accurate than those previously available. Using

Johannes Kepler (1571–1630)

321

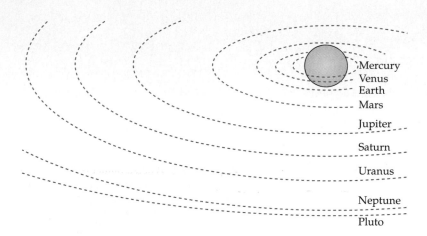

Figure 11-1 Orbits of the planets around the sun.

Brahe's data, Johannes Kepler discovered that the paths of the planets about the sun are ellipses (Figure 11-1). He also showed that the planets move faster when their orbit brings them closer to the sun and slower when their orbit takes them farther away. Finally, Kepler developed a precise mathematical relation between the period of a planet and its average distance from the sun. Kepler stated his results in three empirical laws of planetary motion. Ultimately, these laws provided the basis for Newton's discovery of the law of gravity. Kepler's three laws are

> Law 1. All planets move in elliptical orbits with the sun at one focus.
> Law 2. A line joining any planet to the sun sweeps out equal areas in equal times.
> Law 3. The square of the period of any planet is proportional to the cube of the semimajor axis of its orbit.

An ellipse is the locus of points for which the sum of the distances from two foci F is constant, as shown in Figure 11-2. Figure 11-3 shows a planet following an elliptical path with the sun at one focus. The earth's orbit is nearly circular, with the distance to the sun at perihelion (closest point) being 1.48×10^{11} m, and at aphelion (farthest point) being 1.52×10^{11} m. The semimajor axis equals the average of these distances, which is 1.50×10^{11} m (93 million miles) for the earth's orbit. The mean earth–sun distance defines the astronomical unit (AU):

$$1 \text{ AU} = 1.50 \times 10^{11} \text{ m} = 93.0 \times 10^6 \text{ mi} \qquad 11\text{-}1$$

The AU is used frequently in problems dealing with the solar system.

Figure 11-4 illustrates Kepler's second law, the law of equal areas. A planet moves faster when it is closer to the sun than when it is farther away, so that the area swept out by the radius vector in a given time interval is the same throughout the orbit. The law of equal areas is related to the conservation of angular momentum, as we will see in the next section.

Kepler's third law relates the period of any planet to its mean distance from the sun, which equals the semimajor axis of its elliptical path. In algebraic form, if r is the mean distance

Figure 11-2 An ellipse is the locus of points for which $r_1 + r_2 = $ constant. The distance a is called the *semimajor* axis, and b is the *semiminor* axis. You can draw an ellipse with a piece of string by fixing each end at a focus F and using it to guide the pencil. Circles are special cases in which the two foci coincide.

Figure 11-3 Elliptical path of a planet with the sun at one focus. Point P, where the planet is closest to the sun, is called the perihelion, and point A, where it is farthest, is called the aphelion. The average distance between the planet and the sun is equal to the semimajor axis.

between a planet and the sun and T is the planet's period of revolution, Kepler's third law states that

$$T^2 = Cr^3 \qquad\qquad 11\text{-}2$$

where the constant C has the same value for all the planets. This law is a consequence of the fact that the force exerted by the sun on a planet varies inversely with the square of the distance from the sun to the planet. We will demonstrate this in the next section for the special case of a circular orbit.

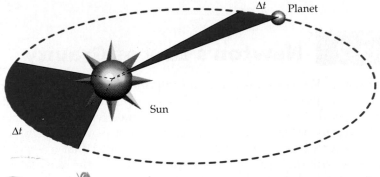

Figure 11-4 When a planet is close to the sun, it moves faster than when it is farther away. The areas swept out by the radius vector in a given time interval are equal.

Example 11-1

The mean distance from the sun to Jupiter is 5.20 AU. What is the period of Jupiter's orbit around the sun?

Picture the Problem We use Kepler's third law to relate the period of Jupiter to its mean distance from the sun. The constant C can be obtained from the known mean distance and period of the earth. Let $T_E = 1$ y be the period of the earth and let $r_E = 1$ AU be the mean distance from the earth to the sun. Let T_J and $r_J = 5.20$ AU be the period and mean distance for Jupiter.

1. Kepler's third law relates Jupiter's period T_J and mean distance r_J:

$$T_J^2 = C r_J^3$$

2. Apply Kepler's third law to the earth to find C in terms of T_E and r_E:

$$T_E^2 = C r_E^3 \quad \text{or} \quad C = \frac{T_E^2}{r_E^3}$$

3. Substitute this value of C and solve for T_J:

$$T_J^2 = C r_J^3 = \frac{T_E^2}{r_E^3} r_J^3$$

$$T_J = \left(\frac{r_J}{R_E}\right)^{3/2} T_E = \left(\frac{5.20\ \text{AU}}{1\ \text{AU}}\right)^{3/2} (1\ \text{y}) = 11.9\ \text{y}$$

Exercise The period of Neptune is 164.8 y. Calculate its mean distance from the sun. (*Answer* 30.1 AU)

Remark Figure 11-5 shows the periods of the planets Earth, Jupiter, and Neptune as functions of their mean distances from the sun. In (*a*), periods are plotted on an arithmetic scale. The same data plotted on a log–log scale (*b*) fall on the straight line $\log T = \frac{1}{2}\log C + \frac{3}{2}\log R$.

Figure 11-5

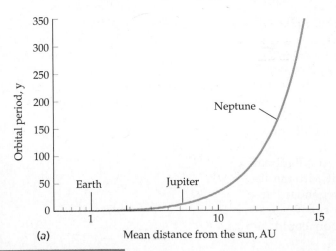

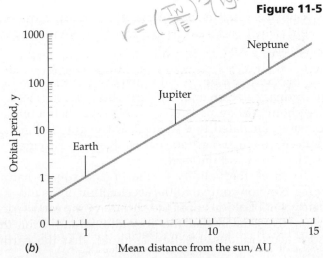

11-2 Newton's Law of Gravity

Although Kepler's laws were an important first step in understanding the motion of planets, they were still just empirical rules obtained from the astronomical observations of Brahe. It remained for Newton to take the next giant step by attributing the acceleration of a planet in its orbit to a specific force exerted on it by the sun. Newton proved that a force that varies inversely with the square of the distance between the sun and a planet results in elliptical orbits, as observed by Kepler. He then made the bold assumption that this force acts between any two objects in the universe. Before Newton, it was not even generally accepted that the laws of physics observed on earth were applicable to the heavenly bodies. **Newton's law of gravity** postulates that there is a force of attraction between each pair of objects that is proportional to the product of the masses of the objects and inversely proportional to the square of the distance separating them. The magnitude of the gravitational force exerted by a particle of mass m_1 on another particle of mass m_2 a distance r away is thus given by

$$F = \frac{Gm_1m_2}{r^2} \qquad \text{11-3}$$

magnitude of g. force

where G is the **universal gravitational constant**, which has the value

$$G = 6.67 \times 10^{-11}\ \text{N·m}^2/\text{kg}^2 \qquad \text{11-4}$$

Newton published his theory of gravitation in 1686, but it was not until a century later that an accurate experimental determination of G was made by Cavendish, whose findings will be discussed in the next section. If m_1 is at position $\vec{r}_1$ and m_2 is at $\vec{r}_2$ (Figure 11-6a), the force $\vec{F}_{1,2}$ exerted by mass m_1 on m_2 is

$$\vec{F}_{1,2} = -\frac{Gm_1m_2}{r_{1,2}^2}\hat{r}_{1,2} \qquad \text{11-5}$$

$F_{12} = F_{21}$

Newton's law of gravity

where $\vec{r}_{1,2}$ is the vector pointing from mass m_1 to m_2 and $\hat{r}_{1,2} = \vec{r}_{1,2}/r_{1,2}$ is a unit vector point from m_1 to m_2. The force $\vec{F}_{2,1}$ exerted by m_2 on m_1 is the negative of $\vec{F}_{1,2}$, according to Newton's third law (Figure 11-6b).

We can use the known value of G to compute the gravitational attraction between two ordinary objects.

Exercise Find the gravitational force that attracts a 65-kg boy to a 50-kg girl when they are 0.5 m apart. Assume that they are point masses. (*Answer* $8.67 \times 10^{-7}\ \text{N}$)

This exercise demonstrates that the gravitational force exerted by an object of ordinary size on another such object is extremely small. For example, the weight of a 50-kg person is 491 N, about half a billion times the force of attraction calculated in the exercise. The gravitational attraction can be easily noticed only if at least one of the objects is extremely massive, as with an ordinary object and the earth.

To check the validity of the inverse-square nature of the gravitational force, Newton compared the acceleration of the moon in its orbit with the acceleration of objects near the surface of the earth (such as the legendary apple). He assumed that the gravitational attraction due to the earth causes both accelerations. He first assumed that the earth and moon could be

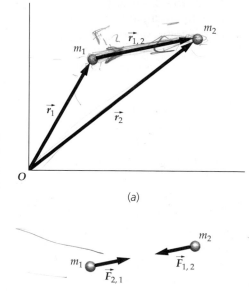

(a)

(b)

Figure 11-6 (a) Particles at $\vec{r}_1$ and $\vec{r}_2$. (b) The particles exert equal and opposite forces on each other.

treated as point particles with their total masses concentrated at their centers. The force on a particle of mass m a distance r from the center of the earth is

$$F = \frac{Gm_E m}{r^2}$$ 11-6

From Newton's second law, the acceleration is

$$a = \frac{F}{m} = \frac{GM_E}{r^2}$$ 11-7

For objects near the surface of the earth, $r = R_E$ and the acceleration is g:

$$g = \frac{GM_E}{R_E^2}$$ 11-8

Since the distance to the moon is about 60 times the radius of the earth, the acceleration of objects near the surface of the earth ($g = 9.81$ m/s²) should be $60^2 = 3600$ times the acceleration of the moon. The moon's centripetal acceleration can be calculated from its known distance from the center of the earth $r = 3.84 \times 10^8$ m, and its period $T = 27.3$ days $= 2.36 \times 10^6$ s:

$$a_m = \frac{v^2}{r} = \frac{(2\pi r/T)^2}{r} = \frac{4\pi^2 r}{T^2} = \frac{4\pi^2(3.84 \times 10^8 \text{ m})}{(2.36 \times 10^6 \text{ s})^2} = 2.72 \times 10^{-3} \text{ m/s}^2$$

Then

$$\frac{g}{a_m} = \frac{9.81 \text{ m/s}^2}{2.72 \times 10^{-3} \text{ m/s}^2} = 3607 \approx 60^2$$

In Newton's words, "I thereby compared the force requisite to keep the Moon in her orb with the force of gravity at the surface of the Earth, and found them answer pretty nearly."

The assumption that the earth and moon can be treated as point particles in the calculation of the force on the moon is reasonable because the moon is far from the earth compared with the radius of either the earth or the moon, but such an assumption is certainly questionable when applied to an object near the earth's surface. After considerable effort, Newton was able to prove that the force exerted by any spherically symmetric object on a point mass either on or outside its surface is the same as if all the mass of the object were concentrated at its center. The proof involves integral calculus, which Newton developed to solve this problem.

Since $g = 9.81$ m/s² is easily measured and the radius of the earth is known, Equation 11-8 can be used to determine either the constant G or the mass of the earth M_E if one of these quantities is known. Newton estimated the value of G from an approximation of the mass of the earth. When Cavendish determined G some 100 years later by measuring the force between small spheres of known mass and separation, he called his experiment "weighing the earth."

Cavendish used the apparatus shown in Figure 11-7. His measurement of G has been repeated by other experimenters with various improvements and refinements. All measurements of G are difficult because of the extreme weakness of the gravitational attraction. Consequently, the value of G is known today only to about 1 part in 10,000. Although G was one of the first physical constants ever measured, it remains one of the least accurately known.

Gravitational torsion balance used in student labs for the measurement of G. A tiny angular deflection of the balance results in a large angular deflection of the laser beam that reflects from a mirror on the balance.

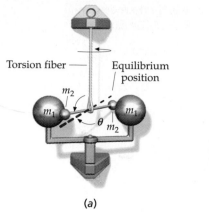

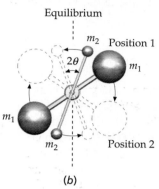

(a)   (b)

Figure 11-7 (a)Two small spheres, each of mass m_2, are at the ends of a light rod that is suspended by a fine fiber. Careful measurements determine the torque required to turn the fiber through a given angle. Two large spheres, each of mass m_1, are then placed near the small spheres. Because of the gravitational attraction of the large spheres of mass m_1 for the small spheres, the fiber is turned through a very small angle θ from its equilibrium position. (b) The apparatus as seen from above. After the apparatus comes to rest, the positions of the large spheres are reversed, as shown by the dashed lines, so that they are at the same distance from the equilibrium position of the balance but on the other side. If the apparatus is again allowed to come to rest, the fiber will turn through angle 2θ in response to the reversal of the torque. Once the torsion constant has been determined, the forces between the masses m_1 and m_2 can be determined from the measurement of this angle. Since the masses and their separations are known, G can be calculated. Cavendish obtained a value for G within about 1% of the presently accepted value given by Equation 11-4.

Example 11-2

What is the free-fall acceleration of an object at the altitude of the space shuttle's orbit, about 400 km above the earth's surface?

Picture the Problem The force is given by Equation 11-6 with $r = R_E +$ 400 km.

1. The acceleration is given by $a = F/m$, where F is given by Newton's law of gravity:
$$a = \frac{F}{m} = \frac{GmM_E/r^2}{m} = \frac{GM_E}{r^2}$$

2. The distance r is related to the radius of the earth R_E and the altitude h:
$$r = R_E + h = 6370 \text{ km} + 400 \text{ km} = 6770 \text{ km}$$

3. The acceleration is then:
$$a = \frac{GM_E}{r^2} = \frac{(6.67 \times 10^{-11} \text{ N·m}^2/\text{kg}^2)(5.98 \times 10^{24} \text{ kg})}{(6770 \text{ km})^2}$$
$$= 8.70 \text{ m/s}^2$$

Remark This is also the acceleration of the "weightless" shuttle astronauts as they accelerate in their circular orbit.

The calculation in Example 11-2 could have been simplified by using Equation 11-8 to write

$$GM_E = gR_E^2 \qquad\qquad\qquad 11\text{-}9$$

Then the acceleration at a distance r is

$$a = \frac{F}{m} = \frac{GM_E}{r^2} = g\frac{R_E^2}{r^2} \qquad\qquad\qquad 11\text{-}10$$

Exercise At what distance h above the surface of the earth is the acceleration of gravity half its value at sea level? (*Answer* $r = \sqrt{2}\, R_E = R_E + h$, $h = (\sqrt{2} - 1)R_E = 2637$ km)

Derivation of Kepler's Laws

Newton showed that when an object such as a planet or comet moves around a $1/r^2$ force center such as the sun, the object's path is an ellipse, a parabola, or a hyperbola.* The parabolic and hyperbolic paths apply to objects that make one pass by the sun and never return. Such orbits are not closed. The only closed orbits in an inverse-square force field are ellipses. Thus, Kepler's first law is a direct consequence of Newton's law of gravity. Kepler's second law, the law of equal areas, follows from the fact that the force exerted by the sun on a planet is directed toward the sun. Such a force is called a **central force.** Figure 11-8 shows a planet moving in an elliptical orbit around the sun. In time dt, the planet moves a distance $v\,dt$ and sweeps out the area indicated in the figure. This is half the area of the parallelogram formed by the vectors $\vec{r}$ and $\vec{v}\,dt$, which is $|\vec{r} \times \vec{v}\,dt|$. Thus, the area dA swept out by the radius vector $\vec{r}$ in time dt is

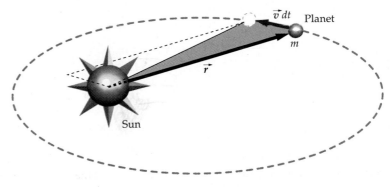

Figure 11-8 Area swept out by the radius vector of a planet orbiting the sun is proportional to the angular momentum of the planet around the sun.

$$dA = \frac{1}{2}|\vec{r} \times \vec{v}\,dt| = \frac{1}{2m}|\vec{r} \times m\vec{v}|\,dt$$

or

$$dA = \frac{1}{2m}L\,dt \qquad\qquad 11\text{-}11$$

where $\vec{L} = \vec{r} \times m\vec{v}$ is the angular momentum of the planet relative to the sun. The area swept out in a given time interval dt is therefore proportional to the angular momentum L. Since the force on a planet is along the line from the planet to the sun, it has no torque about the sun. Thus, the angular momentum of the planet is conserved, that is, L is constant. Therefore, the area swept out in a given time interval dt is the same for all parts of the orbit, which is Kepler's second law.

We will now show that Newton's law of gravity implies Kepler's third law for the special case of a circular orbit. Consider a planet moving with speed v in a circular orbit of radius r about the sun. The gravitational force of attraction between the sun and the planet provides the centripetal acceleration v^2/r. Newton's second law gives

$$F = m_\text{p}a$$

$$\frac{GM_\text{s}m_\text{p}}{r^2} = m_\text{p}\frac{v^2}{r} \qquad\qquad 11\text{-}12$$

where M_s is the mass of the sun and m_p is that of the planet. Solving for v^2, we find

$$v^2 = \frac{GM_\text{s}}{r} \qquad\qquad 11\text{-}13$$

Since the planet moves a distance $2\pi r$ in time T, its speed is related to the period by

$$v = \frac{2\pi r}{T} \qquad\qquad 11\text{-}14$$

*These are paths produced by slicing a cone and are therefore called "conic sections." A circle is a special case of an ellipse.

Substituting this expression for v in Equation 11-13, we obtain

$$v^2 = \frac{4\pi^2 r^2}{T^2} = \frac{GM_s}{r}$$

or

$$T^2 = \frac{4\pi^2}{GM_s} r^3 \qquad\qquad \text{11-15}$$

Kepler's third law

Equation 11-15 is Kepler's third law, which is the same as Equation 11-2 with $C = 4\pi^2/GM_s$.

For the more general case of elliptical orbits, the proof is more difficult. In such cases, the distance r is the mean distance from the sun, which also equals the semimajor axis a. Equation 11-15 also applies to the orbits of the satellites of any planet if we replace the mass of the sun M_s with the mass of the planet.* Finally, since G is known, we can determine the mass of a planet by measuring the period T and the mean orbital radius r of a moon orbiting it.

Example 11-3

A satellite travels in a circular orbit around the earth. Find its period if (a) the satellite is just above the surface of the earth, and (b) the satellite is at the space shuttle's altitude of 400 km. (Assume that air resistance can be neglected.)

Picture the Problem We use Kepler's third law with M_s in Equation 11-15 replaced by the mass of the earth M_E. The numerical calculation is simplified by using $GM_E = R_E^2 g$ from Equation 11-9.

(a)1. Apply Kepler's third law to the satellite:

$$T^2 = \frac{4\pi^2}{GM_E} r^3$$

2. Substitute $r = R_E$ for a satellite just above the earth's surface:

$$T^2 = \frac{4\pi^2}{GM_E} R_E^3$$

3. Use $GM_E = R_E^2 g$ to write T in terms of g:

$$T^2 = \frac{4\pi^2}{GM_E} R_E^3 = \frac{4\pi^2}{R_E^2 g} R_E^3 = \frac{4\pi^2 R_E}{g}$$

$$T = 2\pi\sqrt{\frac{R_E}{g}} = 2\pi\sqrt{\frac{6.37 \times 10^6 \text{ m}}{9.81 \text{ m/s}^2}} = 84.4 \text{ min}$$

(b) At an altitude $h = 400$ km, $r = R_E + h = 6770$ km. We find the period at this altitude by noting that T is proportional to $r^{3/2}$:

$$T = (84.4 \text{ min})\left(\frac{r}{R_E}\right)^{3/2}$$

$$= (84.4 \text{ min})\left(\frac{6.77 \times 10^6 \text{ m}}{6.37 \times 10^6 \text{ m}}\right)^{3/2} = 92.5 \text{ min}$$

Exercise Find the radius of the circular orbit of a satellite that orbits the earth with a period of one day. (*Answer* $r = 6.63R_E = 4.22 \times 10^7$ m $= 26{,}200$ mi. If such a satellite is in orbit over the equator and moves in the same direction as the rotation of the earth, it appears stationary relative to the earth. Most satellites are parked in such an orbit, which is called a geosynchronous orbit.)

---

* For example, it applies to the earth's moon and to all the artificial satellites orbiting the earth if the sun's mass M_s is replaced by the earth's mass M_E.

exploring

Gravitational and Inertial Mass

The property of an object responsible for the gravitational force it exerts on another object is its *gravitational* mass, whereas the property of an object that measures its resistance to acceleration is its *inertial* mass. We have used the same symbol m for these two properties because, experimentally, they are equal. The fact that the gravitational force exerted on an object is proportional to its inertial mass is a characteristic unique to the force of gravity. One consequence is that all objects near the surface of the earth fall with the same acceleration if air resistance is neglected. This fact has seemed surprising to all since it was discovered. The famous story of Galileo demonstrating it by dropping objects from the Tower of Pisa is just one example of the excitement this discovery aroused in the sixteenth century.

We could easily imagine that the gravitational and inertial masses of an object were not the same. Suppose we write m_G for the gravitational mass and m for the inertial mass. The force exerted by the earth on an object near its surface would then be

$$F = \frac{GM_E m_G}{R_E^2} \qquad 1$$

where M_E is the gravitational mass of the earth. The free-fall acceleration of the object near the earth's surface would then be

$$a = \frac{F}{m} = \left(\frac{GM_E}{R_E^2}\right)\frac{m_G}{m} \qquad 2$$

If gravity were just another property of matter, like color or hardness, it might be reasonable to expect that the ratio m_G/m would depend on such things as the chemical composition of the object or its temperature. The free-fall acceleration would then be different for different objects. The experimental fact, however, is that a is the same for all objects. Thus, we need not maintain the distinction between m_G and m and can set $m_G = m$. We must keep in mind, however, that the equivalence of gravitational and inertial mass is an experimental law, one that is limited by the accuracy of experiment. Experiments testing this equivalence were carried out by Simon Stevin in the 1580s. Galileo publicized the law widely, and his contemporaries made considerable improvements in the experimental accuracy with which the law was established.

The most precise early comparisons of gravitational and inertial mass were made by Newton. From experiments using simple pendulums rather than falling bodies, Newton was able to establish the equivalence between gravitational and inertial mass to an accuracy of about 1 part in 1000. Experiments comparing gravitational and inertial mass have improved steadily over the years. Their equivalence is now established to about 1 part in 10^{12}. The equivalence of gravitational and inertial mass is therefore one of the most well established of all physical laws. It is the basis for the principle of equivalence, which is the foundation of Einstein's general theory of relativity.

11-3   Gravitational Potential Energy

Near the surface of the earth, the gravitational force exerted by the earth on an object is constant because the distance to the center of the earth $r = R_E + h$ is always approximately R_E for $h \ll R_E$. The potential energy of an object near the earth's surface is $mg(r - R_E) = mgh$, where we have chosen $U = 0$ at the earth's surface, $r = R_E$. When we are far from the surface of the earth, we must take into account the fact that the gravitational force exerted by the earth is not uniform but decreases as $1/r^2$. The general definition of potential energy (Equation 5–20b) gives

$$dU = -\vec{F} \cdot d\vec{s}$$

where $\vec{F}$ is the force on a particle and $d\vec{s}$ is a general displacement of the particle. For the radial gravitational force $\vec{F}$ given by Equation 11-6 we have

$$dU = -\vec{F} \cdot d\vec{s} = -F_r \, dr = -\left(-\frac{GM_E m}{r^2} \right) dr = +\frac{GM_E m}{r^2} \, dr \qquad 11\text{-}16$$

Integrating both sides of this equation we obtain

$$U = -\frac{GM_E m}{r} + U_0 \qquad 11\text{-}17$$

where U_0 is a constant of integration. Since only changes in potential energy are important, we can choose the potential energy to be zero at any position. The earth's surface is a good choice for many everyday problems, but it is not always a convenient choice. For example, when considering the potential energy associated with a planet and the sun, there is no reason to want the potential energy to be zero at the surface of the sun. In fact, it is nearly always more convenient to choose the gravitational potential energy of a two-object system to be zero when the separation of the objects is infinite. Thus, $U_0 = 0$ is often a convenient choice. Then

$$U(r) = -\frac{GMm}{r}, \qquad U = 0 \text{ at } r = \infty \qquad 11\text{-}18$$

Gravitational potential energy with U = 0 at infinite separation

Figure 11-9 is a plot of $U(r)$ versus r for this choice of $U = 0$ at $r = \infty$ for an object of mass m and the earth of mass M_E. This function begins at the negative value $U = -GM_E m/R_E = -mgR_E$ at the earth's surface and increases as r increases, approaching zero at infinite r.

Escape Speed

In the past few decades, the idea of escaping from the earth's gravity has changed from fantasy to reality. Space probes have been sent out to the far reaches of the solar system. Some of these probes are expected to orbit the sun, others will leave the solar system and drift on into outer space. We will see that there is a minimum initial speed, called the **escape speed**, that is required for an object to escape from the earth.

If we project an object upward from the earth with some initial kinetic energy, the kinetic energy decreases and the poten-

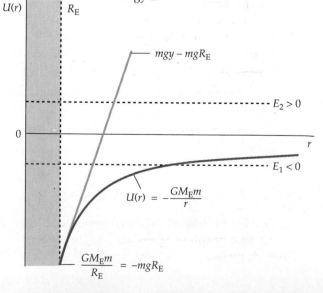

Figure 11-9 Gravitational potential energy $U(r)$ versus distance r from the center of mass. For r just slightly greater than R_E, $U(r)$ is approximately $mgy - mgR_E$, where y is the distance above the earth's surface. Horizontal dashed lines show positive and negative values for the total energy E.

tial energy increases as the object rises. The maximum increase in potential energy is $GM_E m/R_E$, as can be seen from Figure 11-10. Therefore, this is the most that the kinetic energy can decrease. If the initial kinetic energy is greater than $GM_E m/R_E$, the total energy E will be greater than zero (E_2 in Figure 11-10), and the object will still have some kinetic energy when r is very great (or even when r is infinite). Thus, the object will escape from the earth if the initial kinetic energy is greater than $GM_E m/R_E$. Since the potential energy at the earth's surface is $-GM_E m/R_E$, the total energy $E = K + U$ must be greater than or equal to zero for the object to escape. The speed near the earth's surface corresponding to zero total energy is called the escape speed v_e. It is found by setting the total energy at the surface of the earth equal to zero.

$$E = K + U = 0$$

$$\frac{1}{2} mv_e^2 - \frac{GM_E m}{R_E} = 0$$

$$v_e = \sqrt{\frac{2GM_E}{R_E}} = \sqrt{2gR_E} \qquad \text{11-19}$$

Escape speed

Using $g = 9.81 \text{ m/s}^2$ and $R_E = 6.37 \times 10^6$ m, we obtain

$$v_e = \sqrt{2(9.81 \text{ m/s}^2)(6.37 \times 10^6 \text{ m})} = 11.2 \text{ km/s}$$

This is about 6.95 mi/s or 25,000 mi/h. An object with this speed will just escape the earth. (However, it will not escape the solar system, because we have neglected the gravitational attraction of the sun and other planets; see Problem 57.)

The escape speed for a planet or moon relative to the thermal speeds of gas molecules determines the kind of atmosphere a planet or moon can have. The average kinetic energy of gas molecules $(\frac{1}{2}mv^2)_{av}$, is proportional to the absolute temperature T (Chapter 18). At the surface of the earth, the speeds of nearly all of the oxygen and nitrogen molecules are much lower than the escape speed, so these gases are retained in our atmosphere. For the lighter molecules hydrogen and helium, however, a considerable fraction of the molecules have speeds greater than the escape speed. Hydrogen and helium gases are therefore not found in our atmosphere. The escape speed at the surface of the moon is 2.3 km/s, which can be calculated from Equation 11-19 with the mass and radius of the moon replacing M_E and R_E. This is considerably smaller than the escape speed for earth, and in fact is too small for any atmosphere to exist.

Exercise Find the escape speed at the surface of Mercury, which has a mass $M = 3.31 \times 10^{23}$ kg and a radius $R = 2440$ km. (*Answer* $v_e = \sqrt{2GM/R} = 4.25 \text{ km/s}$)

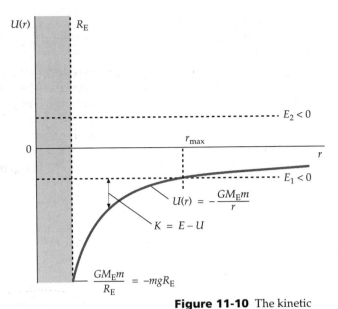

Figure 11-10 The kinetic energy of an object at a distance r from the center of the earth is $E - U(r)$. When the total energy is less than zero (E_1 in the figure), the kinetic energy is zero at $r = r_{max}$ and the object is bound to the earth. When the total energy is greater than zero (E_2 in the figure), the object can escape the earth.

Earth as seen from Apollo 11 orbiting the moon on July 31, 1969.

Classification of Orbits by Energy

In Figures 11-9 and 11-10, two possible values for the total energy E are indicated on a graph of U versus r: E_1, which is negative, and E_2, which is positive. A negative total energy merely means that the kinetic energy at the earth's surface is less than $GM_E m/R_E$, so that $K + U$ is never greater than zero. From these figures, we see that if the total energy is negative, the total-energy line intersects the potential-energy curve at some maximum separation r_{max} and the system is bound. On the other hand, if the total energy is zero or positive, there is no such intersection and the system is unbound. The criteria for a bound or unbound system are simply stated.

> If $E < 0$, the system is bound.
> If $E \geq 0$, the system is unbound.

When E is negative, its absolute value $|E|$ is called the binding energy. The binding energy is the energy that must be added to the system to bring the total energy up to zero.

The potential energy of an object such as a planet or comet of mass m at a distance r from the sun is

$$U(r) = -\frac{GM_s m}{r}$$

11-20

where M_s is the mass of the sun. The kinetic energy of the object is $\frac{1}{2}mv^2$. If the total energy, kinetic plus potential, is less than zero, the orbit will be an ellipse (or a circle), and the object will be bound to the sun. That is, it cannot escape from the sun. On the other hand, if the total energy is positive, the orbit will be a hyperbola, and the object will make one swing around the sun and leave, never to return again. If the total energy is exactly zero, the orbit will be a parabola, and again the object will escape. That is, when the total energy is zero or positive the object is not bound to the sun.

Example 11-4

A projectile is fired straight up from the surface of the earth with an initial speed $v_i = 8$ km/s. Find the maximum height the projectile reaches, neglecting air resistance.

Picture the Problem The maximum height is found using energy conservation. We take the surface of the earth as the initial point, with $U_i = -GM_E m/R_E$ and $K_i = \frac{1}{2}mv_i^2$. At the greatest height, $K_f = 0$.

1. Apply conservation of mechanical energy:

$$U_i + K_i = U_f + K_f$$

$$-\frac{GM_E m}{R_E} + \frac{1}{2}mv_i^2 = -\frac{GM_E m}{r} + 0$$

2. Cancel the common term m, use $g = GM_E/R_E^2$, and solve for r:

$$\frac{1}{2}v_i^2 = \frac{GM_E}{R_E}\left(1 - \frac{R_E}{r}\right) = gR_E\left(1 - \frac{R_E}{r}\right)$$

$$1 - \frac{R_E}{r} = \frac{v_i^2}{2gR_E}$$

$$r = \frac{R_E}{1 - v_i^2/2gR_E}$$

3. Substitute numerical values to find r and $h = r - R_E$:

$$\frac{v_i^2}{2gR_E} = \frac{(8000 \text{ m/s})^2}{2(9.81 \text{ m/s}^2)(6.37 \times 10^6 \text{ m})} = 0.512$$

$$r = \frac{R_E}{1 - 0.512} = 2.05R_E$$

$$h = r - R_E = 1.05R_E$$

Example **11-5**	*try it yourself*

A projectile is fired straight up from the surface of the earth with an initial speed v_i = 15 km/s. Find the speed of the projectile when it is very far from the earth, neglecting air resistance.

Picture the Problem   The initial speed is greater than the escape speed, so the total energy of the projectile is positive, and the projectile will escape the earth with some final kinetic energy. Use conservation of mechanical energy to find this kinetic energy and then solve for the final speed.

Cover the column to the right and try these on your own before looking at the answers.

Steps	**Answers**
1. Apply conservation of mechanical energy, noting that $r_f = \infty$, so $U_f = 0$.	$U_i + K_i = U_f + K_f - \dfrac{GM_E m}{R_E} + \dfrac{1}{2}mv_i^2 = 0 + \dfrac{1}{2}mv_f^2$
2. Solve for v_f^2 using $GM_E/R_E^2 = g$ to simplify.	$v_f^2 = v_i^2 - \dfrac{2GM_E}{R_E} = v_i^2 - 2gR_E$
3. Substitute known values for g and R_E to calculate v_f.	$v_f^2 = 10^8 \text{ m}^2/\text{s}^2$ $v_f = 10^4 \text{ m/s} = 10 \text{ km/s}$

Remark   In Figure 11-11, the speed of the projectile in kilometers per second is plotted versus h/R_E, where h is the height above the earth's surface. At very large values of h/R_E, the speed approaches the dashed line $v_f = 10$ km/s.

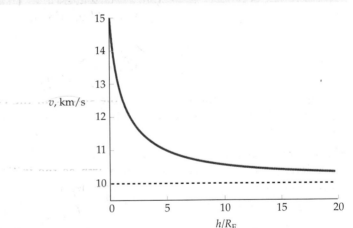

Figure 11-11

Example **11-6**	*try it yourself*

Show that the total energy of a satellite in a circular orbit is half its potential energy.

Picture the Problem   The total energy of a satellite is the sum of its potential and kinetic energy, $E = U + K$. The kinetic energy depends on the satellite's speed, which can be determined by equating the gravitational force with the centripetal force needed for the circular orbit. Assume that the mass of the earth is much greater than that of the satellite so the center of mass is essentially at the center of the earth.

Cover the column to the right and try these on your own before looking at the answers.

Steps	**Answers**
1. Write the potential energy U of the satellite in terms of the separation distance r.	$U = -\dfrac{GM_E m}{r}$
2. Write the kinetic energy K in terms of the mass of the satellite and its velocity v.	$K = \dfrac{1}{2}mv^2$

3. Set the gravitational force on the satellite equal to its mass times its centripetal acceleration.

$$\frac{mv^2}{r} = \frac{GM_E m}{r^2}$$

4. Substitute mv^2 from step 3 into the expression for K in step 2. This will give you an expression for K in terms of the distance r.

$$K = \frac{GM_E m}{2r}$$

5. Write the total energy $E = K + U$ as a function of r. Compare it with U in step 1.

$$E = U + K = -\frac{1}{2}\frac{GM_E m}{r} = \frac{1}{2}U$$

Exercise A satellite of mass 450 kg orbits the earth in a circular orbit at 6830 km above the earth's surface. Find (a) the potential energy, (b) the kinetic energy, and (c) the total energy of the satellite. (*Answers* Note that $r = R_E + h = 13,200$ km. (a) $U = -13.6$ GJ, (b) $K = 6.80$ GJ, (c) $E = -6.80$ GJ)

11-4   The Gravitational Field $\vec{g}$

The gravitational force exerted by a point mass m_1 on a second mass m_2 a distance $r_{1,2}$ away is given by

$$\vec{F}_{1,2} = -\frac{Gm_1 m_2}{r_{1,2}^2}\hat{r}_{1,2}$$

where $\hat{r}_{1,2} = \vec{r}_{1,2}/r_{1,2}$ is a unit vector pointing from m_1 to m_2. The gravitational force on a small test mass m divided by m is called the **gravitational field $\vec{g}$**.

$$\vec{g} = \frac{\vec{F}}{m} \qquad\qquad\qquad 11\text{-}21$$

Definition—Gravitational field

The gravitational field at a point due to a set of point masses is the vector sum of the fields due to the individual masses at that point.

$$\vec{g} = \sum_i \vec{g}_i \qquad\qquad\qquad 11\text{-}22a$$

To find the gravitational field at a point due to a continuous object, we find the field $d\vec{g}$ due to a small mass element dm, assuming it to be a point mass, and integrate over the entire object.

$$\vec{g} = \int d\vec{g} \qquad\qquad\qquad 11\text{-}22b$$

The gravitational field of the earth at a distance $r \geq R_E$ points toward the earth and is given by

$$\vec{g} = -\frac{GM_E}{r^2}\hat{r} \qquad\qquad\qquad 11\text{-}23$$

Gravitational field of the earth

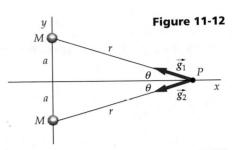

Figure 11-12

Example 11-7

Two particles each of mass M are fixed on the y axis at $y = +a$ and $y = -a$ (Figure 11-12). Find the gravitational field at a point P on the x axis.

Picture the Problem Two particles of mass M produce a gravitational field at point P. The distance between P and either particle is $r = \sqrt{x^2 + a^2}$. The field is the vector sum of the fields $\vec{g}_1$ and $\vec{g}_2$ due to each mass.

1. Calculate the magnitude of either $\vec{g}_1$ or $\vec{g}_2$:

$$g_1 = g_2 = \frac{GM}{r^2} = \frac{GM}{x^2 + a^2}$$

2. By symmetry, the y component of the resultant field is zero. The x component is the sum of g_{1x} and g_{2x}:

$$g_x = g_{1x} + g_{2x} = 2g_{1x} = -2g_1 \cos \theta$$

3. Express $\cos \theta$ in terms of x and a from the figure:

$$\cos \theta = \frac{x}{r} = \frac{x}{\sqrt{x^2 + a^2}}$$

4. Combining the last two results yields $\vec{g}$:

$$\vec{g} = g_x \hat{i} = -\frac{2GMx}{(x^2 + a^2)^{3/2}} \hat{i}$$

Check the Result If $x = 0$, we find that $\vec{g} = 0$; the fields due to m_1 and m_2 are equal and opposite at $x = 0$, and hence they cancel. For $x \gg a$, $\vec{g} \approx (2GM/x^2)\hat{i}$. The field is the same as if a single mass of $2M$ were at the origin.

Example 11-8

A uniform stick of mass M and length L is centered on the origin and lies along the x axis. Find the gravitational field due to the stick at a point x_0 on the x axis, where $x_0 > L/2$.

Picture the Problem We choose a mass element dm at a distance dx (Figure 11-13). All such elements produce a gravitational field at P that points toward the origin. Thus, we can calculate the total field by integrating the magnitude of the field produced by dm from $x = -L/2$ to $x = +L/2$.

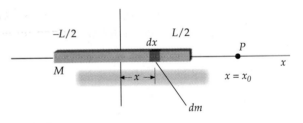

Figure 11-13

1. Find the magnitude of the field at P due to the element dm:

$$dg = \frac{G \, dm}{r^2}$$

2. The mass dm is proportional to the size of the element dx:

$$dm = \frac{M}{L} dx$$

3. Write the distance r between dm and point P in terms of x and x_0:

$$r = x_0 - x$$

4. Substitute these results to express dg in terms of x:

$$dg = \frac{G \, dm}{r^2} = \frac{G(M/L) \, dx}{(x_0 - x)^2}$$

5. Integrate to find the total field:

$$g = \int dg = \frac{GM}{L} \int_{-L/2}^{L/2} \frac{dx}{(x_0 - x)^2}$$

$$= \frac{GM}{L} \left[\frac{1}{x_0 - x} \right]_{-L/2}^{L/2}$$

$$= \frac{GM}{L} \left(\frac{1}{x_0 - L/2} - \frac{1}{x_0 + L/2} \right) = \frac{GM}{x_0^2 - L^2/4}$$

6. Express the resultant field as a vector that points toward the origin:

$$\vec{g} = -\frac{GM}{x_0^2 - L^2/4}\,\hat{i}$$

Check the Result For $x_0 \gg L/2$, the field approaches the field of a point mass $\vec{g} = -(GM/x_0^2)\hat{i}$.

Figure 11-14 A uniform spherical shell of mass M and radius R.

$\vec{g}$ of a Spherical Shell and of a Solid Sphere

One of Newton's motivations for developing calculus was to prove that the gravitational field outside a solid sphere is the same as if all the mass of the sphere were concentrated at its center. In the next section, we will show that the gravitational field at a distance r from the center of a uniform spherical shell of mass M and radius R (Figure 11-14) is given by

$$\vec{g} = -\frac{GM}{r^2}\,\hat{r} \qquad \text{for } r > R \qquad\qquad \text{11-24}a$$

$$\vec{g} = 0 \qquad\qquad \text{for } r < R \qquad\qquad \text{11-24}b$$

Gravitational field of a spherical shell

We can understand the result that $\vec{g} = 0$ inside the shell from Figure 11-15, which shows a point mass m_0 inside a spherical shell. In this figure, the masses of the shell segments m_1 and m_2 are related by

$$m_2 = m_1\left(\frac{r_2^2}{r_1^2}\right)$$

Since the force due to each mass is proportional to $1/r^2$, the force due to the smaller mass on the left is exactly balanced by that due to the more distant larger mass on the right.

The gravitational field outside a solid sphere is a simple extension of Equation 11-24a. We merely consider the solid sphere to consist of a continuous set of spherical shells. Since the field due to each shell is the same as if its mass were concentrated at the center of the shell, the field due to the entire sphere is the same as if the entire mass of the sphere were concentrated at its center:

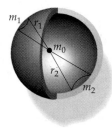

Figure 11-15 A point mass m_0 inside a uniform spherical shell feels no net force.

$$\vec{g} = -\frac{GM}{r^2}\,\hat{r} \qquad \text{for } r > R \qquad\qquad \text{11-25}$$

This result holds whether or not the sphere has a constant density, as long as the density depends only on r so that spherical symmetry is maintained.

$\vec{g}$ Inside a Solid Sphere

We now use Equations 11-24a and 11-24b to find the gravitational field inside of a solid sphere of constant density at a point a distance r from the center, where r is less than the radius R of the sphere. This would apply, for example, to finding the weight of an object at the bottom of a deep mine shaft. As we have seen, the field inside a spherical shell is zero. Thus, in Figure 11-16, the mass of the sphere outside r exerts no force at or inside r. Therefore, only the mass M' within the radius r contributes to the gravitational field at r. This mass produces a field equal to that of a point mass M' at the center of the sphere. The fraction of the total mass of the sphere within r is equal to the ratio of the volume of a sphere of radius r to

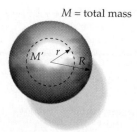

$M = $ total mass

Figure 11-16 A uniform solid sphere of radius R and mass M. Only the mass M', which is inside the sphere of radius r, contributes to the gravitational field at the distance r.

that of a sphere of radius R. Thus, for a uniform mass distribution, if M is the total mass of the sphere, M' is given by

$$M' = \frac{\frac{4}{3}\pi r^3}{\frac{4}{3}\pi R^3} M = \frac{r^3}{R^3} M \qquad \text{11-26}$$

The gravitational field at the distance r is thus

$$g_r = -\frac{GM'}{r^2} = -\frac{GMr^3/R^3}{r^2}$$

or

$$\vec{g} = -\frac{GM}{R^3}\vec{r} \qquad \text{for } r < R \qquad \text{11-27}$$

The magnitude of the field increases with distance r inside the sphere. Figure 11-17 shows a plot of the field g_r as a function of r for a solid sphere of constant mass density.

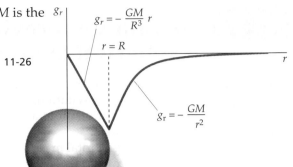

Figure 11-17 Plot of g_r versus r for a uniform solid sphere of mass M. The magnitude of the field increases with r inside the sphere and decreases as $1/r^2$ outside the sphere.

In the figure: $g_r = -\dfrac{GM}{R^3}r$, $r = R$, $g_r = -\dfrac{GM}{r^2}$

Example 11-9   *try it yourself*

A solid sphere of radius R and mass M is spherically symmetric but not uniform. Its density ρ, defined as its mass per unit volume, is proportional to the distance from the center r for $r \leq R$. That is, $\rho = Cr$ for $r \leq R$, and $\rho = 0$ for $r > R$, where C is a constant. (*a*) Find C. (*b*) Find g_r for $r \geq R$. (*c*) Find g_r at $r = R/2$.

Picture the Problem (*a*) You can find C by integrating the density over the volume of the sphere and setting the result equal to M. For a volume element, take a spherical shell of radius r and thickness dr. Its volume is $4\pi r^2\, dr$, and its mass is $dm = \rho\, dV = Cr\,(4\pi r^2\, dr)$. (*b*) The field at $r \geq R$ is the same as if the total mass M were at the center of the sphere. (*c*) The field at $r = R/2$ is the same as if mass M' were at the center of the sphere, where M' is the amount of mass within the sphere of radius $R/2$. The mass between $r = R/2$ and $r = R$ produces zero field at $r = R/2$.

Cover the column to the right and try these on your own before looking at the answers.

Steps	Answers
(*a*)1. Integrate dm from $r = 0$ to $r = R$.	$\displaystyle\int_0^R \rho\, dV = C\pi R^4$
2. Set your result equal to M and solve for C in terms of the given quantities M and R.	$C = \dfrac{M}{\pi R^4}$
(*b*) Write an expression for the field outside the sphere in terms of the mass M and the distance $r \geq R$.	$g_r = \dfrac{GM}{r^2}$
(*c*)1. Compute the mass M' that is within the radius $R/2$ by integrating dm from $r = 0$ to $r = R/2$ and use the value of C found in step 2.	$M' = \displaystyle\int_0^{R/2} \rho\, dV = \dfrac{C\pi R^4}{16} = \dfrac{M}{16}$
2. Write an expression for the field at $r = R/2$ in terms of M and R.	$g_r = \dfrac{GM'}{(R/2)^2} = \dfrac{GM}{4R^2}$

Check the Result The units for C are kg/m^4, so the units for ρ are kg/m^3, which is mass per volume.

*e*xploring

Tidal Forces and the Roche Limit

Because the gravitational field of a spherical object is not uniform, but varies as $1/r^2$, the force exerted on an extended object varies across the object. For example, the force exerted by the moon is stronger on the parts of the earth nearest the moon than on the parts farthest away. Figure 1 shows the earth a distance r from the moon.

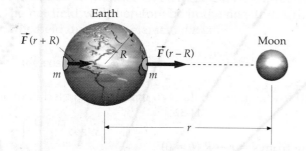

Figure 1

The difference between the force exerted by the moon on a mass m on the near side of the earth at $r - R$ and that exerted on the far side at $r + R$ is

$$\Delta F = F(r - R) - F(r + R)$$

$$= \frac{GMm}{(r - R)^2} - \frac{GMm}{(r + R)^2}$$

$$= \frac{GMm[(r + R)^2 - (r - R)^2]}{(r - R)^2(r + R)^2}$$

$$= \frac{4GMmrR}{(r^2 - R^2)^2} \approx \frac{4GMmR}{r^3} \qquad 1$$

where we have neglected R^2 compared with r^2 in the denominator. Although the sun exerts a much greater force on the earth's oceans than does the moon (see Problem 111), the *difference* in the force exerted by the moon when the ocean is closest compared to when it is farthest is much greater

than the corresponding differential force exerted by the sun. Because this differential force is responsible for the observed tides, it is called a tidal force.

Most large astronomical objects are held together by gravity. If the tidal force on such an object is greater than the gravitational forces holding the object together, the object will fly apart. Consider a planet of mass M. Because the tidal forces exerted by the planet vary as M/r^3, there is a minimum distance r_m at which a satellite can exist. This minimum distance is called the Roche limit after the French scientist Edouard Roche, who investigated this problem in 1848. We can estimate the Roche limit from a simple calculation. We consider an object of mass $2m$ consisting of two uniform spheres each of mass m and radius a (Figure 2). These objects exert an attractive force on each other as if each were a point mass a distance $2a$ from the other. We consider the force of attraction of these spheres $Gmm/(2a)^2$ to be the force that keeps the object together. When this object is at a distance r from a large object of mass M, as shown in Figure 1, the tidal force is given by Equation 1 with $R = a$. At the Roche limit $r = r_m$, the tidal force equals the force of attraction:

$$\frac{4GMma}{r_m^3} = \frac{Gm^2}{(2a)^2}$$

$$r_m^3 = \frac{16Ma^3}{m}$$

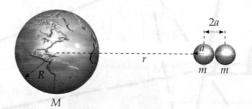

Figure 2

Let ρ_o be the density of the large object, whose radius is R, and ρ_s be the density of the satellite. Then $M = \frac{4}{3}\rho_o R^3$, $m = \frac{4}{3}\rho_s a^3$, and the Roche limit is

$$r_m = \left(\frac{16\rho_o}{\rho_s}\right)^{1/3} R \qquad 2$$

If the densities are equal, the Roche limit is about 2.5 times the radius of the planet. Natural satellites can exist only outside the Roche limit of a planet. Around Saturn, we find that inside the

Roche limit are rings of small particles that cannot form a satellite held together by gravity. Artificial satellites can, of course, exist within the Roche limit of a planet because they are held together by nuts and bolts rather than by gravitational attraction.

If the central object in our system is a black hole, its density ρ_o will be much greater than that of ordinary objects. The Roche limit for a black hole is many times greater than the radius R of the black hole.

The rings and three moons of Saturn. Close to the planet the tidal forces are too great for a satellite to exist. Instead, we find rings of small particles orbiting close to Saturn. The three moons shown orbit at distances greater than the Roche limit, where the tidal forces are much smaller. (Planetary rings are not uncommon. Faint rings have also been observed around Jupiter, Uranus, and Neptune.)

4.   Gravitational Field

1. Due to a set of objects	Find $\vec{g}_i$ due to each object separately from Newton's law of gravitation and sum the vectors. **Example 11-7**
2. Due to a continuous object	Find $d\vec{g}$ due to a mass element dm and integrate. **Example 11-8**
3. At some point r' inside a spherically symmetric mass distribution	Use $g_r = -GM'/r^2$ where M' is the total mass inside $r \leq r'$. The mass M' is found by integrating $\rho\, dV$ where ρ is the mass per unit volume and $dV = 4\pi r^2\, dr$ is the volume of a shell of radius r and thickness dr. **Example 11-9**

Problems

Conceptual Problems

Problems from Optional and Exploring sections

In a few problems, you are given more data than you actually need; in a few other problems, you are required to supply data from your general knowledge, outside sources, or informed estimates.

•   Single-concept, single-step, relatively easy
••   Intermediate-level, may require synthesis of concepts
••• Challenging, for advanced students

Take $g = 9.81$ N/kg $= 9.81$ m/s^2 and neglect friction and air resistance in all problems unless otherwise stated.

Kepler's Laws

1   •   True or false:

(a) Kepler's law of equal areas implies that gravity varies inversely with the square of the distance.
(b) The planet closest to the sun, on the average, has the shortest period of revolution about the sun.

2   •   If the mass of a satellite is doubled, the radius of its orbit can remain constant if the speed of the satellite

(a) increases by a factor of 8.
(b) increases by a factor of 2.
(c) does not change.
(d) is reduced by a factor of 8.
(e) is reduced by a factor of 2.

3   •   One night, Lucy picked up a strange message on her ham radio. "Help! We ran away from earth to live in peace and serenity, and we got disoriented. All we know is that we are orbiting the sun with a period of 5 years. Where are we?" Lucy did some calculations and told the travelers their mean distance from the sun. What is it?

4   •   Halley's comet has a period of about 76 y. What is its mean distance from the sun?

5   •   A comet has a period estimated to be about 4210 y. What is its mean distance from the sun? (4210 y was the estimated period of the comet Hale–Bopp, which was seen in the Northern Hemisphere in early 1997. Gravitational interactions with the major planets that occurred during this apparition of the comet greatly changed its period, which is now expected to be about 2380 y.)

6   •   The radius of the earth's orbit is 1.496×10^{11} m and that of Uranus is 2.87×10^{12} m. What is the period of Uranus?

7   •   The asteroid Hektor, discovered in 1907, is in a nearly circular orbit of radius 5.16 AU about the sun. Determine the period of this asteroid.

8   ••   The asteroid Icarus, discovered in 1949, was so named because its highly eccentric elliptical orbit brings it close to the sun at perihelion. The eccentricity e of an ellipse is defined by the relation $d_p = a(1 - e)$, where d_p is the perihelion distance and a is the semimajor axis. Icarus has an eccentricity of 0.83. The period of Icarus is 1.1 years. (a) Determine the semimajor axis of the orbit of Icarus. (b) Find the perihelion and aphelion distances of the orbit of Icarus.

Newton's Law of Gravity

9   •   Why don't you feel the gravitational attraction of a large building when you walk near it?

10   •   Astronauts orbiting in a satellite 300 km above the surface of the earth feel weightless. Why? Is the force of gravity exerted by the earth on them negligible at this height?

11   ••   The distance from the center of the earth to a point where the acceleration due to gravity is $g/4$ is

(a) R_E.
(b) $4R_E$.
(c) $\frac{1}{2}R_E$.
(d) $2R_E$.
(e) none of the above.

12   ••   At the surface of the moon, the acceleration due to the gravity of the moon is a. At a distance from the center of the moon equal to four times the radius of the moon, the acceleration due to the gravity of the moon is

(a) $16a$.
(b) $a/4$.
(c) $a/3$.
(d) $a/16$.
(e) None of the above.

13 • One of Jupiter's moons, Io, has a mean orbital radius of 4.22×10^8 m and a period of 1.53×10^5 s. (a) Find the mean orbital radius of another of Jupiter's moons, Callisto, whose period is 1.44×10^6 s. (b) Use the known value of G to compute the mass of Jupiter.

14 • The mass of Saturn is 5.69×10^{26} kg. (a) Find the period of its moon Mimas, whose mean orbital radius is 1.86×10^8 m. (b) Find the mean orbital radius of its moon Titan, whose period is 1.38×10^6 s.

15 • Calculate the mass of the earth from the period of the moon $T = 27.3$ d, its mean orbital radius $r_m = 3.84 \times 10^8$ m, and the known value of G.

16 • Use the period of the earth (1 y), its mean orbital radius (1.496×10^{11} m), and the value of G to calculate the mass of the sun.

17 • An object is dropped from a height of 6.37×10^6 m above the surface of the earth. What is its initial acceleration?

18 • Suppose you leave the solar system and arrive at a planet that has the same mass per unit volume as the earth but has 10 times the earth's radius. What would you weigh on this planet compared with what you weigh on earth?

19 • Suppose that the earth retained its present mass but was somehow compressed to half its present radius. What would be the value of g, the acceleration due to gravity, at the surface of this new, compact planet?

20 • A planet moves around a massive sun with constant angular momentum. When the planet is at perihelion, it has a speed of 5×10^4 m/s and is 1.0×10^{15} m from the sun. The orbital radius increases to 2.2×10^{15} m at aphelion. What is the planet's speed at aphelion?

21 • A comet orbits the sun with constant angular momentum. It has a maximum radius of 150 AU, and at aphelion its speed is 7×10^3 m/s. The comet's closest approach to the sun is 0.4 AU. What is its speed at perihelion?

22 •• The speed of an asteroid is 20 km/s at perihelion and 14 km/s at aphelion. Determine the ratio of the aphelion to perihelion distance.

23 •• A satellite with a mass of 300 kg moves in a circular orbit 5×10^7 m above the earth's surface. (a) What is the gravitational force on the satellite? (b) What is the speed of the satellite? (c) What is the period of the satellite?

24 •• At the airport, a physics student weighs 800 N. The student boards a jet plane that rises to an altitude of 9500 m. What is the student's loss in weight?

25 •• Suppose that Kepler had found that the period of a planet's circular orbit is proportional to the square of the orbit radius. What conclusion would Newton have drawn concerning the dependence of the gravitational attraction on distance between two masses?

26 •• A superconducting gravity meter can measure changes in gravity of the order $\Delta g/g = 10^{-11}$. (a) Estimate the maximum range at which an 80-kg person can be detected by this gravity meter. Assume that the gravity meter is stationary, and that the person's mass can be considered to be concentrated at his or her center of gravity. (b) What vertical change in the position of the gravity meter in the earth's gravitational field is detectable?

27 •• During a solar eclipse, when the moon is between the earth and the sun, the gravitational pull of the moon and the sun on a student are in the same direction. (a) If the pull of the earth on the student is 800 N, what is the force of the moon on the student? (b) What is the force of the sun on the student? (c) What percentage correction due to the sun and moon when they are directly overhead should be applied to the reading of a very accurate scale to obtain the student's weight?

28 •• Suppose that the attractive interaction between a star of mass M and a planet of mass $m \ll M$ were of the form $F = KMm/r$, where K is the gravitational constant. What would be the relation between the radius of the planet's circular orbit and its period?

29 •• The mass of the earth is 5.97×10^{24} kg and its radius is 6370 km. The radius of the moon is 1738 km. The acceleration of gravity at the surface of the moon is 1.62 m/s². What is the ratio of the average density of the moon to that of the earth?

30 ••• A plumb bob near a large mountain is slightly deflected from the vertical by the gravitational attraction of the mountain. Estimate the order of magnitude of the angle of deflection using any assumptions you like.

Measurement of G

31 • Why is G so difficult to measure?

32 • The masses in a Cavendish apparatus are $m_1 = 10$ kg and $m_2 = 10$ g, the separation of their centers is 6 cm, and the rod separating the two small masses is 20 cm long. (a) What is the force of attraction between the large and small masses? (b) What torque must be exerted by the suspension to balance these forces?

33 • The masses in a Cavendish apparatus are $m_1 = 12$ kg and $m_2 = 15$ g, and the separation of their centers is 7 cm. (a) What is the force of attraction between these two masses? (b) If the rod separating the two small masses is 18 cm long, what torque must be exerted by the suspension to balance the torque exerted by gravity?

Exploring . . . Gravitational and Inertial Mass

34 •• How would everyday life change if gravitational and inertial mass were not identical?

35 •• If gravitational and inertial mass were not identical, what would change for

(a) an offensive lineman on a football team?
(b) a car?
(c) a paperweight?

36 • A standard object defined as having a mass of exactly 1 kg is given an acceleration of 2.6587 m/s² when a certain force is applied to it. A second object of unknown mass acquires an acceleration of 1.1705 m/s² when the same force is applied to it. (a) What is the mass of the second object? (b)

Is the mass that you determined in part (*a*) gravitational or inertial mass?

37 • The weight of a standard object defined as having a mass of exactly 1 kg is measured to be 9.81 N. In the same laboratory, a second object weighs 56.6 N. (*a*) What is the mass of the second object? (*b*) Is the mass you determined in part (*a*) gravitational or inertial mass?

Gravitational Potential Energy

38 • (*a*) Taking the potential energy to be zero at infinite separation, find the potential energy of a 100-kg object at the surface of the earth. (Use 6.37×10^6 m for the earth's radius.) (*b*) Find the potential energy of the same object at a height above the earth's surface equal to the earth's radius. (*c*) Find the escape speed for a body projected from this height.

39 • A point mass m_0 is initially at the surface of a large sphere of mass M and radius R. How much work is needed to remove it to a very large distance away from the large sphere?

40 • Suppose that in space there is a duplicate earth, except that it has no atmosphere, is not rotating, and is not in motion around any sun. What initial velocity must a spacecraft on its surface have to travel vertically upward a distance above the surface of the planet equal to one earth radius?

41 •• An object is dropped from rest from a height of 4×10^6 m above the surface of the earth. If there is no air resistance, what is its speed when it strikes the earth?

42 •• An object is projected upward from the surface of the earth with an initial speed of 4 km/s. Find the maximum height it reaches.

43 •• A spherical shell has a radius R and a mass M. (*a*) Write expressions for the force exerted by the shell on a point mass m_0 when m_0 is outside the shell and when it is inside the shell. (*b*) What is the potential-energy function $U(r)$ for this system when the mass m_0 is at a distance r ($r \geq R$) if $U = 0$ at $r = \infty$? Evaluate this function at $r = R$. (*c*) Using the general relation for $dU = -\vec{F} \cdot d\vec{r} = -F_r \, dr$, show that U is constant everywhere inside the shell. (*d*) Using the fact that U is continuous everywhere, including at $r = R$, find the value of the constant U inside the shell. (*e*) Sketch $U(r)$ versus r for all possible values of r.

44 ••• Our galaxy can be considered to be a large disk of radius R and mass M of approximately uniform mass density. (*a*) Consider a ring element of radius r and thickness dr of such a disk. Find the gravitational potential energy of a 1-kg mass on the axis of this element a distance x from its center. (*b*) Integrate your result for part (*a*) to find the total gravitational potential energy of a 1-kg mass at a distance x due to the disk. (*c*) From $F_x = -dU/dx$ and your result for part (*b*), find the gravitational field g_x on the axis of the disk.

45 ••• The assumption of uniform mass density in Problem 44 is rather unrealistic. For most galaxies, the mass density increases greatly toward the center of the galaxy. Repeat Problem 44 using a surface mass density of the form $\sigma(r) = C/r$, where $\sigma(r)$ is the mass per unit area of the disk at a

distance r from the center. First determine the constant C in terms of R and M; then proceed as in Problem 44.

Escape Speed

46 • What is the effect of air resistance on the escape speed near the earth's surface?

47 • Would it be possible in principle for the earth to escape from the solar system?

48 • If the mass of a planet is doubled with no increase in its size, the escape speed for that planet will be

(*a*) increased by a factor of 1.4.
(*b*) increased by a factor of 2.
(*c*) unchanged.
(*d*) reduced by a factor of 1.4.
(*e*) reduced by a factor of 2.

49 • The planet Saturn has a mass 95.2 times that of the earth and a radius 9.47 times that of the earth. Find the escape speed for objects near the surface of Saturn.

50 • Find the escape speed for a rocket leaving the moon. The acceleration of gravity on the moon is 0.166 times that on earth, and the moon's radius is $0.273R_E$.

51 •• A particle is projected from the surface of the earth with a speed equal to twice the escape speed. When it is very far from the earth, what is its speed?

52 •• What initial speed should a particle be given if it is to have a final speed when it is very far from the earth equal to its escape speed?

53 •• A space probe launched from the earth with an initial speed v_i is to have a speed of 60 km/s when it is very far from the earth. What is v_i?

54 •• (*a*) Calculate the energy in joules necessary to launch a 1-kg mass from the earth at escape speed. (*b*) Convert this energy to kilowatt-hours. (*c*) If energy can be obtained at 10 cents per kilowatt-hour, what is the minimum cost of giving an 80-kg astronaut enough energy to escape the earth's gravitational field?

55 •• Show that the escape speed from a planet is related to the speed of a circular orbit just above the surface of the planet by $v_e = \sqrt{2} \, v_c$, where v_c is the speed of an object in the circular orbit.

56 •• Find the speed of the earth v_c as it orbits the sun, assuming a circular orbit. Use this and the result of Problem 55 to calculate the speed v_{eS} needed by the earth to escape from the sun.

57 •• If an object has just enough energy to escape from the earth, it will not escape from the solar system because of the attraction of the sun. Use Equation 11-19 with M_S replacing M_E and the distance to the sun r_S replacing R_E to calculate the speed v_{eS} needed to escape from the sun's gravitational field for an object at the surface of the earth. Neglect the attraction of the earth. Compare your answer with that in Problem 56. Show that if v_e is the speed needed to escape from the earth, neglecting the sun, then the speed of an object at the earth's surface needed to escape from the solar system is given by $v_{e,solar}^2 = v_e^2 + v_{eS}^2$, and calculate $v_{e,solar}$.

58 •• Why is it reasonable to neglect the other planets in calculating the speed needed to escape from the solar system? Would you expect the actual value of this speed to be greater or less than that calculated in Problem 57?

59 •• An object is projected vertically from the surface of the earth. Show that the maximum height reached by the object is $H = R_E H'/(R_E - H')$, where H' is the height that it would reach if the gravitational field were constant.

Orbits

60 •• An object (say, a newly discovered comet) enters the solar system and makes a pass around the sun. How can we tell if the object will return many years later, or if it will never return?

61 •• A spacecraft of 100 kg mass is in a circular orbit about the earth at a height $h = 2R_E$. (a) What is the period of the spacecraft's orbit about the earth? (b) What is the spacecraft's kinetic energy? (c) Express the angular momentum L of the spacecraft about the earth in terms of its kinetic energy K and find its numerical value.

62 •• Many satellites orbit the earth about 1000 km above the earth's surface. Geosynchronous satellites orbit at a distance of 4.22×10^7 m from the center of the earth. How much more energy is required to launch a 500-kg satellite into a geosynchronous orbit than into an orbit 1000 km above the surface of the earth?

63 •• It is theoretically possible to place a satellite at a position between the earth and the sun on the line joining them, where the gravitational forces of the sun and the earth on the satellite combine in such a way that the satellite will execute a circular orbit around the sun that is synchronous with the earth's orbit around the sun. (In other words, the satellite and the earth have the same orbital period about the sun, even though they are at different distances from the sun. The satellite always remains on the line joining the earth and the sun.) Write an expression that relates the appropriate circular orbital speed v of a satellite in such a situation to its distance r from the sun. Your expression may also contain quantities shown in Figure 11-20 plus the gravitational constant G.

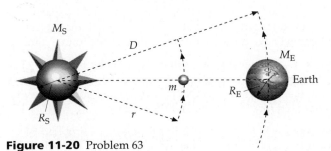

Figure 11-20 Problem 63

The Gravitational Field $\vec{g}$

64 • A 3-kg mass experiences a gravitational force of $12\,\text{N}\,\hat{\imath}$ at some point P. What is the gravitational field at that point?

65 • The gravitational field at some point is given by $\vec{g} = 2.5 \times 10^{-6}\,\text{N/kg}\,\hat{\jmath}$. What is the gravitational force on a mass of 4 g at that point?

66 •• A point mass m is on the x axis at $x = L$ and a second equal point mass m is on the y axis at $y = L$. (a) Find the gravitational field at the origin. (b) What is the magnitude of this field?

67 •• Five equal masses M are equally spaced on the arc of a semicircle of radius R as in Figure 11-21. A mass m is located at the center of curvature of the arc. (a) If M is 3 kg, m is 2 kg, and R is 10 cm, what is the force on m due to the five masses? (b) If m is removed, what is the gravitational field at the center of curvature of the arc?

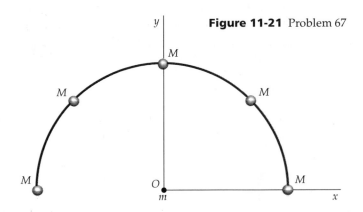

Figure 11-21 Problem 67

68 •• A point mass $m_1 = 2$ kg is at the origin and a second point mass $m_2 = 4$ kg is on the x axis at $x = 6$ m. Find the gravitational field at (a) $x = 2$ m, and (b) $x = 12$ m. (c) Find the point on the x axis for which $g = 0$.

69 •• (a) Show that the gravitational field of a ring of uniform mass is zero at the center of the ring.
(b) Figure 11-22 shows a point P in the plane of the ring but not at its center. Consider two elements of the ring of length s_1 and s_2 at distances of r_1 and r_2, respectively.
   1. What is the ratio of the masses of these elements?
   2. Which produces the greater gravitational field at point P?
   3. What is the direction of the field at point P due to these elements?
(c) What is the direction of the gravitational field at point P due to the entire ring?
(d) Suppose that the gravitational field varied as $1/r$ rather than $1/r^2$. What would be the net gravitational field at point P due to the two elements?
(e) How would your answers to parts (b) and (c) differ if point P were inside a spherical shell of uniform mass rather than inside a plane circular ring?

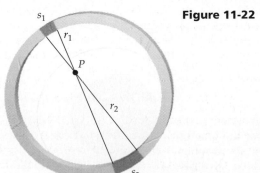

Figure 11-22 Problem 69

70 •• Show that the maximum value of $|g_x|$ for the field of Example 11-7 occurs at the points $x = \pm a / \sqrt{2}$.

71 •• A nonuniform stick of length L lies on the x axis with one end at the origin. Its mass density λ (mass per unit length) varies as $\lambda = Cx$, where C is a constant. (Thus, an element of the stick has mass $dm = \lambda \, dx$.) (a) What is the total mass of the stick? (b) Find the gravitational field due to the stick at a point $x_0 > L$.

72 ••• A uniform rod of mass M and length L lies along the x axis with its center at the origin. Consider an element of length dx at a distance x from the origin. (a) Show that this element produces a gravitational field at a point x_0 on the x axis ($x_0 > \frac{1}{2}L$) given by

$$dg_x = -\frac{GM}{L(x_0 - x)^2} \, dx$$

(b) Integrate this result over the length of the rod to find the total gravitational field at the point x_0 due to the rod. (c) What is the force on an object of mass m_0 at x_0? (d) Show that for $x_0 \gg L$, the field is approximately equal to that of a point mass M.

$\vec{g}$ due to Spherical Objects

73 •• Explain why the gravitational field increases with r rather than decreasing as $1/r^2$ as one moves out from the center inside a solid sphere of uniform mass.

74 • A spherical shell has a radius of 2 m and a mass of 300 kg. What is the gravitational field at the following distances from the center of the shell: (a) 0.5 m; (b) 1.9 m; (c) 2.5 m?

75 • A spherical shell has a radius of 2 m and a mass of 300 kg, and its center is located at the origin of a coordinate system. Another spherical shell with a radius of 1 m and mass 150 kg is inside the larger shell with its center at 0.6 m on the x axis. What is the gravitational force of attraction between the two shells?

76 • Two spheres, S_1 and S_2, have equal radii R and equal masses M. The density of sphere S_1 is constant, whereas that of sphere S_2 depends on the radial distance according to $p(r) = C/r$. If the acceleration of gravity at the surface of sphere S_1 is g_1, what is the acceleration of gravity at the surface of sphere S_2?

77 •• Two homogeneous spheres, S_1 and S_2, have equal masses but different radii, R_1 and R_2. If the acceleration of gravity on the surface of sphere S is g_1, what is the acceleration of gravity on the surface of sphere S_2?

78 •• Two concentric uniform spherical shells have masses M_1 and M_2 and radii a and $2a$ as in Figure 11-23. What is the magnitude of the gravitational force on a point mass m located (a) a distance $3a$ from the center of

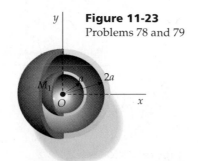

Figure 11-23
Problems 78 and 79

the shells? (b) a distance $1.9a$ from the center of the shells? (c) a distance $0.9a$ from the center of the shells?

79 •• The inner spherical shell in Problem 78 is shifted such that its center is now at $x = 0.8a$. The points $3a$, $1.9a$, and $0.9a$ lie along the same radial line from the center of the larger spherical shell. (a) What is the force on m at $x = 3a$? (b) What is the force on m at $x = 1.9a$? (c) What is the force on m at $x = 0.9a$?

$\vec{g}$ Inside Solid Spheres

80 •• Suppose the earth were a sphere of uniform mass. If there were a deep elevator shaft going 15,000 m into the earth, what would be the loss in weight at the bottom of this deep shaft for a student who weighs 800 N at the surface of the earth?

81 •• A sphere of radius R has its center at the origin. It has a uniform mass density ρ_0, except that there is a spherical cavity in it of radius $r = \frac{1}{2}R$ centered at $x = \frac{1}{2}R$ as in Figure 11-24. Find the gravitational field at points on the x axis for $|x| > R$. (Hint: The cavity may be thought of as a sphere of mass $m = \frac{4}{3}\pi r^3 \rho_0$ plus a sphere of mass $-m$.)

82 ••• For the sphere with the cavity in Problem 81, show that the gravitational field inside the cavity is uniform, and find its magnitude and direction.

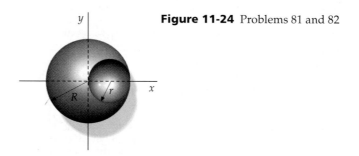

Figure 11-24 Problems 81 and 82

83 ••• A straight, smooth tunnel is dug through a spherical planet whose mass density ρ_0 is constant. The tunnel passes through the center of the planet and is perpendicular to the planet's axis of rotation, which is fixed in space. The planet rotates with an angular velocity ω such that objects in the tunnel have no acceleration relative to the tunnel. Find ω.

84 ••• The density of a sphere is given by $p(r) = C/r$. The sphere has a radius of 5 m and a mass of 1011 kg. (a) Determine the constant C. (b) Obtain expressions for the gravitational field for (1) $r > 5$ m, and (2) $r < 5$ m.

85 ••• A hole is drilled into the sphere of Problem 84 toward the center of the sphere to a depth of 2 km below the sphere's surface. A small mass is dropped from the surface into the hole. Determine the speed of the small mass as it strikes the bottom of the hole.

86 ••• The solid surface of the earth has a density of about 3000 kg/m³. A spherical deposit of heavy metals with a density of 8000 kg/m³ and a radius of 1000 m is centered 2000 m below the surface. Find $\Delta g/g$ at the surface directly above this deposit, where Δg is the increase in the gravitational field due to the deposit.

87 ••• Two identical spherical hollows are made in a lead sphere of radius R. The hollows have a radius $R/2$. They touch the outside surface of the sphere and its center as in Figure 11-25. The mass of the lead sphere before hollowing was M. (a) Find the force of attraction of a small sphere of mass m to the lead sphere at the position shown in the figure below. (b) What is the attractive force if m is located right at the surface of the lead sphere?

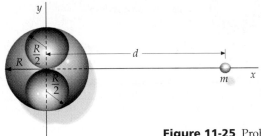

Figure 11-25 Problem 87

General Problems

88 • If K is the kinetic energy of the moon in its orbit around the earth, and U is the potential energy of the earth–moon system, what is the relationship between K and U?

89 •• A woman whose weight on earth is 500 N is lifted to a height two earth radii above the surface of the earth. Her weight will

(a) decrease to one-half of the original amount.
(b) decrease to one-quarter of the original amount.
(c) decrease to one-third of the original amount.
(d) decrease to one-ninth of the original amount.

90 • The mean distance of Pluto from the sun is 39.5 AU. Find the period of Pluto.

91 • The semimajor axis of Ganymede, a moon of Jupiter discovered by Galileo, is 1.07×10^6 km, and its period is 7.155 days. Determine the mass of Jupiter.

92 • Calculate the mass of the earth using the known values of G, g, and R_E.

93 • Uranus has a moon, Umbriel, whose mean orbital radius is 2.67×10^8 m and whose period is 3.58×10^5 s. (a) Find the period of another of Uranus's moons, Oberon, whose mean orbital radius is 5.86×10^8 m. (b) Use the known value of G to find the mass of Uranus.

94 •• Joe and Sally learn that there is a point between the earth and the moon where the gravitational effects of the two bodies balance each other. Being of a New Age bent, they decide to try to conceive a child free from the bondage of gravity, so they book an earth-to-moon trip. How far from the center of the earth should they try to conceive Zerog, the first zero-gravity baby?

95 •• The force exerted by the earth on a particle of mass m a distance r from the center of the earth has the magnitude $GM_Em/r^2 = mgR_E^2/r^2$. (a) Calculate the work you must do against gravity to move the particle from a distance r_1 to r_2. (b) Show that when $r_1 = R_E$ and $r_2 = R + h$, the result can be written

$$W = mgR_E^2 \left(\frac{1}{R_E} - \frac{1}{R_E + h} \right)$$

(c) Show that when $h \ll R_E$, the work is given approximately by $W = mgh$.

96 •• Suppose that the gravitational force of attraction depended not on $1/r^2$ but was proportional to the distance between the two masses, like the force of a spring. In a planetary system like the solar system, what would then be the relation between the period of a planet and its orbit radius, assuming all orbits were circular?

97 •• A uniform sphere of radius 100 m and density 2000 kg/m³ is in free space far from other massive objects. (a) Find the gravitational field outside of the sphere as a function of r. (b) Find the gravitational field inside the sphere as a function of r.

98 •• Two spherical planets have identical mass densities. Planet P_1 has a radius R_1, and planet P_2 has a radius R_2. If the acceleration of gravity at the surface of planet P_1 is g_1, what is the acceleration of gravity at the surface of planet P_2?

99 •• Jupiter has a mass 320 times that of Earth and a volume 1320 times that of Earth. A "day" on Jupiter is 9 h 50 min long. Find the height h above Jupiter at which a satellite must be revolving to have a period equal to 9 h 50 min.

100 •• The average density of the moon is $\rho = 3340$ kg/m³. Find the minimum possible period T of a spacecraft orbiting the moon.

101 •• A satellite is circling around the moon (radius 1700 km) close to the surface at a speed v. A projectile is launched from the moon vertically up at the same initial speed v. How high will it rise?

102 •• Two space colonies of equal mass orbit a star (Figure 11-26). The Yangs in m_1 move in a circular orbit of radius 10^{11} m with a period of 2 y. The Yins in m_2 move in an elliptical orbit with a closest distance $r_1 = 10^{11}$ m and a farthest distance $r_2 = 1.8 \times 10^{11}$ m. (a) Using the fact that the mean radius of an elliptical orbit is the length of the semimajor axis, find the length of the Yin year. (b) What is the mass of the star? (c) Which colony moves faster at point P in Figure 11-26? (d) Which colony has the greater total energy? (e) How does the speed of the Yins at point P compare with their speed at point A?

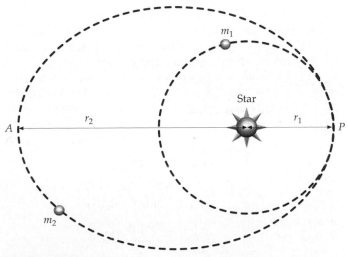

Figure 11-26 Problem 102

103 •• In a binary star system, two stars orbit about their common center of mass. If the stars have masses m_1 and m_2 and are separated by a distance r, show that the period of rotation is related to r by

$$T^2 = \frac{4\pi^2}{G(m_1 + m_2)} r^3$$

104 •• Two particles of mass m_1 and m_2 are released from rest with infinite separation. Find their speeds v_1 and v_2 when their separation distance is r.

105 •• A hole is drilled from the surface of the earth to its center as in Figure 11-27. Ignore the earth's rotation and air resistance. (*a*) How much work is required to lift a particle of mass m from the center of the earth to the earth's surface? (*b*) If the particle is dropped from rest at the surface of the earth, what is its speed when it reaches the center of the earth? (*c*) What is the escape speed for a particle projected from the center of the earth? Express your answers in terms of m, g, and R_E.

Figure 11-27 Problem 105

106 •• A thick spherical shell of mass M and uniform density has an inner radius R_1 and an outer radius R_2. Find the gravitational field g_r as a function of r for all possible values of r. Sketch a graph of g_r versus r.

107 •• (*a*) Sketch a plot of the gravitational field g_x versus x due to a uniform ring of mass M and radius R whose axis is the x axis. (*b*) At what points is the magnitude of g_x maximum?

108 ••• In this problem, you are to find the gravitational potential energy of the stick in Example 11-8 and a point mass m_0 that is on the x axis at x_0. (*a*) Show that the potential energy of an element of the stick dm and m_0 is given by

$$dU = -\frac{Gm_0\, dm}{x_0 - x} = \frac{GMm_0}{L(x_0 - x)} dx$$

where $U = 0$ at $x_0 = \infty$. (*b*) Integrate your result for part (*a*) over the length of the rod to find the total potential energy for the system. Write your result as a general function $U(x)$ by setting x_0 equal to a general point x. (*c*) Compute the force on m_0 at a general point x from $F_x = -dU/dx$ and compare your result with m_0g, where g is the field at x_0 calculated in Example 11-8.

109 ••• A uniform sphere of mass M is located near a thin, uniform rod of mass m and length L as in Figure 11-28. Find the gravitational force of attraction exerted by the sphere on the rod. (See Problem 72.)

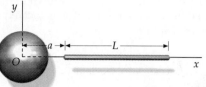

Figure 11-28 Problem 109

110 ••• A uniform rod of mass $M = 20$ kg and length $L = 5$ m is bent into a semicircle. What is the gravitational force exerted by the rod on a point mass $m = 0.1$ kg located at the center of the circular arc?

111 ••• Both the sun and the moon exert gravitational forces on the oceans of the earth, causing tides. (*a*) Show that the ratio of the force exerted by the sun to that exerted by the moon is $M_s r_m^2/M_m r_s^2$, where M_s and M_m are the masses of the sun and moon and r_s and r_m are the distances from the earth to the sun and to the moon. Evaluate this ratio. (*b*) Even though the sun exerts a much greater force on the oceans than the moon does, the moon has a greater effect on the tides because it is the difference in the force from one side of the earth to the other that is important. Differentiate the expression $F = Gm_1m_2/r^2$ to calculate the change in F due to a small change in r. Show that $dF/F = (-2\, dr)/r$. (*c*) During one full day, the rotation of the earth can cause the distance from the sun or moon to an ocean to change by at most the diameter of the earth. Show that for a small change in distance, the change in the force exerted by the sun is related to the change in the force exerted by the moon by

$$\frac{\Delta F_s}{\Delta F_m} \approx \frac{M_s r_m^3}{M_m r_s^3}$$

and calculate this ratio.

CHAPTER

Static Equilibrium and Elasticity

If an object is stationary and remains stationary, it is said to be in static equilibrium. Being able to determine the forces acting on an object in static equilibrium has many important applications. For example, the forces exerted by the cables of a suspension bridge must be known so that the cables can be designed to be strong enough to support the bridge. Similarly, cranes must be designed so that they do not topple over when lifting a weight.

The forces exerted by the cables and beams in a structure are called elastic forces. They are the result of slight deformations—the stretching or compression of solid objects under stress from bearing loads. We will first study the equilibrium of a rigid body, an ideal object whose deformation can be neglected. We then briefly consider the deformations and elastic forces that arise when real solids are under stress.

This gymnast, standing on one hand on a balance beam, is in static equilibrium.

12-1 Conditions for Equilibrium

A necessary condition for a particle at rest to remain at rest is that the net force acting on it must be zero. Similarly, the center of mass of a rigid object remains at rest if the net force acting on the object is zero. However, even if its center of mass is at rest, an object may still rotate. If there is rotation about any point, the object is not in static equilibrium. Therefore, for static equilibrium to exist, the net torque acting on an object in equilibrium must be zero about *any* point. This condition gives us the freedom to choose any point when calculating torques, which is often useful in solving problems.

The two necessary conditions for a rigid body to be in static equilibrium are therefore

1. The net external force acting on the body must be zero:

$$\sum_i \vec{F}_i = 0 \qquad\qquad 12\text{-}1$$

2. The net external torque about any point must be zero:

$$\sum_i \vec{\tau}_i = 0 \qquad\qquad 12\text{-}2$$

Conditions for equilibrium

As we have seen, we can describe the vector nature of rotation about a fixed axis as being positive or negative. We will choose counterclockwise torques* to be positive and clockwise torques to be negative.

---

* A counterclockwise torque is one that tends to produce rotation in the counterclockwise sense.

Example 12-1

A board of length $L = 3$ m and mass $M = 2$ kg is supported by scales on either end, as in Figure 12-1. A 6-kg mass m rests on the board a distance $x_1 = 2.5$ m from the left end and $x_2 = 0.5$ m from the right end. Find the readings on the scales.

Picture the Problem Let $\vec{F}_1$ and $\vec{F}_2$ be the forces exerted by the scales on the left and right ends of the board, respectively (Figure 12-2). Since the board exerts an equal but opposite force on each scale, the magnitudes of $\vec{F}_1$ and $\vec{F}_2$ are the readings on the scales. To find F_1 and F_2, we apply the two conditions for equilibrium. We take upward to be positive.

Figure 12-1

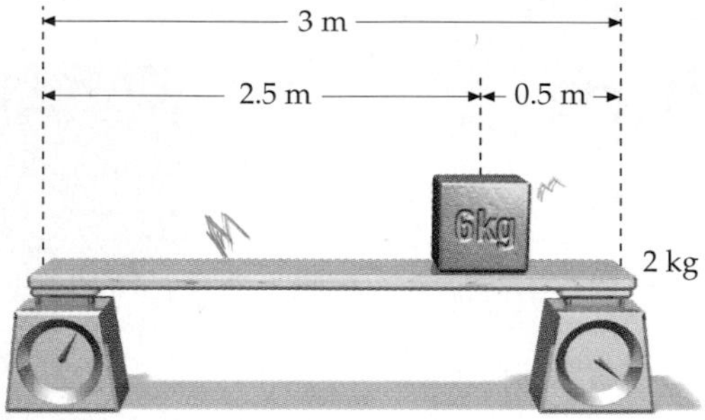

Figure 12-2

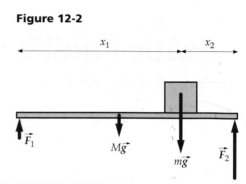

1. Set the net force equal to zero:

$$F_1 + F_2 - Mg - mg = 0$$

2. Set the net torque about the right scale equal to zero:

$$F_1 L - Mg\frac{L}{2} - mgx_2 + F_2 \times (0) = 0$$

3. The torque equation yields F_1:

$$F_1 = \frac{1}{2}Mg + \frac{x_2}{L}mg$$

4. Substitute this result for F_1 into step 1:

$$F_2 = Mg + mg - F_1 = \frac{1}{2}Mg + mg - \frac{x_2}{L}mg$$

5. Substitute numerical values to obtain F_1 and F_2:

$$F_1 = 19.6 \text{ N}$$
$$F_2 = 58.9 \text{ N}$$

Remarks Note that the right scale supports the greater weight, as expected.

Exercise Repeat this example choosing the left scale to be the point about which you set the net torque equal to zero. (*Answer*   In this case, the torque equation is $F_2 L - MgL/2 - mgx_1 = 0$. Again, we find that $F_1 = 19.6$ N and $F_2 = 58.9$ N.)

Example 12-1 can be solved using a pivot point located at the mass m, but in this case both F_1 and F_2 occur in the torque equation, hence the algebra is a bit more complex. In general, a statics problem can be simplified by computing the torques about a point on the line of action of one of the unknown forces, as when we chose the left or right scale above.

12-2   The Center of Gravity

Figure 12-3 shows an object divided into many smaller objects, which we can consider to be particles. The weight of each particle is $\vec{w}_i$, and the total weight of the object is $\vec{W} = \Sigma \vec{w}_i$. We can imagine this total weight concentrated at a single point in the object such that if the object were supported at that point, it would be in static equilibrium. This point is the **center of gravity**, defined so that the torque produced by $\vec{W}$ about any point is the same as that produced by the weights of the particles. If X_{cg} is the x coordinate of the center of gravity relative to some origin O, the magnitude of the torque about O is

$$X_{cg}W = \sum_i w_i x_i \qquad \text{12-3}$$

Center of gravity defined

If the acceleration of gravity is constant over the object (as is nearly always the case), we can write $w_i = m_i g$ and $W = Mg$ and cancel the common factor g. Then

$$X_{cg}Mg = \sum_i m_i g x_i$$

or

$$MX_{cg} = \sum_i m_i x_i \qquad \text{12-4}$$

This is the same as Equation 8-3, which gives the x coordinate for the center of mass. Thus, the center of gravity and the center of mass coincide when the gravitational field is uniform.

If we choose our origin to be at the center of gravity, $X_{cg} = 0$, then

$$X_{cg}W = \sum_i w_i x_i = 0$$

The center of gravity is that point about which the forces of gravity acting on all the particles of an object produce zero torque. We can use the methods discussed in Chapter 8 for locating the center of mass to locate the center of gravity (see Equation 8-6 and the discussion following). For example, the center of gravity of a stick is the point at which it balances on a pivot.

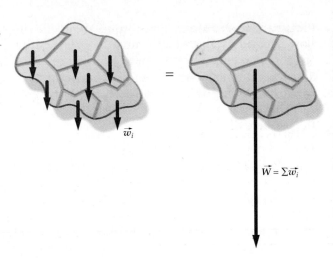

Figure 12-3 The weights of all the particles of an object can be replaced by the total weight $\vec{W}$ of the object acting at the center of gravity.

12-3 Some Examples of Static Equilibrium

Example 12-2

A 60-N weight is held in the hand with the forearm making a 90° angle with the upper arm, as in Figure 12-4. The biceps muscle exerts a force $\vec{F}_m$ that acts 3.4 cm from the pivot point O at the elbow joint. Neglecting the weight of the arm and hand, (*a*) find the magnitude of $\vec{F}_m$ if the distance from the weight to the pivot point is 30 cm, and (*b*) find the force exerted on the elbow joint by the upper arm.

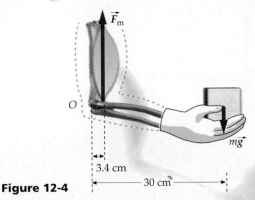

Figure 12-4

Picture the Problem The torque about the elbow exerted by the weight must be balanced by the torque exerted by the force $\vec{F}_m$ (Figure 12-5). The force $\vec{F}_{ua}$ exerted by the upper arm at O is found by setting the net force on the hand and forearm equal to zero. We choose upward to be the positive direction.

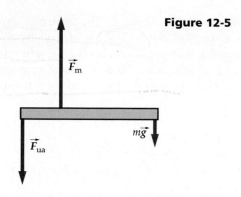

Figure 12-5

(a) $\Sigma \vec{\tau} = 0$ about point O gives F_m:

$$F_{ua} \times (0) + F_m \times (3.4 \text{ cm}) - (60 \text{ N})(30 \text{ cm}) = 0$$

$$F_m = \frac{(30 \text{ cm})(60 \text{ N})}{3.4 \text{ cm}} = 529 \text{ N}$$

(b) $\Sigma \vec{F} = 0$ gives F_{ua}:

$$-F_{ua} + F_m - 60 \text{ N} = 0$$

$$F_{ua} = F_m - 60 \text{ N} = 469 \text{ N}$$

Remarks The force that must be exerted by the muscle is 8.8 times the weight of the object! In addition, as the muscle pulls upward, the upper arm must push downward to keep the forearm in equilibrium. The force exerted by the upper arm is also several times greater than the object's weight.

Exercise Show that F_{ua} can be found in one step by choosing the pivot point to be where the biceps attaches to the forearm. (*Answer* Setting net torque equal to zero gives $(3.4 \text{ cm})F_{ua} + (0)F_m - (30 \text{ cm} - 3.4 \text{ cm})(60 \text{ N}) = 0$. This yields $F_{ua} = (60 \text{ N})(26.6 \text{ cm}/3.4 \text{ cm}) = 469 \text{ N}$.)

Remarks This example and this exercise show that we can choose the pivot point wherever it is convenient for our calculation.

Example 12-3   *try it yourself*

A sign of mass 20 kg hangs from the end of a rod of length 2 m and mass 4 kg. A wire is attached to the end of the rod and to a point 1 m above point O (Figure 12-6). Find the tension $\vec{T}$ in the wire and the force $\vec{F}$ exerted by the wall on the rod at point O.

Picture the Problem We have three unknowns: T, and the components F_x and F_y of the force exerted by the wall on the rod. We can find T_y by setting the net torque about O equal to zero. Then T_x is found from $\tan \theta = T_y/T_x = \frac{1}{2}$. F_x and F_y can then be found by applying the zero-force condition to both the x and y directions.

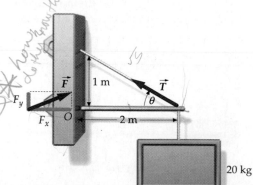

Figure 12-6

Cover the column to the right and try these on your own before looking at the answers.

Steps

Answers

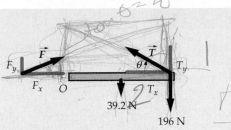

1. Draw a free-body diagram for the rod.

Figure 12-7

2. Set $\Sigma \vec{\tau} = 0$ about point O.

$T_y (2\ \text{m}) - 39.2\ \text{N}(1\ \text{m}) - 196\ \text{N}(2\ \text{m}) = 0$

3. Solve the torque equation for T_y.

$T_y = 216\ \text{N}$

4. Using your result for T_y and the value of $\tan \theta$ from the figure, find T_x.

$T_x = \dfrac{-T_y}{\tan \theta} = -2T_y = -432\ \text{N}$

5. Set $\Sigma F_x = 0$ and $\Sigma F_y = 0$.

$F_x + T_x = 0$

$F_y + T_y - 39.2\ \text{N} - 196\ \text{N} = 0$

6. Use your results for T_x and T_y to find the force components F_x and F_y in step 1.

$F_x = 432\ \text{N}, \qquad F_y = 19.2\ \text{N}$

Example 12-4   *try it yourself*

A wheel of mass M and radius R rests on a horizontal surface against a step of height h ($h < R$). The wheel is to be raised over the step by a horizontal force $\vec{F}$ applied to the axle of the wheel as shown in Figure 12-8. Find the force $\vec{F}$ necessary to raise the wheel over the step.

Picture the Problem   When F is very small, the forces exerted on the wheel are $\vec{F}$, the upward normal force $\vec{F}_n$ exerted by the surface at the bottom of the wheel, its weight $M\vec{g}$, and the force $\vec{F'}$ exerted at the corner in contact with the wheel. As F is increased, F_n decreases. Take torques about the corner to eliminate $\vec{F'}$.

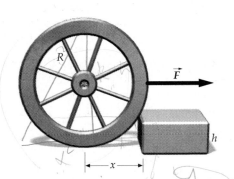

Figure 12-8

Cover the column to the right and try these on your own before looking at the answers.

Steps

Answers

1. Express the clockwise torque exerted by $\vec{F}$ about the corner in terms of F, h, and R.

$\tau_1 = F(R - h)$

2. Express the counterclockwise torque exerted by the weight about the corner in terms of x, the horizontal distance from the center of the wheel to the corner.

$\tau_2 = Mgx$

3. Use trigonometry to express x in terms of h and R.

$x^2 + (R - h)^2 = R^2$

$x = \sqrt{h(2R - h)}$

4. Set the magnitudes of the torques equal to each other and solve for F.

$F = \dfrac{Mgx}{R - h} = \dfrac{Mg\sqrt{h(2R - h)}}{R - h}$

Example 12-5

A uniform, 5-m ladder weighing 60 N leans against a frictionless vertical wall. The foot of the ladder is 3 m from the wall (Figure 12-9). What is the minimum coefficient of static friction necessary between the ladder and the floor if the ladder is not to slip?

Figure 12-9

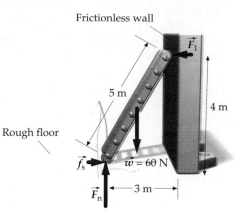

Frictionless wall

Rough floor

Picture the Problem There are three conditions for the ladder to be in equilibrium; $F_{net,x} = 0$, $F_{net,y} = 0$, and $\vec{\tau}_{net} = 0$. The forces acting on the ladder are the force $\vec{w}$ due to gravity acting downward at the ladder's center of gravity, the force $\vec{F}_1$ exerted horizontally by the wall (since the wall is frictionless, it exerts only a normal force), and the force exerted by the floor, which consists of a normal force $\vec{F}_n$ and a horizontal force of static friction $\vec{f}_s$. Thus, our three conditions determine $F_1, f_s,$ and F_n. We choose the foot of the ladder for our pivot point so that the torque equation contains only one unknown, F_1.

1. The coefficient of static friction is related to the frictional force f_s and normal force F_n:

$$f_s \le \mu_s F_n \quad \text{or} \quad \mu_s \ge \frac{f_s}{F_n}$$

2. Set $\Sigma F_x = 0$ and $\Sigma F_y = 0$:

$$F_{net,x} = 0 = f_s - F_1 \quad \text{and} \quad F_{net,y} = 0 = F_n - w$$

3. Solve for f_s and F_n:

$$f_s = F_1 \quad \text{and} \quad F_n = w = 60 \text{ N}$$

4. Set $\Sigma \vec{\tau} = 0$ about the foot of the ladder:

$$\tau_{net} = F_1(4 \text{ m}) - w(1.5 \text{ m}) = 0$$

5. Solve for the force F_1:

$$F_1 = \frac{w(15 \text{ m})}{4 \text{ m}} = \frac{(60 \text{ N})(1.5 \text{ m})}{4 \text{ m}} = 22.5 \text{ N}$$

6. Use this result for F_1 and $f_s = F_1$ from step 3 to find f_s:

$$f_s = F_1 = 22.5 \text{ N}$$

7. Using these results for f_s and F_n, we obtain the minimum value of μ_s from step 1:

$$\mu_s \ge \frac{f_s}{F_n} = \frac{22.5 \text{ N}}{60 \text{ N}} = 0.375$$

Remarks There is another way to look at this problem. Whenever an object is in static equilibrium under the influence of three nonparallel forces, the lines of action of the forces must intersect at one point. In the free-body diagram for the ladder shown in Figure 12-10, the lines of action of the weight $\vec{w}$ and the force $\vec{F}_1$ exerted by the wall intersect at point P. The line of action of the resultant force exerted by the ground $\vec{f} + \vec{F}_n$ must also go through point P or there would be an unbalanced torque about this point. The cotangent of the angle made by this resultant force equals 1.5 m/4 m $= 0.375 = f/F_n$.

Figure 12-10

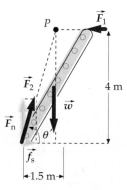

12-4   Couples

The forces $\vec{F}_n$ and $\vec{w}$ in Figures 12-9 and 12-10 of Example 12-5 are equal and opposite. Such a pair of forces, called a couple, tends to produce rotation, but its net force is zero. The forces $\vec{f}_s$ and $\vec{F}_1$ in those figures also constitute a couple. Figure 12-11 shows a couple consisting of forces $\vec{F}_1$ and

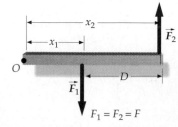

Figure 12-11 Two equal and opposite forces constitute a couple. The torque exerted by a couple has the same value FD about any point in space.

$\vec{F_2}$ a distance D apart. The torque produced by this couple about an arbitrary point O is

$$\tau = Fx_2 - Fx_1 = F(x_2 - x_1) = FD$$

where F is the magnitude of either force and D is the distance between them. This result does not depend on the choice of the point O.

> The torque produced by a couple is the same about all points in space.

12-5   Static Equilibrium in an Accelerated Frame

If we consider an object at rest in an accelerated frame of reference, the net force is not zero. If the object is to be at rest relative to the accelerated frame, the object must have the same acceleration as the frame. The two conditions for an object to be in static equilibrium in an accelerated reference frame are

1. $\Sigma\vec{F} = m\vec{a}_{cm}$

   where $\vec{a}_{cm}$ is the acceleration of the center of mass, which is the acceleration of the reference frame.

2. $\Sigma\vec{\tau}_{cm} = 0$

   The sum of the torques about the center of mass must be zero.

The second condition follows from the fact that Newton's second law for rotation, $\Sigma\vec{\tau}_{cm} = I_{cm}\vec{\alpha}$, holds for torques about the center of mass whether or not the center of mass is accelerating.* (Newton's second law for rotation does not hold for any other point that is accelerating.)

---

* See the discussion surrounding Equation 9-32.

Example 12-6

A truck carries a uniform box of mass m, height h, and square cross section of side L (Figure 12-12). What is the greatest acceleration the truck can have without tipping over the box? Assume that the box tips before it slides.

Picture the Problem The acceleration of the box is caused by the frictional force, as shown in Figure 12-13. This force exerts a counterclockwise torque about the center of mass of the box. The only other force that exerts a torque about the center of mass of the box is the normal force. When the box is not accelerating, $f = 0$, and the normal force is exerted through the center of the box. As the acceleration increases, the normal force moves to the left to provide a balance torque about the center of mass. The greatest balancing torque this force can exert is when it is at the edge of the box.

Figure 12-13

Figure 12-12

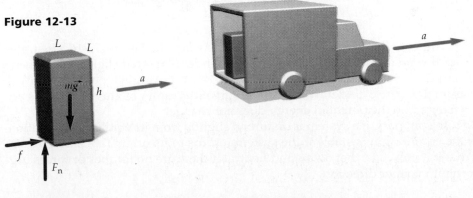

1. Apply $\Sigma \vec{F} = m\vec{a}$ to the box:

$$F_{\text{net},x} = f = ma$$

$$F_{\text{net},y} = F_n - mg = 0 \quad \text{or} \quad F_n = mg$$

2. Apply $\Sigma \vec{\tau}_{\text{cm}} = I_{\text{cm}}\vec{\alpha} = 0$:

$$f\frac{h}{2} - F_n\frac{L}{2} = 0$$

3. Substitute $f = ma$ and $F_N = mg$ and solve for a:

$$mah - mgL = 0$$

$$a = \frac{L}{h}g$$

Remarks The maximum acceleration is proportional to L/h. This maximum acceleration is small for a tall, narrow box (L/h small) and large for a short, wide box (L/h large). Thus, a short, wide box is more stable.

12-6 Stability of Rotational Equilibrium

There are three categories of rotational equilibrium for an object: stable, unstable, or neutral. **Stable rotational equilibrium** occurs when the torques that arise from a small angular displacement of the object urge the object back toward its equilibrium position. Stable equilibrium is illustrated in Figure 12-14. When the box is rotated slightly about one end, the resulting torque about the pivot point tends to restore the box to its original position. Note that this slight rotation lifts the center of gravity, increasing the potential energy of the box.

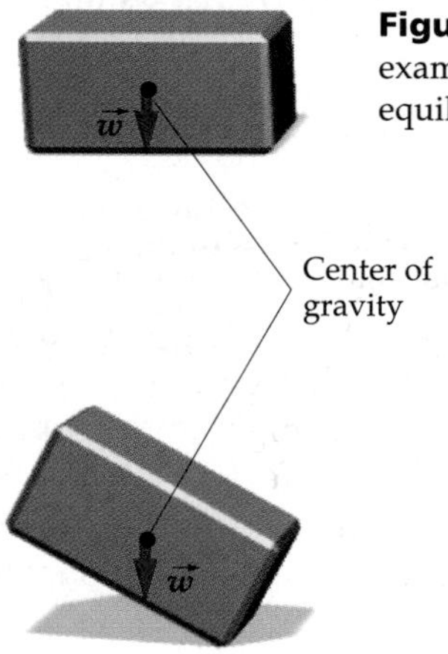

Unstable rotational equilibrium, illustrated in Figure 12-15, occurs when the torques that arise from a small angular displacement of the object urge the object away from its equilibrium position. A slight rotation of the narrow stick causes it to fall over because the torque due to its weight urges it away from its original position. Here the rotation lowers the center of gravity and decreases the potential energy of the stick.

Figure 12-14 An example of stable equilibrium.

Center of gravity

Figure 12-15 An example of unstable equilibrium.

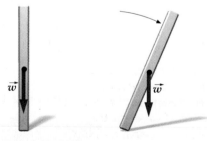

The cylinder resting on a horizontal surface in Figure 12-16 illustrates **neutral rotational equilibrium.** If the cylinder is rotated slightly, there is no torque or force that urges it either back toward its original position or away from it. As the cylinder rotates, the height of the center of gravity remains unchanged, so the potential energy does not change.

In summary, if a system is disturbed slightly from its equilibrium position, the equilibrium is stable if the system returns to its original position, unstable if it moves farther away, and neutral if there are no torques or forces urging it in either direction.

Figure 12-16 An example of neutral equilibrium.

Since "disturbed slightly" is a relative term, stability is also relative. One example of equilibrium may be more or less stable than another. Figure 12-17*a* shows a stick balanced on end that is not as narrow as that in Figure 12-15. Here, if the disturbance is very small (Figure 12-17*b*), the stick will move back toward its original position, but if the disturbance is great enough so that the center of gravity no longer lies over the base of support (Figure 12-17*c*), the stick will fall.

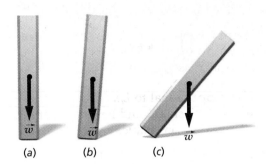

Figure 12-17 Stability of equilibrium is relative. If the stick in (*a*) is rotated slightly, as in (*b*), it returns to its original equilibrium position as long as the center of gravity lies over the base of support. (*c*) If the rotation is too great, the center of gravity is no longer over the base of support, and the stick falls over.

(a)   (b)   (c)

We can improve the stability of a system by either lowering the center of gravity or widening the base of support. Figure 12-18 shows a nonuniform stick that is loaded so that its center of gravity is near one end. If it stands on its heavy end so that the center of gravity is low (Figure 12-18*a*), it is much more stable than if it stands on the other end so that the center of gravity is high (Figure 12-18*b*).

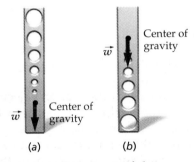

(a)   (b)

Figure 12-18 When a nonuniform stick rests on its heavy end with its center of gravity low (*a*), the equilibrium is more stable than when its center of gravity is high (*b*).

In Figure 12-19 the center of gravity lies below the point of support of the system. This system is stable for any displacement because the resulting torque always rotates the system back toward its equilibrium position.

Standing or walking upright presents a challenge for humans because the center of gravity is high and must be kept in balance over a relatively small base of support, the feet. Human infants take about a year to learn to walk. A four-footed creature has a much easier time because its base of support is larger and its center of gravity is lower. Newborn kittens can walk almost immediately.

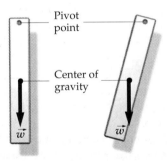

Figure 12-19 When a stick is pivoted so that its center of gravity is below the pivot point, the equilibrium is stable no matter how far the stick is displaced from equilibrium.

12-7   Indeterminate Problems

When objects are not rigid, but deformable, we need more information to determine the forces required for equilibrium. Consider a car resting on a horizontal surface. Suppose there is a very heavy object on one side of the trunk. We wish to find the vertical support force exerted by the road on each tire. Let the road be in the *xy* plane. If we choose one of the tires as our origin, the torque exerted by all the forces about that point has *x* and *y* components, but no *z* component because there are no horizontal forces. We thus obtain two equations by setting the net torque equal to zero, and a third equation by setting the net vertical force equal to zero. We need another equation to find the force exerted by the road on each of the four tires. If we let air out of one of the tires and pump up another tire to a greater pressure, the car remains in equilibrium, but the force exerted on each tire changes. Clearly, the forces on the tires in this problem are not determined by the information given. The tires are not rigid bodies. To some extent every object is deformable, a concept that is the foundation of the next section.

optional

12-8   Stress and Strain

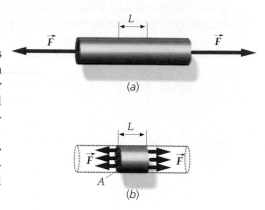

If a solid object is subjected to forces that tend to stretch, shear, or compress the object, its shape changes. If the object returns to its original shape when the forces are removed, it is said to be **elastic.** Most objects are elastic for forces up to a certain maximum, called the **elastic limit.** If the forces exceed the elastic limit, the object does not return to its original shape but is permanently deformed.

Figure 12-20 shows a solid bar of length L subjected to a stretching or **tensile force** F acting equally to the right and to the left. The bar is in equilibrium, but the forces acting on it tend to increase its length. The fractional change in the length of the bar $\Delta L/L$ is called the **strain:**

$$\text{Strain} = \frac{\Delta L}{L} \qquad\qquad 12\text{-}5$$

The ratio of the force F to the cross-sectional area A is called the **tensile stress:**

$$\text{Stress} = \frac{F}{A} \qquad\qquad 12\text{-}6$$

Figure 12-20 (a) A solid bar subjected to stretching forces $\vec{F}$ acting on each end. (b) A small section of the bar of length L. The elements of the bar to the left and right of this section exert forces on this section. If the section is not too close to the end, these forces are distributed equally over the cross-sectional area. The force per unit area is the stress.

Figure 12-21 shows a graph of strain versus stress for a typical solid bar. The graph is linear until point A. Up to this point, known as the proportional limit, the strain is proportional to the stress. The result that strain varies linearly with stress is known as Hooke's law.* Point B in Figure 12-21 is the elastic limit of the material. If the bar is stretched beyond this point, it is permanently deformed. If an even greater stress is applied, the material eventually breaks, shown happening at point C. The ratio of stress to strain in the linear region of the graph is a constant called **Young's modulus** Y:

$$Y = \frac{\text{stress}}{\text{strain}} = \frac{F/A}{\Delta L/L} \qquad\qquad 12\text{-}7$$

Definition—Young's modulus

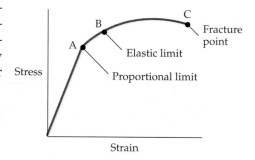

Figure 12-21 A graph of stress versus strain. Up to point A, the strain is proportional to the stress. Beyond the elastic limit at point B, the bar will not return to its original length when the stress is removed. At point C, the bar fractures.

The units of Young's modulus are newtons per square meter (or pounds per square inch). Approximate values of Young's modulus for various materials are listed in Table 12-1.

> **Exercise** A certain person's biceps muscle has a maximum cross-sectional area of 12 cm² = 1.2×10^{-3} m². What is the stress in the muscle if it exerts a force of 300 N? (*Answer*   Stress = $F/A = 2.5 \times 10^{5}$ N/m². The maximum stress that can be exerted is approximately the same for all human muscles. Greater forces can be exerted by muscles with greater cross-sectional areas.)

If a bar is subjected to forces that tend to compress it rather than stretch it, the stress is called **compressive stress.** For many materials, Young's modulus for compressive stress is the same as that for tensile stress. Note that ΔL in Equation 12-7 is then taken to be the *decrease* in the length of the bar. If the tensile or compressive stress is too great, the bar breaks. The stress at which breakage occurs is called the **tensile strength,** or in the case of compression, the **compressive strength.** Approximate values of the tensile and compressive strengths of various materials are listed in Table 12-1. Note that Young's modulus for bone is quite different for compressive and tensile stress, unlike

---

* This is the same behavior as that of a coiled spring for small stretching.

many other materials. This fact has biological significance, because the major job of bone is to resist the compressive load exerted by contracting muscles.

Table 12-1

Young's Modulus Y and Strengths of Various Materials*

Material	Y, GN/m²†	Tensile strength, MN/m²	Compressive strength, MN/m²
Aluminum	70	90	
Bone			
Tensile	16	200	
Compressive	9		270
Brass	90	370	
Concrete	23	2	17
Copper	110	230	
Iron (wrought)	190	390	
Lead	16	12	
Steel	200	520	520

*These values are representative. Actual values for particular samples may differ.
†1 GN = 10^3 MN = 10^9 N.

Example 12-7

A 500-kg mass is hung from a 3-m steel wire with a cross-sectional area of 0.15 cm². How much does the wire stretch?

Picturing the Problem L is the unstretched length of the wire, F is the force acting on it, and A is its cross-sectional area. The stretch in the wire ΔL is related to Young's modulus by $Y = (F/A)/(\Delta L/L)$. From Table 12-1 we find the numerical value of Young's modulus for steel, $Y = 2.0 \times 10^{11}$ N/m².

1. The amount the wire is stretched, ΔL, is found from Young's modulus:

$$Y = \frac{F/A}{\Delta L/L}$$

$$\Delta L = L\frac{F/A}{Y}$$

2. The force acting on the wire is the weight of the 500-kg mass:

$$F = mg = (500 \text{ kg})(9.81 \text{ m/s}^2) = 4.90 \times 10^3 \text{ N}$$

3. Convert the area to m²:

$$A = 0.15 \text{ cm}^2 \times \frac{10^{-4} \text{ m}^2}{1 \text{ cm}^2} = 1.5 \times 10^{-5} \text{ m}^2$$

4. Substituting numerical values yields ΔL:

$$\Delta L = L\frac{F/A}{Y}$$

$$= (3 \text{ m})\frac{(4.9 \times 10^3 \text{ N})/(1.5 \times 10^{-5} \text{ m}^2)}{2.0 \times 10^{11} \text{ N/m}^2}$$

$$= 0.49 \text{ cm}$$

Exercise A wire 1.5 m long has a cross-sectional area of 2.4 mm². It is hung vertically and stretches 0.32 mm when a 10-kg block is attached to it. Find (a) the stress, (b) the strain, and (c) Young's modulus for the wire. (Answers (a) 4.09×10^7 N/m², (b) 2.13×10^{-4}, (c) 192 GN/m²)

In Figure 12-22, a force $\vec{F}_s$ is applied tangentially to the top of a book. Such a force is called a **shear force.** The ratio of the shear force F_s to the area A is called the **shear stress:**

$$\text{Shear stress} = \frac{F_s}{A} \qquad \text{12-8}$$

A shear stress tends to deform an object, as shown in Figure 12-22. The ratio $\Delta X / L$ is called the **shear strain:**

$$\text{Shear strain} = \frac{\Delta X}{L} = \tan \theta \qquad \text{12-9}$$

where θ is the shear angle shown in the figure. The ratio of the shear stress to the shear strain is called the **shear modulus M_s:**

$$M_s = \frac{\text{shear stress}}{\text{shear strain}} = \frac{F_s/A}{\Delta X/L} = \frac{F_s/A}{\tan \theta} \qquad \text{12-10}$$

Definition—Shear modulus

The shear modulus is also known as the **torsion modulus.** The torsion modulus is approximately constant for small stresses, which implies that the shear strain varies linearly with the shear stress. This observation is known as Hooke's law for torsional stress. In a torsion balance, such as that used in Cavendish's apparatus for measuring the universal gravitational constant G, the torque (which is related to the stress) is proportional to the angle of twist (which equals the strain for small angles). Approximate values of the shear modulus for various materials are listed in Table 12-2.

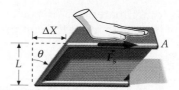

Figure 12-22 The application of the horizontal force F_s to the book causes a shear stress defined as the force per unit area. The ratio $\Delta X / L = \tan \theta$ is the shear strain.

Table 12-2

Approximate Values of the Shear Modulus M_s of Various Materials

Material	M_s, GN/m²
Aluminum	30
Brass	36
Copper	42
Iron	70
Lead	5.6
Steel	84
Tungsten	150

Summary

Topic	Remarks and Relevant Equations

1. Equilibrium of a Rigid Object

Conditions for

1. The net external force acting on the body must be zero:

$$\sum_i \vec{F}_i = 0 \qquad \text{12-1}$$

2. The net external torque about any point must be zero:

$$\sum_i \vec{\tau}_i = 0 \qquad \text{12-2}$$

An alternative statement of the second condition is that the sum of the torques that tend to produce clockwise rotation about any point must equal the sum of the torques that tend to produce counterclockwise rotation about that point.

Stability

The equilibrium of an object can be classified as stable, unstable, or neutral. An object resting on some surface will be in equilibrium if its center of gravity lies over its base of support. Stability can be improved by lowering the center of gravity or by increasing the size of the base.

2. Center of Gravity

The force of gravity exerted on the various parts of an object can be replaced by a single force, the total weight of the object W, acting at the center of gravity. The x coordinate of the center of gravity X_{cg} relative to some origin is given by

$$X_{cg}W = \sum_i w_i x_i \qquad \text{12-3}$$

When the acceleration of gravity is the same at all points of an object, the center of gravity coincides with the center of mass.

3.	**Couples**	A pair of equal and opposite forces constitutes a couple. The torque produced by a couple is the same about any point in space.

4. Stress and Strain

Strain	$\text{Strain} = \dfrac{\Delta L}{L}$	**12-5**
Stress	$\text{Stress} = \dfrac{F}{A}$	**12-6**
Young's modulus	$Y = \dfrac{\text{stress}}{\text{strain}} = \dfrac{F/A}{\Delta L/L}$	**12-7**
Shear stress	$\text{Shear stress} = \dfrac{F_s}{A}$	**12-8**
Shear strain	$\text{Shear strain} = \dfrac{\Delta X}{L} = \tan\theta$	**12-9**
Shear modulus	$M_s = \dfrac{\text{shear stress}}{\text{shear strain}} = \dfrac{F_s/A}{\Delta X/L} = \dfrac{F_s/A}{\tan\theta}$	**12-10**

Problem-Solving Guide

1. Begin by drawing a neat diagram that includes the important features of the problem.
2. To solve problems involving the equilibrium of rigid bodies, set the net force and net torque equal to zero. A judicious choice for the point about which to calculate the torques can simplify the calculation.

Summary of Worked Examples

Type of Calculation	Procedure and Relevant Examples
1. Equilibrium of Rigid Object	
Find the forces exerted on an object in equilibrium.	Use $\Sigma\vec{F} = 0$ and $\Sigma\vec{\tau} = 0$. Choose a point for the torque equation that is on the line of action of one or more of the unknown forces so those forces do not appear in the equation. **Examples 12-1, 12-2, 12-3, 12-4, 12-5**
Equilibrium in an accelerated frame	Use $\Sigma\vec{F} = m\vec{a}_{cm}$ and $\Sigma\vec{\tau}_{cm} = 0$ where the torques must be computed about the center of mass. A box in an accelerated truck will not tip if the normal force can exert a great enough torque to balance that exerted by static friction. **Example 12-6**
2. Stress and Strain	
Find the amount that a wire stretches under a given load.	Use Young's modulus. **Example 12-7**

Problems

In a few problems, you are given more data than you actually need; in a few other problems, you are required to supply data from your general knowledge, outside sources, or informed estimates.

Conditions for Equilibrium

1 • True or false:

(a) $\Sigma \vec{F} = 0$ is sufficient for static equilibrium to exist.

(b) $\Sigma \vec{F} = 0$ is necessary for static equilibrium to exist.

(c) In static equilibrium, the net torque about any point is zero.

(d) An object is in equilibrium only when there are no forces acting on it.

2 • A seesaw consists of a 4-m board pivoted at the center. A 28-kg child sits on one end of the board. Where should a 40-kg child sit to balance the seesaw?

3 • In Figure 12-23, Misako is about to do a push-up. Her center of gravity lies directly above point P on the floor, which is 0.9 m from her feet and 0.6 m from her hands. If her mass is 54 kg, what is the force exerted by the floor on her hands?

Figure 12-23
Problem 3

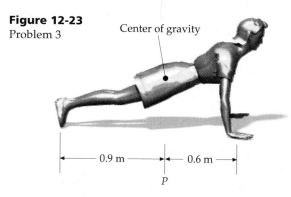

Center of gravity

0.9 m   0.6 m

P

4 • Juan and Bettina are carrying a 60-kg block on a 4-m board as shown in Figure 12-24. The mass of the board is 10 kg. Since Juan spends most of his time reading cookbooks, whereas Bettina regularly does push-ups, they place the block 2.5 m from Juan and 1.5 m from Bettina. Find the force in newtons exerted by each to carry the block.

Figure 12-24 Problem 4

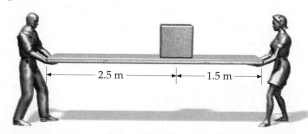

2.5 m   1.5 m

5 • Misako wishes to measure the strength of her biceps muscle by exerting a force on a test strap as shown in Figure 12-25. The strap is 28 cm from the pivot point at the elbow, and her biceps muscle is attached at a point 5 cm from

Figure 12-25 Problem 5

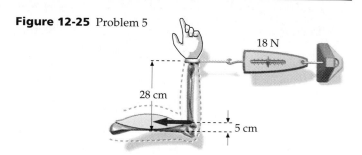

18 N

28 cm

5 cm

the pivot point. If the scale reads 18 N when she exerts her maximum force, what force is exerted by the biceps muscle?

6 • A crutch is pressed against the sidewalk with a force $\vec{F}_c$ along its own direction as in Figure 12-26. This force is balanced by a normal force $\vec{F}_n$ and a frictional force $\vec{f}_s$. (a) Show that when the force of friction is at its maximum value, the coefficient of friction is related to the angle θ by $\mu_s = \tan \theta$. (b) Explain how this result applies to the forces on your foot when you are not using a crutch. (c) Why is it advantageous to take short steps when walking on ice?

Figure 12-26 Problem 6

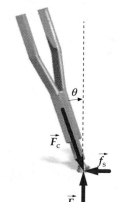

θ

$\vec{F}_c$

$\vec{f}_s$

$\vec{F}_n$

The Center of Gravity

7 • True or false: The center of gravity is always at the geometric center of a body.

8 • Must there be any material at the center of gravity of an object?

9 • If the acceleration of gravity is not constant over an object, is it the center of mass or the center of gravity that is the pivot point when the object is balanced?

10 • Two spheres of radius R rest on a horizontal table with their centers a distance $4R$ apart. One sphere has twice the weight of the other sphere. Where is the center of gravity of this system?

11 • An automobile has 58% of its weight on the front wheels. The front and back wheels are separated by 2 m. Where is the center of gravity located with respect to the front wheels?

12 • Each of the objects shown in Figure 12-27 is suspended from the ceiling by a thread attached to the point

Figure 12-27 Problem 12

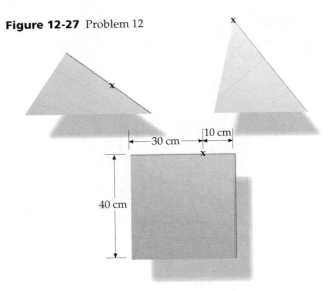

marked ✕ on the object. Describe the orientation of each suspended object with a diagram.

13 ●● A square plate is produced by welding together four smaller square plates, each of side a as shown in Figure 12-28. Plate 1 weighs 40 N; plate 2, 60 N; plate 3, 30 N; and plate 4, 50 N. Find the center of gravity (x_{cg}, y_{cg}).

Figure 12-28
Problem 13

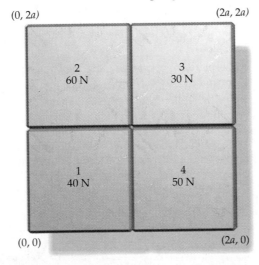

(0, 2a)   (2a, 2a)

| 2 | 3 |
| 60 N | 30 N |

| 1 | 4 |
| 40 N | 50 N |

(0, 0)   (2a, 0)

14 ●● A uniform rectangular plate has a circular section of radius R cut out as shown in Figure 12-29. Find the center of gravity of the system. *Hint:* Do not integrate. Use superposition of a rectangular plate minus a circular plate.

Figure 12-29
Problem 14

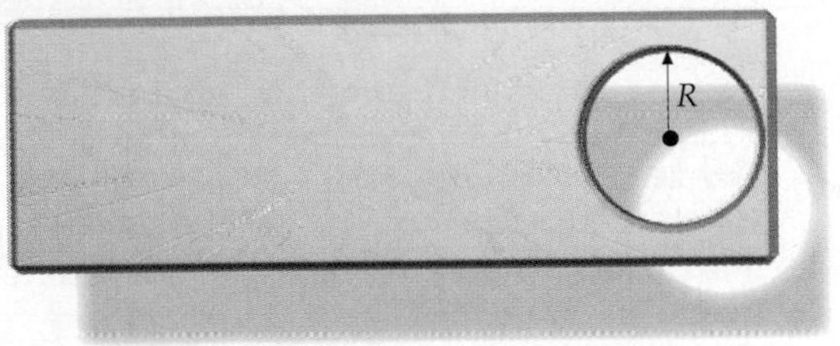

R

Some Examples of Static Equilibrium

15 ● When the tree in front of his house was cut down to widen the road, Jay did not want it to go without ceremony, so he hauled out his electric guitar and amplifier. All that remained was a uniform 10-m log of mass 100 kg resting on two supports, waiting to be cut up and taken away the next day.

One support was 2 m from the left end, and the other was 4 m from the right end. Find the forces exerted on the log by the supports as Jay played his ear-splitting "Requiem for a Fallen Tree."

16 ● Bubba uses a crowbar that is 1 m long to lift a heavy crate off the ground. The crowbar rests on a rigid fulcrum 10 cm from one end as shown in Figure 12-30. (*a*) If Bubba exerts a downward force of 600 N on one end of the crowbar, what is the upward force exerted on the crate by the other end? (*b*) The ratio of the forces at the ends of the crowbar is called the mechanical advantage of the crowbar. What is the mechanical advantage here?

Figure 12-30
Problem 16

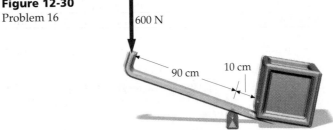

600 N

90 cm   10 cm

17 ● Figure 12-31 shows a 25-foot sloop. The mast is a uniform pole of 120 kg and is supported on the deck and held fore and aft by wires as shown. The tension in the forestay (wire leading to the bow) is 1000 N. Determine the tension in the backstay and the force that the deck exerts on the mast. Is there a tendency for the mast to slide forward or aft? If so, where should a block be placed to prevent the mast from moving?

Figure 12-31
Problem 17

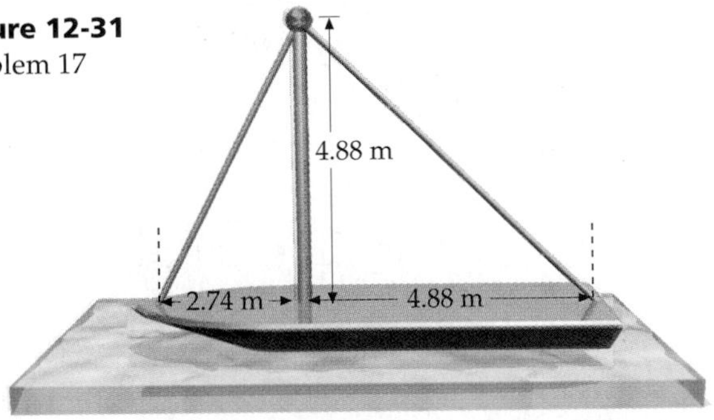

4.88 m

2.74 m   4.88 m

18 ● The sloop in Figure 12-32 is rigged slightly differently from the one in Problem 17. The mass of the mast is 150 kg and the tension in the forestay is again 1000 N. Find the tension in the backstay and the force that the deck exerts on the mast. Is there a tendency for the mast to slide forward or aft? If so, where should a block be placed on the deck to prevent the mast from moving?

Figure 12-32
Problem 18

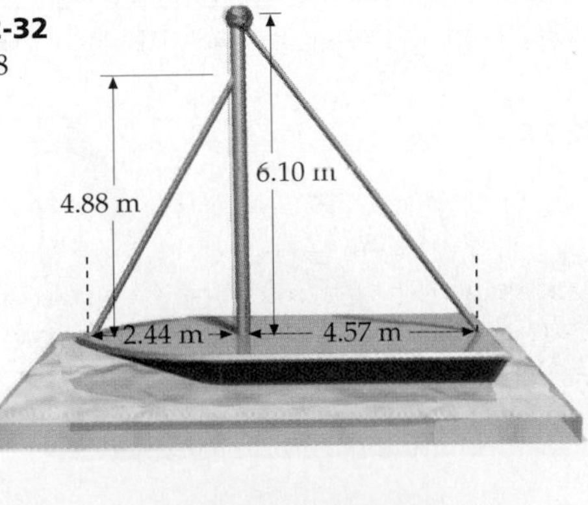

6.10 m

4.88 m

2.44 m   4.57 m

19 •• A 10-m beam of mass 300 kg extends over a ledge as in Figure 12-33. The beam is not attached, but simply rests on the surface. A 60-kg student intends to position the beam so that he can walk to the end of it. How far from the edge of the ledge can the beam extend?

Figure 12-33 Problem 19

20 •• A gravity board for locating the center of gravity of a person consists of a horizontal board supported by a fulcrum at one end and by a scale at the other end. A physics student lies horizontally on the board with the top of his head above the fulcrum point as shown in Figure 12-34. The scale is 2 m from the fulcrum. The student has a mass of 70 kg, and when he is on the gravity board, the scale advances 250 N. Where is the center of gravity of the student?

Figure 12-34
Problem 20

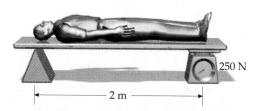

250 N

2 m

21 •• A 3-m board of mass 5 kg is hinged at one end. A force $\vec{F}$ is applied vertically at the other end to lift a 60-kg block, which rests on the board 80 cm from the hinge, as shown in Figure 12-35. (*a*) Find the magnitude of the force needed to hold the board stationary at $\theta = 30°$. (*b*) Find the force exerted by the hinge at this angle. (*c*) Find the magnitude of the force $\vec{F}$ and the force exerted by the hinge if $\vec{F}$ is exerted perpendicular to the board when $\theta = 30°$.

Figure 12-35 Problem 21

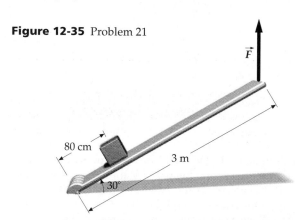

$\vec{F}$

80 cm

3 m

30°

22 •• A cylinder of weight W is supported by a frictionless trough formed by a plane inclined at 30° to the horizontal on the left and one inclined at 60° on the right as shown in Figure 12-36. Find the force exerted by each plane on the cylinder.

Figure 12-36
Problem 22

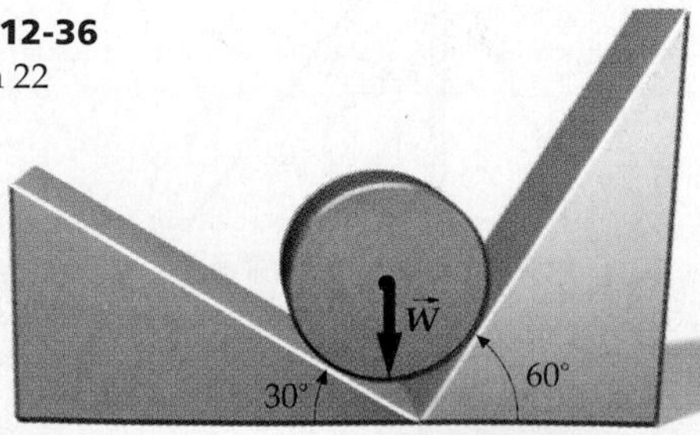

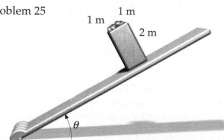

$\vec{W}$

30°   60°

Figure 12-37
Problem 23

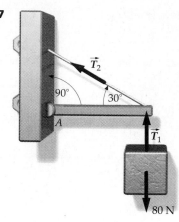

$\vec{T}_2$

90°   30°

A

$\vec{T}_1$

80 N

23 •• An 80-N weight is supported by a cable attached to a strut hinged at point A as in Figure 12-37. The strut is supported by a second cable under tension $\vec{T}_2$. The mass of the strut is negligible. (*a*) What are the three forces acting on the strut? (*b*) Show that the vertical component of the tension $\vec{T}_2$ must equal 80 N. (*c*) Find the force exerted on the strut by the hinge.

24 •• A horizontal board 8.0 m long is used by pirates to make their victims walk the plank. A pirate of mass 105 kg stands on the shipboard end of the plank to prevent it from tipping. Find the maximum distance the plank can overhang for a 63-kg victim to be able to walk to the end if (*a*) the mass of the plank is negligible, and (*b*) the mass of the plank is 25 kg.

25 •• As a farewell prank on their alma mater, Sharika and Chico decide to liberate thousands of marbles in the hallway during final exams. They place a 2-m × 1-m × 1-m box on a hinged board, as in Figure 12-38, and fill it with marbles. When the building is perfectly silent, they slowly lift one end of the plank, increasing θ, the angle of the incline. If the coefficient of static friction is large enough to prevent the box from slipping, at what angle will the box tip? (Assume that the marbles stay in the box until it tips over.)

Figure 12-38 Problem 25

1 m
1 m   2 m

θ

26 •• A uniform 18-kg door that is 2.0 m high by 0.8 m wide is hung from two hinges that are 20 cm from the top and 20 cm from the bottom. If each hinge supports half the weight of the door, find the magnitude and direction of the horizontal components of the forces exerted by the two hinges on the door.

27 •• Find the force exerted by the corner on the wheel in Example 12-4, just as the wheel lifts off the surface.

28 •• Lou is promoting the grand opening of Roswell's, a new nightclub with an alien theme. One end of a uniform 100-kg beam, 10 m long, is hinged to a wall, and the other end sticks out horizontally over the dance floor. A cable connects to the beam 6 m from the wall, as in Figure 12-39. Lou

Figure 12-39
Problems 28 and 32

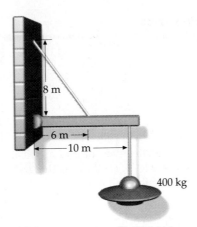

8 m

6 m

10 m

400 kg

sits at the controls of a mock UFO, which hangs from the free end of the beam. From there, he sends down abduction beams, hypnotic light effects, and spaceship noises to the patrons below. If the combined weight of Lou and his UFO is 400 kg, (a) what is the tension in the cable? (b) What is the horizontal force on the hinge? (c) What is the vertical force of the beam on the hinge?

29 •• The diving board shown in Figure 12-40 has a mass of 30 kg. Find the force on the supports when a 70-kg diver stands at the end of the diving board. Give the direction of each support force as a tension or a compression.

Figure 12-40 Problem 29

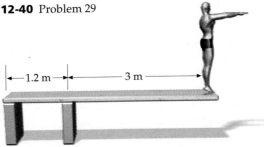

1.2 m

3 m

30 •• Find the force exerted **Figure 12-41** Problem 30 on the strut by the hinge at *A* for the arrangement in Figure 12-41 if (a) the strut is weightless, and (b) the strut weighs 20 N.

31 •• Julie has been hired to help paint the trim of a building, but she is not convinced of the safety of the apparatus. A 5.0-m plank is suspended horizontally from the top of the building by ropes attached at each end. She knows from previous experience, however, that the ropes being used will break if the tension exceeds 1 kN. Her 80-kg boss dismisses Julie's worries and begins painting while standing 1 m from the end of the plank. If Julie's mass is 60 kg and the plank has a mass of 20 kg, then over what range of positions can Julie stand if a colorful plummet is to be avoided?

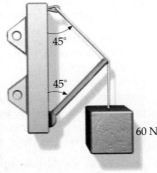

45°

45°

60 N

32 •• The cable in Figure 12-39 must remain attached to the wall 8 m above the hinge, but its length can vary so that it can be connected to the beam at various distances *x* from the wall. How far from the wall should it be attached so that the force on the hinge has no vertical component?

33 •• A cylinder of mass *M* and radius *R* rolls against a step of height *h* as shown in Figure 12-42. When a horizontal force $\vec{F}$ is applied to the top of the cylinder, the cylinder remains at rest. (a) What is the normal force exerted by the floor on the cylinder? (b) What is the horizontal force exerted by the corner of the step on the cylinder? (c) What is the vertical component of the force exerted by the corner on the cylinder?

Figure 12-42
Problems 33 and 34

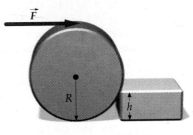

$\vec{F}$

R

h

34 •• For the cylinder in Problem 33, find the minimum horizontal force $\vec{F}$ that will roll the cylinder over the step if the cylinder does not slide on the corner.

35 •• A strong man holds one end of a 3-m rod of mass 5 kg at rest in a horizontal position. (a) What total force does the man exert on the rod? (b) What total torque does the man exert on the rod? (c) If you approximate the effort of the man with two forces that act in opposite directions and are separated by the width of the man's hand, which is taken to be 10 cm, what are the magnitudes and directions of the two forces?

Figure 12-43 Problem 35

36 •• A large gate weighing 200 N is supported by hinges at the top and bottom and is further supported by a wire as shown in Figure 12-44. (a) What must the tension in the wire be for the force on the upper hinge to have no horizontal component? (b) What is the horizontal force on the lower hinge? (c) What are the vertical forces on the hinges?

Figure 12-44
Problem 36

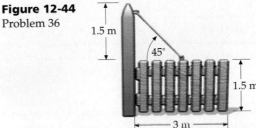

1.5 m

45°

1.5 m

3 m

37 ••• A uniform log with a mass of 100 kg, a length of 4 m, and a radius of 12 cm is held in an inclined position, as shown in Figure 12-45. The coefficient of static friction between the log and the horizontal surface is 0.6. The log is on

Figure 12-45 Problem 37

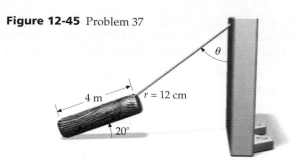

the verge of slipping to the right. Find the tension in the support wire and the angle the wire makes with the vertical wall.

38 ••• A tall, uniform, rectangular block sits on an inclined plane as shown in Figure 12-46. A cord is attached to the top of the block to prevent it from falling down the incline. What is the maximum angle θ for which the block will not slide on the incline? Let b/a be 4 and $\mu_s = 0.8$.

Figure 12-46 Problem 38

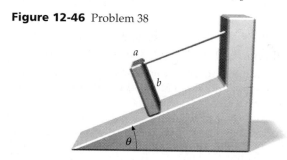

39 ••• A thin rail of length 10 m and mass 20 kg is supported at a 30° incline. One support is 2 m and the other is 6 m from the lower end of the rail. Friction prevents the rail from sliding off the supports. Find the force (magnitude and direction) exerted on the rail by each support.

Couples

40 • Two 80-N forces are applied to opposite corners of a rectangular plate as shown in Figure 12-47. Find the torque produced by this couple.

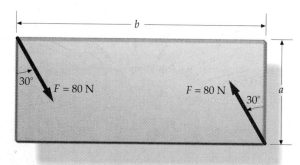

Figure 12-47 Problems 40 and 42

41 • A uniform cube of side a and mass M rests on a horizontal surface. A horizontal force $\vec{F}$ is applied to the top of the cube as in Figure 12-48. This force is not sufficient to move or tip the cube. (a) Show that the force of static friction exerted by the surface and the applied force constitute a couple, and find the torque exerted by the couple. (b) This couple is balanced by the couple consisting of the normal force exerted by the surface and the weight of the cube. Use this fact

Figure 12-48
Problem 41

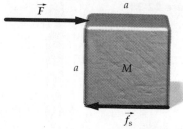

to find the effective point of application of the normal force when $F = Mg/3$. (c) What is the greatest magnitude of $\vec{F}$ for which the cube will not tip?

42 •• Resolve each force in Problem 40 into its horizontal and vertical components, producing two couples. The algebraic sum of the two component couples equals the resultant couple. Use this result to find the perpendicular distance between the lines of action of the two forces.

Ladder Problems

43 • Is it possible to climb a ladder placed against a wall where the ground is frictionless but the wall is not? Explain.

44 •• Romeo takes a uniform 10-m ladder and leans it against the smooth wall of the Capulet residence. The ladder's mass is 22.0 kg, and the bottom rests on the ground 2.8 m from the wall. When Romeo, whose mass is 70 kg, gets 90% of the way to the top, the ladder begins to slip. What is the coefficient of static friction between the ground and the ladder?

45 •• A massless ladder of length L leans against a smooth wall making an angle θ with the horizontal floor. The coefficient of friction between the ladder and the floor is μ_s. A man of mass M climbs the ladder. What height h can he reach before the ladder slips?

46 •• A uniform ladder of length L and mass m leans against a frictionless vertical wall with its lower end on the ground. It makes an angle of 60° with the horizontal ground. The coefficient of static friction between the ladder and ground is 0.45. If your mass is four times that of the ladder, how far up the ladder can you climb before it begins to slip?

47 •• A ladder of mass m and length L leans against a frictionless, vertical wall making an angle θ with the horizontal. The center of mass is at a height h from the floor. A force F pulls horizontally against the ladder at the midpoint. Find the minimum coefficient of static friction μ_s for which the top end of the ladder will separate from the wall while the lower end does not slip.

48 •• A 900-N boy sits on top of a ladder of negligible weight that rests on a frictionless floor as in Figure 12-49. There is a cross brace halfway up the ladder. The angle at the apex is $\theta = 30°$. (a) What is the force exerted by the floor on each leg of the ladder? (b) Find the tension in the cross brace. (c) If the cross brace is moved down toward the bottom of the

Figure 12-49
Problem 48

ladder (maintaining the same angle θ), will its tension be greater or less?

49 •• A ladder rests against a frictionless vertical wall. The coefficient of static friction between the ladder and the floor is 0.3. What is the smallest angle at which the ladder will remain stationary?

50 ••• Having failed in his first attempt, Romeo acquires a new ladder to try once again to get to Juliet's window. This one has a length L and a weight of 200 N. He tries placing it on the other side of the window, where the coefficients of static friction are 0.4 between the ladder and the wall, and 0.7 between the ladder and the ground. Because of bruises suffered in his last fall, Romeo wears heavy padding, which gives him a total mass of 80 kg. Sure enough, when he is $\frac{4}{5}$ of the way up the ladder, it begins to slip. What was the angle between the ladder and the ground when Romeo was making his ascent?

51 ••• A ladder leans against a large smooth sphere of radius R that is fixed in place on a horizontal surface. The ladder makes an angle of 60° with the horizontal surface and has a length $5R/2$. (*a*) What is the force that the sphere exerts on the ladder? (*b*) What is the frictional force that prevents the ladder from slipping? (*c*) What is the normal force that the horizontal surface exerts on the ladder?

Stress and Strain

52 • An aluminum wire and a steel wire of the same length L and diameter d are joined to form a wire of length $2L$. The wire is fastened to the roof and a weight W is attached to the other end. Neglecting the mass of the wires, which of the following statements is true?

(*a*) The aluminum portion will stretch by the same amount as the steel portion.
(*b*) The tensions in the aluminum portion and the steel portion are the same.
(*c*) The tension in the aluminum portion is greater than that in the steel portion.
(*d*) None of the above statements is true.

53 • A 50-kg ball is suspended from a steel wire of length 5 m and radius 2 mm. By how much does the wire stretch?

54 • Copper has a breaking stress of about 3×10^8 N/m². (*a*) What is the maximum load that can be hung from a copper wire of diameter 0.42 mm? (*b*) If half this maximum load is hung from the copper wire, by what percentage of its length will it stretch?

55 • A 4-kg mass is supported by a steel wire of diameter 0.6 mm and length 1.2 m. How much will this wire stretch under this load?

56 • As a runner's foot touches the ground, the shearing force acting on an 8-mm-thick sole is as shown in Figure 12-50. If the force of 25 N is distributed over an area of 15 cm², find the angle of shear θ shown, given that the shear modulus of the sole is 1.9×10^5 N/m².

Figure 12-50
Problem 56

57 •• A steel wire of length 1.5 m and diameter 1 mm is joined to an aluminum wire of identical dimensions to make a composite wire of length 3.0 m. What is the length of the composite wire if it is used to support a mass of 5 kg?

58 •• A force F is applied to a long wire of length L and cross-sectional area A. Show that if the wire is considered to be a spring, the force constant k is given by $k = AY/L$ and the energy stored in the wire is $U = \frac{1}{2}F\,\Delta L$, where Y is Young's modulus and ΔL is the amount the wire has stretched.

59 •• The steel E string of a violin is under a tension of 53 N. The diameter of the string is 0.20 mm, and its length under tension is 35.0 cm. Find (*a*) the unstretched length of this string, and (*b*) the work needed to stretch the string. (See Problem 58.)

60 •• When a rubber strip with a cross section of 3 mm × 1.5 mm is suspended vertically and various masses are attached to it, a student obtains the following data for length versus load:

Load, g	0	100	200	300	300	500
Length, cm	5.0	5.6	6.2	6.9	7.8	10.0

(*a*) Find Young's modulus for the rubber strip for small loads.
(*b*) Find the energy stored in the strip when the load is 150 g. (See Problem 58.)

61 •• A building is to be demolished by a 400-kg steel ball swinging on the end of a 30-m steel wire of diameter 5 cm hanging from a tall crane. As the ball is swung through an arc from side to side, the wire makes an angle of 50° with the vertical at the top of the swing. Find the amount by which the wire is stretched at the bottom of the swing.

62 •• A large mirror is hung from a nail as shown in Figure 12-51. The supporting steel wire has a diameter of 0.2 mm and an unstretched length of 1.7 m. The distance between the points of support at the top of the mirror's frame is 1.5 m. The mass of the mirror is 2.4 kg. What is the distance between the nail and the top of the frame when the mirror is hung?

Figure 12-51
Problems 62 and 91

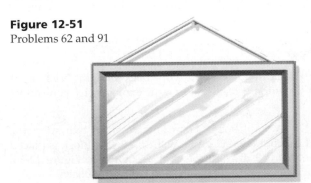

63 •• Two masses, M_1 and M_2, are supported by wires of equal length when unstretched. The wire supporting M_1 is an aluminum wire 0.7 mm in diameter, and the one supporting M_2 is a steel wire 0.5 mm in diameter. What is the ratio M_1/M_2 if the two wires stretch by the same amount?

64 •• A mass of 0.5 kg is attached to an aluminum wire having a diameter of 1.6 mm and an unstretched length of 0.7 m. The other end of the wire is fixed to a post. The mass rotates about the post in a horizontal plane at a rotational speed such that the angle between the wire and the horizontal is 5.0°. Find the tension in the wire and its length.

65 ••• It is apparent from Table 12-2 that the tensile strength of most materials is two to three orders of magnitude smaller than Young's modulus. Consequently, these materials, e.g., aluminum, will break before the strain exceeds even 1%. For nylon, however, the tensile strength and Young's modulus are approximately equal. If a nylon line of unstretched length L_0 and cross section A_0 is subjected to a tension T, the cross section may be substantially less than A_0 before the line breaks. Under these conditions, the tensile stress T/A may be significantly greater than T/A_0. Derive an expression that relates the area A to the tension T, A_0, and Young's modulus Y.

General Problems

66 • If the net torque about some point is zero, must it be zero about any other point? Explain.

67 • The horizontal bar in Figure 12-52 will remain horizontal if

(a) $L_1 = L_2$ and $R_1 = R_2$.   (b) $L_1 = L_2$ and $M_1 = M_2$.
(c) $R_1 = R_2$ and $M_1 = M_2$.   (d) $L_1 M_1 = L_2 M_2$.
(e) $R_1 L_1 = R_2 L_2$.

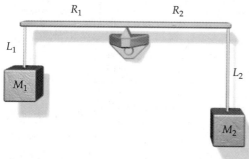

Figure 12-52 Problem 67

68 • Which of the following could not have units of N/m^2?

(a) Young's modulus   (b) Shear modulus
(c) Stress   (d) Strain

69 •• Sit in a chair with your back straight. Now try to stand up without leaning forward. Explain why you cannot do it.

70 • A 90-N board 12 m long rests on two supports, each 1 m from the end of the board. A 360-N block is placed on the board 3 m from one end as shown in Figure 12-53. Find the force exerted by each support on the board.

Figure 12-53
Problem 70

71 • The height of the center of gravity of a man standing erect is determined by weighing the man as he lies on a board of negligible weight supported by two scales as shown in Figure 12-54. If the man's height is 188 cm and the left scale reads 445 N while the right scale reads 400 N, where is his center of gravity relative to his feet?

Figure 12-54 Problem 71

Center of gravity

445 N   400 N

72 • Figure 12-55 shows a mobile consisting of four weights hanging on three rods of negligible mass. Find the value of each of the unknown weights if the mobile is to balance. *Hint:* Find weight w_1 first.

Figure 12-55 Problem 72

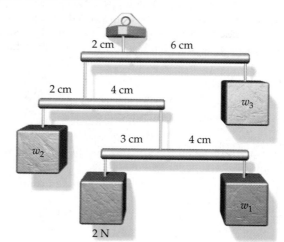

73 • A block and tackle is used to support a mass of 120 kg as shown in Figure 12-56. (a) What is the tension in the rope? (b) What is the mechanical advantage of this device?

Figure 12-56 Problem 73

74 •• A plate in the shape of an equilateral triangle of mass M is suspended from one corner and a mass m is suspended from another corner. What should be the ratio m/M so that the base of the triangle makes an angle of 6.0° with the horizontal?

75 •• A standard six-sided pencil is placed on a paper pad (Figure 12-57). Find the minimum coefficient of static friction μ_s such that it rolls down rather than slides if the pad is inclined.

Figure 12-57
Problem 75

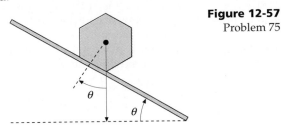

76 •• Having lost his job at the post office, Barry decides to explore the possibility that he might be a brilliant sculptor. Not one to start at the bottom, he borrows the money for a marble slab 3 m × 1 m × 1 m. After loading the marble onto the back of his truck, he drives off with the slab resting on its square end. But on the way home, a confused squirrel runs into his path, and Barry slams on the brakes. What deceleration will cause the uniform slab to tip over?

77 •• A uniform box of mass 8 kg that is twice as tall as it is wide rests on the floor of a truck. What is the maximum coefficient of static friction between the box and floor so that the box will slide toward the rear of the truck rather than tip when the truck accelerates with acceleration $a = 0.6g$ on a level road?

78 •• Barry's art ex- **Figure 12-58**
hibit contains many tiny Problem 78
marble sculptures placed around a central piece called "Politics." The central piece consists of three identical bars, each of length L and mass m, joined as in Figure 12-58. Two bars form a fixed **T**, and the third bar is suspended on a hinge. Asked to explain the name, Barry said, "It swings from left to right with hinges flapping, but no matter where you start it from, you end up in the same place." When the system is in equilibrium, what is the value of θ?

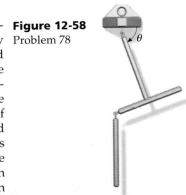

79 •• In the 1996 Olympics, the Russian super-heavyweight weightlifter Andrei Chemerkin broke the world record with a lift of mass 260 kg. Suppose his grip was slightly asymmetrical as shown in Figure 12-59. Find the maximum mass of the barbell Chemerkin could have handled with a symmetrical grip, assuming that his arms are equally strong.

Figure 12-59
Problem 79

|← 0.55 m →|← 0.6 m →|← 0.45 m →|

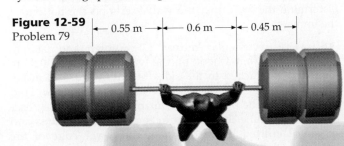

Figure 12-60 Problem 80

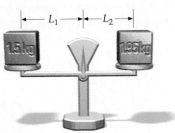

80 •• A balance scale has unequal arms. A 1.5-kg block appears to have a mass of 1.95 kg on the left pan of the scale (Figure 12-60). Find its apparent mass if the block is placed on the right pan.

81 •• A cube of mass M leans against a frictionless wall making an angle θ with the floor as shown in Figure 12-61. Find the minimum coefficient of static friction μ_s between the cube and the floor that allows the cube to stay at rest.

Figure 12-61 Problem 81

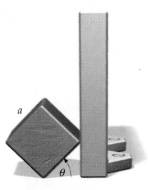

82 •• Figure 12-62 shows a steel meter stick hinged to a vertical wall and supported by a thin wire. The wire and meter stick make angles of 45° with the vertical. The mass of the meter stick is 5.0 kg. When a mass $M = 10.0$ kg is suspended from the midpoint of the meter stick, the tension T in the supporting wire is 52 N. If the wire will break should the tension exceed 75 N, what is the maximum distance along the meter stick at which the 10.0-kg mass can be suspended?

Figure 12-62 Problem 82

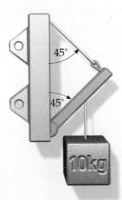

83 •• Figure 12-63 shows a 20-kg ladder leaning against a frictionless wall and resting on a frictionless horizontal surface. To keep the ladder from slipping, the bottom of the ladder is tied to the wall with a thin wire; the tension in the wire is 29.4 N. The wire will break if the tension exceeds 200 N. (*a*) If an 80-kg person climbs halfway up the ladder, what force will be exerted by the ladder against the wall? (*b*) How far up can an 80-kg person climb this ladder?

Figure 12-63 Problem 83

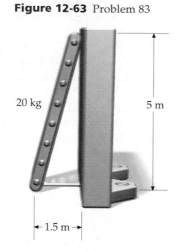

20 kg

5 m

1.5 m

84 •• Suppose that the bar hanging from the end of the T in Problem 78 is of a different length $\ell \neq L$, although its mass per unit length is the same as that of the bars of the T. Find the ratio L/ℓ such that $\theta = 75°$.

85 •• A uniform cube can be moved along a horizontal plane either by pushing the cube so that it slips or by turning it over ("rolling"). What coefficient of static friction μ_k between the cube and the floor makes both ways equal in terms of the work needed?

86 •• A tall, uniform, rectangular block sits on an inclined plane as shown in Figure 12-64. If $\mu_s = 0.4$, does the block slide or fall over as the angle θ is slowly increased?

Figure 12-64 Problem 86

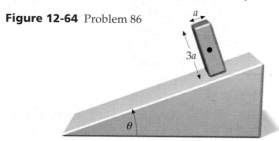

a

$3a$

θ

87 •• A 360-kg mass is supported on a wire attached to a 15-m-long steel bar that is pivoted at a vertical wall and supported by a cable as shown in Figure 12-65. The mass of the bar is 85 kg. (*a*) With the cable attached to the bar 5.0 m from the lower end as shown, find the tension in the cable and the force exerted by the wall on the steel bar. (*b*) Repeat if a somewhat longer cable is attached to the steel bar 5.0 m from its upper end, maintaining the same angle between the bar and the wall.

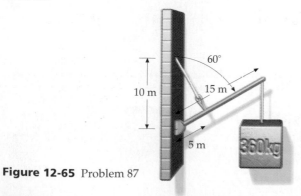

60°

10 m

15 m

5 m

360kg

Figure 12-65 Problem 87

88 •• Repeat Problem 77 if the truck accelerates at $a = 0.6g$ up a hill that makes an angle of 9.0° with the horizontal.

89 •• A thin rod 60 cm long is balanced 20 cm from one end when a mass of $2m + 2$ g is at the end nearest the pivot and a mass of m at the opposite end (Figure 12-66*a*). Balance is again achieved if the mass $2m + 2$ g is replaced by the mass m and no mass is placed at the other end (Figure 12-66*b*). Determine the mass of the rod.

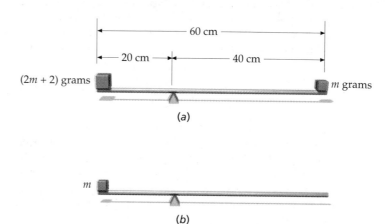

60 cm

20 cm ←→ 40 cm

(2m + 2) grams

m grams

(a)

m

(b)

Figure 12-66 Problem 89

90 •• The planet Mars has two satellites, Phobos and Deimos, in nearly circular orbits. The orbit radii of Phobos and Deimos are 9.38×10^3 km and 23.46×10^3 km, respectively. The mass of Mars is 6.42×10^{23} kg, that of Phobos is 9.63×10^{15} kg, and that of Deimos is 1.93×10^{15} kg. Find the center of gravity and the center of mass of the two-satellite system using the center of Mars as the origin when (*a*) the satellites are in opposition (i.e., on exactly opposite sides of Mars), and (*b*) the satellites are in conjunction (i.e., in line on the same side of Mars).

91 •• When a picture is hung on a smooth vertical wall using a wire and a nail, as in Figure 12-51, the picture almost always tips slightly forward, i.e., the plane of the picture makes a small angle with the vertical. (*a*) Explain why pictures supported in this manner generally do not hang flush against the wall. (*b*) A framed picture 1.5 m wide and 1.2 m high and having a mass of 8.0 kg is hung as in Figure 12-51 using a wire of 1.7 m length. The ends of the wire are fastened to the sides of the frame at the rear and 0.4 m below the top. When the picture is hung, the angle between the plane of the frame and the wall is 5.0°. Determine the force that the wall exerts on the bottom of the frame.

92 •• A rectangular mirror 1.0 m high and 0.60 m wide is hung from a hook on the wall using a wire 0.85 m long that is attached to the sides of the mirror 0.20 m below the top edge. The mass of the mirror is 6.0 kg. (*a*) Find the angle that the plane of the mirror makes with the vertical. (*b*) Determine the tension in the supporting wires and the force exerted by the wall on the lower edge of the mirror.

93 •• If a train travels around a bend in the railbed too fast, the freight cars will tip over. Assume that the cargo portions of the freight cars are regular parallelepipeds of uniform density and 1.5×10^4 kg mass, 10 m long, 3.0 m high, and 2.20 m wide, and that their base is 0.65 m above the rails. The axles are 7.6 m apart, each 1.2 m from the ends of the boxcar. The separation between the rails is 1.55 m. Find the maximum safe speed of the train if the radius of curvature of the bend is (a) 150 m, and (b) 240 m.

94 •• For balance, a tightrope walker uses a thin rod 8 m long and bowed in a circular arc shape. At each end of the rod is a lead mass of 20 kg. The tightrope walker, whose mass is 58 kg and whose center of gravity is 0.90 m above the rope, holds the rod tightly at its center 0.65 m above the rope. What should the radius of curvature of the arc of the rod be so that he will be in stable equilibrium as he slowly makes his way across the rope? Neglect the mass of the rod.

95 •• A large crate weighing 4500 N rests on four 12-cm-high blocks on a horizontal surface (Figure 12-67). The crate is 2 m long, 1.2 m high, and 1.2 m deep. You are asked to lift one end of the crate using a long steel bar. The coefficient of static friction between the blocks and the supporting surface is 0.4. Estimate the length of the steel bar you will need to lift the end of the crate.

Figure 12-67 Problem 95

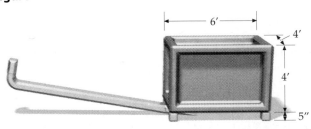

96 ••• Six identical bricks are stacked one on top of the other lengthwise and slightly offset to produce a stepped tower with the maximum offset that will still allow the tower to stand. (a) Starting from the top, give the maximum possible offset for each successive brick. (b) What is the total protrusion or offset of the six bricks?

97 ••• A uniform sphere of radius R and mass M is held at rest on an inclined plane of angle θ by a horizontal string, as shown in Figure 12-68. Let $R = 20$ cm, $M = 3$ kg, and $\theta = 30°$. (a) Find the tension in the string. (b) What is the normal force exerted on the sphere by the inclined plane? (c) What is the frictional force acting on the sphere?

Figure 12-68 Problem 97

98 ••• The legs of a tripod make equal angles of 90° with each other at the apex, where they join together. A 100-kg block hangs from the apex. What are the compressional forces in the three legs?

99 ••• Figure 12-69 shows a 20-cm-long uniform beam resting on a cylinder of 4 cm radius. The mass of the beam is 5.0 kg, and that of the cylinder is 8.0 kg. The coefficient of friction between beam and cylinder is zero. (a) Find the forces that act on the beam and on the cylinder. (b) What must the minimum coefficients of static friction be between beam and floor and between the cylinder and floor to prevent slipping?

Figure 12-69
Problem 99

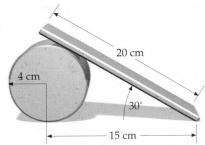

100 ••• Two solid smooth spheres of radius r are placed inside a cylinder of radius R as in Figure 12-70. The mass of each sphere is m. Find the force exerted by the bottom of the cylinder, the force exerted by the wall of the cylinder, and the force exerted by one sphere on the other.

Figure 12-70 Problem 100

101 ••• A solid cube of side length a balanced atop a cylinder of diameter d is in unstable equilibrium if $d \ll a$ and is in stable equilibrium if $d \gg a$ (Figure 12-71). Determine the minimum value of d/a for which the cube is in stable equilibrium.

Figure 12-71
Problem 101

CHAPTER 13

Fluids

Airflow above and below the wing of this Indy race car creates greater pressure above the wing, increasing the effective weight of the car for better control at high speeds. An airplane wing is designed so that the flow creates greater pressure below the wing to lift the plane.

Fluids include both liquids and gases. Liquids flow under gravity until they occupy the lowest possible regions of their containers. Gases expand to fill their containers regardless of the containers' shapes.

In a gas, the average distance between two molecules is large compared with the size of a molecule. The molecules have little influence on one another except during their frequent but brief collisions. In a liquid or solid, the molecules are close together and exert forces on one another that are comparable to the forces that bind atoms into molecules. Molecules in a liquid form temporary short-range bonds that are continually broken due to the thermal kinetic energy of the molecules and then reformed. These bonds hold the liquid together; if the bonds were not present, the liquid would immediately evaporate and the molecules would escape as a vapor. The strength of the bonds in a liquid depends on the type of molecule. For example, the bonds between helium atoms are very weak, and for this reason, helium does not liquefy at atmospheric pressure unless the temperature is 4.2 K or lower.

13-1 Density

An important property of a substance is the ratio of its mass to its volume, which is called its **density**:

$$\text{Density} = \frac{\text{mass}}{\text{volume}}$$

The Greek letter ρ (rho) is usually used to denote density:

$$\rho = \frac{m}{V} \qquad \qquad 13\text{-}1$$

Definition—Density

The gram was originally defined as the mass of one cubic centimeter of water, so the density of water has served as a benchmark. In cgs units, the density of water is 1 g/cm^3. Converting to SI units, we obtain for the density of water

$$\rho_w = \frac{1 \text{ g}}{\text{cm}^3} \times \frac{\text{kg}}{10^3 \text{ g}} \times \left(\frac{100 \text{ cm}}{\text{m}}\right)^3 = 10^3 \text{ kg/m}^3 \qquad 13\text{-}2$$

The densities of most materials, including water, vary with temperature. Equation 13-2 gives the maximum value for the density of water, which occurs at 4°C.

A convenient unit of volume for fluids is the **liter** (L):

$$1 \text{ L} = 10^3 \text{ cm}^3 = 10^{-3} \text{ m}^3$$

In terms of this unit, the density of water at 4°C is 1.00 kg/L. When an object's density is greater than that of water, it sinks in water. When its density is less, it floats. The ratio of the density of a substance to that of water is called the **specific gravity** of the substance. For example, the specific gravity of aluminum is 2.7, meaning that a volume of aluminum has 2.7 times the mass of an equal volume of water. The specific gravities of objects that sink in water range from 1 to about 22.5 (for the densest element, osmium).

Most solids and liquids expand only slightly when heated, and contract slightly when subjected to an increase in external pressure. Since these changes in volume are relatively small, we often treat the densities of solids and liquids as approximately independent of temperature and pressure. The density of a gas, on the other hand, depends strongly on the pressure and temperature, so these variables must be specified when reporting the densities of gases. By convention, **standard conditions** are atmospheric pressure at sea level and a temperature of 0°C. Figure 13-1 gives the densities for a variety of substances under these conditions. Note that the densities of liquids and solids are considerably greater than those of gases. For example, the density of water is about 800 times that of air under standard conditions.

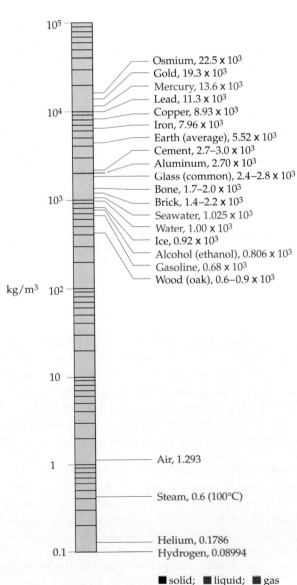

Figure 13-1 Densities of selected substances. Unless otherwise indicated, $t = 0°C$ and $P = 1$ atm.

Example 13-1

A 200-mL flask is filled with water at 4°C. When the flask is heated to 80°C, 6 g of water spill out. What is the density of water at 80°C? (Assume that the expansion of the flask is negligible.)

Picture the Problem The density of water at 80°C is $\rho' = m'/V$, where $V = 200$ mL $= 200$ cm^3 is the volume of the flask, and m' is the mass remaining in the flask after 6 g spills out. We find m' by first finding the mass of water originally in the flask.

1. Calculate the original mass of water in the flask at 4°C using $\rho = 1$ g/cm^3:

$$m = \rho V = (1 \text{ g/cm}^3)(200 \text{ cm}^3) = 200 \text{ g}$$

2. Calculate the mass of water remaining, m', after 6 g spill out:

$$m' = m - 6 \text{ g} = 200 \text{ g} - 6 \text{ g} = 194 \text{ g}$$

3. Use this value of m' to find the density of water at 80°:

$$\rho' = \frac{m'}{V} = \frac{194 \text{ g}}{200 \text{ cm}^3} = 0.97 \text{ g/cm}^3$$

Exercise A solid metal cube 8 cm on a side has a mass of 4.08 kg. (a) What is the density of the cube? (b) If the cube is made from a single element listed in Figure 13-1, what is the element? (*Answers*   (a) 7.97 kg/L, (b) iron. Note that the slight difference between the answer in (a) and Figure 13-1 is due to a difference in the number of significant digits used to derive the two values.)

Exercise A gold brick is 5 cm $\times$ 10 cm $\times$ 20 cm. How much does it weigh? (*Answer*   189 N $\approx$ 42.6 lb)

13-2   Pressure in a Fluid

When a body is submerged in a fluid such as water, the fluid exerts a force perpendicular to the surface of the body at each point on the surface. This force per unit area is called the **pressure** P of the fluid:

$$P = \frac{F}{A} \qquad\qquad\qquad 13\text{-}3$$

Definition—Pressure

The SI unit of pressure is the newton per square meter (N/m^2), which is called the **pascal** (Pa):

$$1 \text{ Pa} = 1 \text{ N/m}^2 \qquad\qquad\qquad 13\text{-}4$$

In the U.S. customary system, pressure is usually given in pounds per square inch (lb/in^2). Another common unit of pressure is the atmosphere (atm), which is approximately the air pressure at sea level. One atmosphere is now defined to be 101.325 kilopascals, which is approximately 14.70 lb/in^2:

$$1 \text{ atm} = 101.325 \text{ kPa} = 14.70 \text{ lb/in}^2 \qquad\qquad 13\text{-}5$$

Other units of pressure in common use are discussed later.

The pressure due to a fluid pressing in on an object tends to compress the object. The ratio of the change in pressure (ΔP) to the fractional decrease in volume ($-\Delta V/V$) is called the **bulk modulus***:

---

* The minus sign in Equation 13-6 is introduced to make B positive since all materials decrease in volume when subjected to external pressure.

$$B = -\frac{\Delta P}{\Delta V / V} \qquad\qquad 13\text{-}6$$

Definition—Bulk modulus

The more difficult a material is to compress, the smaller is $\Delta V/V$ for a given pressure, and hence the greater the bulk modulus. Liquids, gases, and solids all have a bulk modulus. Since liquids and solids are relatively incompressible, they have large values of B, and these values are relatively independent of temperature and pressure. Gases, on the other hand, are easily compressed, and their values for B depend strongly on the pressure and temperature. Figure 13-2 charts values for the bulk modulus of various materials.

Exercise Water is contained in a cylindrical iron container sealed with an iron piston. The pressure on the piston is increased to 100 atm. (*a*) What is the percentage change in the volume of the water? (*b*) What is the percentage change in the volume of the iron? (*Answer* $-\Delta V/V \approx 0.5\%$ for water and 0.01% for iron.)

As any scuba diver knows, the pressure in a lake or ocean increases with depth. Similarly, the pressure of the atmosphere decreases with altitude. For a liquid such as water, whose density is approximately constant throughout, the pressure increases linearly with depth. We can see this by considering a column of liquid of height h and cross-sectional area A, as shown in Figure 13-3. To support the weight of the column, the pressure at the bottom must be greater than the pressure at the top. The weight of this liquid column is

$$w = mg = \rho V g = \rho A h g$$

If P_0 is the pressure at the top and P is the pressure at the bottom, the net upward force exerted by this pressure difference is $PA - P_0 A$. Setting this net upward force equal to the weight of the column, we obtain

$$PA - P_0 A = \rho A h g$$

or

$$P = P_0 + \rho g h \qquad (\rho \text{ constant}) \qquad\qquad 13\text{-}7$$

Exercise Find the pressure at a depth of 10 m below the surface of a lake if the pressure at the surface is 1 atm. (*Answer* With $P_0 = 1$ atm $= 101$ kPa, $\rho = 10^3$ kg/m^3, and $g = 9.81$ N/kg, we have $P = P_0 + \rho g h = 1.97$ atm. The pressure at a depth of 10 m is nearly twice that at the surface.)

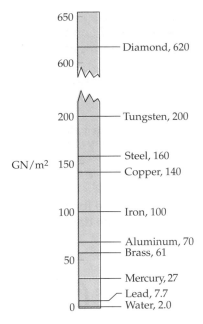

Figure 13-2 Approximate values for the bulk modulus B of various materials.

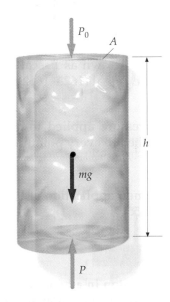

Figure 13-3 Column of water of height h and cross-sectional area A. The pressure P at the bottom must be greater than the pressure P_0 at the top to balance the weight of the water.

Example 13-2

A rectangular dam 30 m wide supports a body of water to a height of 25 m (Figure 13-4). Find the total horizontal force on the dam.

Picture the Problem Because the pressure varies with depth, we cannot merely multiply the pressure times the area of the dam to find the force exerted by the water. We therefore consider the force exerted on a strip of width $L = 30$ m, height dh, and area $dA = L\,dh$ at a depth h, and integrate from $h = 0$ to $h = H = 25$ m. The water pressure at depth h is $P_{\text{atm}} + \rho g h$. We can omit the atmospheric pressure because it is exerted on each side of the wall.

Figure 13-4

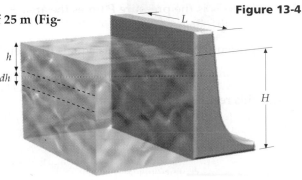

Example 13-4

The average (gauge) pressure in the aorta is about 100 mmHg. Convert this average blood pressure to pascals.

We use the conversion factors implied in Equation 13-9:
$$P = 100 \text{ mmHg}\left(\frac{101.325 \text{ kPa}}{760 \text{ mmHg}}\right) = 13.3 \text{ kPa}$$

Exercise Convert a pressure of 45 kPa to (a) mmHg, and (b) atmospheres. (*Answers* (a) 338 mmHg, (b) 0.444 atm)

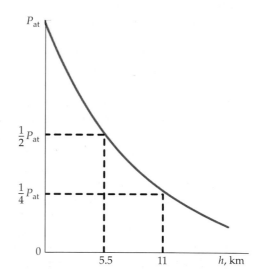

Figure 13-9 Variation in pressure with height above the earth's surface. For each 5.5-km increase in height, the pressure decreases by half.

The relation between pressure and altitude (or depth) is more complicated for a gas than for a liquid because the density of a gas is not constant like that of a liquid, but is approximately proportional to the pressure. As you go up from the surface of the earth, pressure in a column of air decreases, just as the pressure would decrease as you go up from the bottom in a water column. But the decrease in air pressure is not linear with distance. Instead, the air pressure decreases by a constant fraction for a given increase in height, as shown in Figure 13-9. At a height of about 5.5 km (18,000 ft), the air pressure is half its value at sea level. If we go up another 5.5 km to an altitude of 11 km (a typical altitude for airliners), the pressure is again halved so that it is one-fourth its value at sea level, and so on. This example of an *exponential decrease* is called the law of atmospheres. At the high altitudes at which commercial jets fly, the cabins must be pressurized. The density of air is proportional to the pressure, so the density of air decreases with altitude. There is less oxygen available on a mountain than at normal elevations, which makes exercising in the Rockies difficult and climbing in the Himalayas dangerous.

13-3 Buoyancy and Archimedes' Principle

If a heavy object submerged in water is "weighed" by suspending it from a spring scale, the reading on the scale is less than when the object is weighed in air (Figure 13-10a). This is because the water exerts an upward force that partially balances the force of gravity. The force is even more evident when we submerge a piece of cork. When completely submerged, the cork experiences an upward force from the water pressure that is greater than the force of gravity, so it accelerates up toward the surface. The force exerted by a fluid on a body submerged in it is called the **buoyant force.** It is equal to the weight of the fluid displaced by the body.

> A body wholly or partially submerged in a fluid is buoyed up by a force equal to the weight of the displaced fluid.

Archimedes' principle

This result is known as **Archimedes' principle.**

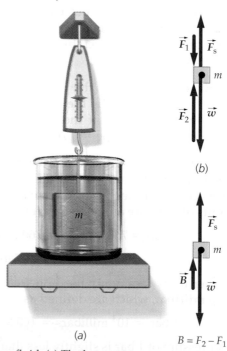

(a)

(b)

(c)

$B = F_2 - F_1$

Figure 13-10 (a) Weighing an object submerged in a fluid. (b) Free-body diagram showing the weight $\vec{w}$, the force $\vec{F}_s$ of the spring, and the forces $\vec{F}_1$ and $\vec{F}_2$ exerted by the surrounding fluid. (c) The buoyant force $B = F_2 - F_1$ is the net force exerted on the object by the fluid.

We can derive Archimedes' principle from Newton's laws by considering the forces acting on a portion of a fluid and noting that in static equilibrium the net force must be zero. Figure 13-10*b* shows the vertical forces acting on an object being weighed while submerged. These are the force of gravity $\vec{w}$ down, the force $\vec{F}_s$ of the spring scale acting up, a force $\vec{F}_1$ acting down because of the fluid pressing on the top surface of the object, and a force $\vec{F}_2$ acting up because of the fluid pressing on the bottom surface of the object. Since the spring scale reads a force less than the weight,

Figure 13-11 Figure 13-10 with the submerged body replaced by an equal volume of fluid. The forces $\vec{F}_1$ and $\vec{F}_2$ due to the pressure of the fluid are the same as in Figure 13-10. The buoyant force is thus equal to the weight w_f of the displaced fluid.

the force $\vec{F}_2$ must be greater in magnitude than the force $\vec{F}_1$. The difference in magnitude of these two forces is the buoyant force $B = F_2 - F_1$. The buoyant force occurs because the pressure of the fluid at the bottom of the object is greater than that at the top.

In Figure 13-11, the spring scale has been eliminated and the submerged object has been replaced by an equal volume of fluid (indicated by the dotted lines). The buoyant force $B = F_2 - F_1$ acting on this volume of fluid is the same as the buoyant force that acted on our original object since the fluid surrounding the space is the same. Because this volume of fluid is in equilibrium, the net force acting on it must be zero. The upward buoyant force thus equals the downward weight of this volume of fluid:

$$B = w_f \qquad\qquad\qquad 13\text{-}11$$

Note that this result does not depend on the shape of the submerged object. If we consider any irregularly shaped portion of fluid, there must be a buoyant force acting on it due to the surrounding fluid that is equal to the weight of that portion. Thus, we have derived Archimedes' principle.

Archimedes (287–212 B.C.) had been given the task of determining whether a crown made for King Hieron II was of pure gold or had been adulterated with some cheaper metal, such as silver. The problem was to determine the density of the irregularly shaped crown without destroying it. As the story goes, Archimedes came upon the solution while sinking himself into a bath and immediately rushed naked through the streets of Syracuse shouting "Eureka!" ("I have found it!"). This flash of insight preceded Newton's laws, which we used to derive Archimedes' principle, by some 1900 years. What Archimedes found was a simple and accurate way to determine the specific gravity of the crown, which he could then compare with the specific gravity of gold.

The specific gravity of an object is the weight of the object in air divided by the weight of an equal volume of water:

$$\text{Specific gravity} = \frac{\text{weight of object in air}}{\text{weight of equal volume of water}} = \frac{w_o}{w_w}$$

But according to Archimedes' principle, the weight of an equal volume of water equals the buoyant force on the object when it is submerged. It is therefore equal to the loss in weight of the object when it is weighed while submerged in water. Thus,

$$\text{Specific gravity} = \frac{\text{weight of object in air}}{\text{weight loss when submerged in water}} = \frac{w_o}{w_{loss}} \qquad 13\text{-}12$$

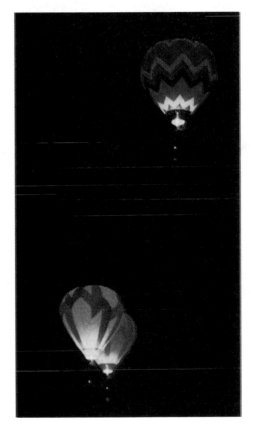

Hot-air balloons rising in the night sky over Albuquerque during a balloon festival.

77 • Suppose that when you are floating in fresh water, 96% of your body is submerged. What is the volume of water your body displaces when it is fully submerged?

78 •• A block of wood of 1.5-kg mass floats on water with 68% of its volume submerged. A lead block is placed on the wood and the wood is then fully submerged. Find the mass of the lead block.

79 •• A Styrofoam cube, 25 cm on a side, is weighed on a simple beam balance. The balance is in equilibrium when a 20-g mass of brass is placed on the opposite pan of the balance. Find the true mass of the Styrofoam cube.

80 •• A spherical shell of copper with an outer diameter of 12 cm floats on water with half its volume above the water surface. Determine the inner diameter of the shell.

81 •• A beaker filled with water is balanced on the left cup of a scale. A cube 4 cm on a side is attached to a string and lowered into the water so that it is completely submerged. The cube is not touching the bottom of the beaker. A weight m is added to the system to retain equilibrium. What is the weight m and on which cup of the balance is it added?

82 •• Crude oil has a viscosity of about 0.8 Pa·s at normal temperature. A 50-km pipeline is to be constructed from an oil field to a tanker terminal. The pipeline is to deliver oil at the terminal at a rate of 500 L/s and the flow through the pipeline is to be laminar to minimize the pressure needed to push the fluid through the pipeline. Estimate the diameter of the pipeline that should be used.

83 •• Water flows through the pipe in Figure 13-32 and exits to the atmosphere at C. The diameter of the pipe is 2.0 cm at A, 1.0 cm at B, and 0.8 cm at C. The gauge pressure in the pipe at A is 1.22 atm and the flow rate is 0.8 L/s. The vertical pipes are open to the air. Find the level of the liquid–air interfaces in the two vertical pipes.

Figure 13-32
Problems 83 and 84

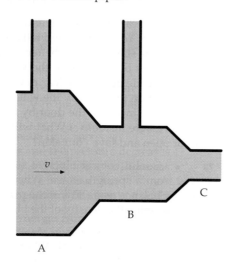

84 •• Repeat Problem 83 with the flow rate reduced to 0.6 L/s and the size of the opening at C reduced so that the pressure in the pipe at A remains unchanged.

85 •• Figure 13-33 is a sketch of an *aspirator*, a simple device that can be used to achieve a partial vacuum in a reservoir connected to the vertical tube at B. An aspirator attached to the end of a garden hose may be used to deliver soap or fertilizer from the reservoir. Suppose that the diameter at A is 2.0 cm and at C, where the water exits to the atmosphere, it is 1.0 cm. If the flow rate is 0.5 L/s and the gauge pressure at A is 0.187 atm, what diameter of the constriction at B will achieve a pressure of 0.1 atm in the container?

Figure 13-33
Problem 85

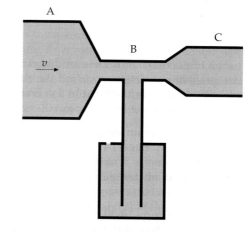

86 •• A cylindrical buoy at the entrance of a harbor has a diameter of 0.9 m and a height of 2.6 m. The mass of the buoy is 600 kg. It is attached to the bottom of the sea with a nylon cable of negligible mass. The specific gravity of the seawater is 1.025. (*a*) How much of the buoy is visible when the cable is slack? (*b*) If a tidal wave completely submerges the buoy, what is the tension in the taut cable? (*c*) If the cable breaks, what is the initial upward acceleration of the buoy?

87 •• Two communicating vessels contain a liquid of density ρ_0 (Figure 13-34). The cross-sectional areas of the vessels are A and $3A$. Find the change in elevation of the liquid level if an object of mass m and density $\rho' = 0.8\rho_0$ is put into one of the vessels.

Figure 13-34 Problem 87

88 •• If an oil-filled manometer ($\rho = 900 \text{ kg/m}^3$) can be read to ±0.05 mm, what is the smallest pressure change that can be detected?

89 •• A rectangular dam 30 m wide supports a body of water to a height of 25 m. (*a*) Neglecting atmospheric pressure, find the total force due to water pressure acting on a thin strip of the dam of height dy located at a depth y. (*b*) Integrate your result in part (*a*) to find the total horizontal force exerted by the water on the dam. (*c*) Why is it reasonable to neglect atmospheric pressure?

90 •• A U-tube is filled with water until the liquid level is 28 cm above the bottom of the tube. An oil of specific gravity 0.78 is now poured into one arm of the U-tube until the level of the water in the other arm of the tube is 34 cm above the bottom of the tube. Find the level of the oil–water and oil–air interfaces in the other arm of the tube.

91 •• A U-tube contains liquid of unknown specific gravity. An oil of density 800 kg/m^3 is poured into one arm of the tube until the oil column is 12 cm high. The oil–air interface is then 5.0 cm above the liquid level in the other arm of the U-tube. Find the specific gravity of the liquid.

92 •• A lead block is suspended from the underside of a 0.5-kg block of wood of specific gravity of 0.7. If the upper surface of the wood is just level with the water, what is the mass of the lead block?

93 •• A helium balloon can just lift a load of 750 N. The skin of the balloon has a mass of 1.5 kg. (a) What is the volume of the balloon? (b) If the volume of the balloon is twice that found in part (a), what is the initial acceleration of the balloon when it carries a load of 900 N?

94 •• A hollow sphere with an inner radius R and an outer radius $2R$ is made of material of density ρ_0 and is floating in a liquid of density $2\rho_0$. The interior is now filled with material of density ρ' so that the sphere just floats completely submerged. Find ρ'.

95 •• A balloon is filled with helium at atmospheric pressure. The skin of the balloon has a mass of 2.8 kg and the volume of the balloon is 16 m^3. What is the greatest weight that this balloon can lift?

96 •• As mentioned in the discussion of *the law of atmospheres*, the fractional decrease in atmospheric pressure is proportional to the change in altitude. Expressed in mathematical terms we have

$$\frac{dP}{P} = -C\, dh$$

where C is a constant. (a) Show that $P(h) = P_0 e^{-Ch}$ is a solution of the differential equation. (b) Show that if $\Delta h \ll h_0$, then $P(h + \Delta h) \approx P(h)(1 - \Delta h/h_0)$, where $h_0 = 1/C$. (c) Given that the pressure at $h = 5.5$ km is half that at sea level, find the constant C.

97 •• A submarine has a total mass of 2.4×10^6 kg, including crew and equipment. The vessel consists of two parts, the pressure hull, which has a volume of 2×10^3 m^3, and the diving tanks, which have a volume of 4×10^2 m^3. When the sub cruises on the surface, the diving tanks are filled with air; when cruising below the surface, seawater is admitted into the tanks. (a) What fraction of the submarine's volume is above the water surface when the tanks are filled with air? (b) How much water must be admitted into the tanks to give the submarine neutral buoyancy? Neglect the mass of air in the tanks and use 1.025 as the specific gravity of seawater.

98 •• A marine salvage crew raises a crate that measures 1.4 m × 0.75 m × 0.5 m. The average density of the empty crate is the same as seawater, 1.025×10^3 kg/m^3, and its mass when empty is 32 kg. The crate contains gold bullion that fills 36% of its volume; the remaining volume is filled with seawater. (a) What is the tension in the cable that raises the crate and bullion while the crate is below the surface of the sea? (b) What is the tension in the cable while the crate is lifted to the deck of the ship if (1) none of the seawater leaks out of the crate, and (2) the crate is lifted so slowly that all of the seawater leaks out of the crate?

99 ••• When the hydrometer in Problem 42 is placed in a liquid whose specific gravity is greater than some minimum value, the device floats with part of the glass tube above the liquid level. Consider a hydrometer that has a spherical bulb 2.4 cm in diameter. The glass tube attached to the bulb is 20 cm long and has a diameter of 7.5 mm. The mass of the hydrometer before lead pellets are dropped into the bulb and the tube is sealed is 7.28 g. (a) What mass of lead should be placed in the bulb so that the hydrometer just floats in a liquid of specific gravity 0.78? (b) If the hydrometer is now placed in water, what is the length of the tube that shows above the surface of the water? (c) The hydrometer is placed in a liquid of unknown specific gravity; the length of the tube above the surface of the liquid is 12.2 cm. Determine the specific gravity of the liquid.

100 ••• A large beer keg of height H and cross-sectional area A_1 is filled with beer. The top is open to atmospheric pressure. At the bottom is a spigot opening of area A_2, which is much smaller than A_1. (a) Show that when the height of the beer is h, the speed of the beer leaving the spigot is approximately $\sqrt{2gh}$. (b) Show that for the approximation $A_2 \ll A_1$, the rate of change of the height h of the beer is given by

$$\frac{dh}{dt} = -\frac{A_2}{A_1}(2gh)^{1/2}$$

(c) Find h as a function of time if $h = H$ at $t = 0$. (d) Find the total time needed to drain the keg if $H = 2$ m, $A_1 = 0.8$ m^2, and $A_2 = (10^{-4})A_1$.

PART II

oscillations and waves

CHAPTER 14

Oscillations

The swaying of the Citicorp Building in New York during high winds is reduced by this tuned-mass damper mounted on an upper floor. It consists of a 400-ton sliding block connected to the building by a spring. The spring constant is chosen so that the natural frequency of the spring–block system is the same as the natural sway frequency of the building. Set into motion by winds, the building and damper oscillate 180° out of phase with each other, thereby significantly reducing the swaying.

Oscillation occurs when a system is disturbed from a position of stable equilibrium. There are many familiar examples: boats bob up and down, clock pendulums swing back and forth, and the strings and reeds of musical instruments vibrate. Other, less familiar examples are the oscillations of air molecules in a sound wave and the oscillations of electric currents in radios and television sets.

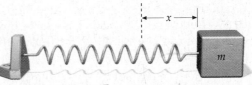

Equilibrium

14-1 Simple Harmonic Motion

A common and very important kind of oscillatory motion is **simple harmonic motion**, such as the motion of an object attached to a spring (Figure 14-1). In equilibrium, the spring exerts no force on the object. When the object is displaced an amount x from its equilibrium position, the spring exerts a force $-kx$, as given by Hooke's law:

$$F_x = -kx \qquad \text{14-1}$$

where k is the force constant of the spring, a measure of the spring's stiffness.

Figure 14-1 An object and spring on a frictionless surface. The displacement x, measured from the equilibrium position, is positive if the spring is stretched and negative if the spring is compressed.

The minus sign indicates that the force is a restoring force; that is, it is oppo-site to the direction of the displacement. Combining Equation 14-1 with Newton's second law, we have

$$F_x = ma_x$$

$$-kx = m\frac{d^2x}{dt^2}$$

or

$$a = \frac{d^2x}{dt^2} = -\frac{k}{m}x \qquad\qquad 14\text{-}2$$

The acceleration is proportional to the displacement and is oppositely di-rected. This is a general characteristic of simple harmonic motion and can be used to identify systems that will exhibit it:

> Whenever the acceleration of an object is proportional to its dis-placement and is oppositely directed, the object will move with simple harmonic motion.

Conditions for simple harmonic motion in terms of acceleration

Since the acceleration is proportional to the net force, whenever the net force on an object is proportional to its displacement and is oppositely directed, the object will move with simple harmonic motion.

The time it takes for a displaced object to make a complete oscillation back and forth about its equilibrium position is called the **period** T. The reciprocal of the period is the **frequency** f, which is the number of oscillations per sec-ond:

$$f = \frac{1}{T} \qquad\qquad 14\text{-}3$$

The unit of frequency is the reciprocal second (s^{-1}), which is called a **hertz** (Hz). For example, if the time for one complete oscillation is 0.25 s, the fre-quency is 4 Hz.

Figure 14-2 shows how we can experimentally obtain x versus t for a mass on a spring. The general equation for such a curve is

$$x = A\cos(\omega t + \delta) \qquad\qquad 14\text{-}4$$

Position in simple harmonic motion

where A, ω, and δ are constants.* The maximum displacement from equilib-rium is called the **amplitude** A. The argument of the cosine function, $\omega t + \delta$, is called the **phase** of the motion, and the constant δ is called the **phase con-stant**. The phase constant depends on the choice of $t = 0$. If we have just one oscillating system, we can always choose $t = 0$ such that $\delta = 0$. If we have two systems oscillating with the same amplitude and frequency but different phase, we can choose $\delta = 0$ for one of them. The equations for the two sys-tems are then

$$x_1 = A\cos(\omega t)$$

and

$$x_2 = A\cos(\omega t + \delta)$$

If the phase difference δ is 0 or an integer times 2π, then $x_2 = x_1$ and the sys-tems are said to be in phase. If the phase difference δ is π or an odd integer times π, then $x_2 = -x_1$ and the systems are said to be out of phase.

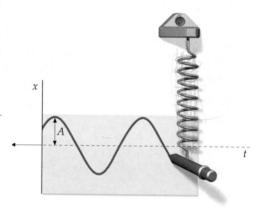

Figure 14-2 A marking pen is attached to a mass on a spring, and the paper is pulled to the left. As the paper moves with con-stant speed, the pen traces out the dis-placement x as a function of time t. (Here we have chosen x to be positive when the spring is compressed.)

---

* Note that $\cos(\omega t + \delta) = \sin(\omega t + \delta + \pi/2)$; thus, whether the equation is expressed as a cosine function or a sine function simply depends on the phase of the oscillation at the moment we designate to be $t = 0$.

We can show that Equation 14-4 is a solution of Equation 14-2 by differentiating x twice with respect to time. The first derivative of x gives the velocity v:

$$v = \frac{dx}{dt} = -A\omega \sin(\omega t + \delta) \qquad \text{14-5}$$

Velocity in simple harmonic motion

Differentiating the velocity with respect to time gives the acceleration:

$$a = \frac{dv}{dt} = \frac{d^2x}{dt^2} = -\omega^2 A \cos(\omega t + \delta) \qquad \text{14-6}$$

or

$$a = -\omega^2 x \qquad \text{14-7}$$

Acceleration in simple harmonic motion

Comparing $a = -\omega^2 x$ with $a = -(k/m)x$ (Equation 14-2), we see that $x = A \cos(\omega t + \delta)$ is a solution of $d^2x/dt^2 = a = -(k/m)x$ if

$$\omega = \sqrt{\frac{k}{m}} \qquad \text{14-8}$$

The amplitude A and the phase constant δ can be determined from the initial position x_0 and the initial velocity v_0 of the system. Setting $t = 0$ in $x = A \cos(\omega t + \delta)$ gives

$$x_0 = A \cos \delta \qquad \text{14-9}$$

Similarly, setting $t = 0$ in $v = dx/dt = -A\omega \sin(\omega t + \delta)$ gives

$$v_0 = -A\omega \sin \delta \qquad \text{14-10}$$

These equations can be solved for A and δ in terms of x_0 and v_0.

The period T is the time after which x repeats. Then

$$x(t) = x(t + T)$$

$$A \cos(\omega t + \delta) = A \cos[\omega(t + T) + \delta] = A \cos(\omega t + \delta + \omega T)$$

The cosine (and sine) function repeats in value when the phase increases by 2π, so

$$\omega T = 2\pi$$

or

$$\omega = \frac{2\pi}{T}$$

The constant ω is called the **angular frequency.** It has units of radians per second and dimensions of inverse time, the same as angular velocity, which is also designated by ω.

The frequency is the reciprocal of the period:

$$f = \frac{1}{T} = \frac{\omega}{2\pi} \qquad \text{14-11}$$

Definition—Frequency, period, and angular frequency

Since $\omega = \sqrt{k/m}$, the frequency and period of an object on a spring are related to the force constant k and the mass m by

$$f = \frac{1}{T} = \frac{1}{2\pi}\sqrt{\frac{k}{m}} \qquad \text{14-12}$$

Frequency and period for an object on a spring

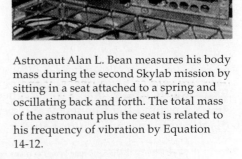

Astronaut Alan L. Bean measures his body mass during the second Skylab mission by sitting in a seat attached to a spring and oscillating back and forth. The total mass of the astronaut plus the seat is related to his frequency of vibration by Equation 14-12.

The frequency increases with increasing k (spring stiffness) and decreases with increasing mass.

Example 14-1

You are on a boat that is bobbing up and down. The boat's vertical displacement y is given by

$$y = (1.2 \text{ m}) \cos\left(\frac{t}{2 \text{ s}} + \frac{\pi}{6}\right)$$

(a) Find the amplitude, angular frequency, phase constant, frequency, and period of the motion. (b) Where is the boat at $t = 1$ s? (c) Find the velocity and acceleration at any time t. (d) Find the initial position, velocity, and acceleration of the boat.

Picture the Problem We find the quantities asked for in (a) by comparing the equation of motion with the standard equation for simple harmonic motion, Equation 14-4. The velocity and acceleration are found by differentiating $y(t)$.

(a) 1. Compare the equation for the boat's vertical displacement with Equation 14-4, $y = A\cos(\omega t + \delta)$, to get A, ω, and δ:

$$y = (1.2 \text{ m}) \cos\left(\frac{t}{2 \text{ s}} + \frac{\pi}{6}\right)$$

$$A = 1.2 \text{ m}, \qquad \omega = 1/2 \text{ rad/s}, \qquad \delta = \pi/6 \text{ rad}$$

2. The frequency and period are found from ω:

$$f = \frac{\omega}{2\pi} = 0.0796 \text{ Hz}, \qquad T = \frac{1}{f} = 12.6 \text{ s}$$

(b) Set $t = 1$ s to find the boat's position:

$$y = (1.2 \text{ m}) \cos\left[\frac{1}{2 \text{ s}}(1 \text{ s}) + \frac{\pi}{6}\right] = 0.624 \text{ m}$$

(c) The velocity and acceleration are obtained from the position by differentiation with respect to time:

$$v_y = \frac{dy}{dt} = -(1.2 \text{ m}) \sin\left(\frac{1}{2 \text{ s}}t + \frac{\pi}{6}\right)\frac{d(t/2 \text{ s})}{dt}$$

$$= -(0.6 \text{ m/s}) \sin\left(\frac{1}{2 \text{ s}}t + \frac{\pi}{6}\right)$$

(d) Set $t = 0$ to find y_0, v_{y0}, and a_{y0}:

$$a_y = \frac{dv_y}{dt} = -(0.6 \text{ m/s}) \cos\left(\frac{1}{2 \text{ s}}t + \frac{\pi}{6}\right)\frac{d(t/2 \text{ s})}{dt}$$

$$= -(0.3 \text{ m/s}^2) \cos\left(\frac{1}{2 \text{ s}}t + \frac{\pi}{6}\right)$$

$$y_0 = (1.2 \text{ m}) \cos\frac{\pi}{6} = 1.04 \text{ m}$$

$$v_{y0} = -(0.6 \text{ m/s}) \sin\frac{\pi}{6} = -0.300 \text{ m/s}$$

$$a_{y0} = -(0.3 \text{ m/s}^2) \cos\frac{\pi}{6} = -0.260 \text{ m/s}^2$$

Exercise A 0.8-kg object is attached to a spring of force constant $k = 400$ N/m. Find the frequency and period of motion of the object when it is displaced from equilibrium. (*Answer* $f = 3.56$ Hz, $T = 0.281$ s)

Figure 14-3 shows two identical masses attached to identical springs and resting on a frictionless surface. One spring is stretched 10 cm and the other 5 cm. If they are released at the same time, which object reaches the equilibrium position first? According to Equation 14-12, the period depends only on k and m and not on the amplitude. Since k and m are the same for both systems, the periods are the same. Thus, the objects reach the equilibrium position at the same time. The second object has twice as far to go to reach equilibrium, but it will also have twice the average speed. Figure 14-4 shows a sketch of the position functions for the two objects. This illustrates an important general property of simple harmonic motion:

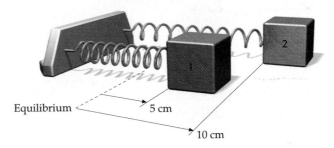

Figure 14-3 Two identical mass–spring systems.

> The frequency and period of simple harmonic motion are independent of the amplitude.

The fact that the frequency in simple harmonic motion is independent of the amplitude has important consequences in many fields. In music, for example, it means that when a note is struck on the piano, the pitch (which corresponds to the frequency) does not depend on how loudly the note is played (which corresponds to the amplitude).* If changes in amplitude had a large effect on the frequency, musical instruments would be unplayable.

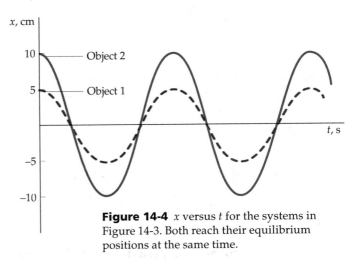

Figure 14-4 x versus t for the systems in Figure 14-3. Both reach their equilibrium positions at the same time.

Example 14-2

An object oscillates with angular frequency $\omega = 8.0$ rad/s. At $t = 0$, the object is at $x_0 = 4$ cm with an initial velocity $v_0 = -25$ cm/s. (a) Find the amplitude and phase constant for the motion. (b) Write x as a function of time.

Picture the Problem The initial position and velocity give us two equations from which to determine the amplitude A and the phase constant δ.

(a)1. The initial position and velocity are related to the amplitude and phase constant:

$$x_0 = A \cos \delta \quad \text{and} \quad v_0 = -\omega A \sin \delta$$

2. Divide these equations to eliminate A:

$$\frac{v_0}{x_0} = \frac{-\omega A \sin \delta}{A \cos \delta} = -\omega \tan \delta$$

3. Substituting numerical values yields δ:

$$\tan \delta = -\frac{v_0}{\omega x_0} = -\frac{-25 \text{ cm/s}}{(8.0 \text{ rad/s})(4 \text{ cm})} = 0.78$$

$$\delta = \tan^{-1}(0.78) = 0.66 \text{ rad}$$

4. The amplitude can be found using either the x_0 or v_0 equation. Here we use x_0:

$$A = \frac{x_0}{\cos \delta} = \frac{4 \text{ cm}}{\cos 0.66} = 5.06 \text{ cm}$$

(b) Comparing with Equation 14-4 yields x:

$$x = (5.06 \text{ cm}) \cos(8.0t + 0.66)$$

---

* For many musical instruments, there is a slight dependence of frequency on amplitude. The vibration of an oboe reed, for example, is not exactly simple harmonic, thus its pitch depends slightly on how hard it is blown. This effect is corrected for by skilled musicians.

When the phase constant is $\delta = 0$, Equations 14-4, 14-5, and 14-6 then become

$$x = A \cos \omega t \qquad \text{14-13a}$$

$$v = -\omega A \sin \omega t \qquad \text{14-13b}$$

and

$$a = -\omega^2 A \cos \omega t \qquad \text{14-13c}$$

These functions are plotted in Figure 14-5.

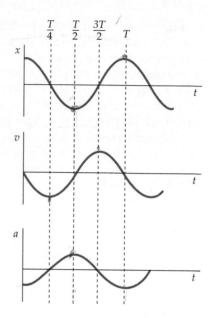

Figure 14-5 Plots of x, v, and a as functions of time t for $\delta = 0$. At $t = 0$, the displacement is maximum, the velocity is zero, and the acceleration is negative and equal to $-\omega^2 A$. The velocity becomes negative as the object moves back toward its equilibrium position. After one quarter-period ($t = T/4$), the object is at equilibrium, $x = 0$, $a = 0$, and the speed has its maximum value of ωA. At $t = T/2$, the displacement is $-A$, the velocity is again zero, and the acceleration is $+\omega^2 A$. At $t = 3T/4$, $x = 0$, $a = 0$, and $v = +\omega A$.

Example 14-3   *try it yourself*

A 2-kg object is attached to a spring as in Figure 14-1. The force constant of the spring is $k = 196$ N/m. The object is held a distance 5 cm from the equilibrium position and is released at $t = 0$. (a) Find the angular frequency ω, the frequency f, and the period T. (b) Write x as a function of time.

Cover the column to the right and try these on your own before looking at the answers.

Steps

Answers

(a)1. Calculate ω from $\omega = \sqrt{k/m}$.

$\omega = 9.90$ rad/s

2. Use your result to find f and T.

$f = 1.58$ Hz, $T = 0.633$ s

3. Find A and δ from the initial conditions.

$A = 5$ cm, $\delta = 0$

(b)   Write $x(t)$ using your results for A, ω, and δ.

$x = (5 \text{ cm}) \cos(9.90 \text{ s}^{-1} t)$

Example 14-4

Consider an object on a spring whose position is given by the equation $x = (5 \text{ cm})\cos(9.90 \text{ s}^{-1} t)$. (a) What is the maximum speed of the object? (b) When does this maximum speed first occur? (c) What is the maximum acceleration of the object? (d) When does maximum acceleration first occur?

Picture the Problem Since the object is released from rest, $\delta = 0$, and the velocity and acceleration are given by Equations 14-13b and c.

(a)1. Equation 14-13b gives the velocity for $\delta = 0$:

$v = -\omega A \sin \omega t$

2. Maximum speed occurs when $\sin \omega t = \pm 1$:

$|v_{\text{max}}| = \omega A = (9.90 \text{ rad/s})(5 \text{ cm}) = 49.5$ cm/s

(b)1. $\sin \omega t = \pm 1$ first occurs when $\omega t = \pi/2$:

$\sin \omega t = \pm 1$

$\omega t = \dfrac{\pi}{2}, \dfrac{3\pi}{2}, \dfrac{5\pi}{2}, \ldots$

2. Solve for t when $\omega t = \pi/2$:

$t = \dfrac{\pi}{2\omega} = \dfrac{\pi}{2(9.90 \text{ s}^{-1})} = 1.59$ s

(c) 1. The acceleration is given by Equation 14-13c: $a = -\omega^2 A \cos \omega t$

    2. Maximum acceleration corresponds to $\cos \omega t = \pm 1$: $|a_{max}| = \omega^2 A = 490 \text{ cm/s}^2 \approx \frac{1}{2} g$

(d) The maximum acceleration occurs first at $t = 0$: $t = 0$

Remark The maximum speed first occurs after one quarter-period,

$$t = \frac{\pi}{2\omega} = \frac{\pi}{2(2\pi/T)} = \frac{1}{4} T$$

Simple Harmonic Motion and Circular Motion

There is a relation between simple harmonic motion and circular motion with constant speed. Imagine a particle moving with constant speed v in a circle of radius A (Figure 14-6). Its angular displacement relative to the positive x axis is

$$\theta = \omega t + \delta$$

where δ is the angular displacement at time $t = 0$ and $\omega = v/A$ is the angular velocity of the particle. The x component of the particle's position is

$$x = A \cos \theta = A \cos(\omega t + \delta)$$

which is the same as Equation 14-4 for simple harmonic motion.

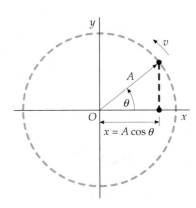

Figure 14-6 When a particle moves in a circular path with constant speed, its x component of position describes simple harmonic motion.

> When a particle moves with constant speed in a circle, its projection on a diameter of the circle moves with simple harmonic motion.

The y component of the particle's position is

$$y = A \sin \theta = A \sin(\omega t + \delta) = A \cos(\omega t + \delta - \pi/2)$$

The y component also describes simple harmonic motion. If we compare the phase of the x and y components, we see that they differ by $\pi/2$ or $90°$. Circular motion is therefore the combination of perpendicular simple harmonic motions having the same amplitude and frequency but with a relative phase difference of $\pi/2$. Figure 14-7 gives two examples demonstrating the relation of circular motion and simple harmonic motion.

Figure 14-7 The relation between circular motion and simple harmonic motion. (a) The projected shadows of a rotating peg and an object on a spring move in unison when the period of the rotating turntable equals that of the oscillating object and the radius of the turntable equals the amplitude of the spring system. (b) Bubbles foaming off the edge of a rotating propeller that is moving through water produce a sinusoidal pattern.

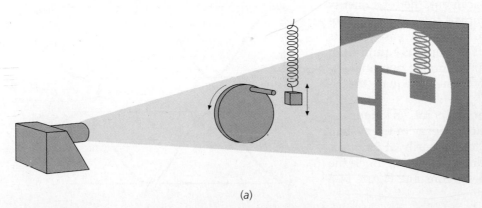

(a)

(b)

14-2   Energy in Simple Harmonic Motion

When an object undergoes simple harmonic motion, the system's potential and kinetic energies vary with time. Their sum, the total energy $E = K + U$, is constant. Consider an object a distance x from equilibrium, acted on by a restoring force $-kx$. The system's potential energy is

$$U = \tfrac{1}{2}kx^2$$

This is Equation 6-23. For simple harmonic motion, $x = A\cos(\omega t + \delta)$. Substituting gives

$$U = \tfrac{1}{2}kA^2 \cos^2(\omega t + \delta) \qquad\qquad\text{14-14}$$

Potential energy in simple harmonic motion

The kinetic energy of the system is

$$K = \tfrac{1}{2}mv^2$$

where m is the object's mass and v is its speed. For simple harmonic motion, $v = -A\omega \sin(\omega t + \delta)$. Substituting gives

$$K = \tfrac{1}{2}mA^2\omega^2 \sin^2(\omega t + \delta)$$

Then using $\omega^2 = k/m$,

$$K = \tfrac{1}{2}kA^2 \sin^2(\omega t + \delta) \qquad\qquad\text{14-15}$$

Kinetic energy in simple harmonic motion

The total energy is the sum of the potential and kinetic energies:

$$E_{\text{total}} = U + K = \tfrac{1}{2}kA^2 \cos^2(\omega t + \delta) + \tfrac{1}{2}kA \sin^2(\omega t + \delta)$$
$$= \tfrac{1}{2}kA^2[\cos^2(\omega t + \delta) + \sin^2(\omega t + \delta)]$$

Since $\sin^2(\omega t + \delta) + \cos^2(\omega t + \delta) = 1$,

$$E_{\text{total}} = \tfrac{1}{2}kA^2 \qquad\qquad\text{14-16}$$

Total energy in simple harmonic motion

This equation reveals an important general property of simple harmonic motion:

> The total energy in simple harmonic motion is proportional to the square of the amplitude.

For an object at its maximum displacement, the total energy is all potential energy. As the object moves toward its equilibrium position, the kinetic energy of the system increases and its potential energy decreases. As it moves through its equilibrium position, the speed of the object is maximum, the potential energy of the system is zero, and the total energy equals the kinetic energy.

As the object moves past the equilibrium point, its kinetic energy begins to decrease, and the potential energy of the system increases until the object again stops momentarily at its maximum displacement (now in the other direction). At all times, the sum of the potential and kinetic energies is constant. Figure 14-8

Figure 14-8 Plots of U and K versus t.

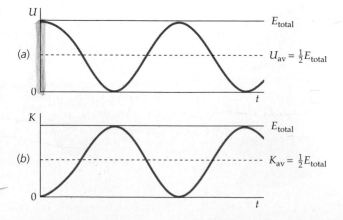

shows plots of U and K versus time. These curves have the same shape except that one is zero when the other is maximum. Their average values over one or more cycles are equal, and since $U + K = E$, their average values are given by

$$U_{av} = K_{av} = \tfrac{1}{2}E \hspace{4cm} \text{14-17}$$

In Figure 14-9, the potential energy U is graphed as a function of x. The total energy E_{total} is constant and is therefore plotted as a horizontal line. This line intersects the potential-energy curve at $x = A$ and $x = -A$, called the **turning points,** since these are the points at which oscillating objects reverse direction and head back toward the equilibrium position.

Figure 14-9 The potential-energy function $U = \tfrac{1}{2}kx^2$ for an object of mass m on a (massless) spring of force constant k. The horizontal blue line represents the total energy E_{total} for an amplitude of A. The kinetic energy K is represented by the vertical distance $K = E_{total} - U$. Since $E_{total} \geq U$, the motion is restricted to $-A \leq x \leq +A$.

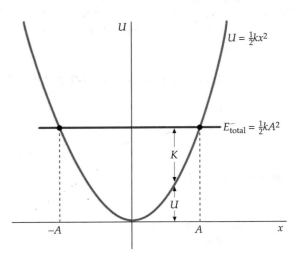

A 3-kg object attached to a spring oscillates with an amplitude of 4 cm and a period of 2 s. (*a*) What is the total energy? (*b*) What is the maximum speed of the object? (*c*) At what position x_1 is the speed equal to half its maximum value?

Picture the Problem (*a*) The total energy can be found from the amplitude and force constant, which can be found from the mass and period. (*b*) The maximum speed occurs when the kinetic energy equals the total energy. (*c*) We can relate the position to the speed by using conservation of energy.

(*a*)1. Write the total energy E in terms of the force constant k and amplitude A:

$$E = \tfrac{1}{2}kA^2$$

2. The force constant is related to the period and mass:

$$k = m\omega^2 = \frac{4\pi^2 m}{T^2} = \frac{4\pi^2 (3\text{ kg})}{(2\text{ s})^2} = 29.6\text{ N/m}$$

3. Substitute $k = 29.6$ N/m, and $A = 0.04$ m to find E:

$$E = \tfrac{1}{2}kA^2 = \tfrac{1}{2}(29.6\text{ N/m})(0.04\text{ m})^2 = 2.37 \times 10^{-2}\text{ J}$$

(*b*) To find v_{max}, set the kinetic energy equal to the total energy and solve for v:

$$\tfrac{1}{2}mv_{max}^2 = E$$

$$v_{max} = \sqrt{\frac{2E}{m}} = \sqrt{\frac{2(2.37 \times 10^{-2}\text{ J})}{3\text{ kg}}} = 0.126\text{ m/s}$$

(*c*)1. Conservation of energy relates the position x to the speed v:

$$E = \tfrac{1}{2}mv^2 + \tfrac{1}{2}kx^2$$

2. Substitute $v = \tfrac{1}{2}v_{max}$ and solve for x_1. It is convenient to find x in terms of E and then write $E = \tfrac{1}{2}kA^2$ to obtain an expression for x in terms of A:

$$E = \tfrac{1}{2}m(\tfrac{1}{2}v_{max})^2 + \tfrac{1}{2}kx_1^2 = \tfrac{1}{4}(\tfrac{1}{2}mv_{max}^2) + \tfrac{1}{2}kx_1^2$$

$$E = \tfrac{1}{4}E + \tfrac{1}{2}kx_1^2$$

$$\tfrac{1}{2}kx_1^2 = E - \tfrac{1}{4}E = \tfrac{3}{4}E = \tfrac{3}{4}(\tfrac{1}{2}kA^2)$$

$$x_1 = \frac{\sqrt{3}}{2}A = \frac{\sqrt{3}}{2}(4\text{ cm}) = 3.46\text{ cm}$$

Exercise Calculate ω for this example and find v_{max} from $v_{max} = \omega A$. (*Answer* $\omega = 3.14$ rad/s, $v_{max} = 0.126$ m/s)

Exercise An object of mass 2 kg is attached to a spring of force constant 40 N/m. The object is moving at 25 cm/s when it is at its equilibrium position. (*a*) What is the total energy of the object? (*b*) What is the amplitude of the motion? (*Answers* (*a*) $E_{total} = \frac{1}{2}mv^2_{max} = 0.0625$ J, (*b*) $A = \sqrt{2E_{total}/k} = 5.59$ cm).

General Motion Near Equilibrium

Simple harmonic motion is important because it occurs whenever a particle is displaced slightly from a position of stable equilibrium. Figure 14-10 is a graph of the potential energy U versus x for a force that has a position of stable equilibrium and a position of unstable equilibrium. As discussed in Chapter 6, the maximum at x_2 on Figure 14-10 corresponds to unstable equilibrium, whereas the minimum at x_1 corresponds to stable equilibrium. Any smooth curve that has a minimum like the one in Figure 14-10 can be approximated near the minimum by a parabola. The dashed curve in this figure is a parabolic curve that approximately fits U near the stable equilibrium point. The general equation for a parabola that has a minimum at point x_1 can be written

$$U = A + B(x - x_1)^2 \qquad \text{14-18}$$

where A and B are constants. The constant A is the value of U at the equilibrium position $x = x_1$. The force is related to the potential energy curve by $F_x = -dU/dx$. Then

$$F_x = -\frac{dU}{dx} = -2B(x - x_1)$$

If we set $2B = k$, this equation reduces to

$$F_x = -\frac{dU}{dx} = -k(x - x_1) \qquad \text{14-19}$$

According to Equation 14-19, the force is proportional to the displacement and oppositely directed, so the motion will be simple harmonic. Figure 14-11 shows a real potential energy function that has a position r_0 of stable equilibrium—$U(r)$ versus separation r for two hydrogen atoms.

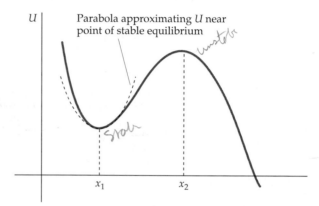

Figure 14-10 Plot of U versus x for a force that has a position of stable equilibrium (x_1) and a position of unstable equilibrium (x_2).

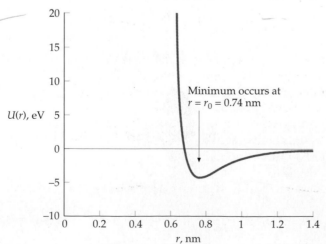

Figure 14-11 Potential energy function $U(r)$ versus separation r for two hydrogen atoms. The minimum value of the potential energy $U = -4.48$ eV occurs at the equilibrium separation $r = r_0 = 0.74$ nm. The atoms in the H_2 molecule oscillate about this point. If the energy of oscillation is not too great, so that $|r - r_0|$ is small, the oscillation is simple harmonic.

14-3 Some Oscillating Systems

Object on a Vertical Spring

When an object hangs from a vertical spring, there is a force mg downward in addition to the force of the spring (Figure 14-12). If we choose the downward direction to be positive, the spring's force on the object is $F_s = -ky$, where y is the difference in position between the end of the spring when it is unstretched and when it is stretched by the weight of the object. Then Newton's second law gives

$$m\frac{d^2y}{dt^2} = -ky + mg \qquad 14\text{-}20$$

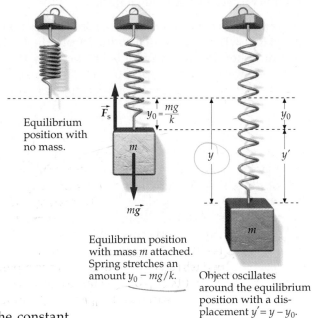

Figure 14-12 The problem of a mass on a vertical spring is simplified if the displacement (y') is measured from the equilibrium position of the spring with the mass attached.

Equilibrium position with no mass.

$y_0 = \dfrac{mg}{k}$

$\vec{F}_s$

mg

y

y_0

y'

Equilibrium position with mass m attached. Spring stretches an amount $y_0 - mg/k$.

Object oscillates around the equilibrium position with a displacement $y' = y - y_0$.

Equation 14-20 differs from Equation 14-2 by the addition of the constant term mg. We handle this extra term by changing to a new variable $y' = y - y_0$, where $y_0 = mg/k$ is the amount the spring is stretched when the object is in equilibrium. Substituting $y = y' + y_0$ into Equation 14-20 gives

$$m\frac{d^2(y' + y_0)}{dt^2} = -k(y' + y_0) + mg = -ky' - ky_0 + mg$$

But $ky_0 = mg$. Using this and the fact that the derivative of a constant is zero, we have

$$m\frac{d^2y'}{dt^2} = -ky'$$

which has the familiar solution

$$y' = A\cos(\omega t + \delta)$$

Thus, the effect of the gravitational force mg is merely to shift the equilibrium position from $y = 0$ to $y' = 0$. When the object is displaced from this equilibrium position by the amount y', the unbalanced force is $-ky'$. The object oscillates about this equilibrium position with an angular frequency $\omega = \sqrt{k/m}$, the same as that for an object on a horizontal spring.

When a mass hangs on a vertical spring, there is both gravitational potential energy U_g and spring potential energy U_s. At the equilibrium point, the spring is stretched and has potential energy $\frac{1}{2}ky_0^2$ relative to its unstretched position $(y = 0)$, and the gravitational potential energy is $-mgy_0$ relative to $y = 0$. We will show below that if we choose the total potential energy (including gravitational potential energy) to be zero at the equilibrium position $y' = 0$, the total potential energy can be written

$$U = U_s + U_g = \frac{1}{2}ky'^2 \qquad (U = 0 \text{ at } y' = 0) \qquad 14\text{-}21$$

So if we measure the displacement from the equilibrium position, we can forget about the effect of gravity.

Example 14-6

A 3-kg object stretches a spring 16 cm when it hangs vertically in equilibrium. The spring is then stretched from its equilibrium position and the object is released. (*a*) Find the frequency of the motion. (*b*) Find the frequency if the 3-kg object is replaced with a 6-kg object.

Picture the Problem (*a*) The frequency depends on the force constant for the spring, which can be determined from the position of the spring at equilibrium, when the weight m_1g is balanced by the spring force ky_0. (*b*) Since the frequency is inversely proportional to the square root of the mass (Equation 14-12), the frequency for a 6-kg object will be $1/\sqrt{2}$ times that for the 3-kg object.

(*a*)1. Write the frequency in terms of the force constant k and the mass m_1:

$$f = \frac{\omega}{2\pi} = \frac{1}{2\pi}\sqrt{\frac{k}{m_1}}$$

2. Set ky_0 equal to m_1g and solve for k:

$$ky_0 = m_1g$$

$$k = \frac{m_1g}{y_0} = \frac{(3\text{ kg})(9.81\text{ N/kg})}{0.16\text{ m}} = 184\text{ N/m}$$

3. Substitute this value for k and the known values for m_1, g, and y_0:

$$f = \frac{1}{2\pi}\sqrt{\frac{k}{m_1}} = \frac{1}{2\pi}\sqrt{\frac{m_1g/y_0}{m_1}} = \frac{1}{2\pi}\sqrt{\frac{g}{y_0}}$$

$$= \frac{1}{2\pi}\sqrt{\frac{9.81\text{ m/s}^2}{0.16\text{ m}}} = 1.25\text{ Hz}$$

(*b*)   Replace m_1 by $m_2 = 2m_1$:

$$f = \frac{1}{2\pi}\sqrt{\frac{k}{m_2}} = \frac{1}{2\pi}\sqrt{\frac{k}{2m_1}} = \frac{1}{\sqrt{2}}\left(\frac{1}{2\pi}\sqrt{\frac{k}{m_1}}\right)$$

$$= \frac{1}{\sqrt{2}}f_1 = \frac{1}{\sqrt{2}}\,1.25\text{ Hz} = 0.884\text{ Hz}$$

Remark Note that in (*a*) we did not need to use the value of m or k because the force constant is mg/y_0, and k/m is therefore equal to g/y_0.

Example 14-7

A block rests on a spring and oscillates vertically with a frequency of 4 Hz and an amplitude of 7 cm. A tiny bead is placed on top of the oscillating block just as it reaches its lowest point (Figure 14-13). Assume that the bead has no effect on the oscillation. (*a*) At what distance from the block's equilibrium position does the bead lose contact with the block? (*b*) What is the speed of the bead when it leaves the block?

Picture the Problem (*a*) Let y' be positive upward with $y' = 0$ at the equilibrium position of the block. The equation of motion for the block is $y' = -A\cos\omega t$ with $A = 0.07$ m and $\omega = 2\pi f = 8\pi$. The forces on the bead are its weight mg downward and the upward normal force exerted by the block. As the block moves upward *from equilibrium*, its acceleration and that of the bead are *downward* and increasing in magnitude. When the acceleration reaches $-g$, the normal force on the bead is zero and the bead leaves the block. (*b*) The velocity of the bead when it leaves the block is the same as that of the block at this time.

Figure 14-13

(a)1. Write the equation of motion for the block: $y' = -A \cos \omega t$

2. Compute the acceleration of the block and set it equal to $-g$:

$$a = -\omega^2 y' = +A\omega^2 \cos \omega t = -g$$

$$\cos \omega t = \frac{-g}{\omega^2 A}$$

3. Find the displacement y at this time:

$$y' = -A \cos \omega t = \frac{g}{\omega^2} = \frac{9.81 \text{ m/s}^2}{(8\pi \text{ rad/s})^2} = 0.0155 \text{ m} = 1.55 \text{ cm}$$

(b)1. Find the velocity of the block (and bead) at any time:

$$v = \frac{dy'}{dt} = +\omega A \sin \omega t = \omega A \sqrt{1 - \cos^2 \omega t}$$

2. Compute v when $y' = 1.55$ cm:

$$\cos \omega t = -\frac{y'}{A} = -\frac{1.55 \text{ cm}}{7 \text{ cm}} = -0.221$$

$$v = \omega A \sqrt{1 - \cos^2 \omega t}$$

$$= (8\pi \text{ rad/s})(0.07 \text{ m})\sqrt{1 - (0.221)^2}$$

$$= 1.72 \text{ m/s}$$

Remark The bead leaves when y' is positive, as expected.

Example 14-8

The 3-kg object in Example 14-6 stretches a spring 16 cm when it hangs vertically in equilibrium. The spring is then stretched 5 cm from its equilibrium position and the object is released. Find the total energy and the potential energy of the spring when the mass is at its maximum displacement.

Picture the Problem The total energy including gravitational energy is $\frac{1}{2}kA^2$. The potential energy of the spring at maximum displacment is the total energy minus the gravitational potential energy, which is $-mgA$. The value of k was calculated to be 184 N/m in Example 14-6.

1. The total energy at maximum displacement is the potential energy given by Equation 14-21 with $y' = A = 0.05$ m:

$$E = U_{max} = \frac{1}{2}kA^2$$

$$= \frac{1}{2}(184 \text{ N/m})(0.05 \text{ m})^2 = 0.23 \text{ N·m} = 0.23 \text{ J}$$

2. To find the potential energy of the spring, subtract the gravitational potential energy:

$$U_s = U_{total} - U_g = 0.23 \text{ J} - (-m_1 gA)$$

$$= 0.23 \text{ J} + (3 \text{ kg})(9.81 \text{ N/kg})(0.05 \text{ m})$$

$$= 0.23 \text{ J} + 1.47 \text{ J} = 1.70 \text{ J}$$

Check the Result When the spring is stretched by 16 cm, its potential energy is $\frac{1}{2}ky_0^2 = \frac{1}{2}(184 \text{ N/m})(0.16 \text{ m})^2 = 2.36 \text{ J}$ *relative to its unstretched position*. When the mass is at the maximum displacement of 5 cm from equilibrium, the spring is stretched by 5 cm + 16 cm = 21 cm and the potential energy of the spring *relative to its unstretched position* is $\frac{1}{2}ky^2 = \frac{1}{2}(184 \text{ N/m})(0.21 \text{ m})^2 = 4.06 \text{ J}$. The potential energy of the spring *relative to its equilibrium position* is then 4.06 J − 2.36 J = 1.70 J.

Derivation of Potential Energy for a Vertical Spring   At the equilibrium point $y = y_0$, the gravitational potential energy is $-mgy_0$ relative to the unstretched length of the spring ($y = 0$), and the potential energy of the mass–spring system is $\frac{1}{2}ky_0^2$. Let us now choose $U_s = U_g = 0$ at the *equilibrium* position $y' = y - y_0 = 0$. Then at a general point y, the potential energy of the spring is $\frac{1}{2}ky^2 - \frac{1}{2}ky_0^2$ and the gravitational potential energy is $-mgy + mgy_0 = -mgy'$. The total potential energy is then

$$U = U_s + U_g = (\tfrac{1}{2}ky^2 - \tfrac{1}{2}ky_0^2) - mgy'$$

$$= \tfrac{1}{2}k(y' + y_0)^2 - \tfrac{1}{2}ky_0^2 - mgy'$$

$$= (\tfrac{1}{2}ky'^2 + ky'y_0 + \tfrac{1}{2}ky_0^2) - \tfrac{1}{2}ky_0^2 - mgy'$$

$$= \tfrac{1}{2}ky'^2 + ky'y_0 - mgy'$$

The terms $+ky'y_0$ and $-mgy'$ cancel because $ky_0 = mg$. The total potential energy is therefore

$$U = U_s + U_g = \tfrac{1}{2}ky'^2$$

which is Equation 14-21.

The Simple Pendulum

A simple pendulum consists of a string of length L and a bob of mass m. When the bob is released from an initial angle ϕ_0 with the vertical, it swings back and forth with some period T. We wish to find the period T.

Exercise in Dimensional Analysis   We might expect the period of a simple pendulum to depend on the mass m of a pendulum bob, the length L of the pendulum, the acceleration g due to gravity, and the initial angle ϕ_0. Find a simple combination of these quantities that gives the correct dimensions for the period. (*Answer* $\sqrt{L/g}$. The units of length, mass, and g are m, kg, and m/s², respectively. The angle ϕ_0 is dimensionless. If we divide L by g, the meters cancel and we are left with seconds squared, suggesting the form $\sqrt{L/g}$. If the formula for period contained the mass, the unit kg must be canceled by some other quantity. But there is no combination of L and g that can cancel mass units. So the period cannot depend on the mass of the bob. Since the initial angle ϕ_0 is dimensionless, we cannot tell whether or not it is a factor in the period. We will see below that for small ϕ_0, the period is given by $T = 2\pi\sqrt{L/g}$.)

The forces on the bob are its weight $m\vec{g}$ and the string tension $\vec{T}$ (Figure 14-14). At an angle ϕ with the vertical, the weight has components $mg\cos\phi$ along the string and $mg\sin\phi$ tangential to the circular arc in the direction of decreasing ϕ. Let s be the arc length measured from the bottom of the circle. Then

$$s = L\phi \qquad\qquad\qquad \text{14-22}$$

where ϕ is in radians. The tangential component of Newton's second law gives

$$\sum F_t = -mg\sin\phi = m\frac{d^2s}{dt^2} = mL\frac{d^2\phi}{dt^2}$$

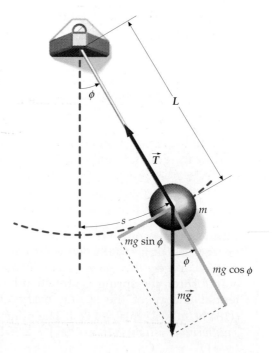

Figure 14-14
Forces on a pendulum bob.

Motion of a simple pendulum. The bob is shown at equal time intervals. It moves faster at the bottom, as shown by the greater spacing of the images.

or

$$\frac{d^2\phi}{dt^2} = -\frac{g}{L}\sin\phi \qquad\qquad 14\text{-}23$$

Note that the mass m does not appear in Equation 14-23—the motion of a pendulum does not depend on its mass. For small ϕ, $\sin\phi \approx \phi$, and

$$\frac{d^2\phi}{dt^2} = -\frac{g}{L}\phi \qquad\qquad 14\text{-}24$$

Equation 14-24 is of the same form as Equation 14-2 for an object on a spring. The motion of a pendulum is thus approximately simple harmonic motion for small angular displacements.

Equation 14-24 can be written

$$\frac{d^2\phi}{dt^2} = -\omega^2\phi \qquad\qquad 14\text{-}25$$

where

$$\omega^2 = \frac{g}{L} \qquad\qquad 14\text{-}26$$

The period of the motion is thus

$$T = \frac{2\pi}{\omega} = 2\pi\sqrt{\frac{L}{g}} \qquad\qquad 14\text{-}27$$

Period of a simple pendulum

The solution of Equation 14-25 is

$$\phi - \phi_0\cos(\omega t + \delta)$$

where ϕ_0 is the maximum angular displacement.

According to Equation 14-27, the greater the length of a pendulum, the greater the period, which is consistent with experimental observation. The period, and therefore the frequency, are independent of the amplitude of oscillation (as long as the amplitude is small), a general feature of simple harmonic motion. Galileo observed this by timing the period of a swinging lamp while in church. (He went back the next Sunday and found that the period also doesn't depend on the mass!)

Exercise Find the period of a pendulum of length 1 m. (*Answer* $T = 2\pi\sqrt{L/g} = 2\pi\sqrt{(1\text{ m})/(9.81\text{ m/s}^2)} = 2.01$ s. The validity of this result can be easily demonstrated by swinging a weight on a 1-m string and timing it.)

The acceleration due to gravity can be measured using a simple pendulum. You need only measure its length L and period T, and using Equation 14-27, solve for g.*

---

* When finding T, one usually measures the time for n oscillations and then divides by n, which helps minimize measurement error.

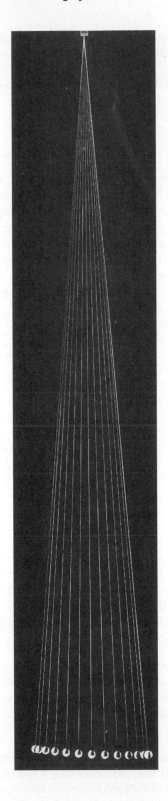

Pendulum in an Accelerated Reference Frame Figure 14-15*a* shows a simple pendulum suspended from the ceiling of a boxcar that has acceleration $\vec{a}_0$ to the right. Relative to a nonaccelerated inertial frame, the bob has a horizontal acceleration $\vec{a}_0$. The vertical and horizontal components of Newton's law for the bob are

$$\sum F_x = T \sin \theta = ma_0$$

$$\sum F_y = T \cos \theta - mg = 0$$

The equilibrium angle is thus given by $\tan \theta = a_0/g$.

Relative to the boxcar, the bob appears to be acted on by a horizontal pseudoforce $-ma_0$ to the left in addition to the downward force of gravity mg. This pseudoforce, like the real gravitational force, is proportional to the mass of the bob. Relative to the boxcar, all objects will fall at an angle θ to the vertical with acceleration $\vec{g}' = \vec{g} - \vec{a}_0$. We can use Newton's laws relative to the accelerating boxcar if we add a pseudoforce $-m\vec{a}_0$ that acts on each object of mass m. This is equivalent to replacing the acceleration due to gravity $\vec{g}$ by $\vec{g}' = \vec{g} - \vec{a}_0$ (Figure 14-15*b*).

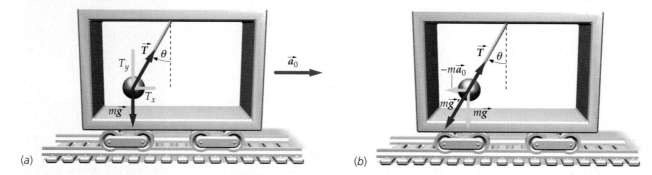

(a)   (b)

Figure 14-15 (a) Simple pendulum in apparent equilibrium in an accelerating boxcar. Forces are those as seen from a separate, stationary frame. (b) Forces on the bob as seen in the accelerated frame. Adding the pseudoforce $-ma_0$ is equivalent to replacing $\vec{g}$ by $\vec{g}'$.

If the bob is displaced slightly from equilibrium, it will oscillate with a period T given by Equation 14-27 with g replaced by g'.

Exercise A simple pendulum of length 1 m is in a boxcar that is accelerating horizontally with an acceleration $a_0 = 3$ m/s². Find g' and the period T. (*Answer* $g' = 10.3$ m/s², $T = 1.96$ s)

Large-Amplitude Oscillations When the amplitude of a pendulum becomes large, its motion continues to be periodic, but it is no longer simple harmonic. A slight dependence on the amplitude must be accounted for when determining the period. For a general angular amplitude ϕ_0, the period can be shown to be

$$T = T_0\left[1 + \frac{1}{2^2}\sin^2 \frac{1}{2}\phi_0 + \frac{1}{2^2}\left(\frac{3}{4}\right)^2 \sin^4 \frac{1}{2}\phi_0 + \cdots\right] \quad \text{14-28}$$

Period for large-amplitude oscillations

where $T_0 = 2\pi\sqrt{L/g}$ is the period for very small amplitudes. Figure 14-16 shows T/T_0 as a function of amplitude ϕ_0.

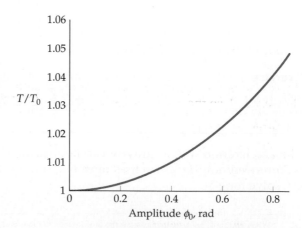

Figure 14-16 T/T_0 as a function of amplitude ϕ_0.

Example 14-9   *try it yourself*

A simple pendulum clock is calibrated to keep accurate time at an angular amplitude of $\phi_0 = 10°$. When the amplitude has decreased to the point that it is very small, does the clock gain or lose time? How much time does the clock gain or lose in one day?

Cover the column to the right and try these on your own before looking at the answers.

Steps	Answers
1. Answer the first question by finding if the period increases or decreases.	T decreases as ϕ decreases so the clock gains time.
2. Use Equation 14-28 to find the percentage change $[(T - T_0)/T_0] \times 100\%$ for $\phi = 10°$. Use only the first correction term.	0.190%.
3. Find the number of minutes in a day.	There are 1440 minutes in a day.
4. Combine steps 2 and 3 to find the change in the number of minutes in a day.	The gain is 2.74 minutes per day.

The Physical Pendulum

A rigid object pivoted about a point other than its center of mass will oscillate when displaced from equilibrium. Such a system is called a **physical pendulum**. Consider a plane figure pivoted about a point a distance D from its center of mass and displaced from equilibrium by the angle ϕ (Figure 14-17). The torque about the pivot has a magnitude $MgD \sin \phi$ and tends to decrease ϕ. Newton's second law applied to rotation is

$$\tau = I\alpha = I\frac{d^2\phi}{dt^2}$$

where α is the angular acceleration, and I is the moment of inertia about the pivot point. Substituting $-MgD \sin \phi$ for the net torque, we have

$$-MgD \sin \phi = I\frac{d^2\phi}{dt^2}$$

or

$$\frac{d^2\phi}{dt^2} = -\frac{MgD}{I}\sin \phi \qquad\qquad 14\text{-}29$$

Again, the motion is approximately simple harmonic if the angular displacements are small, so the approximation $\sin \phi \approx \phi$ holds. In this case, we have

$$\frac{d^2\phi}{dt^2} = -\frac{MgD}{I}\phi = -\omega^2\phi \qquad\qquad 14\text{-}30$$

where $\omega^2 = MgD/I$. The period is therefore

$$T = \frac{2\pi}{\omega} = 2\pi\sqrt{\frac{I}{MgD}} \qquad\qquad 14\text{-}31$$

Period of a physical pendulum

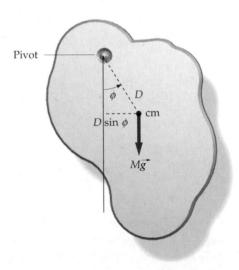

Figure 14-17 A physical pendulum.

optional

For large amplitudes, the period is given by Equation 14-28, with T_0 given by Equation 14-31. For a simple pendulum of length L, the moment of inertia is $I = ML^2$ and $D = L$. Then Equation 14-31 gives $T = 2\pi\sqrt{ML^2/MgL} = 2\pi\sqrt{L/g}$, the same as Equation 14-27.

Example 14-10

A uniform stick of mass M and length L is pivoted at one end. (*a*) Find the period of oscillation for small angular displacements. (*b*) Find the period of oscillation if the stick is pivoted about point P a distance x from the center of mass.

Picture the Problem (*a*) The period is given by $T = 2\pi\sqrt{I/MgD}$. The center of mass is at the center of the stick, so the distance from the center of mass to the pivot is $D = L/2$ (Figure 14-18*a*). The moment of inertia of a uniform stick about one end is $I = \frac{1}{3}ML^2$ (Table 9-1). (*b*) For a pivoting motion around point P, the distance D is given as x (Figure 14-18*b*), and the moment of inertia can be found from the parallel-axis theorem $I = I_{cm} + MD^2$, where $I_{cm} = \frac{1}{12}ML^2$.

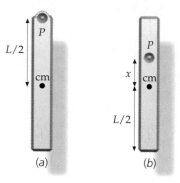

Figure 14-18

(*a*)1. The period is given by Equation 14-31:

$$T = 2\pi\sqrt{\dfrac{I}{MgD}}$$

2. I about the end is known, and D is known:

$$I = \tfrac{1}{3}ML^2$$

$$D = \dfrac{L}{2}$$

3. Substitute the values for I and D to find T:

$$T = 2\pi\sqrt{\dfrac{\frac{1}{3}ML^2}{Mg\frac{1}{2}L}} = 2\pi\sqrt{\dfrac{2L}{3g}}$$

(*b*)1. About point P, $D = x$, and the moment of inertia is given by the parallel axis theorem:

$$D = x$$

$$I = I_{cm} + MD^2 = \tfrac{1}{12}ML^2 + Mx^2$$

2. Substitute these values to find T:

$$T = 2\pi\sqrt{\dfrac{I}{MgD}} = 2\pi\sqrt{\dfrac{\frac{1}{12}ML^2 + Mx^2}{Mgx}}$$

$$= 2\pi\sqrt{\dfrac{\frac{1}{12}L^2 + x^2}{gx}}$$

Check the Result When $x = 0$, the stick is pivoted about its center of mass and the period is infinite. When $x = L/2$, we get the same result as found in (*a*).

Exercise What is the period of oscillation for small angular displacements of a meter stick pivoted about one end? (*Answer* $T = 1.64$ s. Note that this is a smaller period than for a simple pendulum of length $L = 1$ m. The period of the simple pendulum is greater because its moment of inertia is mL^2 rather than $\frac{1}{3}mL^2$.)

Exercise Show that when $x = L/6$, the period is the same as when $x = L/2$.

Remark The period versus distance x from the center of mass for a stick of length 1 m is shown in Figure 14-19.

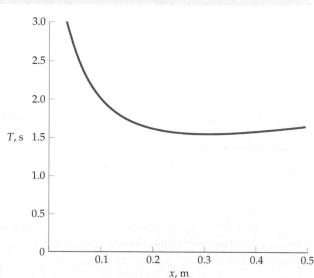

Figure 14-19

| Example 14-11 | *try it yourself* |

Find the value of x in Example 14-10 such that the period is a minimum.

Picture the Problem At the value of x for which T is a minimum, $dT/dx = 0$. To simplify the calculation, let $Z = (L^2 + 12x^2)/x = x^{-1}(L^2 + 12x^2)$; then $T = (2\pi/\sqrt{12g})Z^{1/2}$ and $dT/dx = 0$ when $dZ/dx = 0$.

Cover the column to the right and try these on your own before looking at the answers.

Steps	*Answers*

1. Show that if $dZ/dx = 0$, then $dT/dx = 0$.

$$\frac{dT}{dx} = \frac{2\pi}{\sqrt{12g}}\frac{1}{2}Z^{-1/2}\frac{dZ}{dx} = 0, \qquad \text{only if} \quad \frac{dZ}{dx} = 0$$

2. Compute dZ/dx.

$$\frac{dZ}{dx} = -x^{-2}(L^2 + 12x^2) + x^{-1}(24x)$$

$$= -\frac{L^2}{x^2} + 12$$

3. Set $dZ/dx = 0$ and solve for x.

$$x = \frac{L}{\sqrt{12}}$$

14-4   Damped Oscillations

Left to itself, a spring or a pendulum eventually stops oscillating because the mechanical energy is dissipated by frictional forces. Such motion is said to be **damped.** If damping is small, the system oscillates with an amplitude that decreases slowly with time (Figure 14-20).

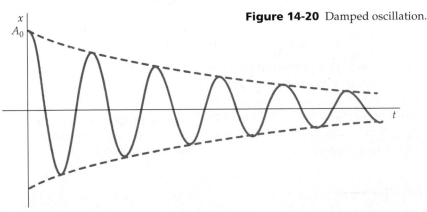

Figure 14-20 Damped oscillation.

Both the amplitude and the energy, which is proportional to the square of the amplitude, decrease by a constant percentage in a given time interval. This type of decrease is called exponential decrease. The force exerted on a damped oscillator such as the one shown in Figure 14-21 can be represented by the empirical expression

$$\vec{F}_d = -b\vec{v}$$

where b is a constant. Since the damping force is opposite the direction of motion, it does negative work and causes the mechanical energy of the system to decrease. The motion of a damped system can be obtained from Newton's second law. For an object of mass m on a spring of force constant k the

Figure 14-21 A damped oscillator. The motion is damped by the plunger immersed in the liquid.

net force is $-kx - b\,dx/dt$. Setting the net force equal to the mass times the acceleration d^2x/dt^2, we obtain

$$F_x = ma_x$$

$$-kx - b\frac{dx}{dt} = m\frac{d^2x}{dt^2} \qquad \text{14-32}$$

Differential equation for a damped oscillator

The exact solution of this equation can be found using standard methods for solving differential equations.* The solution for the case of small damping is

$$x = A_0 e^{-(b/2m)t}\cos(\omega't + \delta) = A_0 e^{-t/2\tau}\cos(\omega't + \delta) \qquad \text{14-33}$$

where A_0 is the maximum amplitude, and

$$\tau = \frac{m}{b} \qquad \text{14-34}$$

is called the **decay time** or **time constant.** The frequency ω' is given by

$$\omega' = \sqrt{\omega_0^2 - (b/2m)^2} = \omega_0\sqrt{1 - (b/2m\omega_0)^2} \qquad \text{14-35}$$

where ω_0 is the frequency with no damping ($\omega_0 = \sqrt{k/m}$ for a mass on a spring). For small damping, $b/2m\omega_0 \ll 1$ and ω' is nearly equal to ω_0. The dashed curves in Figure 14-20 correspond to $x = A$ and $x = -A$ where A is given by

$$A = A_0 e^{-(b/2m)t} = A_0 e^{-t/2\tau} \qquad \text{14-36}$$

If the damping constant b is gradually increased, the angular frequency ω' decreases until it becomes zero at the critical value

$$b_c = 2m\omega_0 \qquad \text{14-37}$$

When b is greater than or equal to b_c, the system does not oscillate. If $b = b_c$, the system is said to be **critically damped;** it returns to equilibrium with no oscillation in the shortest time possible. When b is greater than b_c, the system is **overdamped**. Figure 14-22 shows plots of the displacement versus time for a critically damped and an overdamped oscillator. We often use critical damping when we wish to have a system avoid oscillations and yet return to equilibrium quickly. One example is the use of shock absorbers to damp the oscillations of an automobile on its springs. You can test the damping of your shock absorbers by pushing down on the front or back of your car and releasing it. If the car returns to equilibrium with no oscillation, the system is critically damped or overdamped. Usually, you will note one or two oscillations, indicating that the damping is just under the critical value.

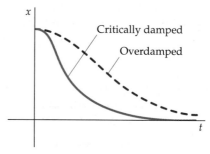

Figure 14-22 Plots of displacement versus time for a critically damped or overdamped oscillator.

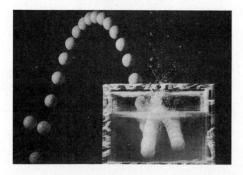

The ball's motion is damped—the energy of each bounce is less than that of the previous bounce because of the resistance of the water.

---

* A solution to this equation using complex numbers is given in the "Exploring . . ." section on page 428.

Because the energy of an oscillator is proportional to the square of the amplitude, the energy of an underdamped oscillator (averaged over a cycle) also decreases exponentially with time:

$$E = \tfrac{1}{2}m\omega^2 A^2 = \tfrac{1}{2}m\omega^2 (A_0 e^{-t/2\tau})^2 = E_0 e^{-t/\tau} \qquad \text{14-38}$$

where $E_0 = \tfrac{1}{2}m\omega^2 A_0^2$. We see that the decay time is the time for the energy to decrease to e^{-1} times its original value.

A damped oscillator is often described by its Q factor (for quality factor),

$$Q = \omega_0 \tau = \frac{\omega_0 m}{b} \qquad \text{14-39}$$

Definition—Q factor

We can relate Q to the fractional energy loss per cycle. Differentiating Equation 14-38 gives

$$dE = -\frac{1}{\tau} E_0 e^{-t/\tau}\, dt = -\frac{1}{\tau} E\, dt$$

If the energy loss per period is small, we can replace dE by ΔE, and dt by the period T. Then $|\Delta E|/E$ in one period is given by

$$\frac{|\Delta E|}{E} = \frac{T}{\tau} = \frac{2\pi}{\omega_0 \tau} = \frac{2\pi}{Q} \qquad \text{14-40}$$

or

$$Q = \frac{2\pi}{(|\Delta E|/E)_{\text{cycle}}} \qquad \text{14-41}$$

Physical interpretation of Q for small damping

Q is thus inversely proportional to the fractional energy loss per cycle.

Shock absorbers (yellow cylinders) are used to damp the oscillations of this truck.

Example 14-12

When middle C on the piano (frequency 262 Hz) is struck, it loses half its energy after 4 s. (a) What is the decay time τ? (b) What is the Q factor for this piano wire? (c) What is the fractional energy loss per cycle?

Picture the Problem   (a) We use $E = E_0 e^{-t/\tau}$ and set E equal to $\tfrac{1}{2}E_0$. (b) The Q value can then be found from the decay time and the frequency.

(a)1. Set the energy at time $t = 4$ s equal to half the original energy:

$$E = E_0 e^{-t/\tau} = E_0 e^{-4\,\text{s}/\tau} = \tfrac{1}{2}E_0$$
$$e^{4\,\text{s}/\tau} = 2$$

2. Solve for the time t by taking the natural logarithm:

$$\frac{4\,\text{s}}{\tau} = \ln 2$$

$$\tau = \frac{4\,\text{s}}{\ln 2} = 5.77\ \text{s}$$

(b)   Calculate Q from τ and ω_0:

$$Q = \omega_0 \tau = (2\pi f)\tau$$
$$= 2\pi(262\ \text{Hz})(5.77\ \text{s}) = 9.50 \times 10^3$$

(c)   The fractional energy loss in a period is given by Equation 14-40:

$$\frac{|\Delta E|}{E} = \frac{T}{\tau} = \frac{1}{f\tau} = \frac{1}{(262\ \text{Hz})(5.77\ \text{s})} = 6.61 \times 10^{-4}$$

Check the Result Q can also be calculated from $Q = 2\pi/(\Delta E/E)_{cycle} = 2\pi/(6.61 \times 10^{-4}) = 9.50 \times 10^3$. Note that the fractional energy loss after 4 s is not just the number of cycles (4×262) times the fractional energy loss per cycle because the energy decrease is exponential, not constant.

Remarks Figure 14-23 shows amplitude versus time for the oscillation of a piano string after middle C is struck. After 4 s, the amplitude has decreased to about 0.7 times its initial value, and the energy, which is proportional to the amplitude squared, drops to about half its initial value.

Note that Q is quite large. You can estimate τ and Q of various oscillating systems. Tap a crystal wine glass and see how long it rings. The longer it rings, the greater the value of τ and Q, and the lower the damping. Glass beakers from your laboratory may also have a high Q. Now tap a plastic cup.

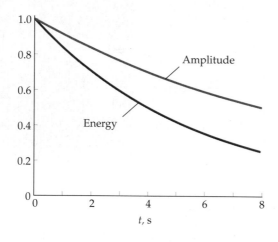

Figure 14-23

In terms of Q, the exact frequency of an underdamped oscillator is

$$\omega' = \omega_0 \sqrt{1 - \left(\frac{b}{2m\omega_0}\right)^2} = \omega_0 \sqrt{1 - \frac{1}{4Q^2}}$$ 14-42

Since Q is quite large for a slightly damped oscillator (Example 14-12), we see that ω' is nearly equal to ω_0.

We can understand the behavior of a damped oscillator qualitatively by considering its energy. The power dissipated by the damping force equals the instantaneous rate of change of the total mechanical energy

$$P = \frac{dE}{dt} = \vec{F}_d \cdot \vec{v} = -bv^2$$ 14-43

For a slightly damped oscillator, the total mechanical energy decreases slowly with time. The average kinetic energy equals half the total energy

$$(\tfrac{1}{2}mv^2)_{av} = \tfrac{1}{2}E$$

If we substitute $(v^2)_{av} = E/m$ for v^2 in Equation 14-43, we have

$$\frac{dE}{dt} = -bv^2 \approx -b(v^2)_{av} = -\frac{b}{m}E$$ 14-44

Equation 14-44 may be solved by direct integration. Its solution is

$$E = E_0 e^{-(b/m)t} = E_0 e^{-t/\tau}$$

which is Equation 14-38.

14-5 Driven Oscillations and Resonance

To keep a damped system going, energy must be put into the system. When this is done, the oscillator is said to be driven or forced. When you keep a swing going by "pumping," that is, by moving your body and legs, you are driving an oscillator. If you put energy into the system faster than it is dissipated, the energy increases with time, and the amplitude increases. If you put energy in at the same rate it is being dissipated, the amplitude remains constant over time.

Figure 14-24 shows a system consisting of an object on a spring that is being driven by moving the point of support up and down with simple harmonic motion of frequency ω. At first the motion is complicated, but eventually a steady state is reached in which the system oscillates with the same frequency as that of the driver and with a constant amplitude and, therefore, a constant energy. In the steady state, the energy put into the system per cycle by the driving force equals the energy dissipated per cycle because of the damping.

The amplitude, and therefore the energy, of a system in the steady state depends not only on the amplitude of the driver, but also on its frequency. The **natural frequency** of an oscillator, ω_0, is the frequency when no driving or damping forces are present. (In the case of a spring, for example, $\omega_0 = \sqrt{k/m}$.) If the driving frequency is approximately equal to the natural frequency of the system, the system will oscillate with a very large amplitude. For example, if the support in Figure 14-24 oscillates with the natural frequency of the mass–spring system, the mass will oscillate with a much greater amplitude than that of the support. This phenomenon is called **resonance**. When the driving frequency equals the natural frequency of the oscillator, the energy absorbed by the oscillator is maximum. The natural frequency of the system is thus called the **resonance frequency** of the system.* The average rate at which energy is absorbed equals the average power delivered by the driving force. Figure 14-25 shows plots of the average power delivered to an oscillator as a function of the driving frequency for two different values of damping. These curves are called **resonance curves**. When the damping is small (large Q), the oscillator absorbs much more energy from the driving force at or near the resonance frequency than it does at other frequencies. The width of the peak of the resonance curve is correspondingly narrow, and we speak of the resonance as being sharp. When the damping is large, the resonance curve is broad. The width $\Delta\omega$ of each resonance curve is indicated in the figure. For relatively small damping, the ratio of the width of the resonance to the frequency can be shown to be equal to the reciprocal of the Q factor (see Problem 130):

$$\frac{\Delta\omega}{\omega_0} = \frac{\Delta f}{f_0} = \frac{1}{Q} \qquad \text{14-45}$$

Resonance width for small damping

Thus, the Q factor is a direct measure of the sharpness of resonance.

You can do a simple experiment to demonstrate resonance. Hold a meter stick at one end so that it acts like a pendulum and then move your hand back and forth horizontally to drive it. Intuitively, you will move your hand back and forth with the natural frequency of the stick, and the amplitude of the oscillations of the stick will be much greater than the amplitude of oscillations of your hand. Now move your hand back and forth at much greater frequency and note the decrease in amplitude of the oscillating stick.

Figure 14-24 An object on a vertical spring can be driven by moving the support up and down.

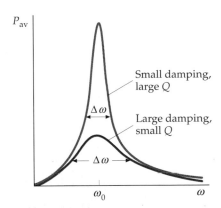

Figure 14-25 Resonance curves. Shown are plots of the average power delivered to an oscillator by a sinusoidal driving force versus the driving angular frequency ω for different values of damping. Resonance occurs when the (angular) frequency of the driving force equals the natural (angular) frequency of the system, ω_0. The resonance is sharp when the damping is small.

*Mathematically, the angular frequency ω is more convenient than the frequency $f = \omega/2\pi$. Since ω and f are proportional, most statements concerning angular frequency also hold for frequency. In verbal descriptions, we usually omit the word angular when the omission will not cause confusion.

There are many familiar examples of resonance. When you sit on a swing, you learn intuitively to pump with the same frequency as the
natural frequency of the swing. Many machines
vibrate because they have rotating parts that are
not in perfect balance. (Observe a washing machine in the spin cycle for an example.) If such a
machine is attached to a structure that can vibrate,
the structure becomes a driven oscillatory system
that is set in motion by the machine. Engineers
pay great attention to balancing the rotary parts of
such machines, damping their vibrations, and iso-
lating them from building supports.

A glass with low damping can be broken by an
intense sound wave at a frequency equal to or
very nearly equal to the natural frequency of vibration of the glass. This is often done in physics
demonstrations using an audio oscillator and an
amplifier.

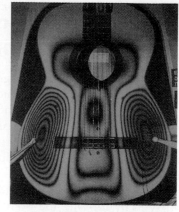

268 Hz ($Q = 52$)   553 Hz ($Q = 66$)

Extended objects have more than one
resonance frequency. When plucked, a
guitar string transmits its energy to the
body of the guitar. The body's oscillations,
coupled to those of the air mass it
encloses, produce the resonance patterns
shown.

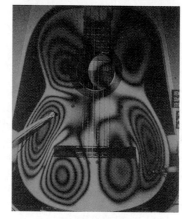
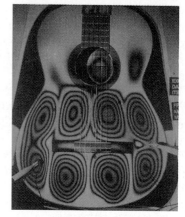

672 Hz ($Q = 61$)   1010 Hz ($Q = 80$)

Mathematical Treatment of Resonance

We can treat a driven oscillator mathematically by assuming that, in addition
to the restoring force and a damping force, the oscillator is subject to an external driving force that varies harmonically with time:

$$F_{ext} = F_0 \cos \omega t \qquad\qquad \textbf{14-46}$$

where ω is the angular frequency of the driving force. This frequency is generally not related to the natural angular frequency of the system ω_0.

Newton's second law applied to an object of mass m attached to a spring
of force constant k and subject to a damping force $-bv$ and an external force
$F_0 \cos \omega t$ gives

$$\sum F = ma = m\frac{dv}{dt}$$

$$-kx - bv + F_0 \cos \omega t = m\frac{dv}{dt}$$

or

$$m\frac{d^2x}{dt^2} + b\frac{dx}{dt} + m\omega_0^2 x = F_0 \cos \omega t \qquad\qquad \textbf{14-47}$$

Differential equation for a driven oscillator

where we have used $k = m\omega_0^2$ and $dv/dt = d^2x/dt^2$.

We will discuss the general solution of Equation 14-47 qualitatively. It consists of two parts, the **transient solution** and the **steady-state solution.** The
transient part of the solution is identical to that for a damped oscillator given

in Equation 14-33. The constants in this part of the solution depend on the initial conditions. Over time, this part of the solution becomes negligible because of the exponential decrease of the amplitude. We are then left with the steady-state solution, which can be written as*

$$x = A \cos(\omega t - \delta)$$
14-48

Position for a driven oscillator

where the angular frequency ω is the same as that of the driving force, and the amplitude A and phase constant δ are given by

$$A = \frac{F_0}{\sqrt{m^2(\omega_0^2 - \omega^2)^2 + b^2\omega^2}}$$
14-49

Amplitude for a driven oscillator

and

$$\tan \delta = \frac{b\omega}{m(\omega_0^2 - \omega^2)}$$
14-50

Phase constant for a driven oscillator

The steady-state solution does not depend on the initial conditions (Figure 14-26). Comparing Equations 14-46 and 14-48, we can see that the displacement and the driving force oscillate with the same frequency, but they differ in phase by δ. When the driving frequency ω is much less than the natural frequency ω_0, $\delta \approx 0$, as can be seen from Equation 14-50. At resonance, $\delta = \pi/2$. When ω is much greater than ω_0, $\delta \approx \pi$. In your simple experiment of driving a meter stick by moving your hand back and forth, you should be able to note that, at resonance, the oscillation of your hand is not in phase with the oscillation of the stick.

The velocity of the object in the steady state is obtained by differentiating x with respect to t:

$$v = \frac{dx}{dt} = -A\omega \sin(\omega t - \delta)$$

At resonance, $\delta = \pi/2$, and the velocity is in phase with the driving force:

$$v = -A\omega \sin\left(\omega t - \frac{\pi}{2}\right) = +A\omega \cos \omega t$$

Thus, at resonance, the object is always moving in the direction of the driving force, as would be expected for maximum power input. The speed is maximum at $\omega = \omega_0$.

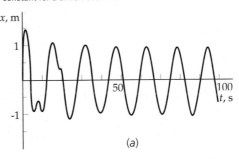

(a)

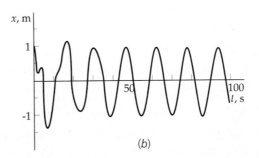

(b)

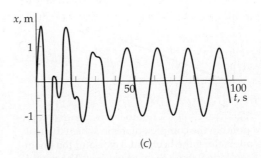

(c)

Figure 14-26 Three solutions of Equation 14-47 for the same driven oscillator under different initial conditions. For each solution, $\omega_0 = 1$ rad/s, $\omega = 0.4$ rad/s, and $Q = 8$. Note that the transient solutions are very different, but the steady-state solutions are all the same.

*The negative sign in the phase of Equation 14-48 is introduced so that the phase constant δ is positive.

*e*xploring

Using Complex Numbers to Solve the Oscillator Equations

Equations 14-32 and 14-47 for the damped and driven oscillator can be solved using complex numbers. A general complex number z can be written

$$z = \alpha + \beta i \qquad 1$$

where α and β are real numbers and $i = \sqrt{-1}$. The number α is called the real part of z, whereas β is called the imaginary part. An important relation for complex numbers is*

$$e^{i\phi} = \cos\phi + i\sin\phi \qquad 2$$

Using Equation 2 and referring to Figure 1, a general complex number can be written in polar form:

$$z = \alpha + \beta i = r\cos\phi + ir\sin\phi$$
$$= r(\cos\phi + i\sin\phi) = re^{i\phi} \qquad 3$$

where $r = |z| = \sqrt{\alpha^2 + \beta^2}$, and $\tan\phi = \beta/\alpha$.

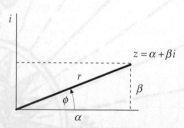

Figure 1 A general complex number $z = \alpha + \beta i$ is represented as a point in the complex plane in which the real part α is plotted along the horizontal axis and the imaginary part β is plotted along the vertical axis. The magnitude $|z| = r$ and the angle ϕ are then the polar coordinates of the point (α, β).

Damped Oscillator

Using $\omega_0^2 = k/m$, and replacing x with z, Equation 14-32 for the damped oscillator can be written

$$\frac{d^2z}{dt^2} + \frac{b}{m}\frac{dz}{dt} + \omega_0^2 z = 0 \qquad 4$$

Consider the complex function $z = x + yi$. Equation 4 is linear, so if the complex function z satisfies the equation, then the real part x (and imaginary part y) must also satisfy the equation. We try the complex function

$$z = Be^{i\omega t} \qquad 5$$

Differentiating with respect to t we obtain

$$\frac{dz}{dt} = i\omega Be^{i\omega t} = i\omega z \qquad 6$$

$$\frac{d^2z}{dt^2} = \frac{d}{dt}\frac{dz}{dt} = (i\omega)^2 Be^{i\omega t} = -\omega^2 z$$

Substituting these results into Equation 4 gives

$$-\omega^2 z + \frac{ib\omega}{m}z + \omega_0^2 z = 0$$

Canceling the common factor z and rearranging, we obtain

$$\omega^2 - \frac{ib}{m}\omega - \omega_0^2 = 0 \qquad 7$$

Equation 7 can be solved using the quadratic formula:

$$\omega = \frac{b}{2m}i \pm \frac{1}{2}\sqrt{\left(\frac{ib}{m}\right)^2 + 4\omega_0^2}$$

$$= \frac{b}{2m}i \pm \omega_0\sqrt{1 - \left(\frac{b}{2m\omega_0}\right)^2} = \frac{b}{2m}i \pm \omega' \qquad 8$$

where

$$\omega' = \omega_0\sqrt{1 - \left(\frac{b}{2m\omega_0}\right)^2}$$

Now substitute this value of ω into Equation 5:

$$z = Be^{i\omega t} = Be^{i(ib/2m \pm \omega')t} = Be^{-(b/2m)t}e^{\pm i\omega't} \qquad 9$$

The real part of z is

$$x = \text{Re}(z) = Be^{-(b/2m)t}\cos\omega't$$

which is the same as Equation 14-33 except for the phase constant δ, which is arbitrary.†

---

* See Appendix D for a brief discussion of complex numbers.

† We could have easily included the arbitrary phase constant by writing B as $B = Ae^{i\delta}$ where A is a real number. Then $z = Be^{i\omega t} = Ae^{i\delta}e^{i\omega t} = Ae^{i\omega t + \delta}$.

Driven Oscillator

To find the steady-state solution for the driven oscillator, we note that the right side of Equation 14-47 is the real part of $F_0 e^{i\omega t}$. We then find the complex function z that satisfies

$$m \frac{d^2 z}{dt^2} + b \frac{dz}{dt} + m\omega_0^2 z = F_0 e^{i\omega t} \qquad 10$$

Then the real part of z must satisfy Equation 14-47. From physical considerations we expect that in the steady state, x will oscillate with frequency ω, so we again try

$$z = Be^{i\omega t} \qquad 11$$

Here, the frequency ω is given, and we wish to determine B. Computing the derivatives as before and substituting them into Equation 10 gives

$$-m\omega^2 z + ib\omega z + m\omega_0^2 z = F_0 e^{i\omega t} = \frac{F_0}{B} z \qquad 12$$

Dividing out the common term z and solving for B, we obtain

$$B = \frac{F_0}{m(\omega_0^2 - \omega^2) + ib\omega} \qquad 13$$

Equation 13 is most easily solved by putting the denominator in polar form.

$$\alpha + \beta i = \sqrt{\alpha^2 + \beta^2}\, e^{i\delta}$$

where $\tan \delta = \beta / \alpha$. Then

$$m(\omega_0^2 - \omega^2) + ib\omega$$
$$= \sqrt{m^2(\omega_0^2 - \omega^2)^2 + b^2\omega^2}\, e^{i\delta}$$

where

$$\tan \delta = \frac{b\omega}{m(\omega_0^2 - \omega^2)} \qquad 14$$

We can then write B as

$$B = \frac{F_0}{\sqrt{m^2(\omega_0^2 - \omega^2)^2 + b^2\omega^2}\, e^{i\delta}}$$

$$= \frac{F_0 e^{-i\delta}}{\sqrt{m^2(\omega_0^2 - \omega^2)^2 + b^2\omega^2}}$$

The complex solution z is then

$$z = Be^{i\omega t} = \frac{F_0 e^{-i\delta}\, e^{i\omega t}}{\sqrt{m^2(\omega_0^2 - \omega^2)^2 + b^2\omega^2}}$$

$$= \frac{F_0 e^{i(\omega t - \delta)}}{\sqrt{m^2(\omega_0^2 - \omega^2)^2 + b^2\omega^2}}$$

or

$$z = Ae^{i(\omega t - \delta)} \qquad 15$$

with

$$A = \frac{F_0}{\sqrt{m^2(\omega_0^2 - \omega^2)^2 + b^2\omega^2}} \qquad 16$$

The real part of Equation 15 is then

$$x = A \cos(\omega t - \delta)$$

with A given by Equation 16 and δ given by Equation 14.

Example 14-13   *try it yourself*

An object of mass 1.5 kg on a spring of force constant 600 N/m loses 3% of its energy in each cycle. The system is driven by a sinusoidal force with a maximum value of $F_0 = 0.5$ N. (*a*) What is Q for this system? (*b*) What is the resonance (angular) frequency? (*c*) If the driving frequency is varied, what is the width $\Delta\omega$ of the resonance? (*d*) What is the amplitude at resonance? (*e*) What is the amplitude if the driving frequency is $\omega = 19$ rad/s?

Picture the Problem We can find Q from $Q = 2\pi/(\Delta E/E)_{\text{cycle}}$ and then use this result to find the width of the resonance $\Delta\omega = \omega_0/Q$. The resonance frequency is the natural frequency. The amplitude can be found from Equation 14-49 both at resonance and off resonance, with the damping constant calculated from Q using Equation 14-41 in the form $b = \omega_0 m/Q$.

Cover the column to the right and try these on your own before looking at the answers.

Steps   **Answers**

(*a*)1. Relate Q to the fractional energy loss. $Q = 2\pi \dfrac{E}{|\Delta E|}$

2. Substitute $\Delta E/E = 3\%$ and calculate Q. $Q = 2\pi \dfrac{100}{3} = 209$

(*b*)   Relate the resonance frequency to the natural
        frequency of the system. $w_0 = \sqrt{k/m} = 20$ rad/s

(*c*)   Relate the width of the resonance $\Delta\omega$ to Q.

                                                             $\Delta w = w_0/Q = 0.0957$ rad/s

(*d*)1. Write an expression for the amplitude A for
        any driving frequency ω. $A = \dfrac{F_0}{\sqrt{m^2(\omega_0^2 - \omega^2)^2 + b^2\omega^2}}$

2. Substitute $\omega = \omega_0$ to calculate A at resonance. $A = \dfrac{F_0}{bw_0}$

3. Use Equation 14-39 to relate the damping
   constant b to Q. $b = \dfrac{m\omega_0}{Q} = \dfrac{(1.5 \text{ kg})(20 \text{ Hz})}{209} = 0.144$ kg/s

4. Use the results of the previous two steps to
   calculate the amplitude at resonance. $A = \dfrac{F_0}{b\omega_0} = \dfrac{0.5 \text{ N}}{(0.144 \text{ kg/s})(20 \text{ Hz})} = 0.174 \text{ m} = 17.4 \text{ cm}$

(*e*)   Calculate the amplitude for $\omega = 19$ Hz. (We
        can omit the units to simplify the equation.
        Since all quantities are in SI units, A will be $A = \dfrac{F_0}{\sqrt{m^2(\omega_0^2 - \omega^2)^2 + b^2\omega^2}}$
        in meters.)
                                                             $= 8.54 \times 10^{-3}$ m $= 0.854$ cm

Remarks At just 1 rad/s off resonance, the amplitude drops by a factor of 20. This is not surprising, because the width of the resonance is only 0.0957 rad/s. Note that off resonance the term $b^2\omega^2$ is negligible compared with the other term in the denominator of the expression for A. When $\omega - \omega_0$ is more than several times the half width $\Delta\omega$, we can neglect that term and calculate A from $A \approx F_0/m(\omega_0^2 - \omega^2)$. Figure 14-27 shows the amplitude versus driving frequency ω.

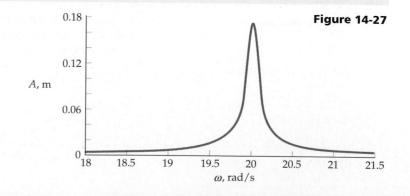

Figure 14-27

Summary

1. Simple harmonic motion occurs whenever the restoring force is proportional to the displacement from equilibrium. It has wide application in the study of oscillations, waves, and electrical circuits.

2. Resonance is an important phenomenon in many areas of physics. It occurs when the frequency of the driving force equals the natural frequency of the oscillating system.

Topic	Remarks and Relevant Equations	
1. Simple Harmonic Motion	In simple harmonic motion, the net force and acceleration are proportional to the displacement and oppositely directed.	
Position function	$x = A \cos(\omega t + \delta)$	14-4
Velocity	$v = -\omega A \sin(\omega t + \delta)$	14-5
Acceleration	$a_x = -\omega^2 x = -\omega^2 A \cos(\omega t + \delta)$	14-6, 14-7
Angular frequency	$\omega = 2\pi f = \dfrac{2\pi}{T}$	14-11
Total energy	$E_{total} = K + U = \frac{1}{2}kA^2$	14-16
Average kinetic or potential energy	$K_{av} = U_{av} = \frac{1}{2}E_{total}$	14-17
Circular motion	When a particle moves in a circle with constant speed, the x and y components of its position vary with simple harmonic motion.	
General motion near equilibrium	When an object is given a small displacement from any position of stable equilibrium, it oscillates about this position with simple harmonic motion	
2. Period for Various Systems		
Mass on a spring	$T = 2\pi\sqrt{\dfrac{m}{k}}$	14-12
Simple pendulum	$T = 2\pi\sqrt{\dfrac{L}{g}}$	14-27
Physical pendulum (optional)	$T = 2\pi\sqrt{\dfrac{I}{MgD}}$	14-31
	where D is the distance of the center of mass to the pivot point and I is the moment of inertia about the pivot point.	
3. Damped Oscillations	In the oscillations of real systems, the motion is damped because of dissipative forces. If the damping is greater than some critical value, the system does not oscillate when disturbed but merely returns to its equilibrium position. The motion of a slightly damped system is nearly simple harmonic with an amplitude that decreases exponentially with time.	
Frequency	$\omega' = \omega_0\sqrt{1 - \left(\dfrac{b}{2m\omega_0}\right)^2} = \omega_0\sqrt{1 - \dfrac{1}{4Q^2}}$	14-35, 14-42
Amplitude	$A = A_0 e^{-(b/2m)t} = A_0 e^{-t/2\tau}$	14-36

Energy	$E = E_0 e^{-(b/m)t} = E_0 e^{-t/\tau}$	14-38
Decay time	$\tau = \dfrac{m}{b}$	14-34
Q Factor	$Q = \omega_0 \tau = \dfrac{\omega_0 m}{b} = \dfrac{2\pi}{(|\Delta E|/E)_{\text{cycle}}}$	14-39, 14-41

4. Driven Oscillations

When a slightly damped system is driven by an external sinusoidal force $F_{\text{ext}} = F_0 \cos \omega t$, the system oscillates with a frequency ω equal to the driving frequency and an amplitude A that depends on the driving frequency.

Resonance frequency	$\omega = \omega_0$	
Resonance width	$\dfrac{\Delta \omega}{\omega_0} = \dfrac{\Delta f}{f_0} = \dfrac{1}{Q}$	14-45
Position function (optional)	$x = A \cos(\omega t - \delta)$	14-48
Amplitude (optional)	$A = \dfrac{F_0}{\sqrt{m^2(\omega_0^2 - \omega^2)^2 + b^2\omega^2}}$	14-49
Phase constant (optional)	$\tan \delta = \dfrac{b\omega}{m(\omega_0^2 - \omega^2)}$	14-50

Problem-Solving Guide

Begin by drawing a neat diagram that includes the important features of the problem.

Summary of Worked Examples

Type of Calculation	Procedure and Relevant Examples	
1. Simple Harmonic Motion		
Find amplitude, phase constant, angular frequency, frequency, and period of simple harmonic motion given the equation for x.	Compare the equation with the standard form $x = A \cos(\omega t + \delta)$.	Example 14-1
Find the velocity and acceleration given the position function x.	Compute the derivatives $v = dx/dt$, and $a = dv/dt = d^2x/dt^2$.	Example 14-1
Find the phase constant and amplitude given the initial position and velocity.	Use $x_0 = A \cos \delta$ and $v_0 = -\omega A \sin \delta$.	Example 14-2
Find the maximum speed and the time that it occurs given $x(t)$.	Use $v = -\omega A \sin \omega t$ and solve for t such that $\sin \omega t = \pm 1$.	Example 14-4
Find x for some given v.	Use conservation of energy.	Example 14-5
2. Object on a Vertical Spring		
Find the frequency of oscillation given the equilibrium stretching of the spring when an object hangs from it.	Calculate $\omega = \sqrt{k/m}$ from $k \Delta y = mg$.	Example 14-6

3.   Pendulums

Find the time lost or gained by a pendulum clock.	If the change is due to an amplitude change, find the percentage change in the period using Equation 14-28.	**Example 14-9**
Find the period of oscillation of a rigid body about some point. (optional)	Use $T = \dfrac{2\pi}{\omega} = 2\pi\sqrt{\dfrac{I}{MgD}}$.	**Example 14-10**

4.   Damped and Driven Oscillators

Find the Q value for an oscillator that loses some fraction of its energy per cycle.	Use $Q = 2\pi E / |\Delta E|$.	**Examples 14-12, 14-13**
Find the amplitude of a driven oscillator. (optional)	Use Equation 14-49.	**Example 14-13**

Problems

In a few problems, you are given more data than you actually need; in a few other problems, you are required to supply data from your general knowledge, outside sources, or informed estimates.

Conceptual Problems

Problems from Optional and Exploring sections

•   Single-concept, single-step, relatively easy
•• Intermediate-level, may require synthesis of concepts
••• Challenging, for advanced students

Simple Harmonic Motion

1   •   Deezo the Clown slept in again. As he roller-skates out the door at breakneck speed on his way to a lunchtime birthday party, his superelastic suspenders catch on a fence post, and he flies back and forth, oscillating with an amplitude A. What distance does he move in one period? What is his displacement over one period?

2   •   A neighbor takes a picture of the oscillating Deezo (from Problem 1) at a moment when his speed is zero. What is his displacement from the fence post at that time?

3   •   What is the magnitude of the acceleration of an oscillator of amplitude A and frequency f when its speed is maximum? When its displacement is maximum?

4   •   Can the acceleration and the displacement of a simple harmonic oscillator ever be in the same direction? The acceleration and the velocity? The velocity and the displacement? Explain.

5   •   True or false:
(a) In simple harmonic motion, the period is proportional to the square of the amplitude.
(b) In simple harmonic motion, the frequency does not depend on the amplitude.
(c) If the acceleration of a particle is proportional to the displacement and oppositely directed, the motion is simple harmonic.

6   •   The position of a particle is given by $x = (7 \text{ cm}) \times \cos 6\pi t$, where t is in seconds. What is (a) the frequency, (b) the period, and (c) the amplitude of the particle's motion? (d) What is the first time after $t = 0$ that the particle is at its equilibrium position? In what direction is it moving at that time?

7   •   (a) What is the maximum speed of the particle in Problem 6? (b) What is its maximum acceleration?

8   •   What is the phase constant δ in Equation 14-4 if the position of the oscillating particle at time $t = 0$ is (a) 0, (b) $-A$, (c) A, (d) $A/2$?

9   •   A particle of mass m begins at rest from $x = +25$ cm and oscillates about its equilibrium position at $x = 0$ with a period of 1.5 s. Write equations for (a) the position x versus the time t, (b) the velocity v versus t, and (c) the acceleration a versus t.

10   •   Find (a) the maximum speed, and (b) the maximum acceleration of the particle in Problem 6. (c) What is the first time that the particle is at $x = 0$ and moving to the right?

11   ••   Work Problem 9 with the particle initially at $x = 25$ cm and moving with velocity $v_0 = +50$ cm/s.

12   ••   The period of an oscillating particle is 8 s, and its amplitude is 12 cm. At $t = 0$, it is at its equilibrium position. Find the distance traveled during the interval (a) $t = 0$ to $t = 2$ s, (b) $t = 2$ s to $t = 4$ s, (c) $t = 0$ to $t = 1$ s, and (d) $t = 1$ s to $t = 2$ s.

13   ••   The period of an oscillating particle is 8 s. At $t = 0$, the particle is at rest at $x = A = 10$ cm. (a) Sketch x as a function of t. (b) Find the distance traveled in the first second, the next second, the third second, and the fourth second after $t = 0$.

14   ••   Military specifications often call for electronic devices to be able to withstand accelerations of $10g = 98.1$ m/s². To make sure that their products meet this specification, manufacturers test them using a shaking table that can vibrate a device at various specified frequencies and amplitudes. If a device is given a vibration of amplitude 1.5 cm, what should its frequency be in order to test for compliance with the 10g military specification?

15 •• The position of a particle is given by $x = 2.5 \cos \pi t$, where x is in meters and t is in seconds. (*a*) Find the maximum speed and maximum acceleration of the particle. (*b*) Find the speed and acceleration of the particle when $x = 1.5$ m.

16 •• The bow of a destroyer undergoes a simple harmonic vertical pitching motion with a period of 8.0 s and an amplitude of 2.0 m. (*a*) What is the maximum vertical velocity of the destroyer's bow? (*b*) What is its maximum acceleration? (*c*) An 80-kg sailor is standing on a scale in the bunkroom in the bow. What are the maximum and minimum readings on the scale in newtons?

Simple Harmonic Motion and Circular Motion

17 • A particle moves in a circle of radius 40 cm with a constant speed of 80 cm/s. Find (*a*) the frequency of the motion, and (*b*) the period of the motion. (*c*) Write an equation for the x component of the position of the particle as a function of time t, assuming that the particle is on the positive x axis at time $t = 0$.

18 • A particle moves in a circle of radius 15 cm, making 1 revolution every 3 s. (*a*) What is the speed of the particle? (*b*) What is its angular velocity ω? (*c*) Write an equation for the x component of the position of the particle as a function of time t, assuming that the particle is on the positive x axis at time $t = 0$.

Energy in Simple Harmonic Motion

19 • If the amplitude of a simple harmonic oscillator is tripled, by what factor is the energy changed?

20 •• An object attached to a spring has simple harmonic motion with an amplitude of 4.0 cm. When the object is 2.0 cm from the equilibrium position, what fraction of its total energy is potential energy?

(*a*) one-quarter   (*b*) one-third   (*c*) one-half
(*d*) two-thirds   (*e*) three-quarters

21 • A 2.4-kg object is attached to a horizontal spring of force constant $k = 4.5$ kN/m. The spring is stretched 10 cm from equilibrium and released. Find its total energy.

22 • Find the total energy of a 3-kg object oscillating on a horizontal spring with an amplitude of 10 cm and a frequency of 2.4 Hz.

23 • A 1.5-kg object oscillates with simple harmonic motion on a spring of force constant $k = 500$ N/m. Its maximum speed is 70 cm/s. (*a*) What is the total energy? (*b*) What is the amplitude of the oscillation?

24 • A 3-kg object oscillating on a spring of force constant 2 kN/m has a total energy of 0.9 J. (*a*) What is the amplitude of the motion? (*b*) What is the maximum speed?

25 • An object oscillates on a spring with an amplitude of 4.5 cm. Its total energy is 1.4 J. What is the force constant of the spring?

26 •• A 3-kg object oscillates on a spring with an amplitude of 8 cm. Its maximum acceleration is 3.50 m/s^2. Find the total energy.

Springs

27 • True or false:
(*a*) For a given object on a given spring, the period is the same if the spring is vertical or horizontal.
(*b*) For a given object oscillating with amplitude A on a given spring, the maximum speed is the same if the spring is vertical or horizontal.

28 • Herb plans to ring in the new year by playing trombone while oscillating up and down on a spring that hangs down from a building at Times Square in New York City. He intends to oscillate with a period of one second to synchronize with the crowd as it counts down to midnight. If he uses a spring with a spring constant of 3000 N/m, Herb must be sure that the total of his vibrating mass adds up to

(*a*) 3000 kg.   (*b*) $\sqrt{3000}$ kg.   (*c*) $4\pi^2(3000)$ kg.
(*d*) $3000/4\pi^2$ kg.   (*e*) none of the above.

29 • A 2.4-kg object is attached to a horizontal spring of force constant $k = 4.5$ kN/m. The spring is stretched 10 cm from equilibrium and released. Find (*a*) the frequency of the motion, (*b*) the period, (*c*) the amplitude, (*d*) the maximum speed, and (*e*) the maximum acceleration. (*f*) When does the object first reach its equilibrium position? What is its acceleration at this time?

30 • Answer the questions in Problem 29 for a 5-kg object attached to a spring of force constant $k = 700$ N/m when the spring is initially stretched 8 cm from equilibrium.

31 • A 3-kg object attached to a horizontal spring oscillates with an amplitude $A = 10$ cm and a frequency $f = 2.4$ Hz. (*a*) What is the force constant of the spring? (*b*) What is the period of the motion? (*c*) What is the maximum speed of the object? (*d*) What is the maximum acceleration of the object?

32 • An 85-kg person steps into a car of mass 2400 kg, causing it to sink 2.35 cm on its springs. Assuming no damping, with what frequency will the car and passenger vibrate on the springs?

33 • A 4.5-kg object oscillates on a horizontal spring with an amplitude of 3.8 cm. Its maximum acceleration is 26 m/s^2. Find (*a*) the force constant k, (*b*) the frequency, and (*c*) the period of the motion.

34 • An object oscillates with an amplitude of 5.8 cm on a horizontal spring of force constant 1.8 kN/m. Its maximum speed is 2.20 m/s. Find (*a*) the mass of the object, (*b*) the frequency of the motion, and (*c*) the period of the motion.

35 •• A 0.4-kg block attached to a spring of force constant 12 N/m oscillates with an amplitude of 8 cm. Find (*a*) the maximum speed of the block, (*b*) the speed and acceleration of the block when it is at $x = 4$ cm from the equilibrium position, and (*c*) the time it takes the block to move from $x = 0$ to $x = 4$ cm.

36 •• An object of mass m is supported by a vertical spring of force constant 1800 N/m. When pulled down 2.5 cm from equilibrium and released from rest, the object oscillates at 5.5 Hz. (*a*) Find m. (*b*) Find the amount the spring is stretched from its natural length when the object is in equilibrium. (*c*) Write expressions for the displacement x, the velocity v, and the acceleration a as functions of time t.

37 •• An object of unknown mass is hung on the end of an unstretched spring and is released from rest. If the object falls 3.42 cm before first coming to rest, find the period of the motion.

38 •• A spring of force constant $k = 250$ N/m is suspended from a rigid support. An object of mass 1 kg is attached to the unstretched spring and the object is released from rest. (*a*) How far below the starting point is the equilibrium position for the object? (*b*) How far down does the object move before it starts up again? (*c*) What is the period of oscillation? (*d*) What is the speed of the object when it first reaches its equilibrium position? (*e*) When does it first reach its equilibrium position?

39 •• The St. Louis Arch has a height of 192 m. Suppose a stunt woman of mass 60 kg jumps off the top of the arch with an elastic band attached to her feet. She reaches the ground at zero speed. Find her kinetic energy K after 2.00 s of the flight. (Assume that the elastic band obeys Hooke's law, and neglect its length when relaxed.)

40 •• A 0.12-kg block is suspended from a spring. When a small stone of mass 30 g is placed on the block, the spring stretches an additional 5 cm. With the stone on the block, the spring oscillates with an amplitude of 12 cm. (*a*) What is the frequency of the motion? (*b*) How long does the block take to travel from its lowest point to its highest point? (*c*) What is the net force of the stone when it is at a point of maximum upward displacement?

41 •• In Problem 40, find the maximum amplitude of oscillation such that the stone will remain on the block.

42 •• An object of mass 2.0 kg is attached to the top of a vertical spring that is anchored to the floor. The uncompressed length of the spring is 8.0 cm, and the equilibrium position of the object on the spring is 5.0 cm from the floor. When the object is resting at its equilibrium position, it is given a downward impulse with a hammer such that its initial speed is 0.3 m/s. (*a*) To what maximum height above the floor does the object eventually rise? (*b*) How long does it take for the object to reach its maximum height the first time? (*c*) Does the spring ever become uncompressed? What minimum initial velocity must be given to the object for the spring to be uncompressed at some time?

43 •• Lou has devised a new kiddie ride and is testing it for safety. A child is placed on a large block that is attached to a horizontal spring. When pulled back and released, the child and block oscillate with a period of 2 s. (*a*) If the coefficient of static friction between the child and the block is 0.25, will an amplitude of 1 m cause the child to slip? (*b*) What is the maximum amplitude that will avoid slipping?

Energy of an Object on a Vertical Spring

44 •• A 2.5-kg object hanging from a vertical spring of force constant 600 N/m oscillates with an amplitude of 3 cm. When the object is at its maximum downward displacement, find (*a*) the total energy of the system, (*b*) the gravitational potential energy, and (*c*) the potential energy in the spring. (*d*) What is the maximum kinetic energy of the object? Choose $U = 0$ when the object is in equilibrium.

45 •• A 1.5-kg object that stretches a spring 2.8 cm from its natural length when hanging at rest oscillates with an amplitude of 2.2 cm. (*a*) Find the total energy of the system. (*b*) Find the gravitational potential energy at maximum downward displacement. (*c*) Find the potential energy in the spring at maximum downward displacement. (*d*) What is the maximum kinetic energy of the object? (Choose $U = 0$ when the object is in equilibrium.)

46 •• A 1.2-kg object hanging from a spring of force constant 300 N/m oscillates with a maximum speed of 30 cm/s. (*a*) What is its maximum displacement? When the object is at its maximum displacement, find (*b*) the total energy of the system, (*c*) the gravitational potential energy, and (*d*) the potential energy in the spring. (Choose $U = 0$ when the object is in equilibrium.)

Simple Pendulums

47 • True or false: The motion of a simple pendulum is simple harmonic for any initial angular displacement.

48 • True or false: The motion of a simple pendulum is periodic for any initial angular displacement.

49 •• The length of the string or wire supporting a pendulum increases slightly when its temperature is raised. How would this affect a clock operated by a simple pendulum?

50 • Find the length of a simple pendulum if the period is 5 s at a point where $g = 9.81$ m/s^2.

51 • What would be the period of the pendulum in Problem 50 if the pendulum were on the moon, where the acceleration due to gravity is one-sixth that on earth?

52 • If the period of a pendulum 70 cm long is 1.68 s, what is the value of g at the location of the pendulum?

53 • A pendulum set up in the stairwell of a 10-story building consists of a heavy weight suspended on a 34.0-m wire. If $g = 9.81$ m/s^2, what is the period of oscillation?

54 •• Show that the total energy of a simple pendulum undergoing oscillations of small amplitude ϕ_0 is approximately $E \approx 1/2mgL\phi_0^2$ (*Hint:* Use the approximation $\cos \phi \approx 1 - \phi^2/2$ for small ϕ.)

55 •• A simple pendulum of length L is attached to a cart that slides without friction down a plane inclined at angle θ with the horizontal as shown (Figure 14-28). Find the period of oscillation of the pendulum on the sliding cart.

Figure 14-28 Problem 55

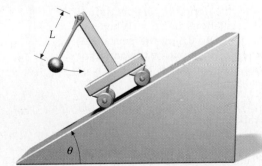

56 •• A simple pendulum of length L is released from rest from an angle ϕ_0. (a) Assuming that the pendulum undergoes simple harmonic motion, find its speed as it passes through $\phi = 0$. (b) Using the conservation of energy, find this speed exactly. (c) Show that your results for (a) and (b) are the same when ϕ_0 is small. (d) Find the difference in your results for $\phi_0 = 0.20$ rad and $L = 1$ m.

Physical Pendulums (optional)

57 • A thin disk of mass 5 kg and radius 20 cm is suspended by a horizontal axis perpendicular to the disk through its rim. The disk is displaced slightly from equilibrium and released. Find the period of the subsequent simple harmonic motion.

58 • A circular hoop of radius 50 cm is hung on a narrow horizontal rod and allowed to swing in the plane of the hoop. What is the period of its oscillation, assuming that the amplitude is small?

59 • A 3-kg plane figure is suspended at a point 10 cm from its center of mass. When it is oscillating with small amplitude, the period of oscillation is 2.6 s. Find the moment of inertia I about an axis perpendicular to the plane of the figure through the pivot point.

60 •• Figure 14-29 shows a dumbbell with two equal masses (to be considered as point masses) attached to a very thin (massless) rod of length L. (a) Show that the period of this pendulum is a minimum when the pivot point P is at one of the masses. (b) Find the period of this physical pendulum if the distance between P and the upper mass is $L/4$.

Figure 14-29
Problem 60

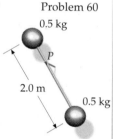

61 •• Suppose the rod in Problem 60 has a mass of $2m$ (Figure 14-30). Determine the distance between the upper mass and the pivot point P such that the period of this physical pendulum is a minimum.

Figure 14-30
Problem 61

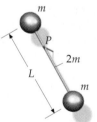

62 •• You are given a meter stick and asked to drill a hole in it so that when pivoted about the hole the period of the pendulum will be a minimum. Where should you drill the hole?

63 •• An irregularly shaped plane object of mass 3.2 kg is suspended by a thin rod of adjustable length and is free to swing in the plane of the object (Figure 14-31). When the length of the supporting rod is 1.0 m, the period of this pendulum for small oscillations is 2.6 s. When the rod is shortened to 0.8 m, the period decreases to 2.5 s. What will be the period of this physical pendulum if the length of the rod is 0.5 m?

Figure 14-31
Problem 63

64 •• When a short person and a tall person walk together at the same speed, the short person will take more steps. Consider the leg to be a physical pendulum

that swings about the hip joint. Estimate the natural frequency of this pendulum for a person of your height, and compare the result with the rate at which you take steps when walking in a leisurely manner.

Figure 14-32
Problem 65

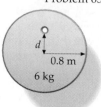

65 •• Figure 14-32 shows a uniform disk of radius $R = 0.8$ m and a 6-kg mass with a small hole a distance d from the disk's center that can serve as a pivot point. (a) What should be the distance d so that the period of this physical pendulum is 2.5 s? (b) What should be the distance d so that this physical pendulum will have the shortest possible period? What is this period?

66 ••• A plane object has moment of inertia I about its center of mass. When pivoted at point P_1, as shown in Figure 14-33, it oscillates about the pivot with a period T. There is a second point P_2 on the opposite side of the center of mass about which the object can be pivoted so that the period of oscillation is also T. Show that $h_1 + h_2 = gT^2/4\pi^2$.

Figure 14-33
Problem 66

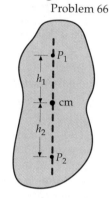

67 ••• A physical pendulum consists of a spherical bob of radius r and mass m suspended from a string (Figure 14-34). The distance from the center of the sphere to the point of support is L. When r is much less than L, such a pendulum is often treated as a simple pendulum of length L. (a) Show that the period for small oscillations is given by

Figure 14-34
Problem 67

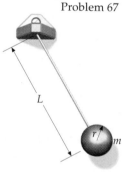

$$T = T_0\sqrt{1 + \frac{2r^2}{5L^2}}$$

where $T_0 = 2\pi\sqrt{L/g}$ is the period of a simple pendulum of length L. (b) Show that when r is much smaller than L, the period is approximately $T \approx T_0(1 + r^2/5L^2)$. (c) If $L = 1$ m and $r = 2$ cm, find the error when the approximation $T = T_0$ is used for this pendulum. How large must the radius of the bob be for the error to be 1%?

68 ••• Figure 14-35 shows the pendulum of a clock. The uniform rod of length $L = 2.0$ m has a mass $m = 0.8$ kg. Attached to the rod is a disk of mass $M = 1.2$ kg and radius 0.15 m. The clock is constructed to keep perfect time if the period of the pendulum is exactly 3.50 s. (a) What should be the distance d so that the period of this pendulum is 2.50 s? (b) Suppose that the pendulum clock loses 5.0 min per day. How far and in what direction should the disk be moved to ensure that the clock will keep perfect time?

Figure 14-35
Problem 68

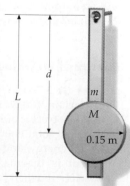

Clocks

69 •• Two clocks have simple pendulums of identical lengths L. The pendulum of clock A swings through an arc of 10°; that of clock B swings through an arc of 5°. When the two clocks are compared one will find that

(a) A runs slow compared to B.
(b) A runs fast compared to B.
(c) both clocks keep the same time.
(d) the answer depends on the length L.

70 •• A simple-pendulum clock keeps accurate time when its length is L. If the length is increased a small amount, how will the accuracy of the clock be affected?

(a) The clock will lose time.
(b) The clock will gain time.
(c) The clock will continue to keep accurate time.
(d) The answer cannot be determined without knowing the original length of the pendulum.
(e) The answer cannot be determined without knowing the percent increase in the length of the pendulum.

71 •• A pendulum clock loses 48 s per day when the amplitude of the pendulum is 8.4°. What should be the amplitude of the pendulum so that the clock keeps perfect time?

72 •• A pendulum clock that has run down to a very small amplitude gains 5 min each day. What angular amplitude should the pendulum have to keep the correct time?

Damped Oscillations

73 • True or false: The energy of a damped, undriven oscillator decreases exponentially with time.

74 • Show that the dampening constant, b, has units of kg/s.

75 • An oscillator has a Q factor of 200. By what percentage does its energy decrease during one period?

76 • A 2-kg object oscillates with an initial amplitude of 3 cm on a spring of force constant $k = 400$ N/m. Find (a) the period, and (b) the total initial energy. (c) If the energy decreases by 1% per period, find the damping constant b and the Q factor.

77 •• Show that the ratio of the amplitudes for two successive oscillations is constant for a damped oscillator.

78 •• An oscillator has a period of 3 s. Its amplitude decreases by 5% during each cycle. (a) By how much does its energy decrease during each cycle? (b) What is the time constant τ? (c) What is the Q factor?

79 •• An oscillator has a Q factor of 20. (a) By what fraction does the energy decrease during each cycle? (b) Use Equation 14-35 to find the percentage difference between ω' and ω_0. (Hint: Use the approximation $(1 + x)^{1/2} \approx 1 + \frac{1}{2}x$ for small x.)

80 •• For a child on a swing, the amplitude drops by a factor of $1/e$ in about eight periods if no energy is fed in. Estimate the Q factor for this system.

81 •• A damped mass–spring system oscillates at 200 Hz. The time constant of the system is 2.0 s. At $t = 0$, the amplitude of oscillation is 6.0 cm and the energy of the oscillating system is then 60 J. (a) What are the amplitudes of oscillation at $t = 2.0$ s and at $t = 4.0$ s? (b) How much energy is dissipated in the first 2-s interval and in the second 2-s interval?

82 •• It has been stated that the vibrating earth has a resonance period of 54 min and a Q factor of about 400 and that after a large earthquake, the earth "rings" (continues to vibrate) for about 2 months. (a) Find the percentage of the energy of vibration lost to damping forces during each cycle. (b) Show that after n periods, the energy is $E_n = (0.984)^n E_0$, where E_0 is the original energy. (c) If the original energy of vibration of an earthquake is E_0, what is the energy after 2 days?

83 ••• A 3-kg sphere dropped through air has a terminal speed of 25 m/s. (Assume that the drag force is $-bv$.) Now suppose the sphere is attached to a spring of force constant $k = 400$ N/m, and that it oscillates with an initial amplitude of 20 cm. (a) What is the time constant τ? (b) When will the amplitude be 10 cm? (c) How much energy will have been lost when the amplitude is 10 cm?

Driven Oscillations and Resonance

84 • True or false:

(a) Resonance occurs when the driving frequency equals the natural frequency.
(b) If the Q value is high, the resonance is sharp.

85 • Give some examples of common systems that can be considered to be driven oscillators.

86 • A crystal wineglass shattered by an intense sound is an example of

(a) resonance.
(b) critical damping.
(c) an exponential decrease in energy.
(d) overdamping.

87 • Find the resonance frequency for each of the three systems shown in Figure 14-36.

Figure 14-36 Problem 87

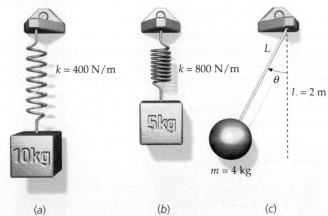

(a) (b) (c)

88 • A damped oscillator loses 2% of its energy during each cycle. (*a*) What is its Q factor? (*b*) If its resonance frequency is 300 Hz, what is the width of the resonance curve $\Delta\omega$ when the oscillator is driven?

89 •• A 2-kg object oscillates on a spring of force constant $k = 400$ N/m. The damping constant has a value of $b = 2.00$ kg/s. The system is driven by a sinusoidal force of maximum value 10 N and angular frequency $\omega = 10$ rad/s. (*a*) What is the amplitude of the oscillations? (*b*) If the driving frequency is varied, at what frequency will resonance occur? (*c*) What is the amplitude of oscillation at resonance? (*d*) What is the width of the resonance curve $\Delta\omega$?

90 •• A damped oscillator loses 3.5% of its energy during each cycle. (*a*) How many cycles elapse before half of its original energy is dissipated? (*b*) What is its Q factor? (*c*) If the natural frequency is 100 Hz, what is the width of the resonance curve when the oscillator is driven?

91 •• Tarzan is depressed again. He ties a vine to his ankle and swings upside-down with a period of 3 s as he contemplates his troubles. Cheetah the chimpanzee pushes him so that the amplitude remains constant. Tarzan's mass is 90 kg and his speed at the bottom of the swing is 2.0 m/s. (*a*) What is Tarzan's total energy? (*b*) If $Q = 20$, how much energy is dissipated during each oscillation? (*c*) What is Cheetah's power input? (*Note:* Pushing a swing is usually not done sinusoidally. However, to maintain a steady amplitude, the energy lost per cycle due to damping must be replaced by an external energy source.)

Collisions

92 •• Peter lays his jack-in-the-box on its side with the lid open, so that Jack, a painted 0.4-kg clown, sticks out horizontally at the end of a spring. Peter then takes a 0.6-kg wad of putty, places it in his favorite slingshot, and fires it at the top of Jack's head. The putty sticks to the clown's head, and the clown and putty oscillate with an amplitude of 16 cm and a frequency of 0.38 Hz. Assuming that the box remains immobile, determine (*a*) the putty's speed before the collision, and (*b*) the spring constant.

93 ••• Figure 14-37 shows a vibrating mass–spring system supported on a frictionless surface and a second equal mass that is moving toward the vibrating mass with velocity v. The motion of the vibrating mass is given by

$$x(t) = (0.1 \text{ m}) \cos(40 \text{ s}^{-1}t)$$

where x is the displacement of the mass from its equilibrium position. The two masses collide elastically just as the vibrating mass passes through its equilibrium position traveling to the right. (*a*) What should be the velocity v of the second mass so that the mass–spring system is at rest following the elastic collision? (*b*) What is the velocity of the second mass after the elastic collision?

Figure 14-37 Problem 93

94 ••• Following the elastic collision in Problem 93, the energy of the recoiling mass is 8.0 J. Find the masses m and the spring constant k.

95 ••• An object of mass 2 kg resting on a frictionless horizontal surface is attached to a spring of force constant 600 N/m. A second object of mass 1 kg slides along the surface toward the first object at 6 m/s. (*a*) Find the amplitude of oscillation if the objects make a perfectly inelastic collision and remain together on the spring. What is the period of oscillation? (*b*) Find the amplitude and period of oscillation if the collision is elastic. (*c*) For each type of collision, write an expression for the position x as a function of time t for the object attached to the spring, assuming that the collision occurs at time $t = 0$.

General Problems

96 • The effect of the mass of a spring on the motion of an object attached to it is usually neglected. Describe qualitatively its effect when it is not neglected.

97 •• A lamp hanging from the ceiling of the club car in a train oscillates with period T_0 when the train is at rest. The period will be (match left and right columns)

1. greater than T_0 when   A. the train moves horizontally with constant velocity.

2. less than T_0 when   B. the train rounds a curve of radius R with speed v.

3. equal to T_0 when   C. the train climbs a hill of inclination θ at constant speed.

D. the train goes over a hill of radius of curvature R with constant speed.

98 •• Two mass–spring systems oscillate at frequencies f_A and f_B. If $f_A = 2f_B$ and the spring constants of the two springs are equal, it follows that the masses are related by
(*a*) $M_A = 4M_B$.   (*b*) $M_A = M_B/\sqrt{2}$.
(*c*) $M_A = M_B/2$.   (*d*) $M_A = M_B/4$.

99 •• Two mass–spring systems A and B oscillate so that their energies are equal. If $M_A = 2M_B$, then which formula below relates the amplitudes of oscillation?
(*a*) $A_A = A_B/4$ (*b*) $A_A = A_B/\sqrt{2}$ (*c*) $A_A = A_B$
(*d*) Not enough information is given to determine the ratio of the amplitudes.

100 •• Two mass–spring systems A and B oscillate so that their energies are equal. If $k_A = 2k_B$, then which formula below relates the amplitudes of oscillation?
(*a*) $A_A = A_B/4$ (*b*) $A_A = A_B/\sqrt{2}$ (*c*) $A_A = A_B$
(*d*) Not enough information is given to determine the ratio of the amplitudes.

101 •• Pendulum A has a bob of mass M_A and a length L_A; pendulum B has a bob of mass M_B and a length L_B. If the period of A is twice that of B, then
(*a*) $L_A = 2L_B$ and $M_A = 2M_B$.
(*b*) $L_A = 4L_B$ and $M_A = M_B$.
(*c*) $L_A = 4L_B$ whatever the ratio M_A/M_B.
(*d*) $L_A = \sqrt{2}L_B$ whatever the ratio M_A/M_B.

102 • A particle has a displacement $x = 0.4 \cos(3t + \pi/4)$, where x is in meters and t is in seconds. (a) Find the frequency f and period T of the motion. (b) Where is the particle at $t = 0$? (c) Where is the particle at $t = 0.5$ s?

103 • (a) Find an expression for the velocity of the particle whose position is given in Problem 102. (b) What is the velocity at time $t = 0$? (c) What is the maximum velocity? (d) At what time after $t = 0$ does this maximum velocity first occur?

104 • An object on a horizontal spring oscillates with a period of 4.5 s. If the object is suspended from the spring vertically, by how much is the spring stretched from its natural length when the object is in equilibrium?

105 •• A small particle of mass m slides without friction in a spherical bowl of radius r. (a) Show that the motion of the particle is the same as if it were attached to a string of length r. (b) Figure 14-38 shows a particle of mass m_1 that is displaced a small distance s_1 from the bottom of the bowl, where s_1 is much smaller than r. A second particle of mass m_2 is displaced in the opposite direction a distance $s_2 = 3s_1$, where s_2 is also much smaller than r. If the particles are released at the same time, where do they meet? Explain.

Figure 14-38 Problem 105

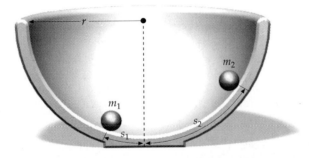

106 •• As your jet plane speeds down the runway on takeoff, you measure its acceleration by suspending your yo-yo as a simple pendulum and noting that when the bob (mass 40 g) is at rest relative to you, the string (length 70 cm) makes an angle of 22° with the vertical. Find the period T for small oscillations of this pendulum.

107 •• Two identical blocks placed one on top of the other rest on a frictionless horizontal air track. The lower block is attached to a spring of spring constant $k = 600$ N/m. When displaced slightly from its equilibrium position, the system oscillates with a frequency of 1.8 Hz. When the amplitude of oscillation exceeds 5 cm, the upper block starts to slide relative to the lower one. (a) What are the masses of the two blocks? (b) What is the coefficient of static friction between the two blocks?

108 •• Two atoms are bound together in a molecule. The potential energy U resulting from their interaction is shown in Figure 14-39. The variable r is the distance between the atom centers, and E_0 is the lowest (ground-state) energy.

(a) As a result of a collision, the molecule acquires a kinetic energy of vibration whose maximum value is 1.0 eV. With this kinetic energy, over what range of separation distance will the molecule vibrate?

(b) Give an approximate value for the force $F(r)$ between the two atoms at $r = 0.4$ nm. Express your answer in units appropriate to those used on the graph.

(c) Calculate the force in (b) in newtons. Is this force attractive or repulsive?

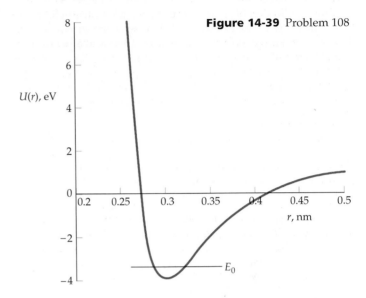

Figure 14-39 Problem 108

109 •• A wooden cube with edge a and mass m floats in water with one of its faces parallel to the water surface. The density of the water is ρ. Find the period of oscillation in the vertical direction if it is pushed down slightly.

110 •• A spider of mass 0.36 g sits in the middle of its horizontal web, which sags 3.00 mm under its weight. Estimate the frequency of vertical vibration for this system.

111 •• A clock with a pendulum keeps perfect time on the earth's surface. In which case will the error be greater: if the clock is placed in a mine of depth h or if the clock is elevated to a height h? Assume that $h \ll R_E$.

112 •• Figure 14-40 shows a pendulum of length L with a bob of mass M. The bob is attached to a spring of spring constant k as shown. When the bob is directly below the pendulum support, the spring is at its equilibrium length. (a) Derive an expression for the period of this oscillating system for small amplitude vibrations. (b) Suppose that $M = 1$ kg and L is such that in the absence of the spring the period is 2.0 s. What is the spring constant k if the period of the oscillating system is 1.0 s?

Figure 14-40
Problem 112

113 •• An object of mass m_1 sliding on a frictionless horizontal surface is attached to a spring of force constant k and oscillates with an amplitude A. When the spring is at its greatest extension and the mass is instantaneously at rest, a second object of mass m_2 is placed on top of it. (a) What is the smallest value for the coefficient of static friction μ_s such that the second object does not slip on the first? (b) Explain how the total energy E, the amplitude A, the angular frequency ω, and the period T of the system are changed by placing m_2 on m_1, assuming that the friction is great enough so that there is no slippage.

114 •• The acceleration due to gravity g varies with geographical location because of the earth's rotation and because the earth is not exactly spherical. This was first discovered in the seventeenth century, when it was noted that a pendulum clock carefully adjusted to keep correct time in Paris lost about 90 s per day near the equator. (a) Show that a small change in the acceleration of gravity Δg produces a small change in the period ΔT of a pendulum given by

$$\frac{\Delta T}{T} \approx -\frac{1}{2}\frac{\Delta g}{g}$$

(Use differentials to approximate ΔT and Δg.) (b) How great a change in g is needed to account for a change in the period of 90 s per day?

115 •• Figure 14-41 shows two equal masses of 0.6 kg glued to each other and connected to a spring of spring constant $k = 240$ N/m. The masses, which rest on a frictionless horizontal surface, are displaced 0.6 m from their equilibrium position and released. Before being released, a few drops of solvent are deposited on the glue. (a) Find the frequency of vibration and total energy of the vibrating system before the glue has dissolved. (b) Find the frequency, amplitude of vibration, and energy of the vibrating system if the glue dissolves when the spring is (1) at maximum compression and (2) at maximum extension.

Figure 14-41
Problem 115

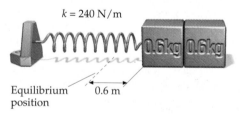

$k = 240$ N/m

0.6kg 0.6kg

Equilibrium position   0.6 m

116 •• Show that for the situations in Figures 14-42a and b, the object oscillates with a frequency $f = [1/(2\pi)]\sqrt{k_{eff}/m}$, where k_{eff} is given by (a) $k_{eff} = k_1 + k_2$ and (b) $1/k_{eff} = 1/k_1 + 1/k_2$. (*Hint:* Find the net force F on the object for a small displacement x and write $F = -k_{eff}x$. Note that in (b) the springs stretch by different amounts, the sum of which is x.)

Figure 14-42 Problem 116

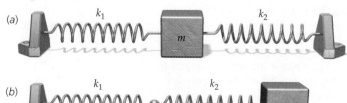

(a) k_1 m k_2

(b) k_1 k_2 m

117 •• A small block of mass m_1 rests on a piston that is vibrating vertically with simple harmonic motion given by $y = A \sin \omega t$. (a) Show that the block will leave the piston if $\omega^2 A > g$. (b) If $\omega^2 A = 3g$ and $A = 15$ cm, at what time will the block leave the piston?

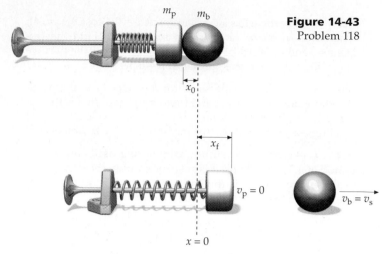

Figure 14-43
Problem 118

m_p m_b

x_0

x_f

$v_p = 0$ $v_b = v_s$

$x = 0$

118 •• The plunger of a pinball machine has mass m_p and is attached to a spring of force constant k (Figure 14-43). The spring is compressed a distance x_0 from its equilibrium position $x = 0$ and released. A ball of mass m_b is next to the plunger. (a) Where does the ball leave the plunger? (b) What is the speed v_s of the ball when it separates? (c) At what distance x_f does the plunger come to rest momentarily? (Assume that the surface is horizontal and frictionless so that the ball slides rather than rolls.)

119 •• A level platform vibrates horizontally with simple harmonic motion with a period of 0.8 s. (a) A box on the platform starts to slide when the amplitude of vibration reaches 40 cm; what is the coefficient of static friction between the body and the platform? (b) If the coefficient of friction between the box and platform were 0.40, what would be the maximum amplitude of vibration before the box would slip?

120 ••• The potential energy of a mass m as a function of position is given by $U(x) = U_0(\alpha + 1/\alpha)$, where $\alpha = x/a$ and a is a constant. (a) Plot $U(x)$ versus x for $0.1a < x < 3a$. (b) Find the value of $x = x_0$ at stable equilibrium. (c) Write the potential energy $U(x)$ for $x = x_0 + \varepsilon$, where ε is a small displacement from the equilibrium position x_0. (d) Approximate the $1/x$ term using the binomial expansion

$$(1 + r)^n = 1 + nr + \frac{n(n-1)}{(2)(1)}r^2 + \frac{n(n-1)(n-2)}{(3)(2)(1)}r^3 + \cdots$$

with $r = \varepsilon/x_0 \ll 1$ and discarding all terms of power greater than r^2. (e) Compare your result with the potential for a simple harmonic oscillator. Show that the mass will undergo simple harmonic motion for small displacements from equilibrium and determine the frequency of this motion.

121 ••• Do Problem 120 with $U(x) = U_0(\alpha^2 + 1/\alpha^2)$.

122 ••• A solid cylindrical drum of mass 6.0 kg and diameter 0.06 m rolls without slipping on a horizontal surface (Figure 14-44). The axle of the drum is attached to a spring of spring constant $k = 4000$ N/m as shown. (a) Determine the frequency of oscillation of this system for small displacements from equilibrium. (b) What is the minimum value of the coefficient of static friction such that the drum will not slip when the vibrational energy is 5.0 J?

Figure 14-44 Problem 122

123 ••• Figure 14-45 shows a solid half-cylinder of mass M and radius R resting on a horizontal surface. If one side of this cylinder is pushed down slightly and then released, the object will oscillate about its equilibrium position. Determine the period of this oscillation.

124 ••• Repeat Problem 123 replacing the half-cylinder with a half-sphere.

Figure 14-45
Problems 123 and 124

Figure 14-46 Problem 125

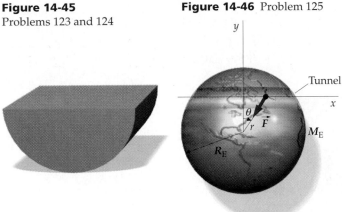

125 ••• A straight tunnel is dug through the earth as shown in Figure 14-46. Assume that the walls of the tunnel are frictionless. (a) The gravitational force exerted by the earth on a particle of mass m at a distance r from the center of the earth when $r < R_E$ is $F_r = -(GmM_E/R_E^3)r$, where M_E is the mass of the earth and R_E is its radius. Show that the net force on a particle of mass m at a distance x from the middle of the tunnel is given by $F_x = -(GmM_E/R_E^3)x$, and that the motion of the particle is therefore simple harmonic motion. (b) Show that the period of the motion is given by $T = 2\pi\sqrt{R_E/g}$ and find its value in minutes. (This is the same period as that of a satellite orbiting near the surface of the earth and is independent of the length of the tunnel.)

126 ••• A damped oscillator has a frequency ω' that is 10% less than its undamped frequency. (a) By what factor is the amplitude of the oscillator decreased during each oscillation? (b) By what factor is its energy reduced during each oscillation?

127 ••• Show by direct substitution that Equation 14-48 is a solution of Equation 14-47.

128 ••• A block of mass m on a horizontal table is attached to a spring of force constant k as shown in Figure 14-47. The coefficient of kinetic friction between the block and the table is μ_k. The spring is stretched a distance A and released. (a) Apply Newton's second law to the block to obtain an equation for its acceleration d^2x/dt^2 for the first half-cycle, during which the block is moving to the left. Show that the resulting equation can be written $d^2x'/dt^2 = -\omega^2x'$, where $x = 0$ at the equilibrium position of the spring, and $x' = x - x_0$, with $x_0 = \mu_k mg/k = \mu_k g/\omega^2$. (b) Repeat part (a) for the second half-cycle as the block moves to the right, and show that $d^2x''/dt^2 = -\omega^2x''$, where $x'' = x + x_0$ and x_0 has the same value. (c) Sketch the first few cycles for $A = 10x_0$.

Figure 14-47 Problem 128

129 ••• In this problem, you will derive the expression for the average power delivered by a driving force to a driven oscillator (Figure 14-25).

(a) Show that the instantaneous power input of the driving force is given by

$$P = Fv = -A\omega F_0 \cos \omega t \sin(\omega t - \delta)$$

(b) Use the trigonometric identity $\sin(\theta_1 - \theta_2) = \sin\theta_1 \cos\theta_2 - \cos\theta_1 \sin\theta_2$ to show that the equation in (a) can be written

$$P = A\omega F_0 \sin\delta \cos^2\omega t - A\omega F_0 \cos\delta \cos\omega t \sin\omega t$$

(c) Show that the average value of the second term in your result for (b) over one or more periods is zero and that therefore

$$P_{av} = \frac{1}{2}A\omega F_0 \sin\delta$$

(d) From Equation 14-50 for $\tan\delta$, construct a right triangle in which the side opposite the angle δ is $b\omega$ and the side adjacent is $m(\omega_0^2 - \omega^2)$, and use this triangle to show that

$$\sin\delta = \frac{b\omega}{\sqrt{m^2(\omega_0^2 - \omega^2)^2 + b^2\omega^2}} = \frac{b\omega A}{F_0}$$

(e) Use your result for (d) to eliminate ωA so that the average power input can be written

$$P_{av} = \frac{1}{2}\frac{F_0^2}{b}\sin^2\delta = \frac{1}{2}\left[\frac{b\omega^2 F_0^2}{m^2(\omega_0^2 - \omega^2)^2 + b^2\omega^2}\right] \qquad \text{14-51}$$

130 ••• In this problem, you are to use the result of Problem 129 to derive Equation 14-45, which relates the width of the resonance curve to the Q value when the resonance is sharp. At resonance, the denominator of the fraction in brackets in Equation 14-51 is $b^2\omega_0^2$ and P_{av} has its maximum value. For a sharp resonance, the variation in ω in the numerator in Equation 14-51 can be neglected. Then the power input will be half its maximum value at the values of ω, for which the denominator is $2b^2\omega_0^2$.

(a) Show that ω then satisfies

$$m^2(\omega - \omega_0)^2(\omega + \omega_0)^2 = b^2\omega_0^2$$

(b) Using the approximation $\omega + \omega_0 \approx 2\omega_0$, show that

$$\omega - \omega_0 \approx \pm\frac{b}{2m}$$

(c) Express b in terms of Q.

(d) Combine the results of (b) and (c) to show that there are two values of ω for which the power input is half that at resonance, and that they are given by

$$\omega_1 = \omega_0 - \frac{\omega_0}{2Q} \quad \text{and} \quad \omega_2 = \omega_0 + \frac{\omega_0}{2Q}$$

Therefore, $\omega_2 - \omega_1 = \Delta\omega = \omega_0/Q$, which is equivalent to Equation 14-45.

CHAPTER 15

Wave Motion

Simple Wave Motion

Transverse and Longitudinal Waves

Waves transport energy and momentum through space without transporting matter. In mechanical waves this happens via a disturbance in a medium. When a string under tension is given a flip, the bump that is produced travels down the string as a wave pulse. The disturbance in this case is the change in shape of the string from its equilibrium shape. Its propagation

Figure 15-1 (*a*) Transverse wave pulse on a spring. The disturbance is perpendicular to the direction of the motion of the wave. (*b*) Three successive drawings of a transverse wave on a string traveling to the right. An element of the string moves up and down.

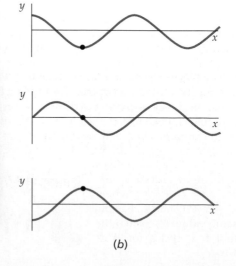

(*a*)

(*b*)

arises from the interaction of each string segment with the adjacent segments. The segments of the string (the medium) move in the direction perpendicular to the string as the pulse propagates down the string. Waves such as those on a string, in which the disturbance is perpendicular to the direction of propagation, are called **transverse** (Figure 15-1). Waves in which the disturbance is parallel to the propagation are called **longitudinal** (Figure 15-2). Sound waves are examples of longitudinal waves—the molecules of a gas, liquid, or solid through which sound travels move back and forth along the line of propagation, alternately compressing and rarefying the medium.

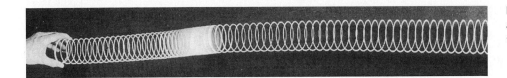

Figure 15-2 Longitudinal wave pulse on a spring. The disturbance is in the direction of the motion of the wave.

Wave Pulses

Figure 15-3 shows a pulse on a string at time $t = 0$. The shape of the string at this time can be represented by some function $y = f(x)$. At some later time, the pulse is farther down the string. In a new coordinate system with origin O' that moves with the speed of the pulse, the pulse is stationary. The string is described in this frame by $f(x')$ for all times. The coordinates of the two reference frames are related by

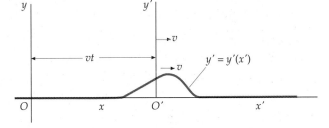

$$x = x' + vt$$

Thus, the shape of the string in the original frame is

$$y = f(x - vt), \qquad \text{wave moving right} \qquad \text{15-1}$$

The same line of reasoning for a pulse moving to the left leads to

$$y = f(x + vt), \qquad \text{wave moving left} \qquad \text{15-2}$$

In both expressions, v is the speed of propagation of the wave. The function $y = f(x - vt)$ is called the **wave function**. For waves on a string, the wave function represents the transverse displacement of the string. For sound waves in air, the wave function can be the longitudinal displacement of the air molecules or the pressure of the air. These wave functions are solutions of a differential equation called the wave equation that can be derived from Newton's laws.

Figure 15-3 A wave pulse moving without change in shape in the positive x direction with a speed v relative to the origin O. In the primed coordinate system moving with the pulse, the wave function is $y' = f(x')$ at all times. In the original, unprimed system, the wave function is $y = f(x - vt)$.

Speed of Waves

A general property of waves is that their speed depends on the properties of the medium but is independent of the motion of the source of the waves. For example, the speed of a sound from a car horn depends only on the properties of air and not on the motion of the car. For wave pulses on a rope, we can easily demonstrate that the greater the tension, the faster the waves propagate. Furthermore, waves propagate faster in a light rope than a heavy rope under the same tension. We show below that if F is the tension* and μ is the linear mass density (mass per unit length), then the wave speed is

$$v = \sqrt{\frac{F}{\mu}} \qquad \text{15-3}$$

Speed of waves on a string

---

*We use F for tension rather than T because we use T for the period.

Example 15-1

The tension in a string is provided by hanging an object of mass $m = 3$ kg at one end as shown in Figure 15-4. The length of the string is $L = 2.5$ m and its mass is $m_s = 50$ g. What is the speed of waves on the string?

Figure 15-4

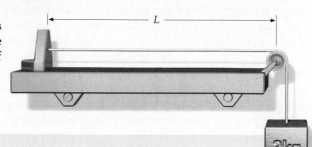

1. The speed is related to the tension F and mass density μ:

$$v = \sqrt{\frac{F}{\mu}}$$

2. Calculate the mass density and tension from the information given:

$$\mu = \frac{m_s}{L} = \frac{0.05 \text{ kg}}{2.5 \text{ m}} = 0.02 \text{ kg/m}$$

$$F = mg = (3 \text{ kg})(9.81 \text{ N/kg}) = 29.4 \text{ N}$$

3. Substitute these values to calculate the speed:

$$v = \sqrt{\frac{F}{\mu}} = \sqrt{\frac{29.4 \text{ N}}{0.02 \text{ kg/m}}} = 38.3 \text{ m/s}$$

Exercise If the 3-kg mass is replaced with a 6-kg mass, what is the speed of waves on the string? (*Answer* 54.2 m/s)

Exercise Show that the units of $\sqrt{F/\mu}$ are m/s when F is in newtons and μ is in kg/m.

For sound waves in a fluid such as air or water, the speed v can be shown to be given by

$$v = \sqrt{\frac{B}{\rho}} \qquad\qquad 15\text{-}4$$

where ρ is the equilibrium density of the medium and B is the bulk modulus (Equation 13-6). Comparing Equations 15-3 and 15-4, we can see that, in general, the speed of waves depends on an elastic property of the medium (the tension for string waves and the bulk modulus for sound waves) and on an inertial property of the medium (the linear mass density or the volume mass density).

For sound waves in a gas such as air, the bulk modulus is proportional to the pressure, which in turn is proportional to the density ρ and to the absolute temperature T of the gas. The ratio B/ρ is thus independent of density and is merely proportional to the absolute temperature T. In Chapter 19, we show that, in this case, Equation 15-4 is equivalent to

$$v = \sqrt{\frac{\gamma RT}{M}} \qquad\qquad 15\text{-}5$$

Speed of sound in a gas

In this equation T is the absolute temperature measured in kelvins (K), which is related to the Celsius temperature t_C by

$$T = t_C + 273 \qquad\qquad 15\text{-}6$$

The constant γ depends on the kind of gas. For diatomic molecules, such as O_2 and N_2, γ has the value 1.4, and, since O_2 and N_2 comprise 98% of the atmosphere, that is the value for air. (For monatomic molecules such as He, γ has the value 1.67.) The constant R is the universal gas constant,

$$R = 8.314 \text{ J/mol·K} \qquad\qquad 15\text{-}7$$

and M is the molar mass of the gas (that is, the mass of 1 mol of the gas), which for air is

$$M = 29 \times 10^{-3} \text{ kg/mol}$$

Example 15-2   *try it yourself*

Calculate the speed of sound in air at (a) 0°C and (b) 20°C.

Cover the column to the right and try these on your own before looking at the answers.

Steps	**Answers**
(a)1. Convert 0°C to kelvins.	$T = 273$ K
2. Substitute your result for T into Equation 15-5 to find v.	$v = 331$ m/s
(b)1. Convert 20°C to kelvins.	$T = 293$ K
2. Use the fact that v is proportional to $\sqrt{T}$ to write an expression for the ratio of the speed at 293 K to the speed at 273 K.	$\dfrac{v_{293}}{v_{273}} = \dfrac{\sqrt{293}}{\sqrt{273}} = 1.15$
3. Calculate v at 293 K.	$v = 343$ m/s

Remarks We see from this example that the speed of sound in air is about 340 m/s at normal temperatures.

Exercise For helium, $M = 4 \times 10^{-3}$ kg/mol and $\gamma = 1.67$. What is the speed of sound waves in helium at 20°C? (*Answer*   1.01 km/s)

Derivation of v for Waves on a String   Equation 15-3 can be obtained from Newton's laws. Consider a pulse traveling along a string with a speed v to the right (Figure 15-5a). If the amplitude of the pulse is small compared to the length of the string, the tension F will be approximately constant along the string. In a reference frame moving with speed v to the right, the pulse is stationary and the string moves with a speed v to the left. Figure 15-5b shows a small segment of the string of length Δs. The segment forms part of a circular arc of radius R. Instantaneously the segment is moving with speed v in a circular path, so it has a centripetal acceleration v^2/R. The forces acting on the segment are the tension F at each end. The horizontal components of these forces are equal and opposite and thus cancel. The vertical components of these forces point radially inward toward the center of the circular arc. These radial forces provide the centripetal acceleration.

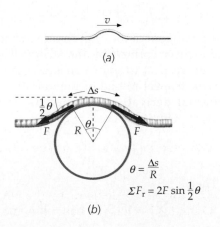

Figure 15-5 (*a*) Wave pulse moving with a speed v along a string. (*b*) In a frame in which the wave pulse of (*a*) is at rest, the string is moving with a speed v to the left. A small segment of the string of length Δs is moving in a circular arc of radius R. The centripetal acceleration of the segment is provided by the radial components of the tension.

Let the angle subtended by the string be θ: The net radial force acting on the segment is

$$\sum F_r = 2F \sin\tfrac{1}{2}\theta \approx 2F(\tfrac{1}{2}\theta) = F\theta$$

where we have used the approximation $\sin\tfrac{1}{2}\theta \approx \tfrac{1}{2}\theta$ for small θ. If μ is the mass per unit length of the string, the mass of a segment of length Δs is $m = \mu\,\Delta s$. The angle θ is related to Δs by

$$\theta = \frac{\Delta s}{R}$$

The mass of the element is thus

$$m = \mu\,\Delta s = \mu R\theta$$

Setting the net radial force equal to the mass times the centripetal acceleration gives

$$F\theta = \mu R\theta\,\frac{v^2}{R}$$

Solving for v, we obtain $v = \sqrt{F/\mu}$. Since v is independent of R and θ, this result holds for all segments of the string. However, the derivation depends on θ being small, which is true if the height of the pulse is small compared to its length.

In the original frame, the string is fixed, and the pulse moves with speed $v = \sqrt{F/\mu}$, which is Equation 15-3.

The Wave Equation

We can apply Newton's laws to a segment of the string to derive a differential equation known as the wave equation, which relates the spatial derivatives of $y(x, t)$ to its time derivatives. Figure 15-6 shows one segment of a string. We consider only small vertical displacements. Then the length of the segment is approximately Δx and its mass is $m = \mu\,\Delta x$, where μ is the string's mass per unit length. The segment moves vertically, and the net force in this direction is

$$\sum F = F \sin\theta_2 - F \sin\theta_1$$

where θ_2 and θ_1 are the angles shown, and F is the tension in the string. Since the angles are assumed to be small, we may approximate $\sin\theta$ by $\tan\theta$. Then the net vertical force on the string segment can be written

$$\sum F = F(\sin\theta_2 - \sin\theta_1) \approx F(\tan\theta_2 - \tan\theta_1)$$

The tangent of the angle made by the string with the horizontal is the slope of the curve formed by the string. The slope S is the first derivative of $y(x, t)$ with respect to x for constant t. A derivative of a function of two variables with respect to one of the variables with the other held constant is called a **partial derivative.** The partial derivative of y with respect to x is written $\partial y/\partial x$. Thus, we have

$$S = \tan\theta = \frac{\partial y}{\partial x}$$

Then

$$\sum F = F(S_2 - S_1) = F\,\Delta S$$

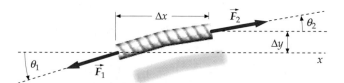

Figure 15-6 Segment of a stretched string used for the derivation of the wave equation. The net vertical force on the segment is $F \sin\theta_2 - F \sin\theta_1$, where F is the tension in the string. The wave equation is derived by applying Newton's second law to the segment.

where S_1 and S_2 are the slopes of either end of the string segment, and ΔS is the change in the slope. Setting this net force equal to the mass $\mu \, \Delta x$ times the acceleration $\partial^2 y / \partial t^2$ gives

$$F \, \Delta S = \mu \, \Delta x \, \frac{\partial^2 y}{\partial t^2}$$

or

$$F \frac{\Delta S}{\Delta x} = \mu \frac{\partial^2 y}{\partial t^2} \qquad \text{15-8}$$

In the limit $\Delta x \to 0$, we have

$$\lim_{\Delta x \to 0} \frac{\Delta S}{\Delta x} = \frac{\partial S}{\partial x} = \frac{\partial}{\partial x} \frac{\partial y}{\partial x} = \frac{\partial^2 y}{\partial x^2}$$

Thus, Equation 15-8 becomes

$$\frac{\partial^2 y}{\partial x^2} = \frac{\mu}{F} \frac{\partial^2 y}{\partial t^2} \qquad \text{15-9a}$$

Wave equation for a stretched string

Equation 15-9a is the **wave equation** for a stretched string.

We now show that the wave equation is satisfied by any function $y(x - vt)$. Let $\alpha = x - vt$ and consider any wave function

$$y = y(x - vt) = y(\alpha)$$

We will use y' for the derivative of y with respect to α. Then, by the chain rule for derivatives,

$$\frac{\partial y}{\partial x} = \frac{\partial y}{\partial \alpha} \frac{\partial \alpha}{\partial x} = y' \frac{\partial \alpha}{\partial x}$$

and

$$\frac{\partial y}{\partial t} = \frac{\partial y}{\partial \alpha} \frac{\partial \alpha}{\partial t} = y' \frac{\partial \alpha}{\partial t}$$

Using $\partial \alpha / \partial x = 1$ and $\partial \alpha / \partial t = -v$ gives

$$\frac{\partial y}{\partial x} = y' \quad \text{and} \quad \frac{\partial y}{\partial t} = -vy'$$

Taking the second derivatives, we obtain

$$\frac{\partial^2 y}{\partial x^2} = y''$$

$$\frac{\partial^2 y}{\partial t^2} = -v \frac{\partial y'}{\partial t} = -v \frac{\partial y'}{\partial \alpha} \frac{\partial \alpha}{\partial t} = +v^2 y''$$

Thus,

$$\frac{\partial^2 y}{\partial x^2} = \frac{1}{v^2} \frac{\partial^2 y}{\partial t^2} \qquad \text{15-9b}$$

General wave equation

The same result can be obtained for any function of $x + vt$. Comparing Equations 15-9a and 15-9b, we see that the speed of propagation of the wave is $v = \sqrt{F/\mu}$, which is Equation 15-3.

optional

Example 15-3

Show by explicitly calculating the derivatives that the function $y(x,t) = A \sin(kx - \omega t)$ satisfies Equation 15-9b.

1. Take two partial derivatives of y with respect to x:

$$\frac{\partial y}{\partial x} = \frac{\partial}{\partial x}[A \sin(kx - \omega t)] = A \cos(kx - \omega t) \frac{\partial(kx)}{\partial x} = kA \cos(kx - \omega t)$$

$$\frac{\partial^2 y}{\partial x^2} = -k^2 A \sin(kx - \omega t)$$

2. Similarly, the two partial derivatives with respect to t are:

$$\frac{\partial y}{\partial t} = \frac{\partial}{\partial t}[A \sin(kx - \omega t)] = A \cos(kx - \omega t) \frac{\partial(-\omega t)}{\partial t} = -\omega A \cos(kx - \omega t)$$

$$\frac{\partial^2 y}{\partial t^2} = -\omega^2 A \sin(kx - \omega t)$$

3. Substitute these results in Equation 15-9b. The equation is satisfied provided that $v = \omega/k$:

$$-k^2 A \sin(kx - \omega t) = \frac{1}{v^2}[-\omega^2 A \sin(kx - \omega t)]$$

Remarks We have shown that the function $y = A \sin(kx - \omega t)$ is a solution to the wave equation. It describes a sinusoidal wave with speed $v = \omega/k$.

Exercise Show that any function $y(x + vt)$ satisfies Equation 15-9b.

A wave equation for sound waves can also be derived using Newton's laws. In one dimension, this equation is

$$\frac{\partial^2 s}{\partial x^2} = \frac{1}{v_s^2} \frac{\partial^2 s}{\partial t^2}$$

where s is the displacement of the medium in the x direction and v_s is the speed of sound.

15-2   Harmonic Waves

Harmonic Waves on a String

If one end of a string is attached to a vibrating tuning fork that is moving up and down with simple harmonic motion, a sinusoidal wave propagates along the string. The shape of the string at some instant in time is that of a sine function,* as shown in Figure 15-7. A sinusoidal wave such as that shown is called a **harmonic wave**. The distance after which the wave repeats (the distance between crests, for example) is the **wavelength** λ.

As the wave propagates, each point on the string moves up and down—perpendicular to the direction of propagation—in simple harmonic motion

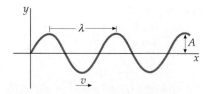

Figure 15-7 Harmonic wave at some instant in time. A is the amplitude and λ is the wavelength. For waves on a string, this figure can be obtained by taking a snapshot of the string.

---

* Whether this is a sine function or a cosine function depends on where the origin is chosen on the x axis.

with the frequency f of the tuning fork. During one period $T = 1/f$, the wave moves a distance of one wavelength, so its speed is given by

$$v = \frac{\lambda}{T} = f\lambda \qquad\qquad\qquad 15\text{-}10$$

Since this relation arises from the definitions of wavelength and frequency, it applies to all harmonic waves.

A harmonic wave has a single frequency and wavelength. Other waves, such as wave pulses, can be considered to be made up of many harmonic waves of different frequencies. The superposition of harmonic waves is discussed in Chapter 16.

The sine function that describes the displacement in Figure 15-7 is

$$y(x) = A \sin(kx + \delta) \qquad\qquad\qquad 15\text{-}11$$

where A is the **amplitude,** k is a constant called the **wave number,** and δ is a phase constant that depends on the choice of the origin $x = 0$. When dealing with a single harmonic wave, we are free to choose the origin anywhere, so we usually choose it to make $\delta = 0$.

Consider a point x_1 and another, x_2, one wavelength away, such that $x_2 = x_1 + \lambda$. The displacement at each point is the same: $y(x_1) = y(x_2)$. So,

$$\sin kx_1 = \sin kx_2 = \sin k(x_1 + \lambda) = \sin(kx_1 + k\lambda)$$

Therefore,

$$k\lambda = 2\pi$$

or

$$k = \frac{2\pi}{\lambda} \qquad\qquad\qquad 15\text{-}12$$

Note that k has dimensions of m^{-1}. (Because the angle must be in radians, we sometimes write the units of k as rad/m.) Since $1/\lambda$ is the number of waves in a length of 1 meter, $k = 2\pi/\lambda$ is the number of waves in a distance of 2π meters.

For a wave traveling to the right with speed v, replace x in Equation 15-11 with $x - vt$ (see "Wave Pulses" in Section 15-1). With δ chosen to be zero, this gives

$$y(x, t) = A \sin k(x - vt) = A \sin(kx - kvt)$$

or

$$y(x, t) = A \sin(kx - \omega t) \qquad\qquad\qquad 15\text{-}13$$

Harmonic wave function

where

$$\omega = kv \qquad\qquad\qquad 15\text{-}14$$

is the angular frequency, which is related to the frequency f and period T by

$$\omega = 2\pi f = \frac{2\pi}{T} \qquad\qquad\qquad 15\text{-}15$$

Substituting $\omega = 2\pi f$ into Equation 15-14 and using $k = 2\pi/\lambda$, we obtain

$$2\pi f = kv = \frac{2\pi}{\lambda} v$$

or $v = f\lambda$, which is Equation 15-10.

The rate of increase of energy is the power passing into the shell. The average incident power is

$$P_{av} = \frac{(\Delta E)_{av}}{\Delta t} = \eta_{av} A v$$

and the intensity of the wave is

$$I = \frac{P_{av}}{A} = \eta_{av} v \qquad\qquad 15\text{-}27$$

Thus, the intensity equals the product of the wave speed v and the average energy density η_{av}. This result applies to all waves. Substituting $\eta_{av} = \frac{1}{2}\rho\omega^2 s_0^2$ from Equation 15-24 for the energy density in a sound wave, we obtain

$$I = \eta_{av} v = \frac{1}{2}\rho\omega^2 s_0^2 v = \frac{1}{2}\frac{p_0^2}{\rho v} \qquad\qquad 15\text{-}28$$

where we have used $s_0 = p_0/\rho\omega v$ from Equation 15-22. This result—that the intensity of a sound wave is proportional to the square of the amplitude—is a general property of harmonic waves.

The human ear can accommodate a large range of sound-wave intensities, from about 10^{-12} W/m^2 (which is usually taken to be the threshold of hearing) to about 1 W/m^2 (a volume so loud it produces pain in most people). The pressure variations that correspond to these extreme intensities are about 3×10^{-5} Pa for the hearing threshold and 30 Pa for the pain threshold. (Recall that a pascal is a newton per square meter.) These very small pressure variations add or subtract to the normal atmospheric pressure of about 101,000 Pa.

Example 15-6

A loudspeaker diaphragm 30 cm in diameter is vibrating at 1 kHz with an amplitude of 0.020 mm. Assuming that the air molecules in the vicinity have this same amplitude of vibration, find (*a*) the pressure amplitude immediately in front of the diaphragm, (*b*) the sound intensity in front of the diaphragm, and (*c*) the acoustic power being radiated. (*d*) If the sound is radiated uniformly into the forward hemisphere, find the intensity at 5 m from the loudspeaker.

Picture the Problem (*a* and *b*) The pressure amplitude is calculated directly from $p_0 = \rho\omega v s_0$ (Equation 15-22), and the intensity from $I = \frac{1}{2}\rho\omega^2 s_0^2 v$ (Equation 15-28). (*c*) The power radiated is the intensity times the area of the diaphragm. (*d*) The area of a hemisphere of radius r is $2\pi r^2$. We can use Equation 15-26 if we replace the area $4\pi r^2$ by $2\pi r^2$.

(*a*) Equation 15-22 relates the pressure amplitude to the displacement amplitude, frequency, wave velocity, and air density:

$$p_0 = \rho\omega v s_0 = (1.29 \text{ kg/m}^3)2\pi (10^3 \text{ Hz})(340 \text{ m/s})(2 \times 10^{-5} \text{ m})$$
$$= 55.1 \text{ N/m}^2$$

(*b*) Equation 15-28 relates the intensity to these same known quantities:

$$I = \frac{1}{2}\rho\omega^2 s_0^2 v$$
$$= (0.5)(1.29 \text{ kg/m}^3)[2\pi (10^3 \text{ Hz})]^2(2 \times 10^5 \text{ m})^2(340 \text{ m/s})$$
$$= 3.46 \text{ W/m}^2$$

(*c*) The power is the intensity times the area of the diaphragm:

$$P = IA = (3.46 \text{ W/m}^2)\pi (0.15 \text{ m})^2 = 0.245 \text{ W}$$

(d) Calculate the intensity at $r = 5$ m assuming uniform radiation into the forward hemisphere:

$$I = \frac{P_{av}}{2\pi r^2} = \frac{0.245\ \text{W}}{2\pi\,(5\ \text{m})^2} = 1.56 \times 10^{-3}\ \text{W/m}^2$$

Remarks The assumption of uniform radiation in the forward hemisphere is not very good because the wavelength in this case ($\lambda = v/f = (340\ \text{m/s})/(1000\ \text{s}^{-1}) = 34$ cm) is not large compared with the speaker diameter. There is also some radiation in the backward direction as can be observed if you stand behind a loudspeaker.

Loudspeakers at a rock concert may put out more than 100 times as much power as the speaker in this example.

Intensity Level and Loudness   The psychological sensation of loudness varies approximately logarithmically rather than directly with intensity. We therefore use a logarithmic scale to describe the **intensity level** of a sound wave β, which is measured in **decibels** (dB) and defined by

$$\beta = 10 \log \frac{I}{I_0} \qquad\qquad 15\text{-}29$$

Definition—Intensity level in dB

Here I is the intensity of the sound and I_0 is a reference level, which we take to be the threshold of hearing:

$$I_0 = 10^{-12}\ \text{W/m}^2 \qquad\qquad 15\text{-}30$$

On this scale, the threshold of hearing is $\beta = 10 \log(I_0/I_0) = 0$ dB and the pain threshold ($I = 1\ \text{W/m}^2$) is $\beta = 10 \log(1/10^{-12}) = 10 \log 10^{12} = 120$ dB. Thus, the range of sound intensities from $10^{-12}\ \text{W/m}^2$ to $1\ \text{W/m}^2$ corresponds to a range of intensity levels from 0 dB to 120 dB. Table 15-1 lists the intensity levels of some common sounds.

Table 15-1

Intensity and Intensity Level of Some Common Sounds ($I_0 = 10^{-12}\ \text{W/m}^2$)

Source	I/I_0	dB	Description
	10^0	0	Hearing threshold
Normal breathing	10^1	10	Barely audible
Rustling leaves	10^2	20	
Soft whisper (at 5 m)	10^3	30	Very quiet
Library	10^4	40	
Quiet office	10^5	50	Quiet
Normal conversation (at 1 m)	10^6	60	
Busy traffic	10^7	70	
Noisy office with machines; average factory	10^8	80	
Heavy truck (at 15 m); Niagara Falls	10^9	90	Constant exposure endangers hearing
Old subway train	10^{10}	100	
Construction noise (at 3 m)	10^{11}	110	
Rock concert with amplifiers (at 2 m); jet takeoff (at 60 m)	10^{12}	120	Pain threshold
Pneumatic riveter; machine gun	10^{13}	130	
Jet takeoff (nearby)	10^{15}	150	
Large rocket engine (nearby)	10^{18}	180	

optional

Example 15-7

A sound absorber attenuates the sound level by 30 dB. By what factor is the intensity decreased?

From Table 15-1, we can see that for every 10-dB drop in the intensity level, the intensity decreases by a factor of 10. Thus, if the sound level drops 30 dB then the intensity drops by a factor of $10^3 = 1000$.

The sensation of loudness depends on the frequency as well as the intensity of a sound. Figure 15-16 is a plot of intensity level versus frequency for sounds of equal loudness to the human ear. (In this figure, the frequency is plotted on a logarithmic scale to display the wide range of frequencies from 20 Hz to 10 kHz.) We note from this figure that the human ear is most sensitive at about 4 kHz for all intensity levels.

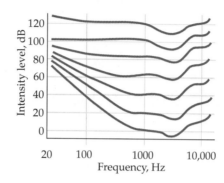

Figure 15-16 Intensity level versus frequency for sounds perceived to be of equal loudness. The lowest curve is below the threshold for hearing of all but about 1% of the population. The second lowest curve is approximately the hearing threshold for about 50% of the population.

Example 15-8

A barking dog delivers about 1 mW of power. (*a*) If this power is uniformly distributed in all directions, what is the sound intensity level at a distance of 5 m? (*b*) What would be the intensity level of two dogs barking at the same time if each delivered 1 mW of power?

Picture the Problem The intensity level is found from the intensity, which is found from $I = P/(4\pi r^2)$. For two dogs, the intensities add.

(*a*)1. The intensity level is related to the intensity:

$$\beta = 10 \log \frac{I}{I_0}$$

2. Calculate the intensity at $r = 5$ m:

$$I = \frac{P}{4\pi r^2} = \frac{10^{-3}\,\text{W}}{4\pi\,(5\,\text{m})^2} = 3.18 \times 10^{-6}\,\text{W/m}^2$$

3. Use your result to find the intensity level at 5 m:

$$\beta = 10 \log \frac{I}{I_0} = 10 \log \frac{3.18 \times 10^{-6}}{10^{-12}} = 10 \log(3.18 \times 10^6)$$

$$= 10(\log 3.18 + \log 10^6) = 10(0.50 + 6) = 65.0\,\text{dB}$$

(*b*) If I_1 is the intensity for one dog, the intensity for two dogs is $I_2 = 2I_1$. The intensity level for two dogs is then:

$$\beta_2 = 10 \log \frac{I_2}{I_0} = 10 \log \frac{2I_1}{I_0} = 10 \log 2 + 10 \log \frac{I_1}{I_0}$$

$$= 10 \log 2 + \beta_1 = 3.01 + 65.0 = 68.0\,\text{dB}$$

Remark We can see from this example that whenever the intensity is doubled, the intensity level increases by 3 dB.

15-4 Waves Encountering Barriers

Reflection and Refraction

When a wave is incident on a boundary that separates two regions of differing wave speed, part of the wave is reflected and part is transmitted. Figure 15-17a shows a pulse on a light string that is attached to a heavier string. In this case, the pulse reflected at the boundary is inverted. If the second string is lighter than the first (Figure 15-14b), the reflected pulse is not inverted. In either case, the transmitted pulse is not inverted. If the string is tied to a fixed point the pulse is reflected and inverted.

Figure 15-17 (*a*) A wave pulse traveling on a light string attached to a heavier string in which the wave speed is smaller. The reflected pulse is inverted, whereas the transmitted pulse is not. (*b*) Photograph of a similar pulse on a light spring attached to a heavier spring. (*c*) A wave pulse traveling on a heavy string attached to a light string in which the wave speed is greater. In this case, the reflected pulse is not inverted. (*d*) Photograph of a similar pulse on a heavy spring attached to a lighter spring.

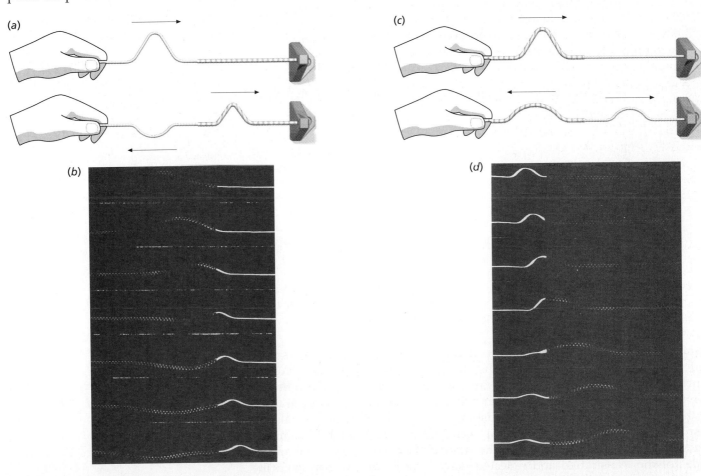

Example 15-9

Two wires of different linear mass densities are soldered together end to end and then stretched under a tension F (the tension is the same in both wires). The wave speed in the first wire is twice that in the second wire. When a harmonic wave traveling in the first wire is reflected at the junction of the wires, the reflected wave has half the amplitude of the transmitted wave. (*a*) If the amplitude of the incident wave is A, what are the amplitudes of the reflected and transmitted waves? (*b*) Assuming no loss in the wire, what fraction of the incident power is reflected at the junction and what fraction is transmitted?

Picture the Problem By conservation of energy, the power incident on the junction equals the power reflected plus the power transmitted. Each power is expressed in Equation 15-19 as a function of the density μ, amplitude A, frequency ω, and wave speed v (Figure 15-18). The angular frequencies of all the waves are equal. Since the reflected wave and incident wave are in the same medium, they have the same wave speed v_1. We are given that the speed in the second wire is $v_2 = \frac{1}{2}v_1$. Then $\mu_2 = F/v_2^2 = 4\mu_1$.

Figure 15-18

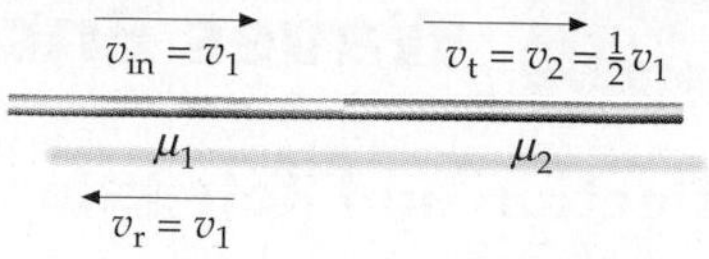

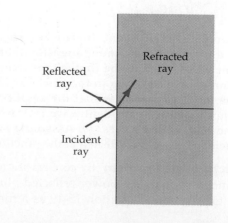

$$v_{in} = v_1 \qquad\qquad v_t = v_2 = \tfrac{1}{2}v_1$$
$$\mu_1 \qquad\qquad \mu_2$$
$$v_r = v_1$$

(a)1. By conservation of energy, the incident power equals the transmitted power plus the reflected power:

$$P_{in} = P_t + P_r$$

2. Express the incident, reflected, and transmitted power in terms of the mass density, amplitude, angular frequency, and wave speed:

$$\tfrac{1}{2}\mu_1\omega^2 A_{in}^2 v_1 = \tfrac{1}{2}\mu_2\omega^2 A_t^2 v_2 + \tfrac{1}{2}\mu_1\omega^2 A_r^2 v_1$$

3. Express the mass densities in terms of the wave speeds, and cancel the common terms:

$$\tfrac{1}{2}(F/v_1^2)\omega^2 A_{in}^2 v_1 = \tfrac{1}{2}(F/v_2^2)\omega^2 A_t^2 v_2 + \tfrac{1}{2}(F/v_1^2)\omega^2 A_r^2 v_1$$
$$A_{in}^2/v_1 = A_t^2/v_2 + A_r^2/v_1$$

4. The following relations are given:

$$v_2 = \tfrac{1}{2}v_1, \qquad A_r = \tfrac{1}{2}A_t$$

5. Substitute the known relations, and eliminate A_t:

$$A_{in}^2/v_1 = (2A_r)^2/(v_1/2) + A_r^2/v_1$$
$$A_{in}^2 = 8\,A_r^2 + A_r^2 = 9A_r^2$$

6. Solve for A_r:

$$A_r = \tfrac{1}{3}A_{in}$$

7. Use this result to find A_t:

$$A_t = 2A_r = \tfrac{2}{3}A_{in}$$

(b)1. Write the reflected power in terms of the incident power:

$$P_r = \tfrac{1}{2}\mu_1\omega^2 A_r^2 v_1 = \tfrac{1}{2}\mu_1\omega^2(\tfrac{1}{3}A_{in})^2 v_1 = \tfrac{1}{9}P_{in}$$

2. Write the transmitted power in terms of the incident power:

$$P_t = \tfrac{1}{2}\mu_2\omega^2 A_t^2 v_2 = \tfrac{1}{2}(4\mu_1)\omega^2(\tfrac{2}{3}A_{in})^2 \tfrac{1}{2}v_1 = \tfrac{8}{9}P_{in}$$

Remarks The reflected wave is inverted relative to the incident wave, so it is 180° out of phase with it. When the displacement of the wire to the left of the junction is y_1 due to the incident wave, it is $-(y_1/3)$ due to the reflected wave. These add (according to the principle of superposition, to be studied in the next chapter) giving a total displacement of $2y_1/3$, which equals the displacement that occurs to the right of the junction due to the transmitted wave. It can be shown that, given the ratio of the wave speeds, the amplitudes of the transmitted and reflected waves can be determined from the conditions that the displacement and slope of the wire must be continuous at the junction.

In three dimensions, a boundary between two regions of differing wave speed is a surface. Figure 15-19 shows a ray incident on such a boundary surface. This example could be a sound wave in air striking a solid or liquid surface. The reflected ray makes an angle with the normal to the surface equal to that of the incident ray, as shown.

Figure 15-19 A wave striking a boundary surface between two media in which the wave speed differs. Part of the wave is reflected and part is transmitted. The change in direction of the transmitted ray is called refraction.

Refracted ray

Reflected ray

Incident ray

The transmitted ray is bent toward or away from the normal—depending on whether the wave speed in the second medium is less or greater than that in the incident medium. The bending of the transmitted ray is called **refraction**. When the wave speed in the second medium is greater than that in the incident medium (as occurs when a light wave in glass or water is refracted into the air), the ray describing the direction of propagation is bent away from the normal, as shown in Figure 15-20. As the angle of incidence is increased, the angle of refraction increases, until a critical angle of incidence is reached for which the angle of refraction is 90°. For incident angles greater than the critical angle, there is no refracted ray, a phenomenon known as **total internal reflection**.

The amount of energy reflected from a surface depends on the surface. Flat walls, floors, and ceilings make good reflectors for sound waves, whereas less rigid and porous materials, such as cloth in draperies and furniture coverings, absorb much of the incident sound. The reflection of sound waves plays an important role in the design of a lecture hall, a library, or a music auditorium. If a lecture hall has many flat reflecting surfaces, speech is difficult to understand because of the many echoes that arrive at different times at the listener's ear. Absorbent material is often placed on the walls and ceiling to reduce such reflections. In a concert hall, a reflecting shell is placed behind the orchestra, and reflecting panels are hung from the ceiling to reflect and direct the sound back toward the listeners.

Tunneling   In total internal reflection, the wave function does not immediately drop to zero at the surface but instead decreases exponentially and becomes negligible within a few wavelengths of the surface. In Figure 15-21, light is totally reflected at the right surface of the glass. When another piece of glass is brought near the surface, some of the light is transmitted across the barrier. This is called **barrier penetration** or **tunneling**. Figure 15-22 shows a barrier penetration by water waves in a ripple tank.

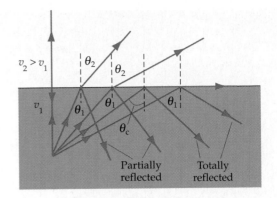

Figure 15-20 Light from a source in the water is bent away from the normal when it enters air. For angles of incidence above a critical angle, there is no transmitted ray, a condition known as total internal reflection.

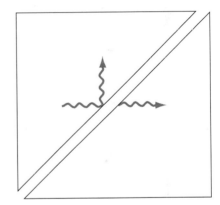

Figure 15-21 Penetration of optical barrier. If the pieces of glass are close enough, some light gets through the barrier.

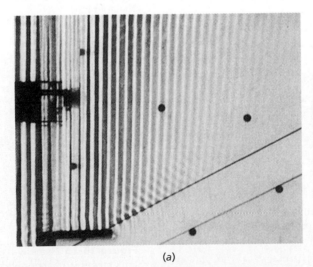

(a)

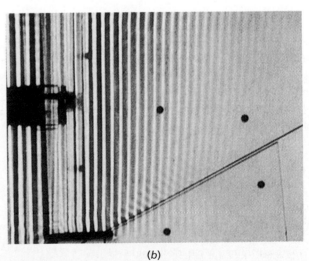

(b)

Figure 15-22 Penetration of a barrier by water waves in a ripple tank. In (a), the waves are totally reflected from a large gap in deeper water. When the gap is very narrow, as in (b), a transmitted wave appears.

(Left) This anechoic chamber at the Naval Research Laboratory is used in the testing of electronic equipment. *(Right)* Davies Symphony Hall in San Francisco. The plastic reflectors above the orchestra reflect the sound waves out toward the audience.

Diffraction

A wave encountering an obstacle tends to bend around the obstacle. When the wave encounters a barrier with a small aperture, the wave bends and spreads out as a spherical or circular wave (Figure 15-23). This bending of the wavefront is called **diffraction**. In contrast, a beam of particles falling upon a barrier with an aperture either is halted by the barrier or passes through cleanly with no change in the direction of the particles (Figure 15-24). Diffraction is one of the key characteristics that distinguishes waves from particles.*

Though waves encountering an obstacle or aperture always bend, or diffract, to some extent, the amount of diffraction depends on whether the wavelength is small or large relative to the size of the obstacle or aperture. If the wavelength is large relative to the aperture, as in Figure 15-23, the diffraction effects are large. There the waves spread out as they pass through the aperture—as if the waves were originating from a point source. On the other hand, if the wavelength is small relative to the aperture, the effect of diffraction is small, as shown in Figure 15-25. Near the edges of the aperture the

Figure 15-23 Plane waves in a ripple tank meeting a barrier with an opening that is small compared to the wavelength λ. To the right of the barrier are circular waves that are concentric about the opening, just as if there were a point source at the opening.

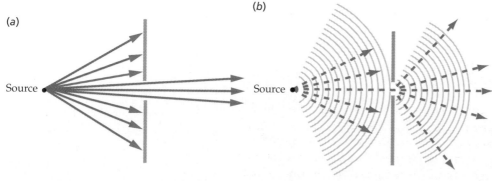

(a)

(b)

Source

Source

Figure 15-24 Comparison of particles and waves passing through a narrow opening in a barrier. (*a*) Transmitted particles are confined to a narrow angle.

(*b*) Transmitted waves radiate widely from the aperture, which acts like a point source of circular waves.

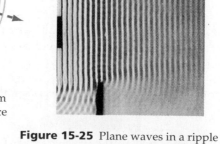

Figure 15-25 Plane waves in a ripple tank meeting a barrier with an opening that is large compared to λ. The barrier has a noticeable effect only near the edges of the opening.

---

* Diffraction is closely related to interference, which we discuss in Chapter 16. We show how diffraction arises when we study the interference and diffraction of light in Chapter 35.

wavefronts are distorted and the waves appear to bend slightly. For the most part, however, the wavefronts are not affected and the waves propagate in straight lines in the direction of the rays, much like a beam of particles. The approximation that these waves propagate in straight lines in the direction of the rays with no diffraction is known as the **ray approximation.**

Because the wavelengths of audible sound (which range from a few centimeters to several meters) are generally large compared with apertures and obstacles (doors or windows, for example), diffraction of sound waves is a common phenomenon. On the other hand, the wavelengths of visible light (4×10^{-7} to 7×10^{-7} m) are so small compared with the size of ordinary objects and apertures that the diffraction of light is not easily noticed; light appears to travel in straight lines. Nevertheless, the diffraction of light is an important phenomenon, which we study in detail in Chapter 35.

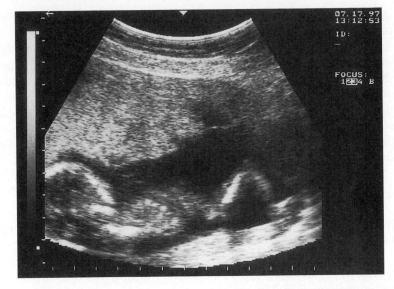

Sonogram of a pregnant woman showing the fetus in the womb.

Diffraction places a limitation on how accurately small objects can be located by reflecting waves off of them, and on how well details of the objects can be resolved. Waves are not reflected appreciably from objects smaller than the wavelength, so detail cannot be observed on a scale smaller than the wavelength used. If waves of wavelength λ are used to locate an object, its position can be known only to within $\pm\lambda$. Since the smallest wavelength of audible sound is about 2 cm, the location of an object cannot be fixed to better than ±2 cm using audible sound.

Sound waves with frequencies above 20,000 Hz are called **ultrasonic waves.** Because of their very small wavelengths, narrow beams of ultrasonic waves can be sent out and reflected from small objects. Bats can emit and detect frequencies up to about 120,000 Hz, corresponding to a wavelength of 2.8 mm, which they use to locate small prey such as moths. Ships use a device called sonar (from *sound navigation and ranging*) to detect the outlines of submarines and other submerged objects with ultrasonic waves. In medicine, ultrasonic waves are used for diagnostic purposes. To create a sonogram, ultrasonic waves are passed through the body and information about the frequency and intensity of the transmitted and reflected waves is processed to construct a three-dimensional picture of the body's interior.

15-5 The Doppler Effect

When a wave source and a receiver are moving relative to each other, the frequency observed is not the same as that emitted. When they are moving toward each other, the observed frequency is greater than the source frequency; when they are moving away from each other, the observed frequency is less than the source frequency. This is called the **Doppler effect.** A familiar example is the change in pitch of a car horn as the car approaches or recedes.

The change in frequency of a sound wave is slightly different depending on whether the source or the receiver moves relative to the medium. When the source moves, the wavelength changes, and the new frequency f' is found by first finding the new wavelength λ' and then computing $f' = v/\lambda'$. When the source is stationary and the receiver moves, the wavelength is unchanged, and the frequency is different simply because the receiver moves past more or fewer waves in a given time.

Consider a source of frequency f_0 moving with speed u_s relative to the medium. The waves in front of the source are compressed, whereas behind the source, they are farther apart, as shown in Figure 15-26. Let v be the speed of the waves relative to the medium. This speed depends only on the properties of the medium and not on the motion of the source. In some time Δt, the source emits a number of waves $N = f_0 \Delta t$. The first wavefront moves a distance $v \Delta t$, while the source moves a distance $u_s \Delta t$. The wavelength λ' is the distance $(v \pm u_s)\Delta t$ divided by the number of waves:

$$\lambda' = \frac{(v \pm u_s)\Delta t}{N} = \frac{(v \pm u_s)\Delta t}{f_0 \Delta t} = \frac{v \pm u_s}{f_0} \qquad \text{15-31}$$

In front of the source, the wavelength decreases, so the minus sign in Equation 15-31 applies. Behind the source, the plus sign applies.

The number of waves that pass a receiver in time Δt is the number of waves in the distance $v_r \Delta t$, where v_r is the speed of the waves relative to the receiver (Figure 15-27):

$$N = \frac{v_r \Delta t}{\lambda'} = \frac{(v \pm u_r)\Delta t}{\lambda'}$$

The frequency observed is this number of waves divided by the time interval:

$$f' = \frac{N}{\Delta t} = \frac{v \pm u_r}{\lambda'} \qquad \text{15-32}$$

If the receiver is stationary, $u_r = 0$, and the frequency is

$$f' = \frac{v}{\lambda'} = \frac{v}{v \pm u_s} f_0 = \frac{1}{1 \pm u_s/v} f_0 \qquad \text{(moving source)} \qquad \text{15-33}$$

When the source is moving toward the receiver, the frequency increases so the minus sign applies.

If the source is stationary, $\lambda' = \lambda_0 = v/f_0$, and the observed frequency is

$$f' = \frac{v \pm u_r}{v/f_0} = \left(1 \pm \frac{u_r}{v}\right)f_0 \qquad \text{(moving receiver)} \qquad \text{15-34}$$

We can combine Equations 15-33 and 15-34 to cover the general case of either source, receiver, or both moving:

$$f' = \frac{v \pm u_r}{\lambda'} = \frac{v \pm u_r}{v \pm u_s} f_0 = \frac{1 \pm u_r/v}{1 \pm u_s/v} f_0 \qquad \text{15-35}$$

The correct choices for the plus or minus signs are most easily determined by remembering that the frequency increases when the source and receiver are moving toward each other, whereas it decreases when they are moving away from each other. Thus, for example, if the source is moving toward the receiver and the receiver is moving toward the source, the plus sign is used in the numerator and the minus sign is used in the denominator.

It can be shown (see Problem 80) that if both u_s and u_r are much smaller than the wave speed v, then the shift in frequency is given approximately by

$$\frac{\Delta f}{f_0} \approx \pm \frac{u}{v} \qquad (u \ll v) \qquad \text{15-36}$$

where $u = u_s \pm u_r$ is the relative speed of the source and receiver.

If the medium is moving (for example, if air is the medium and there is a wind blowing), the wave speed v is replaced by $v' = v \pm u_w$, where u_w is the speed of the wind.

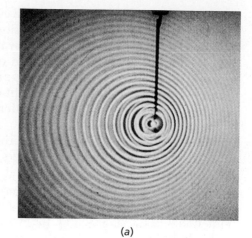

(a)

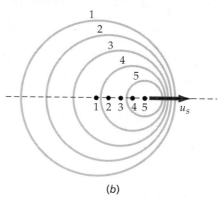

(b)

Figure 15-26 (a) Waves in a ripple tank produced by a point source moving to the right. The wavefronts are closer together in front of the source and farther apart behind the source. (b) Successive wavefronts emitted by a point source moving with speed u_s to the right. The numbers of the wavefronts correspond to the positions of the source when the wave was emitted.

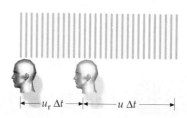

Figure 15-27 The number of waves passing a stationary receiver in time Δt is the number in the distance $v \Delta t$, where v is the wave speed. If the receiver moves toward the source with speed u_r, he passes the additional number of waves in the distance $u_r \Delta t$.

A familiar example of the Doppler effect is the radar used by police to measure the speed of a car. Electromagnetic waves emitted by the radar transmitter strike the moving car. The car acts as both a moving receiver and a moving source as the waves reflect off it back to the radar receiver. Since electromagnetic waves travel at the speed of light, $c = 3 \times 10^8$ m/s, the approximation $u \ll v$ is always valid and Equation 15-36 can be used to calculate the Doppler shift.

Example 15-10

The frequency of a car horn is 400 Hz. Find (a) the wavelength of the sound, and (b) the frequency observed if the car moves with a speed $u_s = 34$ m/s (about 122 km/h) through still air toward a stationary receiver. Take the speed of sound in air to be 340 m/s. (c) Find the frequency observed if the car is stationary and a receiver moves with a speed $u_s = 34$ m/s toward the car.

Picture the Problem (a) The waves in front of the source are compressed, so we use the minus sign in Equation 15-30. (b) We calculate the frequency from $f' = v/\lambda'$. (c) For a moving receiver, the wavelength does not change; the receiver merely passes more waves. We use the plus sign in Equation 15-34.

(a) Calculate the wavelength in front of the car, λ':

$$\lambda' = \frac{v - u_s}{f_0} = \frac{340 \text{ m/s} - 34 \text{ m/s}}{400 \text{ s}^{-1}} = 0.765 \text{ m}$$

(b) Use your result to find the observed frequency:

$$f = \frac{v}{\lambda'} = \frac{340 \text{ m/s}}{0.765 \text{ m}} = 444 \text{ Hz}$$

(c) For a moving receiver, the observed frequency is given by Equation 15-34:

$$f' = f_0\left(1 + \frac{u_r}{v}\right) = f_0\left(1 + \frac{34}{340}\right) = f_0(1.10) = 440 \text{ Hz}$$

Exercise As a train moving at 90 km/h is approaching a stationary listener, it blows its horn, which has a frequency of 630 Hz. (a) What is the wavelength of the sound waves in front of the train? (b) What frequency is heard by the listener? (Use 340 m/s for the speed of sound.) (*Answers* (a) $\lambda_f = 0.5$ m, (b) $f' = 680$ Hz)

Example 15-11   *try it yourself*

The ratio of the frequency of a note to the frequency of the semitone above it on the diatonic scale is about 15:16. How fast is a car going if its horn drops a semitone as it passes you? (Use $v = 340$ m/s for the speed of sound.)

Picture the Problem Let u be the speed of the car and f_0 be the original frequency. The frequency observed as the car approaches f' is greater than f_0, and the frequency observed as the car recedes f'' is less than f_0. Set the ratio $f''/f' = 15/16$ and solve for u.

Cover the column to the right and try these on your own before looking at the answers.

Steps   *Answers*

1. Write the frequency observed as the car ap- $f' = f_0(1 + u/v)$
   proaches in terms of f_0.

2. Write the frequency observed as the car recedes in terms of f_0. $f'' = f_0(1 - u/v)$

3. Set the ratio f''/f' equal to 15/16.

$$\frac{f''}{f'} = \frac{f_0(1 - u/v)}{f_0(1 + u/v)} = \frac{15}{16}$$

4. Solve for u.

$$u = \frac{v}{31} = 11.0 \text{ m/s} = 39.6 \text{ km/h}$$

Example 15-12   *try it yourself*

The radar unit in a stationary police car sends out electromagnetic waves of frequency f_0 that travel at the speed of light c. The waves reflect from a speeding car moving at speed u away from the police car. A frequency difference Δf between the emitted radar and the waves reflected from the speeding car is detected at the police car.* Find u in terms of f_0 and Δf.

Picture the Problem   The frequency of the radar that strikes the speeding car f' is less than f_0 because of the Doppler shift given by Equation 15-36. The car then acts as a moving source emitting waves of frequency f'. The police unit detects waves of frequency $f'' < f'$ because of the Doppler shift due to the moving source. The frequency difference is $f_0 - f''$.

Cover the column to the right and try these on your own before looking at the answers.

Steps   **Answers**

1. Write the frequency f' received by the moving car in terms of f_0, u, and c. $f' = (1 - u/c)f_0$

2. Write the frequency f'' received by the police car in terms of f', u, and c. $f'' = (1 - u/c)f'$

3. Use your result in step 1 to eliminate f'. Simplify using the fact that u is much less than c.

$$f'' = (1 - u/c)^2 f_0 = (1 - 2u/c + u^2/c^2)f_0$$
$$\approx (1 - 2u/c)f_0$$

4. Calculate Δf and solve for u.

$$\Delta f = f_0 - f'' = \frac{2u}{c}f_0, \quad u = \frac{\Delta f}{2f_0}c$$

Exercise   Calculate Δf if $f_0 = 1.5 \times 10^9$ Hz, $c = 3 \times 10^8$ m/s, and $u = 50$ m/s. (*Answer* $\Delta f = 500$ Hz)

*The difference in frequency between two waves of nearly equal frequency is easy to detect because the two waves interfere to produce a wave whose amplitude oscillates with frequency Δf, which is called the beat frequency. Interference and beats are discussed in Chapter 16.

The Doppler Shift and Relativity   We see from Example 15-10 (and Equations 15-33, 15-34, and 15-35) that the magnitude of the Doppler shift in frequency depends on whether it is the source or the receiver that is moving relative to the medium. For sound, these two situations are physically different. For example, if you move relative to still air, you feel air rushing past you. In your reference frame, there is a wind. For sound waves in air, therefore, we can tell whether the source or receiver is moving by noting if there is a wind in the reference frame of the source or the receiver. However, light and other electromagnetic waves propagate through empty space in which there is no medium. There is no "wind" to tell us whether the source or receiver is moving. According to Einstein's theory of relativity, absolute motion cannot be detected, and all observers measure the same speed c for light independent of their motion relative to the source. Thus, Equation 15-35 cannot be correct for the Doppler shift for light. Two modifications must be made in calculating the relativistic Doppler effect for light. First, the speed of waves passing a receiver is c independent of the motion of the receiver. Second, the time interval between the emission of the first wave and the Nth wave, which is $\Delta t = N/f_0$ in the reference frame of the source, is different in the reference frame of the receiver when they are in relative motion because of relativistic time dilation. (We discuss time dilation and the relativistic Doppler effect in Chapter 39.) The result is that the frequency received depends only on the relative speed of approach or recession u, and is related to the frequency emitted by

$$f = \sqrt{\frac{1 + u/c}{1 - u/c}}\, f_0 \qquad \text{(approaching)} \qquad\qquad 15\text{-}37a$$

$$f = \sqrt{\frac{1 - u/c}{1 + u/c}}\, f_0 \qquad \text{(receding)} \qquad\qquad 15\text{-}37b$$

where c is the speed of light. Again, when $u \ll c$, $f/f_0 \approx 1 \pm u/c$, as given by Equation 15-36.

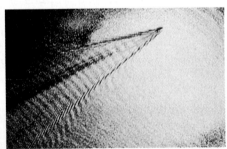

Shock waves from a supersonic airplane.

Bow waves from a boat.

Shock Waves

In our derivations of the Doppler-shift expressions, we assumed that the speed u of the source or receiver was less than the wave speed v. If the receiver moves faster than the wave speed, but toward the source, our derivation is still valid—Equation 15-34 continues to hold for the observed frequency. If the receiver moves faster than the wave speed, but away from the source, then the waves never reach the receiver. If a source moves with speed greater than the wave speed, there will be no waves in front of the source. Instead, the waves pile up behind the source to form a shock wave. In the case of sound waves, this shock wave is heard as a sonic boom when it arrives at the receiver.

Figure 15-28a shows a source originally at point P_1 moving to the right with speed u. After some time t, the wave emitted from point P_1 has traveled

Shock waves produced by a bullet traversing a helium balloon.

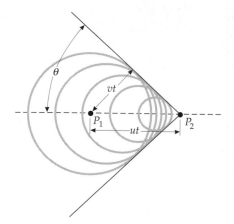

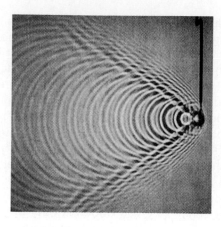

Figure 15-28 (a) Source moving with a speed u that is greater than the wave speed v. The envelope of the wavefronts forms a cone with the source at the apex. (b) Waves in a ripple tank produced by a source moving with a speed $u > v$.

a distance vt. The source has traveled a distance ut and will be at point P_2. The line from this new position of the source to the wavefront emitted when the source was at P_1 makes an angle θ with the path of the source, given by

$$\sin \theta = \frac{vt}{ut} = \frac{v}{u} \qquad \text{15-38}$$

Thus, the shock wave is confined to a cone that narrows as u increases. The ratio of the source speed u to the wave speed v is called the Mach number:

$$\text{Mach number} = \frac{u}{v} \qquad \text{15-39}$$

Equation 15-38 also applies to the electromagnetic radiation called Cerenkov radiation, which is given off when a charged particle moves in a medium with speed u that is greater than the speed of light v in that medium.* The blue glow surrounding the fuel elements used in nuclear reactors is an example of Cerenkov radiation.

Glow produced by Cerenkov radiation in the Waste Encapsulation and Storage Facility at the Hanford weapons complex.

Example 15-13 *try it yourself*

At time $t = 0$, a supersonic plane is directly over point P flying due east at an altitude of 15 km. The sonic boom is heard at point P when the plane is 22 km east of point P (Figure 15-29). What is the speed of the supersonic plane?

Figure 15-29

Cover the column to the right and try these on your own before looking at the answers.

Steps

Answers

1. Sketch the position of the plane when the sonic boom is heard at point P, and calculate $\tan \theta$ from the altitude and distance of the plane; then solve for θ.

$$\tan \theta = \frac{v}{u} = \frac{vt}{ut} = \frac{15 \text{ km}}{22 \text{ km}}$$

$$\theta = 34.3°$$

2. From your result and Equation 15-38, calculate u.

$$u = \frac{v}{\sin \theta} = 604 \text{ m/s}$$

---

* According to the special theory of relativity, it is impossible for a particle to move faster than c, the speed of light in vacuum. In a medium such as glass, however, electrons and other particles can move faster than the speed of light in that medium.

Summary

1. In wave motion, energy and momentum are transported from one point in space to another without the transport of matter.

2. The relation $v = f\lambda$ holds for all harmonic waves.

Topic	Remarks and Relevant Equations	
1. Transverse and Longitudinal Waves	In transverse waves, such as waves on a string, the disturbance is perpendicular to the direction of propagation. In longitudinal waves, such as sound waves, the disturbance is along the direction of propagation.	
2. Speed of Waves	The speed of a wave depends on the density and elastic properties of the medium. It is independent of the motion of the wave source.	
Waves on a string	$v = \sqrt{F/\mu}$	15-3
Sound waves	$v = \sqrt{B/\rho}$	15-4
Sound waves in a gas	$v = \sqrt{\gamma RT/M}$	15-5
	where T is the absolute temperature,	
	$T = t_C + 273$	15-6
	and R is the universal gas constant,	
	$R = 8.314 \text{ J/mol·K}$	15-7
	M is the molar mass of the gas, which for air is 29×10^{-3} kg/mol, and γ is a constant that depends on the kind of gas. For a diatomic gas such as air, $\gamma = 1.4$.	
Electromagnetic waves	The speed of electromagnetic waves such as light in vacuum is a universal constant. $c \approx 3 \times 10^8$ m/s.	
3. Wave Equation (optional)	$\dfrac{\partial^2 y}{\partial x^2} = \dfrac{1}{v^2}\dfrac{\partial^2 y}{\partial t^2}$	15-9*b*
4. Harmonic Waves		
Wave function	$y(x, t) = A\sin(kx - \omega t),$ wave traveling to the right $y(x, t) = A\sin(kx + \omega t),$ wave traveling to the left where A is the amplitude, k is the wave number, and ω is the angular frequency.	15-13
Wave number	$k = \dfrac{2\pi}{\lambda}$	15-12
Angular frequency	$\omega = 2\pi f = \dfrac{2\pi}{T}$	15-15
Speed	$v = f\lambda = \dfrac{\omega}{\kappa}$	15-10, 15-14
Energy	The energy in a harmonic wave is proportional to the square of the amplitude.	
Energy of waves on a string	$\Delta E_{av} = \dfrac{1}{2}\mu\omega^2 y_0^2\,\Delta x$	15-18

Find the change in intensity from the change in intensity level. (optional)	Use Table 15-1. A change of $(n \times 10)$ dB is equivalent to an intensity change by a factor of 10^n.
	Example 15-7

4.   Reflection and Refraction

Find the fraction of power transmitted and reflected given information about the amplitudes, speeds, and densities.	Use $P = \frac{1}{2}\mu\omega^2 A^2 v$ and conservation of energy.
	Example 15-9

5.   The Doppler Effect

Find the frequency of waves from a moving source.	Find the change in wavelength from $\lambda' = (v \pm u_s)/f_0$ and then find the frequency from $f' = v/\lambda'$.
	Examples 15-10, 15-11
Find the frequency observed by a moving receiver.	Use $f' = f_0(1 \pm u_r/v)$.
	Example 15-10
Find the frequency when $u_s, u_r \ll v$.	Use $\Delta f \approx \pm(u/v)f_0$, where $u = u_s \pm u_r$ is the relative speed of the source and receiver.
	Example 15-12

Problems

	Conceptual Problems
	Problems from Optional and Exploring sections

In a few problems, you are given more data than you actually need; in a few other problems, you are required to supply data from your general knowledge, outside sources, or informed estimates.

- Single-concept, single-step, relatively easy
- •• Intermediate-level, may require synthesis of concepts
- ••• Challenging, for advanced students

Use $v = 340$ m/s for the speed of sound in air unless otherwise indicated.

Speed of Waves

1   • A rope hangs vertically from the ceiling. Do waves on the rope move faster, slower, or at the same speed as they move from bottom to top? Explain.

2   • (a) The bulk modulus for water is 2.0×10^9 N/m². Use it to find the speed of sound in water. (b) The speed of sound in mercury is 1410 m/s. What is the bulk modulus for mercury ($\rho = 13.6 \times 10^3$ kg/m³)?

3   • Calculate the speed of sound waves in hydrogen gas at $T = 300$ K. (Take $M = 2$ g/mol and $\gamma = 1.4$.)

4   • A steel wire 7 m long has a mass of 100 g. It is under a tension of 900 N. What is the speed of a transverse wave pulse on this wire?

5   • Transverse waves travel at 150 m/s on a wire of length 80 cm that is under a tension of 550 N. What is the mass of the wire?

6   • A wave pulse propagates along a wire in the positive x direction at 20 m/s. What will the pulse velocity be if we (a) double the length of the wire but keep the tension and mass per unit length constant? (b) double the tension while holding the length and mass per unit length constant? (c) double the mass per unit length while holding the other variables constant?

7   • A steel piano wire is 0.7 m long and has a mass of 5 g. It is stretched with a tension of 500 N. (a) What is the speed of transverse waves on the wire? (b) To reduce the wave speed by a factor of 2 without changing the tension, what mass of copper wire would have to be wrapped around the steel wire?

8   • The cable of a ski lift runs 400 m up a mountain and has a mass of 80 kg. When the cable is struck with a transverse blow at one end, the return pulse is detected 12 s later. (a) What is the speed of the wave? (b) What is the tension in the cable?

9   •• A common method for estimating the distance to a lightning flash is to begin counting when the flash is observed and continue until the thunder clap is heard. The number of seconds counted is then divided by 3 to get the distance in kilometers. (a) What is the velocity of sound in kilometers per second? (b) How accurate is this procedure? (c) Is a correction for the time it takes for the light to reach you important? (The speed of light is 3×10^8 m/s.)

10 •• A method for measuring the speed of sound using an ordinary watch with a second hand is to stand some distance from a large flat wall and clap your hands rhythmically in such a way that the echo from the wall is heard halfway between every two claps. (a) Show that the speed of sound is given by $v = 4LN$, where L is the distance to the wall and N is the number of claps per second. (b) What is a reasonable value for L for this experiment to be feasible? (If you have access to a flat wall outdoors somewhere, try this method and compare your result with the standard value for the speed of sound.)

11 •• A man drops a stone from a high bridge and hears it strike the water below exactly 4 s later. (a) Estimate the distance to the water based on the assumption that the travel time for the sound to reach the man is negligible. (b) Improve your estimate by using your result from part (a) for the distance to the water to estimate the time it takes for sound to travel this distance and then calculate the distance the rock falls in 4 s minus this time. (c) Calculate the exact distance and compare your result with your previous estimates.

12 •• (a) Compute the derivative of the speed of a wave on a string with respect to the tension dv/dF, and show that the differentials dv and dF obey $dv/v = \frac{1}{2}dF/F$. (b) A wave moves with a speed of 300 m/s on a wire that is under a tension of 500 N. Using dF to approximate a change in tension, determine how much the tension must be changed to increase the speed to 312 m/s.

13 •• (a) Compute the derivative of the velocity of sound with respect to the absolute temperature, and show that the differentials dv and dT obey $dv/v = \frac{1}{2}dT/T$. (b) Use this result to compute the percentage change in the velocity of sound when the temperature changes from 0 to 27°C. (c) If the speed of sound is 331 m/s at 0°C, what is it (approximately) at 27°C? How does this approximation compare with the result of an exact calculation?

14 ••• In this problem, you will derive a convenient formula for the speed of sound in air at temperature t in Celsius degrees. Begin by writing the temperature as $T = T_0 + \Delta T$, where $T_0 = 273$ K corresponds to 0°C and $\Delta T = t$, the Celsius temperature. The speed of sound is a function of T, $v(T)$. To a first-order approximation, you can write

$$v(T) \approx v(T_0) + (dv/dT)_{T_0} \Delta T$$

where $(dv/dt)_{T_0}$ is the derivative evaluated at $T = T_0$. Compute this derivative, and show that the result leads to

$$v = (331 \text{ m/s}) \left(1 + \frac{t}{2T_0}\right) = (331 + 0.606t)\text{m/s}$$

15 ••• While studying physics in her dorm room, a student is listening to a live radio broadcast of a baseball game. She is 1.6 km due south of the baseball field. Over her radio, the student hears a noise generated by the electromagnetic pulse of a lightning bolt. Two seconds later, she hears over the radio the thunder picked up by the microphone at the baseball field. Four seconds after she hears the noise of the electromagnetic pulse over the radio, thunder rattles her windows. Where, relative to the ballpark, did the lightning bolt occur?

16 ••• A coiled spring, such as a Slinky, is stretched to a length L. It has a force constant k and a mass m. (a) Show that the velocity of longitudinal compression waves along the spring is given by $v = L\sqrt{k/m}$. (b) Show that this is also the velocity of transverse waves along the spring if the natural length of the spring is much less than L.

The Wave Equation (optional)

17 • Show explicitly that the following functions satisfy the wave equation: (a) $y(x, t) = k(x + vt)^3$; (b) $y(x, t) = Ae^{ik(x-vt)}$, where A and k are constants and $i = \sqrt{-1}$; and (c) $y(x, t) = \ln k(x - vt)$.

18 • Show that the function $y = A \sin kx \cos wt$ satisfies the wave equation.

19 ••• Consider the following equation:

$$\frac{\partial^2 y}{\partial x^2} + i\alpha \frac{\partial y}{\partial t} = 0, \qquad i = \sqrt{-1}$$

where α is a constant. Show that $y(x, t) = A \sin(kx - \omega t)$ is not a solution of this equation but that the functions $y(x, t) = Ae^{i(kx - \omega t)}$ and $y(x, t) = Ae^{i(kx + \omega t)}$ do satisfy that equation.

Harmonic Waves on a String

20 • A traveling wave passes a point of observation. At this point, the time between successive crests is 0.2 s. Which of the following is true?

(a) The wavelength is 5 m.
(b) The frequency is 5 Hz.
(c) The velocity of propagation is 5 m/s.
(d) The wavelength is 0.2 m.
(e) There is not enough information to justify any of these statements.

21 • True or false: The energy in a wave is proportional to the square of the amplitude of the wave.

22 • A rope hangs vertically. You shake the bottom back and forth, creating a sinusoidal wave train. Is the wavelength at the top the same as, less than, or greater than the wavelength at the bottom?

23 • One end of a string 6 m long is moved up and down with simple harmonic motion at a frequency of 60 Hz. The waves reach the other end of the string in 0.5 s. Find the wavelength of the waves on the string.

24 • Equation 15-13 expresses the displacement of a harmonic wave as a function of x and t in terms of the wave parameters k and ω. Write the equivalent expressions that contain the following pairs of parameters instead of k and ω: (a) k and v, (b) λ and f, (c) λ and T, (d) λ and v, and (e) f and v.

25 • Equation 15-10 applies to all types of periodic waves, including electromagnetic waves such as light waves and microwaves, which travel at 3×10^8 m/s in a vacuum.

calculations for a stationary piemaker and a pie eater who moves toward the piemaker at 30 m/min.

64 • For the situation described in Problem 63, derive general expressions for the spacing of the pies λ and the frequency f with which they are received by the pie eater in terms of the speed of the belt v, the speed of the sender u_s, the speed of the receiver u_r, and the frequency f_0 with which the piemaker places pies on the belt.

In Problems 65 through 70, a source emits sounds of frequency 200 Hz that travel through still air at 340 m/s.

65 • The sound source described above moves with a speed of 80 m/s relative to still air toward a stationary listener. (a) Find the wavelength of the sound between the source and the listener. (b) Find the frequency heard by the listener.

66 • Consider the situation in Problem 65 from the reference frame in which the source is at rest. In this frame, the listener moves toward the source with a speed of 80 m/s, and there is a wind blowing at 80 m/s from the listener to the source. (a) What is the speed of the sound from the source to the listener in this frame? (b) Find the wavelength of the sound between the source and the listener. (c) Find the frequency heard by the listener.

67 • The source moves away from the stationary listener at 80 m/s. (a) Find the wavelength of the sound waves between the source and the listener. (b) Find the frequency heard by the listener.

68 • The listener moves at 80 m/s relative to still air toward the stationary source. (a) What is the wavelength of the sound between the source and the listener? (b) What is the frequency heard by the listener?

69 • Consider the situation in Problem 68 in a reference frame in which the listener is at rest. (a) What is the wind velocity in this frame? (b) What is the speed of the sound from the source to the listener in this frame, that is, relative to the listener? (c) Find the wavelength of the sound between the source and the listener in this frame. (d) Find the frequency heard by the listener.

70 • The listener moves at 80 m/s relative to the still air away from the stationary source. Find the frequency heard by the listener.

71 • A jet is traveling at Mach 2.5 at an altitude of 5000 m. (a) What is the angle that the shock wave makes with the track of the jet? (Assume that the speed of sound at this altitude is still 340 m/s.) (b) Where is the jet when a person on the ground hears the shock wave?

72 • If you are running at top speed toward a source of sound at 1000 Hz, estimate the frequency of the sound that you hear. Suppose that you can recognize a change in frequency of 3%. Can you use your sense of pitch to estimate your running speed?

73 •• A radar device emits microwaves with a frequency of 2.00 GHz. When the waves are reflected from a car moving directly away from the emitter, a frequency difference of 293 Hz is detected. Find the speed of the car.

74 •• A stationary destroyer is equipped with sonar that sends out pulses of sound at 40 MHz. Reflected pulses are received from a submarine directly below with a time delay of 80 ms at a frequency of 39.958 MHz. If the speed of sound in seawater is 1.54 km/s, find (a) the depth of the submarine, and (b) its vertical speed.

75 •• Two airplanes, one flying due east and the other due west, are on a near collision course separated by 15 km when the pilot of one plane, traveling at 900 km/h, observes the other on his Doppler radar. The radar unit emits electromagnetic waves of frequency 3×10^{10} Hz. The radar readout indicates that the other plane's speed is 750 km/h. Determine the frequency of the signal received by the pilot's radar.

76 •• A police radar unit transmits microwaves of frequency 3×10^{10} Hz. The speed of these waves in air is 3.0×10^8 m/s. Suppose a car is receding from the stationary police car at a speed of 140 km/h. What is the frequency difference between the transmitted signal and the signal received from the receding car?

77 •• Suppose the police car of Problem 76 is moving in the same direction as the other vehicle at a speed of 60 km/h. What then is the difference in frequency between the emitted and the reflected signals?

78 •• At time $t = 0$, a supersonic plane is directly over point P flying due west at an altitude of 12 km and a speed of Mach 1.6. Where is the plane when the sonic boom is heard?

79 •• A small radio of 0.10 kg mass is attached to one end of an air track by a spring. The radio emits a sound of 800 Hz. A listener at the other end of the air track hears a sound whose frequency varies between 797 and 803 Hz. (a) Determine the energy of the vibrating mass–spring system. (b) If the spring constant is 200 N/m, what is the amplitude of vibration of the mass and what is the period of the oscillating system?

80 •• A sound source of frequency f_0 moves with speed u_s relative to still air toward a receiver who is moving with speed u_r relative to still air away from the source. (a) Write an expression for the received frequency f'. (b) Use the result that $(1 - x)^{-1} \approx 1 + x$ to show that if both u_s and u_r are small compared to v, then the received frequency is approximately

$$f' \approx \left(1 + \frac{u_s - u_r}{v}\right)f_0 = \left(1 + \frac{u_{\text{rel}}}{v}\right)f_0$$

where u_{rel} is the relative velocity of the source and receiver.

81 •• Two students with vibrating 440-Hz tuning forks walk away from each other with equal speeds. How fast must they walk so that they each hear a frequency of 438 Hz from the other fork?

82 •• A physics student walks down a long hall carrying a vibrating 512-Hz tuning fork. The end of the hall is closed so that sound reflects from it. The student hears a sound of 516 Hz from the wall. How fast is the student walking?

83 •• A small speaker radiating sound at 1000 Hz is tied to one end of an 0.8-m-long rod that is free to rotate about its other end. The rod rotates in the horizontal plane at 4.0 rad/s. Derive an expression for the frequency heard by a stationary observer far from the rotating speaker.

84 •• You have won a free trip on the *Queen Elizabeth II* and are in mid-Atlantic steaming due east at 45 km/h as the Concorde passes directly overhead flying due west at Mach 1.6 at an altitude of 12,500 m. Where is the Concorde relative to the *QEII* when you hear the sonic boom?

85 •• A balloon driven by a 36-km/h wind emits a sound of 800 Hz as it approaches a tall building. (*a*) What is the frequency of the sound heard by an observer at the window of this building? (*b*) What is the frequency of the reflected sound heard by a person riding in the balloon?

86 •• A car is approaching a reflecting wall. A stationary observer behind the car hears a sound of frequency 745 Hz from the car horn and a sound of frequency 863 Hz from the wall. (*a*) How fast is the car traveling? (*b*) What is the frequency of the car horn? (*c*) What frequency does the car driver hear reflected from the wall?

87 •• The driver of a car traveling at 100 km/h toward a vertical cliff briefly sounds the horn. Exactly one second later she hears the echo and notes that its frequency is 840 Hz. How far from the cliff was the car when the driver sounded the horn and what is the frequency of the horn?

88 •• You are on a transatlantic flight traveling due west at 800 km/h. A Concorde flying at Mach 1.6 and 3 km to the north of your plane is also on an east-to-west course. What is the distance between the two planes when you hear the sonic boom from the Concorde?

89 ••• Astronomers can deduce the existence of a binary star system even if the two stars cannot be visually resolved by noting an alternating Doppler shift of a spectral line. Suppose that an astronomical observation shows that the source of light is eclipsed once every 18 h. The wavelength of the spectral line observed changes from a maximum of 563 nm to a minimum of 539 nm. Assume that the double star system consists of a very massive, dark object and a relatively light star that radiates the observed spectral line. Use the data to determine the separation between the two objects (assume that the light object is in a circular orbit about the massive one) and the mass of the massive object. (Use the approximation $\Delta f/f_0 \approx v/c$.)

90 ••• A physics student drops a vibrating 440-Hz tuning fork down the elevator shaft of a tall building. When the student hears a frequency of 400 Hz, how far has the tuning fork fallen?

General Problems

91 • When a guitar string is plucked, is the wavelength of the wave it produces in air the same as the wavelength of the wave on the string?

92 • A wave pulse travels along a light string that is attached to a heavier string in which the wave speed is smaller. The reflected pulse is _____, and the transmitted pulse is _____.

(*a*) inverted/inverted
(*b*) inverted/not inverted
(*c*) not inverted/not inverted
(*d*) not inverted/inverted
(*e*) nonexistent/not inverted

93 • True or false:

(*a*) Wave pulses on strings are transverse waves.
(*b*) Sound waves in air are transverse waves of compression and rarefaction.
(*c*) The speed of sound at 20°C is twice that at 5°C.

94 • Sound travels at 340 m/s in air and 1500 m/s in water. A sound of 256 Hz is made under water. In the air, the frequency will be

(*a*) the same, but the wavelength will be shorter.
(*b*) higher, but the wavelength will stay the same.
(*c*) lower, but the wavelength will be longer.
(*d*) lower, and the wavelength will be shorter.
(*e*) the same, and the wavelength too will stay the same.

95 •• Figure 15-30 shows a wave pulse at time $t = 0$ moving to the right. At this particular time, which segments of the string are moving up? Which are moving down? Is there any segment of the string at the pulse that is instantaneously at rest? Answer these questions by sketching the pulse at a slightly later time and a slightly earlier time to see how the segments of the string are moving.

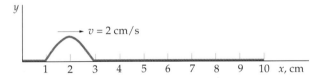

Figure 15–30 Problems 95 and 96

96 •• Make a sketch of the velocity of each string segment versus position for the pulse shown in Figure 15-30.

97 •• Consider a long line of cars equally spaced by one car length and moving slowly with the same speed. One car suddenly slows to avoid a dog and then speeds up until it is again one car length behind the car ahead. Discuss how the space between cars propagates back along the line. How is this like a wave pulse? Is there any transport of energy? What does the speed of propagation depend on?

98 • At time $t = 0$, the shape of a wave pulse on a string is given by the function

$$y(x, 0) = \frac{0.12 \text{ m}^3}{(2.00 \text{ m})^2 + x^2}$$

where x is in meters. (*a*) Sketch $y(x, 0)$ versus x. Give the wave function $y(x, t)$ at a general time t if (*b*) the pulse is moving in the positive x direction with a speed of 10 m/s, and (*c*) the pulse is moving in the negative x direction with a speed of 10 m/s.

99 • A wave with frequency of 1200 Hz propagates along a wire that is under a tension of 800 N. The wavelength of the wave is 24 cm. What will be the wavelength if the tension is decreased to 600 N and the frequency is kept constant?

100 • In a common lecture demonstration of wave pulses, a piece of rubber tubing is tied at one end to a fixed post and is passed over a pulley to a weight hanging at the other end. Suppose that the distance from the fixed support

CHAPTER 16

Superposition and Standing Waves

Water waves on the surface of the ocean diffract as they encounter an aperture formed by the breakwater.

When two waves meet in space, their individual disturbances (represented mathematically by their wave functions) superimpose and add algebraically, creating a new wave. The superposition of harmonic waves is called interference. Interference, like diffraction, is an important wave phenomenon. It was the observation of interference of light by Young in 1801 that led to the understanding that light propagates as a wave motion, not a particle motion as had been proposed by Newton. (It was the inability to observe interference or diffraction of light in Newton's time that led Newton to his particle model of light.) The observation of interference of electron waves by Davisson and Germer in 1927 led to our understanding of the wave nature of electrons and eventually to quantum physics, which we will study in Chapter 17.

In this chapter we begin by studying the superposition of wave pulses on a string and then consider the superposition and interference of harmonic waves. We will examine the phenomenon of beats, which result from the interference of two waves of slightly different frequencies, and we will then study standing waves, which result from the interference of two harmonic waves of the same frequency traveling in opposite directions in a confined space. Finally, we will consider the analysis of complex musical tones in terms of their component harmonic waves, and the inverse problem of the synthesis of harmonic waves to produce complex tones. We will conclude with a qualitative discussion of the extension of harmonic analysis to nonperiodic waves such as wave pulses.

16-1 Superposition of Waves

Figure 16-1 shows small wave pulses moving in opposite directions on a string. The shape of the string when they meet can be found by adding the displacements produced by each pulse separately. The **principle of superposition** is a property of wave motion that states

> When two or more waves combine, the resultant wave is the algebraic sum of the individual waves.

Principle of superposition

Mathematically, when there are two pulses on the string, the total wave function is the algebraic sum of the individual wave functions.

In the special case of two pulses that are identical except that one is inverted relative to the other, as in Figure 16-1b, there will be a moment in time when the pulses exactly overlap and add to zero. At this time the string is horizontal, but it is not at rest. Just to the right of the overlap region the string is moving up whereas just to the left it is moving down. A short time later the pulses emerge, each continuing in its original direction.

Superposition is a characteristic and unique property of wave motion. There is no analogous situation in particle motion; that is, two particles never overlap or add together in this way.

Superposition and the Wave Equation

The principle of superposition follows from the fact that the wave equation (Equation 15-9) is linear for small transverse displacements. That is, the function $y(x, t)$ and its derivatives occur only to the first power. An important property of linear equations is that if y_1 and y_2 are two solutions of the wave equation, the linear combination

$$y_3 = C_1 y_1 + C_2 y_2 \qquad \text{16-1}$$

is also a solution, where C_1 and C_2 are any constants. This can be shown by the direct substitution of y_3 into the wave equation. This result is the mathematical statement of the principle of superposition. If any two waves satisfy a wave equation, their sum also satisfies the same wave equation.

Exercise Show that the function y_3 given by Equation 16-1 satisfies Equation 15-9b if y_1 and y_2 both satisfy Equation 15-9b.

Interference of Harmonic Waves

The result of the superposition of harmonic waves depends on the phase difference between the waves. Let y_1 be the wave function for a harmonic wave traveling to the right with amplitude y_0, angular frequency ω, and wave number k:

$$y_1 = y_0 \sin(kx - \omega t) \qquad \text{16-2}$$

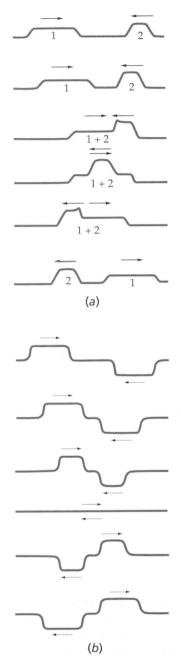

(a)

(b)

Figure 16-1 Wave pulses moving in opposite directions on a string. The shape of the string when the pulses meet is found by adding the displacements of each separate pulse. (a) Superposition of pulses having displacements in the same direction. (b) Superposition of pulses having opposite displacements. Here the algebraic addition of the displacements amounts to the subtraction of the magnitudes.

optional

method of achieving this splitting is by the diffraction of a light beam by two small openings or slits in an opaque barrier (Figure 16-10). The intensity pattern on a screen far from the slits is shown in Figure 16-10b. This method was used by Thomas Young in 1801 to demonstrate that light exhibits interference and is therefore a wave phenomenon, rather than a particle motion as proposed by Newton. The demonstration is now known as Young's experiment. The intensity is maximum when the difference in path from the point on the screen to the two slits is an integer times the wavelength of the light. When this path difference is $\frac{1}{2}\lambda$, $\frac{3}{2}\lambda$, ..., the intensity is zero. We discuss Young's experiment further in Chapter 17.

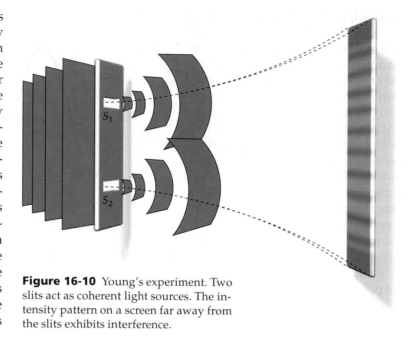

Figure 16-10 Young's experiment. Two slits act as coherent light sources. The intensity pattern on a screen far away from the slits exhibits interference.

16-2 Standing Waves

When waves are confined in space, like the waves on a piano string or sound waves in an organ pipe, reflections at both ends cause the waves to travel in both directions. These waves combine according to the principle of superposition. For a given string or pipe, there are certain frequencies for which superposition results in a stationary vibration pattern called a **standing wave.** Standing waves have important applications in musical instruments and in quantum theory.

Figure 16-11 Standing waves on a string that is fixed at both ends. Points labeled A are antinodes and those labeled N are nodes. In general, the nth harmonic has n antinodes.

String Fixed at Both Ends

If we fix both ends of a string and move a portion of the string up and down with simple harmonic motion of small amplitude, we find that at certain frequencies, standing-wave patterns such as those shown in Figure 16-11 are produced. The frequencies that produce these patterns are called the **resonance frequencies** of the string system. Each such frequency with its accompanying wave function is called a **mode of vibration.** The lowest resonance frequency is called the fundamental frequency f_1. It produces the standing-wave pattern shown in Figure 16-11a, which is called the **fundamental mode** of vibration or the **first harmonic.** The second lowest frequency f_2 produces the pattern shown in Figure 16-11b. This mode of vibration has a frequency twice that of the fundamental frequency and is called the second harmonic.* The third lowest frequency f_3

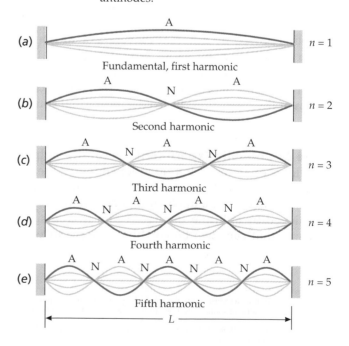

---

*In the terminology often used in music, the second harmonic is called the first **overtone,** the third harmonic is called the second overtone, and so on.

is three times the fundamental frequency, and it produces the third harmonic pattern shown in Figure 16-11c.

We note from Figure 16-11 that for each harmonic there are certain points on the string (the midpoint in Figure 16-11b, for example) that do not move. Such points are called **nodes.** Midway between each pair of nodes is a point of maximum amplitude of vibration called an **antinode.** Both fixed ends of the string are, of course, nodes.* We note that the first harmonic has one antinode, the second harmonic has two antinodes, and so on.

We can relate the resonance frequencies to the wave speed in the string and the length of the string. The length L of the string equals one-half the wavelength in the fundamental mode of vibration (Figure 16-12), and as Figure 16-11 reveals, L equals two half-wavelengths for the second harmonic, $\frac{3}{2}\lambda$ for the third harmonic, and so forth. In general, if λ_n is the wavelength for the nth harmonic, we have

Figure 16-12

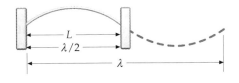

$$L = n\frac{\lambda_n}{2}, \qquad n = 1, 2, 3, \ldots \qquad \text{16-10}$$

Standing-wave condition, both ends fixed

This result is known as the **standing-wave condition.** We can find the frequency of the nth harmonic from the fact that the wave speed v equals the frequency f times the wavelength. Thus,

$$f_n = \frac{v}{\lambda_n} = \frac{v}{2L/n}$$

or

$$f_n = n\frac{v}{2L} = nf_1, \qquad n = 1, 2, 3, \ldots \qquad \text{16-11}$$

Resonance frequencies, both ends fixed

where $f_1 = v/2L$ is the fundamental frequency. You don't need to memorize Equation 16-11. Just sketch Figure 16-11 to remind yourself of the standing-wave condition, $L = n\lambda_n/2$, and then use $f = v/\lambda$.

We can understand standing waves in terms of resonance. Consider a string of length L that is attached at one end to a vibrating tuning fork and is fixed at the other end. The first wave sent out by the tuning fork travels down the string a distance L to the fixed end, where it is reflected and inverted. It then travels back a distance L and is again reflected and inverted at the tuning fork. The total time for the round trip is $2L/v$. If this time equals the period of the vibrating fork, the twice-reflected wave exactly overlaps the second wave produced by the fork, and the two waves interfere constructively, producing a wave with twice the original amplitude. The combined wave travels down the string and back and adds to the third wave produced by the fork, increasing the amplitude threefold, and so on. Thus, the tuning fork is in resonance with the string. Resonance also occurs if the time it takes for the first wave to travel the distance $2L$ is twice the period of the vibrating

A classic Steinway grand piano. The strings vibrate when struck by the hammers, which are controlled by the keys. The longer strings (left) vibrate at lower frequencies than the shorter strings (right).

---

* If one end is attached to a tuning fork rather than being fixed, it will still be approximately a node because the amplitude of the vibration at that end is so much smaller than the amplitude at the antinodes.

Most musical instruments are much more complicated than simple cylindrical tubes. The conical tube, which is the basis for the oboe, bassoon, English horn, and saxophone, has a complete harmonic series with its fundamental wavelength equal to twice the length of the cone. Brass instruments are combinations of cones and cylinders. The analysis of these instruments is extremely complex. The fact that they have nearly harmonic series is a triumph of educated trial and error rather than mathematical calculation.

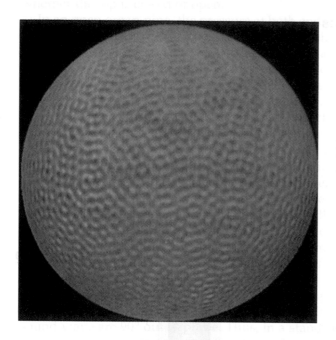

Standing sound waves on the surface of the sun. The surface of the sun is observed to oscillate with a period of about 5 min. Out of the 10 million modes of oscillation known to exist, a combination of approximately 100 modes is illustrated here. The displacements of the surface are exaggerated by a factor of 1000. The period of oscillation for each mode contains information about the structure and dynamics of the solar interior.

Holographic interferograms showing standing waves in a handbell. The "bull's eyes" locate the antinodes.

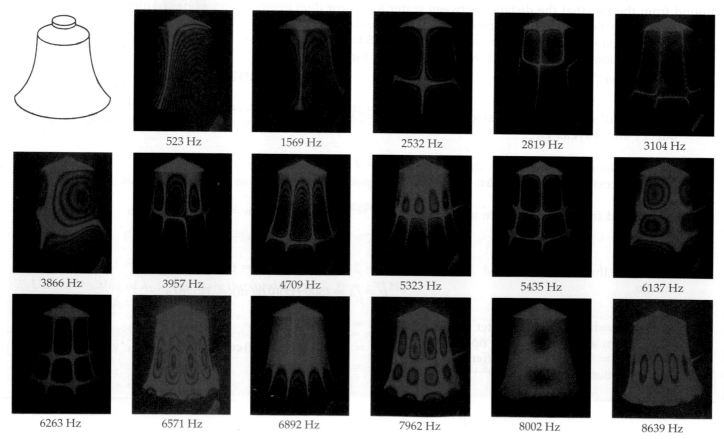

16-3 The Superposition of Standing Waves

We have just seen that there is a set of natural resonance frequencies that produce standing waves for sound waves in air columns or vibrating strings
that are fixed at one or both ends. For example, for a string fixed at both ends,
the frequency of the fundamental mode of vibration is $f_1 = v/2L$, where L is
the length of the string and v is the wave speed, and the wave function is
Equation 16-16:

$$y_1(x,t) = A_1 \sin k_1 x \cos(\omega_1 t + \delta_1)$$

In general, a vibrating system does not vibrate in a single harmonic mode.
Instead, the motion consists of a mixture of the allowed harmonics. The wave
function is a linear combination of the harmonic wave functions:

$$y(x,t) = \sum_n A_n \sin k_n x \cos(\omega_n t + \delta_n) \qquad \text{16-17}$$

where $k_n = 2\pi/\lambda_n$, $\omega_n = 2\pi f_n$, and A_n and δ_n are constants. Since the energy
in a wave is proportional to the square of the amplitude, the quantity A_n^2 describes the fraction of the energy associated with the nth harmonic. The constants A_n and δ_n depend on the initial position and velocity of the string. If a
harp string, for example, is plucked at the center and released, as in Figure
16-20, the initial shape of the string is symmetric about the point $x = \frac{1}{2}L$. The
motion of the string after it has been released will remain symmetric about
this point. Only the odd harmonics, which are also symmetric about $x = \frac{1}{2}L$,
will be excited. The even harmonics, which are antisymmetric about $x = \frac{1}{2}L$,
are not excited; that is, the constant A_n is zero for all even n. The shapes of the
first four harmonics are shown in Figure 16-21. Most of the energy of the
plucked string is associated with the fundamental, but small amounts of energy are associated with the third and higher odd harmonic modes. Figure
16-22 shows an approximation to the initial shape of the string using the superposition of only the first three odd harmonics.

Figure 16-20 A string plucked at the center. When it is released, its vibration is a
linear combination of standing waves.

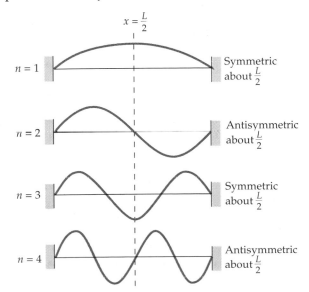

Figure 16-21 The first four harmonics for
a string fixed at both ends. The odd harmonics are symmetrical about the center of
the string, whereas the even harmonics are
not. When a string is plucked at the center,
it vibrates only in its odd harmonics.

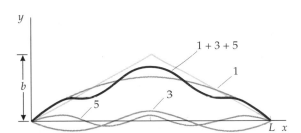

Figure 16-22 Approximating the shape
of a string plucked at the center, as in Figure 16-20, using harmonics. The red line is
an approximation of the original shape of
the string based on the first three odd harmonics. The height of the string is exaggerated in this drawing to show the relative amplitudes of the harmonics. Most of
the energy is associated with the fundamental, but there is some energy in the
third, fifth, and other odd harmonics.

PART III

Two red intersecting laser beams are used here to study the combustion of coal–water slurries in a conventional power conversion device. The test material is injected into the combustion reactor (blue flame), giving off a yellow-orange emission as it ignites and burns. The laser light is used to measure the particle size of combustible material.

thermodynamics

CHAPTER 18

Temperature and the Kinetic Theory of Gases

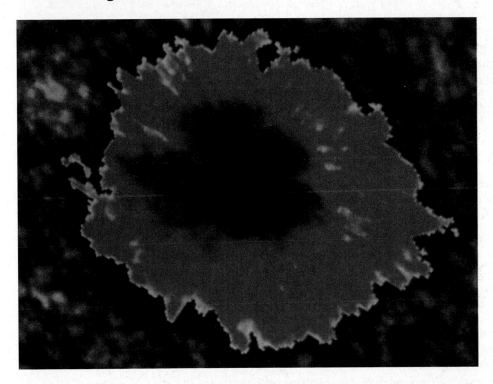

Sunspots appear on the surface of the sun when streams of gases slowly erupt from deep within the star. The solar "flower" is 10,000 miles in diameter. The temperature variation, indicated by computer-enhanced color changes, is not fully understood. The central portion of the sunspot is cooler than the outer regions as indicated by the dark area. The temperature at the sun's core is of the order of 10^7 K, whereas at the surface the temperature is only about 6000 K.

Temperature is familiar to us all as the measure of the hotness or coldness of objects. In this chapter we show that a consistent temperature scale can be defined in terms of the properties of gases at low densities, and that temperature is a measure of the average internal molecular kinetic energy of an object.

18-1 Thermal Equilibrium and Temperature

Our sense of touch can usually tell us if an object is hot or cold. Early in childhood we learn that to make a cold object warmer, we place it in contact with a hot object. To make a warm object cool, we place it in contact with a cold object.

When an object is heated or cooled, some of its physical properties change. Most solids and liquids expand when they are heated. A gas, if permitted, will also expand when it is heated, or, if its volume is kept constant, its pressure will rise. If an electrical conductor is heated, its electrical resistance changes. A physical property that changes with temperature is called a **thermometric property.** A change in a thermometric property indicates a change in the temperature of the object.

Suppose we place a warm copper bar in close contact with a cold iron bar so that the copper bar cools and the iron bar warms. We say the two bars are in **thermal contact**. The copper bar contracts slightly as it cools, and the iron bar expands slightly as it warms. Eventually this process stops and the lengths of the bars remain constant. The two bars are then in **thermal equilibrium** with each other.

Suppose instead we place the warm copper bar in a cool lake. The bar cools until it and the water are in thermal equilibrium. (We assume the lake is large enough so that the warming of its water will be negligible.) Next we place a cold iron bar in the lake far away from the copper bar. The iron bar will warm until it and the lake water are also in thermal equilibrium. If we remove the bars and place them in thermal contact with each other, we find their lengths do not change. They are in thermal equilibrium with each other. Though it is common sense, there is no logical way to deduce this fact, which is called the **zeroth law of thermodynamics** (Figure 18-1):

If two objects are in thermal equilibrium with a third, then they are in thermal equilibrium with each other.

Zeroth law of thermodynamics

Two objects are defined to have the **same temperature** if they are in thermal equilibrium with each other. The zeroth law, as we will see, enables us to define a temperature scale.

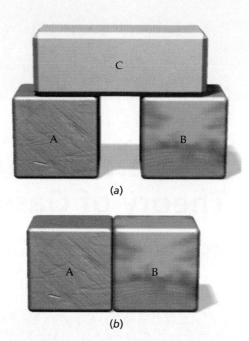

(a)

(b)

Figure 18-1 The zeroth law of thermodynamics. (*a*) Systems A and B are in thermal contact with system C but not with each other. When A and B are each in thermal equilibrium with C, they are in thermal equilibrium with each other, which can be checked by placing them in contact with each other as in (*b*).

18-2   The Celsius and Fahrenheit Temperature Scales

Any thermometric property can be used to establish a temperature scale. The common mercury thermometer consists of a glass bulb and tube containing a fixed amount of mercury. When this thermometer is put in contact with a warmer body, the mercury expands, increasing the length of the mercury column (the glass expands too, but by a negligible amount). We can create a scale along the glass tube as follows. First the thermometer is placed in ice and water in equilibrium* at a pressure of 1 atm. When the thermometer is in thermal equilibrium with the ice water, the position of the mercury column is marked on the glass tube. This is the **ice-point temperature** (also called the **normal freezing point** of water). Next, the thermometer is placed in boiling water at a pressure of 1 atm. When the thermometer is in thermal equilibrium with the boiling water, the new position of the column is marked. This is the **steam-point temperature** (also called the **normal boiling point** of water).

The **Celsius temperature scale** defines the ice-point temperature as zero degrees Celsius (0°C) and the steam-point temperature as 100°C. The space between the 0 and 100° marks is divided into 100 equal intervals (degrees). Degree markings are also extended below and above these points. If L_t is the length of the mercury column, the Celsius temperature t_C is given by

$$t_C = \frac{L_t - L_0}{L_{100} - L_0} \times 100° \qquad 18\text{-}1$$

where L_0 is the length of the mercury column when the thermometer is in an ice bath and L_{100} is its length when the thermometer is in a steam bath. The

---

*Water and ice in equilibrium provide a constant-temperature bath. When ice is placed in warm water, the water cools as some of the ice melts. Eventually, thermal equilibrium is reached and no more ice melts. If the system is heated slightly, some more of the ice melts, but the temperature does not change as long as some ice remains.

normal temperature of the human body measured on the Celsius scale is about 37°C.

The **Fahrenheit temperature scale** (which is used in the United States) defines the ice-point temperature as 32°F and the steam-point temperature as 212°F.* To convert temperatures between Fahrenheit and Celsius, we note there are 100 Celsius degrees and 180 Fahrenheit degrees between the ice and steam points. A temperature change of one Celsius degree therefore equals a change of $1.8 = \frac{9}{5}$ Fahrenheit degrees. To convert a temperature from one scale to the other, we must also take into account the fact that the zero temperatures of the two scales are not the same. The general relation between a Fahrenheit temperature t_F and Celsius temperature t_C is

$$t_C = \frac{5}{9}(t_F - 32°)$$ 18-2

Fahrenheit–Celsius conversion

Example 18-1

(a) **Find the temperature on the Celsius scale equivalent to 41°F.** (b) **Find the temperature on the Fahrenheit scale equivalent to −10°C.**

(a)   Apply Equation 18-2 with $t_F = 41$°F: $t_C = \frac{5}{9}(t_F - 32°) = \frac{5}{9}(41° - 32°) = \frac{5}{9}(9°) = 5°C$

(b)1. Solve Equation 18-2 for t_F in terms of t_C: $t_F = \frac{9}{5}t_C + 32°$

  2. Substitute $t_C = -10$°C: $t_F = \frac{9}{5}(-10°) + 32° = -18° + 32° = 14°F$

Exercise (a) Find the Celsius temperature equivalent to 68°F. (b) Find the Fahrenheit temperature equivalent to −40°C. (*Answers*   (a) 20°C, (b) −40°F)

Other thermometric properties can be used to set up thermometers and construct temperature scales. Figure 18-2 shows a bimetallic strip consisting of two different metals bonded together. When the strip is heated or cooled, it bends to accommodate the difference in the thermal expansion of the two metals. Figure 18-3 shows a thermometer consisting of a bimetallic coil with

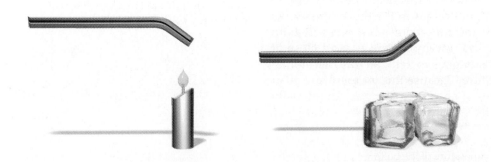

Figure 18-2 A bimetallic strip. When heated or cooled the two metals expand or contract by different amounts, causing the strip to bend.

---

* When the German physicist Daniel Fahrenheit devised his temperature scale, he wanted all measurable temperatures to be positive. Originally, he chose 0°F for the coldest temperature he could obtain with a mixture of ice and saltwater, and 96°F (a convenient number with many factors for subdivision) for the temperature of the human body. He then modified his scale slightly to make the ice-point and steam-point temperatures whole numbers. This resulted in the average temperature of the human body being between 98 and 99°F.

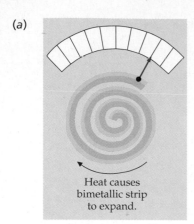

(a)

Heat causes
bimetallic strip
to expand.

(b)

Figure 18-3 (a) A thermometer using a bimetallic strip in the form of a coil. When the coil is heated or cooled, the two different metals expand or shrink at different rates. (b) A home thermostat. The coil on the right controls the air conditioner. When the room gets warmer, the coil expands, the tube mounted on it tilts, and mercury in the tube slides to close an electrical switch, turning on the air conditioning. A slide lever (at the lower right) mounted on the coil is used to set the desired temperature. The circuit will be broken when the cooler air causes the bimetallic coil to contract.

a pointer attached to indicate the temperature. When the thermometer is heated, the coil bends and the pointer moves to the right. Like mercury thermometers, it is calibrated by dividing the interval between the ice point and steam point into 100 Celsius degrees (or 180 Fahrenheit degrees).

18-3 Gas Thermometers and the Absolute Temperature Scale

When different types of thermometers are calibrated in ice water and steam, they agree (by definition) at 0°C and at 100°C, but they give slightly different readings at points in between. Discrepancies increase markedly above the steam point and below the ice point. However, in one group of thermometers, gas thermometers, the measured temperatures agree closely even far from the calibration points. In a **constant-volume gas thermometer**, the gas volume is kept constant, and change in gas pressure is used to measure a change in temperature (Figure 18-4). An ice-point pressure P_0 and steam-point pressure P_{100} are determined by placing the thermometer in ice–water and water–steam baths, and the interval between is divided into 100 equal degrees (for the Celsius scale). If the pressure is P_t in a bath whose temperature is to be determined, that temperature in degrees Celsius is defined to be

$$t_C = \frac{P_t - P_0}{P_{100} - P_0} \times 100°$$

18-3

Figure 18-5 shows the results of measurements of the boiling point of sulfur using constant-volume gas thermometers filled with various gases. The measured temperature is plotted as a function of the steam-point pressure P_{100}, which is varied by changing the amount of gas in the thermometer. As the amount of gas is reduced, its density and the steam-point pressure both decrease. We see that agreement among the thermometers is very close at low gas densities. In the limit as gas density goes to zero, all gas thermometers give the same value for any temperature. Because this temperature measurement is independent of the properties of any particular gas, low-density gas thermometers can be used to define temperature.

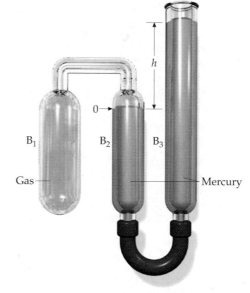

Figure 18-4 A constant-volume gas thermometer. The volume is kept constant by raising or lowering tube B_3 so that the mercury in tube B_2 remains at the zero mark. The temperature is chosen to be proportional to the pressure of the gas in tube B_1, which is indicated by the height h of the mercury column in tube B_3.

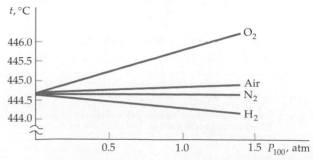

Figure 18-5 Temperature of the boiling point of sulfur measured with constant-volume gas thermometers filled with various gases. The pressure of the steam point of water, P_{100}, is varied by varying the amount of gas in the thermometers. As the amount of gas is reduced, the temperatures measured by all the thermometers approach the value 444.60°C.

Let us now consider a series of temperature measurements with a constant-volume gas thermometer that has a very small but fixed amount of gas. According to Equation 18-3, the pressure P_t in the thermometer varies linearly with the measured temperature t_C. Figure 18-6 shows a plot of measured temperature versus pressure in a constant-volume gas thermometer. When we extrapolate this straight line to zero pressure, the temperature approaches $-273.15°C$. This limit is the same no matter what kind of gas is used.

A reference state that is much more precisely reproducible than either the ice or steam points is the **triple point of water**—the unique temperature and pressure at which water, water vapor, and ice coexist in equilibrium. This equilibrium state occurs at 4.58 mmHg and 0.01°C. The **ideal-gas temperature scale** is defined so that the temperature of the triple point is 273.16 kelvins (K).* The temperature T of any other state is defined to be proportional to the pressure in a constant-volume gas thermometer:

$$T = \frac{273.16 \text{ K}}{P_3} P \qquad\qquad 18\text{-}4$$

Ideal-gas temperature scale

where P is the observed pressure of the gas in the thermometer, and P_3 is the pressure when the thermometer is immersed in a water–ice–vapor bath at its triple point. The value of P_3 depends on the amount of gas in the thermometer. The ideal-gas temperature scale, defined by Equation 18-4, has the advantage that any measured temperature does not depend on the properties of the particular gas that is used, but depends only on the general properties of gases.

The lowest temperature that can be measured with a gas thermometer is about 1 K, and requires helium for the gas. Below this temperature helium liquefies; all other gases liquefy at higher temperatures. In Chapter 20, we will see that the second law of thermodynamics can be used to define the **absolute temperature scale** independent of the properties of any substance, and with no limitations on the range of temperatures that can be measured. Temperatures as low as a millionth of a kelvin have been measured. The absolute scale so defined is identical to that defined by Equation 18-4 for the range of temperatures for which gas thermometers can be used. The symbol T is used when referring to absolute temperature.

Because the Celsius degree and the kelvin are the same size, temperature *differences* are the same on both the Celsius and the absolute temperature scale (also called the **Kelvin scale**). That is, a temperature *change* of 1 K is identical to a temperature *change* of 1 C°.[†] The two scales differ only in the choice of zero temperature. To convert from degrees Celsius to kelvins, we merely add 273.15:[‡]

$$T = t_C + 273.15 \text{ K} \qquad\qquad 18\text{-}5$$

Celsius–Kelvin conversion

Although the Celsius and Fahrenheit scales are convenient for everyday use, the absolute scale is much more convenient for scientific purposes, partly because many formulas are more simply expressed in it, and partly because the absolute temperature can be given a more fundamental interpretation.

Figure 18-6 Plot of pressure versus temperature as measured by a constant-volume gas thermometer. When extrapolated to zero pressure, the plot intersects the temperature axis at the value $-273.15°C$.

H_2O at its triple point. The container is a hollow cylindrical shell that is sealed and evacuated. It contains water, ice, and water vapor in equilibrium. The cylindrical well in the center is filled with water and contains an aluminum bushing for inserting a thermometer. The container, stored at the National Institute of Standards and Technology, is used in an ice-water bath that is just slightly below the triple-point temperature so that the water in the cell freezes very slowly.

---

* The kelvin is a degree unit that is the same size as the Celsius degree.

† We write 1 C° to indicate a *temperature change* of one Celsius degree, in contrast to 1°C, which means a temperature of one degree Celsius.

‡ For most purposes, we can round off the temperature of absolute zero to -273°C.

Example 18-2

What is the Kelvin temperature corresponding to 70°F?

Picture the Problem First convert to degrees Celsius, then to kelvins.

1. Convert to degrees Celsius: $\qquad\qquad\qquad t_C = \frac{5}{9}(70° - 32°) = 21.1°C$

2. To find the Kelvin temperature we add 273: $\qquad T = t_C + 273 = 21.1 + 273 = 294\ \text{K}$

Exercise The "high-temperature" superconductor $YBa_2Cu_3O_7$ becomes superconducting when the temperature is lowered to 92 K. Find this superconducting temperature in degrees Fahrenheit. (*Answer* $-294°F$)

18-4   The Ideal-Gas Law

The properties of gases at low densities allow the definition of the ideal-gas temperature scale. If we compress such a gas while keeping its temperature constant, the pressure increases. Similarly, if a gas expands at constant temperature, its pressure decreases. To a good approximation, the product of the pressure and volume of a low-density gas is constant at a constant temperature. This result was discovered experimentally by Robert Boyle (1627–1691), and is known as **Boyle's law:**

$$PV = \text{constant} \qquad \text{(constant temperature)}$$

A more general law exists that reproduces Boyle's law as a special case. According to Equation 18-4, the absolute temperature of a low-density gas is proportional to its pressure at constant volume. In addition—a result discovered experimentally by Jacques Charles (1746–1823) and Gay–Lussac (1778–1850)—the absolute temperature of a low-density gas is proportional to its volume at constant pressure. We can combine these two results by stating

$$PV = CT \qquad\qquad\qquad \text{18-6}$$

where C is a constant of proportionality. We can see that this constant is proportional to the amount of gas by considering combining two identical containers, each holding the same volume of the same kind of gas at the same temperature. Then twice the amount of gas occupies twice the volume at the same pressure P and temperature T. We therefore write C as a constant k times the number of molecules in the gas N:

$$C = kN$$

Equation 18-6 then becomes

$$PV = NkT \qquad\qquad\qquad \text{18-7}$$

The constant k is called **Boltzmann's constant.** It is found experimentally to have the same value for any kind of gas:

$$k = 1.381 \times 10^{-23}\ \text{J/K} = 8.617 \times 10^{-5}\ \text{eV/K} \qquad \text{18-8}$$

An amount of gas is often expressed in moles. A **mole** (mol) of any substance is the amount of that substance that contains Avogadro's number N_A

of atoms or molecules, defined as the number of carbon atoms in 12 g of ^{12}C:

$$N_A = 6.022 \times 10^{23} \qquad\qquad 18\text{-}9$$

Avogadro's number

If we have n moles of a substance, the number of molecules is then

$$N = nN_A \qquad\qquad 18\text{-}10$$

Equation 18-7 is then

$$PV = nN_A kT = nRT \qquad\qquad 18\text{-}11$$

where $R = N_A k$ is called the **universal gas constant**. Its value, which is the same for all gases, is

$$R = N_A k = 8.314 \text{ J/mol·K} = 0.08206 \text{ L·atm/mol·K} \qquad 18\text{-}12$$

Figure 18-7 shows plots of PV/nT versus the pressure P for several gases. For all gases, PV/nT is nearly constant over a large range of pressures. Even oxygen, which varies the most in this graph, changes by only about 1% between 0 and 5 atm. An **ideal gas** is defined as one for which PV/nT is constant for all pressures. The pressure, volume, and temperature of an ideal gas are related by

$$PV = nRT \qquad\qquad 18\text{-}13$$

Ideal-gas law

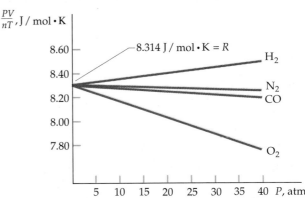

Figure 18-7 Plot of PV/nT versus P for real gases. In these plots, the pressure is varied by varying the amount of gas. As we reduce the density of the gas and thereby the pressure, the ratio PV/nT approaches the same value, 8.314 J/mol·K, for all gases. This value is the universal gas constant R.

The variables P, V, and T describe the **macroscopic state** of a gas at a given instant. (The **microscopic state** of the gas would be described by giving the position and velocity of each of the molecules in the gas.) An equation such as Equation 18-13, which relates these **macroscopic state variables**, is called an **equation of state**. For any gas at any density, there is an equation of state relating P, V, and T for a given amount of gas. Thus, the macroscopic state of a given amount of gas is determined by any two of the three state variables P, V, and T. Equation 18-13 describes the properties of real gases with low densities (and therefore low pressures). At higher densities, corrections must be made to Equation 18-13. In Chapter 21 we discuss another equation of state, the van der Waals equation, that includes such corrections.

<hr>

Example **18-3**

What volume is occupied by 1 mol of gas at a temperature of 0°C and a pressure of 1 atm?

We can find the volume using the ideal-gas law, with $T = 273$ K:

$$V = \frac{nRT}{P} = \frac{(1 \text{ mol})(0.0821 \text{ L·atm/mol·K})(273 \text{ K})}{1 \text{ atm}} = 22.4 \text{ L}$$

Remarks Note that by writing R in L·atm/mol, we could write P in atmospheres to get V in liters.

Exercise Find (a) the number of moles, n, and (b) the number of molecules, N, in 1 cm^3 of a gas at 0°C and 1 atm. (*Answers* (a) $n = 4.46 \times 10^{-5}$ mol, (b) $N - 2.68 \times 10^{19}$ molecules)

The temperature of 0°C (273 K) and the pressure of 1 atm are often referred to as **standard conditions.** We see from Example 18-3 that under standard conditions, 1 mol of an ideal gas occupies a volume of 22.4 L.

Figure 18-8 shows plots of P versus V at several constant temperatures T. These curves are called **isotherms.** The isotherms for an ideal gas are hyperbolas. For a fixed amount of gas, we can see from Equation 18-13 that the quantity PV/T is constant. Using the subscripts 1 for the initial values and 2 for the final values, we have

$$\frac{P_2 V_2}{T_2} = \frac{P_1 V_1}{T_1}$$ 18-14

Ideal-gas law for fixed amount of gas

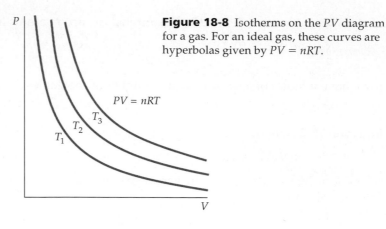

Figure 18-8 Isotherms on the PV diagram for a gas. For an ideal gas, these curves are hyperbolas given by $PV = nRT$.

Example 18-4

A gas has a volume of 2 L, a temperature of 30°C, and a pressure of 1 atm. When the gas is heated to 60°C and compressed to a volume of 1.5 L, what is its new pressure?

Picture the Problem Since the amount of gas is fixed, the pressure can be found using Equation 18-14. Let subscripts 1 and 2 refer to the initial and final state, respectively.

1. Express the pressure P_2 in terms of P_1 and the initial and final volumes and temperatures:

$$\frac{P_1 V_1}{T_1} = \frac{P_2 V_2}{T_2}$$

$$P_2 = \frac{T_2 V_1}{T_1 V_2} P_1$$

2. Calculate the initial and final absolute temperatures:

$T_1 = 273 + 30 = 303$ K

$T_2 = 273 + 60 = 333$ K

3. Substitute numerical values in step 1 to find P_2:

$$P_2 = \frac{(333 \text{ K})(2 \text{ L})}{(303 \text{ K})(1.5 \text{ L})} (1 \text{ atm}) = 1.47 \text{ atm}$$

Exercise How many moles of gas are in the system described in this example? (*Answer* $n = 0.0804$ mol)

The mass per mole of a substance is called its **molar mass** M.* The molar mass of ^{12}C is, by definition, 12 g/mol or 12×10^{-3} kg/mol. Molar masses of the elements are given in the periodic table in Appendix C. The molar mass of a molecule, such as CO_2, is the sum of the molar masses of the elements in the molecule. Since the molar mass of oxygen is 16 g/mol (actually 15.999 g/mol), the molar mass of O_2 is 32 g/mol and that of CO_2 is $12 + 32 = 44$ g/mol.

---

* The terms *molecular weight* and *molecular mass* are also sometimes used.

The mass of n moles of gas is given by

$$m = nM$$

and the density ρ of an ideal gas is

$$\rho = \frac{m}{V} = \frac{nM}{V}$$

Using $n/V = P/RT$ from Equation 18-13, we have

$$\rho = \frac{M}{RT}P \qquad\qquad\qquad 18\text{-}15$$

Density of an ideal gas

At a given temperature, the density of an ideal gas is proportional to its pressure.

The molar mass of hydrogen is 1.008 g/mol. What is the mass of one hydrogen atom?

Picture the Problem Let m be the mass of a hydrogen atom. Since there are N_A atoms in a mole, the molar mass M is given by $M = mN_A$. We can use this to solve for m.

The mass of a hydrogen atom is the molar mass divided by Avogadro's number:

$$m = \frac{M}{N_A} = \frac{1.008 \text{ g/mol}}{6.022 \times 10^{23} \text{ atoms/mol}}$$

$$= 1.67 \times 10^{-24} \text{ g/atom}$$

Remark Note that Avogadro's number is approximately the reciprocal of the mass of the hydrogen atom measured in grams.

try it yourself

One hundred grams of CO_2 occupies a volume of 55 L at a pressure of 1 atm. (*a*) What is the temperature? (*b*) If the volume is increased to 80 L and the temperature is kept constant, what is the new pressure?

Picture the Problem Both questions can be answered using the ideal-gas law (Equation 18-13) if we first find the number of moles, n.

Cover the column to the right and try these on your own before looking at the answers.

Steps	Answers
(*a*)1. Use $M = 44$ g/mol to find the number of moles of CO_2.	$n = m/M = 2.27$ mol
2. Find the temperature T from the ideal-gas law.	$T = PV/nR = 295$ K
(*b*) Use $PV = $ constant to find the new pressure for $V = 80$ L.	$P_2 = 0.688$ atm

18-5   The Kinetic Theory of Gases

The description of the behavior of a gas in terms of the macroscopic state variables P, V, and T can be related to simple averages of microscopic quantities such as the mass and speed of the molecules in the gas. The resulting theory is called **the kinetic theory of gases.**

From the point of view of kinetic theory, a gas consists of a large number of molecules making elastic collisions with each other and with the walls of a container. In the absence of external forces (we may neglect gravity), there is no preferred position for a molecule in the container,* and no preferred direction for its velocity vector. The molecules are separated, on the average, by distances that are large compared with their diameters, and they exert no forces on each other except when they collide. (This final assumption is equivalent to assuming a very low gas density, which as we saw in the last section is the same as assuming that the gas is an ideal gas. Since momentum is conserved, the collisions the molecules make with each other have no effect on the total momentum in any direction—thus such collisions can be neglected.)

Calculating the Pressure Exerted by a Gas   The pressure that a gas exerts on its container is due to collisions between gas molecules and the container walls. This pressure is a force per unit area and, by Newton's second law, this force is the rate of change of momentum of the gas molecules colliding with the wall.

Consider a rectangular container of volume V containing N molecules, each of mass m moving with a speed v. Let us calculate the force exerted by these molecules on the right wall, which is perpendicular to the x axis and has area A. The number of molecules hitting this wall in a time interval Δt is the number that are within distance $v_x \Delta t$ of the wall (Figure 18-9) and are moving to the right. This is the number of molecules per unit volume N/V times the volume $v_x \Delta t\, A$ times $\frac{1}{2}$ because, on the average, only half the molecules are moving to the right. Thus, the number that hit the wall in Δt is

$$\text{Molecules that hit the wall} = \frac{1}{2}\frac{N}{V} v_x \Delta t\, A$$

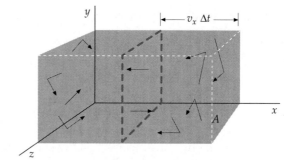

Figure 18-9 Gas molecules in a rectangular container. In a time interval Δt, the molecules at a distance $v_x \Delta t$ from the right wall will hit the right wall if they are moving to the right.

The x component of momentum of a molecule is $+mv_x$ before it hits the wall, and $-mv_x$ after an elastic collision with the wall. The change in momentum has the magnitude $2mv_x$. The magnitude of the total change in momentum Δp of all molecules during a time interval Δt is $2mv_x$ times the number of molecules that hit the wall during this interval:

$$\Delta p = (2mv_x) \times \left(\frac{1}{2}\frac{N}{V} v_x \Delta t\, A\right) = \frac{N}{V} mv_x^2 A\, \Delta t \qquad \text{18-16}$$

The force exerted by the wall on the molecules, and by the molecules on the wall, is this change in momentum divided by the time interval Δt. The pressure is this force divided by the area:

$$P = \frac{F}{A} = \frac{1}{A}\frac{\Delta p}{\Delta t} = \frac{N}{V} mv_x^2$$

---

* Because of gravity, the density of molecules at the bottom of the container is slightly greater than at the top. As discussed in Chapter 13, the density of air decreases by half at a height of about 5.5 km, so the variation over a normal-sized container is negligible.

or

$$PV = Nmv_x^2 \qquad\qquad \text{18-17}$$

To allow for the fact that all the molecules in a container do not have the same speed, we merely replace v_x^2 with the average value $(v_x^2)_{av}$. Then, writing Equation 18-17 in terms of the kinetic energy $\frac{1}{2}mv_x^2$ associated with motion along the x axis, we have

$$PV = 2N(\tfrac{1}{2}mv_x^2)_{av} \qquad\qquad \text{18-18}$$

The Molecular Interpretation of Temperature   Comparing Equation 18-18 with Equation 18-7, which was obtained experimentally for any gas at very low densities, we can see that

$$PV = NkT = 2N(\tfrac{1}{2}mv_x^2)_{av}$$

or

$$(\tfrac{1}{2}mv_x^2)_{av} = \tfrac{1}{2}kT \qquad\qquad \text{18-19}$$

The average energy associated with motion in the x direction

Thus, the average kinetic energy associated with motion along the x axis is $\frac{1}{2}kT$. But there is nothing special about the x direction. On the average,

$$(v_x^2)_{av} = (v_y^2)_{av} = (v_z^2)_{av} \qquad\qquad \text{18-20}$$

and

$$(v^2)_{av} = (v_x^2)_{av} + (v_y^2)_{av} + (v_z^2)_{av} = 3(v_x^2)_{av}$$

Writing $(v_x^2)_{av} = \frac{1}{3}(v^2)_{av}$ and K_{av} for the average kinetic energy of the molecules, Equation 18-19 becomes

$$K_{av} = (\tfrac{1}{2}mv^2)_{av} = \tfrac{3}{2}kT \qquad\qquad \text{18-21}$$

Average kinetic energy of a molecule

The absolute temperature is thus a measure of the average translational kinetic energy of the molecules.* The total translational kinetic energy of n moles of a gas containing N molecules is

$$K = N(\tfrac{1}{2}mv^2)_{av} = \tfrac{3}{2}NkT = \tfrac{3}{2}nRT \qquad\qquad \text{18-22}$$

Kinetic energy of translation for n moles of a gas

where we've used $Nk = nN_A k = nR$. Thus, the translational kinetic energy is $\frac{3}{2}kT$ per molecule and $\frac{3}{2}RT$ per mole.

We can use these results to estimate the order of magnitude of the speeds of the molecules in a gas. The average value of v^2 is, by Equation 18-21,

$$(v^2)_{av} = \frac{3kT}{m} = \frac{3N_A kT}{N_A m} = \frac{3RT}{M}$$

where $M = N_A m$ is the molar mass. The square root of $(v^2)_{av}$ is referred to as the **root mean square** (rms) speed:

$$v_{rms} = \sqrt{(v^2)_{av}} = \sqrt{\frac{3kT}{m}} = \sqrt{\frac{3RT}{M}} \qquad\qquad \text{18-23}$$

---

* We include the word "translational" because the molecules may also have rotational or vibrational kinetic energy. Only the translational kinetic energy is relevant to the calculation of the pressure exerted by a gas on the walls of its container.

Note that Equation 18-23 is similar to Equation 15-5 for the speed of sound in a gas:

$$v_{sound} = \sqrt{\frac{\gamma RT}{M}}$$

18-24

where $\gamma = 1.4$ for air. This is not surprising since a sound wave in air is a pressure disturbance propagated by collisions between air molecules.

Example 18-7

Oxygen gas (O_2) has a molar mass of about 32 g/mol and hydrogen gas (H_2) has a molar mass of about 2 g/mol. Calculate (*a*) the rms speed of an oxygen molecule when the temperature is 300 K, and (*b*) the rms speed of a hydrogen molecule at the same temperature.

Picture the Problem (*a*) We find v_{rms} using Equation 18-23. For the units to work out right we use $R = 8.31$ J/mol·K and we express the molecular mass of O_2 in kg/mol. (*b*) Since v_{rms} is proportional to $1\sqrt{M}$, and the molar mass of hydrogen is one-sixteenth that of oxygen, the rms speed of hydrogen is 4 times that of oxygen.

(*a*) Substitute the given values into Equation 18-23:

$$v_{rms}(O_2) = \sqrt{\frac{3RT}{M}} = \sqrt{\frac{3(8.31 \text{ J/mol·K})(300 \text{ K})}{32 \times 10^{-3} \text{ kg/mol}}}$$

$$= 483 \text{ m/s}$$

(*b*) Use $v_{rms} \propto 1/\sqrt{M}$ to calculate v_{rms} for hydrogen:

$$v_{rms}(H_2) = \frac{\sqrt{M_{O_2}}}{\sqrt{M_{H_2}}} v_{rms}(O_2) = \sqrt{\frac{32 \text{ g/mol}}{2 \text{ g/mol}}} (483 \text{ m/s})$$

$$= 1.93 \text{ km/s}$$

Remarks The rms speed of oxygen molecules is of the same order of magnitude as the speed of sound in air, which at 300 K is about 347 m/s.

Exercise Find the rms speed of a nitrogen molecule ($M = 28$ g/mol) at 300 K. (*Answer* 516 m/s)

The Equipartition Theorem

We have seen that the average kinetic energy associated with translational motion in any direction is $\frac{1}{2}kT$ per molecule (Equation 18-19) (or, equivalently, $\frac{1}{2}RT$ per mole), where k is Boltzmann's constant. If the energy of a molecule associated with its motion in one direction is momentarily increased, say, by a collision between the molecule and a moving piston during a compression, collisions between that molecule and other molecules will quickly redistribute the additional energy. When the gas is again in equilibrium, the energy will be equally partitioned among the translational kinetic energies associated with motion in the x, y, and z directions. This sharing of the energy equally between the three terms in the translational kinetic energy is a special case of the **equipartition theorem**, a result that follows from classical statistical mechanics. Each component of position and momentum (including angular position and angular momentum) that appears as a squared term in the expression for the energy of the system is called a **degree of freedom**. Typical degrees of freedom are associated with the kinetic energy of translation, rotation, and vibration, and with the potential energy of vibration. The equipartition theorem states that

> When a substance is in equilibrium, there is an average energy of $\frac{1}{2}kT$ per molecule or $\frac{1}{2}RT$ per mole associated with each degree of freedom.

Equipartition theorem

In Chapter 19, we use the equipartition theorem to relate the measured heat capacities of gases to their molecular structure.

Mean Free Path

The average speed of molecules in a gas at normal pressures is several hundred meters per second, yet if somebody across a room from you opens a perfume bottle, you don't detect the odor instantly. In fact, if it were not for the bulk flow of air in the room, you wouldn't detect the odor for weeks. The transmission of the odor is slow because perfume molecules, speedy as they are, do not travel directly toward you, but instead travel a zigzag path, as often back as forward, due to collisions with air molecules. The average distance traveled by a molecule between collisions λ is called its **mean free path**.

The mean free path of a molecule is related to its size, the size of the surrounding gas molecules, and the density of the gas. Consider one molecule of radius r_1 moving with speed v through a region of stationary molecules (Figure 18-10). The molecule will collide with another molecule of radius r_2 if the centers of the two molecules come within a distance $d = r_1 + r_2$ from each other. (If all the molecules are the same type, d is the molecular diameter.) As the molecule moves, it will collide with any molecule whose center is in a circle of radius d (Figure 18-11). In some time t, the molecule moves a distance vt and collides with every molecule in the cylindrical volume $\pi d^2 vt$. The number of molecules in this volume is $n_v \pi d^2 vt$, where $n_v = N/V$ is the number of molecules per unit volume. (After each collision, the direction of the molecule changes, so the path actually zigs and zags.) The total path length divided by the number of collisions is the mean free path:

$$\lambda = \frac{vt}{n_v \pi d^2 vt} = \frac{1}{n_v \pi d^2}$$

This calculation of the mean free path assumes that all but one of the gas molecules are stationary. When the motion of all the molecules is taken into account, the correct expression for the mean free path is given by

$$\lambda = \frac{1}{\sqrt{2}\, n_v \pi d^2} \qquad\qquad 18\text{-}25$$

The average time between collisions is called the collision time τ. The reciprocal of the collision time, $1/\tau$, is approximately equal to the average number of collisions per second, or the collision frequency. If v_{av} is the average speed, the average distance traveled between collisions is

$$\lambda = v_{av}\tau \qquad\qquad 18\text{-}26$$

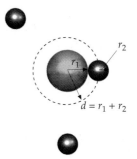

Figure 18-10 Model of a molecule (center sphere) moving in a gas. The molecule of radius r_1 will collide with any molecule of radius r_2 if their centers are a distance $d = r_1 + r_2$ apart, which is any molecule whose center is in a circle of radius $d = r_1 + r_2$ centered about the molecule.

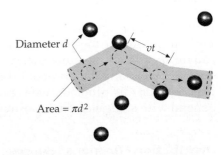

Figure 18-11 Model of a molecule moving with speed v in a gas of similar molecules. In time t the molecule of diameter d will collide with any similar molecule whose center is in a cylinder of volume $\pi d^2 vt$. In this picture, all the molecules but one are assumed to be at rest.

Example 18-8   *try it yourself*

The diameter of a nitrogen molecule is about 3.75×10^{-10} m, and the molar mass of nitrogen is 28 g/mol. (*a*) Calculate the mean free path of a nitrogen molecule at 300 K and 1 atm. (*b*) Estimate the time between collisions, assuming that the average speed equals v_{rms}.

Picture the Problem (*a*) Since d is given, you need only n_v to calculate λ from $\lambda = 1/(\sqrt{2}n_v\pi d^2)$ (Equation 18-25). You can use the ideal-gas law to find $n_v = N/V$. (Be sure to convert 1 atm to pascals so the units come out right.) (*b*) Use Equation 18-26 to relate τ to λ and v_{av}, and estimate v_{av} by $v_{av} \approx v_{rms} = \sqrt{3RT/M}$ (Equation 18-23).

Cover the column to the right and try these on your own before looking at the answers.

Steps	Answers
(*a*)1. Write λ in terms of the number density n_v and the molecular diameter d.	$\lambda = \dfrac{1}{\sqrt{2}n_v\pi d^2}$
2. Use the equation $PV = NkT$ to calculate $n_v = N/V$.	$n_v = 2.45 \times 10^{25}$ molecules/m^3
3. Substitute this value of n_v and the given value of d to calculate λ.	$\lambda = 6.53 \times 10^{-8}$ m
(*b*)1. Write τ in terms of the mean free path λ.	$\tau = \dfrac{\lambda}{v_{av}}$
2. Estimate v_{av} by calculating v_{rms}.	$v_{rms} = 517$ m/s
3. Use $v_{av} \approx v_{rms}$ to estimate τ.	$\tau \approx 1.3 \times 10^{-10}$ s

Remark Note that the mean free path is about 2000 times the diameter of the molecule, and that the collision frequency is about $1/\tau \approx 8 \times 10^9$ collisions per second.

The Distribution of Molecular Speeds

We would not expect all of the molecules in a gas to have the same speed. The calculation of the pressure of a gas allows us to calculate the average square speed and therefore the average energy of molecules in a gas, but it does not yield any details about the *distribution* of molecular speeds. Before we consider this problem, we discuss the idea of distribution functions in general with some elementary examples from common experience.

Distribution Functions Suppose a teacher gave a 25-point quiz to a large number N of students. To describe the results, the teacher might give the average score, but this would not be a complete description. If all the students received a score of 12.5, for example, that would be quite different from half the students receiving 25 and the other half zero, but the average score would be the same in both cases. A complete description of the results would be to give the number n_i of students that received a score s_i for all the scores received. Alternatively, one could give the fraction $f_i = n_i/N$ of the students that received the score s_i. Both n_i and f_i, which are functions of the variable s, are called **distribution functions**. The fractional distribution is somewhat more convenient to use. The probability that one of the N students selected at random received the score s_i equals the total number of students that received that score n_i divided by N, that is, the probability equals f_i. Note that

$$\sum_i f_i = \sum_i \frac{n_i}{N} = \frac{1}{N} \sum_i n_i$$

and since $\Sigma n_i = N$,

$$\sum_i f_i = 1 \qquad\qquad 18\text{-}27$$

Equation 18-27 is called the **normalization condition** for fractional distributions.

To find the average score, we add all the scores and divide by N. Since each score s_i was obtained by $n_i = Nf_i$ students, this is equivalent to

$$s_{av} = \frac{1}{N} \sum_i n_i s_i = \sum_i s_i f_i \qquad\qquad 18\text{-}28$$

Similarly, the average of any function $g(s)$ is defined by

$$g(s)_{av} = \frac{1}{N} \sum_i g(s_i) n_i = \sum_i g(s_i) f_i \qquad\qquad 18\text{-}29$$

In particular, the average square score is

$$(s^2)_{av} = \frac{1}{N} \sum_i s_i^2 n_i = \sum_i s_i^2 f_i \qquad\qquad 18\text{-}30$$

The square root of $(s^2)_{av}$ is called the **root mean square score,** or rms score. A possible distribution function is shown in Figure 18-12. For this distribution, the most probable score (that obtained by the most students) is 16, the average score is 14.2, and the rms score is 14.9.

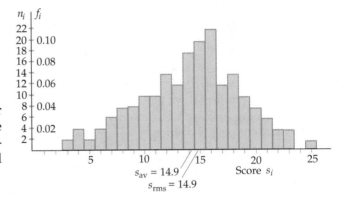

Figure 18-12 Grade distribution for a 25-point quiz given to 200 students. n_i is the number of students receiving grade s_i, and $f_i = n_i/N$ is the fractional distribution function.

Example 18-9

Fifteen students took a 25-point quiz. Their scores were 25, 22, 22, 20, 20, 20, 18, 18, 18, 18, 18, 15, 15, 15, and 10. Find the average score and the rms score.

Picture the Problem The distribution function for this problem is $n_{25} = 1$, $n_{22} = 2$, $n_{20} = 3$, $n_{18} = 5$, $n_{15} = 3$, and $n_{10} = 1$. To find the average score, we use Equation 18-28. To find the rms score, we use Equation 18-30 and then take the square root.

(a) By definition, s_{av} is:

$$s_{av} = \frac{1}{N} \sum_i n_i v_i$$

$$= \tfrac{1}{15}(1(25) + 2(22) + 3(20) + 5(18) + 3(15) + 1(10))$$

$$= \tfrac{1}{15}(274) = 18.3$$

(b) 1. To calculate v_{rms}, first find the average of s^2:

$$(s^2)_{av} = \frac{1}{N} \sum_i n_i s_i^2$$

$$= \tfrac{1}{15}(1(25)^2 + 2(22)^2 + 3(20)^2 + 5(18)^2 + 3(15)^2 + 1(10)^2)$$

$$= \tfrac{1}{15}(5188) = 346$$

2. Take the square root of $(s^2)_{av}$:

$$s_{rms} = \sqrt{(s^2)_{av}} = 18.6 \text{ m/s}$$

Now consider the case of a continuous distribution, for example, the distribution of heights in a population. For any finite number N, the number of people that are exactly 2 m tall is zero. If we assume that height can be determined to any desired accuracy, there are an infinite number of possible heights, so the chance that anybody has a particular exact height is zero. We therefore divide the heights into intervals Δh (for example, Δh might be 1 cm or 0.5 cm) and ask what fraction of people have heights that fall in any particular interval. For very large N, this number is proportional to the size of the interval. We define the distribution function $f(h)$ as the fraction of the number of people with heights in the interval between h and $h + \Delta h$. Then for N people, $Nf(h)\,\Delta h$ is the number of people whose height is between h and $h + \Delta h$. Figure 18-13 shows a possible height distribution.

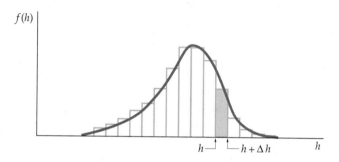

Figure 18-13 A possible height distribution function. The fraction of the number of heights between h and $h + \Delta h$ equals the shaded area $f(h)\,\Delta h$. The histogram can be approximated by a continuous curve as shown.

The fraction of people with heights in a given interval Δh is the area $f(h)\,\Delta h$. If N is very large, we can choose Δh to be very small, and the histogram will approximate a continuous curve. We can therefore consider the distribution function $f(h)$ to be a continuous function, write the interval as dh, and replace the sums in Equations 18-27 through 18-30 with integrals:

$$\int f(h)\,dh = 1 \tag{18-31}$$

$$h_{\text{av}} = \int hf(h)\,dh \tag{18-32}$$

$$g(h)_{\text{av}} = \int g(h)f(h)\,dh \tag{18-33}$$

$$(h^2)_{\text{av}} = \int h^2 f(h)\,dh \tag{18-34}$$

The probability of a person selected at random having a height between h and $h + dh$ is $f(h)\,dh$. We have already encountered a continuous distribution function in Chapter 17, where the probability of measuring the position of an electron described by the wave function $\psi(x)$ is $\psi^2(x)\,dx$.

A useful quantity characterizing a distribution is the **standard deviation** σ defined by

$$\sigma^2 = [(x - x_{\text{av}})^2]_{\text{av}} \tag{18-35a}$$

Expanding the square on the right, we obtain

$$\sigma^2 = (x^2 - 2xx_{\text{av}} + x_{\text{av}}^2)_{\text{av}} = (x^2)_{\text{av}} - 2x_{\text{av}}x_{\text{av}} + x_{\text{av}}^2$$

or

$$\sigma^2 = (x^2)_{\text{av}} - x_{\text{av}}^2 \tag{18-35b}$$

The standard deviation measures the spread of the values about the average value. For most distributions there will be few values that differ from x_{av} by more than a few multiples of σ. For the familiar bell-shaped distribution (called a normal distribution), two-thirds of the values are expected to fall within $x_{\text{av}} \pm \sigma$.

In Example 18-7, we found that the rms value was greater than the average value. This is a general feature for any distribution (unless all the values are identical, in which case $x_{rms} = x_{av}$). According to Equation 18-35b, the square of the rms value $(x^2)_{av}$ minus the square of the average value (x_{av}^2) is σ^2, which is by definition positive.

(a)

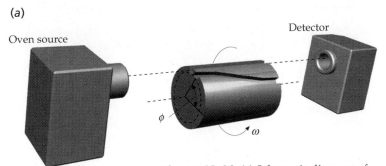

The Maxwell–Boltzmann Distribution The distribution of the molecular speeds of a gas can be measured directly using the apparatus illustrated in Figure 18-14a. In Figure 18-14b, these speeds are shown for two different temperatures. The quantity $f(v)$ in Figure 18-14b is called the **Maxwell–Boltzmann speed distribution function.** In a gas of N molecules, the number that have speeds in the range between v and $v + dv$ is dN, given by

$$dN = N f(v)\, dv \qquad\qquad 18\text{-}36$$

(b)

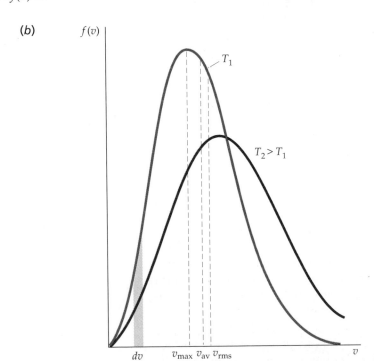

Figure 18-14 (a) Schematic diagram of the apparatus for determining the speed distribution of the molecules of a gas. A substance is vaporized in an oven, and the vapor molecules are allowed to escape through a hole in the oven wall into a vacuum chamber. The molecules are collimated into a narrow beam by a series of slits (not shown). The beam is aimed at a detector that counts the number of molecules that are incident on it in a given period of time. Most of the beam is stopped by a rotating cylinder. Small helical slits in the cylinder (only one of which is depicted here) allow the passage of molecules that have a narrow range of speeds that is determined by the angular velocity of rotation of the cylinder. The number of molecules in each range of speeds is measured by varying the angular velocity of the cylinder and counting the number of molecules that reach the detector for each angular velocity. (b) Distributions of molecular speeds in a gas at two temperatures, T_1 and $T_2 > T_1$. The shaded area $f(v)\, dv$ equals the fraction of the number of molecules having a particular speed in a narrow range of speeds dv. The mean speed v_{av} and the rms speed v_{rms} are both slightly greater than the most probable speed v_{max}.

The fraction $dN/N = f(v)\, dv$ in a particular range dv is illustrated by the shaded region in the figure. The Maxwell–Boltzmann speed distribution function can be derived using statistical mechanics. The result is

$$f(v) = \frac{4}{\sqrt{\pi}}\left(\frac{m}{2kT}\right)^{3/2} v^2 e^{-mv^2/2kT} \qquad\qquad 18\text{-}37$$

Maxwell–Boltzmann speed distribution function

The most probable speed v_{max} is that speed for which $f(v)$ is maximum. It is left as a problem to show that

$$v_{max} = \sqrt{\frac{2kT}{m}} = \sqrt{\frac{2RT}{M}} \qquad\qquad 18\text{-}38$$

Comparing Equation 18-38 with Equation 18-23, we see that the most probable speed is slightly less than the rms speed.

optional

Example 18-10

Use the Maxwell–Boltzmann distribution function to calculate the average value of v^2 for the molecules in a gas.

Picture the Problem The average value of v^2 is calculated from Equation 18-34 with v replacing h and $f(v)$ given by Equation 18-37.

1. By definition, $(v^2)_{av}$ is:

$$(v^2)_{av} = \int_0^\infty v^2 f(v)\, dv$$

2. Use Equation 18-37 for $f(v)$:

$$(v^2)_{av} = \int_0^\infty v^2 \frac{4}{\sqrt{\pi}}\left(\frac{m}{2kT}\right)^{3/2} v^2 e^{-mv^2/2kT}\, dv$$

$$= \frac{4}{\sqrt{\pi}}\left(\frac{m}{2kT}\right)^{3/2} \int_0^\infty v^4 e^{-mv^2/2kT}\, dv$$

3. The integral in step 2 can be found in standard integral tables:

$$\int_0^\infty v^4 e^{-mv^2/2kT}\, dv = \frac{3}{8}\sqrt{\pi}\left(\frac{2kT}{m}\right)^{5/2}$$

4. Use this result to calculate $(v^2)_{av}$:

$$(v^2)_{av} = \frac{4}{\sqrt{\pi}}\left(\frac{m}{2kT}\right)^{3/2}\frac{3}{8}\sqrt{\pi}\left(\frac{2kT}{m}\right)^{5/2} = \frac{3kT}{m}$$

Remarks Note that our result agrees with $v_{rms} = \sqrt{3kT/m}$ from Equation 18-23.

In Example 18-6, we found that the rms speed of hydrogen molecules is about 1.93 km/s. This is about one-sixth of the escape speed at the surface of the earth, which we found to be 11.2 km/s in Section 11-3. So why is there no free hydrogen in the earth's atmosphere? As we can see from Figure 18-14b, a considerable fraction of the molecules of a gas in equilibrium have speeds greater than the rms speed. When the rms speed of the molecules of a particular gas is as great as 15 to 20% of the escape speed for a planet, enough of the molecules have speeds greater than the escape speed so that the gas cannot exist in the atmosphere of that planet. Thus, no hydrogen. The rms speed of oxygen molecules, on the other hand, is about one-fourth that of hydrogen molecules, which makes it only about 4% of the escape speed at the surface of the earth. Therefore, few oxygen molecules have speeds greater than the escape speed, and oxygen is found in the earth's atmosphere.

The Energy Distribution   The Maxwell–Boltzmann speed distribution as given by Equation 18-37 can also be written as an energy distribution. We write the number of molecules with energy E in the range between E and $E + dE$ as

$$dN = NF(E)\, dE$$

where $F(E)$ is the energy distribution function. This will be the same number as given by Equation 18-37 if the energy E is related to the speed v by $E = \frac{1}{2}mv^2$. Then

$$dE = mv\, dv$$

and

$$Nf(v)\, dv = NF(E)\, dE$$

We can write

$$f(v)\,dv = Cv^2 e^{-mv^2/2kT}\,dv = Cve^{-E/kT}\,v\,dv = C\left(\frac{2E}{m}\right)^{1/2}e^{-E/kT}\frac{dE}{m}$$

where $C = (4/\sqrt{\pi})(m/2kT)^{3/2}$ from Equation 18-37. The energy distribution function $F(E)$ is thus given by

$$F(E) = \frac{4}{\sqrt{\pi}}\left(\frac{m}{2kT}\right)^{3/2}\left(\frac{2}{m}\right)^{1/2}\frac{1}{m}E^{1/2}e^{-E/kT}$$

Simplifying, we obtain the **Maxwell–Boltzmann energy distribution function:**

$$F(E) = \frac{2}{\sqrt{\pi}}\left(\frac{1}{kT}\right)^{3/2}E^{1/2}e^{-E/kT} \tag{18-39}$$

Maxwell–Boltzmann energy distribution function

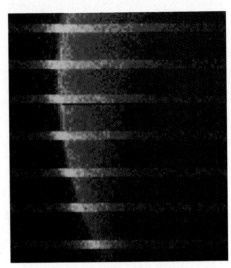

In the language of statistical mechanics, the energy distribution is considered to be the product of two factors: one, called the **density of states,** is proportional to $E^{1/2}$; the other is the probability of a state being occupied, which is $e^{-E/kT}$ and is called the **Boltzmann factor**.

Jupiter as seen from about 12 million miles. Because the escape speed at the surface of Jupiter is about 600 km/s, Jupiter easily retains hydrogen in its atmosphere.

The atmosphere of Venus is almost entirely CO_2. However, measurements by the Pioneer Venus Orbiter show an atomic hydrogen cloud surrounding Venus. The crescent-shaped image shows atomic oxygen, and the bars show hydrogen extending far above the atmosphere. Since the escape speed on Venus is 10.3 km/s, slightly smaller than the escape speed on earth, and since the Venusian atmosphere is considerably warmer than earth's atmosphere, all of the hydrogen in the atmosphere at the time of the formation of Venus should have escaped by now.

Summary

Topic	Remarks and Relevant Equations
1. Celsius and Fahrenheit Scales	On the Celsius scale, the ice point is defined to be 0°C and the steam point is 100°C. On the Fahrenheit scale, the ice point is 32°F and the steam point is 212°F. Temperatures on the Fahrenheit and Celsius scales are related by $$t_C = \tfrac{5}{9}(t_F - 32°) \tag{18-2}$$
2. Gas Thermometers	Gas thermometers have the property that they all agree with each other in the measurement of any temperature as long as the density of the gas is very low. The ideal-gas temperature T is defined by $$T = \frac{273.16\text{ K}}{P_3}P \tag{18-4}$$ where P is the observed pressure of the gas in the thermometer, and P_3 is the pressure when the thermometer is immersed in a water–ice–vapor bath at its triple point.

3.	**Kelvin Temperature Scale**	The absolute temperature or temperature in kelvins is related to the Celsius temperature by

$$T = t_C + 273.15 \text{ K}$$

18-5

4.	**Ideal Gas**	

Equation of state

At low densities, all gases obey the ideal-gas law:

$$PV = nRT$$

18-13

Universal gas constant

$$R = kN_A$$
$$= 8.314 \text{ J/mol·K} = 0.08206 \text{ L·atm/mol·K}$$

18-12

Boltzmann's constant

$$k = 1.381 \times 10^{-23} \text{ J/K} = 8.617 \times 10^{-5} \text{ eV/K}$$

18-8

Avogadro's number

$$N_A = 6.022 \times 10^{23}$$

18-9

Equation for a fixed amount of gas

A form of the ideal-gas law that is useful for solving problems involving a fixed amount of gas is

$$\frac{P_2 V_2}{T_2} = \frac{P_1 V_1}{T_1}$$

18-14

5.	**Kinetic Theory of Gases**	

Molecular interpretation of temperature

The absolute temperature T is a measure of the average molecular energy.

Equipartition theorem

When a system is in equilibrium, there is an average energy of $\frac{1}{2}kT$ per molecule or $\frac{1}{2}RT$ per mole associated with each degree of freedom.

Average kinetic energy

For an ideal gas, the average translational kinetic energy of the molecules is

$$K_{av} = (\tfrac{1}{2}mv^2)_{av} = \tfrac{3}{2}kT$$

18-21

Total kinetic energy

The total translational kinetic energy of n moles of a gas containing N molecules is given by

$$K = N(\tfrac{1}{2}mv^2)_{av} = \tfrac{3}{2}NkT = \tfrac{3}{2}nRT$$

18-22

rms speed of molecules

The rms speed of a molecule of a gas is related to the absolute temperature by

$$v_{rms} = \sqrt{(v^2)_{av}} = \sqrt{\frac{3kT}{m}} = \sqrt{\frac{3RT}{M}}$$

18-23

where m is the mass of the molecule and M is the molar mass.

Mean free path

The mean free path of a molecule is related to its diameter d and the number of molecules per unit volume n_v by

$$\lambda = \frac{1}{\sqrt{2}\, n_v \pi d^2}$$

18-25

6.	**Maxwell–Boltzmann Distribution Functions** (optional)	

Speed distribution function

$$f(v) = \frac{4}{\sqrt{\pi}} \left(\frac{m}{2kT}\right)^{3/2} v^2 e^{-mv^2/2kT}$$

18-37

Energy distribution function

$$F(E) = \frac{2}{\sqrt{\pi}} \left(\frac{1}{kT}\right)^{3/2} E^{1/2} e^{-E/kT}$$

18-39

Problem-Solving Guide

Summary of Worked Examples

Type of Calculation	Procedure and Relevant Examples

1. Temperature

Convert temperatures from one scale to another.

To convert between °F and °C, use $t_C = \frac{5}{9}(t_F - 32°)$. To convert to kelvins, first convert to Celsius, then use

$$T = t_C + 273$$

Examples 18-1, 18-2

2. Ideal gas

Find the volume, pressure, or temperature of an ideal gas.

Use $PV = nRT$ or $\dfrac{P_1V_1}{T_1} = \dfrac{P_2V_2}{T_2}$

Examples 18-3, 18-5

3. Molecules

Find the mass of a molecule given the molar mass.

Use $M = mN_A$

Example 18-4

4. Kinetic Theory

Find the rms speed of molecules.

For a gas at temperature T use

$$v_{rms} = \sqrt{\frac{3kT}{m}} = \sqrt{\frac{3RT}{M}}$$

Example 18-7

Find the mean free path of molecules.

Use $\lambda = \dfrac{1}{\sqrt{2}n_v\pi d^2}$ and calculate $n_v = N/V$ from $PV = NkT$

Example 18-8

Find v_{av} and v_{rms} (optional).

For a small number of objects with given speeds, use $Nv_{av} = \Sigma n_i v_i$ and $v_{rms} = \sqrt{(v^2)_{av}}$ where $N(v^2)_{av} = \Sigma n_i v_i^2$

Example 18-9

For a gas at temperature T use

$$v_{av} = \int v f(v)\, dv$$

and

$$(v^2)_{av} = \int v^2 f(v)\, dv$$

where $f(v)$ is the Maxwell–Boltzmann distribution function.

Example 18-10

Problems

 Conceptual Problems

Problems from Optional and Exploring sections

In a few problems, you are given more data than you actually need; in a few other problems, you are required to supply data from your general knowledge, outside sources, or informed estimates.

- • Single-concept, single-step, relatively easy
- •• Intermediate-level, may require synthesis of concepts
- ••• Challenging, for advanced students

Temperature Scales

1   • True or false:

(a) Two objects in thermal equilibrium with each other must be in thermal equilibrium with a third object.

(b) The Fahrenheit and Celsius temperature scales differ only in the choice of the zero temperature.

(c) The kelvin is the same size as the Celsius degree.

(d) All thermometers give the same result when measuring the temperature of a particular system.

2 • How can you determine if two bodies are in thermal equilibrium with each other if it is impossible to put them into thermal contact with each other?

3 • Which is greater, an increase in temperature of 1 C° or of 1 F°?

4 • "One day I woke up and it was 20°F in my bedroom," said Mert to his old friend Mort. "That's nothing" replied Mort. "My room was once −5°C." Which room was colder?

5 • A certain ski wax is rated for use between −12 and −7°C. What is this temperature range on the Fahrenheit scale?

6 • The melting point of gold (Au) is 1945.4°F. Express this temperature in degrees Celsius.

7 • The highest and lowest temperatures ever recorded in the United States are 134°F (in California in 1913) and −80°F (in Alaska in 1971). Express these temperatures using the Celsius scale.

8 • What is the Celsius temperature corresponding to the normal temperature of the human body, 98.6°F?

9 • The length of the column of mercury in a thermometer is 4.0 cm when the thermometer is immersed in ice water and 24.0 cm when the thermometer is immersed in boiling water. (*a*) What should the length be at room temperature, 22.0°C? (*b*) If the mercury column is 25.4 cm long when the thermometer is immersed in a chemical solution, what is the temperature of the solution?

10 • The temperature of the interior of the sun is about 10^7 K. What is this temperature on (*a*) the Celsius scale, and (*b*) the Fahrenheit scale?

11 • The boiling point of nitrogen N_2 is 77.35 K. Express this temperature in degrees Fahrenheit.

12 • The pressure of a constant-volume gas thermometer is 0.400 atm at the ice point and 0.546 atm at the steam point. (*a*) When the pressure is 0.100 atm, what is the temperature? (*b*) What is the pressure at 444.6°C, the boiling point of sulfur?

13 • A constant-volume gas thermometer reads 50 torr at the triple point of water. (*a*) What will the pressure be when the thermometer measures a temperature of 300 K? (*b*) What ideal-gas temperature corresponds to a pressure of 678 torr?

14 • A constant-volume gas thermometer has a pressure of 30 torr when it reads a temperature of 373 K. (*a*) What is its triple-point pressure P_3? (*b*) What temperature corresponds to a pressure of 0.175 torr?

15 • At what temperature do the Fahrenheit and Celsius temperature scales give the same reading?

16 • Sodium melts at 371 K. What is the melting point of sodium on the Celsius and Fahrenheit temperature scales?

17 • The boiling point of oxygen at one atmosphere is 90.2 K. What is the boiling point of oxygen on the Celsius and Fahrenheit scales?

18 •• On the Réaumur temperature scale, the melting point of ice is 0° R and the boiling point of water is 80° R. Derive expressions for converting temperatures on the Réaumur scale to the Celsius and Fahrenheit scales.

19 ••• A thermistor is a solid-state device whose resistance varies greatly with temperature. Its temperature dependence is given approximately by $R = R_0 e^{B/T}$, where R is in ohms (Ω), T is in kelvins, and R_0 and B are constants that can be determined by measuring R at calibration points such as the ice point and the steam point. (*a*) If $R = 7360$ Ω at the ice point and 153 Ω at the steam point, find R_0 and B. (*b*) What is the resistance of the thermistor at $t = 98.6$°F? (*c*) What is the rate of change of the resistance with temperature (dR/dT) at the ice point and the steam point? (*d*) At which temperature is the thermistor most sensitive?

The Ideal-Gas Law

20 •• Two identical vessels contain different ideal gases at the same pressure and temperature. It follows that

(*a*) the number of gas molecules is the same in both vessels.
(*b*) the total mass of gas is the same in both vessels.
(*c*) the average speed of the gas molecules is the same in both vessels.
(*d*) none of the above is correct.

21 •• Figure 18-15 shows a plot of volume versus temperature for a process that takes an ideal gas from point A to point B. What happens to the pressure of the gas?

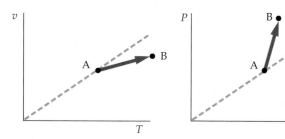

Figure 18-15 Problem 21   **Figure 18-16** Problem 22

22 •• Figure 18-16 shows a plot of pressure versus temperature for a process that takes an ideal gas from point A to point B. What happens to the volume of the gas?

23 • A gas is kept at constant pressure. If its temperature is changed from 50 to 100°C, by what factor does the volume change?

24 • A 10-L vessel contains gas at a temperature of 0°C and a pressure of 4 atm. How many moles of gas are in the vessel? How many molecules?

25 •• A pressure as low as 1×10^{-8} torr can be achieved using an oil diffusion pump. How many molecules are there in 1 cm^3 of a gas at this pressure if its temperature is 300 K?

26 •• A motorist inflates the tires of her car to a pressure of 180 kPa on a day when the temperature is −8.0°C. When she arrives at her destination, the tire pressure has increased to 245 kPa. What is the temperature of the tires if we assume that (*a*) the tires do not expand, or (*b*) that the tires expand by 7%?

27 •• A room is 6 m by 5 m by 3 m. (*a*) If the air pressure in the room is 1 atm and the temperature is 300 K, find the number of moles of air in the room. (*b*) If the temperature rises by 5 K and the pressure remains constant, how many moles of air leave the room?

28 •• A seafood restaurant hires Lou to run its advertising campaign. Lou figures that snorklers are a great pool of potential customers for seafood, so he prints ads on Mylar balloons that he ties to the coral of an underwater reef. Each balloon has a volume of 4 L and is filled with air at 20°C. At 15 m below the ocean surface, the volume has diminished to 1.60 L. What is the temperature of the water at this depth?

29 •• The boiling point of helium at one atmosphere is 4.2 K. What is the volume occupied by helium gas due to evaporation of 10 g of liquid helium at 1 atm pressure and a temperature of (*a*) 4.2 K, and (*b*) 293 K?

30 •• A container with a volume of 6.0 L holds 10 g of liquid helium. As the container warms to room temperature, what is the pressure exerted by the gas on its walls?

31 •• An automobile tire is filled to a gauge pressure of 200 kPa when its temperature is 20°C. (Gauge pressure is the difference between the actual pressure and atmospheric pressure.) After the car has been driven at high speeds, the tire temperature increases to 50°C. (*a*) Assuming that the volume of the tire does not change, and that air behaves as an ideal gas, find the gauge pressure of the air in the tire. (*b*) Calculate the gauge pressure if the volume of the tire expands by 10%.

32 •• A scuba diver is 40 m below the surface of a lake, where the temperature is 5°C. He releases an air bubble with a volume of 15 cm^3. The bubble rises to the surface, where the temperature is 25°C. What is the volume of the bubble right before it breaks the surface? *Hint:* Remember that the pressure also changes.

33 ••• A helium balloon is used to lift a load of 110 N. The weight of the balloon's skin is 50 N, and the volume of the balloon when fully inflated is 32 m^3. The temperature of the air is 0°C and the atmospheric pressure is 1 atm. The balloon is inflated with sufficient helium gas so that the net upward force on the balloon and its load is 30 N. Neglect changes of temperature with altitude.

(*a*) How many moles of helium gas are contained in the balloon?
(*b*) At what altitude will the balloon be fully inflated?
(*c*) Does the balloon ever reach the altitude at which it is fully inflated?
(*d*) If the answer to (*c*) is affirmative, what is the maximum altitude attained by the balloon?

Kinetic Theory of Gases

34 • True or false: The absolute temperature of a gas is a measure of the average translational kinetic energy of the gas molecules.

35 • By what factor must the absolute temperature of a gas be increased to double the rms speed of its molecules?

36 • How does the average translational kinetic energy of a molecule of a gas change if the pressure is doubled while the volume is kept constant? If the volume is doubled while the pressure is kept constant?

37 • A mole of He molecules is in one container and a mole of CH$_4$ molecules is in a second container, both at standard conditions. Which molecules have the greater mean free path?

38 •• A vessel holds an equal number of moles of helium and methane, CH$_4$. The ratio of the rms speeds of the helium atoms to the CH$_4$ molecules is _____ .

(*a*) 1
(*b*) 2
(*c*) 4
(*d*) 16

39 • (*a*) Find v_{rms} for an argon atom if 1 mol of the gas is confined to a 1-L container at a pressure of 10 atm. (For argon, $M = 40 \times 10^{-3}$ kg/mol.) (*b*) Compare this with v_{rms} for a helium atom under the same conditions. (For helium, $M = 4 \times 10^{-3}$ kg/mol.)

40 • Find the total translational kinetic energy of 1 L of oxygen gas held at a temperature of 0°C and a pressure of 1 atm.

41 • Find the rms speed and the average kinetic energy of a hydrogen atom at a temperature of 10^7 K. (At this temperature, which is of the order of the temperature in the interior of a star, the hydrogen is ionized and consists of a single proton.)

42 • In one model of a solid, the material is assumed to consist of a regular array of atoms in which each atom has a fixed equilibrium position and is connected by springs to its neighbors. Each atom can vibrate in the *x*, *y*, and *z* directions. The total energy of an atom in this model is

$$E = \tfrac{1}{2}mv_x^2 + \tfrac{1}{2}mv_y^2 + \tfrac{1}{2}mv_z^2 + \tfrac{1}{2}Kx^2 + \tfrac{1}{2}Ky^2 + Kz^2$$

What is the average energy of an atom in the solid when the temperature is *T*? What is the total energy of one mole of such a solid?

43 • Show that the mean free path for a molecule in an ideal gas at temperature *T* and pressure *P* is given by

$$\lambda = \frac{kT}{\sqrt{2}P\pi d^2}$$

44 •• The escape velocity on Mars is 5.0 km/s, and the surface temperature is typically 0°C. Calculate the rms speeds for (*a*) H$_2$, (*b*) O$_2$, and (*c*) CO$_2$ at this temperature. (*d*) If the rms speed of a gas is greater than about 15 to 20% of the escape velocity of a planet, virtually all of the molecules of that gas will escape the atmosphere of the planet. Based on this criterion, are H$_2$, O$_2$, and CO$_2$ likely to be found in Mars's atmosphere?

45 •• Repeat Problem 44 for Jupiter, whose escape velocity is 60 km/s and whose temperature is typically −150°C.

46 •• A pressure as low as $P = 7 \times 10^{-11}$ Pa has been obtained. Suppose a chamber contains helium at this pressure and at room temperature (300 K). Estimate the mean free path λ and the collision time τ for helium in the chamber. Take the diameter of a helium molecule to be 10^{-10} m.

47 •• Oxygen (O_2) is confined to a cubic container 15 cm on a side at a temperature of 300 K. Compare the average kinetic energy of a molecule of the gas to the change in its gravitational potential energy if it falls from the top of the container to the bottom.

The Distribution of Molecular Speeds

48 •• The class in Room 101 prepares their traditional greeting for a substitute teacher. Ten toy cars are wound up and released as the teacher arrives. The cars have the following speeds:

Speed, m/s	2	5	6	8
Number of cars	3	3	3	1

Calculate (a) the average speed, and (b) the rms speed of the cars.

49 •• Show that $f(v)$ given by Equation 18–37 is maximum when $v = \sqrt{2kT/m}$. *Hint:* Set $df/dv = 0$ and solve for v.

50 •• Since $f(v)\,dv$ gives the fraction of molecules that have speeds in the range dv, the integral of $f(v)\,dv$ over all the possible ranges of speeds must equal 1. Given the integral

$$\int_0^\infty v^2 e^{-av^2}\,dv = \frac{\sqrt{\pi}}{4} a^{-3/2}$$

show that $\int_0^\infty f(v)\,dv = 1$, where $f(v)$ is given by Equation 18–37.

51 •• Given the integral

$$\int_0^\infty v^3 e^{-av^2}\,dv = \frac{a^{-2}}{2}$$

calculate the average speed v_{av} of molecules in a gas using the Maxwell–Boltzmann distribution function.

52 ••• In Chapter 11, we found that the escape speed at the surface of a planet of radius R is $v_e = \sqrt{2gR}$, where g is the acceleration due to gravity. (a) At what temperature is v_{rms} for O_2 equal to the escape speed for the earth? (b) At what temperature is v_{rms} for H_2 equal to the escape speed for the earth? (c) Temperatures in the upper atmosphere reach 1000 K. How does this account for the low abundance of hydrogen in the earth's atmosphere? (d) Compute the temperatures for which the rms speeds of O_2 and H_2 are equal to the escape velocity at the surface of the moon, where g is about one-sixth of its value on earth and $R = 1738$ km. How does this account for the absence of an atmosphere on the moon?

General Problems

53 • True or false: If the pressure of a gas increases, the temperature must increase.

54 • What is the difference between 1°C and 1 C°?

55 • Why might the Celsius and Fahrenheit scales be more convenient than the absolute scale for ordinary, nonscientific purposes?

56 • The temperature of the interior of the sun is said to be about 10^7 degrees. Do you think that this is degrees Celsius or kelvins, or does it matter?

57 • If the temperature of an ideal gas is doubled while maintaining constant pressure, the average speed of the molecules

(a) remains constant.
(b) increases by a factor of 4.
(c) increases by a factor of 2.
(d) increases by a factor of $\sqrt{2}$.

58 • If both temperature and volume of an ideal gas are halved, the pressure

(a) diminishes by a factor of 2.
(b) remains constant.
(c) increases by a factor of 2.
(d) diminishes by a factor of $\sqrt{2}$.

59 • The average translational kinetic energy of the molecules of an ideal gas depends on

(a) the number of moles of the gas and its temperature.
(b) the pressure of the gas and its temperature.
(c) the pressure of the gas only.
(d) the temperature of the gas only.

60 • If a vessel contains equal amounts, by weight, of helium and argon, which of the following are true?

(a) The pressure exerted by the two gases on the walls of the container is the same.
(b) The average speed of a helium atom is the same as that of an argon atom.
(c) The number of helium atoms and argon atoms in the vessel are equal.
(d) None of the above statements is correct.

61 • Two different gases are at the same temperature. What can you say about the rms speeds of the gas molecules? What can you say about the average kinetic energies of the molecules?

62 •• Explain in terms of molecular motion why the pressure on the walls of a container increases when a gas is heated at constant volume.

63 •• Explain in terms of molecular motion why the pressure on the walls of a container increases when the volume of a gas is reduced at constant temperature.

64 •• Oxygen has a molar mass of 32 g/mol, and nitrogen has a molar mass of 28 g/mol. The oxygen and nitrogen molecules in a room have

(a) equal average kinetic energies, but the oxygen molecules are faster.
(b) equal average kinetic energies, but the oxygen molecules are slower.
(c) equal average kinetic energies and speeds.
(d) equal average speeds, but the oxygen molecules have a higher average kinetic energy.
(e) equal average speeds, but the oxygen molecules have a lower average kinetic energy.
(f) None of the above is correct.

65 • At what temperature will the rms speed of an H_2 molecule equal 331 m/s?

66 • A solid-state temperature transducer is essentially a linear amplifier whose amplification is linearly temperature dependent. If the amplification is 25 times at 20°C and 60 times at 70°C, what is the temperature when the amplification is 45 times?

67 •• (a) If 1 mol of a gas in a container occupies a volume of 10 L at a pressure of 1 atm, what is the temperature of the gas in kelvins? (b) The container is fitted with a piston so that the volume can change. When the gas is heated at constant pressure, it expands to a volume of 20 L. What is the temperature of the gas in kelvins? (c) The volume is fixed at 20 L, and the gas is heated at constant volume until its temperature is 350 K. What is the pressure of the gas?

68 •• A cubic metal box with sides of 20 cm contains air at a pressure of 1 atm and a temperature of 300 K. The box is sealed so that the volume is constant, and it is heated to a temperature of 400 K. Find the net force on each wall of the box.

69 •• Water, H_2O, can be converted into H_2 and O_2 gas by electrolysis. How many moles of these gases result from the electrolysis of 2 L of water?

70 •• A massless cylinder 40 cm long rests on a horizontal frictionless table. The cylinder is divided into two equal sections by a membrane. One section contains nitrogen and the other contains oxygen. The pressure of the nitrogen is twice that of the oxygen. How far will the cylinder move if the membrane is removed?

71 •• A cylinder contains a mixture of nitrogen gas (N_2) and hydrogen gas (H_2). At a temperature T_1 the nitrogen is completely dissociated but the hydrogen does not dissociate at all, and the pressure is P_1. If the temperature is doubled to $T_2 = 2T_1$, the pressure is tripled due to complete dissociation of hydrogen. If the mass of hydrogen is m_H, find the mass of nitrogen m_N.

72 •• A vertical closed cylinder is divided into two equal parts by a heavy insulating movable piston of mass m_p. The top part contains nitrogen at a temperature T_1 and pressure P_1, and the bottom part is filled with oxygen at a temperature $2T_1$. The cylinder is turned upside-down. To keep the piston in the middle, the oxygen must be cooled to $T_2 = T_1/3$, with the temperature of the nitrogen remaining at T_1. Find the initial pressure of oxygen P_i.

73 •• Three insulated vessels of equal volumes V are connected by thin tubes that can transfer gas but do not transfer heat. Initially all vessels are filled with the same type of gas at a temperature T_0 and pressure P_0. Then the temperature in the first vessel is doubled and the temperature in the second vessel is tripled. The temperature in the third vessel remains unchanged. Find the final pressure P' in the system in terms of the initial pressure P_0.

74 •• At the surface of the sun, the temperature is about 6000 K, and all the substances present are gaseous. From data given by the light spectrum of the sun, it is known that most elements are present. (a) What is the average kinetic energy of translation of an atom at the surface of the sun? (b) What is the range of rms speeds at the surface of the sun if the atoms present range from hydrogen ($M = 1$ g/mol) to uranium ($M = 238$ g/mol)?

75 •• A constant-volume gas thermometer with a triple-point pressure $P_3 = 500$ torr is used to measure the boiling point of some substance. When the thermometer is placed in thermal contact with the boiling substance, its pressure is 734 torr. Some of the gas in the thermometer is then allowed to escape so that its triple-point pressure is 200 torr. When it is again placed in thermal contact with the boiling substance, its pressure is 293.4 torr. Again, some of the gas is removed from the thermometer so that its triple-point pressure is 100 torr. When the thermometer is placed in thermal contact with the boiling substance once again, its pressure is 146.65 torr. Find the ideal-gas temperature of the boiling substance.

76 ••• A cylinder 2.4 m tall is filled with 0.1 mol of an ideal gas at standard temperature and pressure (Figure 18-17). The top of the cylinder is then closed with a tight-fitting piston whose mass is 1.4 kg and the piston is allowed to drop until it is in equilibrium. (a) Find the height of the piston, assuming that the temperature of the gas does not change as it is compressed. (b) Suppose that the piston is pushed down below its equilibrium position by a small amount and then released. Assuming that the temperature of the gas remains constant, find the frequency of vibration of the piston.

Figure 18-17 Problem 76

1.4 kg

Heat and the First Law of Thermodynamics

Steel ingots in a twin-tube tunnel furnace. The three 53-cm diameter carbon steel ingots seen here have been heated for about 7 hours to approximately 1340°C. Each 3200-kg ingot sits on a furnace car that transports it through the 81-m furnace, which is divided into twelve separate heating zones so that the temperature of the ingot is increased gradually to prevent cracking. The ingots, glowing a yellow-whitish color, exit the furnace to be milled into large, heavy-walled pipes.

Heat is energy that is transferred from one system to another because of a difference in temperature. In the seventeenth century, Galileo, Newton, and other scientists generally supported the theory of the ancient Greek atomists who considered heat to be a manifestation of molecular motion. In the next century, methods were developed for making quantitative measurements of the amount of heat that leaves or enters an object, and it was found that when objects are in thermal contact, the amount of heat that leaves one object equals the amount that enters the other. This discovery led to the caloric theory of heat as a conserved material substance. In this theory, an invisible fluid called "caloric" flowed out of one object and into another and could be neither created nor destroyed.

The caloric theory reigned until the nineteenth century, when it was found that friction between objects could generate an unlimited amount of heat, deposing the idea that caloric was a substance present in a fixed amount. The modern theory of heat did not emerge until the 1840s, when James Joule (1818–1889) showed that the gain or loss of a given amount of heat was accompanied by the disappearance or appearance of an equivalent quantity of mechanical energy. Heat, therefore, is not itself conserved. Instead, heat is a form of energy, and it is energy that is conserved.

19-1 Heat Capacity and Specific Heat

When heat energy flows into a substance, the temperature of the substance usually rises.* The amount of heat energy Q needed to raise the temperature of a substance is proportional to the temperature change and to the mass of the substance:

$$Q = C \, \Delta T = mc \, \Delta T \qquad \text{19-1}$$

where C is the **heat capacity,** which is defined as the heat energy needed to raise the temperature of a substance by one degree, and c is the **specific heat**, the heat capacity per unit mass:

$$c = \frac{C}{m} \qquad \text{19-2}$$

The historical unit of heat energy, the **calorie,** was originally defined to be the amount of heat energy needed to raise the temperature of one gram of water one Celsius degree.[†] Since we now recognize that heat is just another form of energy, we do not need any special units for it. The calorie is now defined in terms of the SI unit of energy, the joule:

$$1 \text{ cal} = 4.184 \text{ J} \qquad \text{19-3}$$

The U.S. customary unit of heat is the **Btu** (for British thermal unit), which was originally defined to be the amount of energy needed to raise the temperature of one pound of water by one Fahrenheit degree. The Btu is related to the calorie and to the joule by

$$1 \text{ Btu} = 252 \text{ cal} = 1.054 \text{ kJ} \qquad \text{19-4}$$

The original definition of the calorie implies that the specific heat of water is[‡]

$$c_{\text{water}} = 1 \text{ cal/g} \cdot \text{C}° = 1 \text{ kcal/kg} \cdot \text{C}°$$

$$= 1 \text{ kcal/kg} \cdot \text{K} = 4.184 \text{ kJ/kg} \cdot \text{K} \qquad \text{19-5}a$$

Similarly, from the definition of the Btu, the specific heat of water in U.S. customary units is

$$c_{\text{water}} = 1 \text{ Btu/lb} \cdot \text{F}° \qquad \text{19-5}b$$

The heat capacity per mole is called the **molar specific heat** c',

$$c' = \frac{C}{n}$$

where n is the number of moles. Since $C = mc$, the molar specific heat c' and specific heat c are related by

$$c' = \frac{C}{n} = \frac{mc}{n} = Mc \qquad \text{19-6}$$

where $M = m/n$ is the molar mass. Table 19-1 lists the specific heats and molar specific heats of some solids and liquids. Note

Table 19-1

Specific Heats and Molar Specific Heats of Some Solids and Liquids

Substance	c, kJ/kg·K	c, kcal/kg·K or Btu/lb·F°	c', J/mol·K
Aluminum	0.900	0.215	24.3
Bismuth	0.123	0.0294	25.7
Copper	0.386	0.0923	24.5
Glass	0.840	0.20	
Gold	0.126	0.0301	25.6
Ice (−10°C)	2.05	0.49	36.9
Lead	0.128	0.0305	26.4
Silver	0.233	0.0558	24.9
Tungsten	0.134	0.0321	24.8
Zinc	0.387	0.0925	25.2
Alcohol (ethyl)	2.4	0.58	111
Mercury	0.140	0.033	28.3
Water	4.18	1.00	75.2

---

* An exception occurs during a change in phase, as when water freezes or evaporates. Changes of phase are discussed in Section 19-2.

† The kilocalorie is then the amount of heat energy needed to raise the temperature of one kilogram of water by one Celsius degree. The "calorie" used in measuring the energy equivalent of foods is actually the kilocalorie.

‡ Careful measurement shows that the specific heat of water varies by about 1% over the temperature range from 0 to 100°C. We will usually neglect this small variation.

that the molar heats of all the metals are about the same. We discuss the significance of this in Section 19-7.

How much heat is needed to raise the temperature of 3 kg of copper by 20 C°?

The required heat is given by Equation 19-1 with $c = 0.386$ kJ/kg·K from Table 19-1:

$$Q = mc\,\Delta T = (3\text{ kg})(0.386\text{ kJ/kg·K})(20\text{ K})$$
$$= 23.2\text{ kJ}$$

Remarks Note that we use $\Delta T = 20$ C° $= 20$ K. Alternatively, we could express the specific heat as 0.386 kJ/kg·C° and write the temperature change as 20 C°.

Exercise A 2-kg aluminum block is originally at 10°C. If 36 kJ of heat energy are added to the block, what is its final temperature? (*Answer*   30°C)

We see from Table 19-1 that the specific heat of water is considerably larger than that of the other substances. Thus, water is an excellent material for storing thermal energy, as in a solar heating system. It is also an excellent coolant, as in a car engine. Large bodies of water, such as lakes or oceans, tend to moderate variations of temperature nearby because they can absorb or release large quantities of thermal energy while undergoing only very small changes in temperature.

Calorimetry

The specific heat of an object can be conveniently measured by heating the object to some temperature, placing it in a water bath of known mass and temperature, and measuring the final equilibrium temperature. If the system is isolated from its surroundings, the heat leaving the object equals the heat entering the water and its container. This procedure is called **calorimetry,** and the insulated water container is called a **calorimeter.**

Let m be the mass of the object, let c be its specific heat, and let T_{io} be the object's initial temperature. If T_f is the final temperature of the object in its water bath, the heat leaving the object is

$$Q_{out} = mc(T_{io} - T_f)$$

Similarly, if T_{iw} is the initial temperature of the water and container, and T_f is their final equilibrium temperature, the heat absorbed by the water and container is

$$Q_{in} = m_w c_w(T_f - T_{iw}) + m_c c_c(T_f - T_{iw})$$

where m_w and $c_w = 4.18$ kJ/kg·K are the mass and specific heat of the water, and m_c and c_c are the mass and specific heat of the container. (Note that we have chosen the temperature differences so that the heat in and heat out are both positive quantities.) Setting these amounts of heat equal yields the specific heat c of the object:

$$Q_{out} = Q_{in}$$

$$mc(T_{io} - T_f) = m_w c_w(T_f - T_{iw}) + m_c c_c(T_f - T_{iw}) \qquad 19\text{-}7$$

Since only temperature differences occur in Equation 19-7, and since the kelvin and Celsius degree are the same size, it doesn't matter whether kelvins or Celsius degrees are used.

Example 19-2

To measure the specific heat of lead, you heat 600 g of lead shot to 100°C and place it in an aluminum calorimeter of mass 200 g that contains 500 g of water initially at 17.3°C. If the final temperature of the mixture is 20.0°C, what is the specific heat of lead? (The specific heat of the aluminum container is 0.900 kJ/kg·K.)

Picture the Problem We set the heat out of the lead equal to the heat into the water and container and solve for the specific heat of lead c_{Pb}.

1. Write the heat given off by the lead in terms of its specific heat:

$$Q_{Pb} = mc_{Pb}(T_{io} - T_f) = (0.6 \text{ kg})c_{Pb}(100°C - 20°C)$$
$$= (0.6 \text{ kg})c_{Pb}(80 \text{ K})$$

2. Find the heat absorbed by the water:

$$Q_w = m_w c_w \Delta T_w = m_w c_w(20.0°C - 17.3°C)$$
$$= (0.5 \text{ kg})(4.18 \text{ kJ/kg·K})(2.7 \text{ K}) = 5.64 \text{ kJ}$$

3. Find the heat absorbed by the container:

$$Q_c = m_c c_c \Delta T_c = (0.2 \text{ kg})(0.900 \text{ kJ/kg·K})(2.7 \text{ K})$$
$$= 0.486 \text{ kJ}$$

4. Set the heat out equal to the heat in:

$$Q_{Pb} = Q_w + Q_c$$
$$(0.6 \text{ kg})c_{Pb}(80 \text{ K}) = 5.64 \text{ kJ} + 0.486 \text{ kJ} = 6.13 \text{ kJ}$$

5. Solve for c_{Pb}:

$$c_{Pb} = \frac{6.13 \text{ kJ}}{(0.6 \text{ kg})(80.0 \text{ K})} = 0.128 \text{ kJ/kg·K}$$

Remarks Note that the specific heat of lead is considerably less than that of water.

19-2   Change of Phase and Latent Heat

When heat is added to ice at 0°C, the temperature of the ice does not change. Instead, the ice melts. This is an example of a **phase change**. Common types of phase changes include fusion (liquid to solid), melting (solid to liquid), vaporization (liquid to vapor or gas), condensation (gas or vapor to liquid), and sublimation (solid directly to vapor, as when solid carbon dioxide (dry ice) changes to vapor). There are other types of phase changes as well, such as the change of a solid from one crystalline form to another. Carbon under intense pressure, for example, becomes diamond.

The fact that the temperature remains constant during a phase change can be understood in terms of molecular theory. The molecules in a liquid are close together and exert attractive forces on each other. The molecules in a gas are far apart. Changing the substance from a liquid to a vapor requires energy to overcome the attraction between the molecules of the liquid. The energy put into the liquid to vaporize it thus increases the potential energy of the molecules, not their kinetic energy. Since temperature measures the kinetic energy of the molecules, the temperature remains the same during a phase change.

For a pure substance, a change in phase at a given pressure occurs only at a particular temperature. For example, pure water at a pressure of 1 atm changes from solid to liquid at 0°C (the normal melting point of water) and from liquid to gas at 100°C (the normal boiling point of water).

The heat energy required to melt a substance of mass m with no change in its temperature is proportional to the mass of the substance:

$$Q_f = mL_f \qquad\qquad 19\text{-}8$$

where L_f is called the **latent heat of fusion** of the substance. At a pressure of 1 atm, the latent heat of fusion for water is 333.5 kJ/kg = 79.7 kcal/kg. When the phase change is from liquid to gas, the heat required is

$$Q_v = mL_v \qquad\qquad 19\text{-}9$$

where L_v is the **latent heat of vaporization**. For water at a pressure of 1 atm, the latent heat of vaporization is 2.26 MJ/kg = 540 kcal/kg. Table 19-2 gives the normal melting and boiling points, and the latent heats of fusion and vaporization at 1 atm, for various substances.

Table 19-2

Normal Melting Point (MP), Latent Heat of Fusion L_f, Normal Boiling Point (BP), and Latent Heat of Vaporization L_v for Various Substances at 1 atm

Substance	MP, K	L_f, kJ/kg	BP, K	L_v, kJ/kg
Alcohol, ethyl	159	109	351	879
Bromine	266	67.4	332	369
Carbon dioxide	—	—	194.6[a]	573[a]
Copper	1356	205	2839	4726
Gold	1336	62.8	3081	1701
Helium	—	—	4.2	21
Lead	600	24.7	2023	858
Mercury	234	11.3	630	296
Nitrogen	63	25.7	77.35	199
Oxygen	54.4	13.8	90.2	213
Silver	1234	105	2436	2323
Sulfur	388	38.5	717.75	287
Water	273.15	333.5	373.15	2257
Zinc	692	102	1184	1768

[a]These values are for sublimation. Carbon dioxide does not have a liquid state at 1 atm.

Example 19-3   *try it yourself*

How much heat do you need to heat 1.5 kg of ice at a pressure of 1 atm from −20°C until all the ice has been changed into steam?

Picture the Problem The heat required consists of four parts: Q_1, the heat needed to warm the ice from −20 to 0°C; Q_2, the heat needed to melt the ice; Q_3, the heat needed to warm the water from 0 to 100°C; and Q_4, the heat needed to vaporize the water. In calculating Q_1 and Q_3, we will assume that the specific heats are constant, with the values 2.05 kJ/kg·K for ice and 4.18 kJ/kg·K for water.

Cover the column to the right and try these on your own before looking at the answers.

Steps

Answers

1. Use $Q_1 = mc\,\Delta T$ to find the heat needed to warm the ice to 0°C.

$Q_1 = 61.5$ kJ $= 0.0615$ MJ

2. Use L_f from Table 19-2 to find the heat Q_2 needed to melt the ice.

$Q_2 = 500$ kJ $= 0.500$ MJ

3. Find the heat Q_3 needed to warm the water from 0 to 100°C.

$Q_3 = 627$ kJ $= 0.627$ MJ

4. Use L_v from Table 19-2 to find the heat Q_4 needed to vaporize the water.

$Q_4 = 3.39$ MJ

5. Sum your results to find the total heat Q.

$Q = 4.58$ MJ

Remarks Notice that most of the heat was needed to vaporize the water, and that the amount needed to melt the ice was almost as much as that needed to raise the temperature of the water by 100 C°. Figure 19-1 shows a graph of temperature versus time for the case in which the heat is added at a constant rate of 1.5 kJ/s. Note that it takes considerably longer to vaporize the water than it does to melt the ice or to raise the temperature of the water. When all of the water has vaporized, the temperature again rises as heat is added.

Exercise An 830-g piece of lead is heated to its melting point of 600 K. How much additional heat energy must be added to melt the lead? (*Answer* 20.5 kJ)

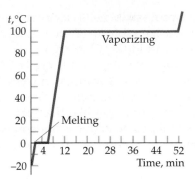

Figure 19-1

Example 19-4

Lemonade has been sitting on the picnic table all day at 33°C. You pour 0.24 kg into a Styrofoam cup and add 2 ice cubes (each 0.025 kg at 0°C). (*a*) Assuming no heat lost to the surroundings, what is the final temperature of the lemonade? (*b*) What is the final temperature if you add 6 ice cubes? Assume that the lemonade has the same heat capacity as water.

Picture the Problem We set the heat lost by the lemonade equal to that gained by the ice cubes. Let t_f be the final temperature of the lemonade and water.

(*a*)1. Write the heat lost by the lemonade in terms of the final temperature t_f:

$$Q_{out} = m_L c\,\Delta t = (0.24\text{ kg})(4.18\text{ kJ/kg·C°})(33°C - t_f)$$
$$= 33\text{ kJ} - (1.00\text{ kJ/C°})t_f$$

2. Write the heat gained by the ice cubes and resulting water in terms of the final temperature:

$$Q_{in} = L_f m_{ice} + m_{ice} c\,\Delta t$$
$$= (0.05\text{ kg})(333.5\text{ kJ/kg})$$
$$\quad + (0.05\text{ kg})(4.18\text{ kJ/kg·C°})(t_f - 0)$$
$$= 16.7\text{ kJ} + (0.209\text{ kJ/C°})t_f$$

3. Set the heat lost equal to the heat gained and solve for t_f:

$$33\text{ kJ} - (1.00\text{ kJ/C°})t_f = 16.7\text{ kJ} + (0.209\text{ kJ/C°})t_f$$
$$t_f = 13.5°C$$

(*b*)1. For 6 ice cubes, $m_{ice} = 0.15$ kg. Step 1 is the same. Find the heat gained by the ice as in step 2 of part (*a*):

$$Q_{in} = L_f m_{ice} + m_{ice} c\,\Delta t$$
$$= (0.15\text{ kg})(333.5\text{ kJ/kg})$$
$$\quad + (0.15\text{ kg})(4.18\text{ kJ/kg·C°})(t_f - 0)$$
$$= 50.0\text{ kJ} + (0.627\text{ kJ/C°})t_f$$

2. Set the heat lost equal to the heat gained and solve for t_f:

$$33\text{ kJ} - (1.00\text{ kJ/C°})t_f = 50.0\text{ kJ} + (0.627\text{ kJ/C°})t_f$$
$$t_f = -10°C$$

But this cannot be correct! We know that if we add ice at 0°C to lemonade at 33°C that the final temperature of the mixture cannot be −10°C. What's wrong? The heat given off by the lemonade as it cools from 33 to 0°C is not enough to melt all of the ice, contrary to our assumption that all of the ice melts. The final temperature is thus

$$t_f = 0°C$$

Check the Result Let's calculate how much ice is melted in part (*b*). For the lemonade to cool from 33 to 0°C, it must give off heat in the amount $Q_{out} =$ (0.24 kg)(4.18 kJ/kg·C°)(33°C) = 33.1 kJ. The mass of ice that this amount of heat will melt is $m_{ice} = Q_{in}/L_f =$ 33.1 kJ/(333.5 kJ/kg) = 0.10 kg. This is about 4 ice cubes. Adding more than 4 ice cubes does not lower the temperature below 0°C. It merely increases the amount of ice in the ice–lemonade mixture. In problems like this one, we should first find out how much ice must be melted to reduce the temperature of the liquid to 0°C. If less than that amount is added, we can proceed as in part (*a*). If more ice is added, the final temperature is 0°C.

19-3   Joule's Experiment and the First Law of Thermodynamics

We can raise the temperature of a system by adding heat, but we can also raise its temperature by just doing work on it. Figure 19-2*a* is a diagram of the apparatus Joule used in his famous experiment for determining the amount of work needed to raise the temperature of one gram of water by one Celsius degree. Joule's machine converts the potential energy of falling weights into work done on the water. Joule found that he could raise the temperature of his water sample by 1 Fahrenheit

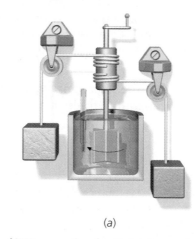

(a)   (b)

degree when he drove his machine by the fall of 772 pounds of weight through one foot. In modern units, Joule found that it takes about 4.184 J (the energy units adopted by the scientific community in 1948) to raise the temperature of 1 g of water by 1 C°. The result that 4.184 J of mechanical energy is equivalent to 1 cal of heat energy is known as the mechanical equivalent of heat.

Figure 19-2 (*a*) Schematic diagram for Joule's experiment. Water is enclosed by insulating walls to prevent heat transfer. As the weights fall at constant speed, they turn a paddle wheel, which does work on the water. If friction is negligible, the work done by the paddle wheel against the water equals the loss of mechanical energy of the weights, which is determined by calculating the loss in the potential energy of the weights. (*b*) Photograph of the apparatus for Joule's experiment.

There are other ways of doing work on this system. For example, we could drop the insulated container of water from some height *h*, letting the system make an inelastic collision with the ground, or we could do mechanical work to generate electricity and then use the electricity to heat the water (Figure 19-3). In all such experiments, the same amount of work is required to produce a given temperature change. By the conservation of energy, the work done must go into an increase in internal energy of the system.

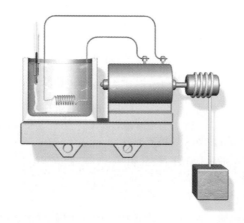

Figure 19-3 Another method of doing work on a thermally insulated container of water. Electrical work is done on the system by the generator, which is driven by the falling weight.

Example 19-5

To demonstrate the equivalence of heat and energy, you drop a thermally insulated container of water from a height *h* to the ground. If the collision is perfectly inelastic and all of the energy lost goes into internal energy of the water, what must *h* be for the temperature of the water to increase by 1 C°?

Picture the Problem The kinetic energy of the water just before it hits the ground equals its original potential energy mgh. During the collision, this energy is converted into thermal energy Q, which in turn causes a rise in temperature given by $Q = mc\,\Delta T$.

1. Set the potential energy equal to the thermal energy:

$$mgh = mc\,\Delta T$$

2. Solve for the height h:

$$h = c\,\Delta T/g$$

3. Substitute $c = 4.18$ kJ/kg·K and $\Delta T = 1\,$C° $= 1$ K:

$$h = \frac{(4.18\ \text{kJ/kg·K})(1\ \text{K})}{9.81\ \text{N/kg}} = 0.426\ \text{km} = 426\ \text{m}$$

Remarks Note that h is independent of the mass of the water. It is also rather large, which illustrates one of the difficulties with Joule's experiment—a large amount of work must be done to produce a measurable change in the temperature of the water.

Now suppose we perform Joule's experiment but replace the insulating walls of the container with conducting walls. We find that the work needed to produce a given change in the temperature of the system depends on how much heat is added to or subtracted from the system by conduction through the walls. However, if we sum the work done on the system and the net heat added to or subtracted from the system, the result is always the same for a given temperature change. That is, the sum of the heat added and the work done on the system equals the change in the internal energy of the system. This is the **first law of thermodynamics**, which is simply a statement of the conservation of energy.

It is customary to write W for the work done *by* the system on its surroundings.* Then $-W$ is the work done *on* the system. For example, if a gas expands against a piston, doing work on the surroundings, W is positive. The heat Q is taken to be positive if it is put *into* the system, and negative if it is taken *out of* the system (Figure 19-4). Using these conventions, and denoting the internal energy by U, the first law of thermodynamics is written[†]

$$Q = \Delta U + W \qquad\qquad 19\text{-}10$$

The heat added to a system equals the change in the internal energy of the system plus the work done by the system.

First law of thermodynamics

Equation 19-10 is the same as the work–energy theorem, $W_{\text{ext}} = \Delta E_{\text{sys}}$ of Chapter 7 (Equation 7-9), except that we have added the heat term Q, changed the sign convention for W, and called the energy of the system U.

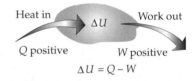

Figure 19-4 Sign convention for the first law of thermodynamics.

*We choose this so that the work done by an expanding gas is positive, and the work done by a heat engine, which we study in the next chapter, is positive.

†The symbol U, which we used in previous chapters to denote potential energy, is now used for the internal energy of a system, which may include both kinetic and potential energy of the molecules in the system.

Example 19-6

You do 25 kJ of work on a system consisting of 3 kg of water by stirring it with a paddle wheel. During this time, 15 kcal of heat is removed. What is the change in the internal energy of the system?

Picture the Problem We express all energies in joules and apply the first law of thermodynamics.

1. ΔU is found from the first law of thermody- $Q = \Delta U + W$
   namics:

2. Heat is *removed* from the system, thus the heat $Q = -15 \text{ kcal} = -(15 \text{ kcal})\left(\dfrac{4.18 \text{ kJ}}{1 \text{ kcal}}\right) = -62.7 \text{ kJ}$
   added is negative:

3. Work is done *on* the system, thus the work done $W = -25 \text{ kJ}$
   by the system is negative:

4. Substitute these quantities and solve for ΔU: $\Delta U = Q - W = (-62.7 \text{ kJ}) - (-25 \text{ kJ}) = -37.7 \text{ kJ}$

Remark The internal energy decreases because the system loses more energy as heat than it gains from the work done on it.

It is important to understand that the internal energy U is a function of the state of the system, just as P, V, and T are functions of the state of the system. Consider a gas in some initial state (P_1, V_1). The temperature T_1 is determined by the equation of state. For example, if the gas is ideal, $T_1 = P_1 V_1/nR$. The internal energy U_1 also depends only on the state of the gas, which is determined by any two state variables, such as P and V. If we compress the gas or let it expand, add or remove heat from it, do work on it or let it do work, the gas will move through a sequence of states; that is, it will have different values of the state functions P, V, T, and U. If the gas is then returned to its original state (P_1, V_1), the temperature T and the internal energy U must equal their original values.

On the other hand, the net heat input Q and the work W done by the gas are not functions of the state of the system. There are no functions Q or W associated with any particular state of the gas. We could take the gas through a sequence of states beginning and ending at state (P_1, V_1) during which the gas did positive work and absorbed an equal amount of heat. Or we could take it through a different sequence of states such that work was done on the gas and heat was removed from the gas. It is correct, then, to say that a system has a large amount of internal energy, but it is not correct to say that a system has a large amount of heat or a large amount of work. Heat is not something that is contained in a system. Rather, it is a measure of the energy that flows from one system to another because of a difference in temperature.

For very small amounts of heat added, work done, or changes in internal energy, it is customary to write Equation 19-10 as

$$dQ = dU + dW \qquad\qquad \text{19-11}$$

In this equation, dU is the differential of the internal-energy function. However, neither dQ nor dW is a differential of any function. Instead, dQ merely represents a small amount of heat added to the system, and dW represents a small amount of work done by the system.

19-4   The Internal Energy of an Ideal Gas

The translational kinetic energy K of the molecules in an *ideal* gas is related to the absolute temperature T by Equation 18-22:

$$K = \tfrac{3}{2}nRT$$

where n is the number of moles of gas and R is the universal gas constant. If the internal energy of a gas is just this translational kinetic energy, then $U = K$, and

$$U = \tfrac{3}{2}nRT \qquad\qquad\qquad 19\text{-}12$$

Then the internal energy will depend only on the temperature of the gas and not on its volume or pressure. If the molecules have other types of energy in addition to translational energy, such as energy of rotation, the internal energy will be greater than that given by Equation 19-12. But according to the equipartition theorem, the average energy associated with any degree of freedom will be $\tfrac{1}{2}kT$ per molecule or $\tfrac{1}{2}RT$ per mole, so again, the internal energy will depend only on the temperature and not on the volume or pressure.

We can imagine that the internal energy of a *real* gas might include other kinds of energy that depend on the pressure and volume of the gas. Suppose, for example, that nearby gas molecules exert attractive forces on each other. Work is then required to increase the separation of the molecules. Then, when the average distance between the molecules is increased, the potential energy associated with the molecular attraction will increase. The internal energy of the gas will then depend on the volume of the gas as well as on its temperature.

Joule, using an apparatus like that shown in Figure 19-5, performed an interesting experiment to determine whether the internal energy of a gas depends on its volume. Initially, the compartment on the left in Figure 19-5 contains a gas and the compartment on the right is evacuated. The compartments are connected by a stopcock that is closed. The whole system is thermally insulated from its surroundings—no heat can go into or out of the system and no work can be done. When the stopcock is opened, the gas rushes into the evacuated chamber. This process is called a **free expansion.** Eventually, the gas reaches thermal equilibrium with itself. Since no work has been done and no heat has been transferred, the final internal energy of the gas must equal its initial internal energy. If the gas molecules exert attractive forces on one another, the potential energy associated with these forces will increase when the volume increases. Since energy is conserved, the kinetic energy of translation must therefore decrease, which will result in a decrease in the temperature of the gas.

When Joule did this experiment, he found the final temperature to be equal to the initial temperature. Subsequent experiments verified this result for low gas densities. This implies that for a gas at low density—that is, for an ideal gas—the temperature depends only on the internal energy, or as we usually think of it, the internal energy depends only on the temperature. However, when the experiment is done with a large amount of gas initially in the left compartment so that the density is high, the temperature after expansion is slightly lower than that before the expansion. This indicates that there is a small attraction between the gas molecules of a real gas.

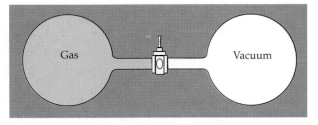

Figure 19-5 Free expansion of a gas. When the stopcock is opened, the gas expands rapidly into the evacuated chamber. Since no work is done and the whole system is thermally insulated, the initial and final internal energies of the gas are equal.

19-5   Work and the *PV* Diagram for a Gas

In many types of engines, work is done by a gas expanding against a movable piston. For example, in a steam engine, water is heated in a boiler to produce steam. The steam then does work as it expands and drives a piston. In an automobile engine, a mixture of gasoline vapor and air is ignited, causing it to explode. The resulting high temperatures and pressures cause the gas to expand rapidly, driving a piston and doing work.

Quasi-static Processes

Figure 19-6 shows an ideal gas confined in a container with a tightly fitting piston that we assume to be frictionless. When the piston moves, the volume of the gas changes. The temperature or pressure or both must also change since these three variables are related by the equation of state $PV = nRT$. If we suddenly push the piston in to compress the gas, the pressure will initially be greater near the piston than far from it. Eventually the gas will settle down to a new equilibrium pressure and temperature. Until equilibrium is restored in the gas, we cannot determine such macroscopic variables as T, P, or U for the entire gas system. However, if we move the piston slowly in small steps and allow equilibrium to be reestablished after each step, we can compress or expand the gas in such a way that the gas is never far from an equilibrium state. In this kind of process, called a **quasi-static process,** the gas moves through a series of equilibrium states. In practice, it is possible to approximate quasi-static processes fairly well.

Let us begin with a gas at a high pressure, and let it expand quasi-statically. The force exerted by the gas on the piston is PA, where A is the area of the piston and P is the gas pressure. If the piston moves a small distance dx, the work done by the gas on the piston is

$$dW = F\,dx = PA\,dx = P\,dV \qquad\qquad 19\text{-}13$$

where $dV = A\,dx$ is the increase in the volume of the gas. To calculate the work done by the gas during an expansion from a volume of V_1 to a volume of V_2, we need to know how the pressure varies during the expansion.

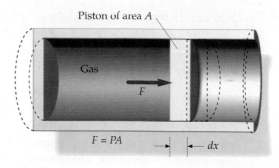

Figure 19-6 Gas confined in a thermally insulated cylinder with a movable piston. When the piston moves a distance dx, the volume of the gas changes by $dV = A\,dx$. The work done by the gas is $PA\,dx = P\,dV$, where P is the pressure.

PV Diagrams

We can represent the states of a gas on a diagram of P versus V. Each point on the PV diagram indicates a particular state of the gas. Figure 19-7 shows a PV diagram with a horizontal line representing a series of states that all have the same value of P. This line represents an expansion at constant pressure. Such a process is called an **isobaric expansion.** For a volume change of ΔV, the work done is $P\,\Delta V$, which is equal to the shaded area under the curve in the figure. In general, the work done by the gas is equal to the area under the P-versus-V curve:

$$W = \int P\,dV = \text{area under the } P\text{-versus-}V \text{ curve} \qquad 19\text{-}14$$

Work done by a gas

Since pressures are often given in atmospheres and volumes are often given in liters, it is convenient to have a conversion factor between liter-atmospheres and joules:

$$1\ \text{L}\cdot\text{atm} = (10^{-3}\,\text{m}^3)(101.3 \times 10^3\,\text{N/m}^2) = 101.3\ \text{J} \qquad 19\text{-}15$$

Exercise If 3 L of an ideal gas at a pressure of 2 atm is heated so that it expands at constant pressure until its volume is 5 L, what is the work done by the gas? (*Answer*   405.2 J)

Figure 19-8 shows three different possible paths on a PV diagram for a gas that is initially in state (P_1, V_1) and is finally in state (P_2, V_2). We assume that the gas is ideal and have chosen the original and final states to have the same temperature so that $P_1 V_1 = P_2 V_2 = nRT$. Since the internal energy depends only on the temperature, the initial and final internal energies are also the same.

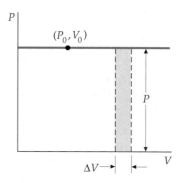

Figure 19-7 Each point on a PV diagram, such as (P_0, V_0), represents a particular state of the gas. The horizontal line represents states with a constant pressure P_0. The work done by a gas as it expands an amount ΔV is represented by the shaded area, $P_0\,\Delta V$.

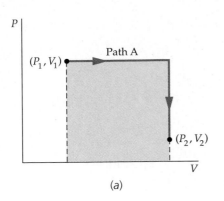

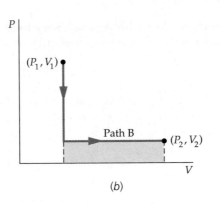

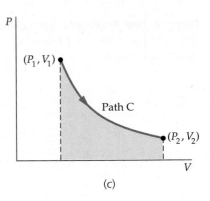

(a)   (b)   (c)

Figure 19-8 Three paths on *PV* diagrams connecting an initial state (P_1, V_1) and a final state (P_2, V_2). The work done along each path is indicated by the shaded area.

In Figure 19-8*a*, the gas is heated at constant pressure until its volume is V_2, after which it is cooled at constant volume until its pressure is P_2. The work done along path A is $P_1(V_2 - V_1)$ for the horizontal part of the path and zero for the constant-volume part.

In Figure 19-8*b*, the gas is first cooled at constant volume until its pressure is P_2, after which it is heated at constant pressure until its volume is V_2. The work done along path B is $P_2(V_2 - V_1)$, which is much less than that done along path A as can be seen by comparing the shaded regions in Figure 19-8*a* and *b*.

In Figure 19-8*c*, path C represents an **isothermal expansion**, meaning that the temperature remains constant. We can calculate the work done along path C by using $P = nRT/V$.

$$dW = P \, dV = \frac{nRT}{V} \, dV$$

Hence, the work done by the gas as it expands from V_1 to V_2 is

$$W = \int_{V_1}^{V_2} P \, dV = \int_{V_1}^{V_2} \frac{nRT}{V} \, dV$$

Since T is constant for an isothermal process, we can remove it from the integral. We then have

$$W_{\text{isothermal}} = nRT \int_{V_1}^{V_2} \frac{dV}{V} = nRT \ln \frac{V_2}{V_1} \qquad \text{19-16}$$

We see that the amount of work done by the gas is different for each process illustrated. Since $U_2 = U_1$ for these states, the net amount of heat added must also be different for each of the processes. This discussion illustrates the fact that both the work done and the heat added depend on just how a system moves from one state to another, but the change in the internal energy of the system does not.

Example 19-7

An ideal gas undergoes a cyclic process from point A to point B to point C to point D and back to point A as shown in Figure 19-9. The gas begins at a volume of 1 L and a pressure of 2 atm and expands at constant pressure until the volume is 2.5 L, after which it is cooled at constant volume until its pressure is 1 atm. It is then compressed at constant pressure until its volume is again 1 L, after which it is heated at constant volume until it is back in its original state. Find the total work done by the gas and the total heat added during the cycle.

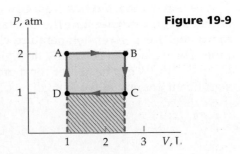

Figure 19-9

Picture the Problem We calculate the work done during each step. Since $\Delta U = 0$ for any complete cycle, the total heat added equals the total work done by the gas.

1. From A to B the pressure is constant. The work done by the gas equals the shaded area under the curve AB in the figure:

$$W_{AB} = P(V_B - V_A) = (2\text{ atm})(2.5\text{ L} - 1\text{ L})$$
$$= 3\text{ L·atm}$$

2. Convert the units to joules:

$$W_{AB} = 3\text{ L·atm} \times \frac{101.3\text{ J}}{1\text{ L·atm}} = 304\text{ J}$$

3. From B to C the gas cools at constant volume so the work done is zero:

$$W_{BC} = 0$$

4. As the gas is compressed at constant pressure from point C to point D, it does negative work. The magnitude of the work is the area under the CD curve indicated by cross hatching:

$$W_{CD} = P(V_D - V_C) = (1\text{ atm})(1\text{ L} - 2.5\text{ L})$$
$$= -1.5\text{ L·atm} = -152\text{ J}$$

5. As the gas is heated back to its original state A, the volume is again constant, so no work is done:

$$W_{DA} = 0$$

6. The total work done by the gas is the sum of the work done along each step:

$$W_{total} = W_{AB} + W_{BC} + W_{CD} + W_{DA}$$
$$= 304\text{ J} + 0 + (-152\text{ J}) + 0 = 152\text{ J}$$

7. Since the gas is back in its original state, the total change in internal energy is zero:

$$\Delta U = 0$$

8. The heat added is found from the first law:

$$Q = \Delta U + W = W = 152\text{ J}$$

Remarks The net work done by the gas is represented by the shaded area enclosed by the cycle in Figure 19-9 (without cross hatching). Such cyclic processes have important applications for heat engines, as we will see in the next chapter.

19-6   Heat Capacities of Gases

The determination of the heat capacity of a substance provides information about its internal energy, which is related to its molecular structure. For all substances that expand when heated, the heat capacity at constant pressure C_p is greater than the heat capacity at constant volume C_v. When heat is added at constant pressure, the substance expands and does work so it takes more heat for a given temperature change than if heated at constant volume. The expansion is usually negligible for solids and liquids, so for them $C_p \approx C_v$. But a gas heated at constant pressure readily expands and does a significant amount of work, so $C_p - C_v$ is not negligible.

When heat is added to a gas at constant volume, no work is done, so the heat added equals the increase in the internal energy of the gas. Writing Q_v for the heat added at constant volume, we have

$$Q_v = C_v \, \Delta T$$

Since $W = 0$, we have from the first law of thermodynamics

$$Q_v = \Delta U + W = \Delta U$$

Thus,

$$\Delta U = C_v \, \Delta T$$

Taking the limit as ΔT approaches zero, we obtain

$$dU = C_v \, dT \qquad\qquad \text{19-17}$$

and

$$C_v = \frac{dU}{dT} \qquad\qquad \text{19-18}$$

The heat capacity at constant volume is the rate of change of the internal energy with temperature. Since U and T are state functions, Equations 19-17 and 19-18 hold for any process.

Now let's calculate the difference $C_p - C_v$ for an ideal gas. From the definition of C_p, the heat added at constant pressure is

$$Q_p = C_p \, \Delta T$$

From the first law of thermodynamics,

$$Q_p = \Delta U + W = \Delta U + P \, \Delta V$$

Then

$$C_p \, \Delta T = \Delta U + P \, \Delta V$$

For infinitesimal changes, this becomes

$$C_p \, dT = dU + P \, dV$$

Using Equation 19-17 for dU, we obtain

$$C_p \, dT = C_v dT + P \, dV \qquad\qquad \text{19-19}$$

The pressure, volume, and temperature for an ideal gas are related by

$$PV = nRT$$

Take the differentials of both sides of the ideal-gas law, with $dP = 0$ for constant pressure.

$$P \, dV + V \, dP = P \, dV = nR \, dT$$

Substituting this into Equation 19-19 gives

$$C_p \, dT = C_v \, dT + nR \, dT$$

Therefore,

$$C_p = C_v + nR \qquad\qquad \text{19-20}$$

For an ideal gas, the heat capacity at constant pressure is greater than that at constant volume by the amount nR.

Table 19-3

Molar Heat Capacities (J/mol·K) of Various Gases at 25°C

Gas	c_p'	c_v'	c_v'/R	$c_p' - c_v'$	$(c_p' - c_v')/R$
Monatomic					
He	20.79	12.52	1.51	8.27	0.99
Ne	20.79	12.68	1.52	8.11	0.98
Ar	20.79	12.45	1.50	8.34	1.00
Kr	20.79	12.45	1.50	8.34	1.00
Xe	20.79	12.52	1.51	8.27	0.99
Diatomic					
N_2	29.12	20.80	2.50	8.32	1.00
H_2	28.82	20.44	2.46	8.38	1.01
O_2	29.37	20.98	2.52	8.39	1.01
CO	29.04	20.74	2.49	8.30	1.00
Polyatomic					
CO_2	36.62	28.17	3.39	8.45	1.02
N_2O	36.90	28.39	3.41	8.51	1.02
H_2S	36.12	27.36	3.29	8.76	1.05

Table 19-3 lists measured molar heat capacities c_p' and c_v' for several gases. We note from this table that the ideal gas prediction, $c_p' - c_v' = R$, holds quite well for all gases. The table also shows that c_v' is approximately $1.5R$ for all monatomic gases, $2.5R$ for all diatomic gases, and greater than $2.5R$ for gases consisting of more complex molecules. We can understand these results by considering the molecular model of a gas discussed in Chapter 18. The total translational kinetic energy of n moles of a gas is $K = \frac{3}{2}nRT$ (Equation 18-22). Thus, if the internal energy of a gas consists of translational kinetic energy only, we have

$$U = \tfrac{3}{2}nRT \qquad\qquad\qquad 19\text{-}21$$

The heat capacities are then

$$C_v = \frac{dU}{dT} = \tfrac{3}{2}nR \qquad\qquad\qquad 19\text{-}22$$

C_v for an ideal monatomic gas

and

$$C_p = C_v + nR = \tfrac{5}{2}nR \qquad\qquad\qquad 19\text{-}23$$

C_p for an ideal monatomic gas

The results in Table 19-3 agree well with these predictions for monatomic gases, but for other gases, the heat capacities are greater than those predicted by Equations 19-22 and 19-23. The internal energy for a gas consisting of diatomic or more complicated molecules is evidently greater than $\frac{3}{2}nRT$. The reason is that such molecules can have other types of energy, such as rotational or vibrational energy, in addition to translational kinetic energy.

Example **19-8**	*try it yourself*

A system consisting of 0.32 mol of a monatomic ideal gas with $c_v' = \frac{3}{2}RT$ occupies a volume of 2.2 L at a pressure of 2.4 atm, as represented by point A in Figure 19-10. The system is carried through a cycle consisting of three processes:
1. The gas is heated at constant pressure until its volume is 4.4 L at point B.
2. The gas is cooled at constant volume until the pressure decreases to 1.2 atm (point C).
3. The gas undergoes an isothermal compression back to point A.
(*a*) What is the temperature at points A, B, and C? (*b*) Find W, Q, and ΔU for each process and for the entire cycle.

Picture the Problem You can find the temperatures at all points from the ideal-gas law. You can find the work for each process by finding the area under the curve, and the heat exchanged from the given heat capacity and the initial and final temperatures for each process. In process 3, T is constant, so $\Delta U = 0$ and the heat input equals the work done.

Figure 19-10

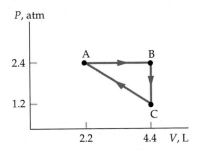

Cover the column to the right and try these on your own before looking at the answers.

Steps

Answers

(*a*)   Find the temperatures at points A, B, and C using the ideal-gas law.

$T_A = T_C = 201$ K, $T_B = 402$ K

(*b*)1. For process 1, use $W_1 = P_C \, \Delta V$ to calculate the work, and $C_p = \frac{5}{2} nR$ to calculate the heat Q_1. Then use W_1 and Q_1 to calculate ΔU_2.

$W_1 = 5.28$ L·atm $= 535$ J, $Q_1 = 1337$ J

$\Delta U_1 = Q_1 - W_1 = 802$ J

2. For process 2, use $C_v = \frac{3}{2} nR$ and $T_C - T_B$ from step 1 to find Q_2. Then, since $W_2 = 0$, $\Delta U = Q_2$.

$W_2 = 0$, $Q_2 = -802$ J, $\Delta U_2 = -802$ J

3. Calculate W_1 from $W = nRT \, \ln(V_A/V_C)$ in the isothermal compression. Then, since $\Delta U_1 = 0$, $Q_3 = W_3$.

$W_3 = -371$ J, $Q_2 = -371$ J, $\Delta U_3 = 0$

5. Find the total work W, the total heat Q, and the total change ΔU by summing the quantities found in steps 2, 3, and 4.

$W_{\text{total}} = W_1 + W_2 + W_3 = 535$ J $+ 0 + (-371$ J$) = 164$ J

$Q_{\text{total}} = Q_1 + Q_2 + Q_3 = 1337$ J $+ (-802$ J$) + (-371$ J$) = 164$ J

$\Delta U_{\text{total}} = \Delta U_1 + \Delta U_2 + \Delta U_3 = 802$ J $+ (-802$ J$) + 0 = 0$

Remarks The total change in internal energy is zero, as it must be for a cyclic process. The work done by the system equals the heat absorbed. This work equals the area under the AB curve minus the area under the CA curve, which equals the area enclosed by the complete figure.

Heat Capacities and the Equipartition Theorem

According to the equipartition theorem stated in Chapter 18, the internal energy of n moles of a gas should equal $\frac{1}{2}nRT$ for each degree of freedom of the gas molecule. The heat capacity at constant volume of a gas should then be $\frac{1}{2}nR$ times the number of degrees of freedom of the molecule. From Table 19-2, nitrogen, oxygen, hydrogen, and carbon monoxide all have molar heat capacities at constant volume of about $\frac{5}{2}R$. Thus, the molecules in each of these gases have five degrees of freedom (Figure 19-11). About 1880, Clausius speculated that these gases must consist of diatomic molecules that can rotate about two axes, giving them two additional degrees of freedom. The two degrees of freedom besides the three for translation are now known to be associated with their rotation about each of the two axes, x' and y', perpendicular to the line joining the atoms. The kinetic energy of a diatomic molecule is therefore

$$K = \tfrac{1}{2}mv_x^2 + \tfrac{1}{2}mv_y^2 + \tfrac{1}{2}mv_z^2 + \tfrac{1}{2}I_{x'}\omega_{x'}^2 + \tfrac{1}{2}I_{y'}\omega_{y'}^2$$

The total internal energy of n moles of such a gas is then

$$U = 5 \times (\tfrac{1}{2}nRT) = \tfrac{5}{2}nRT \qquad\qquad\text{19-24}$$

and the heat capacity at constant volume is

$$C_V = \tfrac{5}{2}nR \qquad\qquad\text{19-25}$$

Apparently, diatomic gases do not rotate about the line joining the two atoms—if they did, there would be six degrees of freedom and C_V would be $\frac{6}{2}nR = 3nR$, contrary to experiment. Furthermore, monatomic gases apparently do not rotate at all. We will see in Section 19-8 that these puzzling facts are easily explained when we take into account the quantization of energy.

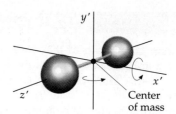

Figure 19-11 Rigid-dumbbell model of a diatomic molecule.

Example 19-9

One mole of oxygen gas is heated from a temperature of 20°C and a pressure of 1 atm to a temperature of 100°C. Assume that oxygen is an ideal gas. (*a*) How much heat must be supplied if the volume is kept constant during the heating? (*b*) How much heat must be supplied if the pressure is kept constant? (*c*) How much work does the gas do in part (*b*)?

Picture the Problem The heat needed for constant-volume heating is $Q_v = C_v\,\Delta T$, where $C_v = \frac{5}{2}nR = \frac{5}{2}R$ since oxygen is a diatomic gas, and $n = 1$ mole. For constant-pressure heating, $Q_p = C_p\,\Delta T$, where $C_p = C_v + R$. Finally, the amount of work done can be found from $W = Q - \Delta U$, or from $W = P\,\Delta V$.

(*a*)1. Write the heat needed for constant volume in terms of C_v and ΔT:

$$Q_v = C_v\,\Delta T$$

2. Calculate the heat capacity at constant volume:

$$C_v = \tfrac{5}{2}nR = \tfrac{5}{2}(1\text{ mol})(8.31\text{ J/mol·K}) = 20.8\text{ J/K}$$

3. Calculate the heat for $\Delta T = 80\ C° = 80$ K:

$$Q_v = C_v\,\Delta T = (20.8\text{ J/K})(80\text{ K}) = 1.66\text{ kJ}$$

(*b*)1. Write the heat needed for constant pressure in terms of C_p and ΔT:

$$Q_p = C_p\,\Delta T$$

2. Calculate the heat capacity at constant pressure:

$$C_p = C_v + nR = 20.8\text{ J/K} + (1\text{ mol})(8.31\text{ J/mol·K}) = 29.1\text{ J/K}$$

3. Calculate the heat added at constant pressure for $\Delta T = 80$ K:

$$Q_p = C_p\,\Delta T = (29.1\text{ J/K})(80\text{ K}) = 2.33\text{ kJ}$$

(*c*)1. The work W can be found from the first law of thermodynamics:

$$W = Q - \Delta U$$

2. The internal energy change equals the heat added at constant volume in (a), since there was no work done:

$$\Delta U = Q_V = C_V \, \Delta T = 1.66 \text{ kJ}$$

3. The work done at constant pressure is then:

$$W = Q_P - \Delta U = 2.33 \text{ kJ} - 1.66 \text{ kJ} = 0.67 \text{ kJ}$$

Remarks Note that the change in internal energy is the same independent of the process. It depends only on the initial and final states.

Exercise Find the initial and final volumes of this gas from the ideal-gas law, and use them to calculate the work done when the heat is added at constant pressure from $W = P \, \Delta V$. (*Answers* $V_i = 24.0$ L, $V_f = 30.6$ L, $W = 6.6$ L·atm $= 0.67$ kJ)

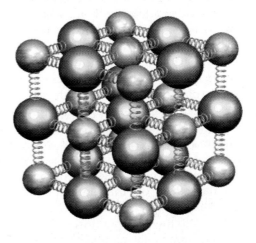

Figure 19-12 Model of a solid in which the atoms are connected to each other by springs. The internal energy of the molecule consists of the kinetic and potential energies of vibration.

19-7 Heat Capacities of Solids

In Section 19-1, we noted that all of the metals listed in Table 19-1 have approximately equal molar specific heats. Experimentally, most solids have molar heat capacities approximately equal to $3R$:

$$c' = 3R = 24.9 \text{ J/mol·K} \qquad \text{19-26}$$

This result is known as the **Dulong–Petit law.** We can understand it by applying the equipartition theorem to the simple model for a solid shown in Figure 19-12. According to this model, a solid consists of a regular array of atoms in which each of the atoms has a fixed equilibrium position and is connected by springs to its neighbors. Each atom can vibrate in the x, y, and z directions. The total energy of an atom in a solid is thus

$$E = \tfrac{1}{2}mv_x^2 + \tfrac{1}{2}mv_y^2 + \tfrac{1}{2}mv_z^2 + \tfrac{1}{2}k_{\text{eff}}x^2 + \tfrac{1}{2}k_{\text{eff}}y^2 + \tfrac{1}{2}k_{\text{eff}}z^2$$

where k_{eff} is the effective force constant of the hypothetical springs. Each atom thus has six degrees of freedom. The equipartition theorem states that a substance in equilibrium has an average energy of $\tfrac{1}{2}RT$ per mole for each degree of freedom. Thus, the internal energy of a mole of a solid is

$$U_m = 6 \times \tfrac{1}{2}RT = 3RT \qquad \text{19-27}$$

which means that c' is equal to $3R$.

Example 19-10

The molar mass of copper is 63.5 g/mol. Use the Dulong–Petit law to calculate the specific heat of copper.

Picture the Problem The Dulong–Petit law gives the molar specific heat of a solid, c'. The specific heat is then $c = c'/M$ (Equation 19-6), where M is the molar mass.

1. The Dulong–Petit law gives c':

$$c' = 3R = 3(8.31 \text{ J/mol·K}) = 24.9 \text{ J/mol·K}$$

2. Using $M = 63.5$ g/mol for copper, the specific heat is:

$$c = \frac{c'}{M} = \frac{24.9 \text{ J/mol·K}}{63.5 \text{ g/mol}} = 0.392 \text{ J/g·K} = 0.392 \text{ kJ/kg·K}$$

Remarks This solution agrees fairly closely with the measured value of 0.386 kJ/kg·K given in Table 19-1.

Exercise The specific heat of a certain metal is measured to be 1.02 kJ/kg·K. (*a*) Calculate the molar mass of this metal, assuming that the metal obeys the Dulong–Petit law. (*b*) What is the metal? (*Answers*   (*a*) $M = 24.4$ g/mol. (*b*) The metal must be magnesium, which has a molar mass of 24.31 g/mol.)

19-8   Failure of the Equipartition Theorem

Although the equipartition theorem had spectacular successes in explaining the heat capacities of gases and solids, it had equally spectacular failures. For example, if a diatomic gas molecule like the one in Figure 19-11 rotates about the line joining the atoms, there should be an additional degree of freedom. Similarly, if a diatomic molecule is not rigid, the two atoms should vibrate along the line joining them. We would then have two more degrees of freedom corresponding to kinetic and potential energies of vibration. According to the measured values of the molar heat capacities in Table 19-2, however, diatomic gases apparently do not rotate about the line joining them, nor do they vibrate. The equipartition theorem gives no explanation for this, nor for the fact that monatomic atoms apparently do not rotate about any of the three possible perpendicular axes in space. Furthermore, heat capacities are found to depend on temperature, contrary to the predictions of the equipartition theorem. The most spectacular case of the temperature dependence of heat capacity is that of H_2, shown in Figure 19-13. At low temperatures, H_2 behaves like a monatomic molecule that does not rotate. At very high temperatures, H_2 begins to vibrate, but the molecule dissociates before c_v' reaches $\frac{7}{2}R$. Finally, the equipartition theorem predicts a constant value of $3R$ for the heat capacity of solids. This result holds for many solids at high temperatures, although not all; however, it does not hold at very low temperatures.

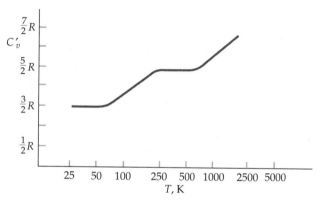

Figure 19-13 Temperature dependence of the molar heat capacity of H_2.

The equipartition theorem fails because energy is quantized. A molecule can have only certain values of energy, as illustrated schematically by the energy-level diagram in Figure 19-14. The molecule can gain or lose energy only if the gain or loss takes it to another allowed level. For example, in a gas the energy that can be transferred between the molecules in collisions is of the order of kT, the typical thermal energy of a molecule. The validity of the equipartition theorem depends on the relative size of kT and the spacing of the energy levels.

Figure 19-14 Energy-level diagram. A system can have only certain discrete energies.

> If the spacing of the levels is large compared with kT, energy cannot be transferred by collisions and the classical equipartition theorem is not valid. If the spacing of the levels is much smaller than kT, energy quantization will not be noticed and the equipartition theorem will hold.

Conditions for the validity of the equipartition theorem

Consider the rotation of a molecule. The energy of rotation is

$$E = \frac{1}{2} I\omega^2 = \frac{(I\omega)^2}{2I} = \frac{L^2}{2I}$$ 19-28

where I is the moment of inertia of the molecule, ω is its angular velocity, and $L = I\omega$ is its angular momentum. In Section 10-5, we mentioned that angular momentum is quantized, and its magnitude is restricted to

$$L = \sqrt{\ell(\ell + 1)}\,\hbar, \qquad \ell = 0, 1, 2, \dots$$ 19-29

where $\hbar = h/2\pi$, and h is Planck's constant. The energy of a rotating molecule is therefore quantized to the values

$$E = \frac{L^2}{2I} = \frac{\ell(\ell + 1)\hbar^2}{2I} = \ell(\ell + 1)E_{0r}$$ 19-30

where

$$E_{0r} = \frac{\hbar^2}{2I}$$ 19-31

is characteristic of the energy gap between levels. If this energy is much less than kT, we expect classical physics and the equipartition theorem to hold. Let us define a critical temperature T_c by

$$kT_c = E_{0r} = \frac{\hbar^2}{2I}$$ 19-32

When T is much greater than this critical temperature, kT will be much greater than the spacing of the energy levels, which is of the order of kT_c, and we expect classical physics and the equipartition theorem to be valid. When T is less than or of the order of T_c, kT will not be much greater than the energy-level spacing, and we expect classical physics and the equipartition theorem to break down. We will estimate T_c for some cases of interest.

1. *Rotation of H_2 about an axis perpendicular to the line joining the H atoms and through the center of mass (see Figure 19-11):* The moment of inertia of H_2 is

$$I_H = 2M_H \left(\frac{r_s}{2}\right)^2 = \frac{1}{2} M_H r_s^2$$

where M_H is the mass of a H atom, and r_s is the separation distance. For hydrogen, $M_H = 1.67 \times 10^{-27}$ kg, and $r_s \approx 8 \times 10^{-11}$ m. The critical temperature is then

$$T_c = \frac{\hbar^2}{2kI_H} = \frac{\hbar^2}{kM_H r_s^2}$$

$$\approx \frac{(1.05 \times 10^{-34}\,\text{J·s})^2}{(1.38 \times 10^{-23}\,\text{J/K})(1.67 \times 10^{-27}\,\text{kg})(8 \times 10^{-11}\text{m})^2} \approx 75\,\text{K}$$

As can be seen from Figure 19-13, this is approximately the temperature below which the rotational energy does not contribute to the heat capacity.

2. O_2: Since the mass of O_2 is about 16 times that of H_2, and the separation is about the same, the critical temperature for O_2 should be about $74/16 \approx 4.7$ K. For all temperatures for which O_2 exists as a gas, $T \gg T_c$, so kT is much greater than the energy level spacing, and we expect the equipartition theorem of classical physics to apply.

3. *Rotation of a monatomic gas:* We consider the He atom. The mass of the electron is about 2000 times smaller than that of the nucleus. But the radius of the nucleus is about 100,000 times smaller than the distance to the electron. So, the moment of inertia of the atom is almost entirely due to its electrons. The distance from the nucleus to the two electrons in He is about half the separation distance of the H atoms in H_2. Thus, using $m_e = M_H/2000$ and $r = r_s/2$, the moment of inertia of the two electrons in He is roughly

$$I_{He} = 2m_e r^2 \approx 2\left(\frac{M_H}{2000}\right)\left(\frac{r_s}{2}\right)^2 = \frac{I_H}{2000}$$

The critical temperature for He is thus about 2000 times that of H_2 or about 150,000 K. This is much higher than the dissociation temperature (the temperature at which electrons are stripped from their nuclei) for helium. So the gap between allowed energy levels is always much greater than kT and the He molecules cannot be induced to rotate by collisions occurring in the gas. Other monatomic gases have slightly greater moments of inertia because they have more electrons, but their critical temperatures are still tens of thousands of kelvins, so their molecules also cannot be induced to rotate by collisions occurring in the gas.

4. *Rotation of a diatomic gas about the axis joining the atoms:* From our discussion of monatomic gases, we see that the moment of inertia for the diatomic case will also be due mainly to the electrons and will be of the same order of magnitude as for a monatomic gas. Again, the critical temperature T_c calculated in order for this rotation to occur due to collisions between molecules in the gas exceeds the gas's dissociation temperature, making rotation under those circumstances impossible.

It is interesting to note that the successes of the equipartition theorem in explaining the measured heat capacities of gases and solids led to the first real understanding of molecular structure in the nineteenth century, whereas its failures played an important role in the development of quantum mechanics in the twentieth century.

Example 19-11

(*a*) **Estimate the lowest (nonzero) rotational energy for the hydrogen atom and compare it to kT at room temperature, $T = 300$ K. (*b*) Calculate the critical temperature T_c.**

Picture the Problem From Equation 19-29, the lowest rotational energy is for $\ell = 1$. Then $E = 2E_{0r} = \hbar^2/I$. Since we can neglect the moment of inertia of the nucleus (because its radius is 100,000 times smaller than the radius of the atom), the moment of inertia for the atom is essentially the moment of inertia of the electron. Then $I = m_e r^2$, where $r \approx 5 \times 10^{-11}$ m is the distance to the electron.

(*a*)1. The lowest energy greater than zero occurs for $\ell = 1$:	$E = \dfrac{\ell(\ell + 1)\hbar^2}{2I} = \dfrac{1(2)\hbar^2}{2I} = \dfrac{\hbar^2}{m_e r^2}$

2. The numerical values are:

$$\hbar = 1.05 \times 10^{-34} \, \text{J·s}$$

$$m_e = 9.11 \times 10^{-31} \, \text{kg}$$

$$r = 5 \times 10^{-11} \, \text{m}$$

3. Substitute the numerical values:

$$E_1 = \frac{\hbar^2}{m_e r^2} = \frac{(1.05 \times 10^{-34} \, \text{J·s})^2}{(9.11 \times 10^{-31} \, \text{kg})(5 \times 10^{-11} \, \text{m})^2}$$

$$= 4.8 \times 10^{-18} \, \text{J}$$

4. The value of kT at $T = 300$ K is:

$$kT = (1.38 \times 10^{-23} \, \text{J/K})(300 \, \text{K}) = 4.1 \times 10^{-21} \, \text{J}$$

(b) Set $kT_c = E_1$ and solve for T_c:

$$kT_c = E_1 = 4.8 \times 10^{-18} \, \text{J}$$

$$T_c = \frac{E_1}{k} = \frac{4.8 \times 10^{-18} \, \text{J}}{1.38 \times 10^{-23} \, \text{J/K}} = 3.48 \times 10^5 \, \text{K}$$

Remarks The lowest rotational energy is about 1000 times kT at room temperature. The critical temperature of a hydrogen atom is so high that the atom would be ionized well before the critical temperature could be reached.

19-9   The Quasi-static Adiabatic Expansion of a Gas

A process in which no heat flows into or out of a system is called an **adiabatic process.** Consider the quasi-static adiabatic expansion of a gas in which the gas in a thermally insulated container expands slowly against a piston, doing work on it. Since no heat enters or leaves the gas, the work done by the gas equals the decrease in the internal energy of the gas, and the temperature of the gas decreases. The curve representing this process on a PV diagram is shown in Figure 19-15.

We can find the equation for the adiabatic curve for an ideal gas using the equation of state and the first law of thermodynamics. We have

$$dQ = dU + dW = C_v \, dT + P \, dV = 0 \qquad \text{19-33}$$

where we have used $dU = C_v \, dT$ from Equation 19-11. Then, using $P = nRT/V$,

$$C_v dT + nRT \frac{dV}{V} = 0$$

Rearranging gives

$$\frac{dT}{T} + \frac{nR}{C_v} \frac{dV}{V} = 0 \qquad \text{19-34}$$

Equation 19-34 can be simplified by noting that $C_p - C_v = nR$, so

$$\frac{nR}{C_v} = \frac{C_p - C_v}{C_v} = \frac{C_p}{C_v} - 1 = \gamma - 1$$

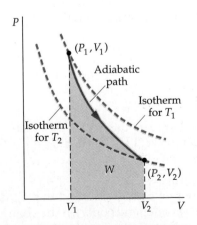

Figure 19-15 Quasi-static adiabatic expansion of an ideal gas. The dashed lines are the isotherms for the initial and final temperatures. The curve connecting the initial and final states of the adiabatic expansion is steeper than the isotherms because the temperature drops during the expansion.

where γ is the ratio of the heat capacities:

$$\gamma = \frac{C_p}{C_v} \qquad \text{19-35}$$

We then have

$$\frac{dT}{T} + (\gamma - 1)\frac{dV}{V} = 0$$

Integration gives

$$\ln T + (\gamma - 1)\ln V = \text{constant}$$

Using the properties of logarithms (Appendix D), we obtain

$$\ln(TV^{\gamma-1}) = \text{constant}$$

or

$$TV^{\gamma-1} = \text{constant} \qquad \text{19-36}$$

Quasi-static adiabatic process

Clouds form when rising moist air cools due to adiabatic expansion of the air and then condenses into liquid droplets.

where the constants in the two preceding equations are not the same. We can eliminate T from Equation 19-36 using $PV = nRT$. We then have

$$\frac{PV}{nR} V^{\gamma-1} = \text{constant}$$

or

$$PV^{\gamma} = \text{constant} \qquad \text{19-37}$$

Quasi-static adiabatic process

Equation 19-37 relates P and V for adiabatic expansions and compressions.

The work done by the gas in an adiabatic expansion can be calculated from the first law of thermodynamics:

$$dQ = dU + dW = dU + p\,dV$$

Since $dQ = 0$, we have

$$dW = -dU = -C_v\,dT$$

Then

$$W_{\text{adiabatic}} = \int dW = \int -C_v\,dT = -C_v\,\Delta T \qquad \text{19-38}$$

Adiabatic work

where we have assumed that C_v is constant.* We note that the work done by the gas depends only on the change in the absolute temperature of the gas. In an adiabatic expansion, the gas does work, and its internal energy and temperature decrease. In an adiabatic *compression*, work is done *on* the gas, and the internal energy and temperature increase.

---

* For an ideal gas, U is proportional to the absolute temperature, and therefore $C_v = dU/dT$ is a constant.

We can use the ideal-gas law to write Equation 19-38 in terms of the initial and final values of the pressure and volume. If T_1 is the initial temperature and T_2 is the final temperature, we have for the work done

$$W_{\text{adiabatic}} = -C_v \Delta T = -C_v(T_2 - T_1) = C_v(T_1 - T_2)$$

Using $PV = nRT$, we obtain

$$W_{\text{adiabatic}} = C_v\left(\frac{P_1V_1}{nR} - \frac{P_2V_2}{nR}\right) = \frac{C_v}{C_p - C_v}(P_1V_1 - P_2V_2)$$

where we have used $nR = C_p - C_v$. Dividing the numerator and denominator by C_v and writing γ for C_p/C_v, we obtain

$$W_{\text{adiabatic}} = \frac{P_1V_1 - P_2V_2}{\gamma - 1} \qquad\qquad\qquad\qquad \text{19-39}$$

Adiabatic work

Example 19-12

A quantity of air ($\gamma = 1.4$) expands adiabatically and quasi-statically from an initial pressure of 2 atm and volume of 2 L at temperature 20°C to twice its original volume. Find (*a*) the final pressure, (*b*) the final temperature, and (*c*) the work done by the gas.

Picture the Problem Since the process is adiabatic, we know that $PV^\gamma =$ constant, and $TV^{\gamma-1} =$ constant. These relations yield the final pressure and final temperature, respectively. The work done is found from Equation 19-39. Let subscript 1 refer to initial values, and subscript 2 to final values. Then $P_1 = 2$ atm, $V_1 = 2$ L, $V_2 = 4$ L, and $T_1 = 20°C = 293$ K.

(*a*)1. Write $PV^\gamma =$ constant in terms of initial and final values:	$P_1V_1^\gamma = P_2V_2^\gamma$
2. Solve for P_2:	$P_2 = P_1\left(\dfrac{V_1}{V_2}\right)^\gamma = (2\text{ atm})\left(\dfrac{2\text{ L}}{4\text{ L}}\right)^{1.4} = 0.758\text{ atm}$
(*b*)1. Write $TV^{\gamma-1} =$ constant in terms of initial and final values:	$T_1V_1^{\gamma-1} = T_2V_2^{\gamma-1}$
2. Solve for T_2:	$T_2 = T_1\left(\dfrac{V_1}{V_2}\right)^{\gamma-1} = (293\text{ K})\left(\dfrac{2\text{ L}}{4\text{ L}}\right)^{0.4} = 222\text{ K} = -51°C$
(*c*) Equation 19-39 gives the work done:	$W_{\text{adiabatic}} = \dfrac{P_1V_1 - P_2V_2}{\gamma - 1} = \dfrac{(2\text{ atm})(2\text{ L}) - (0.758\text{ atm})(4\text{ L})}{1.4 - 1}$
	$= 2.42\text{ L·atm} = 245\text{ J}$

Speed of Sound Waves   Let's use Equation 19-37 to calculate the adiabatic bulk modulus of an ideal gas, which is related to the speed of sound waves in air. We first compute the differential of both sides of Equation 19-37:

$$P\, d(V^{\gamma}) + V^{\gamma}\, dP = 0$$

or

$$P\gamma V^{\gamma-1}\, dV + V^{\gamma}\, dP = 0$$

Then

$$dP = -\frac{\gamma P\, dV}{V}$$

Referring to Equation 13-6, the adiabatic bulk modulus* is then:

$$B_{adiab} = -\frac{dP}{dV/V} = \gamma P \qquad\qquad\text{19-40}$$

The speed of sound (Equation 15-4) is given by

$$v = \sqrt{\frac{B_{adiab}}{\rho}}$$

where the mass density ρ is related to the number of moles n and the molecular mass M by $\rho = m/V = nM/V$. Using the ideal gas law, $PV = nRT$, we can eliminate V from the density:

$$\rho = \frac{nM}{V} = \frac{nM}{nRT/P} = \frac{MP}{RT}$$

Using this result and γP for B_{adiab}, we obtain

$$v = \sqrt{\frac{B_{adiab}}{r}} = \sqrt{\frac{\gamma P}{MP/RT}} = \sqrt{\frac{\gamma RT}{M}}$$

which is Equation 15-5 for the speed of sound in a gas.

*The bulk modulus is the negative ratio of the pressure change to the fractional change in volume, $B = -\Delta P/(\Delta V/V)$ (Chapter 13). The isothermal bulk modulus, which describes changes that occur at constant temperature, differs from the adiabatic bulk modulus, which describes changes with no heat transfer. For sound waves at audible frequencies the changes occur too rapidly for appreciable heat flow, so the appropriate bulk modulus is the adiabatic bulk modulus.

Summary

1. The first law of thermodynamics is a statement of the conservation of energy.

2. The equipartition theorem is a fundamental law of classical physics. It breaks down when the typical thermal energy kT is small compared to the spacing of quantized energy levels.

Topic	Remarks and Relevant Equations
1. Heat	Heat is energy that is transferred from one object to another because of a temperature difference.
Calorie	The calorie, originally defined as the heat necessary to raise the temperature of one gram of water by one Celsius degree, is now defined to be 4.184 joules.
2. Heat Capacity	Heat capacity is the amount of heat needed to raise the temperature of a substance by one degree.

$$c = \frac{Q}{\Delta T} \qquad\qquad\qquad\text{19-1}$$

At constant volume	$C_v = \dfrac{Q_v}{\Delta T}$	
At constant pressure	$C_p = \dfrac{Q_p}{\Delta T}$	
Specific heat (heat capacity per unit mass)	$c' = \dfrac{C}{m}$	19-2
Molar specific heat (heat capacity per mole)	$c' = \dfrac{C}{n}$	19-6
Heat capacity related to internal energy	$C_v = \dfrac{dU}{dT}$	19-18
Of ideal gas	$C_p - C_v = +nR$	19-20
Of monatomic ideal gas	$C_v = \frac{3}{2}nR$	19-22
Of diatomic ideal gas	$C_v = \frac{5}{2}nR$	19-25

3. Fusion and Vaporization

During melting and vaporization, the temperature does not change.

Latent heat of fusion

The heat needed to melt a substance is the product of the mass of the substance and its latent heat of fusion L_f:

$$Q_f = mL_f \qquad \text{19-8}$$

L_f of water

333.5 kJ/kg

Latent heat of vaporization

The heat needed to vaporize a liquid is the product of the mass of the liquid and its latent heat of vaporization L_v:

$$Q_v = mL_v \qquad \text{19-9}$$

L_v of water

2257 kJ/kg

4. First Law of Thermodynamics

The net heat added to a system equals the change in the internal energy of the system plus the work done by the system:

$$Q = \Delta U + W \qquad \text{19-10}$$

5. Internal Energy U

The internal energy of a system is a property of the state of the system, as are the pressure, volume, and temperature. Heat and work are not properties of state.

Ideal gas

U depends only on the temperature T.

Monatomic ideal gas

$$U = \frac{3}{2}nRT \qquad \text{19-12}$$

Internal energy related to heat capacity

$$dU = C_v\, dT \qquad \text{19-18}$$

6. Quasi-static Process

A quasi-static process is one that occurs slowly so that the system moves through a series of equilibrium states

Isobaric

$P = \text{constant}$

Isothermal

$T = \text{constant}$

Adiabatic	$Q = 0$	
Adiabatic, ideal gas	$TV^{\gamma-1} = \text{constant}$ or	**19-36**
	$PV^{\gamma} = \text{constant}$	**19-37**
	where	
	$\gamma = \dfrac{C_p}{C_v}$	**19-35**

7. Work Done by a Gas in a Quasi-static Process

$$W = \int P\, dV \qquad \qquad \textbf{19-14}$$

Isothermal

$$W_{\text{isothermal}} = nRT \ln \frac{V_2}{V_1} \qquad \qquad \textbf{19-16}$$

Adiabatic

$$W_{\text{adiabatic}} = -C_v\, \Delta T = \frac{P_1 V_1 - P_2 V_2}{\gamma - 1} \qquad \qquad \textbf{19-38, 19-39}$$

8. Equipartition Theorem

The equipartition theorem states that when a system is in equilibrium, there is an average energy of $\frac{1}{2}kT$ per molecule or $\frac{1}{2}RT$ per mole associated with each degree of freedom.

Failure of the equipartition theorem

The equipartition theorem fails when the thermal energy ($\sim kT$) that can be transferred in collisions is smaller than the energy gap ΔE between quantized energy levels. For example, monatomic gases cannot rotate because the first nonzero energy permitted is much greater than kT.

9. Dulong–Petit Law

The molar specific heat of most solids is $3R$. This is predicted by the equipartition theorem, assuming that a solid atom has six degrees of freedom.

Problem-Solving Guide

Summary of Worked Examples

Type of Calculation	**Procedure and Relevant Examples**
1. Heat Capacity and Specific Heat	
Find the heat needed to raise the temperature of an object.	Use $Q = C\,\Delta T = mc\,\Delta T$. **Example 19-1**
Calorimeter problems	Use $Q_{\text{in}} = Q_{\text{out}}$. **Examples 19-2, 19-4**
Find the heat needed to change the temperature of an ideal gas.	Use $C_v = \frac{3}{2}nR$ for a monatomic gas, $C_v = \frac{5}{2}nR$ for a diatomic gas, and $C_p = C_v + nR$. **Examples 19-3, 19-7**
Find the specific heat of a metal, or find the molar mass given the specific heat.	Use the Dulong–Petit law, $C_v = 6nR$. **Example 19-6**

2. Latent Heat

Find the heat needed to melt a solid or vaporize a liquid.	Use $Q_f = mL_f$ or $Q_v = mL_v$.	Example 19-3

3. First Law of Thermodynamics

Find the work needed for a given temperature change.	Set the work equal to the thermal energy needed, $mc\,\Delta T$.	Example 19-4
Find the internal energy change of a system.	Use the first law, $\Delta U = Q - W$.	Example 19-5
Find the work done and the heat added during a complete cycle.	Calculate the work from $dW = p\,dV$. Since $\Delta U = 0$, $Q = W$.	Example 19-6
Find W, Q, and ΔU for a quasi-static process.	Use $dW = p\,dV$, $Q = C\,\Delta T$, and $dQ = \Delta U + dW$.	Example 19-3

4. Calculate Work

Find the work done by a gas during expansion or compression.	Use $W = \int p\,dV$. For isothermal expansion of a gas, $W = nRT \ln(V_2/V_1)$.	Example 19-8
	For adiabatic expansion, $$W = \frac{P_1V_1 - P_2V_2}{\gamma - 1} = -C_v\,\Delta T$$	Example 19-9

5. Equipartition Theorem

Determine if the equipartition theorem holds for a system at a certain temperature.	Compare the spacing of the quantized energy levels with kT.	Example 19-11

6. Quasi-static Adiabatic Expansion

Find P, V, or T after an adiabatic expansion.	Use $P_1V_1^\gamma = P_2V_2^\gamma$ to relate P and V. Use $T_1V_1^{\gamma-1} = T_2V_2^{\gamma-1}$ to relate T and V.	Example 19-10

Problems

Conceptual Problems

Problems from Optional and Exploring sections

In a few problems, you are given more data than you actually need; in a few other problems, you are required to supply data from your general knowledge, outside sources, or informed estimates.

- • Single-concept, single-step, relatively easy
- •• Intermediate-level, may require synthesis of concepts
- ••• Challenging, for advanced students

Heat Capacity; Specific Heat; Latent Heat

1 • Body A has twice the mass and twice the specific heat of body B. If they are supplied with equal amounts of heat, how do the subsequent changes in their temperatures compare?

2 • The temperature change of two blocks of masses M_A and M_B is the same when they absorb equal amounts of heat. It follows that the specific heats are related by

(a) $c_A = (M_A/M_B)c_B$.
(b) $c_A = (M_B/M_A)c_B$.
(c) $c_A = c_B$.
(d) none of the above.

In Problems 39 through 42, the initial state of 1 mol of an ideal gas is $P_1 = 3$ atm, $V_1 = 1$ L, and $U_1 = 456$ J, and its final state is $P_2 = 2$ atm, $V_2 = 3$ L, and $U_2 = 912$ J.

39 • The gas is allowed to expand at constant pressure to a volume of 3 L. It is then cooled at constant volume until its pressure is 2 atm. (a) Show this process on a PV diagram, and calculate the work done by the gas. (b) Find the heat added during this process.

40 • The gas is first cooled at constant volume until its pressure is 2 atm. It is then allowed to expand at constant pressure until its volume is 3 L. (a) Show this process on a PV diagram, and calculate the work done by the gas. (b) Find the heat added during this process.

41 •• The gas is allowed to expand isothermally until its volume is 3 L and its pressure is 1 atm. It is then heated at constant volume until its pressure is 2 atm. (a) Show this process on a PV diagram, and calculate the work done by the gas. (b) Find the heat added during this process.

42 •• The gas is heated and is allowed to expand such that it follows a straight-line path on a PV diagram from its initial state to its final state. (a) Show this process on a PV diagram, and calculate the work done by the gas. (b) Find the heat added during this process.

43 •• One mole of the ideal gas is initially in the state $P_0 = 1$ atm, $V_0 = 25$ L. As the gas is slowly heated, the plot of its state on a PV diagram moves in a straight line to the state $P = 3$ atm, $V = 75$ L. Find the work done by the gas.

44 •• One mole of the ideal gas is heated so that $T = AP^2$, where A is a constant. The temperature changes from T_0 to $4T_0$. Find the work done by the gas.

45 •• One mole of an ideal gas initially at a pressure of 1 atm and a temperature of 0°C is compressed isothermally and quasi-statically until its pressure is 2 atm. Find (a) the work needed to compress the gas, and (b) the heat removed from the gas during the compression.

46 •• An ideal gas initially at 20°C and 200 kPa has a volume of 4 L. It undergoes a quasi-static, isothermal expansion until its pressure is reduced to 100 kPa. Find (a) the work done by the gas, and (b) the heat added to the gas during the expansion.

Heat Capacities of Gases and the Equipartition Theorem

47 • The heat capacity at constant volume of a certain amount of a monatomic gas is 49.8 J/K. (a) Find the number of moles of the gas. (b) What is the internal energy of the gas at $T = 300$ K? (c) What is the heat capacity of the gas at constant pressure?

48 • The Dulong–Petit law was originally used to determine the molecular mass of a substance from its measured heat capacity. The specific heat of a certain solid is measured to be 0.447 kJ/kg·K. (a) Find the molecular mass of the substance. (b) What element is this?

49 •• The specific heat of air at 0°C is listed in a handbook as having the value of 1.00 J/g·K measured at constant

pressure. (a) Assuming that air is an ideal gas with a molar mass $M = 29.0$ g/mol, what is its specific heat at 0°C and constant volume? (b) How much internal energy is there in 1 L of air at 0°C and at 1 atm?

50 •• One mole of an ideal diatomic gas is heated at constant volume from 300 to 600 K. (a) Find the increase in internal energy, the work done, and the heat added. (b) Find the same quantities if this gas is heated from 300 to 600 K at constant pressure. Use the first law of thermodynamics and your results for (a) to calculate the work done. (c) Calculate the work done in (b) directly from $dW = P\,dV$.

51 •• A diatomic gas (molar mass M) is confined to a closed container of volume V at a pressure P_0. What amount of heat Q should be transferred to the gas in order to triple the pressure? (Express your answer in terms of P_0 and V.)

52 •• One mole of air ($c_v = 5R/2$) is confined at atmospheric pressure in a cylinder with a piston at 0°C. The initial volume, occupied by gas, is V. Find the volume of gas V' after the equivalent of 13,200 J of heat is transferred to it.

53 •• The heat capacity of a certain amount of a particular gas at constant pressure is greater than that at constant volume by 29.1 J/K. (a) How many moles of the gas are there? (b) If the gas is monatomic, what are C_v and C_p? (c) If the gas consists of diatomic molecules that rotate but do not vibrate, what are C_v and C_p?

54 •• One mole of a monatomic ideal gas is initially at 273 K and 1 atm. (a) What is its initial internal energy? (b) Find its final internal energy and the work done by the gas when 500 J of heat are added at constant pressure. (c) Find the same quantities when 500 J of heat are added at constant volume.

55 •• A certain molecule has vibrational energy levels that are equally spaced by 0.15 eV. Find the critical temperature T_c such that for $T \gg T_c$ you would expect the equipartition theorem to hold and for $T \ll T_c$ you would expect the equipartition theorem to fail.

Quasi-static Adiabatic Expansion of a Gas

56 • When an ideal gas is subjected to an adiabatic process,

(a) no work is done by the system.
(b) no heat is supplied to the system.
(c) the internal energy remains constant.
(d) the heat supplied to the system equals the work done by the system.

57 • One mole of an ideal gas ($\gamma = \frac{5}{3}$) expands adiabatically and quasi-statically from a pressure of 10 atm and a temperature of 0°C to a pressure of 2 atm. Find (a) the initial and final volumes, (b) the final temperature, and (c) the work done by the gas.

58 • An ideal gas at a temperature of 20°C is compressed quasi-statically and adiabatically to half its original volume. Find its final temperature if (a) $C_v = \frac{3}{2}nR$, and (b) $C_v = \frac{5}{2}nR$.

59 • Two moles of neon gas initially at 20°C and a pressure of 1 atm are compressed adiabatically to one-fourth of their initial volume. Determine the temperature and pressure following compression.

60 •• Half a mole of an ideal monatomic gas at a pressure of 400 kPa and a temperature of 300 K expands until the pressure has diminished to 160 kPa. Find the final temperature and volume, the work done, and the heat absorbed by the gas if the expansion is (*a*) isothermal, and (*b*) adiabatic.

61 •• Repeat Problem 60 for a diatomic gas.

62 •• One-half mole of helium is expanded adiabatically and quasi-statically from an initial pressure of 5 atm and temperature of 500 K to a final pressure of 1 atm. Find (*a*) the final temperature, (*b*) the final volume, (*c*) the work done by the gas, and (*d*) the change in the internal energy of the gas.

63 ••• A hand pump is used to inflate a bicycle tire to a gauge pressure of 482 kPa (about 70 lb/in²). How much work must be done if each stroke of the pump is an adiabatic process? Atmospheric pressure is 1 atm, the air temperature is initially 20°C, and the volume of the air in the tire remains constant at 1 L.

64 ••• An ideal gas at initial volume V_1 and pressure P_1 expands quasi-statically and adiabatically to volume V_2 and pressure P_2. Calculate the work done by the gas directly by integrating $P\,dV$, and show that your result is the same as that given by Equation 19-39.

Cyclic Processes

65 •• One mole of N_2 ($C_v = \frac{5}{2}R$) gas is originally at room temperature (20°C) and a pressure of 5 atm. It is allowed to expand adiabatically and quasi-statically until its pressure equals the room pressure of 1 atm. It is then heated at constant pressure until its temperature is again 20°C. During this heating, the gas expands. After it reaches room temperature, it is heated at constant volume until its pressure is 5 atm. It is then compressed at constant pressure until it is back to its original state. (*a*) Construct an accurate *PV* diagram showing each process in the cycle. (*b*) From your graph, determine the work done by the gas during the complete cycle. (*c*) How much heat is added or subtracted from the gas during the complete cycle? (*d*) Check your graphical determination of the work done by the gas in (*b*) by calculating the work done during each part of the cycle.

66 •• Two moles of an ideal monatomic gas have an initial pressure $P_1 = 2$ atm and an initial volume $V_1 = 2$ L. The gas is taken through the following quasi-static cycle: It is expanded isothermally until it has a volume $V_2 = 4$ L. It is then heated at constant volume until it has a pressure $P_3 = 2$ atm. It is then cooled at constant pressure until it is back to its initial state. (*a*) Show this cycle on a *PV* diagram. (*b*) Calculate the heat added and the work done by the gas during each part of the cycle. (*c*) Find the temperatures T_1, T_2, and T_3.

67 ••• At point D in Figure 19-17 the pressure and temperature of 2 mol of an ideal monatomic gas are 2 atm and 360 K. The volume of the gas at point B on the *PV* diagram is three times that at point D and its pressure is twice that at point C. Paths AB and CD represent isothermal processes. The gas is carried through a complete cycle along the path

Figure 19-17
Problems 67–70

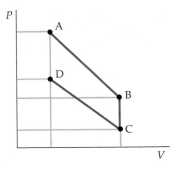

DABCD. Determine the total amount of work done by the gas and the heat supplied to the gas along each portion of the cycle.

68 ••• Repeat Problem 67 with the paths AB and CD representing adiabatic processes.

69 ••• Repeat Problem 67 with a diatomic gas.

70 ••• Repeat Problem 68 with a diatomic gas.

71 ••• An ideal gas of n mol is initially at pressure P_1, volume V_1, and temperature T_h. It expands isothermally until its pressure and volume are P_2 and V_2. It then expands adiabatically until its temperature is T_c and its pressure and volume are P_3 and V_3. It is then compressed isothermally until it is at a pressure P_4 and a volume V_4, which is related to its initial volume V_1 by $T_c V_4^{\gamma-1} = T_h V_1^{\gamma-1}$. The gas is then compressed adiabatically until it is back in its original state. (*a*) Assuming that each process is quasi-static, plot this cycle on a *PV* diagram. (This cycle is known as the Carnot cycle for an ideal gas.) (*b*) Show that the heat Q_h absorbed during the isothermal expansion at T_h is $Q_h = nRT_h \ln(V_2/V_1)$. (*c*) Show that the heat Q_c given off by the gas during the isothermal compression at T_c is $Q_c = nRT_c \ln(V_3/V_4)$. (*d*) Using the result that $TV^{\gamma-1}$ is constant for an adiabatic expansion, show that $V_2/V_1 = V_3/V_4$. (*e*) The efficiency of a Carnot cycle is defined to be the net work done divided by the heat absorbed Q_h. Using the first law of thermodynamics, show that the efficiency is $1 - Q_c/Q_h$. (*f*) Using your results from the previous parts of this problem, show that $Q_c/Q_h = T_c/T_h$.

General Problems

72 • After a potato wrapped in aluminum foil has been baked in an oven, it is taken out and its foil removed. The foil cools much faster than the potato. Why?

73 • True or false:

(*a*) The heat capacity of a body is the amount of heat it can store at a given temperature.

(*b*) When a system goes from state 1 to state 2, the amount of heat added to the system is the same for all processes.

(*c*) When a system goes from state 1 to state 2, the work done on the system is the same for all processes.

(*d*) When a system goes from state 1 to state 2, the change in the internal energy of the system is the same for all processes.

(*e*) The internal energy of a given amount of an ideal gas depends only on its absolute temperature.

(*f*) A quasi-static process is one in which there is no motion.

(*g*) For any material that expands when heated, C_p is greater than C_v.

74 • If a system's volume remains constant while undergoing changes in temperature and pressure, then

CHAPTER 20

The Second Law of Thermodynamics

Solar energy is directed toward the solar oven at the center by this circular array of reflectors at Barstow, California.

We are often asked to conserve energy. But according to the first law of thermodynamics, energy is always conserved. What then does it mean to conserve energy if the total amount of energy in the universe does not change no matter what we do? The first law of thermodynamics does not tell the whole story. Energy is always conserved, but some forms of energy are more useful than others. The possibility or impossibility of putting energy to use is the subject of the second law of thermodynamics. For example, it is easy to convert mechanical work completely into thermal energy, but it is impossible to remove thermal energy from a system and convert it completely into mechanical work with no other changes. This experimental fact is one statement of the **second law of thermodynamics.**

> It is impossible to remove thermal energy from a system at a single temperature and convert it to mechanical work without changing the system or surroundings in some other way.

Second law of thermodynamics: Kelvin statement

We will encounter several other formulations of the same law in this chapter.

A common example of the conversion of mechanical energy into thermal energy is movement with friction. For example, when a block slides along a rough table, the initial mechanical (kinetic) energy of the block is converted into thermal energy as the block and the table are heated. The reverse process never occurs—a block and table that are warm will never spontaneously cool by converting their thermal energy into kinetic energy that sends the block sliding across the table. Yet such an amazing occurrence would not violate the first law of thermodynamics or any other physical laws we have encountered so far. It does, however, violate the second law of thermodynamics. There is thus a lack of symmetry in the roles played by heat and work that is not evident from the first law. This lack of symmetry is related to the fact that some processes are irreversible.

Irreversible processes take many forms, but all are related by the second law. For example, heat conduction is an irreversible process. If we place a hot body in contact with a cold body, heat will flow from the hot body to the cold body until they are at the same temperature. The reverse does not occur. Two bodies in contact at the same temperature remain at the same temperature; heat does not flow from one to the other making one colder and the other warmer. This experimental fact gives us another statement of the second law of thermodynamics.

> There can be no process whose only final result is to transfer thermal energy from a cooler object to a hotter one.

Second law of thermodynamics: Clausius statement

We will show in this chapter that the Kelvin and Clausius statements of the second law are equivalent.

20-1   Heat Engines and the Second Law of Thermodynamics

The study of the efficiency of heat engines gave rise to the first clear statements of the second law. A **heat engine** is a cyclic device whose purpose is to convert as much heat input into work as possible. Heat engines contain a **working substance** (water in a steam engine, air and gasoline vapor in an internal-combustion engine) that absorbs a quantity of heat $Q_{\text{in, h}}$, does work W, and gives off heat $|Q_{\text{out,c}}|$* as it returns to its initial state.

The earliest heat engines were steam engines, invented in the eighteenth century for pumping water from coal mines. Today steam engines are used to generate electricity. In a typical steam engine, water is heated under several hundred atmospheres pressure until it vaporizes at about 500°C (Figure 20-1). This steam expands against a piston, doing work, then exits at a much lower temperature and is further cooled until it condenses. The water is then pumped back into the boiler and heated again.

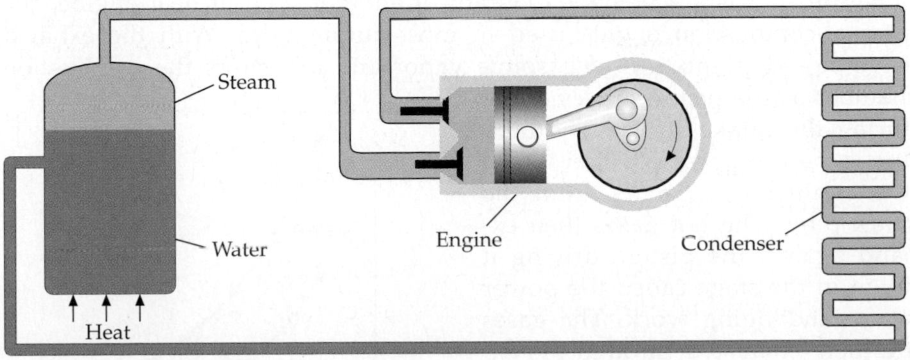

Figure 20-1 Schematic drawing of a steam engine. High-pressure steam does work against the piston.

---

* According to our sign convention for the first law, the heat given off by a system is negative. Since we are interested here only in the magnitudes of the heat absorbed or given off, we will use absolute-value signs where needed.

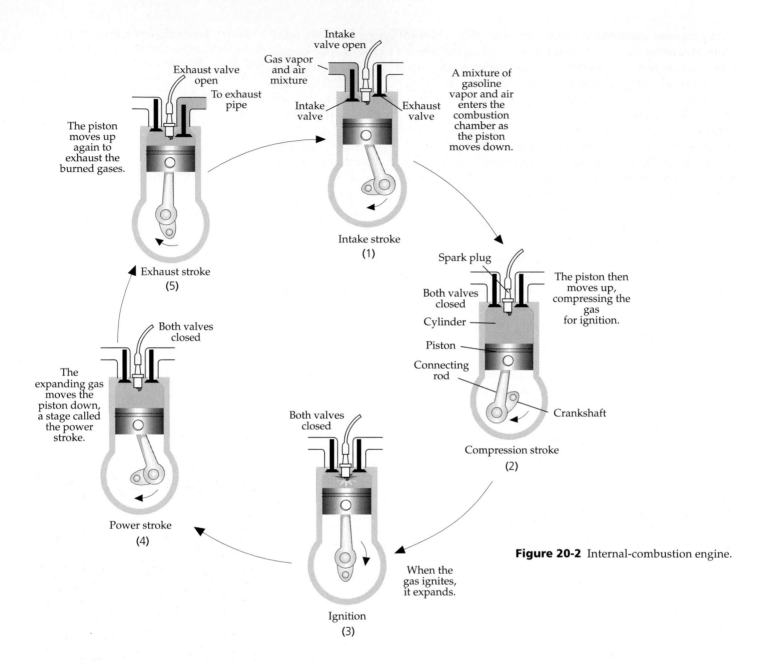

Figure 20-2 Internal-combustion engine.

Figure 20-2 is a schematic diagram of another type of heat engine, the internal-combustion engine used in most automobiles. With the exhaust valve closed, a mixture of gasoline vapor and air enters the combustion chamber as the piston moves down during the intake stroke. The mixture is then compressed, after which it is ignited by a spark from the spark plug. The hot gases then expand against the piston, driving it down in the stage called the power stroke and doing work. The gases are then exhausted through the exhaust valve, and the cycle repeats. An idealized model of the processes in the internal combustion engine is the *Otto cycle* shown in Figure 20-3.

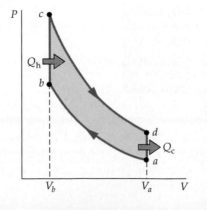

Figure 20-3 Otto cycle, representing the internal-combustion engine. The gasoline–air mixture enters at a and is adiabatically compressed to b. It is then heated (by ignition from the spark plug) at constant volume to c. The power stroke is represented by the adiabatic expansion from c to d. The cooling at constant volume from c to a represents the exhausting of the burned gases and the intake of a fresh gasoline–air mixture.

Figure 20-4 shows a schematic representation of a basic heat engine. The heat input is represented as coming from a **hot reservoir** at temperature T_h, and the exhaust goes into a **cold reservoir** at a lower temperature T_c. A hot or cold reservoir is an idealized body or system that has a very large heat capacity that allows it to absorb or give off thermal energy with no appreciable change in temperature. In practice the surrounding atmosphere or a lake often acts as a thermal reservoir. Since the

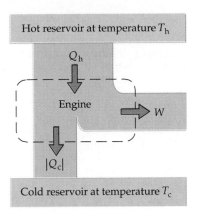

Hot reservoir at temperature T_h

Q_h

Engine → W

$|Q_c|$

Cold reservoir at temperature T_c

Figure 20-4 Schematic representation of a heat engine. The engine removes heat energy Q_h from a hot reservoir at a temperature T_h, does work W, and gives off heat $|Q_c|$ to a cold reservoir at a temperature T_c.

initial and final states of the engine and working substance are the same, the initial and final internal energies of the engine are equal so $\Delta U = 0$. Then, according to the first law of thermodynamics, the work done equals the net heat absorbed:

$$Q = \Delta U + W = W$$

or

$$W = Q_{in,h} - |Q_{out,c}| \qquad \text{20-1}$$

The **efficiency** ε of a heat engine is defined as the ratio of the work done to the heat absorbed from the hot reservoir:

$$\varepsilon = \frac{W}{Q_{in,\,h}} = \frac{Q_{in,\,h} - |Q_{out,\,c}|}{Q_{in,\,h}} = 1 - \frac{|Q_{out,\,c}|}{Q_{in,\,h}} \qquad \text{20-2}$$

Definition—Efficiency of a heat engine

Since the heat Q_{in} is usually produced by burning some fuel like coal or oil that must be paid for, heat engines are designed to have the greatest possible efficiency. The best steam engines operate near 40% efficiency; the best internal-combustion engines operate near 50%. At 100% efficiency ($\varepsilon = 1$), all the thermal energy absorbed from the hot reservoir would be converted into work and no thermal energy would be given off to the cold reservoir. However, *it is impossible to make a heat engine whose efficiency is 100%*. This experimental result is the **heat-engine statement of the second law of thermodynamics,** which is equivalent to the Kelvin statement above:

> It is impossible for a heat engine working in a cycle to produce no other effect than that of extracting thermal energy from a reservoir and performing an equivalent amount of work.

Second law of thermodynamics: Heat-engine statement

The word "cycle" in this statement is important because it *is* possible to convert heat completely into work in a noncyclic process. An ideal gas undergoing an isothermal expansion does just this. But after the expansion, the gas is not in its original state. To bring the gas back to its original state, work must be done on the gas, and some heat will be exhausted.

The second law tells us that to do work with energy extracted from a heat reservoir, we must have a colder reservoir available to receive part of the energy as exhaust. If this were not true, we could design a ship with a heat engine that was powered by simply extracting thermal energy from the ocean. Unfortunately, the lack of a colder reservoir for exhaust makes this enormous reservoir of energy unavailable for such use.* The point is that in order to convert completely disordered thermal energy at a single temperature into the completely ordered energy of an object in motion (with no other changes in the source or object), a separate cold reservoir must be used.

Exhaust manifolds in the 8-cylinder Dusenberg carry heat away from the engine to reduce its temperature.

*It is theoretically possible to run a heat engine between the warmer surface water of the ocean and the colder water at greater depths, but no practical scheme for using this temperature difference has yet emerged.

Example 20-1

A heat engine absorbs 200 J of heat from a hot reservoir, does work, and exhausts 160 J to a cold reservoir. What is the efficiency of the engine?

1. The efficiency is the work done divided by the heat in:
$$\varepsilon = \frac{W}{Q_{in}}$$

2. The heat in is given:
$$Q_{in} = 200 \text{ J}$$

3. The work is found from the first law:
$$W = Q_{in} - |Q_{out}| = 200 \text{ J} - 160 \text{ J} = 40 \text{ J}$$

4. Substitute the values of Q_{in} and W to calculate the efficiency:
$$\varepsilon = \frac{W}{Q_{in}} = \frac{40 \text{ J}}{200 \text{ J}} = 0.20 = 20\%$$

Exercise A heat engine has an efficiency of 35%. (a) How much work does it perform in a cycle if it extracts 150 J of thermal energy from a hot reservoir per cycle? (b) How much thermal energy is exhausted per cycle? (*Answers* (a) 52.5 J, (b) 97.5 J)

Example 20-2   *try it yourself*

(a) Find the efficiency of the Otto cycle shown in Figure 20-3. (b) Express your answer in terms of the ratio of the volumes $r = V_a/V_b = V_d/V_c$.

Picture the Problem (a) To find ε, you need to find Q_{in} and Q_{out}. Heat transfer occurs only during the two constant-volume processes, b to c and d to a. You can thus find Q_{in} and Q_{out} and therefore ε in terms of the temperatures T_a, T_b, T_c, and T_d. (b) The temperatures can be related to the volumes using $TV^{\gamma-1} = $ constant for adiabatic processes.

Cover the column to the right and try these on your own before looking at the answers.

Steps | **Answers**

(a)1. Write the efficiency in terms of Q_{in} and $|Q_{out}|$.
$$\varepsilon = 1 - \frac{|Q_{out}|}{Q_{in}}$$

2. The heat out occurs at constant volume from d to a. Write $|Q_{out}|$ in terms of C_v and the temperatures T_a and T_d.
$$|Q_{out}| = |Q_{da}| = C_v|T_a - T_d| = C_v(T_d - T_a)$$

3. The heat in occurs at constant volume from b to c. Write Q_{in} in terms of C_v and the temperatures T_c and T_b.
$$Q_{in} = Q_{bc} = C_v(T_c - T_b)$$

4. Substitute these values of $|Q_{out}|$ and Q_{in} to find the efficiency in terms of the temperatures T_a, T_b, T_c, and T_d.
$$\varepsilon = 1 - \frac{T_d - T_a}{T_c - T_b}$$

(b)1. Relate T_c to T_d using $TV^{\gamma-1} = $ constant, and $V_d/V_c = r$.

$$T_c V_c{}^{\gamma-1} = T_d V_d{}^{\gamma-1}; \qquad T_c = T_d \frac{V_d{}^{\gamma-1}}{V_c{}^{\gamma-1}} = T_d r^{\gamma-1}$$

2. Relate T_b to T_a as in step 1.

$$T_b = T_a r^{\gamma-1}$$

3. Use these relations to eliminate T_c and T_b from ε in (a) so that ε is expressed in terms of r.

$$\varepsilon = 1 - \frac{T_d - T_a}{T_d r^{\gamma-1} - T_a r^{\gamma-1}} = 1 - \frac{1}{r^{\gamma-1}}$$

Remark The ratio r (volume before compression/volume after compression) is called the compression ratio.

20-2 Refrigerators and the Second Law of Thermodynamics

Figure 20-5 is a schematic representation of a **refrigerator**, which is essentially a heat engine run backward. Work is put into the engine to extract thermal energy from the refrigerator (cold reservoir) and transfer it to the surroundings (hot reservoir). Experience shows that such a transfer always requires some work—a result known as the **refrigerator statement of the second law of thermodynamics**, which is equivalent to the Clausius statement:

It is impossible for a refrigerator working in a cycle to produce no other effect than the transfer of thermal energy from a cold object to a hot object.

Second law of thermodynamics: Refrigerator statement

Were the above statement not true, we could cool our homes in the summer with refrigerators that pumped thermal energy to the outside without using any electricity or other energy.

A measure of a refrigerator's performance is the ratio $Q_{in, c}/W$, where $Q_{in, c}$ is the heat removed from a cold reservoir by the refrigerator doing work W. This ratio is called the **coefficient of performance**, COP:

$$\text{COP} = \frac{Q_c}{W} \qquad \qquad 20\text{-}3$$

Definition—Coefficient of performance

The greater the coefficient of performance, the better the refrigerator. Typical refrigerators have coefficients of performance of about 5 or 6. In terms of this ratio, the refrigerator statement of the second law says that the coefficient of performance of a refrigerator cannot be infinite.

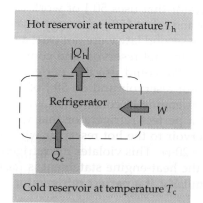

Figure 20-5 Schematic representation of a refrigerator. The refrigerator removes heat energy Q_c from a cold reservoir and gives off heat $|Q_h|$ to a hot reservoir using work W.

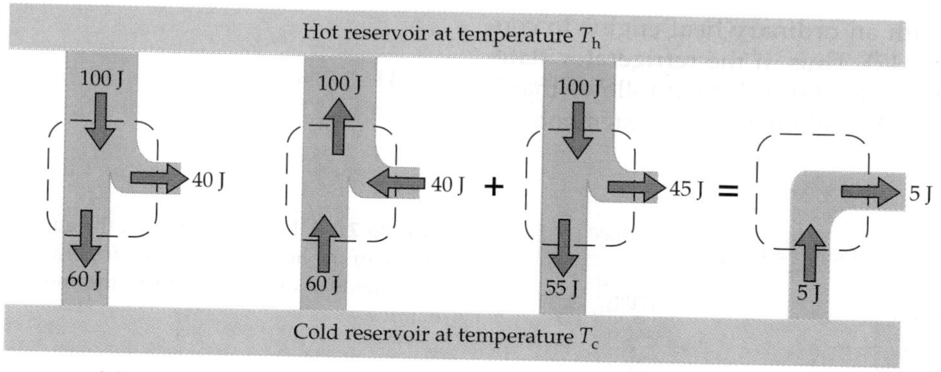

Example 20-3

A refrigerator
needed for this

1. The work is
   heat extracte

2. The total hea
   and to freeze

3. The heat nee
   10 K is:

4. The heat nee

5. Add these h

6. Substitute Q

Exercise A re
heat is exhaust
the cold reserv

Figure 20-7 Demonstration of the Carnot theorem.

(a) A reversible heat engine with 40% efficiency removes 100 J from a hot reservoir, does 40 J of work, and exhausts 60 J to the cold reservoir.

(b) When the same engine runs backward as a refrigerator, 40 J of work are done to remove 60 J from the cold reservoir and exhaust 100 J to the hot reservoir.

(c) This assumed heat engine works between the same two reservoirs with a 45% efficiency, which is greater than that of the reversible engine in (a).

(d) The net effect of running the engine in (c) in conjunction with the refrigerator in (b) is the same as that of a perfect heat engine that removes 5 J from the cold reservoir and converts it completely into work with no other effect, violating the second law of thermodynamics. Thus, the reversible engine in (a) is the most efficient engine that can operate between these two reservoirs.

If no engine can have a greater efficiency than a Carnot engine, it follows that all Carnot engines working between the same two reservoirs have the same efficiency. This efficiency, called the **Carnot efficiency,** must be independent of the working substance of the engine and thus can depend only on the temperatures of the reservoirs.

Let us look at what makes a process reversible or irreversible. According to the second law, the conversion of mechanical energy into heat by friction is *not* reversible; nor is the conduction of heat from a hot object to a cold one. A third type of irreversibility occurs when a system passes through nonequilibrium states, such as when there is turbulence in a gas, or when a gas explodes. For a process to be reversible, we must be able to move the system back through the same equilibrium states in the reverse order.

From these considerations and our statements of the second law of thermodynamics, we can list some conditions that are necessary for a process to be reversible:

1. No work must be done by friction, viscous forces, or other dissipative forces that produce heat.

2. Heat conduction can only occur isothermally.

3. The process must be quasi-static so that the system is always in an equilibrium state (or infinitesimally near an equilibrium state).

Conditions for reversibility

20-3

The heat-engi
statements) of
they are actua
statement is as
merical examp
refrigerator st.

Figure 20-6
move 100 J of
reservoir. If th
fect heat engi
completely in
engine to rem
(Figure 20-6b)
ordinary refri
transfer 100 J
requiring any
tor statement
the refrigerate

Any process that violates any of the above conditions is irreversible. Most processes in nature are irreversible. To have a reversible process, great care must be taken to eliminate frictional and other dissipative forces and to make the process quasi-static. Since this can never be done completely, a reversible process is an idealization, similar to the idealization of motion without friction in mechanics problems. Nevertheless, reversibility can in practice be approximated quite closely.

We can now understand the features of a Carnot cycle, which is a reversible cycle between just two reservoirs. Since all heat transfer must be done isothermally for the process to be reversible, the heat absorbed from the hot reservoir must be absorbed isothermally. The next step must be a quasi-static adiabatic expansion to the lower temperature of the cold reservoir. Then heat is given off isothermally to the cold reservoir. Finally, there is a quasi-static, adiabatic compression to the higher temperature of the hot reservoir. The Carnot cycle thus consists of four reversible steps:

1. A quasi-static isothermal absorption of heat from a hot reservoir

2. A quasi-static adiabatic expansion to a lower temperature

3. A quasi-static isothermal exhaustion of heat to a cold reservoir

4. A quasi-static adiabatic compression back to the original state

Steps in a Carnot cycle

To calculate the efficiency of a Carnot engine, we choose as the working substance a material of which we have some knowledge—an ideal gas—and explicitly calculate the work done by it over a Carnot cycle (Figure 20-8). Since all Carnot cycles have the same efficiency independent of the working substance, our result will be valid in general.

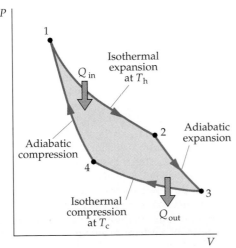

Figure 20-8 Carnot cycle for an ideal gas:
> *Step 1:* Heat is absorbed from a hot reservoir at temperature T_h during an isothermal expansion from state 1 to state 2.
> *Step 2:* The gas expands adiabatically from state 2 to state 3 and its temperature is reduced to T_c.
> *Step 3:* The gas gives off heat to the cold reservoir as it is compressed isothermally at T_c from state 3 to state 4.
> *Step 4:* The gas is compressed adiabatically until its temperature is again T_h.

Work is done on the gas or by the gas during each step. The net work done during the cycle is represented by the shaded area. All processes are reversible. All steps are quasi-static.

The efficiency of this cycle is

$$\varepsilon = 1 - \frac{|Q_{out,c}|}{Q_{in,h}}$$

The heat $Q_{in,h}$ is absorbed during the isothermal expansion from state 1 to state 2. Since $\Delta U = 0$ for an isothermal expansion of an ideal gas, $Q_{in,h}$ equals the work done by the gas.

$$Q_{in,h} = W = \int_1^2 P \, dV = \int_1^2 \frac{nRT_h}{V} \, dV = nRT_h \ln \frac{V_2}{V_1}$$

Similarly, the heat given off to the cold reservoir equals the work done on the gas during the isothermal compression at temperature T_c from state 3 to state 4. This work has the same magnitude as that done by the gas if it expands from state 4 to state 3. The heat rejected is thus

$$|Q_{out,c}| = nRT_c \ln \frac{V_3}{V_4}$$

The ratio of these heats is

$$\frac{|Q_{out,c}|}{Q_{in,h}} = \frac{T_c \ln(V_3/V_4)}{T_h \ln(V_2/V_1)} \qquad \text{20-4}$$

We can relate the volumes V_1, V_2, V_3, and V_4 using Equation 19-22 for a quasi-static adiabatic expansion:

$$TV^{\gamma-1} = \text{constant}$$

For the expansion from state 2 to state 3, we have

$$T_h V_2^{\gamma-1} = T_c V_3^{\gamma-1}$$

Similarly, for the adiabatic compression from state 4 to state 1, we have

$$T_h V_1^{\gamma-1} = T_c V_4^{\gamma-1}$$

Dividing these two equations, we obtain

$$\left(\frac{V_2}{V_1}\right)^{\gamma-1} = \left(\frac{V_3}{V_4}\right)^{\gamma-1}$$

and so $V_2/V_1 = V_3/V_4$. Therefore, $\ln(V_2/V_1) = \ln(V_3/V_4)$ and we can cancel the logarithmic terms in Equation 20-4 to obtain

$$\frac{|Q_{out,c}|}{Q_{in,h}} = \frac{T_c}{T_h} \qquad \text{20-5}$$

(a)

(*a*) Coal-fueled electric generating plant at Four Corners, New Mexico. (*b*) Power plant at Wairakei, New Zealand, that converts geothermal energy into electricity. (*c*) Solar energy is focused and collected individually to produce electricity by these heliostats being tested at Sandia National Laboratory. (*d*) Control rods are inserted into this nuclear reactor at Tihange, Belgium. (*e*) An experimental wind-powered electric generator at Sandia National Laboratory. The propeller is designed for optimum transfer of wind energy to mechanical energy.

The Carnot efficiency ε_C is thus

$$\varepsilon_C = 1 - \frac{T_c}{T_h} \qquad\qquad 20\text{-}6$$

Carnot efficiency

Equation 20-6 demonstrates what we noted earlier, that the Carnot efficiency depends only on the temperatures of the two reservoirs.

(b)

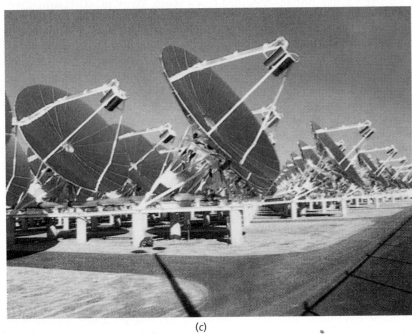

(c)

(d)

(e)

Example 20-4

A steam engine works between a hot reservoir at 100°C = 373 K and a cold reservoir at 0°C = 273 K. (*a*) What is the maximum possible efficiency of this engine? (*b*) If the engine is run backward as a refrigerator, what is its maximum coefficient of performance?

Picture the Problem The maximum efficiency is the Carnot efficiency given by Equation 20-6. To find the coefficient of performance, it is easiest to choose some values for the heat absorbed and given off. If the engine absorbs 100 J from the hot reservoir, it does work $W = \varepsilon(100 \text{ J})$ and gives off heat $Q_c = 100 - \varepsilon(100 \text{ J})$. Then COP = Q_c/W.

(*a*)   The maximum efficiency is the Carnot efficiency:

$$\varepsilon_C = 1 - \frac{T_c}{T_h} = 1 - \frac{273 \text{ K}}{373 \text{ K}} = 0.268 = 26.8\%$$

(*b*)1.   Find the work done by the engine when it absorbs 100 J:

$$W = \varepsilon_C Q_{ih,h} = \ = 0.268(100 \text{ J}) = 26.8 \text{ J}$$

2.   Find the heat given off:

$$Q_{out,c} = Q_{ih,h} - W = 100 \text{ J} - 26.8 \text{ J} = 73.2 \text{ J}$$

3.   Use these results to calculate COP when the engine is run backward:

$$\text{COP} = \frac{Q_{out,c}}{W} = \frac{73.2 \text{ J}}{26.8 \text{ J}} = 2.73$$

Remarks Even though this efficiency seems to be quite low, it is the greatest efficiency possible for any engine working between these temperatures. Real engines will have lower efficiencies because of friction, heat conduction, and other irreversible processes. Similarly, real refrigerators will have a lower coefficient of performance. In Section 20-5 it is shown that the coefficient of performance of a Carnot refrigerator is $T_c/\Delta T$ (Equation 20-9). For these reservoirs, $\Delta T = 100 \text{ K}$ and COP = 273 K/100 K = 2.73.

The Carnot efficiency is useful to know because it gives us an upper limit on possible efficiencies. It tells us, for example, that an engine working between reservoirs at 373 and 273 K with an efficiency of 25% is a very good engine. However much friction and other irreversible losses were reduced, the best efficiency obtainable between those temperatures is 26.8%, as calculated in Example 20-4.

Example 20-5

An engine removes 200 J from a hot reservoir at 373 K, does 48 J of work, and exhausts 152 J to a cold reservoir at 273 K. How much work is "lost" per cycle due to irreversible processes in this engine?

Picture the Problem The difference between maximum amount of work that could be done using a Carnot engine and 48 J is the work lost.

1.   The work lost is the maximum amount of work that could be done minus the work actually done:

$$W_{lost} = W_{max} - W = W_{max} - 48 \text{ J}$$

2. The maximum amount of work that could be done is:

$$W_{max} = \varepsilon_C Q_{in,h}$$

3. The Carnot efficiency was found in Example 20-4:

$$\varepsilon_C = 1 - \frac{T_c}{T_h} = 1 - \frac{273\ K}{373\ K} = 0.268 = 26.8\%$$

4. The maximum amount of work that could be done is then:

$$W_{max} = \varepsilon_C Q_{in,h} = (0.268)(200) = 53.6\ J$$

5. The work lost is then:

$$W_{lost} = W_{max} - W = 53.6\ J - 48\ J = 5.6\ J$$

Remarks The 5.6 J of energy in the answer is not "lost" to the universe—total energy is conserved. That 5.6 J of energy exhausted into the cold reservoir by the nonideal engine of the problem is only lost in that it would have been converted into useful work if an ideal (reversible) engine had been used.

Example 20-6

If 200 J of heat is conducted from a heat reservoir at 373 K to one at 273 K, with no engine between the reservoirs as in Example 20-5, how much work capability is "lost" in this process?

We saw in Example 20-5 that a Carnot engine working between these two reservoirs could do 53.6 J of work if it extracted 200 J from the 373-K reservoir and exhausted to a 273-K reservoir. Thus, if 200 J is conducted directly from the hot reservoir to the cold reservoir without any work being done, 53.6 J of this energy has been "lost" in the sense that it could have been converted into useful work.

Exercise A Carnot engine works between heat reservoirs at 500 and 300 K. (a) What is its efficiency? (b) If it removes 200 kJ of heat from the hot reservoir, how much work does it do? (*Answers* (a) 40%, (b) 80 kJ)

Exercise A real engine works between heat reservoirs at 500 and 300 K. It removes 500 kJ of heat from the hot reservoir and does 150 kJ of work during each cycle. What is its efficiency? (*Answer* 30%)

The Absolute Temperature Scale

In Chapter 18, the ideal-gas temperature scale was defined in terms of the properties of gases at low densities. Since the Carnot efficiency depends only on the temperatures of the two heat reservoirs, it can be used to define the ratio of the temperatures of the reservoirs independent of the properties of any substance. We define the ratio of the absolute temperatures of the hot and cold reservoirs to be

$$\frac{T_c}{T_h} = \frac{|Q_c|}{Q_h} \qquad \text{20-7}$$

where Q_h is the energy removed from the hot reservoir and $|Q_c|$ is the energy exhausted to the cold reservoir by a Carnot engine working between the two reservoirs. Thus, to find the ratio of two reservoir temperatures, we set up a reversible engine operating between them and measure the heat absorbed from or given off by each reservoir during one cycle. The absolute temperature is completely determined by Equation 20-7 and the choice of one fixed point. If the fixed point is defined to be 273.16 K for the triple point of water, then the absolute temperature scale matches the ideal-gas temperature scale for the range of temperatures over which a gas thermometer can be used.

20-5   Heat Pumps

A **heat pump** is essentially a refrigerator that is used to pump thermal energy from a cold reservoir (for example, the cold air outside a house) to a hot reservoir (for example, the hot air inside the house). If work W is done to remove heat Q_c from the cold reservoir and reject heat $|Q_h| = W + Q_c$ to the hot reservoir, the coefficient of performance (Equation 20-3) is

$$\text{COP} = \frac{Q_c}{W}$$

Using $W = |Q_h| - Q_c$, this can be written

$$\text{COP} = \frac{Q_c}{|Q_h| - Q_c} = \frac{Q_c/|Q_h|}{1 - Q_c/|Q_h|} \qquad \text{20-8}$$

The maximum coefficient of performance is obtained using a Carnot heat pump. Then Q_c and Q_h are related by Equation 20-5. Substituting $Q_c/|Q_h| = T_c/T_h$ into Equation 20-8, we obtain for the maximum coefficient of performance

$$\text{COP}_{\text{max}} = \frac{T_c/T_h}{1 - T_c/T_h} = \frac{T_c}{T_h - T_c} = \frac{T_c}{\Delta T} \qquad \text{20-9}$$

where ΔT is the difference in temperature between the hot and cold reservoirs. Real heat pumps and refrigerators have COPs less than the COP_{max} because of losses due to friction, heat conduction, and other irreversible processes.

We are usually interested in the work that must be done to exhaust a given amount of heat $|Q_h|$ into the hot reservoir, which in the case of a home heat pump would be the hot-air supply for the heating fan of a house. Using $|Q_h| = Q_c + W$, we can write Equation 20-3 as

$$\text{COP} = \frac{Q_c}{W} = \frac{|Q_h| - W}{W}$$

or

$$W = \frac{|Q_h|}{1 + \text{COP}} \qquad \text{20-10}$$

Example 20-7   *try it yourself*

An ideal heat pump is used to pump heat from the outside air at −5°C to the hot-air supply for the heating fan in a house, which is at 40°C. How much work is required to pump 1 kJ of heat into the house?

Picture the Problem Use Equation 20-10 with COP calculated from Equation 20-9 for $T_c = -5°C = 268$ K and $\Delta T = 45$ K.

Steps	Answers
1. Calculate the COP from Equation 20-9.	$\text{COP}_{\text{max}} = \dfrac{T_{\text{c}}}{\Delta T} = 5.96$
2. Calculate the work needed from Equation 20-10.	$W = \dfrac{\lvert Q_{\text{c}} \rvert}{1 + \text{COP}} = 0.144 \text{ kJ}$

Remarks Only 0.144 kJ of work is needed to pump 1 kJ of heat into the hot-air supply in the house. We see that the heat pump essentially multiplies the energy needed to run the pump by 1 + COP. If we use 1 kJ to run a heat pump with a COP = 5.96, we can exhaust 6.96 kJ of heat into the house.

20-6   Irreversibility and Disorder

There are many irreversible processes that cannot be described by the heat-engine or refrigerator statements of the second law, such as a glass falling to the floor and breaking. However, all irreversible processes have one thing in common—the system plus its surroundings moves toward a less ordered state.

Suppose a box containing a gas of mass M at a temperature T is moving along a frictionless table with a velocity v_{cm} (Figure 20-9a). The total kinetic energy of the gas has two components: that associated with the movement of the center of mass $\frac{1}{2}Mv_{\text{cm}}^2$, and the energy of the motion of its molecules relative to its center of mass. The center-of-mass energy $\frac{1}{2}Mv_{\text{cm}}^2$ is ordered mechanical energy that could be converted completely into work.* The relative energy is the gas's internal thermal energy, which is related to its temperature T. It is random, nonordered energy that cannot be converted directly into work.

Now, suppose the box hits a fixed wall and stops (Figure 20-9b):

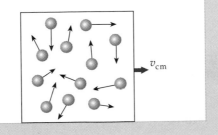

Figure 20-9a

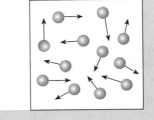

Figure 20-9b

This inelastic collision is clearly an irreversible process. The ordered mechanical energy of the gas is converted into random internal energy and the temperature of the gas rises. The gas still has the same total energy, but now all of it is associated with the random motion of its molecules about its center of mass, which is now at rest. Thus, the gas has become less ordered (or more disordered), and it has lost the ability to do work.

*For example, if a weight were attached with a string to the moving box, this energy could be used to lift the weight.

not freely contract by itself into a smaller volume. This leads us to yet another statement of the second law of thermodynamics:

For any process, the entropy of the universe never decreases.

Example 20-8

Find the entropy change for the free expansion of 0.75 mol of an ideal gas from $V_1 = 1.5$ L to $V_2 = 3$ L.

Picture the Problem The entropy change for this irreversible process is equivalent to the entropy change for an isothermal process from V_1 to V_2. For the isothermal process, we first calculate $Q = W$, then set $\Delta S = Q/T$.

1. The entropy change is the same as for an isothermal expansion from V_1 to V_2:

$$\Delta S = \Delta S_{\text{isothermal}} = \frac{Q}{T}$$

2. The heat Q that would enter the gas during an isothermal expansion at temperature T equals the work done by the gas during the expansion:

$$Q = W = nRT \ln \frac{V_2}{V_1}$$

3. Substitute this value of Q to calculate ΔS:

$$\Delta S = \frac{Q}{T} = nR \ln \frac{V_2}{V_1} = (0.75 \text{ mol})(8.31 \text{ J/mol·K}) \ln 2 = 4.32 \text{ J/K}$$

ΔS for Constant-Pressure Processes When a substance is heated from temperature T_1 to temperature T_2 at constant pressure, the heat absorbed dQ is related to its temperature change dT by

$$dQ = C_{\text{p}} \, dT$$

We can approximate reversible heat conduction if we have a large number of heat reservoirs with temperatures ranging from T_1 to T_2 in very small steps. We then place our substance, whose initial temperature is T_1, in contact with the first reservoir at a temperature just slightly greater than T_1 and let the substance absorb a small amount of heat. Since the heat transfer is approximately isothermal, the process will be approximately reversible. We then place the substance in contact with the next reservoir at a slightly higher temperature, and so on, until the final temperature T_2 is reached. When heat dQ is absorbed reversibly, the entropy change of the substance is

$$dS = \frac{dQ}{T} = C_{\text{p}} \frac{dT}{T}$$

Integrating from T_1 to T_2, we obtain the total entropy change of the substance:

$$\Delta S = C_{\text{p}} \int_{T_1}^{T_2} \frac{dT}{T} = C_{\text{p}} \ln \frac{T_2}{T_1} \qquad 20\text{-}19$$

This result gives the entropy change of a substance that is heated from T_1 to T_2 by any process, reversible or irreversible, as long as the final pressure equals the initial pressure. It also gives the entropy change of a substance that is cooled. In this case, T_2 is less than T_1, and $\ln(T_2/T_1)$ is negative, giving a negative entropy change.

Exercise Find the change in entropy of 1 kg of water that is heated from 0 to 100°C. (*Answer* $\Delta S = 1.31$ kJ/K)

Example **20-9**	*try it yourself*

1 kg of water at temperature $T_1 = 30°C$ is mixed with 2 kg of water at $T_2 = 90°C$ in a calorimeter of negligible heat capacity at a constant pressure of 1 atm. Find the change in entropy of the system.

Picture the Problem When the two amounts of water are mixed, they eventually come to a final equilibrium temperature T_f that can be found by setting the heat lost equal to the heat gained. To calculate the entropy change of each mass of water, we consider a reversible isobaric heating of the 1-kg mass of water from 30°C to T_f and an isobaric cooling of the 2-kg mass from 90°C to T_f using Equation 20-17. The entropy change of the system is the sum of the entropy changes of each part.

Cover the column to the right and try these on your own before looking at the answers.

Steps

Answers

1. Write an expression for the entropy change ΔS_1 of the mass $m_1 = 1$ kg of water going from the state $T = T_1$ to the state $T = T_f$.

$$\Delta S_1 = m_1 c_p \ln \frac{T_f}{T_1}$$

2. Write an expression for the entropy change ΔS_2 of the mass $m_2 = 2$ kg of water going from the state $T = T_2$ to the state $T = T_f$.

$$\Delta S_2 = m_2 c_p \ln \frac{T_f}{T_2}$$

3. Calculate T_f by setting the heat lost equal to the heat gained.

$$m_1 c_p (T_f - 30°C) = m_2 c_p (90°C - T_f)$$

$$T_f = 70°C = 343 \text{ K}$$

4. Use your result for T_f and the data given to calculate ΔS_1 and ΔS_2.

$$\Delta S_1 = m_1 c_p \ln \frac{T_f}{T_1} = 0.518 \text{ kJ/K}$$

$$\Delta S_2 = m_2 c_p \ln \frac{T_f}{T_2} = -0.474 \text{ kJ/K}$$

5. Add ΔS_1 and ΔS_2 to find the total entropy change of the system.

$$\Delta S = \Delta S_1 + \Delta S_2 = +0.044 \text{ kJ/K}$$

Remarks Note that we had to convert the temperatures to the absolute scale to calculate the entropy changes. The entropy change of the universe is positive as expected.

ΔS for an Inelastic Collision   Since mechanical energy is converted into thermal energy in an inelastic collision, such a process is clearly irreversible. The entropy of the universe must therefore increase. Consider a block of mass m falling from a height h and making an inelastic collision with the ground. Let the block, ground, and atmosphere all be at a temperature T, which is not significantly changed by the process. If we consider the block, ground, and atmosphere as our isolated system, there is no heat conducted into or out of the system. The state of the system has been changed because its internal energy has been increased by an amount mgh. This change is the same as if we added heat $Q = mgh$ to the system at constant temperature T.

To calculate the change in entropy of the system, we thus consider a reversible process in which heat $Q_{rev} = mgh$ is added at a constant temperature T. According to Equation 20-11, the change in entropy is then

$$\Delta S = \frac{Q}{T} = \frac{mgh}{T}$$

This positive entropy change is also the entropy change of the universe.

ΔS for Heat Conduction From One Reservoir to Another   Heat conduction is also an irreversible process, so we expect the entropy of the universe to increase. Consider the simple case of heat $|Q|$ conducted from a hot reservoir at a temperature T_h to a cold reservoir at a temperature T_c. The state of a heat reservoir is determined only by its temperature and its internal energy. The change in entropy of a heat reservoir due to a heat exchange is the same whether the heat exchange is reversible or not. If heat $|Q|$ is put into a reservoir at temperature T, the entropy of the reservoir increases by $|Q|/T$. If the heat is removed, the entropy of the reservoir decreases by $-|Q|/T$. In the case of heat conduction, the hot reservoir loses heat, so its entropy change is

$$\Delta S_h = -\frac{|Q|}{T_h}$$

The cold reservoir absorbs heat, so its entropy change is

$$\Delta S_c = +\frac{|Q|}{T_c}$$

The net entropy change of the universe is

$$\Delta S_u = \Delta S_c + \Delta S_h = +\frac{|Q|}{T_c} - \frac{|Q|}{T_h} \qquad\qquad 20\text{-}20$$

Note that since heat always flows from a hot reservoir to a cold reservoir, the change in entropy of the universe is positive.

ΔS for a Carnot Cycle   Since a Carnot cycle is by definition reversible, the entropy change of the universe after a cycle must be zero. We demonstrate this by showing that the entropy change of the reservoirs is zero. (Since a Carnot engine works in a cycle, its entropy change is zero; the entropy change of the universe is just the sum of the entropy changes of the reservoirs.) The entropy change of the hot reservoir is $\Delta S_h = -|Q_h|/T_h$, where $|Q_h|$ is the heat removed. The entropy change of the cold reservoir is $\Delta S_c = +|Q_c|/T_c$, where $|Q_c|$ is the energy exhausted. These energies are related by Equation 20-7,

$$|Q_c| = |Q_h|\frac{T_c}{T_h}$$

The entropy change of the universe is thus*

$$\Delta S_u = \Delta S_h + \Delta S_c = -\frac{|Q_h|}{T_h} + \frac{|Q_c|}{T_c} = -\frac{|Q_h|}{T_h} + \frac{|Q_h|(T_c/T_h)}{T_c} = 0$$

The entropy change of the universe is zero, as expected.

---

* Because the heat *out* of the hot reservoir is the heat *into* the Carnot engine, the sign conventions and the subscripts in and out are confusing here, so we are using absolute values signs for both Q_h and Q_c, and displaying the signs explicitly.

Example 20-10

During each cycle, a Carnot engine removes 100 J of energy from a reservoir at 400 K, does work, and exhausts heat to a reservoir at 300 K. Compute the entropy change of each reservoir for each cycle, and show explicitly that the entropy change of the universe is zero for this reversible process.

Picture the Problem Since the engine works in a cycle, its entropy change is zero. We therefore compute the entropy change of each reservoir and add them to obtain the entropy change of the universe.

1. The entropy change of the universe equals the sum of the entropy changes of the reservoirs:

$$\Delta S_u = \Delta S_{400} + \Delta S_{300}$$

2. Calculate the entropy change of the hot reservoir:

$$\Delta S_{400} = -\frac{|Q_h|}{T_h} = -\frac{100 \text{ J}}{400 \text{ K}} = -0.250 \text{ J/K}$$

3. The entropy change of the cold reservoir is $|Q_c|$ divided by T_c:

$$\Delta S_{300} = \frac{|Q_c|}{T_c} = \frac{|Q_c|}{300 \text{ K}}$$

4. The heat given off $|Q_c|$ is related to the efficiency:

$$\varepsilon_C = 1 - \frac{|Q_c|}{|Q_h|} = 1 - \frac{T_c}{T_h} = 1 - \frac{300 \text{ K}}{400 \text{ K}} = 0.25$$

$$|Q_c| = |Q_h|(1 - 0.25) = (100 \text{ J})(0.75) = 75 \text{ J}$$

5. Calculate the entropy change of the cold reservoir:

$$\Delta S_{300} = \frac{|Q_c|}{T_c} = \frac{75 \text{ J}}{300 \text{ K}} = +0.250 \text{ J/K}$$

6. Substitute these results into step 1 to find the entropy change of the universe:

$$\Delta S_u = \Delta S_{400} + \Delta S_{300}$$
$$= -0.250 \text{ J/K} + 0.250 \text{ J/K} = 0$$

Remarks Suppose an ordinary, nonreversible engine removed 100 J from the hot reservoir. Since its efficiency must be less than that of a Carnot engine, it would do less work and exhaust more heat to the cold reservoir. Then the entropy increase of the cold reservoir would be greater than the entropy decrease of the hot reservoir, and the entropy change of the universe would be positive.

Example 20-11

Since entropy is a state function, thermodynamic processes can be represented as ST, SV, or SP diagrams instead of the PV diagrams we have used so far. Make a sketch of the Carnot cycle on an ST plot.

Picture the Problem The Carnot cycle consists of a reversible isothermal expansion followed by a reversible adiabatic expansion, then a reversible isothermal compression followed by a reversible adiabatic compression. During the isothermal processes, heat is absorbed or expelled at constant temperature, so S increases or decreases at constant T. During the adiabatic processes, the temperature changes, but since $\Delta Q_r = 0$, S is constant.

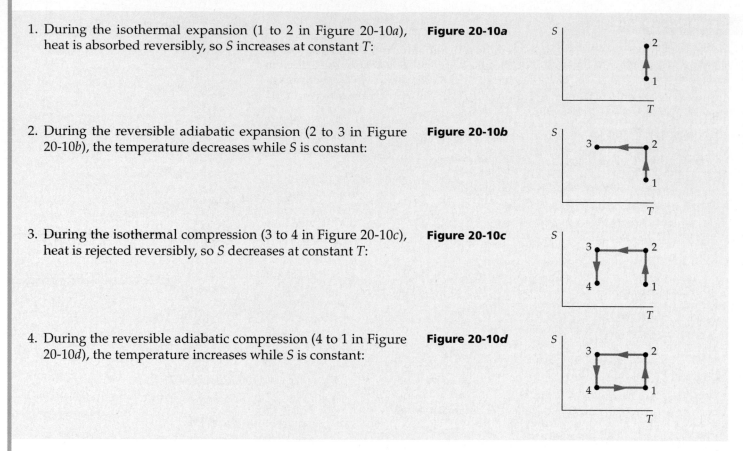

1. During the isothermal expansion (1 to 2 in Figure 20-10*a*), heat is absorbed reversibly, so S increases at constant T:

    Figure 20-10*a*

2. During the reversible adiabatic expansion (2 to 3 in Figure 20-10*b*), the temperature decreases while S is constant:

    Figure 20-10*b*

3. During the isothermal compression (3 to 4 in Figure 20-10*c*), heat is rejected reversibly, so S decreases at constant T:

    Figure 20-10*c*

4. During the reversible adiabatic compression (4 to 1 in Figure 20-10*d*), the temperature increases while S is constant:

    Figure 20-10*d*

Remarks   The Carnot cycle is a rectangle when plotted on an ST diagram.

20-8   Entropy and the Availability of Energy

When an irreversible process occurs, energy is conserved, but some of the energy is "wasted," meaning it becomes unavailable to do work. Consider a block falling to the ground. When the block was at a height h, its potential energy mgh could have been used to do useful work. After the inelastic collision of the block with the ground, this energy is no longer available because it has become the disordered internal energy of the block and its surroundings. The energy that has become unavailable is equal to $mgh = T \Delta S_u$. This is a general result:

> In an irreversible process, energy equal to $T \Delta S_u$ becomes unavailable for doing work, where T is the temperature of the coldest available reservoir.

For simplicity, we will call the energy that becomes unavailable for doing work the "work lost":

$$W_{\text{lost}} = T \Delta S_u \qquad\qquad 20\text{-}21$$

Example 20-12

Suppose that the box discussed in Section 20-5 and shown in Figure 20-9a and b has a mass of 2.4 kg and slides with a speed of $v = 3$ m/s before crashing into a fixed wall and stopping. The temperature T of the block, table, and surroundings is 293 K and does not change appreciably as the block comes to rest. Find the entropy change of the universe.

Picture the Problem The initial mechanical energy of the block, $\frac{1}{2}mv^2$, is converted to thermal energy. The entropy change is equivalent to what would occur if the heat $Q = \frac{1}{2}mv^2$ were added to the system reversibly.

The entropy change of the universe is Q/T:
$$\Delta S_u = \frac{Q}{T} = \frac{\frac{1}{2}mv^2}{T} = \frac{\frac{1}{2}(2.4 \text{ kg})(3 \text{ m/s})^2}{293 \text{ K}} = 0.0369 \text{ J/K}$$

Remark Energy is conserved, but the energy $T \, \Delta S_u = \frac{1}{2}mv^2$ is no longer available to do work.

In the free expansion discussed earlier, the ability to do work was also lost. In that case, the entropy change of the universe was $nR \ln(V_2/V_1)$, so the work lost was $nRT \ln(V_2/V_1)$. This is the amount of work that could have been done if the gas had expanded quasi-statically and isothermally from V_1 to V_2, as given by Equation 19-16.

When heat is conducted from a hot reservoir to a cold reservoir, the change in entropy of the universe is given by Equation 20-20, and the work lost is

$$W_{lost} = T_c \, \Delta S_u = |Q|\left(1 - \frac{T_c}{T_h}\right)$$

We can see that this is just the work that could have been done by a Carnot engine running between these reservoirs, removing heat $|Q|$ from the hot reservoir and doing work $W = \varepsilon_C|Q|$, where $\varepsilon_C = 1 - T_c/T_h$.

20-9   Entropy and Probability

Entropy, which is a measure of the disorder of a system, is related to probability. Essentially, a state of high order has a low probability, whereas a state of low order has a high probability. Thus, in an irreversible process, the universe moves from a state of low probability to one of high probability.

Let us consider a free expansion in which a gas expands from an initial volume V_1 to a final volume $V_2 = 2V_1$. The entropy change of the universe for this process is given by Equation 20-18:

$$\Delta S = nR \ln \frac{V_2}{V_1} = nR \ln 2 \qquad\qquad 20\text{-}22$$

Why is this process irreversible? Why can't the gas compress by itself back into its original volume? Since there is no energy change involved, a compression would not violate the first law of thermodynamics. The reason is merely that such a compression is extremely improbable. To see this, let us begin by assuming that the gas consists of only 10 molecules, and that, initially, these molecules occupy the entire volume of their container. Then the chance that any one particular molecule will be in the left half of the container at any given time is $\frac{1}{2}$. The chance that any two particular molecules will both be in the left half is $\frac{1}{2} \times \frac{1}{2} = \frac{1}{4}$.* The chance that three particular molecules will be in the left half is $\frac{1}{2} \times \frac{1}{2} \times \frac{1}{2} = (\frac{1}{2})^3 = \frac{1}{8}$. The chance that all 10 molecules will be in the left half is $(\frac{1}{2})^{10} = \frac{1}{1024}$. That is, there is 1 chance in 1024 that all 10 molecules will be in the left half of the container at any given time.

Though the probability of all 10 molecules being on one side of the container is small, we would not be completely surprised to see it occur. If we look at the gas once each second, we could expect to see it happen once in every 1024 seconds, or about once every 17 minutes. If we started with the 10 molecules randomly distributed and then found them all in the left half of the original volume, the entropy of the universe would have *decreased* by $nR \ln 2$. However, this decrease is extremely small, since the number of moles n corresponding to 10 molecules is only about 10^{-23}. Still, it would violate the entropy statement of the second law of thermodynamics, which says that for any process, the entropy of the universe never decreases. If we wish to apply the second law to microscopic systems, such as a small number of molecules, we should consider the second law to be a statement of probability.

We can relate the probability of a gas spontaneously compressing itself into a smaller volume to the change in its entropy. If the original volume is V_1, the probability p of finding N molecules in a smaller volume V_2 is

$$p = \left(\frac{V_2}{V_1}\right)^N$$

Taking the natural logarithm of both sides of this equation, we obtain

$$\ln p = N \ln\left(\frac{V_2}{V_1}\right) = nN_A \ln\left(\frac{V_2}{V_1}\right) \qquad \text{20-23}$$

where n is the number of moles and N_A is Avogadro's number. The entropy change of the gas is

$$\Delta S = nR \ln\left(\frac{V_2}{V_1}\right) \qquad \text{20-24}$$

(The entropy change is negative if V_2 is less than V_1.) Comparing Equations 20-23 and 20-24, we see that

$$\Delta S = \frac{R}{N_A} \ln p = k \ln p \qquad \text{20-25}$$

where k is Boltzmann's constant.

It may be disturbing to learn that events such as the spontaneous compression of a gas or the spontaneous conduction of heat from a cold body to a hot body (processes for which $\Delta S_u < 0$) are only improbable, not impossible. But, as we have just seen, there is a reasonable chance that such an event will occur only if the system consists of a very small number of molecules. However, *thermodynamics itself is applicable only to macroscopic systems*, that is, to systems that have a very large number of molecules. Consider trying to measure the pressure of a gas consisting of only 10 molecules. The pressure would vary wildly depending on whether no molecule, 2 molecules, or 10 molecules were colliding with the wall of the container at the time of measurement. The macroscopic variables of pressure and temperature are simply not applicable to a microscopic system with only 10 molecules.

As we increase the number of molecules in a system, the chance of an event occurring for which $\Delta S_u < 0$ decreases dramatically. For example, if we have

---

* This is the same as the chance that a coin flipped twice will come up heads both times.

50 molecules in a container, the chance that they will all be in the left half of the volume is $(\frac{1}{2})^{50} \approx 10^{-15}$. Thus, if we look at the gas once each second, we could expect to see all 50 molecules in the left half of the volume about once in every 10^{15} s or once in every 36 million years! For 1 mol $= 6 \times 10^{23}$ molecules, the chance that all will wind up in half of the volume is vanishingly small, essentially zero. For macroscopic systems, the probability of a process resulting in a decrease in the entropy of the universe is so extremely small that the distinction between improbable and impossible becomes blurred.

Summary

The second law of thermodynamics is a fundamental law of nature.

Topic	Remarks and Relevant Equations
1. Efficiency of a Heat Engine	If the engine removes Q_h from a hot reservoir, does work W, and exhausts heat $\lvert Q_c \rvert$ to a cold reservoir, its efficiency is $$\varepsilon = \frac{W}{Q_h} = 1 - \frac{\lvert Q_c \rvert}{Q_h} \qquad \text{20-2}$$
2. Coefficient of Performance of a Refrigerator	$$\text{COP} = \frac{Q_c}{W} \qquad \text{20-3}$$
3. Equivalent Statements of the Second Law of Thermodynamics	
The Kelvin statement	It is impossible to remove thermal energy from a system at a single temperature and convert it to mechanical work without changing the system or surroundings in some other way.
The heat-engine statement	It is impossible for a heat engine working in a cycle to remove heat from a reservoir and convert it completely into work with no other effects.
The Clausius statement	A process whose only final result is to transfer thermal energy from a cooler object to a hotter one is impossible.
The refrigerator statement	It is impossible for a refrigerator working in a cycle to produce no other effect than the transfer of thermal energy from a cold object to a hot object.
The entropy statement	The entropy of the universe (system plus surroundings) can never decrease.
4. Conditions for a Reversible Process	1. No work must be done by friction, viscous forces, or other dissipative forces that produce heat. 2. Heat conduction can occur only isothermally. 3. The process must be quasi-static so that the system is always in an equilibrium state (or infinitesimally near an equilibrium state).
5. Carnot Engine	A Carnot engine is a reversible engine that works between two reservoirs. It uses a Carnot cycle, which consists of
Carnot cycle	1. A quasi-static isothermal expansion absorbing heat at temperature T_h 2. A quasi-static adiabatic expansion 3. A quasi-static isothermal compression rejecting heat at temperature T_c 4. A quasi-static adiabatic compression back to the original state

Carnot efficiency	$$\varepsilon_C = 1 - \frac{|Q_c|}{Q_h} = 1 - \frac{T_c}{T_h}$$	20-6

6. Absolute Temperature Scale	The ratio of the absolute temperatures of two reservoirs is defined to be the ratio of the heat exhausted to the heat intake of a Carnot engine running between the reservoirs. $$\frac{T_c}{T_h} = \frac{|Q_c|}{Q_h}$$	20-7

7. Entropy	Entropy is a measure of the disorder of a system. The difference in entropy between two nearby states is given by $$dS = \frac{dQ_{rev}}{T}$$ where dQ_{rev} is the heat added in a reversible process connecting the states. The entropy change of a system can be positive or negative.	20-11

Entropy and loss of work capability	During an irreversible process, the entropy of the universe S_u increases and a certain amount of energy $$W_{lost} = T \, \Delta S_u$$ becomes unavailable for doing work.	20-21

Entropy and probability	Entropy is related to probability. A highly ordered system is one of low probability and low entropy. An isolated system moves toward a state of low order, high probability, and high entropy.

Problem-Solving Guide

Summary of Worked Examples

Type of Calculation	Procedure and Relevant Examples	
1. Heat Engines		
Find the efficiency of a heat engine.	Use $\varepsilon = W/Q_{in} = 1 - Q_{out}/Q_{in}$.	**Examples 20-1, 20-2**
Find the maximum efficiency of an engine.	Use $\varepsilon_C = 1 - T_c/T_h$.	**Example 20-3**
Find the work lost due to irreversibility of an engine.	Compare the work done with the maximum work that could be done using a Carnot engine.	**Examples 20-4, 20-5**
2. Refrigerators		
Find the work needed to run a refrigerator.	Use $W = Q_c/\text{COP}$.	**Example 20-2**
3. Entropy		
Find the change in entropy when two objects at different temperatures are put in contact.	First find the final temperature. Then use $\Delta S = mC_p \ln(T_f/T_i)$ for each object.	**Example 20-2**

Find the entropy change due to a free expansion of an ideal gas.	Use $\Delta S = Q/T$ where $Q = W = nRT \ln(V_2/V_1)$ is the heat input for an isothermal expansion to the same final volume. **Example 20-6**
Find the entropy change when mechanical energy ΔE is converted into thermal energy.	Use $\Delta S = \Delta E/T$ where T is the temperature of the surroundings. **Example 20-11**
Find the entropy change of a heat reservoir of temperature T.	Use $\Delta S = \pm Q/T$, where $+Q$ is for heat in and $-Q$ is for heat out. **Example 20-8**
Plot a reversible cyclic process on an S versus T diagram.	During adiabatic expansions or compressions, T changes at constant S. During isothermal expansions or compressions, S changes at constant T. **Example 20-9**

Problems

In a few problems, you are given more data than you actually need; in a few other problems, you are required to supply data from your general knowledge, outside sources, or informed estimates.

- Single-concept, single-step, relatively easy
- Intermediate-level, may require synthesis of concepts
- Challenging, for advanced students

Heat Engines and Refrigerators

1 • Where does the energy come from in an internal-combustion engine? In a steam engine?

2 • How does friction in an engine affect its efficiency?

3 • John is house-sitting for a friend who keeps delicate plants in her kitchen. She warns John not to let the room get too warm or the plants will wilt, but John forgets and leaves the oven on all day after his brownies are baked. As the plants begin to droop, John turns off the oven and opens the refrigerator door, intending to use the refrigerator to cool the kitchen. Explain why this doesn't work.

4 • Why do power-plant designers try to increase the temperature of the steam fed to engines as much as possible?

5 • An engine with 20% efficiency does 100 J of work in each cycle. (a) How much heat is absorbed in each cycle? (b) How much heat is rejected in each cycle?

6 • An engine absorbs 400 J of heat and does 120 J of work in each cycle. (a) What is its efficiency? (b) How much heat is rejected in each cycle?

7 • An engine absorbs 100 J and rejects 60 J in each cycle. (a) What is its efficiency? (b) If each cycle takes 0.5 s, find the power output of this engine in watts.

8 • A refrigerator absorbs 5 kJ of energy from a cold reservoir and rejects 8 kJ to a hot reservoir. (a) Find the coefficient of performance of the refrigerator. (b) The refrigerator is reversible and is run backward as a heat engine between the same two reservoirs. What is its efficiency?

9 •• An engine operates with 1 mol of an ideal gas for which $C_v = \frac{3}{2}R$ and $C_p = \frac{5}{2}R$ as its working substance. The cycle begins at $P_1 = 1$ atm and $V_1 = 24.6$ L. The gas is heated at constant volume to $P_2 = 2$ atm. It then expands at constant pressure until $V_2 = 49.2$ L. During these two steps, heat is absorbed by the gas. The gas is then cooled at constant volume until its pressure is again 1 atm. It is then compressed at constant pressure to its original state. During the last two steps, heat is rejected by the gas. All the steps are quasi-static and reversible. (a) Show this cycle on a PV diagram. Find the work done, the heat added, and the change in the internal energy of the gas for each step of the cycle. (b) Find the efficiency of the cycle.

10 •• An engine using 1 mol of a diatomic ideal gas performs a cycle consisting of three steps: (1) an adiabatic expansion from an initial pressure of 2.64 atm and an initial volume of 10 L to a pressure of 1 atm and a volume of 20 L, (2) a compression at constant pressure to its original volume of 10 L, and (3) heating at constant volume to its original pressure of 2.64 atm. Find the efficiency of this cycle.

11 •• An engine using 1 mol of an ideal gas initially at $V_1 = 24.6$ L and $T = 400$ K performs a cycle consisting of four steps: (1) an isothermal expansion at $T = 400$ K to twice its initial volume, (2) cooling at constant volume to $T = 300$ K, (3) an isothermal compression to its original volume, and (4) heating at constant volume to its original temperature of 400 K. Assume that $C_v = 21$ J/K. Sketch the cycle on a PV diagram and find its efficiency.

12 •• One mole of an ideal monatomic gas at an initial volume $V_1 = 25$ L follows the cycle shown in Figure 20-11. All the processes are quasi-static. Find (a) the temperature of each state of the cycle, (b) the heat flow for each part of the cycle, and (c) the efficiency of the cycle.

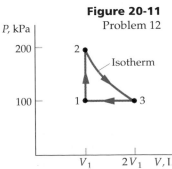

Figure 20-11
Problem 12

79 ••• (a) Show that if the refrigerator statement of the second law of thermodynamics were not true, the entropy of the universe could decrease. (b) Show that if the heat-engine statement of the second law were not true, the entropy of the universe could decrease. (c) An alternative statement of the second law is that the entropy of the universe cannot decrease. Have you just proved that this statement is equivalent to the refrigerator and heat-engine statements?

80 ••• Suppose that two heat engines are connected in series, such that the heat exhaust of the first engine is used as the heat input of the second engine as shown in Figure 20-22. The efficiencies of the engines are ε_1 and ε_2, respectively. Show that the net efficiency of the combination is given by

$$\varepsilon_{net} = \varepsilon_1 + (1 - \varepsilon_1)\varepsilon_2$$

Figure 20-22
Problems 80 and 81

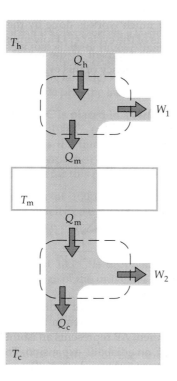

81 ••• Suppose that each engine in Figure 20-22 is an ideal reversible heat engine. Engine 1 operates between temperatures T_h and T_m and engine 2 operates between T_m and T_c, where $T_h > T_m > T_c$. Show that

$$\varepsilon_{net} = 1 - \frac{T_c}{T_h}$$

This means that two reversible heat engines in series are equivalent to one reversible heat engine operating between the hottest and coldest reservoirs.

82 ••• The cooling compartment of a refrigerator and its contents are at 5°C and have an average heat capacity of 84 kJ/K. The refrigerator exhausts heat to the room, which is at 25°C. What minimum power will be required by the motor that runs the refrigerator if the temperature of the cooling compartment and its contents is to be reduced by 1 C° in 1 min?

83 ••• An insulated container is separated into two chambers of equal volume by a thin partition. On one side of the container there are twelve ^{131}Xe atoms, and on the other side there are twelve ^{132}Xe atoms. The partition is then removed. Calculate the change in entropy of the system after equilibrium has been established (that is, when the ^{131}Xe and ^{132}Xe atoms are evenly distributed throughout the total volume).

CHAPTER 21

Thermal Properties and Processes

The temperature in the interior of a furnace is measured by a radiation thermometer.

When an object absorbs thermal energy, various changes in the physical properties of the object may occur. The temperature of the object may rise, accompanied by an expansion or contraction of the object, or the object may liquefy or vaporize, during which the temperature remains constant. In this chapter, we examine some of the thermal properties of matter and some important processes involving thermal energy.

21-1 Thermal Expansion

When the temperature of an object increases, the object usually expands. Consider a long rod of length L at a temperature T. When the temperature changes by ΔT, the change ΔL in length is proportional to ΔT and to the original length L:

$$\Delta L = \alpha L \, \Delta T \qquad\qquad 21\text{-}1$$

where α, called the **coefficient of linear expansion,** is the ratio of the fractional change in length to the change in temperature:

$$\alpha = \frac{\Delta L/L}{\Delta T} \qquad \text{21-2}$$

The units for the coefficient of linear expansion are reciprocal Celsius degrees (1/C°), which are the same as reciprocal kelvins (1/K). The value of α for a solid or liquid doesn't vary much with pressure, but it may vary significantly with temperature. Equation 21-2 gives the average value over the temperature interval ΔT. The coefficient of linear expansion at a particular temperature T is found by taking the limit as ΔT approaches zero:

$$\alpha = \lim_{\Delta T \to 0} \frac{\Delta L/L}{\Delta T} = \frac{1}{L}\frac{dL}{dT} \qquad \text{21-3}$$

The accuracy obtained by using the average value of α over a wide temperature range is sufficient for most purposes.

The **coefficient of volume expansion** β is similarly defined as the ratio of the fractional change in volume to the change in temperature (at constant pressure):

$$\beta = \lim_{\Delta T \to 0} \frac{\Delta V/V}{\Delta T} = \frac{1}{V}\frac{dV}{dT} \qquad \text{21-4}$$

Like α, β does not usually vary with pressure for solids and liquids, but may vary with temperature. Average values for α and β for various substances are given in Figure 21-1.

For a given material, $\beta = 3\alpha$. We can show this by considering a box of dimensions L_1, L_2, and L_3. Its volume at a temperature T is

$$V = L_1 L_2 L_3$$

The rate of change of the volume with respect to temperature is

$$\frac{dV}{dT} = L_1 L_2 \frac{dL_3}{dT} + L_1 L_3 \frac{dL_2}{dT} + L_2 L_3 \frac{dL_1}{dT}$$

Dividing each side of the equation by the volume, we obtain

$$\beta = \frac{1}{V}\frac{dV}{dT} = \frac{1}{L_3}\frac{dL_3}{dT} + \frac{1}{L_2}\frac{dL_2}{dT} + \frac{1}{L_1}\frac{dL_1}{dT}$$

Since each term on the right side of the above equation equals α, we have

$$\beta = 3\alpha \qquad \text{21-5}$$

Similarly, the coefficient of *area* expansion is twice that of *linear* expansion.

The increase in size of any part of a body for a given temperature change is proportional to the original size of that part of the body. Thus, if we increase the temperature of a steel ruler, for example, the effect will be similar to that of a (very) slight) photographic enlargement. The dimensions of the ruler itself will be larger, as will the distance between the equally spaced lines. If the

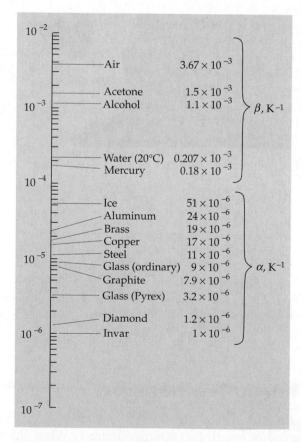

Figure 21-1 Approximate values of the coefficients of thermal expansion for various substances.

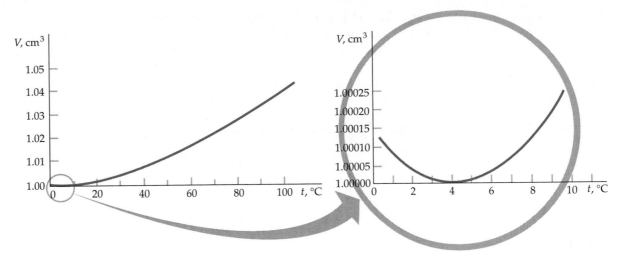

Figure 21-2 Volume of 1 g of water at atmospheric pressure versus temperature. The minimum volume, which corresponds to the maximum density, occurs at 4°C.

ruler has a hole in it, the hole will get larger, just as it would in a photographic enlargement.

Most materials expand when heated. Water, however, presents an important exception. Figure 21-2 shows the volume occupied by 1 g of water as a function of temperature. The volume is minimum, and therefore the density is maximum, at 4°C. Thus, when water is heated from temperatures below 4°C, it contracts rather than expands. This property has important consequences for the ecology of lakes. At temperatures above 4°C, the water in a lake becomes denser as it cools, and thus sinks to the bottom. But as the water cools below 4°C it becomes less dense, and rises to the surface. Ice therefore forms first on the surface of a lake and, being less dense than water, it remains there and acts as a thermal insulator for the water below. If water contracted when it froze, as most substances do, ice would sink and new water would be exposed at the surface to freeze. Lakes would fill with ice from the bottom up and would be much more likely to freeze completely in the winter, killing fish and other aquatic life.

Example 21-1

A steel bridge is 1000 m long. By how much does it expand when the temperature rises from 0 to 30°C?

Picture the Problem Use $\alpha = 11 \times 10^{-6} \, \text{K}^{-1}$ from Figure 21-1 and calculate ΔL from Equation 21-2.

The change in length for a 30-C° = 30-K change in temperature is the product of α, L, and ΔT:

$$\Delta L = \alpha L \, \Delta T = (11 \times 10^{-6} \text{K}^{-1})(1000 \text{ m})(30 \text{ K})$$
$$= 0.33 \text{ m} = 33 \text{ cm}$$

Expansion joints are included in bridges to relieve the enormous stresses that would occur without them. We can calculate the stress that would result in a steel bridge without expansion joints by using Young's modulus (Equation 12-7):

$$Y = \frac{\text{stress}}{\text{strain}} = \frac{F/A}{\Delta L/L}$$

Then

$$\frac{F}{A} = Y\frac{\Delta L}{L} = Y\alpha\,\Delta T$$

For $\Delta T = 30$ K, $\Delta L/L = 0.33$ m$/1000$ m as found in Example 21-1. Then using $Y = 2 \times 10^{11}$ N/m^2 (from Table 12-1),

$$\frac{F}{A} = Y\frac{\Delta L}{L} = (2 \times 10^{11}\,\text{N/m}^2)\,\frac{0.33\,\text{m}}{1000\,\text{m}} = 6.6 \times 10^7\,\text{N/m}^2$$

This stress is about one-third of the breaking stress for steel under compression. A compression stress of this magnitude would cause a steel bridge to buckle and become permanently deformed.

Example 21-2

Your 1-L glass flask is filled to the brim with alcohol at 10°C. If the temperature is raised to 30°C, how much alcohol spills out of the flask?

Picture the Problem The glass flask and the alcohol both expand when heated, but the alcohol expands more, so some spills out. We calculate the amount spilled by finding the changes in volume for $\Delta T = 20$ K using $\Delta V_a = \beta V\,\Delta T$ with $\beta = 1.1 \times 10^{-3}$ K^{-1} for alcohol (from Figure 21-1), and $\Delta V_g = \beta V\,\Delta T = 3\alpha V\,\Delta T$ with $\alpha = 9 \times 10^{-6}$ K^{-1} for glass. The difference in these volume changes equals the volume spilled.

1. The volume of alcohol spilled, ΔV_s, is the difference in the changes in volume of the alcohol and glass:

$$\Delta V_s = \Delta V_a - \Delta V_g$$

2. Find the increase in the volume of the alcohol:

$$\Delta V_a = \beta V\,\Delta T = (1.1 \times 10^{-3}\,\text{K}^{-1})(1\,\text{L})(20\,\text{K})$$
$$= 2.2 \times 10^{-2}\,\text{L} = 22.0\,\text{mL}$$

3. Find the increase in the volume of the glass flask:

$$\Delta V_g = \beta V\,\Delta T = 3\alpha V\,\Delta T$$
$$= 3(9 \times 10^{-6}\,\text{K}^{-1})(1\,\text{L})(20\,\text{K})$$
$$= 5.4 \times 10^{-4}\,\text{L} = 0.54\,\text{mL}$$

4. Subtract to find the amount of alcohol spilled:

$$\Delta V_s = \Delta V_a - \Delta V_g = 22.0\,\text{mL} - 0.54\,\text{mL} = 21.5\,\text{mL}$$

Example 21-3

A copper bar is heated to 300°C and is then clamped rigidly between two fixed points so that it can neither expand nor contract. If the breaking stress of copper is 230 MN/m^2, at what temperature will the bar break as it cools?

Picture the Problem As the bar cools, the change ΔL in length that *would* occur if the bar contracted is offset by an equal stretching which is due to tensile stress in the bar. The stress F/A is related to the stretching ΔL by $Y = (F/A)/(\Delta L/L)$, where Young's modulus for copper is $Y = 110$ GN/m^2 (from Table 12-1). The maximum allowable stretching occurs when F/A equals 230 MN/m^2. We thus find the temperature change that would produce this maximum contraction.

1. Calculate the change ΔL in length that would occur if the bar were unclamped and cooled by ΔT:

$$\Delta L = \alpha L\, \Delta T$$

2. A tensile stress F/A stretches the bar by the same amount ΔL found in step 1:

$$\Delta L = L\frac{F/A}{Y} = \alpha L\, \Delta T$$

3. Solve for ΔT and set the stress equal to the breaking value:

$$\Delta T = \frac{F/A}{\alpha Y} = \frac{230 \times 10^6 \text{ N/m}^2}{(17 \times 10^{-6}\,\text{K}^{-1})(110 \times 10^9 \text{ N/m}^2)}$$
$$= 123 \text{ K} = 123 \text{ C}°$$

4. Subtract this result from the original temperature to find the final temperature at which the bar breaks:

$$300°\text{C} - 123°\text{C} = 177°\text{C}$$

21-2   The van der Waals Equation and Liquid–Vapor Isotherms

Although most gases behave like an ideal gas at ordinary pressures, this ideal behavior breaks down when the pressure is high enough or the temperature is low enough that the density of the gas is high and the molecules are, on average, close together. An equation of state called the **van der Waals equation** describes the behavior of many real gases over a wide range of pressures more accurately than does the ideal-gas equation of state ($PV = nRT$). The van der Waals equation for n moles of gas is

$$\left(P + \frac{an^2}{V^2}\right)(V - bn) = nRT \qquad\qquad 21\text{-}6$$

The van der Waals equation of state

The constant b in this equation arises because the gas molecules are not point particles but objects that have a finite size; therefore, the volume available to each molecule is reduced. The magnitude of b is the volume of one mole of gas molecules. The term an^2/V^2 arises from the attraction of the gas molecules for each other. As a molecule approaches the wall of the container, it is pulled back by the molecules surrounding it with a force that is proportional to the density of those molecules, n/V. Since the number of molecules that hit the wall in a given time is also proportional to the density of the molecules, the decrease in pressure due to the attraction of the molecules for each other is proportional to the square of the density and therefore to n^2/V^2. The constant a depends on the gas and is small for inert gases, which have very weak chemical interactions. The terms bn and an^2/V^2 are both negligible when the volume V is large, so at low densities the van der Waals equation approaches the ideal-gas law. At high densities the van der Waals equation provides a much better description than the ideal-gas law of the behavior of real gases.

Figure 21-3 shows PV isothermal curves for a real substance at various temperatures. For temperatures above some critical temperature T_c, these curves are described quite accurately by the van der Waals equation and can be used to determine the constants a and b. For example, the values of these constants that give the best fit to the experimental curves for nitrogen are $a = 0.14$ Pa·m^6/mol^2 and $b = 39.1$ cm^3/mol. This volume of 39.1 cm^3 per mole is about 0.2% of the volume of 22,400 cm^3 occupied by 1 mol of nitrogen under

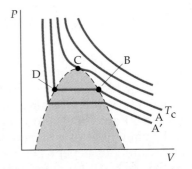

Figure 21-3 Isotherms on the PV diagram for a real substance. For temperatures above the critical temperature T_c, the substance remains a gas at all pressures and is described by the van der Waals equation. The pressure for the horizontal portions of the curves in the shaded region is the vapor pressure, which is the pressure at which the vapor and liquid are in equilibrium. To the left of the shaded region for temperatures below the critical temperature, the substance is a liquid and is nearly incompressible.

standard conditions. Since the molar mass of nitrogen is 28 g/mol, if 1 mol of nitrogen molecules were packed into a volume of 39.1 cm³, the density would be

$$\rho = \frac{M}{V} = \frac{28 \text{ g}}{39.1 \text{ cm}^3} = 0.72 \text{ g/cm}^3$$

which compares favorably with the density of liquid nitrogen, 0.80 g/cm³.

The value of the constant b can be used to estimate the size of a molecule. Since 1 mol = N_A molecules of nitrogen has a volume of 39.1 cm³, the volume of one nitrogen molecule is

$$V = \frac{b}{N_A} = \frac{39.1 \text{ cm}^3/\text{mol}}{6.02 \times 10^{23} \text{ molecules/mol}}$$

$$= 6.50 \times 10^{-23} \text{ cm}^3/\text{molecule}$$

If we assume that each molecule is a sphere of diameter d occupying a cubic volume of side d, we obtain

$$d^3 = 6.50 \times 10^{-23} \text{ cm}^3$$

or

$$d = 4.0 \times 10^{-8} \text{ cm} = 4.0 \times 10^{-10} \text{ m}$$

which is a reasonable estimate for the diameter of a molecule.

At temperatures below T_c, the van der Waals equation describes those portions of the isotherms outside the shaded region in Figure 21-3 but not those portions inside the shaded region. Suppose we have a gas at a temperature below T_c that initially has a low pressure and a large volume. We begin to compress the gas while holding the temperature constant (isotherm A in the figure). At first the pressure rises, but when we reach point B on the dashed curve, the pressure ceases to rise and the gas begins to liquefy at constant pressure. Along the horizontal line BD in the figure, the gas and liquid are in equilibrium. As we continue to compress the gas, more and more gas liquefies until at point D on the dashed curve we have only liquid. Then, if we try to compress the substance further, the pressure rises sharply because a liquid is nearly incompressible.

Now consider putting a liquid such as water in a sealed evacuated container. As some of the water evaporates, water-vapor molecules fill the previously empty space in the container. Some of these molecules will hit the liquid surface and rejoin the liquid water in the process called condensation. Initially, the rate of evaporation will be greater than the rate of condensation, but eventually equilibrium will be reached. The pressure at which a liquid is in equilibrium with its own vapor is called the **vapor pressure.** If we now heat the container slightly, the liquid boils, more liquid will evaporate, and a new equilibrium will be established at a higher vapor pressure. Vapor pressure thus depends on the temperature. We can see this from Figure 21-3. If we had started compressing the gas at a lower temperature, as with isotherm A′ in Figure 21-3, the vapor pressure would be lower, as is indicated by the horizontal constant-pressure line for A′ at a lower value of pressure. The temperature for which the vapor pressure for a substance equals 1 atm is the **normal boiling point** of that substance. For example, the temperature at which the vapor pressure of water is 1 atm is 373 K = 100°C, so this temperature is the normal boiling point of water. At high altitudes, such as on the top of a mountain, the pressure is less than 1 atm, and water then boils at a temperature lower than 373 K. Figure 21-4 gives the vapor pressures of water at various temperatures.

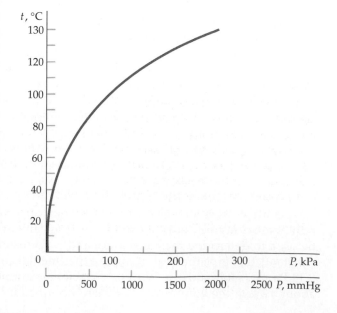

Figure 21-4 Vapor pressure of water versus temperature.

At temperatures greater than the critical temperature T_c, a gas will not liquefy at any pressure. The critical temperature for water vapor is 647 K = 374°C. The point at which the critical isotherm intersects the dashed curve in Figure 21-3 (point C) is called the critical point.

21-3 Phase Diagrams

Figure 21-5 is a plot of pressure versus temperature at a constant volume for water. Such a plot is called a **phase diagram.** The portion of the diagram between points O and C shows the vapor pressure versus the temperature. As we continue to heat the container, the density of the liquid decreases and the density of the vapor increases. At point C on the diagram, these densities are equal. Point C is called the **critical point.** At this point and above it, there is no distinction between the liquid and the gas.* Critical-point temperatures T_c for various substances are listed in Figure 21-6. At temperatures greater than the critical temperature a gas will not liquefy at any pressure.

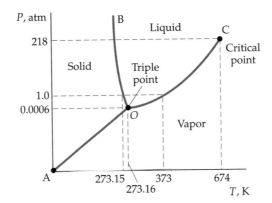

Figure 21-5 Phase diagram for water. The pressure and temperature scales are not linear but are compressed to show the interesting points. Curve OC is the curve of vapor pressure versus temperature. Curve OB is the melting curve, and curve OA is the sublimation curve.

If we now cool our container, some of the vapor condenses into a liquid as we move back down the curve OC until the substance reaches point O in Figure 21-5. At this point, the liquid begins to solidify. Point O is the triple point, that one point at which the vapor, liquid, and solid phases of a substance can coexist in equilibrium. Every substance has a unique triple point at a specific temperature and pressure. The triple-point temperature for water is 273.16 K = 0.01°C and the triple-point pressure is 4.58 mmHg.

At temperatures and pressures below the triple point, the liquid cannot exist. The curve OA in the phase diagram of Figure 21-5 is the locus of pressures and temperatures for which the solid and vapor coexist in equilibrium. The direct change from a solid to a vapor is called **sublimation.** You can observe sublimation if you put ice cubes in the freezer compartment of a refrigerator (especially a self-defrosting refrigerator). The ice cubes will eventually disappear due to sublimation. Because atmospheric pressure is well above the triple-point pressure of water, equilibrium is never established between the ice and water vapor. The triple-point temperature and pressure of carbon dioxide (CO_2) are 216.55 K and 3880 mmHg, which means that liquid CO_2 can exist only at pressures above 3880 mmHg = 5.1 atm. Thus, at ordinary atmospheric pressures, liquid carbon dioxide cannot exist at any temperature. When solid carbon dioxide "melts," it sublimates directly into gaseous CO_2 without going through the liquid phase, hence the name "dry ice."

The curve OB in Figure 21-5 is the melting curve separating the liquid and solid phases. For a substance like water for which the melting temperature decreases as the pressure increases, curve OB slopes upward to the left from the triple point, as in this figure. For most other substances, the melting temperature increases as the pressure increases. For such a substance, curve OB slopes upward to the right from the triple point.

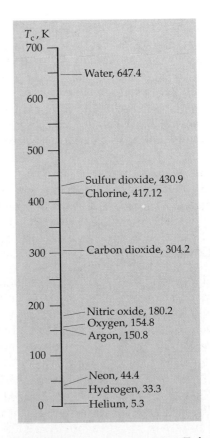

Figure 21-6 Critical temperatures T_c for various substances.

---

* Often the word "vapor" is used if the temperature is below the critical temperature, and the word "gas" is used if the temperature is above the critical temperature, though there is no need for such a distinction.

For a molecule to escape from a liquid, energy is required to break the molecular bonds at the liquid's surface. Vaporization therefore cools the liquid left behind. If water is brought to a boil over heat, the cooling effect keeps the temperature of the liquid constant at the boiling point. This is the reason that the boiling point of a substance can be used to calibrate thermometers. However, water can also be caused to boil without adding heat by evacuating the air above it, thereby lowering the applied pressure. The energy needed for vaporization is then taken from the liquid left behind. As a result, the liquid will cool down, even to the point that ice forms on top of the boiling water.

21-4   The Transfer of Thermal Energy

Thermal energy is transferred from one place to another by three processes: conduction, convection, and radiation. In **conduction,** thermal energy is transferred by interactions among atoms or molecules, though there is no transport of the atoms or molecules themselves. For example, if one end of a solid bar is heated, the atoms in the heated end vibrate with greater energy than those at the cooler end. Because of the interaction of the more energetic atoms with their neighbors, this energy is transported along the bar.*

In **convection,** heat is transported by direct mass transport. For example, warm air in a room expands and rises because of its lower density. Thermal energy is thus transported upward along with the mass of warm air.

In **radiation**, thermal energy is transported through space in the form of electromagnetic waves that move at the speed of light. Thermal radiation, light waves, radio waves, television waves, and X rays are all forms of electromagnetic radiation that differ from one another only in their wavelengths and frequencies.

In all mechanisms of heat transfer, the rate of cooling of a body is approximately proportional to the temperature difference between the body and its surroundings. This result is known as **Newton's law of cooling.**

In many real situations, all three mechanisms for heat transfer occur simultaneously, though one may be more effective than the others. For example, an ordinary space heater uses both radiation and convection. If the heating element is quartz, the main mechanism of heat transference is radiation. If the heating element is metal, which does not radiate as efficiently as quartz, convection is the main mechanism by which heat is transmitted, with the heated air rising to be replaced by cooler air. Often a fan is included in heaters with hot elements to speed the convection process.

Conduction

Figure 21-7 shows a uniform solid bar of cross-sectional area A. If we keep one end of the bar at a high temperature and the other end at a low temperature, thermal energy is conducted down the bar from the hot end to the cold end. In the steady state, the temperature varies uniformly from the hot end to the cold end. The rate of change of the temperature along the bar, $\Delta T/\Delta x$, is called a **temperature gradient**.

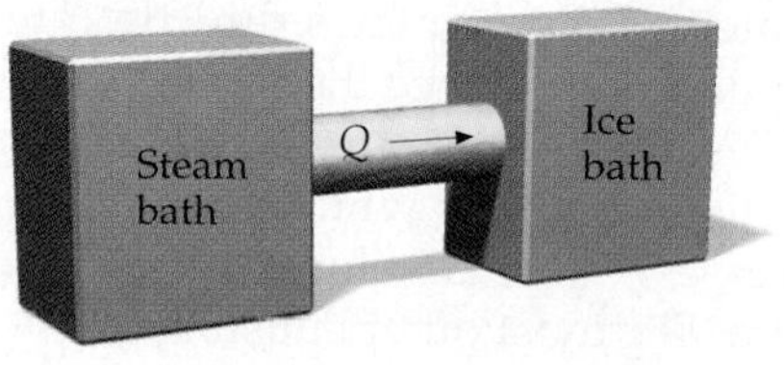

(a)

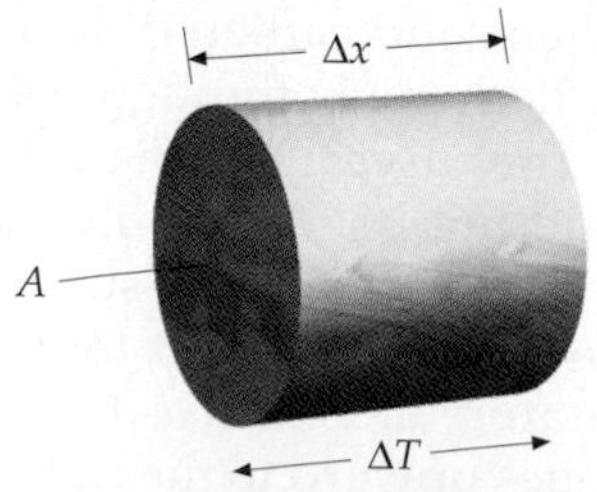

(b)

$$\frac{\Delta Q}{\Delta t} = kA\frac{\Delta T}{\Delta x}$$

Figure 21-7 (*a*) A conducting bar with its ends at two different temperatures. (*b*) A segment of the bar of length Δx. The rate at which thermal energy is conducted across the segment is proportional to the cross-sectional area and the temperature difference and is inversely proportional to the thickness of the segment.

---

* If the solid is a metal, the transport of thermal energy is helped by free electrons, which move throughout the metal.

Let ΔT be the temperature difference across a small segment of thickness Δx (Figure 21-7b). If ΔQ is the amount of thermal energy conducted through the segment in some time Δt, the rate of conduction of thermal energy, $\Delta Q/\Delta t$, is called the thermal current I. Experimentally, it is found that the thermal current is proportional to the temperature gradient* and to the cross-sectional area A:

$$I = \frac{\Delta Q}{\Delta t} = kA\frac{\Delta T}{\Delta x} \qquad \text{21-7}$$

Thermal current

The proportionality constant k, called the **thermal conductivity**, depends on the composition of the bar.† In SI units, thermal current is expressed in watts, and the thermal conductivity has units of watts per meter-kelvin.‡ In practical calculations in the United States, the thermal current is usually expressed in Btu per hour, the area in square feet, the thickness in inches, and the temperature in Fahrenheit degrees. The thermal conductivity is then given in Btu·in/h·ft²·F°. Figure 21-8 lists thermal conductivities for various materials in both SI and U.S. customary units.

If we solve Equation 21-7 for the temperature difference, we obtain

$$\Delta T = I\frac{\Delta x}{kA} \qquad \text{21-8}$$

or

$$\Delta T = IR \qquad \text{21-9}$$

Temperature change versus current

where $\Delta x/kA$ is the **thermal resistance** R:

$$R = \frac{\Delta x}{kA} \qquad \text{21-10}$$

Definition—Thermal resistance

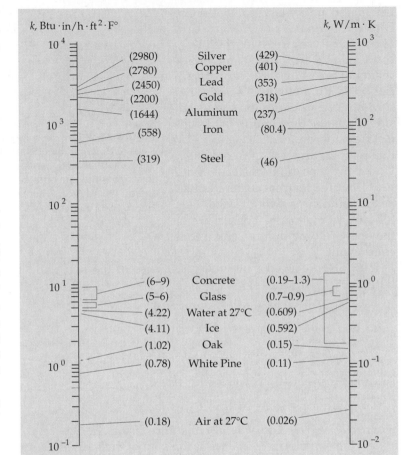

Figure 21-8 Thermal conductivities k for various materials.

Exercise Calculate the thermal resistance of an aluminum slab of cross-sectional area 15 cm² and thickness 2 cm. (*Answer*   0.0563 K/W)

Exercise What thickness of silver would be required to give the same thermal resistance as a 1-cm thickness of air of the same area? (*Answer* $\Delta x = (1\text{ cm})(429)/(0.026) = 16{,}500$ cm $= 165$ m)

In many practical problems, we are interested in the flow of heat through two or more conductors (or insulators) in series. For example, we may wish to know the effect of adding insulating material of a certain thickness and

---

* The thermal current is in the direction of the decrease in temperature.

† Don't confuse the thermal conductivity with Boltzmann's constant, which is also designated by k.

‡ In some tables, the energy may be given in calories or kilocalories and the thickness in centimeters.

thermal conductivity to the space between two layers of plasterboard. Figure 21-9 shows two thermally conducting slabs of the same cross-sectional area but of different materials and different thicknesses. Let T_1 be the temperature on the warm side, T_2 be the temperature at the interface between the slabs, and T_3 be the temperature on the cool side. Under the conditions of steady-state heat flow, the thermal current I must be the same through both slabs. This follows from energy conservation; the energy going in must equal that coming out.

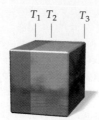

$T_1\ T_2\quad T_3$

Figure 21-9 Two thermally conducting slabs of different materials in series. The equivalent thermal resistance of the slabs in series is the sum of their individual thermal resistances. The thermal current is the same through both slabs.

If R_1 and R_2 are the thermal resistances of the two slabs, we have from Equation 21-9 for each slab,

$$T_1 - T_2 = IR_1$$

and

$$T_2 - T_3 = IR_2$$

Adding these equations gives

$$\Delta T = T_1 - T_3 = I(R_1 + R_2) = IR_{eq} \qquad \text{21-11}$$

where R_{eq} is the **equivalent resistance.** Thus, for thermal resistances in series, the equivalent resistance is the sum of the individual resistances:

$$R_{eq} = R_1 + R_2 + \cdots \qquad \text{21-12}$$

Thermal resistances in series

This result can be applied to any number of resistances in series. In Chapter 26 we will find that the same formula applies to electrical resistances in series.

To calculate the amount of heat leaving a room by conduction in a given time, we need to know how much leaves through the walls, the windows, the floor, and the ceiling. For this type of problem, in which there are several paths for heat flow, the resistances are said to be in parallel. The temperature difference is the same for each path, but the thermal current is different. The total thermal current is the sum of the thermal currents through each of the independent or parallel paths:

$$I_{total} = I_1 + I_2 + \cdots = \frac{\Delta T}{R_1} + \frac{\Delta T}{R_2} + \cdots = \Delta T\left(\frac{1}{R_1} + \frac{1}{R_2} + \cdots\right)$$

or

$$I_{total} = \frac{\Delta T}{R_{eq}} \qquad \text{21-13}$$

where the equivalent thermal resistance is given by

$$\frac{1}{R_{eq}} = \frac{1}{R_1} + \frac{1}{R_2} + \cdots \qquad \text{21-14}$$

Thermal resistances in parallel

We will encounter this equation again when we study electrical conduction through parallel resistances. Note that I is proportional to ΔT, in agreement with Newton's law of cooling.

Example 21-4   *try it yourself*

Two metal bars, each of length 5 cm and rectangular cross section with sides 2 cm and 3 cm, are wedged between two walls, one held at 100°C and the other at 0°C (Figure 21-10). The bars are lead and silver. Find (*a*) the total thermal current through the bars, and (*b*) the temperature at the interface.

Picture the Problem (*a*) You can find the total thermal current I from $\Delta T = IR_{eq}$. The equivalent resistance R_{eq} is the sum of the individual resistances, which you can find from the thermal conductivities given in Figure 21-8. (*b*) You can find the temperature at the interface by applying $\Delta T = IR_1$ to the first cube only, and solving for ΔT in terms of the value for I found in (*a*).

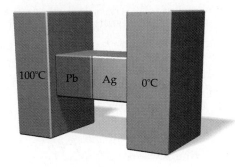

Figure 21-10

Cover the column to the right and try these on your own before looking at the answers.

Steps   **Answers**

(*a*)1. Write the equivalent thermal resistance in terms of the $R_{eq} = R_{Pb} + R_{Ag}$
       thermal resistances of the two cubes.

   2. Calculate the thermal resistance from its definition for $R_{Pb} = 0.236 \text{ K/W};\qquad R_{Ag} = 0.194 \text{ K/W}$
      each cube.

   3. Find the equivalent thermal resistance from your results $R_{eq} = 0.430 \text{ K/W}$
      in step 2.

   4. Substitute R_{eq} and $\Delta T = 100$ K into Equation 21-13 to find $I = 232.6 \text{ W}$
      the thermal current.

(*b*)1. Calculate the temperature difference across the lead cube $\Delta T_{Pb} = IR_{Pb} = 54.9 \text{ K} = 54.9 \text{ C}°$
       using the current and thermal resistance found in (*a*).

   2. Use your result in the previous step to find the tempera- $T_{if} = 100° \text{ C} - \Delta T_{Pb} = 45.1°\text{C}$
      ture at the interface.

   3. Check your answer in (*b*)1 by finding the temperature dif- $\Delta T_{Ag} = IR_{Ag} = 45.1 \text{ C}°$
      ference across the silver cube.

Example 21-5

The metal bars in Example 21-4 are rearranged as shown in Figure 21-11. Find (*a*) the thermal current in each bar, (*b*) the total thermal current, and (*c*) the equivalent thermal resistance of the two-bar system.

Picture the Problem The current in each bar is found from $I = \Delta T/R$, where R is the thermal resistance of the bar found in Example 21-4. The total current is the sum of the currents. The equivalent resistance can be found from Equation 21-14 or from $I_{total} = \Delta T/R_{eq}$.

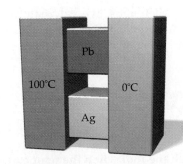

Figure 21-11

(*a*)   Calculate the thermal current for each cube: $I_{Pb} = \dfrac{\Delta T}{R_{Pb}} = \dfrac{100 \text{ K}}{0.236 \text{ K/W}} = 424 \text{ W}$

$$I_{Ag} = \dfrac{\Delta T}{R_{Ag}} = \dfrac{100 \text{ K}}{0.194 \text{ K/W}} = 515 \text{ W}$$

(b)   Add these results to find the total current:

$I_{total} = I_{Pb} + I_{Ag} = 424\ W + 515\ W = 939\ W$

(c) 1. Use Equation 21-14 to calculate the equivalent resistance of the two cubes in parallel:

$$\frac{1}{R_{eq}} = \frac{1}{R_{Pb}} + \frac{1}{R_{Ag}}$$

$$R_{eq} = \frac{R_{Pb}R_{Ag}}{R_{Pb} + R_{Ag}} = \frac{(0.236)(0.194)}{0.236 + 0.194} = 0.106\ K/W$$

2. Check the result, using $I_{total} = \Delta T/R_{eq}$:

$$I_{total} = \frac{\Delta T}{R_{eq}}; \qquad R_{eq} = \frac{\Delta T}{I_{total}} = \frac{100\ K}{939\ W} = 0.106\ K/W$$

Remark Note that the equivalent resistance is less than either of the individual resistances. This is always the case for parallel resistors.

In the building industry, the thermal resistance in U.S. customary units for a square foot of material is called the **R factor,** R_f. The R factor is simply the thickness of the material divided by its thermal conductivity:

$$R_f = \frac{\Delta x}{k} = RA \qquad \text{21-15}$$

Definition—R factor

Table 21-1 lists R factors for several materials. In terms of the R factor, Equation 21-9 for the thermal current is

$$\Delta T = IR = \frac{I}{A}R_f \qquad \text{21-16}$$

For slabs of insulating material of the same area in series, R_f is replaced by the equivalent R factor $R_{f,eq}$

$$R_{f,eq} = R_{f1} + R_{f2} + \cdots$$

For parallel slabs, we calculate the thermal current through each slab and add to obtain the total current.

Table 21-1

R Factors $\Delta x/k$ for Various Building Materials

Material	Thickness, in	R_f, h·ft^2·F°/Btu
Building board		
Gypsum or plasterboard	0.375	0.32
Plywood (Douglas fir)	0.5	0.62
Plywood or wood panels	0.75	0.93
Particle board, medium density	1.0	1.06
Finish flooring materials		
Carpet and fibrous pad	1.0	2.08
Tile		0.5
Wood, hardwood finish	0.75	0.68
Roof insulation	1.0	2.8
Roofing		
Asphalt roll roofing		0.15
Asphalt shingles		0.44
Windows		
Single-pane		0.9
Double-pane		1.8

Example 21-6

A 60 × 20-ft roof is made of 1-in pine board covered with asphalt shingles. (a) If the overlap in the shingles is neglected, at what rate is heat conducted through the roof when the temperature inside is 70°F and the temperature outside is 40°F? (b) Find the rate at which heat is conducted if 2 in of roof insulation is added.

Picture the Problem (a) The rate at which heat is lost, I, is given by $I = \Delta T/R = A\ \Delta T/R_{eq,f}$. For materials in series, the R factors add. R_f for pine board is found from its thermal conductivity, given in Figure 21-8. The R_f for asphalt shingles is given in Table 21-1. (b) The heat loss will be reduced by the ratio of the new R factor to the old R factor.

(a)1. The rate of heat loss is the thermal current: $I = \dfrac{A\,\Delta T}{R_{f,eq}}$

2. The equivalent R factor is the sum of the individual R factors: $R_{f,eq} = R_{f,p} + R_{f,a}$

3. We find the R factor for 1-in pine board using k from Figure 21-8:

$$R_{f,p} = \frac{\Delta x}{k} = \frac{1\text{ in}}{0.78\text{ Btu}\cdot\text{in/h}\cdot\text{ft}^2\cdot\text{F}^\circ}$$

$$= 1.28\,\frac{\text{h}\cdot\text{ft}^2\cdot\text{F}^\circ}{\text{Btu}}$$

4. We find the R factor for asphalt shingles from Table 21-1: $R_{f,a} = 0.44\text{ h}\cdot\text{ft}^2\cdot\text{F}^\circ/\text{Btu}$

5. The equivalent R factor is the sum of $R_{f,p}$ and $R_{f,a}$:

$$R_{f,eq} = R_{f,p} + R_{f,a} = (1.28 + 0.44)\text{h}\cdot\text{ft}^2\cdot\text{F}^\circ/\text{Btu}$$

$$= 1.72\text{ h}\cdot\text{ft}^2\cdot\text{F}^\circ/\text{Btu}$$

6. Calculate the area and temperature: $A = 60\text{ ft} \times 20\text{ ft} = 1200\text{ ft}^2$

and

$$\Delta T = 70^\circ\text{F} - 40^\circ\text{F} = 30\text{ F}^\circ$$

7. Substitute to find the thermal current:

$$I = \frac{A\Delta T}{R_{f,eq}} = \frac{(1200\text{ ft}^2)(30\text{ F}^\circ)}{1.72\text{ h}\cdot\text{ft}^2\cdot\text{F}^\circ/\text{Btu}} = 20{,}900\text{ Btu/h}$$

(b)1. The R factor for roof insulation, given in Table 21-1, is 2.8 h·ft²·F°/Btu for 1 in, twice that for 2 in:

$$R_{f,in} = 2 \times 2.8\text{ h}\cdot\text{ft}^2\cdot\text{F}^\circ/\text{Btu}$$

$$= 5.6\text{ h}\cdot\text{ft}^2\cdot\text{F}^\circ/\text{Btu}$$

2. Add the R factor for the insulation to that for the roof to find the new equivalent R factor:

$$R_{f,eq} = 1.72 + 5.6 = 7.32\text{ h}\cdot\text{ft}^2\cdot\text{F}^\circ/\text{Btu}$$

3. Since the ratio of R factors in (b) and (a) is $7.32/1.72 = 4.26$, the rate of heat loss will be reduced by a factor of 4.26:

$$I_b = I_a\frac{R_{f,eq\,a}}{R_{f,eq\,a}} = (20{,}900\text{ Btu/h})\frac{1.72\text{ h}\cdot\text{ft}^2\cdot\text{F}^\circ/\text{Btu}}{7.32\text{ h}\cdot\text{ft}^2\cdot\text{F}^\circ/\text{Btu}}$$

$$= \frac{20{,}900\text{ Btu/h}}{4.26} = 4910\text{ Btu/h}$$

The thermal conductivity of air is very small compared with that of solid materials, so air is a very good insulator. However, when there is a large air gap—say, between a storm window and the inside window—the insulating efficiency of air is greatly reduced because of convection. Whenever there is a temperature difference between different parts of the air space, convection currents act quickly to equalize the temperature, so the effective conductivity is greatly increased. For storm windows, air gaps of about 1 to 2 cm are optimal. Wider air gaps actually reduce the thermal resistance of a double-pane window because of convection. The insulating properties of air are used most effectively when the air is trapped in small pockets that prevent convection from taking place. This is the principle underlying the excellent insulating properties of both goose down and Styrofoam.

If you touch the inside surface of a glass window when it is cold outside, you will observe that the surface is considerably colder than the inside air. The thermal resistance of windows is due mainly to thin films of insulating air that adhere to either side of the glass surface. The thickness of the glass

optional

R factor	The R factor is the thermal resistance for a unit area of a slab of material: $$R_f = \frac{\Delta x}{k} = RA \qquad \textbf{21-15}$$

7. Thermal Radiation

Rate of power radiated	$$P_r = e\sigma AT^4 \qquad \textbf{21-17}$$ where $\sigma = 5.6703 \times 10^{-8}\,\text{W}/\text{m}^2\cdot\text{K}^4$ is Stefan's constant, and e is the emissivity, a number between 0 and 1 that depends on the composition of the surface of the object. Materials that are good heat absorbers are good heat radiators.
Net power radiated by an object at T to its environment at T_0	$$P_{net} = e\sigma A(T^4 - T_0^4) \qquad \textbf{21-20}$$
Blackbody	A blackbody has an emissivity of 1. It is a perfect radiator, and it absorbs all the radiation incident upon it.
Wien's law	The power spectrum of electromagnetic energy radiated by a blackbody has a maximum at a wavelength λ_{max}, which varies inversely with the absolute temperature of the body: $$\lambda_{max} = \frac{2.898\ \text{mm}\cdot\text{K}}{T} \qquad \textbf{21-21}$$

Problem-Solving Guide

Summary of Worked Examples

Type of Calculation	Procedure and Relevant Examples
1. Thermal Expansion	
Find the expansion due to temperature change.	Use definitions of coefficients α or β, and values from tables. **Examples 21-1, 21-2**
Find the breaking temperature of an object that is constrained	Find ΔL for which stress equals breaking stress. Then find ΔT that would give ΔL if expansion were allowed. **Example 21-3**
2. Thermal Conduction	
Find the thermal current through series conductors.	Use $\Delta T = IR$ with $R = R_1 + R_2 + \cdots$. **Examples 21-4, 21-6**
Find the thermal current through parallel resistors.	Use $I_{total} = \dfrac{\Delta T}{R_{eq}}$ with $\dfrac{1}{R_{eq}} = \dfrac{1}{R_1} + \dfrac{1}{R_2} + \cdots$. **Example 21-5**
Find the temperature at the interface of two series conductors.	First find the total current, then use $\Delta T_1 = IR_1$. **Example 21-4**
3. Blackbody Radiation	
Find the peak wavelength of a blackbody spectrum.	Use $\lambda_{max} = \dfrac{2.898\ \text{mm}\cdot\text{K}}{T}$. **Example 21-7**
Find the net loss in radiated energy of a body.	Use $P_{net} = e\sigma A(T^4 - T_0^4)$. **Example 21-8**

Problems

In a few problems, you are given more data than you actually need; in a few other problems, you are required to supply data from your general knowledge, outside sources, or informed estimates.

Thermal Expansion

1 • Why does the mercury level first decrease slightly when a thermometer is placed in warm water?

2 • A large sheet of metal has a hole cut in the middle of it. When the sheet is heated, the area of the hole will

(a) not change.
(b) always increase.
(c) always decrease.
(d) increase if the hole is not in the exact center of the sheet.
(e) decrease only if the hole is in the exact center of the sheet.

3 • A steel ruler has a length of 30 cm at 20°C. What is its length at 100°C?

4 • A bridge 100 m long is built of steel. If it is built as a single, continuous structure, how much will its length change from the coldest winter days (−30°C) to the hottest summer days (40°C)?

5 •• (a) Define a coefficient of area expansion. (b) Calculate it for a square and a circle, and show that it is 2 times the coefficient of linear expansion.

6 •• The density of aluminum is 2.70×10^3 kg/m^3 at 0°C. What is the density of aluminum at 200°C?

7 •• A copper collar is to fit tightly about a steel shaft whose diameter is 6.0000 cm at 20°C. The inside diameter of the copper collar at that temperature is 5.9800 cm. To what temperature must the copper collar be raised so that it will just slip on the steel shaft, assuming that the steel shaft remains at 20°C?

8 •• Repeat Problem 7 when the temperature of both the steel shaft and copper collar are raised simultaneously.

9 •• A container is filled to the brim with 1.4 L of mercury at 20°C. When the temperature of container and mercury is raised to 60°C, 7.5 mL of mercury spill over the brim of the container. Determine the linear expansion coefficient of the container.

10 •• A hole is drilled in an aluminum plate with a steel drill bit whose diameter at 20°C is 6.245 cm. In the process of drilling, the temperature of the drill bit and of the aluminum plate rise to 168°C. What is the diameter of the hole in the aluminum plate when it has cooled to room temperature?

11 •• Len sells trees that double in price when they are over 2.00 m high. To make a standard, he cuts an aluminum rod 2.00 m in length, as measured by a steel measuring tape. That day, the temperature of both the rod and the tape is 25°C. What will the tape indicate the length of the rod to be when both the tape and the rod are at (a) 0°C and (b) 50°C?

12 •• A rookie crew was left to put in the final 1 km of rail for a stretch of railroad track. When they finished, the temperature was 20°C, and they headed to town for some refreshments with their coworkers. After an hour or two, one of the old-timers noticed that the temperature had gone up to 25°C, so he said, "I hope you left some gaps to allow for expansion." By the look on their faces, he knew that they had not, and they all rushed back to the work site. The rail had buckled into an isosceles triangle. How high was the buckle?

13 •• A car has a 60-L steel gas tank filled to the top with gasoline when the temperature is 10°C. The coefficient of volume expansion of gasoline is $\beta = 0.900 \times 10^{-3}$ K^{-1}. Taking the expansion of the steel tank into account, how much gasoline spills out of the tank when the car is parked in the sun and its temperature rises to 25°C?

14 •• A thermometer has an ordinary glass bulb and thin glass tube filled with 1 mL of mercury. A temperature change of 1 C° changes the level of mercury in the thin tube by 3.0 mm. Find the inside diameter of the thin glass tube.

15 •• A mercury thermometer consists of a 0.4-mm capillary tube connected to a glass bulb. The mercury level rises 7.5 cm as the temperature of the thermometer increases from 35 to 43°C. Find the volume of the thermometer bulb.

16 ••• A grandfather's clock is calibrated at a temperature of 20°C. (a) On a hot day, when the temperature is 30°C, does the clock run fast or slow? (b) How much does it gain or lose in a 24-h period? Assume that the pendulum is a thin brass rod of negligible mass with a heavy bob attached to the end.

17 ••• A steel tube has an outside diameter of 3.000 cm at room temperature (20°C). A brass tube has an inside diameter of 2.997 cm at the same temperature. To what temperature must the ends of the tubes be heated if the steel tube is to be inserted into the brass tube?

18 ••• What is the tensile stress in the copper collar of Problem 7 when its temperature returns to 20°C?

The van der Waals Equation, Liquid–Vapor Isotherms, and Phase Diagrams

19 • Mountaineers say that you cannot hard boil an egg on the top of Mount Rainier. This is true because

(a) the air is too cold to boil water.
(b) the air pressure is too low for stoves to burn.
(c) boiling water is not hot enough to hard boil the egg.
(d) the oxygen content of the air is too low.
(e) the eggs always break in their backpacks.

20 • Which gases in Figure 21-6 cannot be liquefied by applying pressure at 20°C?

21 •• The phase diagram in Figure 21-14 can be interpreted to yield information on how the boiling and melting points of water change with altitude. (*a*) Explain how this information can be obtained. (*b*) How might this information affect cooking procedures in the mountains?

Figure 21-14 Problems 21 and 22

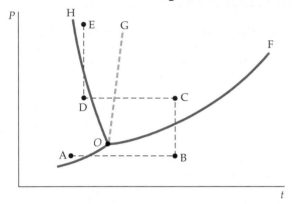

22 •• For the phase diagram given in Figure 21-14, state what changes (if any) occur for each line segment—AB, BC, CD, and DE—in (*a*) volume and (*b*) phase. (*c*) For what type of substance would *OH* be replaced by *OG*? (*d*) What is the significance of point F?

23 • (*a*) Calculate the volume of 1 mol of steam at 100°C and a pressure of 1 atm, assuming that it is an ideal gas. (*b*) Find the temperature at which the steam will occupy the volume found in part (*a*) if it obeys the van der Waals equation with $a = 0.55$ Pa·m^6/mol^2 and $b = 30$ cm^3/mol.

24 •• From Figure 21-4, find (*a*) the temperature at which water boils on a mountain where the atmospheric pressure is 70 kPa, (*b*) the temperature at which water will boil in a container in which the pressure has been reduced to 0.5 atm, and (*c*) the pressure at which water will boil at 115°C.

25 •• The van der Waals constants for helium are $a = 0.03412$ L^2·atm/mol^2 and $b = 0.0237$ L/mol. Use these data to find the volume in cubic centimeters occupied by one helium atom and to estimate the radius of the atom.

26 ••• (*a*) For a van der Waals gas, show that the critical temperature is $8a/27Rb$ and the critical pressure is $a/27b^2$. (*b*) Rewrite the van der Waals equation of state in terms of the reduced variable $V_r = V/V_c$, $P_r = P/P_c$, and $T_r = T/T_c$.

Heat Conduction

27 • A copper bar 2 m long has a circular cross section of radius 1 cm. One end is kept at 100°C and the other end is kept at 0°C. The surface of the bar is insulated so that there is negligible heat loss through it. Find (*a*) the thermal resistance of the bar, (*b*) the thermal current *I*, (*c*) the temperature gradient $\Delta T/\Delta x$, and (*d*) the temperature of the bar 25 cm from the hot end.

28 • A 20 × 30-ft slab of insulation has an *R* factor of 11. How much heat (in Btu per hour) is conducted through the slab if the temperature on one side is 68°F and that on the other side is 30°F?

29 •• Two metal cubes with 3-cm edges, one copper (Cu) and one aluminum (Al), are arranged as shown in Figure 21-15. Find (*a*) the thermal resistance of each cube, (*b*) the thermal resistance of the two-cube system, (*c*) the thermal current *I*, and (*d*) the temperature at the interface of the two cubes.

Figure 21-15 Problem 29

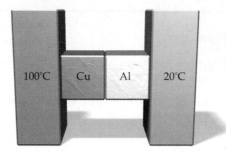

30 •• The cubes in Problem 29 are rearranged in parallel as shown in Figure 21-16. Find (*a*) the thermal current carried by each cube from one side to the other, (*b*) the total thermal current, and (*c*) the equivalent thermal resistance of the two-cube system.

Figure 21-16 Problem 30

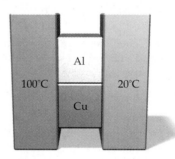

31 •• A spherical shell of thermal conductivity *k* has inside radius r_1 and outside radius r_2 (Figure 21-17). The inside of the shell is held at a temperature T_1, and the outside at temperature T_2. In this problem, you are to show that the thermal current through the shell is given by

$$I = \frac{4\pi k r_1 r_2}{r_2 - r_1}(T_2 - T_1) \qquad \text{21-22}$$

Consider a spherical element of the shell of radius *r* and thickness *dr*. (*a*) Why must the thermal current through each such element be the same? (*b*) Write the thermal current *I* through such a shell element in terms of the area $A = 4\pi r^2$, the thickness *dr*, and the temperature difference *dT* across the element. (*c*) Solve for *dT* in terms of *dr* and integrate from $r = r_1$ to $r = r_2$. (*d*) Show that when r_1 and r_2 are much larger than $r_2 - r_1$, Equation 21-22 is the same as Equation 21-7.

Figure 21-17
Problems 31 and 50

32 •• A group of anthropologists is staying in the high Arctic for a month, and they need accommodation. They are directed to a small company, Inuit Igloos."How thick do you want the walls?" asks Inuk, the head igloo maker. After some conferring, they reply that it should be 20°C inside when the temperature is −20°C outside. After looking the anthropologists over and poking them a bit, Inuk estimates that they would give off 38 MJ of heat per day. If the inside radius of the hemispherical igloo is to be 2 m, and the thermal conductivity of the compacted snow is 0.209 W/m·K, how thick should the walls be? (As an approximation, assume that the inner surface area of the igloo is equal to the outer surface area.)

33 •• For a boiler at a power station, heat must be transferred to boiling water at the rate of 3 GW. The boiling water passes through copper pipes having a wall thickness of 4.0 mm and a surface area of 0.12 m² per meter length of pipe. Find the total length of pipe (actually there are many pipes in parallel) that must pass through the furnace if the steam temperature is 225°C and the external temperature of the pipes is 600°C.

34 ••• A steam pipe of length L is insulated with a layer of material of thermal conductivity k. Find the rate of heat transfer if the temperature outside the insulation is t_1, the temperature inside is t_2, the outside radius of the insulation is r_1, and the inside radius is r_2.

35 ••• Brine at −16°C circulating through copper pipes with walls 1.5 mm thick is used to keep a cold room at 0°C. The diameter of each pipe is very large compared to the thickness of its walls. By what fraction is the transfer of heat reduced when the pipes are coated with a 5-mm layer of ice?

Radiation

36 • If the absolute temperature of an object is tripled, the rate at which it radiates thermal energy

(a) triples.
(b) increases by a factor of 9.
(c) increases by a factor of 27.
(d) increases by a factor of 81.
(e) depends on whether the absolute temperature is above or below zero.

37 • Calculate λ_{max} for a human blackbody radiator, assuming the surface temperature of the skin to be 33°C.

38 •• The heating wires of a 1-kW electric heater are red hot at a temperature of 900°C. Assuming that 100% of the heat output is due to radiation and that the wires act as blackbody radiators, what is the effective area of the radiating surface? (Assume a room temperature of 20°C.)

39 •• A blackened, solid copper sphere of radius 4.0 cm hangs in a vacuum in an enclosure whose walls have a temperature of 20°C. If the sphere is initially at 0°C, find the rate at which its temperature changes, assuming that heat is transferred by radiation only.

40 •• The surface temperature of the filament of an incandescent lamp is 1300°C. If the electric power input is doubled, what will the temperature become? *Hint:* Show that you can neglect the temperature of the surroundings.

41 •• Liquid helium is stored at its boiling point (4.2 K) in a spherical can that is separated by a vacuum space from a surrounding shield that is maintained at the temperature of liquid nitrogen (77 K). If the can is 30 cm in diameter and is blackened on the outside so that it acts as a blackbody, how much helium boils away per hour?

General Problems

42 • In a cool room, a metal or marble table top feels much colder to the touch than a wood surface does even though they are at the same temperature. Why?

43 • True or false:

(a) During a phase change, the temperature of a substance remains constant.
(b) The rate of conduction of thermal energy is proportional to the temperature gradient.
(c) The rate at which an object radiates energy is proportional to the square of its absolute temperature.
(d) All materials expand when they are heated.
(e) The vapor pressure of a liquid depends on the temperature.

44 • Conduction is a method of heat transfer that

(a) can proceed in vacuum.
(b) involves the transfer of mass.
(c) is dominant in solids.
(d) depends on the fourth power of the absolute temperature.

45 • The earth loses heat by

(a) conduction.
(b) convection.
(c) radiation.
(d) all of the above.

46 • Which heat-transfer mechanisms are most important in the warming effect of a fire in a fireplace?

47 • Which heat-transfer mechanism is important in the transfer of energy from the sun to the earth?

48 •• Two cylinders made of materials A and B have the same lengths; their diameters are related by $d_A = 2d_B$. When the same temperature difference is maintained between the ends of the cylinders they conduct heat at the same rate. Their thermal conductivities are related by

(a) $k_A = k_B/4$.
(b) $k_A = k_B/2$.
(c) $k_A = k_B$.
(d) $k_A = 2k_B$.
(e) $k_A = 4k_B$.

49 • A steel tape is placed around the earth at the equator when the temperature is 0°C. What will the clearance between the tape and the ground (assumed to be uniform) be if the temperature of the tape rises to 30°C? Neglect the expansion of the earth.

50 •• Use the result of Problem 31 (Equation 21-22) to calculate the wall thickness of the hemispherical igloo of Problem 32 without assuming that the inner surface area equals the outer surface area.

51 •• Show that change in the density of an isotropic material due to an increase in temperature ΔT is given by $\Delta \rho = -\beta \rho \, \Delta T$.

52 •• The solar constant is the power received from the sun per unit area perpendicular to the sun's rays at the mean distance of the earth from the sun. Its value at the upper atmosphere of the earth is about 1.35 kW/m^2. Calculate the effective temperature of the sun if it radiates like a blackbody. (The radius of the sun is 6.96×10^8 m.)

53 •• Lou has patented a cooking timer, which he is marketing as "Nature's Way: Taking You Back To Simpler Times." The timer consists of a 28-cm copper rod having a 5.0-cm diameter. Just as the lower end is placed in boiling water, an ice cube is placed on the top of the rod. When the ice melts completely, the cooking time is up. A special ice cube tray makes cubes of various sizes to correspond to the boiling time required. What is the cooking time when a 30-g ice cube at $-5.0°C$ is used?

54 •• To determine the R value of insulating material that comes in sheets of $\frac{1}{2}$-in thickness, you construct a cubical box of 12 in per side and place a thermometer and a 100-W heater inside the box. After thermal equilibrium has been attained, the temperature inside the box is 90°C when the external temperature is 20°C. Determine the R value of this material.

55 •• A 2-cm-thick copper sheet is pressed against a sheet of aluminum. What should be the thickness of the aluminum sheet so that the temperature of the copper–aluminum interface is $(T_1 + T_2)/2$, where T_1 and T_2 are the temperatures at the copper–air and aluminum–air interfaces?

56 •• At a temperature of 20°C, a steel bar of radius 2.2 cm and length 60 cm is jammed horizontally perpendicular between two vertical concrete walls. With a blowtorch, the temperature of the bar is raised to 60°C. Find the force exerted by the bar on each wall.

57 •• (a) From the definition of β, the coefficient of volume expansion (at constant pressure), show that $\beta = 1/T$ for an ideal gas. (b) The experimentally determined value of β for N_2 gas at 0°C is 0.003673 K^{-1}. Compare this value with the theoretical value $\beta = 1/T$, assuming that N_2 is an ideal gas.

58 •• One way to construct a device with two points whose separation remains the same in spite of temperature changes is to bolt together one end of two rods having different coefficients of linear expansion as in the arrangement shown in Figure 21-18. (a) Show that the distance L will not change with temperature if the lengths L_A and L_B are chosen such that $L_A/L_B = \alpha_B/\alpha_A$. (b) If material B is steel, material A is brass, and $L_A = 250$ cm at 0°C, what is the value of L?

Figure 21-18 Problem 58

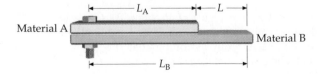

Material A

Material B

L_A L

L_B

59 •• On the average, the temperature of the earth's crust increases 1.0 C° for every 30 m of depth. The average thermal conductivity of the earth's crust is 0.74 J/m·s·K. What is the heat loss of the earth per second due to conduction from the core? How does this heat loss compare with the average power received from the sun? (The solar constant is about 1.35 kW/m^2.)

60 •• A copper-bottomed saucepan containing 0.8 L of boiling water boils dry in 10 min. Assuming that all the heat flows through the flat copper bottom, which has a diameter of 15 cm and a thickness of 3.0 mm, calculate the temperature of the outside of the copper bottom while some water is still in the pan.

61 •• A hot-water tank of cylindrical shape has an inside diameter of 0.55 m and inside height of 1.2 m. The tank is enclosed with a 5-cm-thick insulating layer of glass wool whose thermal conductivity is 0.035 W/m·K. The metallic interior and exterior walls of the container have thermal conductivities that are much greater than that of the glass wool. How much power must be supplied to this tank to maintain the water temperature at 75°C when the external temperature is 1°C?

62 ••• The diameter of a rod is given by $d = d_0(1 + ax)$, where a is a constant and x is the distance from one end. If the thermal conductivity of the material is k what is the thermal resistance of the rod if its length is L?

63 ••• A solid disk of radius R and mass M is spinning in a frictionless environment with angular velocity ω_1 at temperature T_1. The temperature of the disk is then changed to T_2. Express the angular velocity ω_2, rotational kinetic energy E_2, and angular momentum L_2 in terms of their values at the temperature T_1 and the linear expansion coefficient α of the disk.

64 ••• A small pond has a layer of ice 1 cm thick floating on its surface. (a) If the air temperature is $-10°C$, find the rate in centimeters per hour at which ice is added to the bottom of the layer. The density of ice is 0.917 g/cm^3. (b) How long does it take for a 20-cm layer to be built up?

65 ••• A body initially at a temperature T_i cools by convection and radiation in a room where the temperature is T_0. The body obeys Newton's law of cooling, which can be written $dQ/dt = hA(T - T_0)$, where A is the area of the body and h is a constant called the surface coefficient of heat transfer. Show that the temperature T at any time t is given by $T = T_0 + (T_i - T_0)e^{-hAt/mc}$, where m is the mass of the body and c is its specific heat.

66 ••• Two 200-g copper containers, each holding 0.7 L of water, are connected by a 10-cm copper rod of cross-sectional area 1.5 cm^2. Initially, one container is at 60°C; the second is maintained at 0°C. (a) Show that the temperature t_c of the first container changes over time t according to

$$t_c = t_{c0}e^{-t/RC}$$

where t_{c0} is the initial temperature of the first container, R is the thermal resistance of the rod, and C is the total heat capacity of the container plus the water. (b) Evaluate R, C, and the "time constant" RC. (*c*) Show that the total amount of heat Q conducted in time t is

$$Q = Ct_{c0}(1 - e^{-t/RC})$$

(*d*) Find the time it takes for the temperature of the first container to be reduced to 30°C.

67 ••• Liquid helium is stored in containers fitted with 7-cm-thick "superinsulation" consisting of a large number of layers of very thin aluminized Mylar sheets. The rate of evaporation of liquid in a 200-L container is about 0.7 L per day. Assume that the container is spherical and that the external temperature is 20°C. The specific gravity of liquid helium is 0.125 and the latent heat of vaporization is 21 kJ/kg. Estimate the thermal conductivity of superinsulation.

A

SI Units and Conversion Factors

Basic Units

Length	The *meter* (m) is the distance traveled by light in a vacuum in 1/299,792,458 s.
Time	The *second* (s) is the duration of 9,192,631,770 periods of the radiation corresponding to the transition between the two hyperfine levels of the ground state of the ^{133}Cs atom.
Mass	The *kilogram* (kg) is the mass of the international standard body preserved at Sèvres, France.
Current	The *ampere* (A) is that current in two very long parallel wires 1 m apart that gives rise to a magnetic force per unit length of 2×10^{-7} N/m.
Temperature	The *kelvin* (K) is 1/273.16 of the thermodynamic temperature of the triple point of water.
Luminous intensity	The *candela* (cd) is the luminous intensity, in the perpendicular direction, of a surface of area $1/600,000$ m^2 of a blackbody at the temperature of freezing platinum at a pressure of 1 atm.

Derived Units

Force	newton (N)	$1\,N = 1\,kg \cdot m/s^2$
Work, energy	joule (J)	$1\,J = 1\,N \cdot m$
Power	watt (W)	$1\,W = 1\,J/s$
Frequency	hertz (Hz)	$1\,Hz = s^{-1}$
Charge	coulomb (C)	$1\,C = 1\,A \cdot s$
Potential	volt (V)	$1\,V = 1\,J/C$
Resistance	ohm (Ω)	$1\,\Omega = 1\,V/A$
Capacitance	farad (F)	$1\,F = 1\,C/V$
Magnetic field	tesla (T)	$1\,T = 1\,N/A \cdot m$
Magnetic flux	weber (Wb)	$1\,Wb = 1\,T \cdot m^2$
Inductance	henry (H)	$1\,H = 1\,J/A^2$

Conversion Factors

Conversion factors are written as equations for simplicity;
relations marked with an asterisk are exact.

Length

1 km = 0.6215 mi

1 mi = 1.609 km

1 m = 1.0936 yd = 3.281 ft = 39.37 in

*1 in = 2.54 cm

*1 ft = 12 in = 30.48 cm

*1 yd = 3 ft = 91.44 cm

1 lightyear = 1 $c \cdot y$ = 9.461 × 10^{15} m

*1 Å = 0.1 nm

Area

*1 m^2 = 10^4 cm^2

1 km^2 = 0.3861 mi^2 = 247.1 acres

*1 in^2 = 6.4516 cm^2

1 ft^2 = 9.29 × 10^{-2} m^2

1 m^2 = 10.76 ft^2

*1 acre = 43,560 ft^2

1 mi^2 = 640 acres = 2.590 km^2

Volume

*1 m^3 = 10^6 cm^3

*1 L = 1000 cm^3 = 10^{-3} m^3

1 gal = 3.786 L

1 gal = 4 qt = 8 pt = 128 oz = 231 in^3

1 in^3 = 16.39 cm^3

1 ft^3 = 1728 in^3 = 28.32 L = 2.832 × 10^4 cm^3

Time

*1 h = 60 min = 3.6 ks

*1 d = 24 h = 1440 min = 86.4 ks

1 y = 365.24 d = 31.56 Ms

Speed

1 km/h = 0.2778 m/s = 0.6215 mi/h

1 mi/h = 0.4470 m/s = 1.609 km/h

1 mi/h = 1.467 ft/s

Angle and Angular Speed

*π rad = 180°

1 rad = 57.30°

1° = 1.745 × 10^{-2} rad

1 rev/min = 0.1047 rad/s

1 rad/s = 9.549 rev/min

Mass

*1 kg = 1000 g

*1 tonne = 1000 kg = 1 Mg

1 u = 1.6606 × 10^{-27} kg

1 kg = 6.022 × 10^{23} u

1 slug = 14.59 kg

1 kg = 6.852 × 10^{-2} slug

1 u = 931.50 MeV/c^2

Density

*1 g/cm^3 = 1000 kg/m^3 = 1 kg/L

(1 g/cm^3)g = 62.4 lb/ft^3

Force

1 N = 0.2248 lb = 10^5 dyn

1 lb = 4.4482 N

(1 kg)g = 2.2046 lb

Pressure

*1 Pa = 1 N/m^2

*1 atm = 101.325 kPa = 1.01325 bars

1 atm = 14.7 lb/in^2 = 760 mmHg

      = 29.9 inHg = 33.8 ftH$_2$O

1 lb/in^2 = 6.895 kPa

1 torr = 1 mmHg = 133.32 Pa

1 bar = 100 kPa

Energy

*1 kW·h = 3.6 MJ

*1 cal = 4.1840 J

1 ft·lb = 1.356 J = 1.286 × 10^{-3} Btu

*1 L·atm = 101.325 J

1 L·atm = 24.217 cal

1 Btu = 778 ft·lb = 252 cal = 1054.35 J

1 eV = 1.602 × 10^{-19} J

1 u·c^2 = 931.50 MeV

*1 erg = 10^{-7} J

Power

1 horsepower = 550 ft·lb/s = 745.7 W

1 Btu/min = 17.58 W

1 W = 1.341 × 10^{-3} horsepower

    = 0.7376 ft·lb/s

Magnetic Field

*1 G = 10^{-4} T

*1 T = 10^4 G

Thermal Conductivity

1 W/m·K = 6.938 Btu·in/h·ft^2·F°

1 Btu·in/h·ft^2·F° = 0.1441 W/m·K

Numerical Data

Terrestrial Data

Acceleration of gravity g	9.80665 m/s^2
Standard value	32.1740 ft/s^2
At sea level, at equator†	9.7804 m/s^2
At sea level, at poles†	9.8322 m/s^2
Mass of earth M_E	$5.98 \times 10^{24} \text{ kg}$
Radius of earth R_E, mean	$6.37 \times 10^6 \text{ m}; 3960 \text{ mi}$
Escape speed $\sqrt{2R_E g}$	$1.12 \times 10^4 \text{ m/s}; 6.95 \text{ mi/s}$
Solar constant‡	1.35 kW/m^2
Standard temperature and pressure (STP):	
Temperature	273.15 K
Pressure	$101.325 \text{ kPa}; 1.00 \text{ atm}$
Molar mass of air	28.97 g/mol
Density of air (STP), ρ_{air}	1.293 kg/m^3
Speed of sound (STP)	331 m/s
Heat of fusion of H_2O (0°C, 1 atm)	333.5 kJ/kg
Heat of vaporization of H_2O (100°C, 1 atm)	2.257 MJ/kg

†Measured relative to the earth's surface.
‡Average power incident normally on 1 m² outside the earth's atmosphere at the mean distance from the earth to the sun.

Astronomical Data

Earth	
Distance to moon†	$3.844 \times 10^8 \text{ m}; 2.389 \times 10^5 \text{ mi}$
Distance to sun, mean†	$1.496 \times 10^{11} \text{ m}; 9.30 \times 10^7 \text{ mi}; 1.00 \text{ AU}$
Orbital speed, mean	$2.98 \times 10^4 \text{ m/s}$
Moon	
Mass	$7.35 \times 10^{22} \text{ kg}$
Radius	$1.738 \times 10^6 \text{ m}$
Period	27.32 d
Acceleration of gravity at surface	1.62 m/s^2
Sun	
Mass	$1.99 \times 10^{30} \text{ kg}$
Radius	$6.96 \times 10^8 \text{ m}$

† Center to center.

Physical Constants

Gravitational constant	G	$6.672\,6 \times 10^{-11}\,\mathrm{N \cdot m^2/kg^2}$
Speed of light	c	$2.997\,924\,58 \times 10^8\,\mathrm{m/s}$
Electron charge	e	$1.602\,177 \times 10^{-19}\,\mathrm{C}$
Avogadro's number	N_A	$6.022\,137 \times 10^{23}\,\mathrm{particles/mol}$
Gas constant	R	$8.314\,51\,\mathrm{J/mol \cdot K}$
		$1.987\,22\,\mathrm{cal/mol \cdot K}$
		$8.205\,78 \times 10^{-2}\,\mathrm{L \cdot atm/mol \cdot K}$
Boltzmann's constant	$k = R/N_A$	$1.380\,658 \times 10^{-23}\,\mathrm{J/K}$
		$8.617\,385 \times 10^{-5}\,\mathrm{eV/K}$
Unified mass unit	$u = (1/N_A)\,\mathrm{g}$	$1.660\,540 \times 10^{-24}\,\mathrm{g}$
Coulomb constant	$k = 1/4\pi\epsilon_0$	$8.987\,551\,788 \times 10^9\,\mathrm{N \cdot m^2/C^2}$
Permittivity of free space	ϵ_0	$8.854\,187\,817 \times 10^{-12}\,\mathrm{C^2/N \cdot m^2}$
Permeability of free space	μ_0	$4\pi \times 10^{-7}\,\mathrm{N/A^2}$
		$1.256\,637 \times 10^{-6}\,\mathrm{N/A^2}$
Planck's constant	h	$6.626\,076 \times 10^{-34}\,\mathrm{J \cdot s}$
		$4.135\,669 \times 10^{-15}\,\mathrm{eV \cdot s}$
	$\hbar = h/2\pi$	$1.054\,573 \times 10^{-34}\,\mathrm{J \cdot s}$
		$6.582\,122 \times 10^{-16}\,\mathrm{eV \cdot s}$
Mass of electron	m_e	$9.109\,390 \times 10^{-31}\,\mathrm{kg}$
		$510.999\,1\,\mathrm{keV}/c^2$
Mass of proton	m_p	$1.672\,623 \times 10^{-27}\,\mathrm{kg}$
		$938.272\,3\,\mathrm{MeV}/c^2$
Mass of neutron	m_n	$1.674\,929 \times 10^{-27}\,\mathrm{kg}$
		$939.565\,6\,\mathrm{MeV}/c^2$
Bohr magneton	$m_B = e\hbar/2m_e$	$9.274\,015\,4 \times 10^{-24}\,\mathrm{J/T}$
		$5.788\,382\,63 \times 10^{-5}\,\mathrm{eV/T}$
Nuclear magneton	$m_n = e\hbar/2m_p$	$5.050\,786\,6 \times 10^{-27}\,\mathrm{J/T}$
		$3.152\,451\,66 \times 10^{-8}\,\mathrm{eV/T}$
Magnetic flux quantum	$\phi_0 = h/2e$	$2.067\,834\,6 \times 10^{-15}\,\mathrm{T \cdot m^2}$
Quantized Hall resistance	$R_K = h/e^2$	$2.581\,280\,7 \times 10^4\,\Omega$
Rydberg constant	R_H	$1.097\,373\,153\,4 \times 10^7\,\mathrm{m^{-1}}$
Josephson frequency–voltage quotient	$2e/h$	$4.835\,979 \times 10^{14}\,\mathrm{Hz/V}$
Compton wavelength	$\lambda_C = h/m_e c$	$2.426\,310\,58 \times 10^{-12}\,\mathrm{m}$

For additional data, see the last four pages in the book and the following tables in the text.

Periodic Table of Elements

1																	18
1 **H** 1.00797	**2**																**2** **He** 4.003
3 **Li** 6.941	**4** **Be** 9.012											**13** **5** **B** 10.81	**14** **6** **C** 12.011	**15** **7** **N** 14.007	**16** **8** **O** 15.9994	**17** **9** **F** 19.00	**10** **Ne** 20.179
11 **Na** 22.990	**12** **Mg** 24.31	**3**	**4**	**5**	**6**	**7**	**8**	**9**	**10**	**11**	**12**	**13** **Al** 26.98	**14** **Si** 28.09	**15** **P** 30.974	**16** **S** 32.064	**17** **Cl** 35.453	**18** **Ar** 39.948
19 **K** 39.102	**20** **Ca** 40.08	**21** **Sc** 44.96	**22** **Ti** 47.88	**23** **V** 50.94	**24** **Cr** 52.00	**25** **Mn** 54.94	**26** **Fe** 55.85	**27** **Co** 58.93	**28** **Ni** 58.69	**29** **Cu** 63.55	**30** **Zn** 65.38	**31** **Ga** 69.72	**32** **Ge** 72.59	**33** **As** 74.92	**34** **Se** 78.96	**35** **Br** 79.90	**36** **Kr** 83.80
37 **Rb** 85.47	**38** **Sr** 87.62	**39** **Y** 88.906	**40** **Zr** 91.22	**41** **Nb** 92.91	**42** **Mo** 95.94	**43** **Tc** (98)	**44** **Ru** 101.1	**45** **Rh** 102.905	**46** **Pd** 106.4	**47** **Ag** 107.870	**48** **Cd** 112.41	**49** **In** 114.82	**50** **Sn** 118.69	**51** **Sb** 121.75	**52** **Te** 127.60	**53** **I** 126.90	**54** **Xe** 131.29
55 **Cs** 132.905	**56** **Ba** 137.33	**57–71** **Rare Earths**	**72** **Hf** 178.49	**73** **Ta** 180.95	**74** **W** 183.85	**75** **Re** 186.2	**76** **Os** 190.2	**77** **Ir** 192.2	**78** **Pt** 195.09	**79** **Au** 196.97	**80** **Hg** 200.59	**81** **Tl** 204.37	**82** **Pb** 207.19	**83** **Bi** 208.98	**84** **Po** (210)	**85** **At** (210)	**86** **Rn** (222)
87 **Fr** (223)	**88** **Ra** (226)	**89–103** **Actinides**	**104** **Rf** (261)	**105** **Ha** (260)	**106** (263)	**107** (262)	**108** (265)	**109** (266)									

Rare Earths (Lanthanides)	**57** **La** 138.91	**58** **Ce** 140.12	**59** **Pr** 140.91	**60** **Nd** 144.24	**61** **Pm** (147)	**62** **Sm** 150.36	**63** **Eu** 152.0	**64** **Gd** 157.25	**65** **Tb** 158.92	**66** **Dy** 162.50	**67** **Ho** 164.93	**68** **Er** 167.26	**69** **Tm** 168.93	**70** **Yb** 173.04	**71** **Lu** 174.97
Actinides	**89** **Ac** 227.03	**90** **Th** 232.04	**91** **Pa** 231.04	**92** **U** 238.03	**93** **Np** 237.05	**94** **Pu** (244)	**95** **Am** (243)	**96** **Cm** (247)	**97** **Bk** (247)	**98** **Cf** (251)	**99** **Es** (252)	**100** **Fm** (257)	**101** **Md** (258)	**102** **No** (259)	**103** **Lr** (260)

The 1–18 group designation has been recommended by the International Union of Pure and Applied Chemistry (IUPAC).

Atomic Numbers and Atomic Masses

Name	Symbol	Atomic Number	Mass	Name	Symbol	Atomic Number	Mass
Actinium	Ac	89	227.03	Mercury	Hg	80	200.59
Aluminum	Al	13	26.98	Molybdenum	Mo	42	95.94
Americium	Am	95	(243)	Neodymium	Nd	60	144.24
Antimony	Sb	51	121.75	Neon	Ne	10	20.179
Argon	Ar	18	39.948	Neptunium	Np	93	237.05
Arsenic	As	33	74.92	Nickel	Ni	28	58.69
Astatine	At	85	(210)	Niobium	Nb	41	92.91
Barium	Ba	56	137.3	Nitrogen	N	7	14.007
Berkelium	Bk	97	(247)	Nobelium	No	102	(259)
Beryllium	Be	4	9.012	Osmium	Os	76	190.2
Bismuth	Bi	83	208.98	Oxygen	O	8	15.9994
Boron	B	5	10.81	Palladium	Pd	46	106.4
Bromine	Br	35	79.90	Phosphorus	P	15	30.974
Cadmium	Cd	48	112.41	Platinum	Pt	78	195.09
Calcium	Ca	20	40.08	Plutonium	Pu	94	(244)
Californium	Cf	98	(251)	Polonium	Po	84	(210)
Carbon	C	6	12.011	Potassium	K	19	39.102
Cerium	Ce	58	140.12	Praseodymium	Pr	59	140.91
Cesium	Cs	55	132.905	Promethium	Pm	61	(147)
Chlorine	Cl	17	35.453	Protactinium	Pa	91	231.04
Chromium	Cr	24	52.00	Radium	Ra	88	(226)
Cobalt	Co	27	58.93	Radon	Rn	86	(222)
Copper	Cu	29	63.55	Rhenium	Re	75	186.2
Curium	Cm	96	(247)	Rhodium	Rh	45	102.905
Dysprosium	Dy	66	162.50	Rubidium	Rb	37	85.47
Einsteinium	Es	99	(252)	Ruthenium	Ru	44	101.1
Erbium	Er	68	167.26	Rutherfordium	Rf	104	(261)
Europium	Eu	63	152.0	Samarium	Sm	62	150.36
Fermium	Fm	100	(257)	Scandium	Sc	21	44.96
Fluorine	F	9	19.00	Selenium	Se	34	78.96
Francium	Fr	87	(223)	Silicon	Si	14	28.09
Gadolinium	Gd	64	157.25	Silver	Ag	47	107.870
Gallium	Ga	31	69.72	Sodium	Na	11	22.990
Germanium	Ge	32	72.59	Strontium	Sr	38	87.62
Gold	Au	79	196.97	Sulfur	S	16	32.064
Hafnium	Hf	72	178.49	Tantalum	Ta	73	180.95
Hahnium	Ha	105	(260)	Technetium	Tc	43	(98)
Helium	He	2	4.003	Tellurium	Te	52	127.60
Holmium	Ho	67	164.93	Terbium	Tb	65	158.92
Hydrogen	H	1	1.00797	Thallium	Tl	81	204.37
Indium	In	49	114.82	Thorium	Th	90	232.04
Iodine	I	53	126.90	Thulium	Tm	69	168.93
Iridium	Ir	77	192.2	Tin	Sn	50	118.69
Iron	Fe	26	55.85	Titanium	Ti	22	47.88
Krypton	Kr	36	83.80	Tungsten	W	74	183.85
Lanthanum	La	57	138.91	Uranium	U	92	238.03
Lawrencium	Lr	103	(260)	Vanadium	V	23	50.94
Lead	Pb	82	207.19	Xenon	Xe	54	131.29
Lithium	Li	3	6.941	Ytterbium	Yb	70	173.04
Lutetium	Lu	71	174.97	Yttrium	Y	39	88.906
Magnesium	Mg	12	24.31	Zinc	Zn	30	65.38
Manganese	Mn	25	54.94	Zirconium	Zr	40	91.22
Mendelevium	Md	101	(258)				

Review of Mathematics

In this appendix, we will review some of the basic results of algebra, geometry, trigonometry, and calculus. In many cases, we will merely state results without proof. Table D-1 lists some mathematical symbols.

Equations

The following operations can be performed on mathematical equations to facilitate their solution:

1. The same quantity can be added to or subtracted from each side of the equation.

2. Each side of the equation can be multiplied or divided by the same quantity.

3. Each side of the equation can be raised to the same power.

It is important to understand that the preceding rules apply to each *side* of the equation and not to each *term* in the equation.

Table D-1

Mathematical Symbols

$=$	is equal to
$\neq$	is not equal to
$\approx$	is approximately equal to
$\sim$	is of the order of
$\propto$	is proportional to
$>$	is greater than
$\geq$	is greater than or equal to
$\gg$	is much greater than
$<$	is less than
$\leq$	is less than or equal to
$\ll$	is much less than
Δx	change in x
$\|x\|$	absolute value of x
$n!$	$n(n-1)(n-2)\cdots 1$
Σ	sum
lim	limit
$\Delta t \to 0$	Δt approaches zero
$\dfrac{dx}{dt}$	derivative of x with respect to t
$\dfrac{\partial x}{\partial t}$	partial derivative of x with respect to t
$\int$	integral

Example D-1

Solve the following equation for x: $(x - 3)^2 + 7 = 23$.

1. Subtract 7 from each side: $\qquad\qquad\quad (x - 3)^2 = 16$

2. Take the square root of each side: $\quad x - 3 = \pm 4$

3. Add 3 to each side: $\qquad\qquad\quad x = 4 + 3 = 7 \quad$ or $\quad x = -4 + 3 = -1$

Remark Note that in step 2 we do not need to write $\pm(x - 3) = \pm 4$ because all possibilities are included in $x - 3 = \pm 4$.

Check the Result We check our result by substituting each value into the original equation: $(7 - 3)^2 + 7 = 16 + 7 = 23$ and $(-1-3)^2 + 7 = 16 + 7 = 23$.

Example D-2

Solve the following equation for x:

$$\frac{1}{x} + \frac{1}{4} = \frac{1}{3}$$

1. Subtract $\frac{1}{4}$ from each side: $\qquad \dfrac{1}{x} = \dfrac{1}{3} - \dfrac{1}{4} = \dfrac{4}{12} - \dfrac{3}{12} = \dfrac{1}{12}$

2. Multiply each side by $12x$: $\qquad x = 12$

Remark This type of equation occurs both in geometric optics and in analyses of electric circuits. Although it is easy to solve, errors are often made. A typical mistake is to take the reciprocal of each *term*, obtaining $x + 4 = 3$. Taking the reciprocal of each term is not allowed; taking the reciprocal of each *side* of an equation is allowed. Note that multiplying each side by $12x$ in step 2 is equivalent to taking the reciprocal of each side of the equation.

Direct and Inverse Proportion

The relationships of direct proportion and inverse proportion are so important in physics that they deserve special consideration. Often much algebraic manipulation can be avoided through a simple knowledge of these relationships. Suppose, for example, that you work for 5 days at a certain pay rate and earn \$400. How much would you earn at the same pay rate if you worked 8 days? In this problem, the money earned is *directly* proportional to the time worked. We can write an equation relating the money earned M to the time worked t using a constant of proportionality R:

$$M = Rt$$

The constant of proportionality in this case is the pay rate. We can express R in dollars per day. Since \$400 was earned in 5 d, the value of R is \$400/(5 d) = \$80/d. In 8 d, the amount earned is therefore

$$M = (\$80/\text{d})(8 \text{ d}) = \$640$$

Chapter 13

Opener p. 374 Michael Dunn/The Stock Market; **p. 381** Chuck O'Rear/Woodfin Camp and Assoc.; **p. 382 (a)** PAR/NYC, Inc. Archives; **(b)** David Burnett/Woodfin Camp and Assoc.; **p. 384 (a)** Estate of Harold E. Edgerton/Palm Press, Inc.; **(b)** Richard Megna/Fundamental Photographs; **p. 388 (a) and (b)** Office National d'Etudes et de Recherches Areospatiales; **p. 392 (top)** Picker International; **(bottom)** Dr. Owen M. Griffin, Naval Research Laboratory.

Chapter 14

Opener p. 403 Citibank; **p. 405** N.A.S.A. (73-HC-787); **p. 409** Institute for Marine Dynamics; **p. 417** Berenice Abbott/Commerce Graphics Ltd, Inc.; **p. 422** Estate of Harold E. Edgerton/Palm Press Inc.; **p. 423** Monroe Auto Equipment; **p. 426** Royal Swedish Academy of Music.

Chapter 15

Opener p. 442 (top) Four by Five Inc.; **(bottom)** Berenice Abbott/Photo Researchers; **p. 443** Berenice Abbott/Photo Researchers; **p. 454 (Figure 15-11)** *PSSC Physics*, 2nd ed., 1965. D.C. Heath & Co., and Education Development Center Inc., Newton, MA; **(Figure 15-12)** From Winston E. Kock, *Lasers and Holography*, Dover Publications, New York, 1981; **p. 459 (a) and (b)** *PSSC Physics*, 2nd ed., 1965. D.C. Heath & Co., and Education Development Center Inc., Newton, MA; **p. 461 (Figure 15-22)** *PSSC Physics*, 2nd ed., 1965. D.C. Heath & Co., and Education Development Center Inc., Newton, MA; **p. 462 (left)** Naval Research Laboratory; **(right)** Courtesy San Francisco Symphony; **(Figures 15-23 and 15-25)** *PSSC Physics*, 2nd ed., 1965. D.C. Heath & Co., and Education Development Center Inc., Newton, MA; **p. 463** Courtesy of the author; **p. 464** *PSSC Physics*, 2nd ed., 1965. D.C. Heath & Co., and Education Development Center Inc., Newton, MA; **p. 467 (left)** Sandia National Laboratory; **(right)** Robert de Gast/Photo Researchers; **(bottom)** Estate of Harold E. Edgerton/Palm Press Inc.; **p. 468 (Figure 15-28)** *PSSC Physics*, 2nd ed., 1965. D.C. Heath & Co., and Education Development Center Inc., Newton, MA; **(right)** Department of Energy.

Chapter 16

Opener p. 480 Dr. John S. Shelton; **p. 486** Berenice Abbott(8J 1328)/Photo Researchers; **p. 489** David Yost/Steinway & Sons; **p. 490 (bottom)** University of Washington; **p. 492** *PSSC Physics*, 2nd ed., 1965. D.C. Heath & Co., and Education Development Center Inc., Newton, MA; **p. 496 (top)** David Hathaway/N.A.S.A.; **(bottom)** Professor Thomas D. Rossing, Northern Illinois University, DeKalb.

Chapter 17

Opener p. 509 (a–d) Courtesy of Akira Tonomura, Advanced Research Laboratory, Hitachi, Ltd.; **p. 519 (a, b)** *PSSC Physics*, 2nd ed., 1965. D.C. Heath & Co., and Education Development Center Inc., Newton, MA; **(c)** C.G. Shull; **(d)** Claus Jönsson; **(bottom)** Jack Griffith/University of North Carolina.

Chapter 18

Opener p. 541 Lockheed Corporation; **p. 544** Courtesy Honeywell, Inc.; **p. 545** Dr. William Magnum/National Bureau of Standards; **p. 559 (top)** N.A.S.A.; **(bottom)** Jet Propulsion Laboratory/N.A.S.A.

Chapter 19

Opener p. 566 Phoenix Pipe and Tube/Lana Berkovich; **p. 572 (b)** Science Museum, London; **p. 588** Will and Deni McIntyre/Photo Researchers.

Chapter 20

Opener p. 600 Sandia National Laboratory; **p. 603** J.M. Mejuto/FPG International; **p. 610** Michael Collier/Stock, Boston; **p. 611 (b)** Jean-Pierre Horlin/The ImageBank; **(c)** Sandia National Laboratories; **(d)** Peter Miller/The Image Bank; **(e)** Sandia National Laboratories.

Chapter 21

Opener p. 633 Michael Melford/The Image Bank.

ANSWERS

To help you master the techniques in Examples and to solve the intermediate-level problems at the end of the chapter, the maps indicate which Examples and odd-numbered intermediate-level Problems deal with similar material. The Problem answers are calculated used $g = 9.81$ m/s² unless otherwise specified in the Problem. Differences in the last figure can easily result from differences in rounding the input data and are not important.

Chapter 1

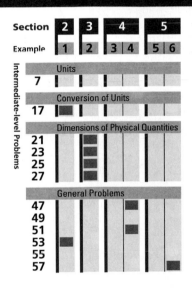

1. (c)

3. (a) 1 MW; (b) 2 mg; (c) 3 μm; (d) 30 ks

5. (a) 1 picoboo; (b) 1 gigalow; (c) 1 microphone; (d) 1 attoboy; (e) 1 megaphone; (f) 1 nanogoat; (g) 1 terabull

7. (a) C_1: ft, C_2: ft/s; (b) C_1: ft/s²; (c) C_1: m/s²; (d) C_1: ft, C_2: s⁻¹; (e) C_1: ft/s, C_2: s⁻¹

9. 2450 km/h, 1520 mi/h

11. (a) 62.1 mi/h; (b) 23.6 in; (c) 91.4 m

13. 1.61 km/mi

15. (a) 3.784 L; (b) 0.1589 m³

17. (a) 0.505 ft³; (b) 0.0143 m³; (c) 14.3 L; (d) 3.78 gal

19. (a) C_1: L, C_2: L/T; (b) C_1: L/T²; (c) C_1: L/T²; (d) C_1: L, C_2: 1/T; (e) C_1: L/T, C_2: 1/T

21. $L^3/(MT^2)$, m³/kg·s²

23. $M(L/T^2)(L/T) = ML^2/T^3$

25. $(ML/T^2)(L/T)$

27. $T = Cr^{3/2}/\sqrt{M_s G}$

29. (d)

31. (c)

33. (a) 30,000; (b) 0.0062; (c) 0.000004; (d) 217,000

35. (a) 1.14×10^5; (b) 2.24×10^{-8}; (c) 8.27×10^3; (d) 6.27×10^2

37. 4×10^6 membranes

39. (a) 1.69×10^3; (b) 4.8; (c) 5.6; (d) 10

41. One advantage of using your arm is that a measure of standard length is always available. The disadvantage is that arm lengths are not uniform. For example, if you wish to purchase a board of "two arm lengths," it may be shorter or longer than you expected, depending on the owner's arm length.

43. (a) True; (b) False; (c) True

45. 31.7 yr

47. 2.0×10^{23}

49. 3.51 Mm

51. 3.86×10^2 mi²

53. (a) 1.8×10^2; (b) 3.4; (c) 2.9×10^8; (d) 0.45

55. (a) 6.6×10^3 electrons/m³; (b) 3.6 protons/m³

57. (a) 1×10^{11} cans/yr; (b) 2×10^9 kg/yr; (c) $2 billion

59. (a) In a plot of T versus m^n, $n = 1/2$ provides the best fit to a straight line.

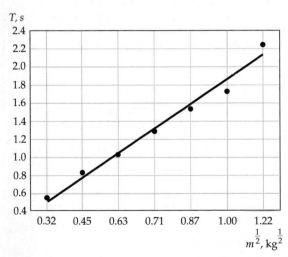

The slope of the line $C \approx 1.77$ s/kg$^{1/2}$. (b) The points with the greatest deviation are $T = 1.75$ s and $T = 2.22$ s.

61. (a) $C\sqrt{L/g}$; (b) Check by using pendulums of lengths 1 m and 0.5 m; the periods should be about 2 s and 1.4 s.

63. (a) 1.6×10^{10} bits; (b) 350 books

Chapter 2

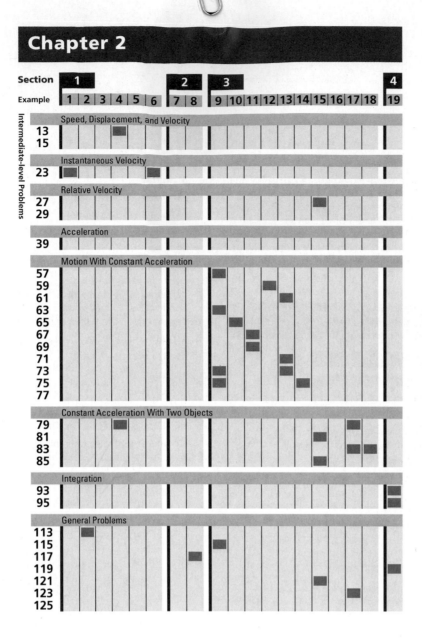

1. Since the cars go round a closed circuit and return nearly to the starting point, the displacement is nearly zero, and the average velocity is zero.

3. Yes. In a round trip, A to B and back to A, the average velocity is zero; the average velocity between A and B is not zero.

5. (a) 4 ns; (b) 4 ks

7. (a) 260 km; (b) 65 km/h

9. (a) 73.3 mi/h; (b) 44.3

11. (a) 434 yr; (b) 4.34×10^6 yr; he need not pay for pizza.

13. (a) 13 m; (b) 2.2 s

15. (a) 7.9×10^4 m/s; (b) 3.16×10^7 m/s; (c) 2×10^{10} yr

17. No, because for constant velocity, the instantaneous velocity and average velocities are equal.

19. (b)

21. (a) 1 m/s; (b) 2 m/s

23. (a) 2 m, 2 m/s; (b) $\Delta x = [(2t - 5)\,\Delta t + \Delta t^2]$ m; (c) $(2t - 5)$ m/s

25. (a)

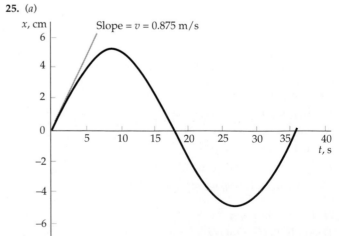

(b) 0.875 m/s; (c) 0.723, 0.835, 0.857, 0.871, 0.874, 0.875; (d) 0.875 cm/s

27. 120 km

29. 6 h

31. (*a*) After getting started, walk in the opposite (negative) direction. Gradually slow the speed of walking, until the other end of the room is reached.

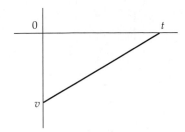

33. Yes. An object tossed up has a constant downward acceleration; at the top its instantaneous velocity is zero.

35. (*a*) $a = 0$; (*b*) $a > 0$; (*c*) $a < 0$; (*d*) $a = 0$

37. (*a*) 2.42 m/s², (*b*) 89.2 km/h

39. (*a*) 8 m/s², 8 m/s²; (*b*) 8 m/s²

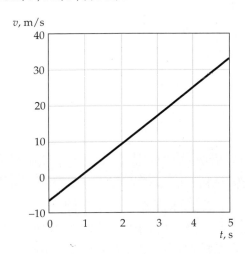

41. The initial downward velocities of the two rocks are not the same.

43. (*b*)

45. (*a*)

47. (*d*)

49. (*c*)

51. (*a*) 80 m/s; (*b*) 400 m; (*c*) 40 m/s

53. 15.6 m/s²

55. 4.59×10^3 m

57. (*a*) 4.08 s; (*b*) 20.4 m; (*c*) 0.99 s or 3.09 s are both acceptable.

59. (*a*) -5.1×10^5 m/s²; (*b*) 0.686 ms

61. 1467 m

63. (*a*)

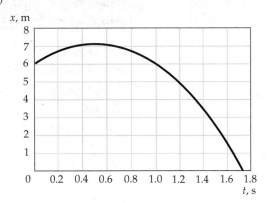

(*b*) 7.27 m; (*c*) 1.73 s; (*d*) 11.9 m/s

65. 1.62 m

67. 43.6 m

69. ±68 m/s; the stone may be thrown either up or down.

71. (*a*) 666 m; (*b*) 13.6 m/s

73. 19.0 km; (*b*) 2 min 18 s; (*c*) 610 m/s

75. (*a*) 2.24 s; (*b*) 6.17 m

77. 145.7 m

79. 15.6 m

81. 100.4 km/h

83. Deceleration ≥ 0.754 m/s², $v_{rel} = 3.77$ m/s, distance = 518 m

85. 4.8 m/s

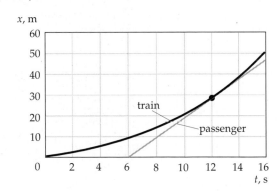

87. $2h/3$

89. 32.4 km

91. (*a*)

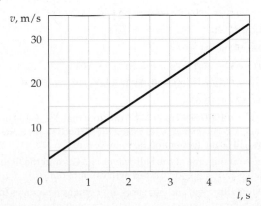

(*b*) 90 m

93. $x(t) = \int (7t^2 - 5)\, dt = (7/3)\, t^3 - 5t + C$

95. (a) 0.25 m/s; (b) 0.9 m/s, 3 m/s, 6 m/s; (c) 6.5 m

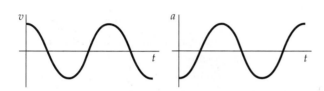

97.

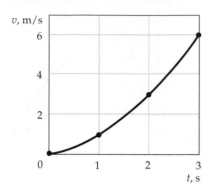

99. (c)

101. (a)

103. (d)

105. (d)

107. (a)

109. (a) a, f, i; (b) c, d; (c) a, d, e, f, h, i; (d) b, c, g. The graphs d and h and the graphs f and i are mutually consistent.

111. (a) 36 m; (b) −36 m; (c) −9 m/s

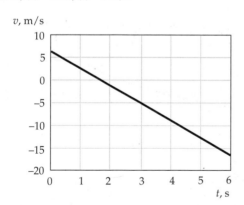

113. 134 m/s

115. 24 m, 1.4 s

117. $v(t) = (50 - 10t)$ m/s, $a = -10$ m/s^2, $x(t) = 50t - 5t^2$

119. (a) $x(t) = Ct^3/6 + At + B$; (b) 37.5 m/s, 62.5 m

121. (a) 6.67 s; (b) 6.67 s

123. (a) $\Delta x_s = 100.3$ m; $\Delta x_p = 232.3$ m; Note that $\Delta x_s + 100$ m $< x_p$, so the cars collide. (b) 5.52 s. (c) With added reaction time, the collision will occur sooner and be more severe.

125. The running time of the ball player is 2.42 s and the flight time of the ball is 2.19 s. A good umpire will call him out!

127. (a) Initially, $v = 0$ and increases as gt, but as v becomes finite, the acceleration diminishes to $g - bv$. Ultimately, the acceleration approaches zero and v remains constant, $v_{term} = g/b$.

(b)

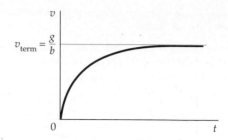

129. $x(t) = x_0\, e^{t - t_0}$.

131. (a)

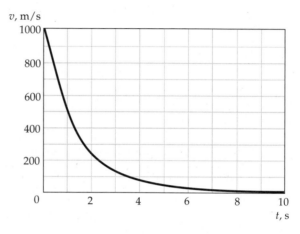

(b) 144.1 m/s

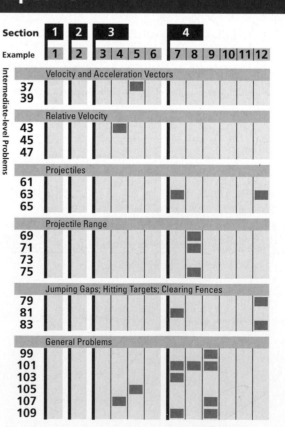

1. The magnitude of the displacement can be less than but never more than the distance traveled.

3. (*e*)

5. $D = (-5\hat{i} + 5\hat{j})$ m

7.

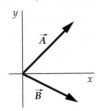

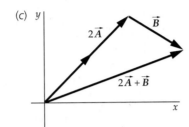

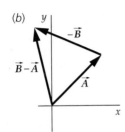

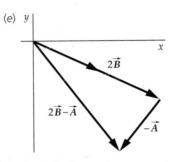

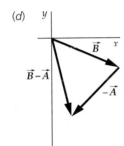

9. No, the magnitudes can be equal if the vector is along the component direction.

11. No, if $\vec{B} = -\vec{A}$, then $\vec{C} = 0$.

13. (*b*)

15. (*a*) $A_x = 8.66$ m, $A_y = 5$ m; (*b*) $A_x = 3.54$ m, $A_y = 3.54$ m; (*c*) $A_x = 3.5$ km, $A_y = 6.06$ km; (*d*) $A_x = 0$ km, $A_y = 5$ km; (*e*) $A_x = -13$ km/s, $A_y = 7.5$ km/s; (*f*) $A_x = -5$ m/s, $A_y = -8.66$ m/s; (*g*) $A_x = 0$, $A_y = -8$ m/s^2

17. (*a*) $A = 5.83$, $\theta = 31°$; (*b*) $B = 12.2$, $\theta = -35°$; (*c*) $C = 5.39$, $\theta = 42°$, $\phi = 236°$

19. (*a*) 5 m/s $\hat{i} + 8.66$ m/s $\hat{j}$; (*b*) -3.54 m $\hat{i} - 3.54$ m $\hat{j}$; (*c*) 14 m $\hat{i} - 6$ m $\hat{j}$

21. $\vec{B} = -1.5\vec{A}$

23. No

25. A particle moving at constant speed in a circular path is accelerating (the direction of the velocity vector is changing). If a particle is moving at constant velocity, it is not accelerating.

27. (*a*) The velocity vector is always tangent to the path.

(*b*)

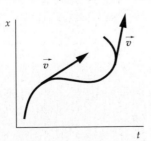

29.

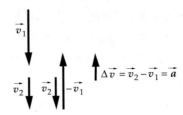

31. 14.1 km/h $\hat{i} - 4.1$ km/h $\hat{j}$

33. (*b*)

35. (*a*) -10 m/s; (*b*) $90°$; (*c*) 40 m/s $\hat{i} + 30$ m/s $\hat{j}$; (*d*) 10 m/s^2 at $37°$ with *x* axis

37. (*a*) 10 m/s $\hat{i} - 3$ m/s $\hat{j}$; (*b*) 44 m $\hat{i} - 9$ m $\hat{j}$; $r = 44.9$ m, $\theta = -11.6°$

39. (*a*) $\vec{v} = [6t\,\hat{i} + 4t\,\hat{j}]$ m/s, $\vec{r} = [(10 + 3t^2)\,\hat{i} + 2t^2\,\hat{j}]$ m

(*b*)

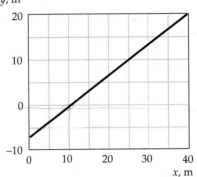

41. (*d*)

43. (*a*) 0.8 m/s; (*b*) 1.79 m/s; (*c*) $30°$ upstream

45. 5.18 km/h

47. (*a*) -120 m $\hat{i} + 4$ m $\hat{j}$; (*b*) -20 m/s $\hat{i} - 12$ m/s $\hat{j}$; (*c*) -2 m/s^2 $\hat{j}$

49. 261.7 km/h, $6.58°$ west of north

51. True

53. (*e*)

55. (*a*) A and E; (*b*) C; (*c*) A and E have the same speed, but not the same velocities.

57. 1.1 m

59. 15 m/s, 9.81 m/s^2 downward

61. 33.8 m/s

63. 20.3 m/s, $36.2°$

65. $69.3°$

67. (*a*) 49.5 s; (*b*) 12.4 km; (*c*) 12.0 km

69. (*a*) 8.14 m/s; (*b*) 23.2 m/s

71. $-63.4°$

73. $d[(v_0^2 \sin 2\theta)/g]/d\theta = (v_0^2/g)\,d(\sin 2\theta)/d\theta = (2v_0^2/g)\cos 2\theta$; set equal to 0; $2\theta = 90°$, $\theta = 45°$

75. $2v^2 \tan \phi/(g \cos \phi)$

77. 18.2 m from wall

79. $-1.93°$

81. 3.19 m/s

83. (*a*) 44.3 m/s, $9.33°$ above horizontal; (*b*) 0.733 s; (*c*) 3.23 m

85. 0.785 m

87. (*a*) 11.0 m; (*b*) $55.8°$; (*c*) 17.8 m/s

89. (*a*) False; (*b*) False; (*c*) True

91.

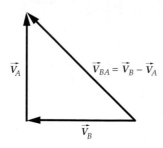

93. (*a*) north, northeast, east, southeast, south; (*b*) AB north, BC southeast, CD $\vec{a} = 0$, DE southwest, EF north; (*c*) equal

95. 4.9 m/s^2 $\hat{i}$ + 8.5 m/s^2 $\hat{j}$

97. (*a*)

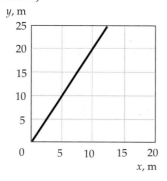

(*b*) (5$\hat{i}$ + 10$\hat{j}$) m/s, 11.2 m/s

99. 4.29 m

101. (*a*) 14 m; (*b*) 64 m; (*c*) 41 m; (*d*) 75 m

103. 14.8 m/s

105. (*a*) ($\hat{i}$ + $\hat{j}$) m/s; (*b*) (2$\hat{i}$ − 3.5$\hat{j}$) m/s^2;
(*c*) [$\hat{i}$ + $\hat{j}$ + (2$\hat{i}$ − 3.5$\hat{j}$)t] m/s;
(*d*) [4$\hat{i}$ + 3$\hat{j}$ + ($\hat{i}$ + $\hat{j}$)t + $\frac{1}{2}$(2$\hat{i}$ − 3.5$\hat{j}$)t^2] m

107. (*a*) (25$\hat{i}$ + 7.07$\hat{j}$ + 7.07$\hat{k}$) m/s; (*b*) (39.9$\hat{i}$ + 11.3$\hat{j}$) m

109. Yes, if it is a high fly.

111. (*a*) 0.98 km/h; (*b*) 39.6 west of north; (*c*) 3 h 19 min

113. $v_y^2 = v_{0y}^2 + 2gh$, regardless of direction (up or down).
$v^2 = v_x^2 + v_y^2$, and since $v_x = v_{0x}$, $v^2 = v_{0x}^2 + v_{0y}^2 + 2gH = v_0^2 + 2gH$,
for any angle. $v = (v_0^2 + 2gh)^{1/2}$

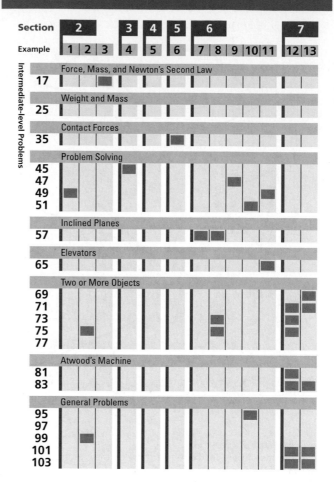

1. If Newton's first law is obeyed, the reference frame is an inertial reference frame.

3. No

5. No

7. (*d*)

9. (*b*)

11. $F_2 = 3F_1$

13. 4.46 m/s^2

15. (*a*) 250 N; (*b*) 400 N

17. (*a*) (1.5$\hat{i}$ − 3.5$\hat{j}$) m/s^2; (*b*) (4.5$\hat{i}$ − 10.5$\hat{j}$) m/s;
(*c*) (6.75$\hat{i}$ − 15.75$\hat{j}$) m

19. To accelerate, she must exert a force proportional to her mass.

21. (*c*)

23. 74.8 kg

25. (*a*) 395 N; (*b*) No

27. (*b*)

29. (*c*)

31.

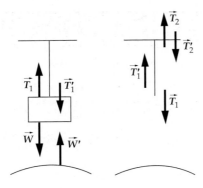

33. (a) 60 N; (b) 57.7 N

35. (a)

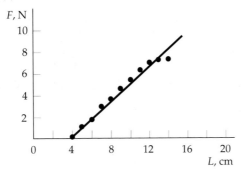

(b) 7.15 N; (c) 5.3 cm

37. No, the tension in the line must have a vertical component to support the weight of the towel.

39. (b)

41. (c)

43. (a) -8.81 m/s^2; (b) -7.81 m/s^2; (c) 10.19 m/s^2

45. (a) $v = -10.8 \text{ m/s}$; (b) $v = -7.7 \text{ m/s}$

47. (a) $T = w/2 \, (\sin \theta)$; $\theta = 90°$, T least; $\theta \to 0°$, T greatest; (b) 19.6 N

49. (a) 11,810 N; (b) 9810 N; (c) 7810 N

51. (a) $T_1 = 60$ N, $T_2 = 52$ N, $M = 5.3$ kg; (b) $T_1 = 46.2$ N, $T_2 = 46.2$ N, $M = 4.71$ kg; (c) $T_1 = T_3 = 34$ N, $T_2 = 58.9$ N, $M = 3.46$ kg

53.

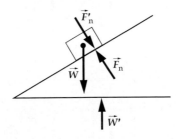

55. (a) and (b) = 98.1 N; (c) 49 N; (d) 49 N

57. (a) 149 N; (b) 3.41 m/s^2

59. (a)

61. high

63. 5.19 m/s^2

65. (a) 19.6 N; (b) 19.6 N; (c) 39.6 N; for $0 \le t \le 2$ s, $F = 19.6$ N; for 2 s $\le t \le 4$ s, $F = 9.62$ N

67. (d)

69. (a) $a = F/(m_1 + m_2)$; $F_c = F \, [m_2/(m_1 + m_2)]$; (b) $a = 0.4 \text{ m/s}^2$; $F_c = 2.4$ N

71. $F_A = 160$ N, $F_B = 161.2$ N, $F_C = 321.6$ N

73. (a) 492 N, 3.71 m/s; (b) 20.7 m/s

75. (a) $m_1 = m_2(g - a)/(a + g \sin \theta) = 48$ kg; (b) 424 N

77. (a) 10 m/s^2; (b) $60 + 8x$ N

79. $a_{10} = 1.12 \text{ m/s}^2$, $a_{20} = -1.12 \text{ m/s}^2$, $T = 44.8$ N

81. $\Sigma F_x = m_1 a = T - m_1 g \sin \theta$, $\Sigma F_y = m_2 a = m_2 g - T$. Add the two equations and solve for a, using $\theta = 90°$. Use the expression for a to solve for T.

83. $2 \, m_1 m/(m_1 + m_2)$

85. (a) $(2L/t^2) \, [(m_1 + m_2)/(m_1 - m_2)]$; (b) Differentiate with respect to t; $dg/dt = -2g/t$ or $dg/g = -2dt/t$; $m_2 = 0.926$ kg or 1.08 kg

87. (a)

89. 3 kg

91. 14.3 m/s^2

93. 20.8 lb

95. (a) The object will swing backward; (b) $T_x = ma$, $T_y = mg$, $T_y/T_x = \tan \theta = g/a$, $a = g \tan \theta$; (c) 1.6 m/s^2, $9.3°$

97. (a) $a = F/(m_1 + m_2)$; (b) $F_{net} = Fm_2/(m_1 + m_2)$; (c) $Fm_1/(m_1 + m_2)$

99. 4000 N

101. (a) 55 g; (b) 2.45 m/s^2

103. (a) $T = (F_2 + 2F_1)/3$; (b) $3T_0/4C$

105. $-g$

Chapter 5

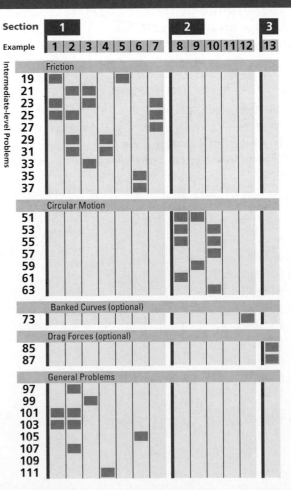

1. The force of friction between the objects and the floor of the truck

3. (*a*) False; (*b*) True; (*c*) True; (*d*) True

5. (*d*)

7. (*e*)

9. (*b*)

11. 0.417

13. (*a*) $\mu_s > \mu_k$; therefore *f* is greater if wheels do not spin; (*b*) $\mu_s = 0.425$

15. (*a*) 49.05 N; (*b*) 123 N

17. (*a*) $\theta > 0$ is preferable; it reduces F_n and therefore f_s; (*b*) 294 N

19. (*b*)

21. (*a*) 0.614 kg, 3.39 kg; (*b*) 9.81 N

23. 83.9 m

25. (*a*) 49.1 m; (*b*) 110 m

27. (*a*) 2.75 m/s^2; (*b*) 10.1 s

29. (*a*) -0.809 m/s^2; (*b*) $T = 0.176$ N

31. (*a*) -0.944 m/s^2; (*b*) -0.425 N

33. (*a*) One expects that *F* will increase with increasing magnitude of the angle since the normal component increases and the tangential component decreases.

(*b*) A force applied at a 0° angle will be most efficient.

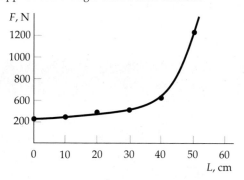

35. (*a*) 0.238; (*b*) 1.4 m/s^2

37. (*a*) 17.7 N; (*b*) 2.95 m/s^2, 5.9 N; (*c*) $a_1 = 1.96$ m/s^2, $a_2 = 7.87$ m/s^2

39. (*a*) $F = \mu_s mg/(\cos\theta + \mu_s \sin\theta)$; (*b*) *F* is a minimum when the denominator is a maximum. Differentiate $(\cos\theta + \mu_s \sin\theta)$ and set to 0. Solve for θ: $\theta = \tan^{-1}\mu_s$. For $\mu_s = 0.6$, $\theta = 31°$

41. $a_{min} = g(\sin\theta - \mu_s \cos\theta)/(\cos\theta + \mu_s \sin\theta)$, $a_{max} = g(\sin\theta + \mu_s \cos\theta)/(\cos\theta - \mu_s \sin\theta)$

43. True, it requires centripetal force.

45. (*e*)

47. 33.4 rpm

49. (*a*) 1.41 m/s; (*b*) 8.5 N

51. (*a*) 4660 N; (*b*) 110 m

53. $R = (m_1/m_2)\,v_2/R$

55. $T_1 = [m_1 L_1 + m_2(L_1 + L_2)](2\pi/T)^2$; $T_2 = m_2(L_1 + L_2)(2\pi/T)^2$

57. 410 N, 53.3°

59. 640 N to 1000 N

61. (*a*) 2.61 N; (*b*) 1.21 m/s

63. 51.6°

65. (*a*) $a_c = \dfrac{v_0^2}{r}\left[\dfrac{1}{1 + (\mu_k v_0/r)t}\right]^2$; (*b*) $-\mu_k a_c$; (*c*) $a = a_c(1 + \mu_k^2)^{1/2}$

67. (*d*)

69. Bonita

71. (*a*) 7.25 m/s; (*b*) 0.536

73. (*a*) 8245 N; (*b*) 1565 N; (*c*) 0.19

75. 176 m

77. The constant *b* should increase with density as more air molecules collide with the object as it falls.

79. (*d*)

81. 3.27×10^{-9} kg/s

83. (*a*) 589 N; (*b*) 0.942 kg/m

85. 88.2 km/h

87. (*a*) 21.8 kg/m; (*b*) 78.48 kN. This initial acceleration would cause internal damage.

89. (*a*) 2.42 cm/s; (*b*) 1.15 h

91. (*d*)

93. (*b*)

95. (*a*) 10.7 m/s; (*b*) 8.0 N

97. (*a*) 0.289; (*b*) 600 N

99. 1486 N

101. (*d*)

103. $\mu_s = 0.577$, $\mu_k = 0.342$

105. $(m_1 + m_2)g\mu_s \cos\theta$

107. 0.433

109. 23.6 rpm

111. 1.22 kg, 0.672

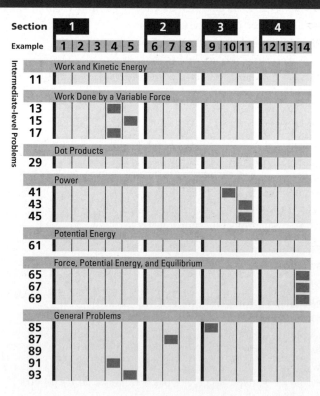

1. (a) False; (b) True; (c) True

3. Yes, if your center of mass is higher when standing than when lying in bed. Otherwise, no.

5. No, $dW = \vec{F} \cdot d\vec{r}$ and here $\vec{F}$ is perpendicular to $d\vec{r}$.

7. (a) 10.8 kJ; (b) 2.7 kJ; (c) 43.2 kJ

9. (a) 240 J; (b) −177 J; (c) 63 J

11. 54.4 kg

13. (a) 6 J; (b) 12 J; (c) 3.46 m/s

15. $19C$ J

17. (a) 2.75 J; (b) 11.4 J; (c) 2.76 m/s; (d) 3.5 J; (e) 2.84 m/s

19. No. Since the force is perpendicular to the motion, it will cause the particle to depart from straight-line motion.

21. (a) $F_{3m} = 417$ N, $F_{4m} = 313$ N, $F_{5m} = 250$ N; (b) 1.25 kJ for each L; (c) Choosing a longer length means one can exert a smaller force.

23. 180°

25. (a) −24; (b) −10; (c) 0

27. (a) 1 J; (b) 0.213 N

29. (a) $(d/dt)(\vec{r} \cdot \vec{r}) = \vec{r} \cdot (d\vec{r}/dt) + (d\vec{r}/dt) \cdot \vec{r} = 2\vec{v} \cdot \vec{r} = 0$. Therefore $\vec{v} \perp \vec{r}$. (b) $(d/dt)(\vec{v} \cdot \vec{v}) = 2\vec{a} \cdot \vec{v} = 0$. Therefore $\vec{a} \perp \vec{v}$. The above implies that the component of $\vec{a}$ in the plane formed by $\vec{r}$ and $\vec{v}$ is colinear with $\vec{r}$. (c) $(d/dt)(\vec{v} \cdot \vec{r}) = \vec{v} \cdot (d\vec{r}/dt) + \vec{r} \cdot (d\vec{v}/dt) = v^2 + \vec{r} \cdot \vec{a} = 0$. Therefore, $a_r = -v^2/r$.

31. (d)

33. (a)

35. F_B

37. (a) 2 m/s; (b) 24 J

39. (a) 24 W; (b) −50 W; (c) 24 W

41. 26.7 kW

43. From Example 6-11, $x = (8P/9m)^{1/2}t^{3/2}$. From $v = (2P/m)^{1/2}t^{1/2}$, solve for $t^{3/2}$ and substitute this into the expression for x. Simplifying, one obtains $x = (m/3P)v^3$.

45. 4.9 m/s, 19.6 m

47. (c)

49. (a)

51. (c)

53. No. The steeper trail requires fewer steps, but more effort per step.

55. 4.71 kJ

57. 879 MW

59. (a) $U(x) = -6(x - x_0)$; (b) $U(x) = 24 - 6x$; (c) $U(x) = 50 - 6x$

61. $m_1 = 5.68$ kg, $m_2 = 4.32$ kg

63. (a) $A, E: F_x < 0$; $B, D, E: F_x = 0$; $C: F_x > 0$; (b) A; (c) B: unstable equilibrium; D: stable equilibrium

65. (a) $F_x = -C/x^2$; (b) directed toward origin; (c) decreases; (d) directed away from origin, increases

67. $U(x) = a/x + U_0$

69. (a) $U(y) = 2Mg[(d^2 + y^2)^{1/2} - d] - mgy$; (b) $y = d(m/2M)/\sqrt{1 - (m/2M)^2}$; (c) $y_0 = d(m/2M)/\sqrt{1 - (m/2M)^2}$

71. Stable equilibrium at $x = 0$, unstable equilibrium at $x = \pm 2$. $U(x)$ is local minimum at $x = 0$, local maximum at $x = \pm 2$.

73. (a) $U(x)$ decreases; (b) $U(x) = 4/x^2$ J;
(c)

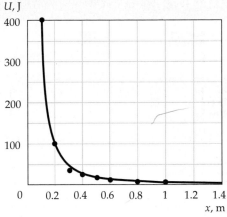

75. (d)

77. (a) $2h$; (b) wh; (c) $F \times 2h = wh$

79. (a) 706 MJ; (b) 11.8 MW

81. 0.5 m

83. 39.3 m

85. (a) 12 MW; (b) 1261 citizens

87. (a) 0; (b) 78 J

89. (a) 9 J; (b) −22 W

91. (a)

x, m	−4	−3	−2	−1	0	1	2	3	4
W, J	−11	−10	−7	−3	0	1	0	−2	−3

(b)

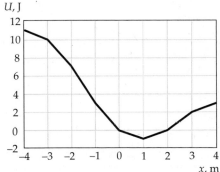

93. (a) μyg; (b) $\mu g(l^2/2)$

95. (a) $W = Tx$; (b) $v(x) = \sqrt{2[(T/m) - g\sin\theta]x}$;
(c) $P = T\sqrt{2[(T/M) - g\sin\theta]x}$

97. (a) $|\vec{F}| = (F_x^2 + F_y^2)^{1/2} = [(F_0^2/r^2)(y^2 + x^2)]^{1/2} = F_0$;
$\vec{F} \cdot \vec{r} = (F_0/r)(y\hat{i} - x\hat{j}) \cdot (x\hat{i} + y\hat{j}) = (F_0/r)(yx - xy) = 0$, thus $\vec{F} \perp \vec{r}$; (b) $W = 10\pi F_0$ (clockwise rotation); the force is not conservative.

Chapter 7

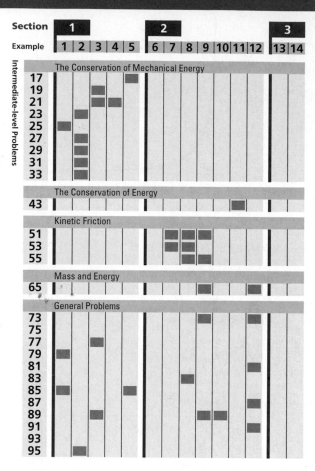

1. Using conservation of mechanical energy is generally simpler, involving only scalars; however, some details can't be obtained (e.g., trajectories).

3. (c)

5. (c)

7. (a) 1.215 J; (b) 3.49 m/s; (c) 2.34 m

9. (a) 0.858 m; (b) The spring will accelerate the mass and it will then retrace the path, rising to a height of 5 m.

11. (a)

13. 26.7 m/s

15. 25.6°

17. 1.40 m/s

19. $[mg(\sin \theta + \mu_s \cos \theta)]^2/2k$

21. 8.0 m/s

23. $6mg$

25. 27.2 m/s

27. (a) 60°; (b) 51.3°

29. (a) 3.52 m/s; (b) 7.89 J; (c) 25.3 N; (d) 49°

31. (a) $(5/2)\,mgL$; (b) 6 mg

33. (a) 20.2°; (b) 6.4 m/s

35. $v = L\theta\sqrt{g/L + (k/m)\theta^2}$

37. The kinetic energy of the man increases at the expense of metabolic (chemical) energy.

39. (b)

41. (a) 0; (b) 560 J; (c) The increase in kinetic energy is at the expense of a loss in metabolic (chemical) energy.

43. (a) 94.2 kJ; (b) The energy comes from metabolic energy. (c) 471 kJ

45. Metabolic (chemical) energy converted to thermal energy is released through friction.

47. (a) 104 J; (b) 70.2 J; (c) 33.8 J; (d) 2.91 m/s

49. (a) 7.67 m/s; (b) 58.9 J; (c) 1/3

51. (a) 13.73y J; (b) −13.73y J; (c) 1.98 m/s

53. (a) $(3/8)\,mv_0^2$; (b) $(3v_0^2)/(16\pi gr)$; (c) 1/3 revolution

55. (a) 2.15 m; (b) 0.63 m/s; (c) 133 J

57. (a) $d = (mg/d)(\sin \theta + \mu_s \cos \theta)$; (b) $\mu_k = \tan \theta - \frac{1}{2}(1 + \mu_s \cot \theta)$

59. (a) 9×10^{13} J; (b) $\$2.5 \times 10^6$; (c) 9×10^{11} s = 28,400 yr

61. 3.56×10^{14} reactions

63. 0.782 MeV

65. 1.05 kg; (b) 3.06×10^9 kg

67. (b)

69. (a) 8.64 MJ; (b) 2065 kcal

71. 5.76 MJ

73. 350 kW

75. 45.1 kW

77. 17.3 m/s

79. (a) 1600 J; (b) 619 J; (c) 16 m/s

81. 18 kW

83. (a) $y = y_0 = 0$; (b) $F = mg - ky$; (c) $y = y_{max} = 2mg/k$; (d) $y_{eq} = mg/k$; (e) $W_f = \frac{1}{2}(m^2g^2/k)$

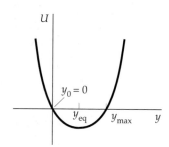

85. (a) 17.34 m; (b) 4905 N; (c) 4.905 m/s²; (d) 13.41 kN; (e) 5461 N; (f) 1440 N

87. (a) 981 N; (b) 29.4 kW; (c) 8.85°; (d) 6.36 km/L

89. (a) 0.989 m; (b) 0.783 m; (c) 1.54 m

91. (a) 115.7 kJ; (b) 90 kJ; (c) 1.58

93. 10.96 m/s, 9.12 s. The pole vaulter uses additional metabolic energy to raise himself on the pole.

95. $\theta_2 = \cos^{-1}\left[1 - \dfrac{L}{L - x}(1 - \cos \theta_1)\right]$

97. (a) $F_{tan} = -mg \sin \theta$; therefore $a = dv/dt = -g \sin \theta$; (b) For circular motion, $v = r\omega = L\,d\theta/dt$; $d\theta/dt = v/L$; (c) $dv/dt = (dv/d\theta)(d\theta/dt) = (v/L)(dv/d\theta)$; (d) $dv/d\theta = (L/v)(dv/dt) = -(L/v)\,g \sin \theta$; $v\,dv = -gL \sin \theta\,d\theta$;

(e) $\displaystyle\int_0^v v\,dv = \int_{\theta_0}^0 -gL \sin \theta\,d\theta; \frac{1}{2}v^2 = gL(1 - \cos \theta_0)$;

Note that $L(1 - \cos \theta_0) = h$; consequently, $v = \sqrt{2gh}$.

Chapter 8

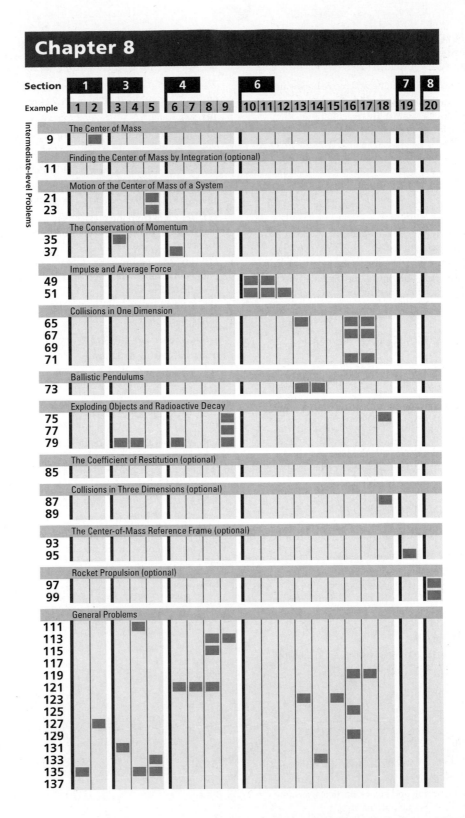

Section	1		3			4				6									7	8
Example	1	2	3	4	5	6	7	8	9	10	11	12	13	14	15	16	17	18	19	20

Intermediate-level Problems

The Center of Mass
9

Finding the Center of Mass by Integration (optional)
11

Motion of the Center of Mass of a System
21
23

The Conservation of Momentum
35
37

Impulse and Average Force
49
51

Collisions in One Dimension
65
67
69
71

Ballistic Pendulums
73

Exploding Objects and Radioactive Decay
75
77
79

The Coefficient of Restitution (optional)
85

Collisions in Three Dimensions (optional)
87
89

The Center-of-Mass Reference Frame (optional)
93
95

Rocket Propulsion (optional)
97
99

General Problems
111
113
115
117
119
121
123
125
127
129
131
133
135
137

1. hollow sphere

3. 15.6 m

5. 3.36 m

7. (2 m, 1.4 m)

9. (1.5 m, 1.36 m)

11. (9/16)L

13. (0, 0, R/2)

15. The external force is the force of the static friction between the tires and the road.

17. (b)

19. 4 m/s east

21. (a) Yes. Initially the scale reads $(M + m)g$. After m drops, the scale reads Mg. (b) $mg/(M + m)$ directed downward; (c) $F_{net} = (M + m)g - (M + m) a_{cm} = Mg$.

23. (a) $[(m_1 - m_2)^2 g]/[(m_1 + m_2)(m_1 + m_2 + m_c)]$; (b) $4gm_1m_2/(m_1 + m_2) + m_c g$; (c) Since $T = 2gm_1m_2/(m_1 + m_2)$, then $F = 2T + m_c g$

25. (a) True; (b) True; (c) True

27. If the man throws something forward, he will move backward.

29. Because of the conservation of momentum, the rocket can move without having something to push against. By expelling fuel, the rocket increases in velocity in the opposite direction.

31. 1.83 m/s to the left

33. (c)

35. 40 m

37. $\sqrt{gh/2}$

39. rolling ball

41. (a) 60 J; (b) 3.75 m/s to the right; (c) v of 3-kg block = 1.25 m/s to the right, v of 5-kg block = 0.75 m/s to the left; (d) 3.75 J; (e) 6.75 J

43. Assume that the ball travels 80 mi/h = 35 m/s and that the ball stops in a distance of 1 cm. The collision time is the distance for which the ball and bat are in contact divided by the average speed. Thus the collision time ≈ 0.02 m/18 m/s = 1 ms.

45. (a) 10.75 N·s; (b) 1344 N

47. $I = 1.81 \times 10^6$ MN·s; $F_{av} = 0.602$ MN

49. 1.15 kN

51. (a) If $d = 0.7$ m, then $F_{av} = 84$ N. (b) Yes, the weight of the ball is much less than the average force.

53. (a) 4.95×10^{-5} N; (b) The weight of the water droplet is about six times the average force due to 10 drops.

55. (a) False; (b) True; (c) True

57. (a) The energy losses are the same. (b) The percentage loss is greater for the two objects with oppositely directed velocities of $v/2$.

59. 0.56 m/s

61. 3.13 m/s

63. $3v$

65. (a) 1.5 m/s to the right; (b) v of 0.3-kg muffin = 2.5 m/s, v of 0.5-kg muffin = 4.5 m/s

67. v of 2-kg mass = 0.8 m/s, v of 3-kg mass = 4.8 m/s

69. (a) 5 m/s; (b) 25 cm; (c) $v_1 = 0$ m/s, $v_2 = 7$ m/s

71. (a) $0.2v_0$; (b) $0.4 v_0$

73. $[2(m_1 + m_2)/m_1] \sqrt{gL}$

75. (a) 9 m/s $\hat{i}$ − 2 m/s $\hat{j}$; (b) 6 m/s $\hat{i}$

77. $v_p = 1.74 \times 10^7$ m/s, $v_\alpha = 4.34 \times 10^6$ m/s

79. (a) $v_x = 312$ m/s, $v_y = 66.6$ m/s; (b) 5610 m; (c) 35.8 kJ

81. 0.913

83. 0.894

85. (a) 1.7 m/s; (b) 0.83

87. (a) $\frac{1}{2}v_0\hat{i} + v_0\hat{j}$; (b) $mv_0^2/16$

89. (a) $v_2 = \sqrt{2}v_0$; (b) $K_i = \frac{1}{2}m(3v_c)^2 = 4.5\,mv_0^2$, $K_f = \frac{1}{2}m(\sqrt{5}v_0)^2 + \frac{1}{2}(2m)(\sqrt{2}v_0)^2 = 4.5mv_0^2$

91. (a) Let m_1 and m_2 be the incoming and struck particles; assume $\vec{v}_0$ is in the x direction. From $\vec{p}_i = \vec{p}_f$ one obtains $m_1v_0 = m_1v \cos \phi + m_2v_2 \cos \theta$ (1) and $m_1v \sin \phi = m_2v_2 \sin \theta$ (2). Rearrange the first equation to $m_1(v_0 - v \cos \phi) = m_2v_2 \cos \phi$ (1a). Divide (2) by (1a) to obtain $\tan \theta = (v \sin \phi)/(v_0 - v \cos \phi)$. (b) Did not use energy conservation; therefore, the result is valid for elastic and inelastic collisions.

93. $K_i = p_1^2/2m_1 + p_1^2/2m_2$; $K_f = p_1'^2/2m_1 + p_1'^2/2m_2$. In an elastic collision, $K_f = K_i$, thus $p_1'^2 = p_1^2$ and $p_1' = \cdot\ p_1$. If $p_1' = +p_1$, the particles do not collide.

95. (a) 3.75 m/s; (b) $u_{3kg} = 1.25$ m/s, $u_{5kg} = -0.75$ m/s; (c) $u'_{3kg} = -1.25$ m/s, $u'_{5kg} = 0.75$ m/s; (d) $v'_{3kg} = 2.5$ m/s, $v'_{5kg} = 4.5$ m/s; (e) $K_i = K_f = 60$ J

97. 15 km/s

99. (a) 3.6×10^5 N; (b) 120 s; (c) 1.72 km/s

101. (e)

103. $N_f = 0.192$ m/s, $K_i = 31.3$ mJ, $K_f = 11.7$ mJ

105. 0.46 m/s

107. (a) -1.1×10^5 kg·km/h $\hat{i}$ + 1.05×10^5 kg·km/h $\hat{j}$; (b) $v_f = 43.4$ km/h, $\theta = 43°$ north of west

109. (a) 81 J; (b) 0 m/s; (c) $K_{cm} = 0$; (d) $K_{rel} = 81$ J

111. (a) 2.5 m from the pier; (b) $K_{total} = 330$ J, $K_{on\ land} = 270$ J; (c) 60 J derives from the chemical energy of the woman. (d) Shot will land in the water.

113. (a) Momentum of the system is not conserved because the rails will exert a vertical reaction force. (b) 4.33 m/s in direction opposite to shell; (c) 346 kJ

115. (a) 2.3 s; (b) The pumpers hit the ground at 24 km/h.

117. $0.75v\hat{i} - v\hat{j}$

119. (a) 29.6 km/s away from planet; (b) 8.1: This energy comes from the slowing of Saturn.

121. (a) -0.6 m/s; (b) 960 N

123. No

125. 8.85 kg

127. $r/14\hat{k}$

129. (a) From Problem 128, the fraction of energy retained after collision is $-\Delta K/K = 4(m/M)/(1 + m/M)^2 = 0.284$. The fraction of energy retained = $1 - 0.284 = 0.716$. After N collisions, $K_f = K_i \times 0.716^N$. (b) 55

131. (a) $v^2t^2/2L$; (b) v^2/L; (c) $Mv^3t/L^2 + Mvtg/L$

133. (a) $(m_p + m_b)g$; (b) $F = g[m_p + m_b \sqrt{(2kh)/g(m_c + m_b)}]$

135. (a) 4670 km; (b) gravitational force of sun and other planets; (c) acceleration is toward the sun; (d) 9330 km

137. -1.5×10^{-22} kg·m/s $\hat{j}$

139. $1.2\sqrt{L}$

Chapter 9

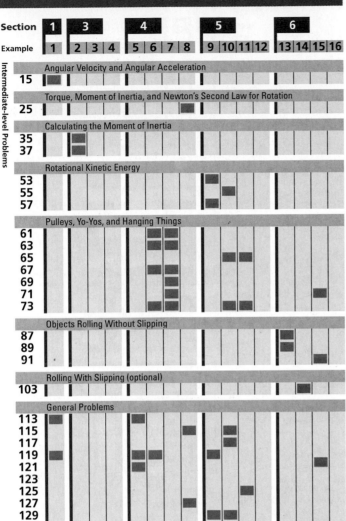

1. The point on the rim moves the greater distance. Both turn through the same angle. The point on the rim has the greater speed. Both have the same angular velocity. Both have zero tangential acceleration. Both have zero angular acceleration. The point on the rim has the greater centripetal acceleration.

3. (c)

5. (a) 15.6 rad/s; (b) 46.8 rad; (c) 7.45 rev; (d) 4.68 m/s, 73 m/s^2

7. (a) 40 rad/s; (b) 0.96 m/s^2; 192 m/s^2

9. (a) 0.233 rad/s; (b) 2.8 m/s; 0.65 m/s^2

11. 7.27×10^{-5} rad/s

13. (a)

15. 1.04 rad/s, 9.93 rev/min

17. (d)

19. No, it may cause a rotating object to come to rest.

21. (b)

23. (a) 1.87 N·m; (b) 124 rad/s^2; (c) 620 rad/s

25. (a) $g \sin \theta$; (b) $mgL \sin \theta$; (c) $g \sin\theta$

27. (a) $2(M/R^2) \mu_k g r^2 \, dr$; (b) $(2/3) MR\mu_k g$; (c) $3Rw/4\mu_k g$

29. 4.66×10^{-5} kg·m^2

31. 28 kg·m^2

33. $(7/5)MR^2$

35. (a) $m_1 x^2 + m_2 (L - x)^2$; (b) $2(m_1 x + m_2 x - m_2 L)$, $dI/dx = 0$ when $x = m_2 L/(m_1 + m_2)$

37. (a) 0.04 kg·m^2/0.04145 kg·m^2 = 0.965; (b) The moment of inertia would increase because I_{cm} of the hollow sphere $> I_{cm}$ of the solid sphere.

39. Let the element of mass be $dm = \rho \, dV = 2\pi\rho h r \, dr$, where h is the height of the cylinder. The mass m of the hollow cylinder is $m = \pi\rho h(R_2^2 - R_1^2)$, so $\rho = m/[\pi h(R_2^2 - R_1^2)]$. The element $dI = r^2 \, dm = 2\pi\rho h r^3 \, dr$. Integrate dI from R_1 to R_2 and obtain $I = \frac{1}{2}\pi\rho h(R_2^4 - R_1^4) = \frac{1}{2}\pi\rho h(R_2^2 + R_1^2)(R_2^2 - R_1^2) = \frac{1}{2}m(R_2^2 + R_1^2)$.

41. (a) $0.509 \, M/R^3$; (b) $0.329 \, MR^2$

43. $(\frac{1}{2})MR^2$

45. $(\frac{1}{2})MR^2$

47. (b)

49. (a) 52 kg·m^2; (b) 20.9 rev/min

51. 155 kW

53. $K_{rot} = 2.6 \times 10^{29}$ J, $K_{orb} = 2.7 \times 10^{33}$ J; $K_{orb} \simeq 10^4 \, K_{rot}$

55. (a) $\sqrt{(8mg/R)/2m + M}$; (b) $mg[1 + 8m/(2m + M)]$

57. 1.95 m

59. $\sqrt{(1 + M/2m)/(1 + M/3m)}$

61. 3.11 m/s^2; $T_1 = 12.4$ N, $T_2 = 13.4$ N

63. 1.56 m/s^2; $T_1 = 16.0$ N, $T_2 = 16.5$ N

65. 8.2 m/s

67. (a) $g/[1 + (2M/5m)]$; (b) $2mMg/(5m + 2M)$

69. (a) 72 kg; (b) 1.37 rad/s^2, $T_1 = 294$ N, $T_2 = 745$ N

71. (a) mg; (b) $2g/R$; (c) $2g$

73. (a) $(g \sin \theta)/(1 + m_1/2m_2)$; (b) $(\frac{1}{2}m_1 g \sin \theta)/(1 + m_1/2m_2)$; (c) $m_2 gh$; (d) $m_2 gh$; (e) $\sqrt{(2gh)/(1 + m_1/2m_2)}$; (f) For $\theta = 0$: $a = T = 0$; for $\theta = 90°$: $a = g/(1 + m_1/2m_2)$, $T = \frac{1}{2}m_1 a$, $v = \sqrt{(2gh)/(1 + m_1/2m_2)}$; for $m_1 = 0$: $a = g \sin \theta$, $T = 0$, $v = \sqrt{2gh}$

75. True

77. (b)

79. (c)

81. Assume the ball is rolling to the right. If f acts to the right along the direction of motion, the center of mass will then accelerate to the right and its linear speed will increase. But the torque produced by f about the center of mass will be counterclockwise tending to decrease the angular speed. But $v = Rw$, so w cannot decrease while v is increasing. If f acts to the left so that v decreases, it produces a clockwise torque that tends to increase w. The only possibility consistent with $v = Rw$ is $f = 0$.

83. 1125 J

85. 45.9 m

87. 3/2

89. 1.09L

91. 0.325°

93. 233 J

95. (a) From the figure it is evident that $x = r_0 \cos \theta$ and $y = r_0 \sin \theta$ relative to the center of the wheel. Therefore, if the coordinates of the center are X and R, those of point P are as stated.
(b) $v_{Px} = d(X + r_0 \cos \theta)/dt = dX/dt - r_0 \sin \theta \, d\theta/dt$. Note that $dX/dt = V$ and $d\theta/dt = -\omega = -V/R$; therefore, $v_{Px} = V + (r_0 V \sin \theta)/R$. $v_{Py} = d(R + r_0 \sin \theta)/dt = r_0 \cos \theta \, d\theta/dt \, (dR/dt = 0)$. Again, $d\theta/dt = -\omega$, so $v_{Py} = -(r_0 V \cos \theta)/R$.

(c) $\vec{v}\cdot\vec{r} = v_{Px}r_x + v_{Py}r_y = (V + r_0V\sin\theta/R)(r_0\cos\theta) - (r_0V\cos\theta/R)(R + r_0\sin\theta) = 0$

(d) $v^2 = v_x^2 + v_y^2 = V^2[1 + (2r_0/R)\sin\theta + r_0^2/R^2]$; $r^2 = r_x^2 + r_y^2 = R^2[1 + (2r_0/R)\sin\theta + r_0^2/R^2]$; so $v/r = V/R = \omega$.

97. $2F/R(M + 3m)$, counterclockwise; (b) $F/(M + 3m)$; (c) $-2F/(M + 3m)$.

99. $54°$

101. (a)

The translational impulse $P_t = F_0\,\Delta t = mv_0$. The rotational impulse about the center of mass is $P_\tau = P_t(h - r) = I\omega_0$.

103. With $I = (2/5)mr^2$ one then obtains $\omega_0 = 5v_0(h - r)/2r^2$.

105. (a) 200 m/s; (b) 8000 rad/s; (c) 257 m/s, 11.6 s

107. (a) $5v_0/3R$; (b) $(5/21)v_0$; (c) $1.056\ mv_0^2$; (d) $-1.016\ mv_0^2$

109. (a) $(2/3)\,v_0$; (b) $5v_0^2/18\mu_k g$; (c) $1/3$

111. 2.7×10^{-6} rad/s

113. (a) 0.0873 rad/s²; (b) 572 N·m; (c) 6550 kg·m²

115. (a) 7.36 m/s²; (b) 14.7 m/s²; (c) 2.43 m/s

117. $(6gL/7)^{1/2}$

119. (a) 780 kJ; (b) 211 N·m, 151 N; (c) 1375 rev

121. (a) 15 m; (b) 15.4 rad/s

123. (a), (b) $I_z = \int r^2\,dm = \int(x^2 + y^2)\,dm = \int x^2\,dm + \int y^2\,dm = I_y + I_x$; (c) Let the z axis be the axis of rotation of the disk. By symmetry, $I_x = I_y$. So $I_x = \frac{1}{2}I_z = (1/4)MR^2$. See Table 9-1.

125. (a) The spool will move down the plane at a constant acceleration, spinning in a counterclockwise direction as the string unwinds. $v = \sqrt{(2MgD\sin\theta)/(M + I/r^2)}$; (b) The direction of the friction force is up along the plane. $f_s = (Mg\sin\theta)/(1 + R/r)$

127. $0.75mg, 0.5mg, 0.25mg, 0, 0, 0$

129. (a) 230.4 N/m; (b) 122.7 J

131. (a) The only force is F; therefore, $a_{cm} = F/M$. The torque about the center of mass is $\tau = FR$ and $I = MR^2/2$. Thus $\alpha = \tau/I = 2F/MR$. If the cylinder rolls without slipping, $a_{cm} = \alpha R$. Here, $\alpha = 2a_{cm}/R$. (b) $f = F/3$, same direction as F; $a_{cm} = 4F/3M$

133. (a) $T(2r/R - 1)/3$; (b) $(2T/3m)(1 + r/R)$; (c) $r > R/2$ (d) Same direction as T.

Chapter 10

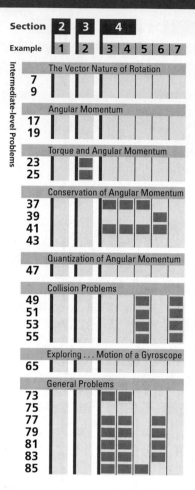

1. (a) True; (b) True; (c) True

3. $\vec{F} = -F\hat{i}, \vec{r} = R\hat{j}, \tau = +FR\hat{k}$

5. (a) $24\hat{k}$; (b) $-24\hat{j}$; (c) $13\hat{k}$

7. (a) Let $\vec{r}$ be in the x-y plane. If $\vec{\omega}$ points in the positive z direction, i.e., $\vec{\omega} = \omega\hat{k}$, the particle's velocity is in the $\hat{j}$ direction when $\vec{r} = r\hat{i}$ (see Figure) and has the magnitude $r\omega$. Thus, $\vec{v} = \vec{\omega}\times\vec{r} = r\omega\hat{j}$.

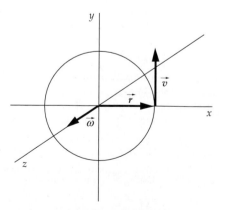

(b) $\vec{a} = d\vec{v}/t = (d\vec{\omega}/dt)\times\vec{r} + \vec{\omega}\times(d\vec{r}/dt) = (d\vec{\omega}/dt)\times\vec{r} + \vec{\omega}\times\vec{v} = \vec{a}_t + \vec{\omega}\times(\vec{\omega}\times\vec{r}) = \vec{a}_t + \vec{a}_c$, where $\vec{a}_t$ and $\vec{a}_c$ are the tangential and centripetal accelerations, respectively.

9. $4\hat{j} + 3\hat{k}$

11. (b)

13. constant

15. (a) 28 kg·m²/s; (b) 32 kg·m²; (c) 0.875 rad/s²

17. The area at $t = t_1$ is $A_1 = \frac{1}{2}br_1 \cos\theta_1 = \frac{1}{2}bx_1$, where θ_1 is the angle between $\vec{r}_1$ and $\vec{v}$ and x_1 is the component of $\vec{r}_1$ in the direction of $\vec{v}$. At $t = t_1 + dt$, $A = A_1 + dA = \frac{1}{2}b(x + dx) = \frac{1}{2}b(x + v\,dt)$. Thus, $dA/dt = \frac{1}{2}bv = $ constant. Note that $r \sin\theta = b$; consequently, $\frac{1}{2}bv = \frac{1}{2}L/m$.

19. $\vec{\tau} = \vec{r}_1 \times \vec{F}_1 + \vec{r}_2 \times \vec{F}_2 = (\vec{r}_1 - \vec{r}_2) \times \vec{F}_1$ since $\vec{F}_2 = -\vec{F}_1$. But $\vec{r}_1 - \vec{r}_2$ points along $-\vec{F}_1$ so $(\vec{r}_1 - \vec{r}_2) \times \vec{F}_1 = 0$.

21. (a) 4 N·m; (b) $0.192t$ rad/s

23. (a) $Rg(m_2 \sin\theta - m_1)$ into the page; (b) $vR(I/R^2 + m_1 + m_2)$; (c) $(m_2 \sin\theta - m_1)g/(I^2/R + m_1 + m_2)$

25. (a) 0.785 N·m; (b) 2.75 kg·m²/s; (c) 0.131 kg·m²/s; (d) 0.317, 0.634, 0.048

27. False

29. (e)

31. No

33. (b)

35. (c)

37. (a) $\omega_i/(1 + mR^2/I_0)$; (b) ω_f

39. (a) $r_0 mv_0$; (b) $\frac{1}{2}mv_0^2$; (c) mv_0^2/r_0, $(3/2) mv_0^2$

41. 0.11 ms or 1.31×10^{-9} days

43. (a) 4 m/s; (b) 3.2 N/m

45. 54.7°

47. (a) 3.46×10^{-47} kg·m²; (b) 2.0 meV, 6.0 meV, 12 meV

49. 14 J·s

51. $L[(M - m)/12m]^{1/2}$

53. 7.75 m/s

55. 0.88

57. 7.36 rad/s; 11.1 J

59. (a) False; (b) True; (c) False

61. (d)

63. First, imagine that $\vec{L}$ points vertically upward and the car is moving directly away from you, rounding the top of a hill. The original $\vec{L}$ is now altered by the addition of a ΔL pointing to the left, hence $\vec{L}$ is rotated counterclockwise and the car rolls to its left. If the original $\vec{L}$ had pointed downward, the same ΔL would have been added, but now it causes a clockwise rotation of $\vec{L}$, and hence the car rolls to the right. Similarly, imagine $\vec{L}$ points forward, the car is moving away from you, and is turning to the left. The torque exerted by the road on the car in this case points upward. This means an upward component of angular momentum is added to $\vec{L}$, thus tending to lift the front of the car. If $\vec{L}$ had pointed backward, the same left turn would add the same angular momentum increment, this time raising the rear of the car.

65. (a) 3.27 rad/s; (b) 0.163 m/s; (c) 0.535 m/s²; (d) 24.5 N, 1.34 N

67. To prevent body of helicopter from rotating. If rear rotor fails, body of helicopter will tend to rotate on its main axis.

69. (e) The ball rotates counterclockwise. The torque about the center of the pole is clockwise and of magnitude RT, where R is the pole's radius and T is the tension. So L must decrease.

71. (a) $-47.7\hat{k}$; (b) $15.9\hat{k}$ N·m

73. 4.2 rev/s with $m_{arm} = 4$ kg

75. (a) 1. Use Newton's laws to determine v; $T \cos\theta = mg$; $T \sin\theta = mv^2/(r \sin\theta)$; $v = (rg \sin\theta \tan\theta)^{1/2} = 2.06$ m/s;
$\vec{v} = d\vec{r}/dt = -2.06 \sin\omega t\,\hat{i} + 2.06 \cos\omega t\,\hat{j}$

2. Find $\vec{L} = m\vec{r} \times \vec{v} = 6.18[\sin 30°(\cos\omega t\,\hat{i} + \sin\omega t\,\hat{j}) - \cos 30°\,\hat{k}] \times (-\sin\omega t\,\hat{i} + \cos\omega t\,\hat{j}) = [(3.09\hat{k} + 5.35(\sin\omega t\,\hat{j} + \cos\omega t\,\hat{i})]$ J·s

(b) $dL/dt = 14.7$ N·m, $\tau = mgr \sin 30° = 14.7$ N·m

77. $\omega_f = [(M + 5 m\ell^2/L^2)/(M + 5m)]\omega$; $K_i = (ML^2 + 5m\ell^2)\omega^2/20$; $K_f = [(ML^2 + 5m\ell^2)^2/(ML^2 + 5mL^2)]\omega^2/20$

79. 30 rad/s; 10.5 rad/s; 176 J, 61.7 J

81. $\omega_i = 30$ rad/s, $\omega_f = 14.13$ rad/s; 93.3 J

83. 6.32 rad/s; 1.54 J

85. 0.55 s

87. $\omega = 12.5$ rad/s in both instances

89. (a) 21.4 rad/s; (b) $0.378\,e^{1.14t}$ J·s

Chapter 11

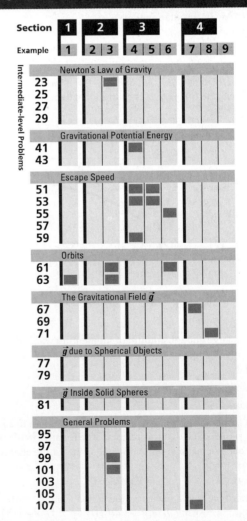

1. (a) False; (b) True

3. 4.39×10^{11} m

5. 3.91×10^{13} m

7. 11.7 y

9. The mass of the building is significant compared to the mass of the earth.

11. (d)

13. (a) 1.88×10^9 m; (b) 1.9×10^{27} kg

15. 6.02×10^{24} kg

17. 2.45 m/s^2

19. 39.2 m/s^2

21. 7.5×10^5 m/s

23. (a) 37.6 N; (b) 2.66 km/s; (c) 1.33×10^5 s

25. $F \propto 1/R^3$

27. (a) 0.0027 N; (b) 0.48 N; (c) -0.061%

29. 0.605

31. Because gravity is weak compared to other forces of nature, either an extremely massive object or an extremely accurate instrument is needed to see its effects.

33. (a) 2.45×10^{-9} N; (b) $\tau = 4.41 \times 10^{-10}$ N·m

35. (a) His effectiveness would depend on his mass rather than his weight. (b) The power requirement would not be determined by the car's weight, but by its mass. (c) No significant effect.

37. (a) 5.77 kg; (b) gravitational mass

39. GMm_0/R

41. 6.95 km/s

43. (a) Outside: $\vec{F} = -(GMm_0/r^2)\hat{r}$. Inside: $\vec{F} = 0$. (b) $U(r) = -GMm_0/r$; $U(R) = -GMm_0/R$. (c) Since $F = 0$ for $r < R$, $dU/dr = 0$ and $U = $ constant. (d) Since U is continuous, then for $r < R$, $U(r) = U(R) = -GMm_0/R$.

(e)

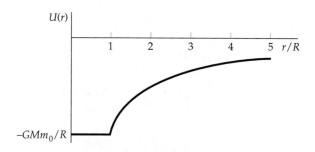

45. (a) $M/2\pi R$; (b) $-\dfrac{GM}{R}\ln\left(\dfrac{R + \sqrt{x^2 + R^2}}{x}\right)$

(c) $GM/(x\sqrt{x^2 + R^2})$

47. Yes, if a large enough amount of momentum was imparted to the earth, say through a collision with a huge comet.

49. 35.5 km/s

51. 19.4 km/s

53. 61.0 km/s

55. From Example 11-6, the kinetic energy of a mass m in a circular orbit is $K = \frac{1}{2}|U|$. Note that this result is true for any circular orbit of a mass m about a massive center. But $|U|$ is the escape energy of $\frac{1}{2}mv_e^2$. Consequently, $K = \frac{1}{2}mv_c^2 = \frac{1}{2}(\frac{1}{2}mv_e^2)$ and $v_c = v_e/\sqrt{2}$.

57. $v_{eS} = 42.2$ km/s, $v_{e,\,solar} = 43.7$ km/s

59. To determine H use energy conservation: $\frac{1}{2}mv^2 = -\Delta U = GMm[1/R_E - 1/(R_E + H)]$ or $v^2 = 2gR_E^2[1/R_E - 1/(R_E + H)]$. So $H' = R_E H/(R_E + H)$ and, solving for H, $H = R_E H'/(R_E - H')$.

61. (a) 7.3 h; (b) 1.04×10^9 J; (c) 8.73×10^{10} J·s

63. $v = \sqrt{GM_S/r - GM_E r/(D - r)^2}$

65. 1×10^{-8} N $\hat{j}$

67. (a) 9.66×10^{-8} N $\hat{j}$; (b) 4.83×10^{-8} N/kg $\hat{j}$

69. (a) g of opposite elements of mass $R\lambda\, d\theta$ cancel. By symmetry, $g = 0$ at center. (b) 1. $m_1/m_2 = r_1/r_2$; 2. $g_1 > g_2$; 3. toward m_1; (c) toward m_1; (d) $g = 0$; (e) 1. $m_1/m_2 = r_1^2/r_2^2$; 2. $g_1 = g_2$; 3. $g = 0$; $g = 0$

71. (a) $CL^2/2$; (b) $-GC\displaystyle\int_0^L \dfrac{x\,dx}{(x_0 - x)^2} = \dfrac{2GM}{L^2}\left[\ln\left(\dfrac{x_0}{x_0 - L}\right) - \left(\dfrac{L}{x_0 - L}\right)\right]\hat{i}$

73. g is proportional to the mass within the sphere and inversely proportional to the radius, i.e., proportional to $r^3/r^2 = r$.

75. Zero. The gravitational field inside the 2-m shell is zero; therefore, it exerts no force on the 1-m shell, and, by Newton's third law, that shell exerts no force on the larger shell.

77. $g_1\,(R_1^2/R_2^2)$

79. (a) $(Gm/a^2)(M_1/4.84 + M_2/9)$; (b) $GmM_1/1.21\,a^2$; (c) 0

81. $G\left(\dfrac{4\pi\rho_0 R_3}{3}\right)\left[\dfrac{1}{x^2} - \dfrac{1}{8(x - \frac{1}{2}R)^2}\right]$

83. $\sqrt{4\pi\rho_0 G/3}$

85. 1.04×10^{-5} m/s

87. (a) $-(GMm/d^2)[1 - (d^3/4)/(d^2 + R^2/4)^{3/2}]\hat{i}$; (b) $-0.821(GMm/R^2)\hat{i}$

89. (d)

91. 1.9×10^{27} kg

93. (a) 1.16×10^6 s; (b) 8.79×10^{25} kg

95. (a) $GM_E m\left(\dfrac{1}{r_1} - \dfrac{1}{r_2}\right)$; (b) In the above expression, replace $GM_E m$ by mgR_E^2, r_1 by R_E, and r_2 by $R_E + h$ to obtain the result given. (c) $[(1/R_E) - 1/(R_E + h)] = h/[R_E(R_E + h)]$; if $h \ll R_E$, the denominator $\approx R_E^2$ and $W = mgh$.

97. (a) $-0.56/r^2$; (b) $-(5.6 \times 10^{-7})r$

99. 8.94×10^7 m

101. 1700 m

103. Take the coordinate origin at the center of mass. Then $r_1 m_1 = r_2 m_2$ and $r = r_1 + r_2$. The force holding m_2 in orbit is $Gm_1 m_2/(r_1 + r_2)^2 = m_2 r_2 \omega^2$. $\omega^2 = Gm_1/r_2(r_1 + r_2)^2$. Now $r_2 = rm_1/(m_1 + m_2)$, so $\omega^2 = 4\pi^2/T^2 = G(m_1 + m_2)/r^3$ and $T^2 = 4\pi^2 r^3/G(m_1 + m_2)$.

105. (a) $gmR_E/2$; (b) $\sqrt{gR_E}$; (c) $\sqrt{3gR_E}$

107. (a)

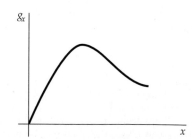

(b) $x = R/\sqrt{2}$

109. $Gm^2/[a(a + L)]$

111. (a) Since force on a mass m is $F = GMm/r^2$, then $F_s F_m = M_s r_m^2/M_m r_s^2 = 179$. (b) $dF/dr = -2Gm_1 m_2/r^3 = -2F/r$; $dF/F = -2dr/r$. (c) Since $\Delta F = -2(F/r)\,\Delta r$, then $\Delta F_s/\Delta F_m = (F_s/F_m)(r_m/r_s) = (M_s r_m^3)/(M_m r_s^3) = 0.46$.

Chapter 12

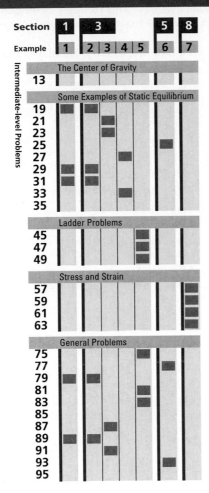

1. (a) False; (b) True; (c) True; (d) False

3. 318 N

5. 101 N

7. False. Location depends on mass distribution.

9. center of gravity

11. 3.4 m

13. $(0.94a, a)$

15. Left support, $F = 245$ N; right support, $F = 736$ N

17. $T_B = 690$ N, $F = 2540$ N, No

19. 4.17 m

21. (a) 182 N; (b) 456 N; (c) $\vec{F} = 157$ N, $\vec{F}_{hinge} = 508$ N

23. (a) $\vec{T}_1, \vec{T}_2, \vec{F}_H$ (the force exerted on the strut by the hinge); (b) $\Sigma\tau$ about hinge = 0, so $T_{2y} = T_1 = 80$ N; (c) 139 N

25. 27°

27. $Mg(2Rh - h^2)^{1/2}(h - R)\hat{i} + Mg\hat{j}$

29. Force on middle support = 2920 N, compression. Force on end support = 1940 N, tension.

31. If her boss is 1 m from the left end, Julie must stand 2.33 m or less from the right end.

33. (a) $Mg - F\sqrt{(2R - h)/h}$; (b) $-F$; (c) $F\sqrt{(2R - h)/h}$

35. (a) 49 N; (b) 73.6 N·m; (c) 736 N, acting upward; (d) 687 N, acting downward

37. 636 N, 21.5°

39. $F_{2m} = 49.1$ N, $F_{6m} = 147$ N; both forces act in a direction 60° above the x axis.

41. (a) Cube is stationary, thus $f_s = -F$. $\tau = Fa$. (b) $x = a/3$ from center of cube. (c) $Mg/2$

43. Yes, the coefficient of static friction between the wall and ladder must be sufficiently great to prevent slipping.

45. $\mu_s L \sin\theta \tan\theta$

47. $2h/L \sin\theta \tan\theta$

49. 59°

51. (a) $0.361mg$; (b) $0.313mg$; (c) $0.820mg$

53. 0.98 mm

55. 0.83 mm

57. 1.81 mm

59. (a) 34.7 cm; (b) 0.0776 J

61. 0.515 mm

63. 0.686

65. $A = A_0 - T/Y$

67. (c)

69. The body's center of gravity must be above the feet.

71. 99 cm

73. (a) 392 N; (b) 3

75. 0.577

77. 0.5

79. 304 kg

81. $(\cot\theta - 1)/2$

83. (a) 147 N; (b) 72.4% of total length

85. 0.207

87. (a) $T = 10,260$ N, $F_x = 5130$ N, $F_y = -4520$ N; (b) $T = 5924$ N, $F_x = 5130$ N, $F_y = 1403$ N

89. $2g$

91. (a) Since the center of gravity of the picture is in front of the wall, the torque due to mg about the nail must be balanced by an opposing torque due to the force of the wall on the picture, acting horizontally. So that $\Sigma F_x = 0$, the tension in the wire must have a horizontal component, and the picture must therefore tilt forward. (b) 3.43 N

93. (a) 83 km/h; (b) 105 km/h

95. 1.5 m

97. (a) 7.9 N; (b) 29.4 N; (c) 7.9 N

99. (a) Forces that act on the beam: weight = 49.1 N; force of the cylinder on the beam, $F_c = 28.3$ N; normal force of the ground, $F_{nb} = 24.5$ N; and friction force, $f_{sb} = 14.2$ N. Forces that act on the cylinder: weight = 78.5 N; force of the beam on the cylinder, $F_b = 28.3$ N; normal force of the ground, $F_{nc} = 103$ N; and friction force, $f_{sc} = 14.2$ N. (b) μ_s(beam–floor) = 0.58; μ_s(cylinder–floor) = 0.14.

101. $d/a = 1$

Chapter 13

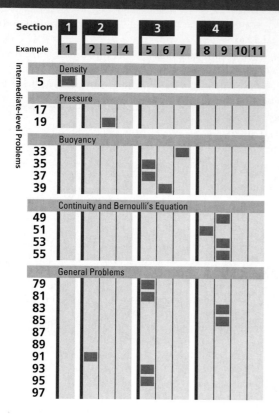

1. 0.673 kg

3. 103 kg

5. 13,621 kg/m^3

7. 29.8 in

9. (a) 1.5 atm; (b) 0.5 atm

11. 230 N

13. 200 atm

15. (a) 14,800 N; (b) 340 g

17. 45 cm

19. $W = Fh$; $F_1 = F_2 (A_1/A_2)$; $h_1A_1 = h_2A_2$; $h_1 = h_2 (A_2/A_1)$; $W_1 = F_1h_1 = F_2 (A_1/A_2)h_2 (A_2/A_1) = F_2h_2 = W_2$

21. (a) $V = 5.89 \times 10^{-3}$ m^3; $w = 57.8$ N; (b) $F = 173$ N

23. (c)

25. Nothing. The fish is in neutral buoyancy, so the upward acceleration of the fish is balanced by the downward acceleration of the displaced water.

27. (b) Volume of copper > volume of lead.

29. 4.36 N

31. (a) 11.1×10^3 kg/m^3; (b) lead

33. $\rho = 800$ kg/m^3; sp. gr. = 1.11

35. 250 kg/m^3

37. 183 m^3

39. 3.89 kg in freshwater

41. 2.06×10^7 kg

43. It blows over the ball, reducing the pressure above the ball below atmospheric pressure.

45. As the water falls, its velocity increases, causing the internal pressure to decrease; the pressure of air pushes it into a smaller stream.

47. (a) 12 m/s; (b) 132.5 kPa; (c) The volume flows are equal.

49. (a) 4.58 L/min; (b) 763 cm^2

51. $2\sqrt{h(H - h)}$

53. 1.85 m/s

55. 241 kPa

57. 1.47 kPa

59. 3.98 mPa·s

61. The water level remains constant.

63. The density of humid air is less than that of dry air.

65. The force acting is the difference in pressure between the wide and narrow parts times the area of the narrow part.

67. (a) Water spills over because the amount displaced is now greater than the volume of the lead block.

69. $F = 6.48 \times 10^4$ N. The table doesn't collapse because the atmosphere also exerts an upward force on the bottom surface of the table.

71. 1061 kg/m^3

73. 5000 kg/m^3

75. 72% of the block will be submerged.

77. 0.06 m^3 if mass = 60 kg

79. 40.2 g

81. The additional weight on the beaker side equals the weight of the displaced water, i.e., 64 g. That is the mass that must be placed on the other cup to maintain balance.

83. $h_A = 12.6$ m; $h_B = 5.3$ m

85. 6.5 mm

87. Since the object floats, the volume of displaced liquid is $m/\rho_0 = 4A\,\Delta h$; $\Delta h = m/4\rho_0 A$

89. (a) $dF = \rho gyL\,dy$; (b) 9.20×10^7 N; (c) Atmospheric pressure is exerted on each side of the dam and can therefore be neglected.

91. 1.37

93. (a) 70 m^3; (b) 6.61 m/s^2

95. 150 N (mass ≈ 15 kg)

97. (a) 2.4%; (b) 59 m^3

99. (a) 5.3 g; (b) 8 cm; (c) 1.17

Chapter 14

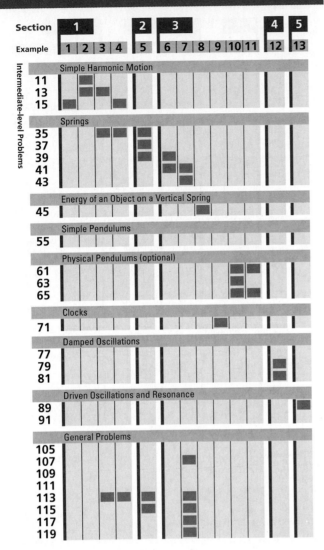

1. $4A, 0$

3. $0, A(2\pi f)^2$

5. (a) False; (b) True; (c) True

7. (a) 1.3 m/s; (b) 25 m/s

9. (a) $x(t) = (0.25 \text{ m}) \cos(4\pi/3)t$;
(b) $v(t) = dx/dt = -(\pi/3 \text{ m/s}) \sin(4\pi/3)t$;
(c) $a(t) = dv/dt = -(4\pi^2/9 \text{ m/s}^2) \cos(4\pi/3)t$

11. (a) $x(t) = (0.277 \text{ m}) \cos(4\pi t/3 - 0.445)$; (b) $v(t) = dx/dt = -(1.16 \text{ m/s}) \sin(4\pi t/3 - 0.445)$; (c) $a(t) = dv/dt = -(4.86 \text{ m/s}^2) \cos(4\pi t/3 - 0.445)$

13. (a) $x(t) = 10 \cos(\pi t/4)$

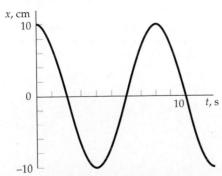

(b) 2.9 cm, 7.1 cm, 7.1 cm, 2.9 cm

15. (a) 2.5π; (b) $2.5\pi^2$

17. (a) $f = 0.32$ Hz, $\omega = 2$ rad/s; (b) 3.1 s; (c) $x(t) = (40 \text{ cm}) \cos(2t)$

19. 9

21. 23 J

23. (a) 0.37 J; (b) 3.8 cm

25. 1.4×10^3 N/m

27. (a) True; (b) True

29. (a) 6.9 Hz; (b) 0.15 s; (c) 0.1 m; (d) 4.3 m/s; (e) 1.9×10^2 m/s^2; (f) at $T/4, a = 0$

31. (a) 680 N/m; (b) 0.42 s; (c) 1.5 m/s; (d) 23 m/s^2

33. (a) 3.1 kN/m; (b) 4.2 Hz; (c) 0.24 s

35. (a) 0.44 m/s; (b) 0.38 m/s, 1.2 m/s^2; (c) 0.095 s

37. 0.26 s

39. 10 kJ

41. 25 cm

43. (a) yes; (b) $A_{\max} = 0.25$ m

45. (a) 0.13 J; (b) -0.32 J; (c) 0.45 J; (d) 0.13 J

47. False, the amplitude must be small.

49. The clock would run too slowly.

51. 12 s

53. 11.7 s

55. $T = 2\pi\sqrt{L/g(1 \sin \theta)}$

57. 1.1 s

59. 0.5 kg·m^2

61. $0.0918L$

63. 2.4 s

65. (a) 0.25 m; (b) 0.57 m, 2.1 s

67. (a) $T = 2\pi\sqrt{I/mgL}$. Since $I = \frac{2}{5} mr^2 + mL^2$,
$T = 2\pi\sqrt{L/g + 2r^2/5gL} = 2\pi\sqrt{L/g} (\sqrt{1 + 2r^2/5L^2}) = T_0\sqrt{1 + 2r^2/5L^2}$; (b) $\sqrt{1 + x} \approx 1 + x/2$ for small x. When $r \ll L$, $T = T_0\sqrt{1 + 2r^25L^2} \approx T_0(1 + r^2/5L^2)$; (c) 0.008%, 22 cm

69. (a)

71. 6.43°

73. True

75. 3.1%

77. $A(t + T)/A(t) = e^{-T/2\tau}$

79. (a) 31%; (b) 0.031%

81. (a) 3.6 cm, 2.2 cm; (b) 38 J, 14 J

83. (a) 2.6 s; (b) 3.6 s; (c) 6.0 J

85. Pendulum of a clock, violin string when bowed

87. (a) $f_0 = 1.01$ Hz; (b) $f_0 = 2.01$ Hz; (c) $f_0 = 0.352$ Hz

89. (a) 0.498 m; (b) 14.1 rad/s; (c) 0.354 m; (d) 1 rad/s

91. 180 J; (b) 57 J; (c) 19 W

93. (a) 0; (b) 4 m/s

95. (a) 0.14 m, 0.44 s; (b) 0.23 m, 0.36 s; (c) For inelastic collisions, $x(t) = (0.14 \text{ m}) \sin[(14 \text{ rad/s})t]$; for elastic collisions, $x(t) = (0.231 \text{ m}) \sin[(17 \text{ rad/s})t]$

97. 1. B; 2. D; 3. A, C

99. (d)

101. (c)

103. (a) $v(t) = (1.2 \text{ m/s}) \sin(3t + \pi/4)$; (b) -0.85 m/s; (c) 1.2 m/s; (d) 0.26 s or 1.3 s

105. (*a*) Since there is no friction, the only forces acting on the particle are *mg* and the normal force acting radially inward; the normal force is identical to the tension in the string that keeps the particle moving in a circular path if attached to a string of length *r*. (*b*) The particles meet at the bottom of the bowl because in simple harmonic motion, the period is independent of amplitude.

107. (*a*) 2.3 kg; (*b*) 0.65

109. $T = (2\pi/a)\sqrt{m/\rho g}$

111. The error is greater when the clock is elevated.

113. (*a*) $\mu_s = kA/(m_1 + m_2)g$; (*b*) *A* and *E* are unchanged, $\omega_f = \sqrt{m_1/(m_1 + m_2)}\,\omega_i$, $T_f = \sqrt{(m_1 + m_2)m_1}\,T_i$

115. (*a*) 2.3 Hz, 43 J; (*b*) maximum compression: 3.2 Hz, 0.60 m, 22 J; maximum extension: 3.2 Hz, 0.60 m, 43 J.

117. (*a*) If $\omega^2 A > g$, then the piston's acceleration is greater than the maximum acceleration of the block. (*b*) 2.4×10^{-2} s

119. (*a*) 2.5; (*b*) 6.4 cm

121. (*a*)

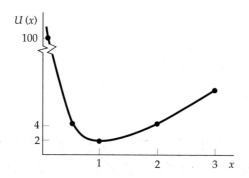

(*b*) $x_0 = 1$; (*c*) $U(x) = (1 - \epsilon)^2 + \dfrac{1}{(1 - \epsilon)^2}$; (*d*) $U(x) = \text{constant} + 4\epsilon^2$;

(*e*) $f = (1/\pi)\sqrt{2/m}$

123. $T = 8.59\sqrt{R/g}$

125. (*a*) $F = -(Gm\,M_E/R_E^3)r$; $F_x = (\sin\theta)F = (x/r)F = -(Gm\,M_E/R_E^3)x$; (*b*) $T = 2\pi\sqrt{m/k}$; $k = GmM_E/R_E^3 = gm/R_E$. Substituting *k*, $T = 2\pi\sqrt{R_E/g}$; $T = 85$ min.

127. Answer given in the problem.

129. Answer given in the problem.

Chapter 15

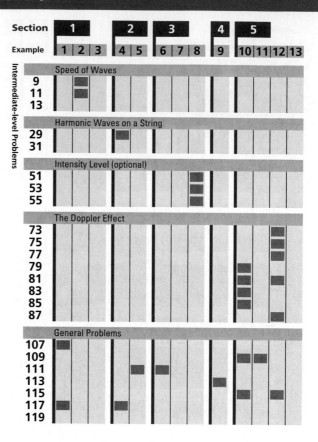

1. They move faster as they move up because the tension increases due to the weight of the rope below.

3. 1.32 km/s

5. 19.6 g

7. (*a*) 265 m/s; (*b*) 15 g

9. (*a*) 0.340 km/s; (*b*) 98% accurate; (*c*) No, because the speed of light is much greater than the speed of sound.

11. (*a*) 78 m; (*b*) 0.23 s, 70 m; (*c*) 70 m

13. (*a*) $v = \sqrt{\gamma RT/M}$; $dv/dt = \frac{1}{2}\sqrt{\gamma R/MT} = \sqrt{\gamma RT/M}/2T = v/2T$. Thus $dv/v = dT/2T$; (*b*) 4.95%; (*c*) $v_{\text{approximate}} = 347.4$ m/s, $v_{\text{exact}} = 347$ m/s

15. The lightning struck 680 m from the ballpark, 58.4° W or E of North.

17. (*a*) $d^2y/dx^2 = 6k(x + vt)$, $d^2y/dt^2 = 6kv^2(x + vt)$; thus $d^2y/dx^2 = (1/v^2)(d^2y/dt^2)$; (*b*) $d^2y/dx^2 = A(ik)^2 e^{ik(x - vt)} = -Ak^2 e^{ik(x - vt)}$, $d^2y/dt^2 = A(ikv)^2 e^{ik(x - vt)} = -Ak^2v^2 e^{ik(x - vt)}$; thus $d^2y/dx^2 = (1/v^2)(d^2y/dt^2)$; (*c*) $d^2y/dx^2 = -k^2/[k(x - vt)]^2 = -1/(x - vt)^2$ $d^2y/dt^2 = -(-kv)^2/[k(x - vt)]^2 = -v^2/(x - vt)^2$; thus $d^2y/dx^2 = (1/v^2)(d^2y/dt^2)$

19. For $y(x,t) = A\sin(kx - \omega t)$: $d^2y/dx^2 + i\,dy/dt = -Ak^2\sin(kx - \omega t) - Ai\omega\cos(kx - \omega t) \neq 0$. For $y(x,t) = Ae^{i(kx - \omega t)}$: $d^2y/dx^2 + i\,dy/dt = -k^2y + \omega y = 0$ if $k^2 = \alpha\omega$.

21. True

23. 20 cm

25. (a) 7.5×10^{14} to 4.3×10^{14} Hz; (b) 10 GHz

27. 27.2 Hz

29. (a) $y(x,t) = (0.025 \text{ m}) \sin(42x - 500t)$; (b) 13 m/s; (c) 6.3 km/s

31. (a) $P_0 = \frac{1}{2}\mu\omega^2 A_0^2 v$; (b) $P(x) = \frac{1}{2}\mu\omega^2 A_0^2 v e^{-2bx}$

33. (a) 0.75 Pa; (b) 4 m; (c) 85 Hz; (d) 340 m/s

35. (a) 36.7×10^{-5} m; (b) 8.27×10^{-2} Pa

37. (a) 0; (b) 3.67×10^{-7} m

39. (a) 138 Pa; (b) 21.7 W/m²; (c) 0.217 W

41. (a) 50 W; (b) 2 m; (c) 4.4×10^{-3} W/m²

43. False

45. (a) 10^{-11} W/m²; (b) 2×10^{-12} W/m²; (c) for (a) 9.4×10^{-5} Pa, for (b) 4.2×10^{-5} Pa

47. (a)

49. 99%

51. (a) 100 m; (b) 0.13 W

53. (a) 100 dB; (b) 50 W; (c) 20 m; (d) 96 dB

55. (a) 81 dB; (b) Eliminating the two least intense sources will not help much in reducing the intensity level of the noise. Most of the intensity comes from the 80-dB source.

57. 88 dB

59. 57 dB

61. (a)

63. (a) 15 m, 20 pies/min; (b) 13.5 m, 22.2 pies/min; (c) 15 m, 22 pies/min

65. (a) 1.3 m; (b) 262 Hz

67. (a) 2.1 m; (b) 162 Hz

69. (a) 80 m/s toward listener; (b) 420 m/s; (c) 1.7 m; (d) 247 Hz

71. (a) 24°; (b) 50 km high and 11.5 km over

73. 22 m/s = 79.2 km/h

75. $\Delta f = 9.2 \times 10^4$ Hz, $f = 30.00092$ GHz

77. 4.45 kHz

79. (a) 8.1×10^{-2} J; (b) 29 cm, 0.14 s

81. Both students must walk 0.75 m/s away from each other.

83. $f' = 1000 + 9.4 \sin 4t$ Hz

85. (a) 823.5 Hz; (b) 848.5 Hz

87. 183 m, 713 Hz.

89. 6.73×10^{10} m, 4.3×10^{34} kg

91. No; the frequencies are the same, but since the propagation speeds differ, so do the wavelengths.

93. (a) True; (b) False; (c) False

95. Segments 2 cm $< x \leq 3$ cm are moving up. Segments 1 cm $\leq x <$ 2 cm are moving down. $x = 2$ cm is instantaneously at rest.

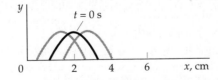

97. Cars behind slow down and then speed up in turn, as each reacts to the reduced distance between them and the car in front of them. This gives rise to a longitudinal wave pulse propagating backwards along the line of cars. No energy is transported. The speed of propagation is proportional to the length of a car and inversely proportional to the average driver's reaction time.

99. 21 cm

101. (a) 34 m/s to the left; (b) 20 m/s to the right; (c) 10 m/s to the right

103. (a) 110 Hz; (b) 1.1×10^4 Hz; (c) 570 Hz, 5.7×10^4 Hz

105. (a) 0.59 Hz; (b) 1.59 Hz

107. 8.0 m from the left end

109. 65 km/h

111. 1.07 kHz

113. (a) $A_r = A/2$, $A_t = 3A/2$; (b) $P_r = P_{in}/4$, $P_t = 3P_{in}/4$; (c) $A_1 = A + A_r = A_t$ so the amplitudes of displacement are the same.

115. (a) 4.5 m/s; (b) 887 Hz; (c) 842 Hz

117. (a) $T = 2.5 \times 10^{-3}$ s, $f = 400$ Hz; (b) 316 m/s; (c) $\lambda = 79$ cm, $k = 7.9$ m^{-1} (d) $y(x,t) = [5 \times 10^{-4} \sin(7.9x - 2.5 \times 10^3 t)]$ m; (e) $v_{max} = 1.26$ m/s, $a_{max} = 3.16$ km/s²; (f) 2.5 W

119. (a) $F = \mu v_0^2$; (b) $v_0 = \sqrt{F/\mu}$; (c) With respect to a fixed point on the chain, the pulse travels through 360°.

121. (a) The speed is given by $v = \sqrt{F/\mu}$. At a distance y from the bottom, $F = \mu g y$. Thus, $v = \sqrt{g y}$. (b) 2.2 s

123. (a) For $\Delta y/\Delta x \ll 1$, $\Delta\ell = \Delta x[1 + \frac{1}{2}(\Delta y/\Delta x)^2]$; so, $\Delta\ell - \Delta x = \frac{1}{2}(\Delta y/\Delta x)^2 \Delta x$ and $\Delta U = \frac{1}{2}F(\Delta y/\Delta x)^2 \Delta x$. (b) $(dy/dx)^2 = A^2 k^2 \cos^2(kx - \omega t)$. So $\Delta U = \frac{1}{2}FA^2 k^2 \Delta x \cos^2(kx - \omega t)$. (c) Replace F by $\mu v^2 = \mu\omega^2/k^2$. This gives Eq. 15-16b.

Chapter 16

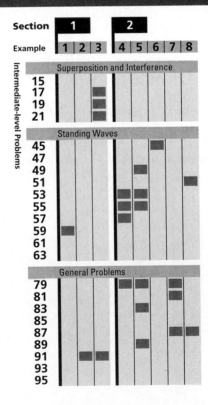

1. (a) False; (b) False; (c) True

3.

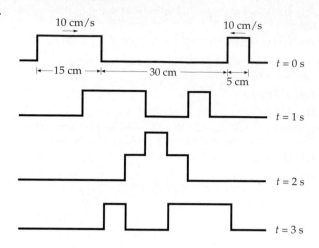

5. (a) 3.86 cm; (b) 3.46 cm

7. 0.071 m

9. (a) $\pi/2$ rad; (b) $\sqrt{2}A$

11. (a) 85 Hz, 255 Hz; (b) Some sound will reflect off walls, thus giving a variety of path differences.

13. (a) $4I_0$; (b) $2I_0$; (c) 0

15. For destructive interference, $\delta = \pi = 2\pi\Delta x/\lambda$. $\Delta x = d \sin \theta \le d$, where d is the source separation. So if $\Delta x < \lambda/2$, $\delta < \pi$ and there is no complete destructive interference in any direction.

17. 0.8π rad

19. 2361 Hz, 7083 Hz

21. 1.81 m; $\theta = 51°$

23. (a) 0.279 m; (b) 1220 Hz; (c) $\theta_3 = 0.432$ rad, $\theta_4 = 0.592$ rad, $\theta_5 = 0.772$ rad, $\theta_6 = 0.992$ rad, $\theta_7 = 1.354$ rad; (d) 0.07 rad

25. 1.98 rad

27. True

29. 4 Hz

31. (a) 496 Hz, 504 Hz; (b) If the beat frequency is diminished, the second fork has a frequency of 496 Hz, whereas if the beat frequency is increased, the frequency of the second fork is 504 Hz.

33. (b)

35. (c)

37. 180 m/s

39. (a) 17 Hz; (b) 8.5 Hz

41. (a) 1.25 Hz; (b) No second harmonic is excited. (c) 3.75 Hz

43. (a) 4.25 m; (b) 8.5 m

45. (a) $\lambda = 31$ cm, $f = 48$ Hz; (b) 15 m/s; (c) 63 cm

47. (a)

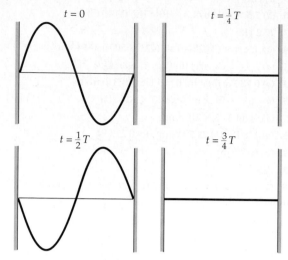

(b) 0.0126 s; (c) All the energy is kinetic.

49. (a) 75 Hz/125 Hz = 3/5, 125 Hz/175 Hz = 5/7; (b) The even harmonics are missing. (c) 25 Hz; (d) 3rd, 5th, 7th harmonics; (e) 4 m

51. (a) 345 m/s; (b) 0.45 cm

53. (a) 0.8 m; (b) 480 N; (c) 9.2 cm from the end

55. (a) 75 Hz; (b) 5th, 6th; (c) 2 m

57. 16.2 Hz

59. (a) The two sounds produce a beat because the third harmonic of 440 Hz equals the second harmonic of 660 Hz, and the original frequency of the E string is slightly greater than 660 Hz. If $f_E = (660 + \Delta f)$ Hz, a beat of $2\Delta f$ will be heard. (b) 661.5 Hz; (c) 79.6 N

61. 12.5 cm

63. (a) $y = 0.03 \sin(3\pi/4)x \cos 200\pi t$; (b) $dK = \frac{1}{2}\mu [6\pi \sin(3\pi/4)x \sin 200\pi t]^2 dx$, $t = 2.5 \times 10^{-3}$ s, straight; (c) $89m$ J

65. (a) $N \approx f_0 \Delta t$; (b) $\lambda \approx \Delta x/N$; (c) $k \approx 2\pi N/\Delta x$; (d) N is uncertain because the waveform dies out gradually, rather than stopping abruptly at some time; hence, where the pulse starts and stops is not well defined. (e) $\Delta k \approx 2\pi \Delta N/\Delta x$. Since $\Delta N = 1$, $\Delta k \approx 2\pi/\Delta x$.

67. No

69. When the edges of the glass vibrate, sound waves are produced in the air in the glass. The resonance frequency of the air columns depends on the length of the air column, which depends on how much water is in the glass.

71. (b)

73. The pitch is determined in large part by the resonant cavity of the mouth. Since $v_{He} > v_{air}$, the resonance frequency is higher if helium is the gas in the cavity.

75. (a) $3400n$ Hz, $n = 1, 3, 5 \ldots$; (b) Frequencies near 3400 Hz will be the most readily perceived.

77. (a) 2270 Hz; (b) 9th harmonic

79. (a) $f_n = n(0.66)$ Hz, $n = 1, 2, 3 \ldots$; (b) $f_n = n(0.33)$ Hz, $n = 1, 3, 5 \ldots$

81. (a) 4 m; (b) $\pi/2$ m^{-1}; (c) 800π rad/s; (d) $y = 0.03[\sin(\pi/2)x][\cos 800\pi t]$

83. (a) $n = 5$; (b) 5.5 Hz, 11.0 Hz, 16.4 Hz

85. (a) 720 N; (b) 2880 N, 6480 N, 11,520 N

87. 338 m/s. This method is not very accurate because the antinode generally does not occur exactly at the end of an open pipe.

89. 9.1 g

91. (a) $I_1 = 1.99 \times 10^{-5}$ W/m^2, $I_2 = 8.84 \times 10^{-6}$ W/m^2; (b) 5.53×10^{-5} W/m^2; (c) 2.21×10^{-6} W/m^2; (d) 2.87×10^{-5} W/m^2

93. (a) Answer given in the problem. (b) 203.4 Hz

95. (a) $K = \frac{1}{4}\mu L \omega_n^2 A_n^2 \sin^2 \omega_n t$; (b) $K_{max} = \frac{1}{4}\mu L \omega_n^2 A_n^2$; (c) $y = 0$; (d) Since $\omega_n = n\omega_1$, $K_{max} \propto n^2 A_n^2$.

97. $y_1 = A_0 \cos(kx - \omega t)$, $y_2 = A_0 \cos(kx - \omega t + k\,\Delta x + \delta_0)$; (b) $y = 2A_0 \cos\frac{1}{2}(k\,\Delta x + \delta_0)\cos\frac{1}{2}[kx - \omega t + \frac{1}{2}(k\,\Delta x + \delta_0)]$; $A = 2A_0 \cos\frac{1}{2}(\delta + \delta_0)$, where $\delta = 2\pi\,\Delta x/\lambda$.

(c)

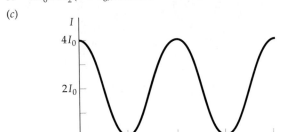

The time-average intensity $= 2I_0$.

(d)

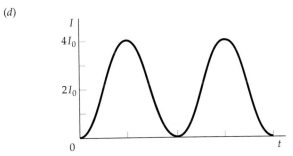

99. (a) 0.02 J; (b) 0; (c) dK_{max} at midpoint: $x = 1$ m; (d) dU_{max} at midpoint: $x = 1$ m.

Chapter 17

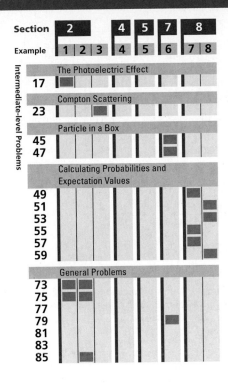

1. (c)

3. (a) 6.63×10^{-26} J, 4.14×10^{-7} eV; (b) 5.96×10^{-28} J, 3.72×10^{-9} eV

5. (a) 2.42×10^{14} Hz; (b) 2.42×10^{17} Hz; (c) 2.42×10^{20} Hz

7. (a) 12 keV; (b) 1.2 GeV

9. (a) True; (b) False; (c) True; (d) True

11. (a)

13. (a) 1.11×10^{15} Hz, 270 nm; (b) 1.62 eV; (c) 0.38 eV

15. (a) 4.73 eV; (b) 2.36 eV

17. 1.52 eV

19. 30°

21. (a) 17.4 keV; (b) 0.0760 nm; (c) 16.3 keV

23. (a) 2.43 pm; (b) 60 keV

25. (a) True; (b) True; (c) True; (d) False

27. (c)

29. 2.9 nm

31. (a) 0.061 eV; (b) 15 keV

33. 2.02×10^{-14} m

35. (a) 8.2×10^{-4} eV; (b) 8.2×10^{8} eV

37. 0.167 nm

39. 4.6 pm

41. Aperture $\approx 1.7 \times 10^{-33}$ m. The size of an atomic nucleus is on the order of 10^{-15}, which is much larger than the size of the aperture.

43. 0.087 nm, atoms in crystals

45. (a) $E_1 = 205$ MeV, $E_2 = 821$ MeV, $E_3 = 1846$ MeV

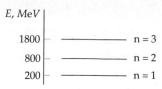

(b) 2.01 fm; (c) 1.21 fm; (d) 0.758 fm

47. (a) 5.6×10^{-55} J, 2.2×10^{-54} J, 5.0×10^{-54} J; (b) 5×10^{-16} J, 3×10^{19}

49. (a) 0; (b) 0.003; (c) 0

51. $L/2$, $L^2/3$ for box between 0 and L

53. (a) $L/2$; (b) $0.328L^2$

55. (a) $A = \sqrt{1/a}$; (b) 0.865

57. (a) 0.5; (b) 0.402; (c) 0.75

59. $\langle x \rangle = 0$, $\langle x^2 \rangle = 0.0327L^2$

61. Yes

63. (a) 0; (b) 1.5

65. Quantum mechanics predicts only probabilities. Change "results" to "probabilities of various results."

67. 3×10^{19}

69. 1.0×10^{22} photons

71. 4.2×10^{-20} nm

73. 6800 km

75. 1.7×10^{-14} W/m^2

77. 1.9 eV

79. 5

81. 1.04 eV, 554 nm

83. (a) $\dfrac{E_{n+1} - E_n}{E_n} = \dfrac{(n+1)^2 - n^2}{n^2} = \dfrac{2}{n} + \dfrac{1}{n^2} \approx \dfrac{2}{n}$ for large n.

(b) 0.2%; (c) For large quantum numbers n, the fractional difference between adjacent energy levels here is proportional to $1/n$, so as n approaches infinity, the quantized energy levels approach a continuum of levels as in classical physics.

85. (a) 6.25×10^{-4} eV/s; (b) 3200 s

Chapter 18

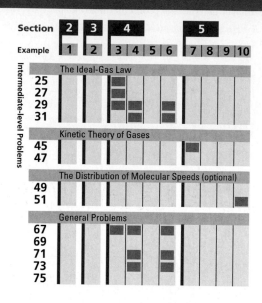

1. (a) False; (b) False; (c) True; (d) False

3. 1 C° > 1 F°

5. 10°F to 19°F

7. 57°C, −62°C

9. (a) 8.4 cm; (b) 107°C

11. −320.4°F

13. (a) 54.9 torr; (b) 3700 K

15. −40°

17. −183°C, −297°F

19. (a) $R_0 = 3.91 \times 10^{-3}$ Ω, $B = 3940$ K; (b) 1310 Ω; (c) −390 Ω/K at ice point, −4.34 Ω/K at steam point; (d) 0 K

21. Pressure increases

23. 1.15

25. 3.2×10^8 molecules

27. (a) 3700 mol; (b) 61 mol

29. 0.86 L; (b) 60 L

31. (a) 230 kPa; (b) 200 kPa

33. (a) 776 mol; (b) 4.86 km; (c) yes; (d) 6.0 km

35. 4

37. helium molecules

39. (a) 276 m/s; (b) 872 m/s

41. 5×10^5 m/s, 1.29 keV

43. $\lambda = 1/\sqrt{2}n_v\pi d^2$; since $n_v = N/V = P/KT$, $\lambda = KT/P\sqrt{2}\pi d^2$

45. (a) 1.24 km/s; (b) 0.31 km/s; (c) 0.26 km/s; (d) All three gases are likely to be present.

47. $K/mgh = 8.0 \times 10^4$

49. Answer given in the problem.

51. $v_{\mathrm{av}} = \sqrt{8kT/\pi m}$

53. False

55. The Fahrenheit and Celsius scales emphasize biologically significant temperatures; 0° is "cold" and 100° is "hot" in everyday experience.

57. (d)

59. (d)

61. $v_2/v_1 = \sqrt{M_1/M_2}$, $K_{2,av}/K_{1,av} = 1$

63. The rate of molecule–wall collisions increases with increased density.

65. 8.79 K

67. (a) 122 K; (b) 244 K; (c) 1.43 atm

69. 110 mol of H_2, 55 mol of O_2

71. $m_N = 7m_H$

73. $(18/11)P_0$

75. 400 K

Chapter 19

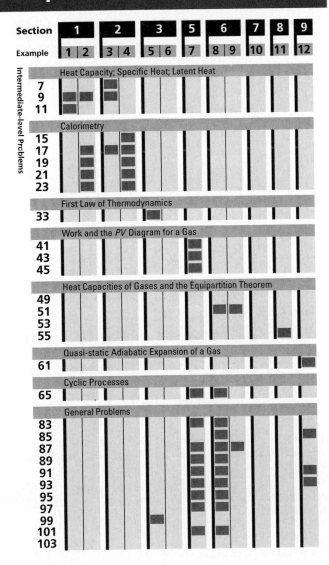

1. $\Delta T_A/\Delta T_B = 1/4$

3. (c)

5. 5×10^5 kJ

7. 402 kJ

9. 100 g

11. 6.25 kg

13. 0.092 cal/g·K

15. (a) 0°C; (b) 125 g

17. (a) 4.9°C; (b) no

19. (a) 3.01°C; (b) 200 g; (c) no

21. 18 ice cubes

23. (a) 28.5°C; (b) 15.5°C

25. Yes

27. 2.2 kJ

29. 176°C

31. 54 J

33. (a) 34 km; (b) yes; (c) The snowflakes reach terminal velocity and hit without melting.

35. For an ideal gas, $\Delta U = 0$ implies $\Delta T = 0$ because U is a function of T alone. A real gas is likely to have small, long-range, attractive interactions, which may introduce a dependence on volume (so that molecular kinetic energies decrease slightly as volume increases).

37. Since $Q = 0$ and $W = 0$, $\Delta U = 0$. Thus the temperature must increase to offset the decrease of potential energy from the increase in average distance between the ions.

39. (a) 608 J; (b) 1060 J

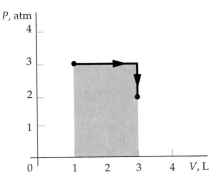

41. (a) 334 J; (b) 790 J

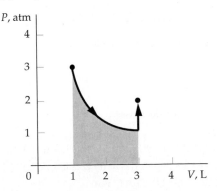

43. 100 L·atm

45. (a) −1.57 kJ; (b) −1.57 kJ

47. (a) 4.00 mol; (b) 14.9 kJ; (c) 83.0 J/K

49. (a) 0.713 J/g·K; (b) 252 J/L

51. $5P_0V$

53. (a) 3.50 mol; (b) $C_v = 43.7$ J/K, $C_p = 72.7$ J/K; (c) $C_v = 72.7$ J/K, $C_p = 102$ J/K

55. 1740 K

57. (a) $V_i = 2.2$ L, $V_f = 5.9$ L; (b) 143 K; (c) 1.62 kJ

59. 739 K, 10.1 atm

61. (a) 300 K, 7.79×10^{-3} m^3, $W = 1.14$ kJ, $Q = 1.14$ J; (b) 231 K, 6.00×10^{-3} m^3, $W = 287$ J, $Q = 0$

63. 575 J

65. (a)

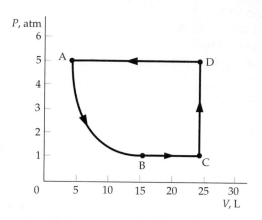

(b) $W \approx -64$ L·atm; (c) -65.1 L·atm; (d) $W = -65.1$ L·atm

67. D to A: $W = 0$, $Q = 89$ L·atm; A to B: $W = 130$ L·atm, $Q = 130$ L·atm; B to C: $W = 0$, $Q = -89$ L·atm; C to D: $W = -65$ L·atm, $Q = -65$ L·atm; Total $W = 65$ L·atm, total $Q = 65$ L·atm

69. D to A: $W = 0$, $Q = 148$ L·atm; A to B: $W = 105$ L·atm, $Q = 0$; B to C: $W = 0$, $Q = -95$ L·atm; C to D: $W = -52$ L·atm, $Q = 0$; Total $W = 53$ L·atm, Total $Q = 53$ L·atm

71. (a)

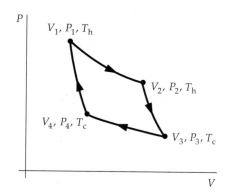

(b)–(f) Answer given in the problem.

73. (a) False; (b) False; (c) False; (d) True; (e) True; (f) False; (g) True

75. (d)

77. The temperature decreases.

79. 43 kcal

81. (a) $T_A = 65.2$ K, $T_C = 81.3$ K; (b) 16.0 L·atm; (c) 21.9 L·atm

83. (a) $T_A = 65.2$ K, $T_C = 81.3$ K; (b) 26.2 L·atm; (c) 32.1 L·atm

85. $W = 21.0$ L·atm, $Q = 26.9$ L·atm

87. $W = -170$ cal, $\Delta U = 0$, $T_f = T_i = 53$ K

89. $W = -170$ cal, $\Delta U = 0$, $T_f = T_i = 53$ K

91. (a) 3.91 kJ; (b) 5.49 kJ

93. (a) 2.74 kJ; (b) 4.91 kJ

95. (a) For an ideal gas, ΔU depends only on the states of the gas (n, T) and not on the path taken to go between the states. (b) Answer given in the problem.

97. (a) 8.6 K; (b) 143 J; (c) 1.03

99. 0.20 g

101. $4RT$

103. 4650 J/mol

105. For adiabatic expansion, $PV^\gamma = $ constant. Taking the derivative of both sides, $V^\gamma dP + \gamma PV^{\gamma-1} dV = 0$. Thus $dP/dV = -\gamma PV^{\gamma-1}/V^\gamma = -\gamma P/V$.

For isothermal expansion, $PV = nRT = $ constant. Taking the derivative, $V\,dP + P\,dV = 0$. Thus $dP/dV = -P/V$. $(-P/V)_{\text{isothermal}} = \gamma(-P/V)_{\text{adiabatic}}$.

107. (a) $T = 7418$ K, $V = 1832$ L; (b) $P = 0.0403$ atm

109. (a) $P = 218.2$ kPa; (b) $T = 3426$ K, $P = 2136$ kPa; (c) $P = 170.2$ kPa. Since N_2 is a diatomic gas, the contribution to specific heat from vibrations is R only, not $3R$ as for the triatomic gases.

111. (a) Answer given in the problem. (b) 4650 J

Chapter 20

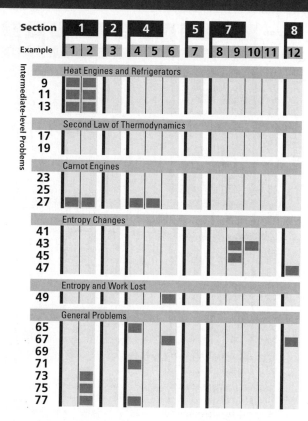

1. In an internal combustion engine, fuel is burned internally to produce hot gases which press a piston. In a steam engine, an external heat source converts water to hot steam, which then presses against a piston.

3. A refrigerator exhausts more heat into the kitchen than it absorbs from its interior.

5. (a) 500 J; (b) 400 J

7. (a) 40%; (b) 80 W

9.

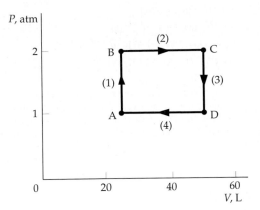

(a)

Step	W, L·atm	Q, L·atm	ΔU, L·atm
1	0	36.9	36.9
2	49.2	123.6	74.4
3	0	−74.4	−74.4
4	−24.6	−61.5	−36.9

(b) 15.3%

11.

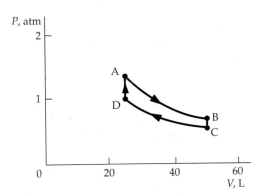

$\varepsilon = 13\%$

13. (a) 600 K, 1800 K, 600 K; (b) 15.4%

15. $\varepsilon = \varepsilon_C/\{1 + C_v\varepsilon_C/[R \ln(V_a/V_b)]\}$, where ε_C = Carnot efficiency = $(T_h - T_c)/T_h$.

17. This engine takes 200 J from a hot reservoir, does 60 J of work, and rejects the remaining 140 J to a cold reservoir. If the refrigerator statement of the second law of thermodynamics is false, then it would be possible to build a perfect refrigerator that would transfer the 140 J of heat from the cold to the hot reservoir with no other effects. The net result is that 60 J of heat is taken from the hot reservoir, 60 J of work is done, and no heat is rejected to the cold reservoir, a violation of the heat-engine statement of the second law.

19. If the proposed cycle is performed in the clockwise sense, then positive net work is done during the cycle. On the other hand, heat is exchanged only during the isothermal expansion. In this process, heat is drawn in from a high-temperature source. Note, however, that no heat is ever rejected to a low-temperature reservoir in the cycle, so the cycle would violate the heat-engine statement of the second law, namely, that it is impossible to have a cycle that uses heat to do work without exhausting waste heat in the process.

21. (a) 13.7; (b) 8.77

23. Answer given in the problem.

25. (a) 33.3%; (b) Answer given in the problem.

27. (a) 373 K; (b) 3.12 kJ, 0, −2.91 kJ; (c) 7%; (d) 35%

29. 5.26; (b) 3.2 kW; (c) 4.8 kW

31. (a) 303 kJ/min; (b) 212 kJ

33. (c)

35. (a) 11.5 J/K; (b) 0

37. 6.05 kJ/K

39. 6.05 kJ/K

41. (a) 0; (b) 267 K

43. (a) 244 kJ/K; (b) −244 kJ/K; (c) 0

45. (a) −24 J/K; (b) 138 J/K; (c) 114 J/K

47. 1.98 kJ/K

49. (a) 0.42 J/K; (b) 125 J

51. (c)

53. (d)

55. The diagram is an Otto cycle, which represents an internal combustion engine.

57.

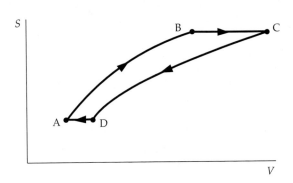

59.

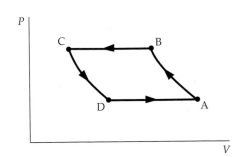

AB, CD are adiabatic; BC, DA are isobaric.

61. The 5-K decrease in the temperature of the cold reservoir

63. (a) 51%; (b) 102 kJ; (c) 98 kJ

65. 60%

67. (a) (1); (b) 7.66 J/K; 0.83 J/K

69. 313 K

71. 10 W

73. (a) 2.5 atm; (b) 462 K; (c) 68.7 L·atm, 26%

75. (a) 2.5 atm; (b) 415 K; (c) 65.3 L·atm, 35%

77. $\varepsilon = \varepsilon_C/\{1 + C_v\varepsilon_C/[R \ln(V_a/V_b)]\}$, which is less than the Carnot efficiency ε_C.

79. (a) Perfect refrigerator = $\Delta S_u = Q(T_h - T_c)/(T_hT_c) < 0$
(b) Perfect heat engine: $\Delta S_u = -Q/T_h < 0$; (c) $\Delta S \geq 0$ implies both statements, but the converse has not been proven here.

81. $\varepsilon = \varepsilon_1 + \varepsilon_2 (1 - \varepsilon_1)$. Substitute $W/Q_h = (1 - T_m/T_h) + (1 - T_c/T_m)(T_m/T_h) = 1 - T_c/T_h$.

83. 2.30×10^{-22} J/K

Chapter 21

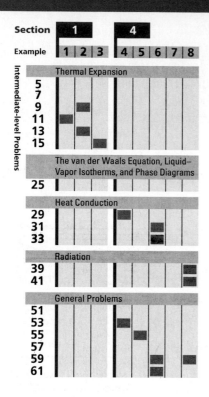

1. The glass of the thermometer expands first, causing an increased volume of space and a corresponding dip in the level of mercury.

3. 30.026 cm

5. (a) $(1/A)(dA/dT)$; (b) 2α in each case

7. 217°C

9. $15 \times 10^{-6}\,\text{K}^{-1}$

11. (a) 1.99935 m; (b) 2.00065 m

13. 0.78 L

15. 6.54 cm³

17. 395°C

19. (c)

21. For boiling, the temperature and pressure values must lie on the OF curve. As the altitude increases, the pressure decreases. The OF curve shows that at lower pressures, lower temperatures are required for boiling to occur. The OH curve shows the temperature and pressure values required for melting ice to water. As the pressure decreases (with increased altitude), higher temperatures are required for melting. (b) Because at higher altitudes water boils at lower temperatures, foods boiled in water would take a longer time to heat thoroughly. Also, more heat would have to be applied to heat foods to a certain temperature.

23. (a) 3.07×10^{-2} m; (b) 376 K

25. 3.94×10^{-23} cm³, 2.1×10^{-8} cm

27. (a) 16 K/W; (b) 6.3 W; (c) 50 K/m; (d) 87.5°C

29. (a) Cu: 0.083 K/W, Al: 0.14 K/W; (b) 0.22 K/W; (c) 360 W; (d) 70°C

31. (a) Because of conservation of energy, the thermal current into an element must equal the thermal current out of the element. (b) $I = 4\pi r^2 k\, dT/dr$; (c) $dT = (I/4\pi k)(dr/r^2)$; integrating from r_1 to r_2 gives $I = 4\pi k r_1 r_2\,(T_2 - T_1)/(r_2 - r_1)$; (d) Answer given in the problem.

33. 665 m

35. 4.4×10^{-4}

37. 9470 nm

39. 0.022 K/s

41. 97 g/h

43. (a) True; (b) True; (c) False; (d) False; (e) True

45. (d) assuming that the atmosphere is part of the earth.

47. radiation

49. 2.1 km

51. Rearranging $\beta = (1/V)\,(\Delta V/\Delta T)$ gives $\Delta V/V = \beta\,\Delta T$; $\rho = m/V$. Since $m = \text{constant}$, $\Delta\rho/\rho = -\Delta V/V = -\beta\,\Delta T$; hence $\Delta\rho = -\rho\beta\,\Delta T$

53. 360 s

55. 1.2 cm

57. (a) $\beta = (1/V)\,(dV/dT)$, where $V = nRT/P$ and $dV/dT = nR/P$. Thus $\beta = (P/nRT)/(nR/P) = 1/T$; (b) $\beta_t = 3.663 \times 10^{-3}\,\text{K}^{-1}$

59. 1.26×10^{13} W; $P_{\text{lost}}/P_{\text{absorbed}} = 7.3 \times 10^{-5}$ (Note that the effective area for absorption of energy from the sun is πR^2 whereas the loss is through area $4\pi R^2$.)

61. 142 W

63. $\omega_2 = \omega_1/(1 + \alpha\,\Delta T)^2$, $K_2 = MR_1^2\,\omega_1^2/4\,(1 + \alpha\,\Delta T)^2$, $L_2 = MR_1^2\,\omega_1/2$

65. Answer given in the problem.

67. 3.1×10^{-6} W/m·K

INDEX

Numbers in **bold** indicate additional display material, such as diagrams; *n* indicates a footnote; AP indicates material in the Appendixes.

Prefixes for Powers of 10	Multiple	Prefix	Abbreviation
	10^{18}	exa	E
	10^{15}	peta	P
	10^{12}	tera	T
	10^9	giga	G
	10^6	mega	M
	10^3	kilo	k
	10^2	hecto	h
	10^1	deka	da
	10^{-1}	deci	d
	10^{-2}	centi	c
	10^{-3}	milli	m
	10^{-6}	micro	μ
	10^{-9}	nano	n
	10^{-12}	pico	p
	10^{-15}	femto	f
	10^{-18}	atto	a

Some Physical Data			
	Acceleration of gravity at earth's surface	g	$9.81 \text{ m/s}^2 = 32.2 \text{ ft/s}^2$
	Radius of earth	R_E	$6370 \text{ km} = 3960 \text{ mi}$
	Mass of earth	M_E	$5.98 \times 10^{24} \text{ kg}$
	Mass of sun		$1.99 \times 10^{30} \text{ kg}$
	Mass of moon		$7.36 \times 10^{22} \text{ kg}$
	Escape speed at earth's surface		$11.2 \text{ km/s} = 6.95 \text{ mi/s}$
	Standard temperature and pressure (STP)		$0°C = 273.15 \text{ K}$ $1 \text{ atm} = 101.3 \text{ kPa}$
	Earth–moon distance		$3.84 \times 10^8 \text{ m} = 2.39 \times 10^5 \text{ mi}$
	Earth–sun distance (mean)		$1.50 \times 10^{11} \text{ m} = 9.30 \times 10^7 \text{ mi}$
	Speed of sound in dry air (at STP)		331 m/s
	Density of air		1.29 kg/m^3
	Density of water		1000 kg/m^3
	Heat of fusion of water	L_f	333.5 kJ/kg
	Heat of vaporization of water	L_v	2.257 MJ/kg

The Greek Alphabet

Alpha	A	α	Iota	I	ι	Rho	P	ρ
Beta	B	β	Kappa	K	κ	Sigma	Σ	σ
Gamma	Γ	γ	Lambda	Λ	λ	Tau	T	τ
Delta	Δ	δ	Mu	M	μ	Upsilon	Y	υ
Epsilon	E	ϵ	Nu	N	ν	Phi	Φ	ϕ
Zeta	Z	ζ	Xi	Ξ	ξ	Chi	X	χ
Eta	H	η	Omicron	O	o	Psi	Ψ	ψ
Theta	Θ	θ	Pi	Π	π	Omega	Ω	ω

Abbreviations for Units

A	ampere	lb	pound
Å	angstrom (10^{-10} m)	L	liter
atm	atmosphere	m	meter
Btu	British thermal unit	MeV	mega-electron volt
Bq	becquerel	Mm	megameter (10^6 m)
C	coulomb	mi	mile
°C	degree Celsius	min	minute
cal	calorie	mm	millimeter
Ci	curie	ms	millisecond
cm	centimeter	N	newton
dyn	dyne	nm	nanometer (10^{-9} m)
eV	electron volt	pt	pint
°F	degree Fahrenheit	qt	quart
fm	femtometer, fermi (10^{-15} m)	rev	revolution
ft	foot	R	roentgen
Gm	gigameter (10^9 m)	Sv	seivert
G	gauss	s	second
Gy	gray	T	tesla
g	gram	u	unified mass unit
H	henry	V	volt
h	hour	W	watt
Hz	hertz	Wb	weber
in	inch	y	year
J	joule	yd	yard
K	kelvin	μm	micrometer (10^{-6} m)
kg	kilogram	μs	microsecond
km	kilometer	μC	microcoulomb
keV	kilo-electron volt	Ω	ohm

Some Conversion Factors

$1 \text{ m} = 39.37 \text{ in} = 3.281 \text{ ft} = 1.094 \text{ yd}$

$1 \text{ m} = 10^{15} \text{ fm} = 10^{10} \text{ Å} = 10^9 \text{ nm}$

$1 \text{ km} = 0.6215 \text{ mi}$

$1 \text{ mi} = 5280 \text{ ft} = 1.609 \text{ km}$

$1 \text{ lightyear} = 1 \ c \cdot y = 9.461 \times 10^{15} \text{ m}$

$1 \text{ in} = 2.540 \text{ cm}$

$1 \text{ L} = 10^3 \text{ cm}^3 = 10^{-3} \text{ m}^3 = 1.057 \text{ qt}$

$1 \text{ h} = 3.6 \text{ ks}$

$1 \text{ y} = 365.24 \text{ d} = 3.156 \times 10^7 \text{ s}$

$1 \text{ km/h} = 0.278 \text{ m/s} = 0.6215 \text{ mi/h}$

$1 \text{ ft/s} = 0.3048 \text{ m/s} = 0.6818 \text{ mi/h}$

$1 \text{ rev} = 2\pi \text{ rad} = 360°$

$1 \text{ rad} = 57.30°$

$1 \text{ rev/min} = 0.1047 \text{ rad/s}$

$1 \text{ slug} = 14.59 \text{ kg}$

$1 \text{ tonne} = 10^3 \text{ kg} = 1 \text{ Mg}$

$1 \text{ atm} = 101.3 \text{ kPa} = 1.013 \text{ bar} = 76.00 \text{ cmHg} = 14.70 \text{ lb/in}^2$

$1 \text{ N} = 10^5 \text{ dyn} = 0.2248 \text{ lb}$

$1 \text{ lb} = 4.448 \text{ N}$

$1 \text{ Pa·s} = 10 \text{ poise}$

$1 \text{ J} = 10^7 \text{ erg} = 0.7373 \text{ ft·lb} = 9.869 \times 10^{-3} \text{ L·atm}$

$1 \text{ kW·h} = 3.6 \text{ MJ}$

$1 \text{ cal} = 4.184 \text{ J} = 4.129 \times 10^{-2} \text{ L·atm}$

$1 \text{ L·atm} = 101.3 \text{ J} = 24.22 \text{ cal}$

$1 \text{ eV} = 1.602 \times 10^{-19} \text{ J}$

$1 \text{ Btu} = 778 \text{ ft·lb} = 252 \text{ cal} = 1054 \text{ J}$

$1 \text{ horsepower} = 550 \text{ ft·lb/s} = 746 \text{ W}$

$1 \text{ W/m·K} = 6.938 \text{ Btu·in/h·ft}^2\text{·°F}$

$1 \text{ T} = 10^4 \text{ G}$

$1 \text{ kg weighs about } 2.205 \text{ lb}$

Some Physical Constants

Avogadro's number	N_A	$6.022\ 137 \times 10^{23}$ particles/mol
Boltzmann's constant	k	$1.380\ 658 \times 10^{-23}$ J/K
Bohr magneton	$m_B = e\hbar/2m_e$	$9.274\ 015\ 4 \times 10^{-24}$ J/T
Coulomb constant	$k = 1/4\pi\epsilon_0$	$8.987\ 551\ 788 \times 10^9$ N·m²/C²
Compton wavelength	$\lambda_C = h/2e$	$2.426\ 310\ 58 \times 10^{-12}$ m
Fundamental charge	e	$1.602\ 177 \times 10^{-19}$ C
Gas constant	$R = N_A k$	$8.314\ 51$ J/mol·K $= 1.987\ 22$ cal/mol·K $= 8.205\ 78 \times 10^{-2}$ L·atm/mol·K
Gravitational constant	G	$6.672\ 6 \times 10^{-11}$ N·m²/kg²
Mass, of electron	m_e	$9.109\ 390 \times 10^{-31}$ kg $= 510.999\ 1$ keV/c^2
of proton	m_p	$1.672\ 623 \times 10^{-27}$ kg $= 938.272\ 3$ MeV/c^2
of neutron	m_n	$1.674\ 929 \times 10^{-27}$ kg $= 939.565\ 6$ MeV/c^2
Permeability of free space	μ_0	$4\pi \times 10^{-7}$ N/A²
Planck's constant	h	$6.626\ 076 \times 10^{-34}$ J·s $= 4.135\ 669 \times 10^{-15}$ eV·s
	$\hbar$	$1.054\ 573 \times 10^{-34}$ J·s $6.582\ 122 \times 10^{-16}$ eV·s
Speed of light	c	$2.997\ 924\ 58 \times 10^8$ m/s
Unified mass unit	u	$1.660\ 540 \times 10^{-27}$ kg $= 931.494\ 32$ MeV/c^2

Mathematical Symbols

$=$	is equal to		Δx	change in x
$\neq$	is not equal to		$\lvert x \rvert$	absolute value of x
$\approx$	is approximately equal to		$n!$	$n(n-1)(n-2)\cdots 1$
$\sim$	is of the order of		Σ	sum
$\propto$	is proportional to		$\lim$	limit
$>$	is greater than		$\Delta t \to 0$	Δt approaches zero
$\geq$	is greater than or equal to		$\dfrac{dx}{dt}$	derivative of x with respect to t
$\gg$	is much greater than			
$<$	is less than		$\dfrac{\partial x}{\partial t}$	partial derivative of x with respect to t
$\leq$	is less than or equal to			
$\ll$	is much less than		$\int$	integral